Silicon Based Unified Memory Devices and Technology

Silicon Based Unified Memory Devices and Technology

Arup Bhattacharyya

CRC Press
Taylor & Francis Group
Boca Raton London New York

CRC Press is an imprint of the
Taylor & Francis Group, an **informa** business

The materials covered here may describe multiple technical innovations owned by the author which are subject of potential patent assignments to third parties. The information provided in this book is in no form whatsoever to be used as an inducement to infringe into any such potential patents.

June 2017   Arup Bhattacharyya

CRC Press
Taylor & Francis Group
6000 Broken Sound Parkway NW, Suite 300
Boca Raton, FL 33487-2742

© 2017 by Taylor & Francis Group, LLC
CRC Press is an imprint of Taylor & Francis Group, an Informa business

No claim to original U.S. Government works

Printed on acid-free paper

International Standard Book Number-13: 978-1-138-03271-2 (Hardback)
International Standard Book Number-13: 978-1-138-74632-9 (Paperback)

Visit the Taylor & Francis Web site at
http://www.taylorandfrancis.com

and the CRC Press Web site at
http://www.crcpress.com

Dedication

This book is dedicated to our two adorable grandchildren:
Kailas Bhattacharyya Kahler (13-years-old) and Ishan Bhattacharyya Kahler
(8-years-old): The budding scientist, mathematician, engineer, and visionary
of tomorrow, who have been my deepest source of inspiration and joy.

Contents

Foreword xxi
Preface.. xxv
Acknowledgments... xxvii
Author xxxi

PART I Conventional Silicon Based NVM Devices

Chapter 1   Silicon Based Digital Volatile and Nonvolatile Memories: An Introductory
Overview.. 3

1.1   Digital Memories and Binary States: Basic Concepts.................................. 3
1.2   Volatile Memories and NVMs 4
        1.2.1   Static Random Access Memory: SRAM.. 5
        1.2.2   Dynamic Random Access Memory: DRAM 6
        1.2.3   Read-Only Memory: ROM 7
        1.2.4   EPROM, EEPROM, and E²PROM.. 7
        1.2.5   Recent NVMs: NROM and NAND Flash Memories 8
1.3   Memory Hierarchy in Digital Systems.. 9
1.4   Fundamental Memory Concept in NVMs... 10
1.5   NVM Device Groupings and Nomenclature 11
References 12

Chapter 2   Historical Progression of NVM Devices 13

2.1   Floating-Gate Devices.. 13
        2.1.1   The FAMOS Device... 13
        2.1.2   The SAMOS Device.. 14
        2.1.3   The SAMOS 8 Kb EAROM... 14
        2.1.4   The SIMOS Device 15
        2.1.5   Chronology of Floating-Gate Device Evolution............................. 15
        2.1.6   The HIMOS Cell... 16
        2.1.7   The ETOX/FLOTOX Cell 16
        2.1.8   The DINOR Cell 17
        2.1.9   The NAND Cell 17
2.2   Conventional Charge-Trapping Devices... 18
        2.2.1   MNOS/MXOS/MONOS/SNOS/SONOS 18
        2.2.2   Historical Evolution of MNOS, MXOS to SONOS CT Devices 18
        2.2.3   Evolution of CT-NVM Cell Designs 19
                 2.2.3.1   Tri-Gate Memory Cell 19
                 2.2.3.2   Pass Gate Memory Cell 20
2.3   Nanocrystal Charge-Trapping NVM Devices... 20
        2.3.1   Early History 20
        2.3.2   Nanocrystal Physics and Charge Trapping 21
        2.3.3   Review of Nanocrystal NVM Devices.. 24
        2.3.4   Nanocrystal Device General Characteristics 25

2.3.5   Silicon Nanocrystal Device Characteristics.....................................26
2.3.6   Metal Nanocrystal Device Characteristics..28
2.3.7   Germanium Nanocrystal Device Characteristics.............................29
2.4   Direct Tunnel Memory 29
References 32

Chapter 3   General Properties of Dielectrics and Interfaces for NVM Devices 35

3.0   The NVM Gate Stack Layers and Interfaces 35
3.1   Attributes of Gate Stacks for NVM Devices.......................................36
       3.1.1   The Energy Band of the Gate Stack Layers 37
3.2   General Properties of Thin Dielectric Films..38
       3.2.1   Physical, Chemical, and Thermal Stability................................38
       3.2.2   Electronic Properties...39
       3.2.3   Bulk and Interface Defects and Charge Trapping.....................41
       3.2.4   Charge Transport...42
       3.2.5   Figure of Merit for Selected Metal-Oxide Dielectric Films 42
                3.2.5.1   Metal Work Function and Electron Affinity 43
3.3   Interfaces, Electrode Compatibility, and Process Sensitivity..........44
3.4   Gate Material for NVM Devices...44
3.5   Dielectric Conductivity Mechanisms 45
       3.5.1   Bulk-Controlled Poole-Frenfel Mechanism 45
       3.5.2   Electrode-Controlled Quantum Mechanical Tunneling Mechanisms 46
       3.5.3   Direct Tunneling and/or Modified Fowler–Nordheim Tunneling......46
                3.5.3.1   Fowler–Nordheim Tunneling...47
                3.5.3.2   Enhanced Fowler–Nordheim Tunneling...........................48
3.6   Carrier Transport Mechanisms for Multilayer Dielectrics............49
References 49

Chapter 4   Dielectric Films for NVM Devices...51

4.0   Conventional Dielectric Films for NVM Devices................................51
4.1   Thermal Oxide: SiO_2 51
       4.1.1   Defect Generation and Oxide Degradation 53
                4.1.1.1   Oxide Reliability..54
4.2   CVD or LPCVD Nitride: Si_3N_4 55
       4.2.1   Nitride Traps, Trap Creation: Process and Stress Sensitivity.............55
4.3   Silicon Oxynitrides: SiONs 56
       4.3.1   Trapping Characteristics of SiON 57
       4.3.2   MIOS Trapping Comparison between Nitride and Oxynitride..........59
       4.3.3   Properties of Oxynitride Films: Co-Relationship with
                Refractive Index and Composition 60
                4.3.3.1   Refractive Index and SiON Films....................................60
                4.3.3.2   Dielectric Constant and SiON Films 62
                4.3.3.3   Breakdown Strength and SiON Films 62
       4.3.4   Advantage of Oxynitride Films..63
       4.3.5   Applications of SiONs in NVM Devices 63
4.4   Silicon-Rich Insulators: (SROs and SRNs) 64
       4.4.1   Single-Phase SRN 65
       4.4.2   Two-Phase SRN 65
                4.4.2.1   Electrical Characteristics...66

4.4.3 Applications in NVM Devices: IN-SRN Films 68
4.4.4 Application of CT-SRN Films in NVM Devices 69
References 70

Chapter 5 NVM Unique Device Properties 71

5.1 Memory Window W 71
5.1.1 Memory Window Generation: Charge Storage and
Trap Density 73
5.1.1.1 Vpp and EOT: Impact on Memory Window 75
5.2 Memory Retention 75
5.2.1 Retention, Charge Transport, and Dielectric Conductivity 76
5.2.2 Charge Loss and Leakage Paths for Floating-Gate Devices 77
5.2.3 Band Diagram-Based Assessment of FG Current Components 78
5.2.4 Tunnel Oxide Charge Loss Characteristics 79
5.2.5 Tunnel Oxide Thickness Limit for FG Device 80
5.2.6 Enhancement of Window and Retention in FG Devices 81
5.2.7 Charge Loss and Leakage Paths for Charge-Trapping Devices 81
5.2.7.1 Band Diagram and Charge Loss Modes for
Floating-Gate NVM Devices 81
5.2.7.2 Charge Loss from Charged Nitride Floating Node to
Substrate: (JeS and JhS) 82
5.2.7.3 Charge Loss from Charged Nitride Floating Node to
the Gate: (JeG and JhG) 82
5.2.7.4 Charge Loss Mechanisms in Conventional CT Devices 83
5.2.8 Comparative Retention Attributes: MNOS/SNOS versus
MONOS/SONOS 84
5.2.8.1 Retention Time Modeling for MONOS/SONOS
Devices 84
5.2.9 Enhanced Charge Retention for MXOS Devices: Correlation
with Band Structures 85
5.3 Memory Endurance and Endurance Traits of FG, CT, NC NVMs 86
5.3.1 Endurance and Peak Field 87
5.3.2 Endurance, Charge Fluence, Dielectric Conductance, and QBD 89
5.3.3 Endurance and Depend Types: Tunnel Layer Dependency 90
5.3.4 Requirement for High Endurance 91
5.3.5 MXOS Device with High Endurance 91
References 93

Chapter 6 NVM Device Stack Design 95

6.1 Fundamentals 95
6.2 FG-NAND Flash Devices: Stack/Cell Design 97
6.2.1 Floating Poly Gate, GCR, and STI in Stack Design 97
6.3 Charge Trapping Devices: Cell/Stack Designs 102
6.3.1 Stack Design Elements and Device Properties: MNOS/SNOS/
MXOS/SONOS/MONOS 103
6.3.1.1 Tunneling Oxide 103
6.3.1.2 Trapping Nitride (Si$_3$N$_4$) and Oxynitride (SiON) 106
6.3.1.3 Blocking Oxide or Blocking OR-SiON 107
6.3.2 ONO Stack Design and SONOS Device Properties 108

6.3.3   Examples of SONOS Stack Designs 110
6.3.4   SONOS Devices with Oxynitride 112
6.3.5   Nano Crystal Charge Trapping Devices: Stack
         Design Elements and Device Properties: SiNC/GeNC/MNC.......... 115
References 116

Chapter 7   NVM Cells, Arrays, and Disturbs.. 119

7.1   Pre-Silicon Gate Technology and NVM Cells.. 119
7.2   The Floating-Gate Flash NVM Cells in Silicon Gate Technology 119
       7.2.1   Floating-Gate Flash Cells and Arrays.. 120
                7.2.1.1   The FLOTOX Cell.. 120
                7.2.1.2   The ACEE Cell 121
                7.2.1.3   The ETOX Cell 122
7.3   The NAND Flash Cell.. 122
       7.3.1   Floating-Gate NAND Cell and Array Design................................ 123
       7.3.2   Erasing, Writing, and Reading Operations of 4 Mb
                NAND EEPROM 124
       7.3.3   Progress in Floating-Gate NAND Cell Design.............................. 126
7.4   Floating-Gate NOR Cells and Arrays 128
       7.4.1   The DINOR Memory Cell and the VGA Architecture
                for NOR Arrays 129
7.5   Charge Trap NVM Cells and Arrays... 130
       7.5.1   NROM Cells and Arrays... 130
       7.5.2   General Operations of NROM Array.. 131
       7.5.3   Advancement in NROM.. 131
       7.5.4   NROM Scaling and Product Progression....................................... 133
       7.5.5   NROM Product Potential and Challenges...................................... 134
7.6   Disturbs and Mitigation in NVM Cells and Arrays 134
       7.6.1   Disturbs in FG Devices and Arrays 134
                7.6.1.1   Read Disturb... 134
                7.6.1.2   Program Disturb 136
       7.6.2   Disturb Minimization or Mitigation 136
                7.6.2.1   Read-Disturb Mitigation... 136
                7.6.2.2   Mitigation or Minimization of Drain Disturb 137
       7.6.3   Disturbs in CT (NROM) Devices and Arrays 138
                7.6.3.1   Read Disturb... 138
                7.6.3.2   Program Disturb or Drain Disturb 138
References 139

Chapter 8   NVM Process Technology and Integration Scheme 141

8.1   Silicon-Gate CMOS Process Technology History 141
8.2   NVM-Unique Technology Integration Features.. 142
       8.2.1   High-Voltage Generation, Transmission, and Circuit
                Requirements.. 142
       8.2.2   Self-Aligned Shallow Trench Isolation: SASTI 143
       8.2.3   NVM Process Flow and Integration Scheme................................. 144
8.3   NROM Process Flow and Integration Scheme.. 144
8.4   NAND Process Steps 145
References 146

Chapter 9   NVM Device Reliability 147

      9.1   Reliability Issues for Gate Stack Dielectric Elements 147

      9.2 SiO_2 Tunnel Dielectric Film 148

           9.2.1   Stress-Induced Traps/Defects in Tunnel Oxide:

                   Floating-Gate Cell Issues 148

                   9.2.1.1   Erratic Bit Phenomenon 148

                   9.2.1.2   Transconductance Degradation 149

                   9.2.1.3   Data Retention Degradation: Stress-Induced

                            Leakage Current, Anomalous Stress-Induced

                            Leakage Current, or "Tail Cell" 149

                   9.2.1.4   Fast Erase Cells 150

                   9.2.1.5   Drain Disturb 150

           9.2.2   CT Cell (SONOS and Nanocrystal) Issues 151

                   9.2.2.1   Ultrathin Tunnel Oxide Cells: (Type a):

                            Transconductance Degradation 151

                   9.2.2.2   Thicker Tunnel Oxide Cells: (Type b):

                            Hot-Hole Effects 153

                   9.2.2.3   Hot-Carrier-Induced CT Cells: (Type c) 153

                   9.2.2.4   Data Retention Loss 153

                   9.2.2.5   Sub-Threshold Slope Degradation (STS) 155

                   9.2.2.6   Drain Disturb 156

      9.3   Radiation-Induced Instability and Radiation Immunity 156

      References 158

Chapter 10   Conventional NVM Challenges 159

      10.1   Scaling Challenges of Conventional NVM 159

           10.1.1   Short-Channel Effects (SCEs) 159

           10.1.2   Few-Electron Effects (FEE) 159

           10.1.3   Vertical Scalability 160

           10.1.4   Limitations and Challenges 160

      References 161

PART II   *Advanced NVM Devices and Technology*

Chapter 11   Voltage Scalability 167

      11.1   Why Voltage Scaling? 167

      11.2   Early History of Voltage Scaling 168

      11.3   Recent Developments in Voltage Scaling 168

           11.3.1   Floating-Gate Devices 168

      11.4   CT Device: SONOS Voltage Scaling 169

      References 170

Chapter 12   High K Dielectric Films for NVM 173

      12.1   Historical Perspective 173

      12.2   General Requirements 174

      12.3   Common High K Dielectric Films for NVM Gate 175

12.4   FET Gate-Insulator Requirements... 176
12.5   Thin Amorphous Films of Al_2O_3, ZrO_2, HfO_2, Ta_2O_5, and TiO_2
          for Gate-Insulator Applications 176
12.6   Alumina (Al_2O_3) 178
12.7   Hafnia (HfO_2) 180
12.8   Zirconia (ZrO_2) 184
12.9   Tantalum Oxide (Ta_2O_5) and Titanium Oxide (TiO_2)..................... 187
          12.9.1 TiO_2 187
                      12.9.1.1 Ta_2O_5.. 188
12.10   Bimetal Oxides and Aluminates of Hafnium and Zirconium 188
12.11   Aluminates of Lanthanides... 188
          12.11.1   Lanthanum Aluminate ($LaAlO_3$)..................................... 189
12.12   Nitrides and Oxynitrides of Hafnium and Zirconium........................... 191
12.13   Hafnium Silicon Oxynitride (HfSiON) 192
12.14   Comparison of Band Diagrams of HfO_2, HfSiON, ZrO_2, and $LaAlO_3$..... 194
12.15   Common High K Films for NVM Applications 195
          12.15.1   Unique Functional Requirements of High K Films for NVM .. 195
          12.15.2   Tunnel Dielectric Film 196
          12.15.3   Trapping Dielectric Film.. 196
          12.15.4   Blocking Dielectric Film.. 197
12.16   Review of High K Dielectric Applications for Current NVM Devices..... 198
          12.16.1   High K Dielectric for Charge Tunneling............................. 198
          12.16.2   High K Dielectric for Charge Trapping 199
          12.16.3   High K Dielectric for Charge Blocking 199
12.17   Applicability of High K Dielectric Films for NVM Gate Stack Design ... 199
          12.17.1   Application Case Exercise 200
                      12.17.1.1   Case Example of the CT-NVM Device Stack
                                   Design and High K Film Applicability 200
12.18   High K Films for Tunneling 201
          12.18.1   Electron Transport and Current Requirements
                      for Window Generation.. 201
12.19   Leakage and Retention for High K Tunnel Dielectric Films................... 203
          12.19.1   Charge Stored during Writing and Internal Potential.............. 203
          12.19.2   Leakage Characteristics for High K Films for the Tunnel
                      Applications 203
12.20   High K Films for Charge Trapping... 204
12.21   High K Films for Charge Blocking... 207
12.22   Device Design Objectives and Dielectric Selection Options.................... 208
12.23   Other Potential Future High K Dielectric Films for NVM Devices 209
12.24   Integration of High K Films in Silicon Based CMOS
          NVM Technology 210
References 210

Chapter 13 Band Engineering for NVM Devices.. 213

          13.1   Revisiting Band Diagram and Band Engineering for
                    Conventional NVMs... 214
                    13.1.1   FG Devices... 214
                    13.1.2   Band Diagram for CT-SONOS Devices................... 216
                    13.1.3   Limitations of Conventional NVM Stack Designs 216

13.2 Band Engineering Using Single and Multilayer Dielectrics 217
    13.2.1 Single-Layer Tunnel Dielectric Films: Conventional SiO_2
        versus High K Films 217
    13.2.2 Multilayer Tunnel Dielectric and Associated Concepts of
        "Band Engineering" 218
    13.2.3 Band Engineering Objectives through Multilayered
        Dielectric Films 219
    13.2.4 Band-Engineering Design Options: Distinction and
        Classification Based on Single and Multilayered Tunneling
        Modes and Dielectric Thickness 220
13.3 Band-Engineering Options for Thicker Multilayer Tunnel Dielectrics 221
    13.3.1 Double-Layer Dielectrics 221
        13.3.1.1 Option A: Low K High Barrier/High K Low
            Barrier (D01/D10) 222
        13.3.1.2 Option B: High K Low Barrier/Low K High
            Barrier (D10/D01) 223
        13.3.1.3 Option C: High K High Barrier/Low K Low
            Barrier (D11/D00) 223
        13.3.1.4 Option D: Low K Low Barrier/High K High
            Barrier (D00/D11) 224
    13.3.2 Current–Voltage Characteristics of Double-Layer Dielectrics 224
        13.3.2.1 Other Examples of Double-Layer Dielectrics 225
    13.3.3 Crested Barrier and Variot Barrier: Triple-Layer Dielectrics 226
        13.3.3.1 Crested Barrier 226
        13.3.3.2 Variot Barrier 226
        13.3.3.3 Variot versus Crested Barrier for Charge
            Tunneling 228
        13.3.3.4 Combining Crested Barrier and High K Dielectrics 229
    13.3.4 Progressive Band Offset (Barrier) 230
13.4 Band-Engineering Options for Direct Tunnel Multilayer
    Tunnel Dielectrics 231
    13.4.1 Double-Layer Direct Tunnel Dielectrics 231
    13.4.2 Modified Double-Layer Direct Tunnel Barrier 232
    13.4.3 Triple-Layer PBO Direct Tunnel Barrier ("DT-PBO") 232
13.5 Applications of Band Engineering for Specific NVM Device Attributes 233
    13.5.1 Vpp Lowering and Associated Reduction in Programming
        Time and Power 233
        13.5.1.1 Floating-Gate and Floating Plate Devices 233
        13.5.1.2 Conventional NAND Flash Devices 234
        13.5.1.3 Two-Layer Tunneling Media with Oxide Replaced
            by High K Dielectric Interfacing Silicon 234
        13.5.1.4 Three-Layer Tunneling with Barrier Engineering
            with Oxide or with High K Interfacing Silicon 235
    13.5.2 CT/NC/DT NVMs 236
    13.5.3 Applications of Band Engineering in Other NVM Property
        Enhancements 237
        13.5.3.1 Programming Speed Enhancement through Band
            Engineering 237

13.5.3.2   Memory Window Enhancements through Band
                        Engineering 237
            13.5.3.3   "Trap Engineering" 240
    13.5.4   Memory Retention Enhancement through Band Engineering 240
            13.5.4.1   The TANOS Device 241
            13.5.4.2   Other TANOS-Follow-on Band-Engineered
                        Devices with Enhanced Retention 242
            13.5.4.3   Band-Engineered Stack Requirements to Achieve
                        Enhanced Retention: Devices Designed to Operate
                        in Fowler–Nordheim Tunneling Mode 243
            13.5.4.4   Multiple Examples of Band-Engineered Stacks
                        with Enhanced Retention 244
            13.5.4.5   Endurance Enhancement through Band
                        Engineering 246
13.6   Band Engineering for Direct Tunnel NVM Devices 247
    13.6.1   The Band-Engineered Multiple Direct-Tunnel High K Device 247
    13.6.2   Double Tunnel Junction DTM 248
    13.6.3   BE-SONOS DTM 249
    13.6.4   Resonant Tunnel Barrier DTM Device 250
    13.6.5   Progressive Band Offset DTM Device 251
13.7   Band Engineering for Multilevel (MLC) and
        Multifunctional (MF) NVMs 253
    13.7.1   Single Polarity Vpp for Both Writing and Erasing 254
    13.7.2   NVMs with Progressive Nonvolatility 255
    13.7.3   Multibit per Cell (MLC) NVM 256
    13.7.4   Multifunctional Cell (MFC) NVM 259
    13.7.5   Reverse Mode and MultiMode NVM Devices 260
            13.7.5.1   High K Reverse Mode Device 260
            13.7.5.2   High-Performance Dual-Mode Device 260
    13.7.6   Multimechanism NVM Devices 261
References 261

Chapter 14 Enhanced Technology Integration for NVM 265

14.1   Functional Integration at Interconnect and Packaging Levels 265
    14.1.1   System-In-Package: SIP 265
    14.1.2   System-On-Chip: SOC 267
    14.1.3   Other Examples of Functional Integration 268
14.2   NVM Integration at Memory Level 269
    14.2.1   Embedded NVMs 269
    14.2.2   Stand-Alone NVMs 269
14.3   Integration at Front-End-Of-Line (FEOL) Level 269
14.4   NVM Technology/Device Integration Schemes 271
    14.4.1   Compatibility with CMOS Platform (Embedded) Technology 271
    14.4.2   Compatibility with Gate Stack Design 271
    14.4.3   Compatible Planar Implementation: Illustration of Integration
            Schemes ("SMLK" Concept) 272
            14.4.3.1   Well and Isolation Schemes 272
            14.4.3.2   Compatible CMOS FET Gate Stack and NVM
                        Gate Stack 273

14.5 NVM Device Transition and Integration Challenges 276

14.6 Addressing NVM Device/Arrays Challenges and Integration 277

 14.6.1 NAND Flash Devices and Arrays.. 277

 14.6.2 NOR Flash and NROM Devices and Arrays 278

References... 279

Chapter 15 Planar Multilevel Storage NVM Devices.. 281

15.1 Multilevel NVM: Early Developmental History 281

15.2 Planar Floating-Gate MLC NAND Flash Devices................................... 282

 15.2.1 FG to FG Capacitive Coupling (Cell-to-Cell Coupling:
    CCC) for MLC NAND Design 282

15.3 GCR for MLC NAND Design.. 284

15.4 FG-MLC Cell Designs for Memory Levels, Stability, Sense Margin,
  and Reliability.. 284

15.5 Advanced Technology: FG-MLC Extendibility Challenges 287

 15.5.1 Sub-Lithographic Patterning.. 288

 15.5.2 Effects of Advanced Technology and Improved Program
    Algorithm on FG-MLC Device Design 289

 15.5.3 STI/Sidewall Fringing Field and Cell Geometry Effects for
    Planar FG Cell.. 289

 15.5.4 Controlling Read-Current Reduction..................................... 290

15.6 Planar and Wrap-Around Floating-Gate Flash Designs:
  Extendibility Issues... 290

 15.6.1 High K IPD and Metal Gate for Planar Conductive FG Cell.... 291

 15.6.2 GCR Reduction Yielding Smaller Memory Window 293

 15.6.3 Parameter Degradation: STI Edge-Fringing Field Effect........ 293

15.7 Current State-of-the-Art in Planar MLC FG-NAND Flash Device
  and Products... 295

 15.7.1 Planar Charge-Trapping MLC/MNSC Flash Devices 296

 15.7.2 Uniform Charge Storage (CT-UNSC or CT-CMLC) and
    Local Charge Storage (CT-MNSC) Cells................................ 297

15.8 Planar Charge-Trapping MLC Devices: Enhanced SONOS/MONOS...... 298

 15.8.1 Stack Design Options for Planar CT-MLC Devices 299

    15.8.1.1 Single Trapping Layer Options 299

    15.8.1.2 Multiple Trapping Layer Options............................ 300

 15.8.2 Advanced CT-MLC Stack Design Concepts with
    Multimode and Multilayer Trapping....................................... 300

 15.8.3 Three-Bit/Cell Storage 302

15.9 Additional Examples of CT-MLC Multilayer NVM Stack Designs 303

 15.9.1 DCSL Cell: 2 Bit/Cell MLC Technology............................... 303

 15.9.2 DSM Cell: 4 Bit/Cell MLC Technology 305

15.10 Planar Charge-Trapping MNSC Devices and Extendibility:
  MNSC-NROM and NAND 306

 15.10.1 Erase.. 307

 15.10.2 Sensing (Reading) 308

 15.10.3 Programming (Writing) 308

 15.10.4 Disturbs 309

 15.10.5 Endurance and Retention 309

15.11   Enhanced CT-MNSC Devices 309
         15.11.1   Band-Engineered MNSC Device 309
         15.11.2   Dual-Gated Enhanced CT-MNSC Device: 2 Bit/Cell
                    and 4 Bit/Cell Designs 310
15.12   Future of Planar MLC NVM Devices 312
         15.12.1   Future of Planar Multilevel Storage NVM Technology,
                    Device, and Products...312
         15.12.2   Multiplanar Stackable NAND Devices and Technology 315
15.13   Addressing Current Planar MLC FG-NAND Flash/SSD Limitations
         and Scalability Issues...316
         15.13.1   MLC Endurance (W/E Cycling)-Related Issues on Data
                    Integrity: ECC and Controller Mitigation 316
         15.13.2   Endurance and Overprovisioning 318
         15.13.3   Other MLC NAND Data Integrity Issues...................318
References 319

Chapter 16 Nonplanar and 3D Devices and Arrays 323

16.1   Vertical Channel and Nonplanar NVM Devices: Historical Evolution.....324
16.2   Nonplanar Multibit/Cell Vertical Channel CT Devices 325
        16.2.1   Trench-Based Vertical Channel CT NROM 325
        16.2.2   Vertical Channel DTM-NROM 327
        16.2.3   Vertical Channel FG-NAND 329
16.3   FinFET and Gate-All-Around (GAA) NV Devices....................329
        16.3.1   FinFET Technology and Devices...........................329
        16.3.2   Gate-All-Around (GAA) and Nano-Wired (NW) FET
                  (TSNWFET) and NVM Devices 336
16.4   Surround Gate NV Devices (SGT) 340
        16.4.1   Perspective of S-SGT and GAA Device Structures.................341
        16.4.2   Extendable S-SGT NV Devices and Arrays 342
        16.4.3   Multibit S-SGT Devices and Arrays 343
                  16.4.3.1   Surround Gate Floating Gate...............343
                  16.4.3.2   Stacked Gate-All-Around or Surround Gate
                             CT NROM (SGAA-NROM OR SGT-NROM)........344
                  16.4.3.3   SGT Multifunctional NROM....................344
                  16.4.3.4   S-SGT Very High-Density (3D) NAND Design......347
16.5   Full 3D NV Devices and Arrays...348
References 352

Chapter 17 Emerging NVMs and Limitations of Current NVM Devices 355

17.1   Device Level and Functional Level Attributes of DRAM, NVMs
        (SSDs), and HDD 356
17.2   The NVM Market Horizon and Driving Factors.....................356
        17.2.1   Observations...356
                  17.2.1.1   Consumer and Portables 356
                  17.2.1.2   Enterprise..359
                  17.2.1.3   Cloud..359
17.3   Emerging Contenders for Conventional Silicon Based
        NVM Memories..360
        17.3.1   PCM/PRAM 362

              17.3.2   STT-MRAM..362
              17.3.3   ReRAM or RRAM...363
       17.4   Requirements of Memory Attributes for Future Applications/Systems 363
       References...364

Chapter 18   Advanced Silicon Based NVM Device Concepts...367
       18.1   Device Parameter Enhancement Consideration 367
              18.1.1   Device Parameter Drivers...367
                        18.1.1.1   Memory Window and Charge Retention...............368
                        18.1.1.2   Charge Retention and Programming Speed..........368
                        18.1.1.3   Power Reduction..368
                        18.1.1.4   Endurance and Cyclic Reliability..........................368
       18.2   Application Parameter Enhancement Consideration 368
       18.3   Application Drivers for Embedded and Stand-Alone NVMs...................369
       18.4   Functional and Architectural Requirements and Grouping of NROMs.......370
       18.5   NVM Embedded Device Types and Options 370
       18.6   NVM Device Integration Options 371
       18.7   NVM Product and Stack Design Basic Consideration 372
       18.8   Advanced NVM Devices and Array Concepts..372
       18.9   Advanced NVM Devices and Arrays: Device Stack and Band Features..374
              18.9.1   Advanced NVM Devices: Explanations and Illustrations
                        of Devices and Band Features..377
       18.10 Scalable and Nonplanar NROMs...382
       18.11 MLS and Dense NROM Design Concepts: Both Planar and Nonplanar..383
              18.11.1   Planar MLS Potentials 383
              18.11.2   NonPlanar Cell Designs and MLS Potentials...........................385
       18.12 Advanced NAND Design Concepts: Planar and Nonplanar 386
              18.12.1   Embedded NAND Flash..386
              18.12.2   Advanced Stand-Alone NAND Flash.......................................389
                        18.12.2.1   Cyclability 389
       18.13 Advanced Nanocrystal Device Concepts...389
       18.14 Other Advanced MLC NVM Device Concepts..390
       18.15 Multifunctional NVM Devices..390
       References...391

PART III   SUM: Silicon Based Unified Memory

Chapter 19   SUM Perspective, Device Concepts, and Potentials...397
       19.1   SUM Device Perspective: Objective, Applicability, and Functionality.....397
              19.1.1   SUM Objective..397
              19.1.2   SUM Applicability 397
              19.1.3   SUM Functionality..398
       19.2   SUM Device Concepts and Classifications 398
              19.2.1   USUM and MSUM/URAM Definitions....................................398
              19.2.2   Historical Background of SUM-RAM Cells 399
       19.3   SUM Devices and Arrays in Memory Hierarchy 400
              19.3.1   USUM 401
              19.3.2   MSUM/URAM...401

19.4   Comparative Attributes of SUM versus Other Memories............................402
19.5   Application Advantages of SUM Devices and Technology.......................403
References...403

Chapter 20 SUM Technology 405

20.1   Consideration and Selection of Dielectric Films for SUM Devices 405
        20.1.1   Dielectric Layer Selection for Silicon Interface: Preferred
                  Film: OR-SiON..405
        20.1.2   Insulator-Metal-Interface (IMI) Layer Selection for Metal
                  Gate Interface: Preferred Film → TiN or TaN 406
        20.1.3   Charge Trapping and Charge Storage Layers for SUM
                  Technology: Preferred Films Are: Si_3N_4, or Si_2ON_2,
                  or GaN or a Combination 406
                  20.1.3.1   Role of I-SRN (IN-SRN) for Charge Storage of
                              SUM Devices 407
        20.1.4   Blocking Layer Selection for SUM Stack and CMOS FET
                  Gate Insulator for SUM Technology 407
        20.1.5   Tunneling Dielectric for SUM Technology 408
                  20.1.5.1   For USUM Device 409
                  20.1.5.2   For MSUM Device...410
20.2   Integration Scheme for SUM Technology..413
        20.2.1   SUM Technology and Device Interplay 413
        20.2.2   Technology for Planar and 3D SUM Devices and Arrays.............413
        20.2.3   Common Elements of USUM and MSUM.......................................414
20.3   Stack Designs for SUM Devices...414
        20.3.1   Guiding Device Concepts for Stack Design for
                  "CT-DTM" USUM 414
        20.3.2   Stack SUM Devices Design Examples of "CT-DTM" USUM 415
        20.3.3   High-Performance Stack Options for CT-DTM USUM
                  Device Design (L1 Applications)..416
        20.3.4   Stack Design Examples of CT-MSUM Devices...........................416
        20.3.5   Key Design Concepts for MSUM Stacks 419
        20.3.6   Parameter Objectives for MSUM Devices 419
References 420

Chapter 21 Band Engineering for SUM Devices...423

21.1   Band Engineering for USUM Devices...423
        21.1.1   PBO-DTM USUM Device 423
        21.1.2   Multimechanism-Carrier-Transport (MMCT) USUM Device......424
21.2   Band Engineering for MSUM Devices...426
        21.2.1   Background and Historical References 426
        21.2.2   Band Diagram Illustrations for MSUM Devices..........................427
        21.2.3   PBO Multifunctional CT-NVM → PBO-CT-MSUM:
                  1.1 of Table 20.3...427
        21.2.4   VARIOT-CT-MSUM 1.3 of Table 20.3 428

21.2.5 CRESTED Barrier CT-MSUM (Table 20.3) →
        CRESTED Barrier-CT-MSUM 2.1 of Table 20.3 429
21.2.6 CRESTED Barrier Multifunctional CT-NVM [40] →
        CRESTED Barrier-CT-MSUM 2.3 of Table 20.3: Dual Carrier
        Source Design 430
References 431

Chapter 22 Uni-Functional SUM: The USUM Cells and Arrays 433

22.1 The FB RAM USUM Cell 433
      22.1.1 The FB RAM Cell: (The Floating Body-Charge Trapping
              USUM RAM Cell) 433
              22.1.1.1 Mode of Body-Charge Generation 434
              22.1.1.2 Reading the Memory Cell 435
              22.1.1.3 Processing of FB RAM Device Structure 435
22.2 The GDRAM USUM Cell 437
      22.2.1 The Diode I-V Characteristics 438
      22.2.2 Cell Operation 440
              22.2.2.1 Memory Cell 22.9 (a): Gated N+-i-P+ Diode 440
              22.2.2.2 Memory Cell 22.9 (b): Gated P+-i-N+ Diode 441
      22.2.3 SOI Technology Implementation of the Diode Charge
              Storage RAM Cell 441
      22.2.4 SOI Diode Charge Storage RAM Cell Operation 443
      22.2.5 Process Considerations for the Gated P-i-N or N-i-P Diode
              Charge Storage USUM RAM Cell Bulk Versions 444
22.3 The GTRAM USUM Cell 444
      22.3.1 High Performance SOI Nonvolatile Two-Device SRAM
              Using Floating Body Charge 444
22.4 The CPRAM USUM Cell 445
      22.4.1 The CPRAM Cell: (The Cross-Point Thyristor-Based Charge
              Storage RAM Cell) 445
      22.4.2 Cell Operation 448
22.5 The FET USUM Cells 449
      22.5.1 Configurations and Operations of Memory Cells and Arrays
              for CT-NVM USUM Devices 450
      22.5.2 Planar Merged Two-Device (Split-Gate) CT-NVM USUM
              NROM Cells 451
              22.5.2.1 Planar/Vertical and Vertical-Channel Single
                        Device and Merged Two-Device (Split-Gate)
                        CT-NVM USUM NROM Cells 452
              22.5.2.2 CT-NROM USUM Array Layout 453
              22.5.2.3 Operational Schemes of CT-NROM USUM Cells 453
      22.5.3 High Density CT-NAND USUM CELLS 455
      22.5.4 Very High Density Vertical-Channel CT-NAND USUM
              Configurations 455
References 456

Chapter 23 Multifunctional SUM: The MSUM Cells and Arrays 459

23.1 Integrated DRAM-NVRAM Multilevel and Multifunctional
MSUM Cell 460
23.1.1 Multifunctionality 461
23.1.2 Operational Attributes 462
23.1.3 Array Layout 463
23.2 The Band-Engineered DTM MSUM Cells and Arrays 463
23.2.1 General Considerations for DTM MSUM Cell Design 463
23.2.2 MSUM Cell Band Characteristics for NROM and NAND
Applications 464
23.2.3 CT-MSUM Cell and Array Configurations 465
23.3 Other CT-MSUM Devices 466
23.3.1 Nanocrystal-Based CT-MSUM Device 466
23.3.1.1 Nanocrystal Size, Trapping Characteristics, and
Estimated Memory Window (Shift in Vth due to
Trap Filling) 468
23.3.1.2 Operational Scheme for Nanocrystal-Based MSUM
Device of Figure 47 and Associated MLC Memory
States 470
23.3.1.3 DRAM MLC States of 11D/10D/01D and
Reference State 00 470
23.3.1.4 NVRAM MLC States of 11NV/10NV/01NV and
Reference State 00 470
23.3.2 Reverse Mode CT-MSUM Devices 471
23.3.2.1 Multilayered Conductive Floating/Floating-Plate
MSUMs 472
23.4 URAM and Other Multifunctional Silicon Based Memories 473
23.4.1 The URAM and "Universal" Devices from KAIST 473
23.4.2 Other MSUM/UNIFIED Memory Concepts 476
References 477

Chapter 24 SUM Functional Integration, Packaging, and Potential Applications 479

24.1 SUM and NVM Integration Perspective: Chip and Packaging 479
24.2 Integration at Silicon Technology/Chip Level 479
24.3 Integration at Packaging Level 480
24.4 Integration at Large System Level 481
24.5 Integration at Functional and Architectural Level 482
24.5.1 SUM and NVM Application Perspective 483
24.6 Current NVM and DRAM Market: Potential SUM Applications 483
24.7 Advanced Applications 484
References 485

Conclusion 487

Appendix: Rare Earth Metal-Based Future Dielectric Thin Films for SUM Devices 489

Index 499

Foreword

Solid-state memories were invented in the 1960s with the rise of the semiconductor industry and grew to replace the magnetic core memories of the 1950s. They are segmented into two main categories: volatile and nonvolatile memories (NVMs).

The volatile category was and still is dominated by dynamic random access memories (DRAMs) and static random access memories (SRAMs). The disadvantage of *volatile* memories is that the information stored in the memories is lost when power is removed from the device. The advantage of volatile memories is that they are capable of reading and writing an infinite number of times.

The SRAM is dominated by a very expensive six-transistor cell, which has a large area and is the highest speed memory in existence. It is typically placed on the same die as the microprocessor and serves as several layers of cache memories tasked with feeding data to the microprocessor as fast as possible. It sits at the peak of the memory hierarchy.

The DRAM serves as the workhorse memory storing large amounts of data actively being manipulated by the microprocessor. It is far less expensive than SRAM but is substantially slower. Typically, overall system performance is handicapped by the performance of the DRAM, and great effort has been exerted to overcome the handicap of DRAM performance. The DRAM sits at the second level of the memory hierarchy tasked with feeding the processor with large volumes of data.

The options for innovation and alternative structures for DRAM and SRAM fundamentally are limited because of the simple structure of these devices. The differences in the implementation of DRAM and SRAM technologies between manufacturers is very small and almost indistinguishable to nonexperts.

In contrast, NVMs can be implemented in a vast spectrum of different ways using many different potential structures. The result has been an explosion of diversity similar to life itself.

NVMs fulfill a vast spectrum of applications from a small amount of memory necessary to make a simple device like a microwave oven work, to provide the memory function to a massive solid-state disk drive capable of storing terabytes of data.

The major challenges of NVMs include the following:

1. It has the ability to achieve high endurance (read and write cycles).
2. It has the ability to retain the data without power for years at high temperatures.
3. Fundamentally, from a thermodynamics viewpoint, nonvolatility requires more energy than a volatile memory to achieve data retention. The challenge to the engineer is to be able to achieve nonvolatility in the most energy-efficient manner.
4. The cost structure of NVMs is a challenge to all manufacturers. The ability to produce a product that can meet the customer's requirements while maintaining profitability is a fundamental requirement to stay in business in a global competitive environment.
5. There is a myriad of second-order problems: read disturb, pattern dependence, disturb, and so on, which have to be exhaustively studied and solved by the NVM engineer. The complexity of these issues is vast and requires an extreme effort to solve.
6. NVM technology has achieved the ability to store two or three bits per cell requiring the memory technology to have the capability of operating with and distinguishing four or eight separate memory states, which complicate all the other NVM issues enormously.

NVM has been segmented into two major categories:

1. NOR memories tend to be of lower density and higher cost per bit. NOR has the advantage of being byte addressable allowing small processors to execute the code directly from the NVM. It fills a very large amount of lower order applications giving intelligence to many everyday electronic devices. Recently, it has been declining in market share as the explosive growth of NAND has cannibalized its market. NOR technology advancement has essentially stagnated because the current planar NOR can no longer be scaled to lower cost structures.

2. NAND memories have been able to achieve enormous cost reductions enabling the technology to store large volumes of data. NAND has enabled the advanced devices that we take for granted (digital cameras, cell phones, tablets, laptop computers, and many more devices). It has become the dominant form of NVM on the planet.

The first part of the book covers a historical perspective on NVM memories. The different types of NVM devices are reviewed. A heavy focus area is the physics of charge trap flash (CTF) memories, which is quite complex. The digital state in a CTF is stored in a thin insulating layer that traps and detraps electrons. This section will enable the reader to grasp these difficult concepts. The materials science of early CTF memories was difficult because the thin films useful for NVM applications and available to the materials scientists were extremely limited. There was basically only silicon dioxide and silicon nitride. The processing technologies were also limited to just a few techniques such as thermal oxidation and chemical vapor deposition. A great deal of experimental work was necessary to successfully develop a useful CTF cell. These early limitations made it difficult to deliver a commercially viable charge trap product, which resulted in the dominance of the floating-gate technology. The floating-gate cell stores electrons on a minute region of polysilicon instead of in an insulating layer.

In recent years, the field of materials science has advanced tremendously. Materials scientists now have most of the useful elements of the periodic table available for engineering films to advance the state of the art of NVM. Atomic layer deposition and other processing techniques now allow layering of films at the atomic level. This allows new film capabilities where superlattices of mixtures of single atomic layers of dielectric materials can be placed in the NVM films. New electrical properties can be created and controlled enabling enhanced performance of the NVM. Recent NAND memories have used high-K dielectrics in their formation, benefiting from the advancement of materials science. Ultra-thin multilayered high-K dielectric films can now be readily integrated within the framework of charge trap memories providing potential solutions for NVM scalability but also addressing solutions for (1) endurance, (2) energy reduction through voltage scaling, (3) enhancement in charge retention, (4) disturb mitigation, (5) avoidance of parasitic coupling and associated disturb, and (6) enhanced Multi-level Cell (MLC) capability through localized trapping over diffusion nodes (interchangeable source and drain nodes) yielding higher sense margins for MLC capabilities.

Charge trap memories are now becoming the main stream as NAND makes a transition from planar to three-dimensional (3D) memories. The simplicity of CTF has advantages in 3D structures over the floating gate. These 3D memories include the formation of atomic layer structure. The charge trap memories provide ease of integration with a scaled baseline CMOS technology in both planar (two-dimensional) and nonplanar (3D) configurations. Direct tunnel transport-based charge trap memory designs exhibit the potential of providing a significant enhancement in energy efficiency and device reliability.

Appropriate combinations of the technical techniques described previously potentially provide future NVMs with both MLC capability and multifunctional capability in some applications combining the functionality of volatile SRAMs and DRAMs as well as conventional NVMs. This potential is suggested in the third part of the book and is broadly termed as silicon based unified memory (SUM).

It may be noted that the book is divided into three parts:

Part I: Historical progression up to the year 2000

Part II: Scalability and extendibility of CMOS technology and NVMs post 2000 (including high-K gate insulator and work function-tuned metal gate FET-CMOS technology, and nonplanar device technology)

Part III: Future SUM-NVM device concepts and possibilities

I hope you enjoy this book and learn from it as I have.

Kirk Prall
MICRON TECHNOLOGY
Boise, Idaho, USA
February 2016

Preface

Silicon based digital electronics have impacted human lives unprecedentedly in human history. Yet it is only half-a-century old. This justifiably called "The Digital Age" is ushered by the rapid development and application of silicon based integrated digital logic and digital memory. Central to this development is the successful miniaturization of the metal–insulator–silicon transistor at the core of digital logic and digital memory. This miniaturization is known as "scaling" and has followed "Moore's Law"[*] for doubling the component capacity in every two years for over four decades. As a result, the capacity for processing and storing information over a piece of silicon increased more than a millionfold. Product cost got drastically reduced vastly expanding applicability and affordability. This phenomenal growth of silicon based digital electronics is expected to continue for at least another two decades.

Digitally stored memories in electronic systems today are hierarchical to optimize speed, power, and functionality of such systems. Both "volatile" and "nonvolatile" memories (NVMs) may coexist even on a "system-on-chip" solution. By definition, volatile memories lose information when the power is not provided, whereas NVMs retain such information. Depending on the speed and capacity requirements of processing information, a unit of digital information (bit) is stored in a variety of memory "cells." These units could be built not only into the processor logic system and in discrete registers, but also in memory arrays. Multiple memory arrays may provide varying performance, density, and volatility all existing within a silicon chip. Alternately, such memory arrays are integrated into multiple silicon memory chips. Large amount of information is also stored in non-silicon medium such as magnetic discs. For processing, information stored is accessed either randomly or serially depending on the characteristics of the memory cell. Randomly accessible memory arrays are significantly faster, whereas serially accessible ones are significantly slower. Key volatile memory arrays are customarily labeled in terms of accessibility of information: static random access memory (SRAM) and dynamic random access memory (DRAM). Key NVM arrays are labeled in terms of logical framework of information stored: NOR read-only memory (NROM) and NOT-AND (NAND) flash memory. NROM is a randomly accessible memory, whereas NAND is a serially accessible memory. Although traditional SRAM memory cell requires six transistor elements, DRAM cell requires a single switching transistor integrated with a storage capacitor. NROM and NAND memory elements could be incorporated within a single transistor. However, NROM is often configured into an integrated form of two-transistor memory cell. Consequently, NAND memory today is the densest with significantly higher storage capacity, whereas NROM, DRAM, and SRAM are typically less dense by factors of 1.5, 2.5, and 12, respectively. Accessibility of information is faster by many orders of magnitude in SRAM and NROM. Accessibility of information in DRAM is also much faster compared to NAND. Alterability of information is faster in SRAM, followed by DRAM, and is unlimited. Alterability of information in NROM and NAND NVMs is limited severely restricting its applicability.

In spite of their inherent limitations in performance and alterability of information, silicon based NVMs had been increasingly incorporated in memory hierarchy. Both NROM and NAND have expanded their application base in every decade since their inception. Several factors have favored enhanced application of such NVMs as well as nonvolatile storage of information. Besides the reduced cost per bit of memory, the other factors are (1) simpler memory cell structure with proven record of scalability and feature size reduction, (2) capacity to store multiple bits of information within a single memory cell, (3) compatibility with baseline digital logic technology, and (4) applications of

---

[*] Gordon E. Moore, Cramming more components onto integrated Circuits, *Electronics Magazine*, April 19, 1965. McGraw-Hill.

logic circuitry and memory system architecture to ensure information integrity and improvement in accessibility. In the year 2000, nearly 70% of the memory revenue worldwide of approximately $100 billion came from DRAM memory alone. DRAM technology drove the scaling of digital electronics in the early decades. In the year 2011, NVM NAND flash revenue exceeded that of DRAM. NAND flash technology drove the technology scaling even earlier. The cost/bit of NAND flash memory was reduced to nearly one-fifth of that of DRAM in 2012. Application of the flash bit grew also by nearly five times during 2012. The majority of the flash memory bits were in the form of NAND flash, which grew 100 times from 3E17 bytes in 2002 to 3E19 bytes in 2012!!

This book was conceived with the objective of extending the momentum gained in the application of silicon based NVM devices and technology in the foreseeable future by reducing or eliminating the current limitations in memory cell design. It treats the evolution of silicon based NVM technology primarily from the basic device concepts, memory cell design, and process technology integration. It does not deal with circuit design and memory system architecture and associated design in any depth. The book is divided into three parts. Part I consisting of 10 chapters provides an in-depth coverage of conventional NVM devices, stack structures from device physics, and historical perspectives as well as identifies limitations of conventional devices. Part II consisting of eight chapters reviews the advances made since 2000 in reducing and/or eliminating existing limitations of NVM device parameters from the standpoint of device scalability, application extendibility, and reliability. Part III consisting of six chapters proposes multiple options of silicon based unified (nonvolatile) memory cell concepts and stack designs (silicon based unified memories) for potential applicability covering the entire memory hierarchy spectrum. The SUM NVM cell concepts and structures are unified from the following standpoints:

1. Compatibility with the current and future silicon based metal gate FET CMOS technology
2. Applicability with both planar and nonplanar or three-dimensional devices and technology integration scheme
3. Extendibility of NVM memory cell design beyond the current limit of 3 bit/cell
4. Expandability of application of NVM devices for both high-performance requirements (replacing SRAMs and DRAMs) and high-capacity requirements (replacing magnetic storage mediums)
5. Providing multifunctionality within the unified cell design
6. Aiding potentials for the development of future memory subsystem architecture as well as digital system architecture for novel applications through circuits and system developments

It should be noted that in recent years, several NVM devices have emerged where the "nonvolatile" element is not FET based. These NVMs are grouped as a whole as "emerging NVMs" in this book. These include MRAM, PCM, RRAM, STTRAM, and molecular memories. These are not treated in any depth in this book and are discussed briefly in the context of silicon based NVMs only.

Arup Bhattacharyya

Acknowledgments

I must admit that this book project, in many ways, is a very personal story spanning nearly five decades, and apparently with multiple forces subtle and not so subtle paving the way. I am keenly aware, with regret, that the acknowledgment would, at best be well intended, and at worst, be incomplete and unconventional.

Reflecting back, the history of evolution of silicon based nonvolatile memory (NVM) is intimately linked to nearly the entire span of my professional life as a technologist. My interest in solid-state memory development led me to join the newly established research and development (R&D) group of IBM Burlington, Vermont, in the late 1960s. Within few years, I found myself involved in setting up a small 6-μm silicon based LSI R&D facility, for the first time, in IBM, Burlington, Vermont, under the inspiring patronage of late Dr. Benjamin Agusta, who organized the advanced R&D group pulling together some of the youngest and brightest device and circuit designers and technologists of our time under one umbrella. He fostered an ambitious mission and vision for us to challenge the roles of IBM's Thomas J. Watson Research Center team and "Ma Bell's" (A T & T Bell Research Laboratory, Murray Hill, N.J.) formidable teams in semiconductor-based memory developments!! My interest in silicon based memory device and technology grew with Dr. Agusta's support and passion for development and his patronage to embark on an R&D effort to develop an "MXOS" version of 8 kb alterable read-only memory (a challenge to the "MNOS" version from Bell Labs) in the early 1970s. Another leader whose patronage helped me shape my lifelong interest in memory technology development was Mr. Jim Webster, who was the laboratory director and subsequently served as General manager for IBM General Technology Division, Burlington. I had the privilege to serve him briefly as his technical assistant. His long-range vision, extraordinary nurturing, and compassionate spirit reflected in his support and encouragement for "off-the-record" innovative efforts have left an everlasting impression on me and motivated me and our team to develop new materials (e.g., silicon-rich nitride) and new NVMs (e.g., first nano-crystal charge-trapping NVRAM device) during the late 1970s and the early 1980s, over and beyond the scheduled and planned items. I take this opportunity to express my immense sense of gratitude to late Dr. Agusta and Mr. Webster for their trust, support, and encouragement in our exploratory missions. I also wish to express my sincere thanks to my immediate manager at that time, Dr. M.L. Joshi, for his prudent support in technology developments in the early 1970s. Although IBM management did not embrace any productization plan for NVMs during the 1970s through the 1990s, their encouragement in CMOS technology scalability helped me and our team to successfully learn and develop deep submicron CMOS technology during the 1980s (many years ahead of the industry) and subsequent follow-ons in the early 1990s (applied initially toward CMOS logic chips). Some of the scalability and integration concepts discussed in the book and applied for advanced nonvolatile technology emerged from such learning. I express my gratitude to two individuals for that: Dr. Wolfgang Liebman, General Manager, Burlington, during the 1980s for his patronage and Dr. Robert Dennard, a friend, colleague, and inventor of dynamic random access memory for his mentoring and involvement in the subject during that time. My interest in NVMs would have stayed dormant if not for a very dear colleague, late Dr. Paul Farrar Sr. from IBM (an earlier Micron Advanced Research Institute [MARI] fellow) insisted in the year 2000 (I was involved with Motorola Austin, Texas at that time to develop the state-of-the-art logic technology) that I accepted an invitation to aid the memory technology for Micron Technology in Boise, Idaho. My interest in NVM resurfaced when I accepted the invitation to be a MARI fellow in 2001, and when Micron Technology was actively involved in developing NAND flash technology. In many ways, I feel indebted (directly or indirectly) to late Dr. Farrar for reviving my interest in NVMs. Under the active patronage of Dr. Mark Durcan, V.P. MICRON Technology, Boise, Idaho (currently, CEO, Micron Technology) and the leadership of Dr. Kirk Prall, director of flash development, I served as a consultant to the flash development team for several years. In addition, as a MARI fellow,

I was provided the opportunity to help develop the intellectual property of Micron Technology in silicon based microelectronics and nanoelectronics, and in particular in the areas of future NVM devices. I am deeply thankful to Drs. Durcan and Prall for inviting me to be associated with Micron Technology, and to present seminars on NVM devices and technology to my Micron Technology colleagues. My sincere appreciation and thanks also go to Miss Kelly Maculley (Brewer) and late Charles Brentley of Micron Technology's legal department for their prompt support in providing me patent-related materials and to help me build the patent portfolio. I express additional special appreciation and thanks to Dr. Prall for his generous support and encouragement in writing this book, for technical reviewing and editing the book, and for graciously agreeing to write the Foreword for this book. Among my other Micron Technology colleagues, I am especially grateful for the suggestion and forthright support for this book I received from Dr. Chandra Mouli, Manager, technology and product reliability, toward many aspects of the book. I am also indebted to my friend and associate, Dr. Perry Pelley, previously manager of memory array design and an active inventor from Motorola Inc. (subsequently associated with the intellectual property (IP) group of Freescale Semiconductor and recently NXP, Austin, Texas), for his suggestion and review of my book at several phases. The book proposal was previewed by several well-known experts in the field of memory in general and NVM in particular, including Prof. Sanjay Banerjee, Cockrell Family Regents chair–professor of electrical and computer engineering, University of Texas at Austin, Austin, Texas; Prof. Joe Brewer, University of Florida at Gainesville, Florida, who is the editor and author of several NVM books published by IEEE; John Wiley; Dr. Howard Kalter, technical and IP consultant, who is a leader in memory chip design previously from IBM Corporation; and also by Dr. Chandra Mouli; and Dr. Perry Pelley. I am especially thankful to Prof. Sanjay Banerjee for his technical review of the manuscript during preparation and Prof. Joe Brewer for his support in the initial phase of manuscript preparation. I express my thanks to Mr. Ken Moore of IEEE publication, for lending his prompt support for granting permissions for multiple scores of IEEE publication materials used in this book.

My original intention was to initiate the book project in 2007 when I completed my involvement with the Micron Technology and to publish the book within the time frame around 2010. I had to postpone the project due to my health problem for several years and started the project in 2012 after my recovery. After the book project got approved, it took me a while to get acquainted with the current world of digital publication. I had no previous experience of the effort it entails coming from an "old school," which effectively took an additional year. I was fortunate to receive education, guidance, and support from John Blackman. He was God-sent to me, and I am very thankful to this young and energetic software application specialist in helping me in every possible way to satisfy the manuscript preparation requirements. Much of the credit in transforming my work into digitized form belongs to him, and I remain indebted to him. I am also very thankful to my friend and longtime neighbor, Bruce Blackman, from Essex, Vermont, and to Joshua Rose from Tennessee for their considerable help in providing me high-resolution digital artwork. My special thanks and warm appreciation are also due to Miss. Nora Konopka, the technical editor of the CRC Press, and Miss. Kyra Lindholm of the technical support team of the CRC Press (Taylor & Francis), for their many faceted support in getting the manuscript off the ground. I cannot overstate Miss. Konopka's prompt support, care, patience, and consideration in helping me to prepare the manuscript in spite of her very busy schedule. She also helped immensely in educating me of all the nuances of the modern world of digital publication. Thanks are also due to Dr. Lipeng Cao of Qualcom, San Diago, California, who also helped me in reviewing and providing me inputs from a reader's perspective. I also thankfully acknowledge the support for expeditiously digitizing multiple scores of hand-drawn and publication figures by Mr. Jakir Pathan of IP Design Tech, Pune, India, and Maureen Besede of Essex, Vermont, who made my task manageable. Thanks are also due to Mr. and Mrs. Dinesh and Indrani Jaiswal, my dear relatives, who helped me in arranging support from IP Design Tech of India. I express my thanks also to a dear friend and a past IBM colleague, Dr. Albert Bergendahl, who helped me provide many relevant technical publications from Essex Junction, Vermont.

Last but not the least, my sincerest gratitude and thanks go to my dear wife and life partner for over past 50 years, Dipa Bhattacharyya. Without her constant vigilance, tough love, exemplary discipline, and support throughout my marital life and especially the past few years, I would have never succeeded to make this book project a reality in this life!

Arup Bhattacharyya
Essex Junction, Vermont

Author

Arup Bhattacharyya was born in Varanasi, India, and graduated with a Bachelor of engineering from Bengal Engineering College of the University of Calcutta, Kolkata, West Bengal, India, in 1960. After a brief training in TISCO, Jamshedpur, India, Dr. Bhattacharyya moved to the University of Pennsylvania, Philadelphia, Pennsylvania, with a student research fellowship to pursue graduate studies at the Laboratory for Research on Structure of Matter of the University of Pennsylvania. After completing PhD, he joined The Foxboro Company, Foxboro, Massachusetts, as a staff physicist in January 1966, where he developed magnetic thin-film sensors for process control applications. Dr. Bhattacharyya joined IBM Advanced R&D Laboratory, Burlington in December 1968, subsequently held many technical and management positions, and pioneered many generations of VLSI and ULSI technologies in IBM, related to microelectronics development. One of his major technological achievements include IBM's successful transition from CMOS-VLSI era to CMOS-ULSI era during mid-1980s, under his technical leadership, which was many years ahead of the VLSI industry at that time. During his tenure at IBM, he also volunteered his service as a United Nations Development Project's (UNDP) Technical Advisor to India and as an associate Dean of Engineering at the Washington State University, Pullman, Washington promoting education in Science and Technology amongst Women and Minorities. He retired from IBM in June 1996 and took a senior technical management position in Motorola, Austin, Texas, where he led the development effort in SOI technology. In addition, he served as a MARI (Micron Advanced Research Institute) fellow, from 2001 to 2007, and as a consultant in Micron Technology, Boise, Idaho, where he helped the development effort of flash memory technology and products. Dr. Bhattacharyya and his wife, Dipa, have been the residents of Essex Junction, Vermont, since 1968. They have two children and two grandchildren, who were born and raised in the United States. He enjoys living in rural green mountains of Vermont, walking in the woods, gardening, hiking, and tinkering with concepts related to "green energy."

Part I

Conventional Silicon Based NVM Devices

INTRODUCTION

The first part of this book provides a basic background and reviews key device elements and characteristics of conventional silicon field-effect-transistor (FET)-based nonvolatile memory (NVM) devices. This part also describes the development and progression of fundamental features, memory cell and memory array design attributes, technology integration schemes, unique properties, and reliability challenges of conventional NVM devices since the inception of such devices through the turn of the century. The part concludes with a brief assessment of extendibility limitations of conventional NVM devices.

Chapter 1 introduces the basic concepts behind current digital memory products: both volatile memories such as registers, SRAMs, and DRAMs, as well as NVMs such as NROMs and NANDs. This chapter covers at elementary level the device concepts, the memory cell and array concepts, and the current memory hierarchy schemes for digital systems. In addition, it also familiarizes the naming schemes (nomenclature) employed to distinguish different types of NVM devices.

The introductory chapter is followed in Chapter 2 by a review of the key historical progression made in different types of FET-based NVMs. The review leads to the development of current "conventional" NVM devices and products. This chapter emphasizes different "tunnel-based" NVM devices which include NROMs, NAND-FLASH, "Charge Trap" (CT) devices, "Nano-Crystal" (NC) devices, and "Direct Tunnel Memory" (DTM) devices.

Chapters 3 and 4 describe respectively the general properties of dielectric films and interfaces for the NVM device stack designs, and the detailed assessment of transport and trapping characteristics of the most widely employed conventional dielectric films in device stacks. Additionally, process sensitivity on relevant dielectric properties for NVM devices and required interface characteristics at both silicon and gate interfaces are discussed in Chapter 4 to enable NVM gate stack designs. These chapters provide the necessary basic background to recognize the significance of properties of dielectric films and interfaces in imparting unique NVM properties from technology integration and device design perspectives.

Chapter 5 describes in detail the unique properties associated with FET-based NVM devices. These include memory window and memory state stability associated with nonvolatility ("retention"), and stress-stability associated with cycling between the memory states ("endurance"). The concept of band diagram in explaining the energetics of electronic charge transport and trapping through various interfaces and dielectric layers was introduced in this chapter and was used to explain NVM device properties.

Chapter 6 explains in detail various elements in NVM device stack designs and applications of band diagrams to optimize device properties. Additionally, array schemes for NAND flash and NROM NVM devices were discussed and parasitic coupling effects between adjacent cells were explained. Nonplanar FinFET NVMs and multi-bit per cell NVM device concepts in CT devices were introduced.

Chapter 7 discusses the evolution and progression of NVM cells, arrays, and associated products and explained the operational characteristics of such device designs. These include floating gate EEPROMS, NAND flash, and NROM NVMs. Origin of various disturb mechanisms in array designs is explained and mitigation solutions are addressed.

Chapter 8 provides brief introduction of CMOS-based process technology evolution and integration highlights for NVM technology.

Chapter 9 briefly covers key reliability issues and radiation immunity considerations for conventional NVMs.

Finally, Chapter 10 addresses fundamental extendibility challenges associated with conventional NVMs.

1 Silicon Based Digital Volatile and Nonvolatile Memories
An Introductory Overview

CHAPTER OUTLINE

An introductory review of currently employed silicon based digital memories, both volatile and nonvolatile types, is provided here. The basic device, memory cell, and array concepts are discussed in this chapter. Memory hierarchy schemes for current digital systems are introduced. Naming conventions for different types of nonvolatile memories (NVMs) are explained.

1.1 DIGITAL MEMORIES AND BINARY STATES: BASIC CONCEPTS

Silicon based digital memories are an information storage device where such information is stored in unit memory cells. Currently, these unit memory cells are built with the complementary metal–oxide–silicon (CMOS) technology wherein each memory cell consists of one to several CMOS transistors. Information is stored within each memory cell in the form of either of the two binary states called bits. These binary states are conventionally called "0" or "on" memory state and "1" or "off" memory state. The binary state of any memory cell is sensed by associated built-in circuitry within the memory device. This mode of sensing the state is called "reading" the memory state of any specific memory bit. Bits of binary memories are stored in many forms of CMOS memory cells generally classified as "registers" and "memory arrays." Registers are not discussed in this book because such memories are integrated into logic components and do not contain addressing and decoding schemes required to specify the memory cell. Memory arrays provide a built-in circuitry to address and decode any specific memory cell within the memory array. Memory cells and memory arrays capable of storing binary bits are the subject matter of this book.

A large amount of information is stored in unit memory cells laid out typically in the form of rectangular planar arrays of memory cells built in the silicon CMOS technology. These memory cells could be accessed either randomly or serially through an appropriate addressing scheme. The process of altering the memory state of a memory cell from "1" to "0" or from "0" to "1" is called "writing" the memory cell or "programming" the memory cell. The process of removing the information contained within a single memory cell or simultaneously within a large segment of memory cells in an array is often called "erasing." It has been implied earlier that the memory arrays and the associated silicon memory chip(s) contain not only the memory cells and arrays but also built-in functional memory circuits for addressing and decoding the memory cells. Additionally, the memory chips also provide a built-in circuitry for sensing and programming the schemes for the memory cells. Memory cells are typically organized in terms of "word length": identified as N, which is defined by powers of 2 (binary digit). Memory address scheme: identified as M is also organized in terms of powers of 2 on the memory chip. Memory functional blocks typically containing memory arrays, circuit blocks for addressing and decoding memory cells, as well as sensing and programming blocks. These are schematically identified for a memory (DRAM) chip in Figure 1.1 [1]. In addition to the typical functional blocks discussed above, the figure also shows the integration of a built-in read-only memory (ROM) block to be discussed in Section 1.2.3 later on. The ROM block was employed for built-in self-test (BIST) for the memory chip.

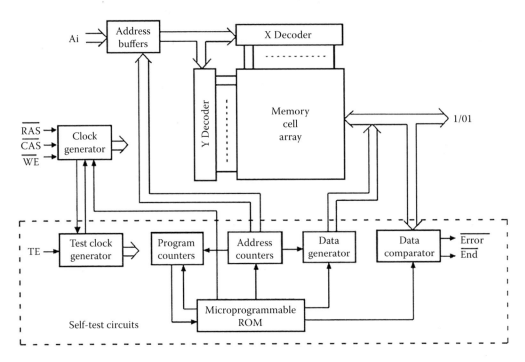

FIGURE 1.1 Functional blocks for a DRAM memory chip which additionally includes BIST and programmable ROM blocks. (From Takeshima, T. et al., A 55 ns 16 Mb DRAM, *IEEE ISSCC*, p. 246 © 1989 IEEE.)

For a specific memory chip containing M bits of addressing, the number of addresses on the chip would be 2^M consisting of N-bit words, providing a chip memory capacity of $N2^M$ bits [2]. Currently, the memory bit capacity of silicon chips could be as high as 256 gigabits (256 billion bits). In multiple chip packages, the memory capacity of silicon based memories has increased up to terabytes (TB or trillion bytes) today (1 byte = 8 bits; a byte is represented as "B," while a bit is represented as "b").

1.2 VOLATILE MEMORIES AND NVMs

Silicon based digital memories have evolved in many forms over the years based on the progression of technology associated with the feature size reduction of the transistor, considered as the core element in a memory cell. The latest of such transistor technology had been the CMOS-field-effect transistor (FET) technology, which dominated the digital memory products since the 1980s. Such memories could be either "volatile" where the binary memory states are lost if the power supply is withdrawn or "nonvolatile" where the memory states could be retained even when the power supply is removed. NVMs, therefore, have an intrinsic advantage over volatile memories in retaining the information in digital systems in case of a power failure. However, as pointed out in the preface of this book that the current NVM products operate significantly slower than the volatile memories thereby restricting system applications if solely designed with NVMs. Therefore, current systems employ both volatile memories and NVMs in a hierarchical order, which will be explained later on in Section 1.3. One primary objective of this book is to reduce the speed limitations of current NVMs, thereby expanding the applications potentials of such memories in future digital systems.

The most prominent volatile silicon based digital memories employed in systems today are static random access memory (SRAM) and dynamic random access memory (DRAM). Despite of the volatility, SRAM and DRAM influenced the design of the digital processors and digital system architectures more profoundly than any other types of memories. A brief introduction of SRAMs and DRAMs is made here, which includes the respective memory cells and arrays. Except for relative comparison with NVM cells and arrays, these memories are not discussed in depth in this book. Readers are encouraged to study in depth such memories because such topics are beyond the scope of this book.

1.2.1 Static Random Access Memory: SRAM

SRAMs are currently provided as embedded memories and integrated into logic systems, register files, microcontrollers, microprocessors, high-performance processors, and virtually all digital systems. The extent of SRAM bits employed is dependent on the specifics of design and application requirements. SRAM cells and arrays are differentiated from those of DRAMs because the state of SRAM cells exhibits stability (remanence) compared to that of DRAM cells, the latter requiring periodic refresh of the memory states—hence the names "static" for SRAM and "dynamic" for DRAM. However, SRAMs are still volatile because the information stored in SRAM cells is eventually lost when the memory is not powered. SRAMs operate significantly faster than DRAMs and are coupled directly to the logic circuit elements of processors and controllers to optimize the system performance. SRAMs are significantly more expensive per bit compared to DRAMs because the memory cells are least dense and require larger number of transistors compared to DRAM. High performance requires more power consumption for SRAM, although SRAM may consume less power during standby compared to DRAM.

The conventional SRAM cell consists of six transistors: two pairs of cross-coupled invertors (four transistors) at the core act as "flip-flops" and define the memory state, while the two other external transistors serve to select the memory cell. An electrical schematic for the conventional six-transistor CMOS SRAM cell is shown in Figure 1.2a and a chip with embedded SRAM

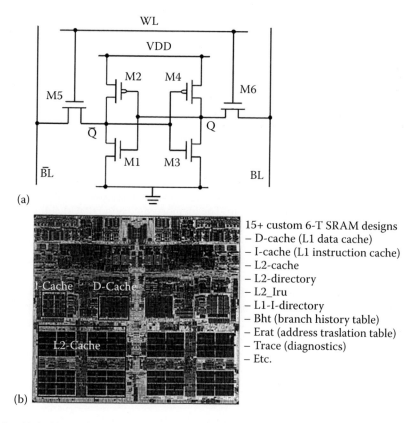

(a)

(b)

15+ custom 6-T SRAM designs
- D-cache (L1 data cache)
- I-cache (L1 instruction cache)
- L2-cache
- L2-directory
- L2_lru
- L1-I-directory
- Bht (branch history table)
- Erat (address traslation table)
- Trace (diagnostics)
- Etc.

FIGURE 1.2 (a) A six-transistor CMOS SRAM cell and (b) a custom-embedded SRAM array depicting the functionality of L1 and L2 data and instruction caches. (From Pilo, H., IEDM SRAM short course, Memory Technology for 45nm and Beyond, *IEEE IEDM* © 2006 IEEE.)

arrays for L1 and L2 (level 1 and level 2 for the memory hierarchy, respectively) cache functions is illustrated in Figure 1.2b [3]. The cross-coupled invertor pairs consisting of the four transistors M1, M2, M3, and M4 store the memory state of either "0" or "1." The word line tied to the gates of transistors M5 and M6 enables the access to the memory cell for reading and writing the cell. These transistors control the complementary bit-line nodes BL and BL—for sensing the memory state as shown in Figure 1.2a.

SRAMs could function synchronously with clock frequency or asynchronously independent of clock frequency. SRAM arrays are employed with varying capacities for a wide range of applications including computers, work stations, routers, and peripheral devices, as well as in scientific and industrial systems and subsystems. Due to their high speed, SRAMs are employed as caches.

1.2.2 Dynamic Random Access Memory: DRAM

DRAMs have been the working memory for the digital systems since their inception. DRAM cell is simple and consists of a transistor with an integrated metal–insulator–silicon (MIS) capacitor at the floating node (source node) of the transistor. The DRAM was invented by R. H. Dennard of IBM Corporation in 1967 [4] and has driven the digital electronics technology scaling and memory revenue ever since except for the last decade.

The binary state of the DRAM cell is defined by charging or discharging the floating capacitor node either to a high potential or to a low potential through the access (switching) transistor. A simple electrical schematic of a DRAM cell is shown in Figure 1.3a with a corresponding cell cross section shown in Figure 1.3b. The gate node of the memory cell transistor is connected to the word line laid out in the X-direction of the memory array, while the drain node of the memory cell transistor is connected to the bit/sense line laid out in the Y-direction of the memory array. The MIS capacitor is integrated at the floating source node of the transistor as shown in Figure 1.3a and b. The capacitor shown in Figure 1.3b is a capacitor fabricated in a vertical trench of the silicon substrate (in the Z-direction). However, options of vertically stacked capacitor (in the X–Z plane) fabricated over the gate of the transistor are also frequently adopted for the DRAM cell design.

DRAM is volatile because the charged state of the floating node changes with time due to a finite charge leakage. Therefore, it is periodically refreshed (rewritten) to maintain the memory state. The process of refreshing all cells in the memory array at regular time intervals is built into the memory chip through appropriate circuitry. This process of storing the data at the floating node is achieved through the bit line of the memory cell and through turning the gate of the cell transistor on (via the word line), thereby charging or discharging the floating node and subsequently turning the transistor off to maintain the memory state. The memory state of a DRAM cell is sensed by first pre-charging the bit line to an appropriate potential level in between those of the memory states at the floating node of the cell. This is followed by turning the cell transistor on via the word line and sensing the rise or fall of the bit line potential due to charge transfer from the floating node capacitor memory state.

Since their inception, an outstanding progress has been made in cost reduction, power reduction, and enhancement in bit/product density of DRAM chips. A significant progress has also been made in the rate of data availability from DRAM over the decades through advancement in memory design and architecture. DRAM organizations evolved from earlier asynchronous designs to synchronous designs (SDRAMs). Currently, DRAM products are readily available with DDR4 SDRAM designs whereby 16 consecutive words (bytes) of data get transferred per internal clock cycle running at >0.5 GHz. DRAM has currently achieved up to a 4 GB chip capacity in DDR4 SDRAM product form.

The 16 MB DRAM memory chip with a functional memory block described in Figure 1.1 is shown in Figure 1.3c. The memory was designed by NEC Corporation to provide data at 55 ns access time.

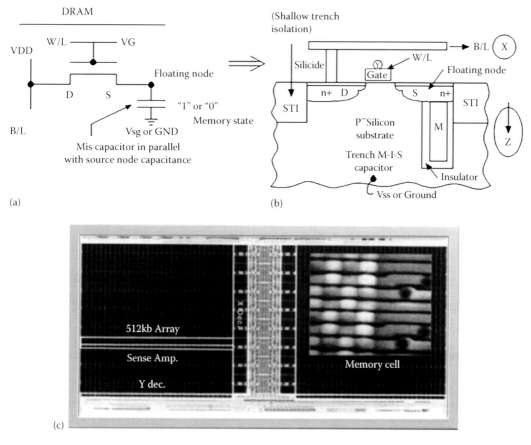

FIGURE 1.3 (a) A simple circuit schematic of a DRAM cell, (b) a cross-sectional schematic of DRAM cell with trench capacitor, and (c) a micrograph of 16 MB DRAM from NEC Corporation with functional blocks shown in Figure 1.1. (From Takeshima, T. et al., A 55 ns 16 Mb DRAM, *IEEE ISSCC*, p. 246 © 1989 IEEE.)

1.2.3 Read-Only Memory: ROM

Historically, read-only memory (ROM) had been the first NVM where each memory cell contained either of the two binary memory states of "0" or "1" through a process called "mask programming." Such memory could only be read or sensed in the random access mode to hold permanent data. Application of ROM has been in "Built-in Operating System" or BIOS and in installing the permanent microcode for digital systems. ROM cells are single transistors with fixed thresholds fabricated to be either "conducting" to reflect the "0" memory state or "nonconducting" to reflect the "1" memory state while being "read" or sensed. ROM arrays are laid out in a similar architecture as those of SRAM or DRAM containing the functional blocks discussed earlier.

1.2.4 EPROM, EEPROM, and E²PROM

The legacy of ROM evolved into a set of NVMs where the memory state of each cell could be kept permanently, yet could be electrically programmed and altered when or if required. This is achieved by "permanently" storing either electrons or holes within the gate stack of the FET memory cell, thereby altering the threshold of the memory cell into two "stable" memory states and subsequently sensing (or reading) the bistable memory states. This process of storing charges was labeled as electrically programming of such memory cell. The label EP stands for "electrically programmable," the label EEP stands for "electrically erasable and programmable," and the label E²PROM also stands for "electrically erasable and programmable." Although the label ROM is misleading from the stricter sense of the term, EPROM, EEPROM, E²PROM cells are limited in "writing" capability from the standpoint of writing speed and number of writing capability contrary to SRAM and DRAM. The label PROM succeeding ROM reflects the application legacy of ROM with on-chip programmable capability. Reading features in these devices are similar to those of ROM. EEPROM is erasable in memory blocks, whereas E²PROM has a byte-erasable capability. Programming mechanisms in these devices evolved over time and have been discussed in detail in a previous IEEE publication [4]. Section 1.2.5 explains some of the recent NVM devices in historical context.

1.2.5   Recent NVMs: NROM and NAND Flash Memories

Recent NVM products are labeled in terms of the logical architecture of the arrays combined with built-in features for erasing capability provided within the chip. When the memory arrays are organized in logical NOR configuration and randomly accessed, these are called NROMs. When a series of bit strings (memory cells) within the array are accessed in serial NAND (logically "not AND") mode with typically a string of 32 or 64 bits or lately 128 bits, such memories are called NAND memories. NROMs that store charges in the floating gate (see the next section) and are block erasable are called flash devices. Similarly, NAND memories with block-erasable features are called NAND flash devices. NROMs that store charges in the dielectric medium and are byte erasable are called EEPROMs. NAND flash memories are most dense and lowest cost/bit among all silicon based volatile and nonvolatile memories. In the history of memory technology. NAND flash memories, for the first time, drove the feature size technology scalability during the last decade. It has already been noted in the preface that, NAND flash revenue exceeded that of DRAM in 2011 and continued growing at an annual compound growth rate of 70% ever since. NAND flash products are offered in the form of single-level cell per bit and multilevel cell per bit (MLC)

FIGURE 1.4   A scanning electron micrograph of a 16 GB (MLC) capacity NAND flash memory chip from Micron Technology produced in 2007 in the 50 nm technology node. (Courtesy of Micron Technology Incorporated, Boise, ID.)

defined and discussed in the later sections. Figure 1.4 illustrates a scanning electron micrograph of a 16 GB MLC NAND flash chip produced by Micron Technology Boise, Idaho, with the 50 nm minimum feature size in 2007. Recently, Micron Technology announced the availability of 128 Gb and (at the time of publication 256 Gb) NAND flash memory produced in the 15-nm technology node. NAND flash memories are discussed in considerable detail in a recently published book by Wiley-IEEE [5]. A further clarification of recent NVMs is discussed in Section 1.5 below on Nomenclature and in Chapter 2.

1.3 MEMORY HIERARCHY IN DIGITAL SYSTEMS

Digital systems employ different types of memories in the form of memory hierarchy to meet the objectives in cost, performance, power, and applicability. A large volume of permanent data is stored in magnetic medium such as magnetic tapes and hard disk drives. These devices contain moving parts and are characterized by lower speed and lower reliability compared with silicon based memories. However, such devices provide lower cost pet bit of storage and larger capacity. Data are accessed serially in such memories. Silicon based digital memories, both volatile and nonvolatile types, together constitute primary working memory requirements to meet the critical system objectives as above. A hierarchy of memories employed in a computer system in 2012 is illustrated in Figure 1.5 below showing the characteristics of different types of memories [6]. It should be noted that NVMs are increasingly employed in digital systems as the cost per bit for such memories is reduced and the functional attributes of such memories are enhanced. This trend is expected to continue for at least the next two decades.

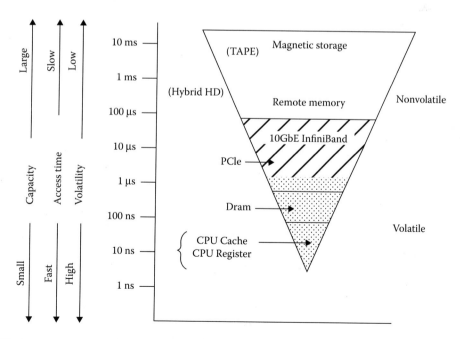

FIGURE 1.5 Hierarchy of memories in a computer system in 2012. (From Dinker, D., Architectural approaches to explaining flash memory, *Flash Memory Summit* © 2012 IEEE.)

1.4   FUNDAMENTAL MEMORY CONCEPT IN NVMs

Silicon based NVM device operates like a metal–oxide–semiconductor FET transistor. Electronic charges are stored within the gate dielectric medium. The process of storing the charges is called the "writing" of the device. The writing puts the device to an altered high threshold state "Vth high" or memory state "1." The process of removing or neutralizing the stored electronic charge is called the "erasing" of the device. The erasing puts the device to a low threshold state "Vth low" or memory state "0." The relative permanency of these binary memory states provides the basis of the nonvolatile characteristics of the NVM device. The NVM device could be sensed or "read" by imposing a gate potential between the erased Vth ("0") and the written Vth ("1"), thereby determining whether some current flows through the device or not. This is schematically shown in Figure 1.6, where the former is a device cross section and the latter is the Id–Vg characteristics of the NVM device. The illustrated memory cell is an NMOS schematic implementable on a p-silicon substrate.

By storing different amounts of charges, multiple levels of written memory states could be obtained and sensed (read). These form the basis of multilevel storage of the NVM device.

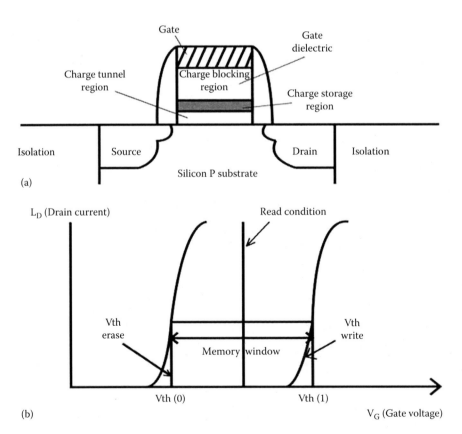

FIGURE 1.6   NVM FET: (a) Conceptual schematic and (b) bistable memory states Vth (0) and Vth (1). (From Bhattacharyya, A., Technology development and optimization of P-channel and N-channel MXOS FET technology for nonvolatile memory application, Part I: P-Channel: 1972, Part II: N-Channel 1973, IBM Internal Technical Report, 1973, Unpublished.)

1.5 NVM DEVICE GROUPINGS AND NOMENCLATURE

In general, NVM devices are differentiated and grouped based on the medium of the charge storage region of the NVM FET gate stack. NVM devices are broadly classified into two groups:

1. A floating (3D) storage medium providing uniform potential across the medium such as polysilicon (doped or undoped) or metal
2. A dielectric medium or nanocrystals (NCs; metallic or semiconductor) embedded in a dielectric medium where charges are discretely trapped in local potential wells

The first type of devices is commonly known as "floating gate" (FG) devices, while the second type of devices is commonly known as "charge-trapping" (CT) devices. The CT devices had been referred to in the past as "discrete traps" or "embedded traps" often interchangeably. They could hold charges at the interface or bulk of a dielectric film with characteristically high density of available "traps" or local "defects." This type of dielectric medium is called a "trapping" dielectric. Silicon nitride is a common example of a trapping dielectric.

Alternately, the CT devices could hold charges in local quantum wells or defects or traps associated with nanoparticles embedded in any dielectric film. This latter type is commonly called an NC charge trapping device to differentiate them from devices with trapping dielectric.

In this book, we will define the commonly known NVM device type called silicon–oxide–nitride–oxide–silicon (SONOS) as the trapping dielectric CT device because in such devices N is associated with charge trapping dielectric, silicon nitride or nitride.

The NC charge trapping devices will be referred to as CT-NC or simply NC or nanodot (ND) devices. Also in this book, the terms NC and ND have been used interchangeably.

The nitride charge trapping or SONOS types may exhibit unique characteristics. Similarly, the NC or ND devices may exhibit distinctive characteristics due to embedding silicon-NC or germanium-NC or different metal NCs (e.g., W, Pt, Au, Ir, and Pd) embedded into either oxide (SiO_2) or other dielectric films. Therefore, in discussing the properties of such devices, the nature of charge trapping medium is identified for specific CT devices. However, the FG devices use silicon film as the floating gate; therefore, unless otherwise stated, polysilicon should be assumed as the floating gate for the FG devices. In some special FG devices, charges can be stored in a semiconducting film consisting of high-density closely spaced silicon NCs or NDs creating an equipotential medium. The medium is created by incorporating high concentration of silicon NDs in nitride. This special form of FG devices has been named by the author as "floating plate" (FP) devices since, such film could be almost two-dimensional being of a single- or double-layer thickness of the NCs. The author developed such material and applied to the NVM devices. The material has been identified as silicon-rich nitride (SRN) with a wide range of silicon concentrations. A detailed discussion of SRN has been provided in Chapter 4. SRN has been classified into two broad groups for NVM applications: (1) I-SRN (or IN-SRN)—SRN of high concentration of injecting silicon-NDs—and (2) T-SRN (or TR-SRN)—silicon-rich insulator containing silicon-NDs for charge trapping. The author used this naming following the naming convention used by D. De Maria et al. [7], who earlier developed similar materials using silicon-NDs in oxide and identified those media as I-SRO and T-SRO.

The two groups of devices, namely, the FG devices (with the special case of FP) and the CT devices, are also differentiated by the mechanisms of incorporating charges into the trapping medium or media. Such mechanisms have profound effects on NVM device characteristics and reliability. Charges could be incorporated or neutralized by injecting from the substrate or from diffusion regions as well as from the gate of the devices. Source of charged carriers in some early devices were electron–hole pairs generated by the ultraviolet radiation. Such devices will not be discussed here. The mechanism of charging and discharging the trapping medium may involve high-energy carriers (e.g., hot electrons and hot holes), or medium energy carriers with sufficient energy to overcome the energy barrier by quantum mechanical tunneling. The mechanism and energetics of charge carrier transport strongly determine the properties of all silicon based NVM devices. With few exceptions,

TABLE 1.1

Naming Conventions for NVM Devices

Gate Stack Vertical Layer Method		Charge and Discharge Combinations	
Device	**Vertical Layers**	**Charge**	**Discharge**
Floating gate	Charge trapping	Hot carrier	Hot carrier
MON-FG-OS	MNOS/MAOS/MXOS/SNOS	Hot carrier	Tunneling
SON-FG-OS	MONOS/SONOS/MO-NC-OS/SO-NC-OS	FN tunneling	FN tunneling
DTM	Both FG and CT types	Direct tunneling	Direct tunneling

M, metal; O, oxide; N, nitride; S, silicon or polysilicon; A, alumina; X, oxynitride; SRN, silicon-rich nitride; SRO, silicon-rich oxide; NC, nanocrystal; FG, floating gate; FP, floating plate.

both FG and CT devices evolved into the current form whereby the quantum mechanical tunneling and related modes are being employed as primary modes of charging and discharging the trapping medium. Therefore, much of the device description in this book will focus on such mechanisms.

Table 1.1 outlines the naming conventions for NVM devices, which are based on gate stack design concepts for NVM devices and also defined by the mechanisms of transporting charges and the medium of storing charges.

The above grouping reflects frequently used convention of defining the types of NVM devices in terms of the dielectric layers employed in the gate stack design of the NVM FET. It also differentiates the gate material employed: for example, M for metal and S for polysilicon. These abbreviations are also followed in the text, which are defined as follows:

MNOS: **metal–nitride** (silicon nitride or Si_3N_4)–**oxide** (silicon dioxide or SiO_2)–**silicon** substrate
MAOS: **metal–alumina** (aluminum oxide or Al_2O_3)–**oxide-silicon** substrate
MXOS: **metal–silicon oxynitride–oxide–silicon** substrate

The design concepts for NVM devices are also defined by the mechanisms of transporting charges and the medium of storing charges.

As discussed previously, the first alphabet M or S implies the gate of the FET device, the second alphabet N stands for nitride trapping layer, and the third alphabet O stands for oxide in the charge tunneling medium interfacing the silicon substrate S. Similarly, metal–oxide–nitride–oxide–semiconductor or SONOS devices include an oxide blocking layer interfacing the metal or silicon gate, respectively.

REFERENCES

1. A. Bhattacharyya, "Technology development and optimization of P-channel and N-channel MXOS FET technology for nonvolatile memory application, Part I: P-Channel: 1972, Part II: N-Channel 1973", IBM Internal Technical Report, 1973, Unpublished.
2. T. Takeshima, M. Takada, H. Koike, H. Watanabe, S. Koshimaru, K. Mitake, W. Kikuchi, T. et al., "A 55 ns 16 Mb DRAM", *IEEE, ISSCC*, pp. 246–247, 1989.
3. H. Pilo, 2006 IEDM SRAM short course, "Memory technology for 45nm and Beyond", *IEEE, IEDM*, 2006.
4. W. D. Brown and J. E. Brewer (Eds.), *Nonvolatile Semiconductor Memory Technology*, IEEE Press, Piscataway, NJ, 1997.
5. J. E. Brewer and M. Gill, *Nonvolatile Memory Technologies with Emphasis on Flash*, IEEE Press, John Wiley & Sons, Hoboken, NJ, 2008.
6. D. Dinker, "Architectural approaches to explaining flash memory", *IEEE, Flash Memory Summit*, Santa Clara, CA, 2012.
7. D. DiMaria et al., "Electrically alterable ROM using silicon-rich-SiO2 injector and polycrystalline silicon storage layer" *Journal of Applied Physics*, 52, 4825–4842, 1982.

2 Historical Progression of NVM Devices

CHAPTER OUTLINE

A variety of silicon field effect transistor (FET)-based nonvolatile memory device concepts evolved since the inception of such devices. These devices may be grouped primarily into several subgroups. The chronological evolution of four types of nonvolatile memory devices is briefly sketched in this chapter. These include floating-gate devices, conventional charge-trapping devices, nanocrystal charge-trapping devices, and direct tunnel memory devices. The first three of the four devices and associated memory cell types are distinguished from each other by the charge-storage mediums employed. The last type is defined by the mechanism of electronic charge-transfer mode in such devices. Some of the prominent devices of the first three types are further discussed with technical details in Chapter 7 of this book.

2.1 FLOATING-GATE DEVICES

Floating gate nonvolatile memory or FG-NVM evolved out of metal–oxide–silicon field effect transistor (MOSFET) technology and followed the development and integration schemes adopted by the general FET technology. The first FG device was published by D. Kahng and S. M. Sze in 1967 [1]. Figure 2.1 shows the cross section of the transistor memory cell. The gate stack consisted of an n-type silicon substrate; a 5 nm tunnel oxide (SiO_2) layer was grown over the substrate; a metal floating gate was deposited over the tunnel oxide and followed by a 100 nm zirconium oxide layer and the metal gate of the FET. A positive gate bias was applied to inject electrons from the silicon substrate through the tunnel oxide to be stored in the insulated metal floating gate. The stored negative charge on the floating gate raised the threshold of the PFET device turning it off. Thus one of the binary memory states termed as the "written" state and defined by the higher threshold of the memory cell was created. The memory FET device would not be conducting at this state when turned on. A large negative potential was, thereafter, applied to the gate to remove the electrons from the metal floating gate by back tunneling. This returns the memory device to the other binary memory state termed as the "erased" state defined by a lower negative threshold thus created. The memory FET device would be conducting at this memory state when turned on. The stable "turned off" and "turned on" characteristics of the memory states thus defined the nonvolatile feature of the memory cell. While the device proved the concept of a floating-gate memory, the combination of thin tunneling oxide and the metal floating gate proved to be leaky and unreliable.

2.1.1 The FAMOS Device

The first polysilicon floating-gate device was introduced by Frohman–Bentchkowsky in 1971 using a PFET device on an n-type silicon substrate [2]. A later version was introduced in 1974 using an NFET device on a p-type silicon substrate [3]. In the original version, a 100 nm tunnel oxide was used to separate a polysilicon floating gate from the silicon substrate, while an overlayer of 1000 nm of oxide was used to prevent charge leakage from the floating gate to the FET gate. Electrons were injected from the silicon substrate by avalanche injection at the drain–substrate depletion layer by

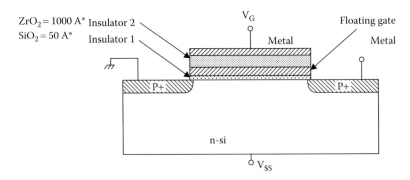

FIGURE 2.1   Schematic cross section of the first floating-gate memory device. (From Kahng, D. and Sze, S. M., *Bell Syst. Tech. J.*, 46, p. 1288 © 1967 IEEE.)

reverse biasing the p+ drain. The drain was biased at 50 V for 5 ms, while the source and substrate were held at ground. The device was erased (conducting state) by the UV radiation.

The above device was called a "floating gate avalanche injection MOS," or the FAMOS device. The FAMOS chip is the first 2048 bit NVM product providing a field-programmable feature, and it replaced read-only memory (ROM) chips for applications such as code and look-up table updates and was also employed for character generation. The 2 Kb FAMOS memory chip was described by Frohman–Bentchkowsky in 1974 [3].

2.1.2   The SAMOS Device

With the introduction of double-layer polysilicon gate MOS technology, a stacked gate avalanche injection scheme for writing into the floating gate was introduced by H. Iizuka et al. [4,5]. The device was called "Stacked Gate Avalanche Injection MOS" or SAMOS. The device featured a control gate on the top of the floating gate separated by an oxide layer. Application of an appropriate potential at the drain of the device caused the breakdown voltage of the drain junction to decrease and created enhanced avalanche injection of electrons. This improved the writing efficiency when compared over the FAMOS devices. Additionally, the hole injection from the top gate was feasible for erasing such devices besides other available erasure mechanisms.

2.1.3   The SAMOS 8 Kb EAROM

In 1977, an 8 Kb electrically alterable ROM (EAROM) chip was designed and fabricated for the first time using the SAMOS memory cell [6]. The chip area was 19.7 mm². The memory chip features and specifications are listed below:

1 K × 8 organization
N-channel technology
One-transistor memory cell
Electrically erasable by block (30 s)
Fully decoded
Standard operating voltages: ±5 V and +12 V
Programming by single pulse: +26 V, 100 ms/byte
Erasure using voltage ramp: +35 V, 30 s
Memory access time: 250 ns

The SAMOS EAROM device was built on the NFET technology for enhanced performance, and it created the framework for subsequent devices. The memory chip employed a block erasure scheme by the tunneling mechanism for the first time. It featured a single-transistor NVM memory cell with stacked gate. Subsequent innovations in the NVM Flash devices followed such ideas. Since the development of the SAMOS device, several variations of SAMOS were proposed during the next decade (1977–1987), which involved hot-electron injection for programming [7–11].

2.1.4   The SIMOS Device

By the mid-1970s, the NFET technology was well established and the channel hot-electron effect was understood and modeled [12,13]. B. Rossler and R. G. Muler proposed a novel floating-gate cell with stacked control gate where channel hot-electron injection was used to inject electrons from silicon to the floating gate for writing. Subsequently, they proposed a dual-polysilicon split-channel cell named as the "Stacked Gate Injection MOS" or the SIMOS memory cell [14]. While the device was written by channel hot-electron injection, the device was erased by tunneling through a thinner 40 nm thick oxide off the channel and over the source side. At 6 μm lithography, the cell size was 841 μm². The channel length was 3.5 μm. The writing condition was 26 V, 100 ms, whereas the erasing was accomplished at −35 V, 30 s. The substrate was back biased at −5 V.

During the same period, several nonplanar floating-gate device concepts evolved that took advantage of enhanced tunneling of electronic charges between polysilicon electrodes separated by an oxide layer. These devices are not discussed in this book. An excellent review of such devices was provided by H. A. R. Wagener and W. Owen in a previous publication [15].

2.1.5   Chronology of Floating-Gate Device Evolution

During the period of 1977–1987, FG memory products grew in density from 8 Kb Flash EEPROM to 256 Kb Flash EEPROM (32× density enhancement). During the same period, the FG cell density increased by nearly 20× (841 μm² cell to 43 μm² cell) by a combination of lithographic scaling and improved cell configuration.

The decade that followed (1987–1997) had seen phenomenal growth in floating-gate memory density, an increase from 256 Kb to 128 Mb resulted in a 500× enhancement. Associated reduction in cell size was from 43 μm² to 0.32 μm², by a factor of 130×. The progress in bit density of EPROM cells during 1980s and 1990s is illustrated in Figure 2.2a [15]. Evolution of cell size of EPROM cells is shown for the same period in Figure 2.2b [15]. This rate of growth has been sustained even today.

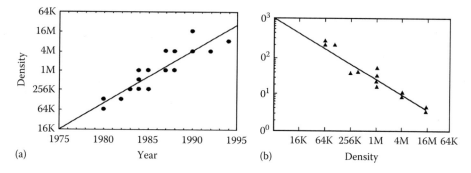

FIGURE 2.2   (a) Evolution of nonvolatile bit density showing new generations at three year intervals. (From Groeseneken, G. et al., Basics of nonvolatile semiconductor memory devices, Chapter 1, in *Nonvolatile Semiconductor Memory Technology*, IEEE Press, p. 27 © 1997 IEEE.) (b) EPROM cell size evolution as a function of bit density. (From Groeseneken, G. et al., Basics of nonvolatile semiconductor memory devices, Chapter 1, in *Nonvolatile Semiconductor Memory Technology*, IEEE Press, p. 26 © 1997 IEEE.)

The chronological development from 1987 to 1994 was well documented and reviewed by Manzur Gill and Stefan Lai [16] and subsequently up to 2003 by A. Bhattacharyya [17]. The second decade gave rise to many innovations in the NVM cell design. Notable among these memory cells are the "HIMOS" cell, the "ETOX/FLOTOX" cell, and the "DINOR" cell. These EPROM and EEPROM cells may be considered as the precursors for the current NROM cells. These cell designs are briefly mentioned below and also further discussed in Chapter 7.

2.1.6 The HIMOS Cell

The HIMOS cell introduced many innovative attributes [16]. This included source-side hot carrier injection, which reduced programming power by an order of magnitude compared to channel hot-electron injection [18–22]. It featured a single-transistor split-channel floating-gate cell. The design allowed the use of a single 5 V power supply. The high-performance HIMOS device [23] made use of an efficient virtual ground array (VGA) architecture, which has been followed since. As mentioned, the HIMOS cell used source-side hot carrier injection for writing, and erasing is done by tunneling through the thin oxide. The HIMOS cell was a larger memory cell and was succeeded by ETOX and FLOTOX memory cells, which outlasted the HIMOS cell.

2.1.7 The ETOX/FLOTOX Cell

The "ETOX" NVM cell was originally proposed in the form of a two-transistor design and subsequently evolved into a "FLOTOX" cell, which is a single-transistor design [24–26]. These memory cells are considered to be the forerunner of the currently used one-transistor NROM Flash device. Due to their historical significance, both the FLOTOX and the ETOX cells are described in greater detail in Chapter 7. The schematic cross section of the two-transistor version of FLOTOX/ETOX cell is shown in Figure 7.2. The cell design employed two levels of polysilicon. The first level of polysilicon was used as the floating gate, which was extended into a thinner tunnel oxide region for electronic charge transfer by tunneling while overlapping the N+ drain region. Two fixed threshold FET elements were placed at the two ends of the memory cell defined by the second level of polysilicon gate (POLY-2). The first element acted as the sense transistor element, while the second one acted as the select transistor element. The POLY-2 gate also acted as the gate for the nonvolatile element. The programming (writing) and the erasing scheme of the EEPROM cells are illustrated in Figure 2.3a and b, respectively. The programming was accomplished by channel hot-electron injection from silicon substrate as shown in Figure 2.3a. The Figure 2.3b illustrates the tunnel erase scheme for the flash device employed for block erasing.

The ETOX was incorporated into a highly successful 256 Kb EEPROM Flash memory chip [26].

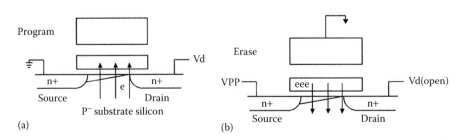

FIGURE 2.3 Schematic representation of: (a) Channel hot-electron injection for programming and (b) Erasing by F–N tunneling. The above scheme was adopted for FLOTOX EEPROM cell.

2.1.8 The DINOR Cell

Another important device development is known as divided bit-line NOR cell or DINOR [27,28]. The DINOR device was a single-transistor stacked gate NOR Flash device based on tunneling for both writing and erasing. The DINOR device reversed the convention of memory state: the high Vth state was the erased state, while the low Vth state was the write state. Since the low-threshold state is defined as the written state, the DINOR programming scheme enjoyed the advantage of allowing tight Vth distribution through employment of the bit-by-bit programming algorithm. During "write" operation, charges tunnel from the floating gate to the drain, while during "erase" operation electrons tunnel from the channel to the floating gate. Energy associated with the writing and erasing operations were further reduced due to the tunneling mechanisms requiring lower energy for carrier transport as well as application of a 3 V power supply for the first time. The DINOR cell paved the way to develop the charge-trapping NOR devices and the current forms of NROM devices. This is further discussed in Chapter 7. An excellent relatively recent review on DINOR Flash memory has been provided by M. Nakashima and N. Ajika [29].

2.1.9 The NAND Cell

Previous devices discussed above were associated with parallel or random access memory architecture. In 1990, R. Kirisawa et al. [30] proposed a serial access NAND flash memory utilizing stacked gate device elements and tunnel-based writing and erasing schemes. Current NAND flash memory was thus born achieving significantly enhanced density by reducing the number of bit-line contacts. A cross-sectional view of an 8-bit string NAND flash design published by M. Monosomy and co-workers is shown in Figure 2.4, illustrating a twin-well technology [16]. NAND cells and arrays as well as associated technology integration schemes are discussed in considerably greater detail in Chapters 7 and 8. The FG Flash has been the work horse for NVM devices ever since.

A synopsis for the evolution of FG Flash devices during the first two decades since inception is shown in Table 2.1, starting as a two-transistor device evolving into a single-transistor NVM cell.

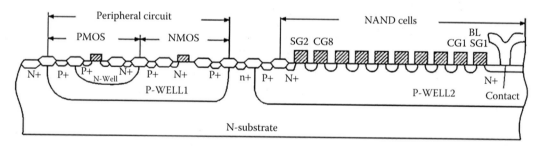

FIGURE 2.4 Cross-sectional view of a first generation 8-bit string NAND EEPROM memory. (From Gill, M., and Lai, S., Floating gate flash memories, Chapter 4, in *Nonvolatile Semiconductor Memory Technology*, IEEE Press, p. 189 © 1997 IEEE.)

TABLE 2.1

Synopsis of Floating-Gate Device Evolution in the First Two Decades

Transistors/Cell	NVM Device	Write	Erase
2	SIMOX	CHE byte	Tunnel block
	FLOTOX	Tunnel byte	Tunnel byte
1	ETOX/HIMOS	CHE byte source side	Tunnel block
	DINOR	Tunnel block	Tunnel block
	NAND	Tunnel byte	Tunnel byte

M. Gill and S. Lai [16] and A. Bhattacharyya [17] had covered more extensively the historical progression of early FG and charge-trapping (CT) devices. The phenomenal progress led to the current NVM products of FG-NAND flash and CT NROM. The evolution of current CT NROM devices is discussed below. Recent developments and associated challenges for the conventional flash NVM devices have been outlined in Part I as well as in Part II of this book. A relatively recent publication from IEEE press reviews the developments until 2008 of developments in nonvolatile memories with emphasis on NAND flash developments [29].

2.2 CONVENTIONAL CHARGE-TRAPPING DEVICES

2.2.1 MNOS/MXOS/MONOS/SNOS/SONOS

Conventional charge-trapping devices (the "CT" NVMs) employ thin nitride film, typically in the range of 50–100 nm as charge-trapping medium and a very thin oxide of 1.5–3.0 nm range as the medium of tunneling of electrons and holes from the silicon substrate to the nitride traps for writing and erasing, respectively. These devices employ either a metal gate or a doped polysilicon gate. The evolution of the gate stack layers for the NV element initially had been PFET MNOS, followed by NFET MNOS, and thereafter, NFET SNOS (silicon replacing metal gate) and finally to NFET SONOS and NFET or PFET MONOS whereby a charge-blocking oxide layer was introduced between the gate electrode and the trapping layer. The random access memory CT cell consisted of a two-transistor cell design: a select transistor with fixed Vth in series with an NV-transistor with bistable memory states. Alternately, such memory cell design was evolved into a "merged" device of "one-and-a-half transistor" cell whereby the select element with fixed Vth was placed in series with the NV element, splitting the channel and sharing the common source and drain with the gates of the elements (select gate and NV control gate) overlapping each other. Such a cell design was electrically erasable and programmable (EEPROM). Eventually the CT cells adopted the NOR-ROM (NROM) architecture of a single-transistor NV-random access memory (RAM) cell similar to the FG NROM Flash cell. A CT memory cell provided the following advantages over the FG memory cell:

1. Higher endurance capability by at least one to two orders of magnitude greater than FG cells (~1E7 write/erase cycle capability).
2. Lower programming and erasing voltages typically nearly half of those required for FG memory devices.
3. Greater compatibility of process fabrication with general CMOS technology.
4. Greater potential bit density. By storing charges discretely and locally in nitride layer above the source and drain regions and sensing the Vth after writing and erasing by altering the roles of source and drain (current flow directions in the channel), a single cell can be used to store 2 memory bits of information, thereby doubling the storage capacity. Additional advantages of CT memory cell include:
5. Stacked design of CT cells is not constrained by capacitive coupling ratio of that of FG cell design. Therefore, CT devices are inherently more scalable.
6. CT devices are more radiation immune due to charge trapping in nitride compared to FG devices where charge is stored within silicon floating gate.

The disadvantage, however, is the larger cell size when compared to the single-transistor implementation of the FG cell and, specifically, when compared with the NAND implementation of the FG cell.

2.2.2 Historical Evolution of MNOS, MXOS to SONOS CT Devices

The very first MNOS CT device was proposed by Wagener et al. In 1967 [31] around the same time, the FG-NVM was conceived [1]. MNOS CT device characteristics were illustrated by D. Frohman

Bentchkowsky [2]. In 1972 and 1973, IBM R&D groups developed successively fully integrated PFET and, thereafter, NFET MNOS technology for 8 Kb "alterable read only storage" (AROS) memory programs [32,33]. The 8 Kb AROS (EAROM or EPROM) technology was a two-transistor NVM RAM cell, which is further described in Figure 7.1. The 8 Kb AROS memory chip disclosed in IBM invention Technical Bulletin in 1975 had similar applicability and features as that of the SAMOS 8 Kb EAROM chip published subsequently (see Section 2.1.4). The memory cell at 6 μm design rule was smaller than the SAMOS floating gate 8 Kb EAROM cell. By 1974–1975, the high-quality CVD silicon oxynitride (SiON) film technology was developed by IBM researchers [34,35]. Nitride was replaced in MNOS NVM structures by multilayered SiON films of different oxygen to nitrogen ratios for tunneling layers, trapping layers, and gate-induced charge injection blocking layers. These devices were called as MXOS CT devices [34,35]. The MXOS devices demonstrated vastly superior device characteristics over MNOS devices in memory window, charge retention, and write–erase cycling ("Endurance") capability with programmability at ±30 V, 10 ms. A limited information was published of MXOS devices [35,36]. The 8 Kb "AROS" technology was qualified but the product program was abandoned in favor of the DRAM product program. The work remained mostly unpublished and subsequently partially published and patented [37]. It was not until a decade later, in 1987, SiON films replaced nitride in CT devices in the industry [38].

With the establishment of the silicon gate NFET technology, a 16 Kb SNOS EEPROM was published in 1980 for the first time, by T. Higiwara et al. [39]. The EEPROM was a two-transistor cell similar to the MNOS cell and the MXOS cell as mentioned above. The cell employed two levels of polysilicon, the first one was used as gate for the fixed Vth select device, while the second polysilicon gate was used for the nonvolatile element.

With the exception of MXOS devices mentioned above, MNOS and SNOS devices suffered from limited memory state window due to undesirable hole injection from metal or silicon gate to the nitride layer during write operation. To overcome this problem, an oxide layer was introduced between the gate and the nitride trapping layer to provide a large energy barrier for the charge injection from the gate [40,41]. SONOS device thus developed became the standard CT-NVM device for the following decade (1985–1995). During this period, the top blocking oxide and the CT nitride layers were optimized to achieve the following device characteristics:

- Reduced hole injection from the gate.
- Maximized hole trapping in nitride while reducing the nitride thickness.
- Scaled nitride thickness for programming voltage reduction and maximized programming window.

Typical optimized ONO stack design consisted of 1.5–2.0 nm of tunnel oxide with 3.0–5.0 nm of nitride trapping layer and oxidized nitride charge-blocking layer of 4–5 nm thickness [42]. The programming voltage for such design was shown to be as low as ±7 V. Programming voltage as low as ±5.0 V was reported in the literature for optimized stack design [43].

2.2.3 Evolution of CT-NVM Cell Designs

A brief outline of cell designs for CT devices from inception through the mid-1990s is discussed here. More detailed description on cell designs had been reviewed by Libsch and White [44–47] and more recently by M. Taguchi [48]. A detailed discussion of selected cells is covered in Chapter 7.

2.2.3.1 Tri-Gate Memory Cell

The very first MNOS memory cell was a PFET device and was published by J. R. Cricchi et al. in 1973 [49]. The cell was called a "tri-gate memory cell" (TMS) and contained the NV-MNOS element at the center of the channel with a single aluminum gate. The source and drain side of the channel consisted of thicker oxide fixed Vth MNOS gate stack elements to address the cell as select device. The TMS memory cell was the precursor for all subsequent split-channel NVM cells. A variation of the TMS (Tri-gate NFET CT NVM) cell design using double-polysilicon gate in technology was published in 1983 by A. Lanchester et al. a decade later [50].

The nonvolatile element for the above Tri-gate device was an SNOS device with a second polysilicon gate overlapping the two first polysilicon-gated MOS elements. The source-side MOS element acted to isolate the transistor channel, while the drain side MOS element acted as a select element for the cell. The memory array was designed within a P-well with an n-type silicon substrate. A 5 V power supply was used and an on-chip −10 V voltage source was generated for programming and erasing. Programming and erasing required ±15 V and were supplied by applying Vcc (5 V) at the control gate and −10 V Vpp at the P-well for programming. The applied potentials were reversed for erasing. The cell was read by applying Vcc (+5 V) to the first polysilicon gates while grounding the P-well and second polysilicon control gate. Channel current flowed when the memory was in the erased state by imposing a drain potential of 1 V. The memory cell demonstrated 10 years of charge retention and >1E6 write/erase cycle ability.

2.2.3.2 Pass Gate Memory Cell

The second type of cell widely implemented from the mid-1980s through mid-1990s for SNOS and SONOS devices was a simple two-transistor implementation consisting of a pass fixed Vth transistor in series with an NVM device and is called the "pass gate memory cell" [51]. The pass gate memory cell typically used first polysilicon gate MOS transistor as the select gate to access the cell and was integrated with a second polysilicon-gated SNOS or SONOS NVM device. The cell was first integrated into a standard NFET technology and later on into a standard CMOS technology. The cell was used both as a stand-alone EEPROM as well as for multiple embedded applications.

The pass-gate memory cell led to the development of the so-called "Split Gate" or "Merged Gate" cell, which is still being used today in the current versions of NROM CT devices. The Split Gate is a merged version of the pass-gate cell, thereby providing enhanced density.

In 1991, T. Nozaki et al. published the first 1 Mb EEPROM SONOS device and array design as well as the operational scheme for the memory [51]. Similar layout and operational scheme has been adopted by subsequent NROM devices and memory arrays described in Section 7.5.1. The above device operated at +9 V, 100 μs for writing and −9 V, 10 ms for erasing. The memory window was 3 V: (+1 Vth and −2 Vth). The device demonstrated 10 years of charge retention and >1E7 write/erase cycle ability. The gate insulator ONO stack consisted of 2 nm of tunnel oxide, 5 nm of trapping nitride, and 4 nm of blocking oxide.

2.3 NANOCRYSTAL CHARGE-TRAPPING NVM DEVICES

2.3.1 Early History

Aside from the conventional nitride-based SONOS type of CT NVM devices, another CT device types emerged in the early 1980s out of the developments of silicon-rich oxides (SROs) and silicon-rich nitrides (SRNs). Both SROs and SRNs deposited by LPCVD processing provided two phase insulating structures containing silicon nanocrystals in SiO_2 and Si_3N_4, respectively, and charge trapping was observed associated with silicon nanocrystals. This has been briefly mentioned earlier. During the mid-1990s, such novel type of CT devices was named as "Nanocrystal" devices, and drew considerable exploratory activities since that time. Although no nanocrystal-based stand-alone flash NVM product has yet been available, considerable understanding of such devices exist as well as future product potential. This section will cover an extended introduction of the historical development and future potentials for such devices.

MIS capacitor investigations by A. Bhattacharyya et al. [52] demonstrated for the first time in 1984 that the efficient charge-storage possibility can be obtained by controlling the concentration of excess silicon and associated silicon nanocrystal size and distribution in LPCVD SRN dielectric

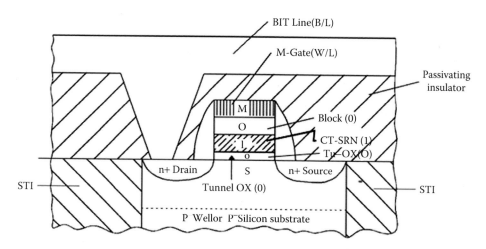

FIGURE 2.5 Cross-section of the first silicon nano-crystal CT-MOIOS NVM cell. (From Bass, R. et al., Nonvolatile memory cell having silicon rich silicon nitride charge trapping layer, U.S. Patent No. 4870470, 1989.)

films. Properties of SROs and SRNs are discussed in Section 4.4. It has been shown that size and distribution of silicon nanocrystals can be altered by processing conditions and postprocessing thermal anneal which, in turn, affects charge trapping. Furthermore, when the excess silicon concentration is between 1.5% and 7%, charge trapping for both electrons and holes could be observed in SRN films. Large reproducible flatband shift was noted in MIS structures in the range of excess silicon concentration between 1.5% and 3.0%. It was further observed that with higher silicon concentration in SRN films, size of silicon nanocrystals was progressively increased, and beyond ~10% of excess silicon concentration, charges could not be stored in silicon nanocrystals and SRN films exhibited very high conductivity.

Subsequently, silicon nanocrystal NVM cells with MOIOS gate stack designs were disclosed for the first time in 1984. The stack design contained a thin tunnel oxide layer with an overlayer of silicon-rich nitride for charge trapping [53,54]. Such devices exhibited significantly larger window (>4 V), high writing and erasing speed and significantly enhanced write–erase cycle-ability (endurance). A cross section of the first silicon nanocrystal CT-NC device is shown in Figure 2.5 [53].

During 1982, D. De Maria et al. [54] also disclosed a floating-gate NVM cell using enhanced carrier transport properties of SROs with very high excess silicon concentration (>14% excess silicon with closely spaced Si nanocrystals without CT characteristics). The cell, known as dual-electron injector structure or DEIS cell, was a floating gate "DEIS" (SRO) cell, which operated on the principle of enhanced carrier transport from the control gate (second polysilicon level) to the floating gate (first polysilicon level). Nevertheless, DEIS cell should be considered as the first application of silicon nanocrystal for NVM. During 1995 and the following year, other IBM researchers revived the interest in nanocrystal-based NVM devices attracting world-wide attention [55,56]. A flurry of R&D activities followed to develop nanocrystal-based CT NVMs. Such activities are still being pursued. It is believed that complexity in process integration and control had been the primary obstacle, which prevented product introduction thus far. A brief review of nanocrystal physics impacting CT device characteristics as well as future possibilities is summarized below.

2.3.2 Nanocrystal Physics and Charge Trapping

Nanocrystals or nanodots (NC or NDs) are considered to be quantum dots ranging in size from <1 to 20 nm in dot diameter. When a nanocrystal dot of a semiconductor or metal element is incorporated into an otherwise homogeneous dielectric medium such as SiO_2 (oxide) or Si_3N_4 (nitride) or Al_2O_3 (alumina), it destroys local bonding and alters local characteristic in lattice potential creating deep

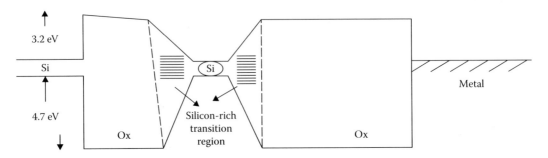

FIGURE 2.6 Schematic energy diagram modification with the presence of silicon nanocrystal in oxide.

quantum potential well or "defects" or trapping centers. Such quantum wells provide energy states where electrons and/or holes could be trapped and remain confined assuming a set of defined levels of potential energy. The defect creation, energy states generation, and trapping could be depicted in the form of modification of the band diagram of an insulator when a nanocrystal is introduced. This is shown for a silicon-rich oxide in Figure 2.6 when a silicon nanocrystal is formed in the oxide insulator. A transition in band energy level is introduced as shown with generation of energy states around the silicon nanocrystal. These energy states would then be available to capture and trap either electrons or holes. When an electron is captured in the potential well created by a nanodot, the nanodot is charged. The energy associated with the charged dot in the defect center is given by:

$$Ed = e2/2C, \text{ and is proportional to the quantum dot diameter, Dnd}$$

where, Ed is the columbic nanodot energy, e is the electronic charge in coulomb, and C is the capacitance of the dielectric medium. If more than one electronic charges are captured in the defect center, the electrostatic energy of the defect center is raised rapidly (e.g., 4× for two electrons and 9× for three electrons) and counters the applied field for programming. This phenomenon is known as "coulomb blockade." The effect of the phenomenon of coulomb blockade on programming the NC-NVM device could be explained in the following way:

Assuming planar and uniform nanocrystal (NC) or nanodot (ND) distribution, the resulting defect centers associated with the nanodots are expected to be uniformly distributed along the nanodot plane. The planar field thus created due to charge trapping in the defect centers counters and effectively reduces the external "writing" field. As a result, write current per nanodot is self-limited by coulomb blockade. Consequently, the nanodot devices display "write saturation," limiting the change in threshold during writing when such devices are programmed.

A second aspect of coulomb blockade is related to stored charge retention. If more than one electronic charge is stored in a nanodot defect center, the second charge is stored at a higher energy state closer to the conduction band edge and, therefore, more amenable for reverse tunneling. For a given density of nanodots, therefore, there exists a trade-off between memory window and charge retention.

A second phenomenon associated with quantum dot-induced trapping or defect centers is known as "quantum confinement" (QC). QC is related to the quantum-well dimension, which in turn is related to the diameter of the quantum dot. QC is inversely proportional to the square of the diameter of the quantum dot. QC is proportional to trap energy depth. The deeper the trap energy depth, the greater is the QC and, consequently, greater would be the charge retention. If the nanodots are distributed uniformly, higher dot density implies smaller dot size and reduced charge retention capability associated with reduced QC. The relationship between quantum dot diameter and trapping energy associated with it can be calculated. For example, for silicon nanodot in SiO_2, the associated trap energy depths for 5, 3, and 2 nm dot diameters are, respectively, 3.05, 2.65, and 2.2 eV. Due to such sensitivity, nanodot size control is critical to achieve desired charge retention for such devices.

Therefore, both coulomb blockade and QC provide limits of NC-CT quantum dot devices in terms of window and retention. Separation between quantum dots needs to be greater than direct tunnel distance to prevent enhanced lateral charge transport. It had been shown that when distance between silicon nanodots in nitride is <3 nm, the conductivity in SRN is increased by many orders of magnitude and the material ceases to have characteristics of a classical dielectric [52]. Due to the close proximity of silicon nanodots in such cases, the trapping medium becomes effectively an equipotential plane and charge-storage characteristics are similar to those of the floating-gate devices without any local defect-induced leakage paths. Therefore, to take unique advantages of the charge trapping due to nanodots, the dots need to be tightly controlled in diameter and should have optimal spacing in the range of 3.5–5 nm. Such requirements could be challenging for manufacturability.

When nanodot size, separation, and charging (single electronic charge per dot for maximizing retention) are optimized for charge trapping, an upper limit is set for the device in terms of charge-storage density and memory window. For example, for nanodot diameter $Dnd = 4.5 \pm 0.5$ nm with separation of nanodots of 5.5 ± 0.5 nm, the total number of nanodot density available for charging in a single plane would be:

$$NNd = 1E12/cm^2$$

Therefore, change in Vth due to charging assuming one electronic charge per nanodot would yield:

$$\text{Delta Vth} = qNNd/Ei(Di + 0.5Ei.Dnd/Esi), \text{ when,}$$

Ei is the permittivity of the dielectric medium where nanodot charge center resides, Esi is the silicon permittivity, Di is the insulator thickness from the nanocrystal plane to the gate (blocking insulator thickness), and Dnd, as previously, the diameter of the quantum nanodot.

As an example, Delta Vth = 0.485 V for Di = 10 nm, and $NNd = 1E12/cm^2$.

It has been experimentally observed that for the above case example Delta Vth has been measured to be higher than the one predicted above. This is explained to be due to the capture of more than one electron per nanocrystal quantum dot in such device.

The effect of available nanodot density on threshold shift (Delta Vth) in different dielectric medium for a dielectric thickness Di of 10 nm for different dielectric medium is shown in Table 2.2, assuming no additional charge trapping could be associated with the trap centers of the bulk dielectric medium. It should be noted, however, that for dielectric medium with high bulk and interface trap density, for example, nitride or HfO_2 (see section on dielectric films), such assumption is not valid. In general, when nanodots are embedded into trapping dielectric medium, charge trapping takes place in multiple energy levels. In such cases, Delta Vth could be significantly greater than that predicted by the above formulation and is dependent on the trap energetics of the trapping dielectric medium and those created by the quantum nanodots. Table 2.2 suggests, however, that under the constraints of a single-planar, single-energy nanodot-induced charge storage within a trap-free dielectric medium with nanodot density optimized for charge retention, the memory window would be limited due to the effects of coulomb blockade (CB) and QC.

TABLE 2.2

Impact of Nanocrystal Density on Threshold Shift for Several 10 nm Blocking Dielectrics

Nanocrystal Density	Computed Threshold Shift (V)			
	SiO_2	Si_3N_4	Al_2O_3	HfO_2
1.00E+12	0.485	0.33	0.25	0.1
2.5E12[a]	1.16	0.785	0.55	0.22

[a] Equivalent to 3 nm nanocrystal diameter with 3.32 nm spacing.

2.3.3 Review of Nanocrystal NVM Devices

Nanocrystal NVM devices employed silicon nanodots, germanium nanodots, a combination of Si and Ge nanodots, and a variety of metal nanodots with different work functions. A major number of investigations incorporated silicon nanodots in oxide since it had been easier to fabricate such devices. It has been stated earlier that with appropriate SRN composition [52] consisting of Si nanodots in nitride, significantly larger memory window could be achieved with equivalent tunnel oxide thickness compared to the MONOS stack design. This is shown in Figure 2.7. For a thicker tunnel oxide device of thickness >3.5 nm, electrons or holes transport is dictated by the Fowler–Nordheim tunneling. This leads to achieving larger memory window and a more stable NVM device. Larger window is associated with multiple charge trapping per nanodot associated with multiple energy states generated as shown previously in Figure 2.6.

Activation energy studies by Y. Liu et al. [57] confirmed the evidence of deep trap energy levels below the Si–Ge conduction band level of 0.33–0.46 eV, in P-MOS capacitors. The structure consisted of 3.0 nm tunnel oxide, 7–10 nm Si–Ge nanodots (16% Ge), and 9.0 nm of blocking oxide over the nanodot trapping layer.

Device window based on nanocrystal-based CT devices depend on the following factors: (a) Dot density, (b) Dot size and dot spacing, (c) Excess silicon concentration if SRIs are employed as source of nanodots, (d) trapping energy levels of dielectric material in which nanodots are embedded (e.g., nitride in SRN), and (e) work function of semiconductor nanodot or metal nanodot incorporated in the device. In general, nanocrystal CT devices have the potential advantages over the floating-gate devices and SONOS types of CT devices due to the unique feature of what may be described as discrete "planar charge trapping" (PCT) characteristics. Most significant outcome of PCT characteristics is the reduction in field requirement for erasing since the electrostatic field created by the electronic PCT enhances effective hole transport, thereby minimizing requirement of external field-imposed "hot holes" during erasing. Because of PCT, hole-induced damage in tunnel oxide is reduced, reducing stress-induced leakage current (SILC),

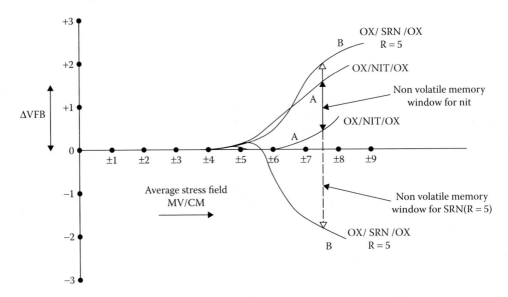

FIGURE 2.7 Example of larger memory window (enhanced flat band shift) for MO-SRN-OS: (Si Nc in Nitride) stack compared to conventional MONOS. (From Bhattacharyya, A. et al., Physical and electrical characteristics of LPCVD silicon-rich-nitride, *ECS Fall meeting*, New Orleans, CA, 1984.)

and thereby enhancing endurance and reliability. This has been further discussed in Chapter 9, Section 9.2. Additional advantages include:

- Faster writing since less electron injection is required for writing limited by CB.
- Reduction in parasitic fringing field effect with neighboring cells ("edge effect," and "disturb," discussed later on), thereby enhancing scalability as cell feature sizes are reduced.
- Reduction in local defect-induced leakage path through tunnel oxide, a common concern in FG devices.

Such potential advantages motivated a flurry of innovations and investigations in nanocrystal memory development during the last decade. Thus far, the R&D activities had been largely limited to capacitors and transistors. Enhancements in programming speed, endurance, and reliability of nanocrystal devices had been demonstrated at the transistor level. Concepts to overcome limitations related to memory window and retention had been suggested. However, nanocrystal memory products are yet to emerge as a mainstream NVM product offering. As stated earlier, critical challenges exist in integration and manufactureability to develop large-scale reproducible nanocrystal single or multiplanar devices with controlled size and distribution of nanodots. Template self-assembly schemes [58,59] to achieve size and distribution uniformity had been demonstrated. However, complexity in technology integration, stability, and manufactureability remains as a challenge.

2.3.4 Nanocrystal Device General Characteristics

Selected references of nanocrystal device characteristics will be reviewed here. A broader yet limited reference of published nanocrystal memory work is added at the end of the section. Most of the studies involved silicon nanodots embedded on top of tunnel dielectric with an overlayer of blocking dielectric. The insulator stack structure could be obtained in at least two ways:

1. After forming the tunnel insulator (e.g., thermal oxidation of silicon substrate to form desired thickness of SiO_2 tunnel oxide), a layer of appropriate composition of silicon-rich insulator (either SRO or SRN) is deposited by the LP CVD process. This is followed by an appropriate blocking insulator (e.g., SiO_2 or Si_3N_4 or SiON) deposited in situ by vapor phase. Such fabrication can be implemented sequentially. After deposition, and appropriate postdeposition anneal, the desired size and distribution of nanocrystals are achieved.
2. After forming the tunnel insulator as above, a layer of silicon nanocrystal could be deposited by either of several means, e-beam evaporation, sputtering deposition, or by vapor phase deposition as above. This may be followed by an appropriate anneal in situ to form the desired size and distribution of silicon nanocrystals in size and distribution followed by the deposition of the blocking insulator by similar means and a postdeposition anneal.

Care is taken in either approaches to ensure the elimination of any possible partial oxidation of silicon nanocrystal by means of ambient control. The blocking dielectric layer could be ONO (oxide-nitride-oxide), similar to SONOS blocking layer, or high K dielectric such as CVD-deposited alumina (Al_2O_3). Either metal gate (e.g., titanium nitride (TiN) or tantalum nitride (TaN)) or polysilicon gate is compatible with the standard CMOS process technology and could be adopted.

Early silicon nanocrystal device investigators fabricated simple device structures of NV capacitors and transistors as outlined above to study the device characteristics. Subsequently, tunnel-engineered stack designs such as NON tunnel barrier or high K tunnel and blocking layers such as HfO_2 or ZrO_2 (replacing SiO_2) were also employed to investigate enhancements in device characteristics.

2.3.5   Silicon Nanocrystal Device Characteristics

General observations of silicon nanocrystal device characteristics are as follows:

1. Devices demonstrated improved writing speed and endurance.
2. Retention was strongly sensitive to tunnel oxide thickness. For 2.5 nm tunnel oxide, retention was nearly 1E4 seconds, while for 4.5 nm tunnel oxide thickness, retention was greater than 1E8 seconds (nearly 10 years).
3. Programming voltage levels in Si nanocrystal devices were similar to those of SONOS devices for equivalent EOT gate stack design. Programming voltage level is also approximately 0.5× those of FG devices similar to those of SONOS devices. This is expected since tunnel oxide thickness for equivalent retention was also approximately 0.5× compared to those of FG Flash NV devices.
4. Equivalent EOT silicon nanocrystal device with similar tunnel oxide structures showed somewhat lower memory window, compared to the SONOS device. For example, memory window for nanocrystal device typically would be in the range of 2.5–3.0 V initially and nearly 1.5 V end-of-life (EOL) while corresponding SONOS device would yield 4 V window initially and EOL window of nearly 2.5 V. As should be anticipated, this property is dependent on size, density, and distribution of nanodots and other factors.
5. Silicon nanocrystal device showed "write–saturation," implying Vth will not increase beyond a certain value on increasing additional write-voltage and pulse duration, thereby reaching a steady state between trapping and detrapping of injected electrons. SONOS devices on the contrary showed "Erase-Saturation," which has been explained earlier due to electron injection from the gate compensating hole injection from the substrate, thereby reaching steady state during erase operation of the SONOS device.

   In recent years, several authors have investigated silicon nanocrystal devices using SRO films. Relatively large window (>4 V) with enhanced write/erase performance (e.g., /E at ±18 V, 200 μs) and 10 years of retention had been reported by K.-H. Joo et al. in 2006 for silicon nanocrystal device consisting of TaN, gate/alumina, charge-blocking dielectric/ silicon nanodots in SRO/ ≥2.5 nm of tunnel oxide NVM stack design [60]. The SRO films were deposited by a HDP-CVD flash deposition technique [61] and consisted of 2–3 nm of silicon nanocrystals in 4–10 nm HD-CVD SRO containing excess silicon concentration ranging 10%–20%. Silicon nanocrystals precipitated from SRO films by postdeposition annealing at 850°C–1050°C in mixed nitrogen–oxygen ambient.

   Investigations during the 1980s have demonstrated that silicon nanodots in SRN are significantly more stable compared to those in SRO films [52,53]. It was shown that post-deposition thermal budget and ambient alters the shape, size, and distribution of silicon nanodots in SRO due to silicon atom interdiffusion, which can be readily suppressed for silicon nanodots in SRN films [52]. Consequently, integration challenges are less severe for silicon nanocrystal devices in SRN compared to those in SRO [54]. Additional advantage of SRN-based devices is that such devices yield larger memory window due to the availability of higher trapping density as explained earlier.

R. Muralidhar et al. [62] have outlined several key integration challenges in fabricating a 6 V, 4 Mb NV NAND memory array using SRO-based silicon nanodot devices. The 4 Mb NVM NAND array was employed for embedded application in a standard CMOS technology by the addition of four non-critical mask over the base-line CMOS logic process (replacing 10 mask additional process requirements for embedding conventional FG-NVM NAND arrays [63]). The gate insulator stack design for the NVM device consisted of 5 nm of tunnel oxide followed by 5 nm of silicon nanodot deposited by CVD technique for a density of 1E12/cm², a thin SiON overlayer to protect the silicon nanodots from oxidation during blocking dielectric deposition, and an ONO blocking dielectric of 12 nm of EOT. Integration challenges addressed by Steimle et al. for the embedded NVM array fabrication included:

1. Nanodot formation
2. Nanodot protection from oxidation and subsequent processing
3. Removal of nanodots outside the array region
4. Removal of nanodots from bit-cell, source/drain, and extension areas in order to ensure bit-cell performance stability.

Device characteristics obtained by the above authors are shown in Figure 2.8a–c. Ten years of retention and 1E6 W/E cycle of endurance were demonstrated in Figure 2.8a and b respectively.

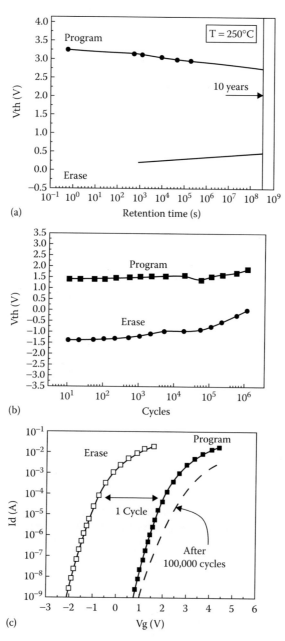

FIGURE 2.8 High temperature charge retention characteristics with ONO blocking dielectric stack, (a) Retention, (b) Endurance and (c) Transconductance post 100,000 W/E cycling. (From Muralidhar, R. et al., A 6V embedded 90 nm silicon nanocrystal nonvolatile memory, *IEDM*, p. 601 © 2003 IEEE.)

Figure 2.8c plots Id versus Vg characteristics after the first W/E cycle and post 100,000 W/E cycles. Transconductance degradation post 100,000 W/E cycles was considered to be acceptable by the authors.

The importance of charge-blocking dielectric to achieve stable device characteristics as well as performance enhancements was investigated by several researchers [62,64]. Replacement of ONO by high K blocking layers such as $O/HfO_2/O$ was shown to enhance writing characteristics. Furthermore, protecting Si nanodots from blocking oxide formation and subsequent processing were shown to be critical for reproducible device characteristics.

Tunnel insulator engineering was also shown to provide device enhancements of silicon nanocrystal devices. S. Baik et al. [64] had demonstrated that significant enhancements in retention, endurance, and programming (writing/erasing) speed could be achieved by replacing SiO_2 tunnel dielectric layer by a NON tunnel barrier for such devices. They reported 10 μs writing and 100 μs erasing performance for silicon nanodot devices using the above tunnel barrier and 10 nm of deposited oxide for charge blocking. The NON tunnel barrier consisted of 1.0 nm of SiN/2.2 nm MTO/2 nm SiN to provide what is known as "rested tunnel barrier" (see Section 2.3). Both speed and retention were enhanced using above tunnel-engineered barrier.

2.3.6 Metal Nanocrystal Device Characteristics

Metal nanocrystals can be readily embedded into dielectric medium by several low-temperature techniques for device fabrications. These techniques are as follows: (1) E-beam evaporation followed by an RTA [65], (2) by D. C. Or R. F. Sputtering [67], by plasma vapor deposition (PVD) followed by nitrogen RTP at temperature ranging 500°C–800°C [66] and (4) by PNL (pulsed nucleation layer) process [67]. Multiple planes of metal nanodots could be embedded incorporating high density of nanodots to enhance the memory window. Nanodot density as high as $2E13/cm^2$ was reported by sputtering of tungsten (W) as well as of cobalt (Co) nanodots [67]. WN nanodot of density of $2E12/cm^2$ was achieved by S. H. Lim et al. using the PNL process [67]. In recent times, ALD techniques were also developed by several investigators to incorporate high density of metal nanodots in dielectric layers. Metal nanodots with varying work functions (WF) were embedded for device studies. These studies include platinum (Pt, WF = 5.3 eV), tungsten (W, WF = 4.55 eV), silver (Ag, WF = 4.3 eV), gold (Au, WF = 4.25 eV), and tantalum (Ta, WF = 4.2 eV). Metal dots ranged in diameter from 2 to 7 nm with density ranging $2E12/cm^2$ to $2E13/cm^2$. Both single-planar and double-planar devices were investigated, and it was generally observed that multiplanar devices demonstrated enhanced memory window and retention. Device retention characteristics of metal nanodots devices are, in general, similar to silicon nanodot devices and primarily driven by the tunnel insulator thickness. However, comparatively superior retention was achieved by employing Pt nanodots with largest work function when compared with silicon nanodot devices. Ten years of retention can be achieved with Pt nanodot devices using tunnel oxide thickness as thin as 4 nm. For direct tunnel devices, however, retention is essentially similar to silicon nanodot devices for equivalent dot dimensions and distribution.

Work function of potential metal and semiconductor elements are shown in Figure 2.9. Future devices may explore other metal or semiconductor nanodot devices with larger work function differences, for example, arsenic (As, WF = 5.11 eV), iridium (Ir, WF = 5.3 eV), rhenium (Re, WF = 5.0 eV), tellurium (Te, WF = 5.0 eV), and palladium (Pd, WF = 5.0 eV). From retention enhancement standpoint, band engineered multilevel direct tunneling layers could be employed (discussed in section on retention) with or without incorporation of nanodots inside the direct tunnel dielectric layers. It should be noted, however, that multiple layers of high density metal nanodots could be incorporated into the device structure to significantly enhance the memory window, especially for multibit per cell storage applications. However, as discussed earlier, process integration issues into the standard scaled CMOS technology, selective incorporation and removal of metal nanodots in

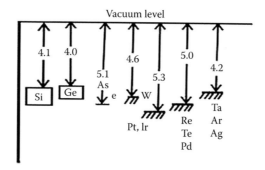

FIGURE 2.9 Work functions of semiconductors and metals.

device structures, thermal budget and postprocessing stability of size and distributions remain critical challenges for metal nanodot devices.

2.3.7 Germanium Nanocrystal Device Characteristics

Along with silicon nanodot devices, germanium (Ge) and germanium–silicon (Ge–Si) nanodot devices were investigated and characterized. In general, Ge nanodot devices provided superior charge retention compared to silicon devices due to its deeper valence band offset against SiO_2, although conduction band offset of Ge and Si is similar. Larger window device structure was reported for a Ge nanodot device stack consisting of nitride/HfO_2 tunnel dielectric with 2.5 nm Ge nanodots embedded by sputtering with blocking dielectric layer of deposited SiO_2 [68]. A method to incorporate high density Ge or Ge–Si nanodots into device structure was disclosed by Bhattacharyya [69] to take advantage of superior retention of Ge nanodot devices. Several advanced nanocrystal-based device could be envisioned if manufacturability issues could be resolved. Such advanced devices are discussed in Part II, Chapter 5 of this book.

2.4 DIRECT TUNNEL MEMORY

Another type of NVM device based on quantum mechanical direct tunneling transport of electronic charges (electrons and/or holes) is known as "direct tunnel memory" or DTM. DTM memories are often considered as a class by themselves wherein charges could be stored optionally either in floating gate, or in trapping planes based on SRIs or embedded nanodots, or in bulk traps associated with CT dielectric such as nitride (Si_3N_4) or oxynitride (SiON). As the name suggests, the DTM device is based on carrier transport by quantum mechanical direct tunneling. Since direct tunneling is energetically the most efficient mode of carrier transport through the dielectric medium, such devices have potential advantages in reliability over any other types of device discussed earlier. A brief historical review of DTM devices is presented here. A primary objective of this book is to propose a unified set of nonvolatile memory devices based primarily on the DTM concept and yet overcoming all the current limitations of silicon based NVMs in the form of "silicon based unified memory (SUM)" devices (Part III). The DTM concept, therefore, has been discussed in several sections in this book: to extend "NVM-unique" device properties, namely, memory window, retention, and endurance; to achieve multifunctionality as well as multilevel storage within the memory cell; to functionally replace current volatile higher performance memories such as SRAMs and DRAMs; to extend applications into high capacity storage arena of less reliable HDDs and magnetic tapes; and to provide evolutionary options against "nonsilicon based" emerging NVMs.

A DTM cell was first proposed by K. Tusunoda et al. [70] in 2004, to functionally replace DRAM. The memory cell was a split-channel FET device with the central part of the channel consisting of

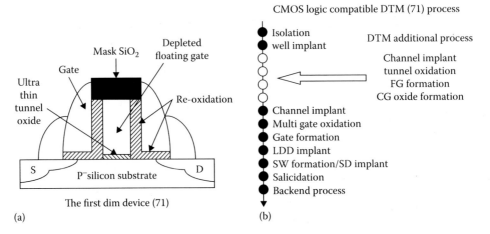

FIGURE 2.10 (a) Schematic cross section of the first DTM device and (b) Process flow. (From Tusunoda, K. et al., Ultra-high speed direct tunnel memory (DTM) for embedded RAM applications, VLSI Technology Digest, p. 133, VLSI Technology Digest Publication © 2004 IEEE.)

a very thin direct tunnel oxide with an overlayer of depleted floating gate. The control gate wrapped around the floating gate while overlapping sidewise the source and drain edges of the channel. The cross section of the device across the channel is shown in Figure 2.10a. The underlying gate insulator of the control gate is a thicker oxide, which sets the low Vth of the memory cell (Vth low = 1.5 V). The control gate capacitively couples the floating gate via the sidewall oxide as shown. The fabrication process integrates readily with the conventional CMOS technology by adding several additional steps at the front end as shown in Figure 2.10b. This makes such device attractive for embedded applications.

The tunnel oxide thickness for the device was 1.55 nm. Memory threshold window of 1.3 V was demonstrated by the authors with programming voltages of ±4.0 V, 40 ns writing and erasing. Endurance and charge retention for the device were respectively >1E11 W/E cycles and 10 seconds. During the same year (2004), several embodiments of DTM devices were proposed, providing significantly enhanced retention, increased coupling between the floating node and the control gate as well as both planar and 3D configurations to further enhance density. These devices are described in greater detail in chapter in Part II of this book.

The DTM device proposed by K. Tusunoda had several limitations. These were: (1) Voltage coupling between the control gate and the floating gate is inefficient due to the low dielectric constant of SiO_2 between the control gate and floating gate, requiring large coupling area and affecting both cell density and voltage scalability; and (2) Limited retention due to enhanced leakage of the floating gate edge to the substrate and diffusion.

A brief description of one of the improved DTM devices is illustrated below.

An improved DTM cell proposed subsequently was a similar split-channel DTM cell called "Embedded Trap DTM" (ETDTM), which improves memory retention by many orders of magnitude while providing high program/erase speed and infinite endurance of the DTM [71]. The tunnel oxide is replaced by a thin layer of scalable oxynitride, while the floating gate is replaced by a thin layer of either an embedded trap CVD oxynitride film or an embedded nanometal-dot insulator (see section on LP CVD SRN film) to provide many orders of magnitude longer memory retention in the deep quantum-well trap thus formed. Additionally, the thin embedded trap layer has a thick overlayer of silicon-rich nitride of high conductivity (injector SRN, see also section on SRN films as above), which acts electrically like a top floating electrode while chemically passivating the

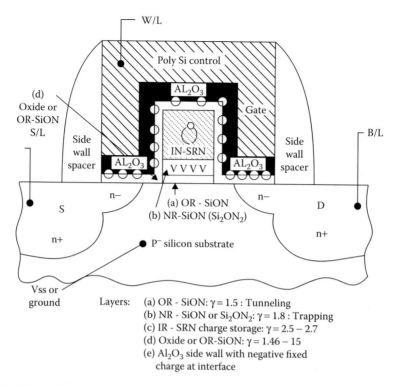

Layers: (a) OR - SiON: $\gamma = 1.5$: Tunneling
(b) NR - SiON or Si_2ON_2: $\gamma = 1.8$: Trapping
(c) IR - SRN charge storage: $\gamma = 2.5 - 2.7$
(d) Oxide or OR-SiON: $\gamma = 1.46 - 15$
(e) Al_2O_3 side wall with negative fixed charge at interface

FIGURE 2.11 Schematic cross section of embedded trap DTM: The split-channel device nonvolatile element consists of oxygen-rich SiON tunnel layer, nitrogen-rich SiON trapping layer combined with injector SRN for charge storage and field enhancement. (From Bhattacharyya, A., Embedded trap direct tunnel NVM, A. Bhattacharyya, ADI Associates Internal Publication, U.S. Patent 7365388 [4/29/08], 2004.)

underlayers of gate insulator dielectric films and the silicon substrate. Both of the above layers (SiON or SRN for trapping as well as SRN-injector for floating node) are deposited by low-temperature CVD or ALD processes. This stack is separated from the wrapped control gate by a high K coupling medium of thin layers of oxide plus Al_2O_3 (or oxide/HfO_2 or oxide/ZrO_2), which provides enhanced coupling between the control gate and the embedded floating-trap layer. This allows enhanced voltage scalability over the earlier DTM device proposed by K. Tusunoda et al. The device is shown in Figure 2.11. The gate insulator stack consisted of 1.5 nm of oxygen-rich SiON (refractive index of 1.55) for tunneling/4.5 nm of a nitrogen-rich layer of SiON (refractive index of 1.8) for the trapping layer/10–15 nm of highly conductive SRN layer (refractive index of 2.5–2.7) to provide enhanced injection and equipotential medium/and a combination of deposited oxide and alumina on all sides including the sidewall as shown in Figure 2.11. The device is further discussed in the section on advanced devices in Part II.

DTM concept has been the basis for many advanced NVM device features to be discussed in Part II and Part III of this book. Direct tunneling mode of charge transport could be combined with other lower energy mode of charge transport, namely, the enhanced Fowler–Nordheim tunneling and source-side charge injection mode of transport to simultaneously achieve DRAM as well as NVM functionality in a single-transistor memory cell. Furthermore, dense DTM memory cell could be envisioned to achieve multibit memory per cell.

These potentials of combining DTM with conventional and/or advanced CT devices provide an important framework for future highly scalable advanced NVM devices including the SUM devices. These will be further elaborated in Part II and Part III of this book.

REFERENCES

1. D. Kahng, and S. M. Sze, A floating gate and its application to memory devices, *Bell Syst. Tech. J.*, 46, 1288–1295, 1967.
2. D. Frofman-Bentchkowsky, Memory behavior in a floating gate avalanche injection MOS (FAMOS) structure, *Appl. Phys. Lett.*, 18, 332–334, 1971.
3. D. Frofman-Bentchkowsky, FAMOS—A new semiconductor charge storage device, *Solid State Electron.*, 17, 517–529, 1974.
4. H. Lizuka, T. Sato, F. Masuoka et al., Stacked gate avalanche injection type MOS (SAMOS) memory, *Proceedings of the 4th Conference on Solid State Device*, Tokyo, Japan, 1972; *J. Japan Soc. Appl. Phys.*, 42, p. 158, 1973.
5. H. Lizuka, F. Masuoka, T. Sato et al., Electrically alterable avalanche injection type MOS read-only-memory with stacked gate structure, *IEEE Trans. Elec. Dev.*, ED-23, 379–387, 1976.
6. R. G. Muller, H. Nietsch, B. Rossler et al., An 8192-bit electrically alterable ROM employing a one transistor cell with floating gate, *IEEE J. Solid State Circuits*, SC 12(5), 507–514, 1977.
7. M. Gill, R. Cleavelin, S. Lin et al., A novel sub-lithographic tunnel diode based 5V flash memory, *IEEE, IEDM Technology Digest*, 1990, pp. 119–122.
8. F. Masuka, M. Asano, H. Iwahashi et al., A new Flash EEPROM cell using triple polysilicon technology, *IEEE, IEDM*, 1984, pp. 464–467.
9. F. Masuoka, M. Asano, H. Iwahashi et al., A 256-kbit flash E^2PROM using triple-polysilicon technology, *IEEE J. Solid State Circuits*, SC-22, 548–552, 1987.
10. S. Mukherjee, T. Chang, R. Pang et al., A single transistor EEPROM cell and its implementation in a 512Kb CMOS EEROM, *IEEE, IEDM*, 1985, pp. 616–619.
11. G. Samachisa, C.-S. Su, Y.-S. Kao et al., A 128 K flash EEPROM using double polysilicon technology, *IEEE, ISSCC Digest of Technical Papers*, 1987, pp. 76–77.
12. T. H. Ning, and H. N. Yu, Optically-induced injection of hot electron into SiO_2, APS, *J. Appl. Phys.*, 45, 5373–5378, 1974.
13. P. E. Cottrell, R. R. Troutman, and T. H. Ning, Hot electron emission in n-channel IGFETs, *IEEE J. Solid State Circuits*, SC 14, 442–455, 1979.
14. B. Rossler, and R. Muller, Electrically erasable and reprogrammable read only memory using n-channel SIMOS one-transistor cell, *IEEE Trans. Elec. Dev.*, Ed-24, 606–610, 1977.
15. G. Groeseneken, H. E. Maes, J. Van Houdt, and J. S. Witters, Basics of nonvolatile semiconductor memory devices, Chapter 1, in *Nonvolatile Semiconductor Memory Technology*, IEEE Press, 1997, pp. 1–88.
16. M. Gill, and S. Lai, Floating gate flash memories, Chapter 4, in *Nonvolatile Semiconductor Memory Technology*, IEEE Press, 1997, pp. 189–308.
17. A. Bhattacharyya, Non volatile memory: Challenges and opportunities, ADI Associates Internal Publication, Micron Technology Seminar Presentation, Micron Confidential, Unpublished, July, 2004.
18. A. T. Wu, T. Y. Chen, P. K. Ko et al., A novel high speed 5V programming EPROM structure with source-side injection, *IEEE, IEDM*, 1986, pp. 584–587.
19. K. Naruke, S. Yamada, E. Obi et al., A new flash-erase EEPROM cell with sidewall select-gate on its source side, *IEEE, IEDM*, 1989, pp. 603–606.
20. Y. Yamauchi, K. Tanaka, H. Shibayama et al., A 5V only virtual ground flash cell with an auxiliary gate for high density high speed applications, *IEEE, IEDM*, 1991, pp. 319–322.
21. Y. Ma, C. S. Pang, J. Pathak et al., A novel high density contact less flash memory array using split-gate source side injection cell for 5V only applications, IEEE & JSAP, *VLSI Technology*, 1994, pp. 49–50.
22. S. Kianian, A. Levi, D. Lee et al., A novel 3V only small sector erase high density flash EEPROM, IEEE & JSAP, *VLSI Technology*, 1994, pp. 71–72.
23. J. Van Houdt, D. Wellekens, L. Haspeslagh et al., A 5V/3.3V compatible flash E^2PROM cell with a 400ns/70us programming time for embedded memory application, *IEEE, Nonvolatile Memory Technology Review*, June, 1993, pp. 54–57.
24. W. Johnson, G. Perlegos, A. Reninger et al., A 16 Kb electrically erasable nonvolatile memory, *IEEE, ISSCC Conference on Digest of Technical Papers*, 1980, pp. 152–153.
25. S. K. Lai, Y. W. Hu, S. Tam et al., Design of an E^2PROM memory cell less than 100 square micron using 1 micron technology, *IEEE, IEDM*, 1984, pp. 468–471.
26. V. N. Kunett, A. Baker, M. Fandrich et al., An in-system reprogrammable 32KX8 CMOS flash memory, *IEEE J. Solid State Circuits*, 83, 1137–1162, 1988.
27. H. Onoda, Y. Kunori, S. Kobayashi et al., A novel cell structure suitable for a 3V operation, sector erase Flash Memory, *IEEE, IEDM Technology Digest*, 1992, pp. 599–602.

28. S. Kobayashi, M. Mihara, Y. Miyawaki at al., A 3.3 V-only 16 Mb DINOR flash memory, *IEEE, ISSCC Digest Technology Papers*, 1995, pp. 122–123.
29. M. Nakashima, and N. Ajika, Dinor flash memory technology, Chapter 7, in *Nonvolatile Memory Technologies with Empkasis on Flash*, edited by J. E. Brewer and M. Gill, pp. 313–336. IEEE Press, 2008.
30. R. Kirisawa, S. Aritome, R. Nakayama et al., A NAND structured cell with a new programming technology for highly reliable 5V only Flash EEPROM, IEEE & JSAP, *VLSI Technology*, 1990, pp. 129–130.
31. H. A. R. Wagener, A. J. Lincoln, H. C. Pao et al., The variable threshold transistor, a new electrically alterable, non-destructive read-only storage device, *IEEE, IEDM Technology Digest*, Washington, DC, 1967.
32. A. Bhattacharyya, Technology development and optimization of P-Channel and N-Channel MXOS FET technology for nonvolatile memory application, Part I: P-Channel: 1972, Part II: N-Channel, IBM Internal Technical Report, 1973 (Unpublished).
33. A. Bhattacharyya, R. Silverman, M. L. Joshi et al., IBM 8Kb AROS memory program: T1 results, IBM Internal Technical Report, 1974 (Unpublished).
34. A. Bhattacharyya, C. T. Kroll, H. W. Mock et al., Properties and applications of silicon oxynitride films, IBM Internal Technical Report, October, 1976 (Unpublished).
35. A. Bhattacharyya, C. T. Kroll, H. W. Mock et al., Processing and characterization of CVD thin silicon oxynitride films, *ECS*, 1976, Las Vegas, CA.
36. A. Bhattacharyya, M. L. Joshi, and C. T. Kroll, FET gate structure for nonvolatile N-channel read mostly memory device, *IBM Tech. Discl. Bull.*, 18(6), 1975, 1768.
37. A. Bhattacharyya, M. L. Joshi, C. T. Kroll et al., (a) Fixed and variable threshold N-channel MNOS FET integration technique, U.S. Patent 3978577, September 7, 1976.
37b. (b) A. Bhattacharyya, T. Horoun., Small system organization for a nonvolatile mass memory, *IBM Tech. Discl. Bull.*, 18(3), 654–655, 1975.
38. H. E. Maes et al., Nonvolatile memory characteristics of polysilicon-oxynitride-oxide-silicon devices and circuits, *Proceedings of ECS*, vol. 87–10, edited by J. Kapoor and K. T. Hankins, p. 28, 1987.
39. T. Hagiwara, Y. Yatsuda, R. Kondo et al., A 16kbit electrically erasable PROM using n-channel Si-gate MNOS technology, *IEEE J. Solid State Circuits*, SC-15, 346–353, 1980.
40. E. Suzuki, H. Hiraishi, K. Ishi et al., A low voltage alterable EEPROM with metal-oxide-nitride-oxide-semiconductor (MONOS) structure, *IEEE Trans. Elect. Dev.*, ED-30, 122, 1983.
41. C. C. Chao, and M. H. White, Characterization of charge injection and trapping in scaled SONOS/MONOS memory devices, *Solid State Electron.*, 30, 307–319, 1987.
42. T. A. Dellin, and P. J. McWhorter, Scaling of MONOS nonvolatile memory transistors, *Proceedings of ECS*, vol. 87–10, p. 3, edited by V. J. Kapoor and K. T. Hankins, 1987.
43. F. R. Libsch, A. Roy, and M. H. White, Amphoteric trap modelling of multidielectric scaled SONOS nonvolatile memory structures, *IEEE, 8th NVSM Workshop*, 1986, Vail, CO.
44. F. R. Libsch, and M. H. White, SONOS nonvolatile semiconductor memory, Chapter 5, in *Non Volatile Semiconductor Memory Technology*, edited by W. D. Brown and J. E. Brewer, p. 309. IEEE Press, 1997.
45. Y. Yatsuda, S. Nabetani, K. Uchida et al., Hi-MNOS II technology for a 64-kbit byte-erasable 5-V-only EEPROM, *IEEE Trans. Elect. Dev.*, 20, 144–151, 1985.
46. M. H. White, and F. R. Libsch, SONOS nonvolatile semiconductor memories, Chapter 5, in *Nonvolatile Semiconductor Memory Technology*, edited by W. D. Brown and J. E. Brewer, p. 325. IEEE Press, 1997.
47. H. E. Maes, G. Groesenehen, H. Lehen, and J. Witters, Trend in semiconductor memories, *IEEE, Microelectron. J.*, 20, 9–58, 1989.
48. M. Taguchi, NOR flash memory technology, 2006 IEDM Short Course; Course Organizer: Rich Liu, IEDM, and IEEE Press., 2006.
49. J. R. Cricchi, F. C. Blaha, M. D. Fitzpatrick et al., The drain-source protected MNOS memory device and memory endurance, *IEEE, IEDM Technology Digest*, 1973, pp. 126–129.
50. A. Lancaster, B. Johnstone, J. Chritz et al., A 5V only EEPROM with internal program/erase control, *IEEE, ISSCC Technology Digest*, 1983, pp. 164–165.
51. T. Nozaki, T. Tanaka, Y. Kijiya et al., A 1 Mb EEPROM with MONOS memory cell for semiconductor disk application, *IEEE J. Solid State Circuits*, 26(4), 497–501, 1991.
52. A. Bhattacharyya, R. Bass, W. Tice et al., Physical and electrical characteristics of LPCVD silicon-rich-nitride, *ECS Fall meeting*, October 1984, New Orleans, CA.
53. R. Bass, A. Bhattacharyya, and G. D. Grise, Nonvolatile memory cell having silicon rich silicon nitride charge trapping layer, U.S. Patent No. 4870470, September 26, 1989.
54. D. J. DiMaria, K. M. DeMeyer, C. M. Serrano et al., Electrically alterable read-only-memory using Si-rich SiO2 injectors and a polycrystalline silicon storage layer, *J. Appl. Phys.*, 52, 4825–4842, 1982.

55. S. Tiwari, F. Rana, K. Chan et al., Volatile and nonvolatile memories in silicon with nano-crystal storage, *IEEE, IEDM*, 1995, pp. 521–524.

56. S. Tiwari, F. Rana, H. Hanafi et al., Volatile and nonvolatile memories in silicon with nano-crystal storage, APS, *Appl. Phys. Lett.*, 68, 1996, 1377–1379.

57. K. W. Guarini, C. T. Black, Y. Zhang et al., Low voltage scalable nanocrystal flash memory fabricated by templated self-assembly, *IEEE, IEDM*, 2003, pp. 541–544.

58. S. Tang, C. Mao, Y. Liu et al., Nanocrystal flash memory fabricated with protein-mediated Assembly, *IEEE, IEDM*, Washington DC, 2005, pp. 181–184.

59. K.-H. Joo, X. Wang, J. H. Han et al., Novel transition layer engineered Si nanocrystal Flash memory with MHSOS structure Featuring Large Vth window and Fast P/E speed, *IEEE, IEDM*, Washington DC, 2005, pp. 868–871.

60. K.-H. Joo, X. Wang, J. H. Han et al., Flash memory with MHSOS structure, Korean Patent P2005-0029290, 2005.

61. R. Muralidhar, R. F. Steimle, M. Sadd et al., A 6V embedded silicon nanocrystal nonvolatile memory for the 90nm technology node operating at 6V, *IEEE Integrated Circuit Design and Technology*, Austin TX, 2004, pp. 31–35.

62. R. Muralidhar, R. F.Steimle, M. Sadd et al., A 6V embedded 90 nm silicon nanocrystal nonvolatile memory, *IEEE, IEDM*, Washington DC, 2003, pp. 601–604.

63. C. Gerardi, B. DeSalvo, S. Lombardo et al., Performance of Si nanocrysral memories obtained by CVD and their potentialities to further scaling of non-volatile memories, *IEEE, International Conference on Integrated Circuit Design and Technology*, 2004, pp. 37–43.

64. S. J. Baik, S. Choi, U.-I. Chung et al., High speed and nonvolatile Si nanocrystal memory for scaled flash technology using highly field sensitive tunnel barrier, *IEEE, IEDM*, Washington DC, 2003, pp. 545–548.

65. C. Lee, A. Gorur-Seetharam, and E. C. Kan et al., Operational and reliability comparison of discrete storage nonvolatile memories: Advantages of single and double layer metal Nanocrystals, *IEEE, IEDM*, Washington DC, 2003, pp. 557–510.

66. M. Takata, S. Kondoh, T. Sakaguchi et al., New nonvolatile memory with extremely high density metal nano-dots, *IEEE, IEDM*, Washington DC, 2003, pp. 553–556.

67. S.-H. Lim, K. H. Joo, J.-H. Park et al., Nonvolatile MOSFET memory based on high density WN nanocrystal layer fabricated by novel PNL (pulse nucleation layer) method, IEEE & JSAP, *Symposium on VLSI technology*, Kyoto, Japan, 2005, pp. 190–191.

68. T. H. Ng, W. K. Chim, W. K. Choi et al., Minimization of germanium penetration, nanocrystal formation, charge storage, and retention in a trilayer memory structure with silicon nitride/hafnium dioxide stack as the tunnel dielectric, APS, *Appl. Phys. Lett.*, 84, 4385–4387, 2004.

69. A. Bhattacharyya, Scalable multi-functional and multi-level nanocrystal NVM, ADI Associates internal Publication, November 2004, U.S. Patent 7759715, issued; 7/30/2010.

70. K. Tusunoda, A. Sato, H. Tashiro et al., Ultra-high speed direct tunneling memory (DTM) for embedded RAM applications, IEEE & JSAP, *VLSI Technology Digest*, Honolulu HI, 2004, pp. 152–153.

71. A. Bhattacharyya, Embedded trap direct tunnel NVM, ADI Associates Internal Publication, 2004, U.S. Patent 7365388 (4/29/08).

3 General Properties of Dielectrics and Interfaces for NVM Devices

CHAPTER OUTLINE

Nonvolatile memory device performs as field effect transistor (FET) exhibiting multiple "stable" threshold states. These states are achieved by injecting charges (electrons and/or holes) from either the silicon substrate and/or the gate electrode, and by "permanently" holding the charges either into a floating storage medium or into a dielectric medium adjacent to the substrate to create appropriate threshold states. The dielectric media between the substrate and the gate and associated interfaces assume an essential and critical role in developing the characteristics of the nonvolatile memory device. This chapter will discuss the basic properties associated with dielectric films in general and related interface characteristics with silicon substrate and other dielectric, semiconductor, and metal films the context of FET gate stack structure. The selection and significance of various dielectric layers and interfaces in developing the characteristics of nonvolatile memories could be appreciated once the fundamental properties of materials and interfaces and roles played are understood.

3.0 THE NVM GATE STACK LAYERS AND INTERFACES

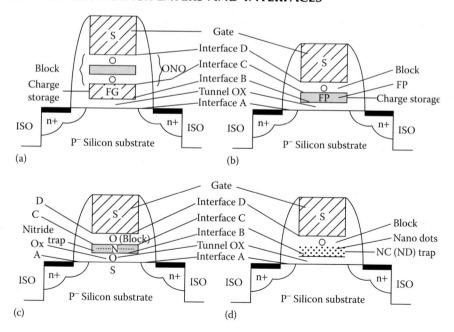

FIGURE 3.1 NVM device stack cross-sections: Dielectrics and charge storage layers and interfaces (A, B, C and D): (a) FG device, (b) FP device, (c) CT device, and (d) NC device.

3.1 ATTRIBUTES OF GATE STACKS FOR NVM DEVICES

The gate stack cross-sections of the conventional NVM-FET element for FG, FP (floating plate), CT, and NC (nanocrystal) devices are shown above in Figure 3.1a–d, respectively. The common elements are:

1. The silicon substrate.
2. *Interface A*: The silicon substrate with source and drain edges interfacing the tunnel dielectric layer defines interface A. This layer is a primary active layer for injection of electrons and holes from the silicon substrate.
3. *Tunnel insulator*: The tunnel dielectric above interface A had conventionally been SiO_2. The tunnel dielectric layer had been recently replaced by either SiON or high dielectric constant metal oxide or metal oxynitride, for example, HfO_2 or ZrO_2 or HfSiON or in some advanced devices, multiple very thin tunnel dielectric layers such as oxide/nitride/oxide (ONO). Advanced device tunneling layers employing higher dielectric constant metal-oxide films as mentioned above have been discussed in this book in Part II Chapter 12.
4. *Interface B*: This interface layer exists between the tunnel dielectric and the charge storage medium. In the conventional FG device, the polycrystalline silicon (doped or undoped) with associated grain edges interface the amorphous tunnel insulator. In the case of FP device, the interface could be an amorphous silicon-rich nitride (SRN) layer with high concentration of silicon nanocrystals at the interface. In the case of CT devices, the interface is typically a transitional oxide–nitride interface, both layers being amorphous. Finally, for the NC devices, the interface could either be similar to CT devices if the nanodots are embedded in nitride (as in SRN or nitride) or could be relatively uniform when the nanodots are embedded into deposited oxide.
5. *Charge storage layer*: The charge storage layer is deposited over the tunnel dielectric layer. For the FG device, this is the polysilicon layer. For the FP device, this is the injector SRN layer, which could be considerably thinner (<5–10 nm) compared to FG thickness, which could be >>10 nm. For the CT device, typical nitride thickness may be around 6–10 nm although thinner nitride layers may also be used for SONOS devices. For NC devices, the charge storage layer is similar in thickness comparable to SONOS nitride layer.
6. *Interface C*: This interface layer exists between the charge storage layer and the blocking dielectric layer. This interface is similar in characteristics to interface B.
7. *Charge blocking layer*: The charge blocking insulator layer is deposited over the charge storage layer. For the FG device, conventionally, with polysilicon control gate (CG), this layer is often called interpoly dielectric or IPD layer, being a single dielectric layer or multiple dielectric layers between the FG and the CG. For the FG device, this is typically an ONO structure with an equivalent effective oxide thickness (EOT) (equivalent oxide thickness, defined later on in the text) of 10 nm. For the FP device, this could be similar to the FG device but thinner. For the CT and NC devices, this could be oxide only with thickness around ~5 nm. For advanced device structures, charge blocking layer is currently often a single or multilayer high K dielectric film, for example, alumina or HfO_2 or ZrO_2 or HfSiON to reduce the stack EOT and thereby the writing and erasing voltage level requirements for the device.
8. *Interface D*: This interface layer exists between the blocking dielectric layer and the CG. Similar to the interface layer A, this layer is also a primary active layer for injection of electrons and holes from the CG.
9. *Control gate*: The CG could be either a heavily doped polysilicon layer ("S") or Metal or Metal Silicide ("M"). In recent years with feature size scaling, the CG is increasingly being changed from polysilicon gate to metal gate and often typically changed from

multigrain interface D associated with polysilicon grain to smoother interface D with metallic nitride such as titanium nitride (TiN) or tantalum nitride (TaN). Some popular SONOS devices are now called TANOS (TaN/alumina/nitride/oxide/silicon) device with associated change in blocking layer and gate material. The TANOS devices will be discussed in Part II Section 13.5.4, in this book.

The sidewalls of the gate insulator stack are isolated along the length and width of the FET channel by thicker oxide and associated interfaces. These regions are considered relatively passive and not taken into consideration from device characteristics point of view. However, as feature size is scaled, parasitic capacitive coupling with neighboring memory cells and associated disturb becomes important in device characteristics, which will be discussed in this book. Additionally, fringing fields due to edge effects associated with device isolation (shallow trench isolation, "STI," not shown, along the device width direction) also affect device characteristics when feature size is scaled. This will also be discussed in the later chapters of this book.

It should be noted, therefore, that all devices have at least four active interfaces and at least three dielectric layers and either metal or silicon gate. The characteristics of the interfaces, gate material, and the dielectric layers have profound impact on the various NVM device properties. This section will review the material properties of these layers and their significance on NVM device characteristics.

3.1.1 The Energy Band of the Gate Stack Layers

The gate stack layer of the NVM device with the associated dielectric layers and interfaces can be depicted in terms of the energy band diagram identifying the different elements in the stack design. As an example, the energy band diagram for the conventional floating-gate device is illustrated in Figure 3.2. On the left, is the silicon substrate with the energy band gap Eb = 1.1 eV. On the right, is the metal or silicon gate with the electronic work function (energy required to remove an electron from the gate to vacuum: WF; for silicon gate, WF = 4.1 eV) with reference to the vacuum. The tunnel dielectric oxide layer, the silicon floating gate charge storage layer, and the oxide-nitride-oxide (ONO) charge blocking layers are identified with their appropriate band gaps. The interfaces are identified with the shaded region. The potential energy barriers (U) or (Φ) [used interchangeably in this book] are identified as: the first suffix for electrons (e) and holes (h) and the second two-letter suffix relate to the function of the dielectric layer: "tu"

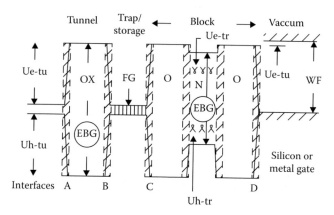

FIGURE 3.2 Energy band diagram for the FG stack: The band gap EBG for silicon, SiO$_2$, and Si$_3$N$_4$ are, respectively, 1.1, 9.0, and 5.1 eV; Ue–tu, and Uh–tu are respectively 3.2 and 4.7 eV, Ue'–tu for silicon gate is 4.3 eV and Ue–tr for nitride is 1.1 eV.

for tunneling, "tr" for trapping, and "bl" for blocking. Therefore, Ue–tu (or Φe–tu) implies the energy barrier for electron tunneling, Ue–tr (or Φe–tr) is the electron trap energy depth, and so on. The band diagrams are extremely useful means to recognize the energetics associated with the interface barriers impacting electrons and hole transport in metal/insulator/semiconductor structures. Band diagrams has been used extensively in this book to qualitatively explain various modes of electronic charge transport through NVM devices and consequently NVM device properties. It should be noted, however, that in many illustrations and especially with complex gate stack designs employing multilayered dielectric films, interface modulations of potentials (and associated energy) are ignored for simplicity.

3.2 GENERAL PROPERTIES OF THIN DIELECTRIC FILMS

Properties of thin dielectric films can be broadly categorized into three groups: (1) Physical, chemical, and thermal stability and associated reliability and breakdown strength; (2) Electronic properties such as dielectric constant, transport properties, and band gaps; and (3) Process-sensitive properties such as compositional uniformity or stoichiometry, scalability, interface properties such as fixed charge density, interface states density, and trap density. Selection of thin dielectric films has been a critical aspect of the evolution of microelectronics.

Since multiple layers of thin dielectric films are employed in the gate stacks of nonvolatile memories, commonly used thin dielectric films and future possibilities will also be discussed in this chapter addressing potential advantages and limitations of potential options.

The above discussion would be limited to amorphous thin dielectric films only, and to those dielectric films maintaining amorphous characteristics during the processing temperatures of 0°C–1000°C (typical of microelectronic device fabrication) without undergoing any structural transformations. Crystalline formations create defects along the grain boundaries and consequently exhibit higher leakage. Therefore, crystalline dielectric films are not suitable for microelectronics and nanoelectronics applications and will not be considered even if such dielectric films exhibit otherwise desirable dielectric characteristics.

3.2.1 Physical, Chemical, and Thermal Stability

Physical, chemical, and thermal stability of amorphous bulk dielectrics as well as thin films is dependent on purity, local, and long-range compositional uniformity, local and long-range bonding, electronic structure (position in periodic table of elements), and electron affinity. Most dielectric compounds exhibit either ionic or covalent bond characteristics, which determine their bond strength. A good indicator of the bond strength and related thermal stability associated with bulk or thin films of any dielectric is its melting point. The higher the melting point, the stronger is the bond strength.

Ionic bonds are established between metal cations and oxygen or silicates or halides to form stable metal oxides or silicates or halides. For oxides, metals exchange one, two, three, or four electrons (dependent on the energy states of the electrons in the outer orbitals, position in periodic table) with oxygen to form corresponding ionic bonds for crystalline dielectric establishing long-range order. In contrast for amorphous solid dielectrics, such bonding exists with short-range order only. Bond strengths in metal and semimetal (semiconductor) oxides are generally related to the electronic structure of the metal (or semimetal) and the number of electrons exchanged with oxygen to form stable inorganic compounds. For example, silicon (Si) and zirconium (Zr) with four electronic exchange form respective oxides of SiO_2 and ZrO_2 with high bond strength. Lanthanides (La, Pr, Sm, etc.) form corresponding oxides: La_2O_3, Pr_2O_3, Sm_2O_3, and so on possessing intermediate bond strength. Whereas metals such as Ba and Sr form BaO and SrO, respectively, with two electronic exchange exhibiting lower bond strength. Exceptions exist for metals such as aluminum, titanium, and tin.

TABLE 3.1
Melting Point and Dielectric Strength of Common Dielectric Materials

	SiO_2	Si_3N_4	Al_2O_3	TiO_2	ZrO_2	HfO_2
Melting point: C	1710	1900	2015	1640	2710	2810
Dielectric strength/cm	>10	10	5–10[a]	low[a]	≥4[a]	≥6[a]

[a] Strongly process sensitive.

In case of Aluminum (AL) with three electron exchange forming Al_2O_3 results in bond strength as nearly as strong as SiO_2. Titanium (Ti) could exchange 3 or 4 electrons forming, respectively, Ti_2O_3 and TiO_2 exhibiting intermediate bond strengths. In contrast, Tin with four electronic exchange forming SnO_2 exhibits lower bond strength. However, it is generally expected that metals with similar electronic structure will exchange electrons with oxygen in a similar manner. Therefore, such exchanges will result in similar characteristics related to thermal and chemical stability as well as short-range order from thermodynamics and reaction kinetics.

Electronegativity (or electron affinity for dielectric) is also an important parameter, which relates to bond formations and stability of dielectrics. While large difference in electronegativity between oxygen and metal results in ionic bonding, elements with similar electronegativity form covalent dielectric. Silicon carbide (SiC) is a good example of covalent stable dielectric and has wide application in electronics. Electron affinity and band gap of dielectrics are related and will be discussed in the section on electronic properties.

Melting point and dielectric strength of some common dielectric materials are shown in Table 3.1.

Breakdown strength and reliability of thin dielectric films will be discussed later on in a separate section. Both breakdown strength and reliability depend on material stability but also on electronic properties and process sensitivity.

3.2.2 Electronic Properties

Key electronic properties of dielectric materials as well as dielectric films for nonvolatile memory applications are: (1) Band gap and electron affinity, (2) Charge holding capacity or dielectric constant, (3) Destructive breakdown strength, (4) Leakage or charge transport characteristics, (5) Reliability under electrical stress, and (6) Interface characteristics and stability with reference to silicon substrate and gate electrode metallurgy, plus interface characteristics between different insulators when multilayer insulators are involved. A significant parameter in nonvolatile devices that are related to the above list is "EOT," and the second one is "scalability." These are also discussed in the section below.

Band gap and electron affinity: Dielectric materials prevent charge movement within the material and have the capacity to hold charges at the meta–dielectric interface. This is related to the energy barrier between electrons and holes called band gap. The larger the band gap, the less conductive the dielectric material under voltage stress. Electron affinity is defined as the energy difference between dielectric conduction band and vacuum. It represents the energy required to remove an electron from the dielectric material. The higher the electron affinity, the more chemically reactive the dielectric material with other elements at the interface in electronic exchanges. The band gap and electron affinity are represented as Eb and Ue, respectively. Dielectric materials with large band gap and low electron affinity provide low leakage and high breakdown strength. Both silicon dioxide (SiO_2) and alumina (Al_2O_3) are characterized by such attributes and are widely used in electronics. Eb and Ue for SiO_2 and Al_2O_3 are, respectively, 9 eV and 0.9 eV for SiO_2 and 8.7 eV and 1.35 eV for Al_2O_3. Both insulators exhibit high breakdown strength: >10E6 V/cm and low leakage. In contrast, insulator such as TiO_2 has a low band gap of Eb = 3.5 eV and large electron affinity of Ue = 4.1 eV. Consequently, TiO_2 films have limited applicability due to high leakage in spite of its high charge holding capacity (dielectric constant of 80).

Charge holding capacity or dielectric constant: Dielectric materials are used to hold charges as capacitors. Capacitor charge density is expressed in terms of capacitance per square micrometer in the form of $fF/\mu m^2$. One way to characterize the charge holding capacity of a dielectric is its dielectric constant (K), which is the ratio of its permittivity to that of vacuum characterizing the dielectric constant of vacuum as K = 1. The higher the dielectric constant, more is the capacity to hold charge. Thin SiO_2 films have a K value equal to 3.9, while a thin Al_2O_3 film could have a K value nearly equal to 9. Therefore, the same thickness of Al_2O_3 can hold ~2.3 times electronic charge when compared to the same thickness of SiO_2 film. For reasons to be discussed later on, SiO_2 films had been the reference dielectric films for the microelectronics industry. For technology scaling purposes, it is convenient to express other dielectric films in reference of "Equivalent SiO_2 oxide thickness" or "EOT." EOT is the ratio of dielectric constant of another dielectric film when compared to that of SiO_2 film. For the same geometry, EOT implies the scalability factor when SiO_2 film is replaced by a different dielectric film to achieve the same value of capacitance. Therefore, EOT of Al_2O_3 would be nearly $3.9/9 = 0.43\times$ of those of SiO_2.

Dielectric breakdown or dielectric strength: When sufficient electrical field is applied across a dielectric capacitor, electrons and holes get injected into the body of the dielectric and get transported across the dielectric exhibiting current flow. With increasing field, current flow increases. Eventually, at a certain critical field, permanent conductive path (or paths) is established and the dielectric breaks down and loses charge holding capacity. The field at which such destructive breakdown takes place is called the breakdown strength of the dielectric and is typically expressed in mega volts/cm or MV/cm. In general, the breakdown strength of a bulk dielectric is intimately related to the band gap and bond strength and chemistry of the dielectric. However, in thin dielectric films, the breakdown strength is influenced by local defects, compositional nonhomogeneity, surface irregularities, and process parameters. For ultrathin dielectric films, surfaces and interfaces play increasingly important role in determining the breakdown strength. Dielectric reliability is connected to the breakdown strength. The higher the breakdown strength, more reliable the dielectric is expected to be. Well processed and relatively defect-free SiO_2 films exhibit breakdown strength greater than 10 MV/cm, while ultrathin SiO_2 films show breakdown strength significantly higher. Table 3.2 shows schematic representation of electronic affinity, band gaps, and dielectric constant of common metal-oxide dielectric films against those of silicon and Ge substrates.

Charge confinement and charge transport characteristics: Amorphous dielectric exhibits the capacity to hold charges in the body of the dielectric, what are known as defect centers or traps. When sufficient potential is applied across the dielectric, both the electrodes as well as the charge centers could be active in providing charges to flow in response to the potential imposed. These phenomena known as charge trapping and charge detrapping and field plus temperature-induced charge flow are critical to the characteristics of FETs, and especially to the attributes of nonvolatile memory devices. These will be elaborated below.

TABLE 3.2

Schematic Representation of Electron Affinity (Ue), Band Gap (Eb), and Dielectric Constant (K) of Common Metal Oxide Films against Ue of Silicon and Germanium

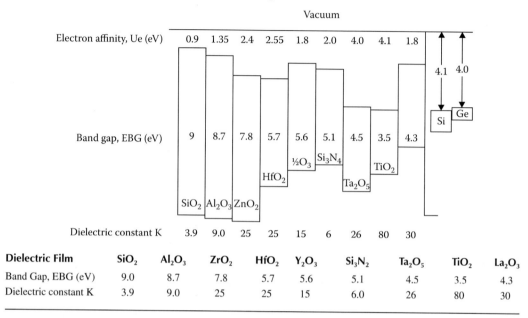

Dielectric Film	SiO_2	Al_2O_3	ZrO_2	HfO_2	Y_2O_3	Si_3N_2	Ta_2O_5	TiO_2	La_2O_3
Band Gap, EBG (eV)	9.0	8.7	7.8	5.7	5.6	5.1	4.5	3.5	4.3
Dielectric constant K	3.9	9.0	25	25	15	6.0	26	80	30

3.2.3 Bulk and Interface Defects and Charge Trapping

Amorphous dielectrics contain varying degrees of compositional nonhomogeneity within the bulk of the material. Such nonhomogeneity manifests at the molecular dimensions into defect centers often called traps. Such traps create local potential well providing local confinement of charges. Traps could be positively charged (with trapped holes), or negatively charged (with trapped electrons), or neutral. Additionally, fluctuations in nuclear potentials give rise to fixed charge centers either positive or negative within the bulk dielectric as well as at the interfaces. Furthermore, energy states are generated within the forbidden band gap due to the presence of impurities and variation in local chemistry. In planar thin dielectric films, surfaces provide broken chemical bonding as well as chemical nonuniformity. In a field effect transistor structure (FET device), a thin dielectric film (e.g., SiO_2) is in contact with the silicon substrate. In such a case, the interface can generate fixed charge centers as well as interface energy states, which adversely affect device properties. In a multi-, dielectric gate insulator stack, the interface created by the two dielectric layers of different chemical composition provides discontinuity in chemical bonding as well as thermomechanical stresses. This may result in increased defect density at the interface and associated trap density and interface states. Similarly, when a dielectric film interfaces a metal gate or a polysilicon gate (doped or undoped or a floating gate, examples being FET or FLASH devices, respectively), the interfaces created are chemically, thermally, and mechanically nonideal. The resulting interfaces give rise to interface defects, traps, fixed charges, and interface energy states. As devices get scaled in dimensions, interfaces become increasingly important in influencing different device characteristics including charge transport.

3.2.4 Charge Transport

In thicker dielectric films, charge carriers, electrons, and holes exist in bulk traps. In thick or thin films, such carriers could also be provided by the electrodes interfacing the dielectric film. For example, in metal–insulator–silicon (MIS) capacitors or devices, charge carrier sources could be metal at one end and silicon at the other end. When a voltage stress is applied across the insulator (dielectric) film, charge carriers could be provided either from the bulk dielectric or from the electrodes, that is, metal or silicon or both. Depending on the polarity of the stress and the magnitude, electrons will flow toward the positive terminal (anode), while holes will flow toward the negative (cathode) terminal. This results in leakage and current conductivity in the dielectric reducing its capacity to store charge. When the charge source is bulk dominated, current flow is bulk controlled reflecting the bulk characteristics of the dielectric film. Whereas when the current source is electrode controlled, either metal or silicon, the current transport is electrode controlled and reflects the interface energy states of the metal (work function) or silicon (band energy or interface energy levels).

Multiple examples of carrier transport exist in dielectric films. Even in a particular dielectric film, transport mechanisms may change depending on the stress level, processing history, chemistry, and so on, and several mechanisms may be operative. The most significant bulk-controlled mechanism is called the Poole–Frankel mechanism, which involves transport of trapped charges released due to thermal excitation. Such mechanism is field enhanced and sensitive to temperature. Another variation of above mechanism prevalent in dielectrics with shallow traps involves hopping by thermal excitation from one shallow trap to another with strong temperature dependence. Another variation which has weak temperature dependence and is present in dielectrics with relatively shallow trap is called resonance tunneling or trap-assisted direct tunneling. All the above mechanisms are bulk controlled.

In contrast to the above mechanism, a significant electrode-controlled mechanism based on quantum mechanical tunneling of charge transport is called the Fowler–Nordheim (FN) tunneling. For thicker dielectric films with carrier sources being electrode controlled, this transport mechanism is prevalent with strong field dependence and weak temperature dependence. For very thin films, direct quantum mechanical tunneling transport of carriers takes place, which has no temperature dependence but is strongly dependent on voltage and dielectric film thickness. A variation of the FN tunneling is called enhanced FN tunneling where transport of carrier is enhanced by local high-field effects at the interface.

Carrier transport mechanisms determine how conductivity of any dielectric film is sensitive to field across the dielectric film and temperature. Since these mechanisms are of considerable interest in nonvolatile memory devices, all above mechanisms will be discussed in greater detail in a separate section on Dielectric Conductivity.

3.2.5 Figure of Merit for Selected Metal-Oxide Dielectric Films

Using oxide (SiO_2) as a reference dielectric, a commonly used figure of merit is often employed to compare potentially applicable metal-oxide dielectric films as a replacement for oxide in FET device and memory applications [1]. This figure of merit combines the charge holding capacity (dielectric constant) with a permissible dielectric leakage level and expresses in terms of EOT providing a useful parameter for device application. Assuming the SiO_2 film reference limit of thickness 3.5 nm and leakage level $\leq$1E-3 A/cm^2 at 1.25 V (field of ~3.6 MV/cm), charge holding capacity of other dielectric films is compared for the same leakage level. This is shown in Table 3.3 for selected metal-oxide dielectric films of interest. The figure of merit demonstrates

TABLE 3.3

Single Metal Oxide Figure of Merit [1]

	SiO$_2$	Al$_2$O$_3$	ZrO$_2$	HfO$_2$	Ta$_2$O$_5$	TiO$_2$
Band gap: EBG, eV	9.0	8.7	7.8	5.7	4.5	3.5
EOT: nm	3.5	1.8	1.7	1.5	3.75	2.8
Thickness: [a] nm	3.5	5.5	8.5	13.6	25.0	58.0
Capacitance: [a] fF/µ^2	10	~20	~20	~17	~9	~12.5
Figure of merit	1	2	2	1.7	0.9	1.25

[a] Ref: SiO$_2$: Leakage limit: 1E-3A/cm^2 at 1.25 V.

that higher K metal oxides such as TiO$_2$ ($K = 80$) and Ta$_2$O$_5$ ($K = 26$) may not be appropriate replacement for SiO$_2$ due to considerably higher leakage. On the other hand, alumina and zirconia films are similar in figure of merits (2× of SiO$_2$) and worthy of consideration even though the latter has the 2.5× enhancement in K value. HfO$_2$ film also exhibits enhanced figure of merit compared to SiO$_2$.

3.2.5.1 Metal Work Function and Electron Affinity

For MIS types of devices (here metal [M] also includes heavily doped silicon electrode), work function (Wf) for metal electrode interfacing the dielectric (I) is an important parameter. While work function is the energy required to remove a "free electron" from the metal electrode to vacuum, the electron affinity, Ue for the dielectric (I) is the corresponding energy required to remove an electron from the conduction band of the dielectric to the vacuum. When the difference between work function and electron affinity (Wf-Ue) is large, charge injection from the metal electrode to the insulator is minimized, thereby minimizing both charge trapping and charge transport. Similarly, larger band gap insulator (I) at the silicon substrate, (S), interface ensures electron or hole injection into the insulator to minimize charge trapping and transport at the substrate side. Figure 3.3 shows metal work functions values in a schematic representation. Corresponding values for silicon and germanium are, respectively, 4.1 eV and 4.0 eV.

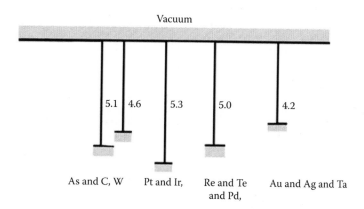

FIGURE 3.3 Work functions of As, C, and metals: Pt/Ir (5.3 eV); Re/Te/Pd (5.0 eV); and Au/Ag/Ta (4.2 eV).

3.3   INTERFACES, ELECTRODE COMPATIBILITY, AND PROCESS SENSITIVITY

We have defined various interfaces for the NVM devices at the beginning of the section (Figure 3.1) and discussed some aspects of the metal–insulator and insulator–silicon interfaces earlier. For multiple dielectric layers, there are additional interfaces created between the dielectric layers as well as between each dielectric and silicon and other interfacing electrode. Since interfaces involve transition of one material chemistry to another, the processing goal is to minimize defect creation at the interfaces, which gives rise to traps, interface states, fixed charges, and so on as mentioned earlier. Interfaces are expected to be free of contaminants, clean, and discrete. Interface properties to consider are of physical, chemical, and electronic in nature. Physical compatibility implies materials selection to reduce thermal mismatch and stresses at the interface and to ensure adhesion and stability during processing and post processing conditions. Chemical compatibility involves bond stability during processing and post processing variables such as temperature, pressure, and chemical environments. Electronic compatibility involves reproducibility and stability of energy states with reference to electrons and holes.

Microelectronics involves extensive applications of MIM (metal–insulator–metal) and MIS (metal–insulator–silicon) elements as passive and active device elements. Therefore, compatibility with electrode material is also extremely important in achieving desirable characteristics. Electrode compatibility involves material selection to simultaneously achieve, (a) large work function difference to minimize charge injection from the electrode to the insulator, (b) reduce inter facial stress and thermal mismatch, (c) physical/chemical/mechanical integrity during processing and post processing conditions of temperature and ambient, and (d) process reproducibility and ease of fabrication of dielectric and electrode materials. From process integration point of view, additional requirements include selective patterning of deposition and etching, thickness uniformity, and control.

Process sensitivity relates to process techniques and variables in process parameters impacting the properties of the dielectric films. It has been well known that dielectric thin-film properties (bulk as well as surface) are process sensitive. Properties impacted include physical properties such as adhesion, stresses, thickness uniformity, and refractive index, chemical properties such as purity, compositional uniformity, chemical bonding, and etch rate, electronic properties such as band gap, dielectric constant, leakage, electron affinity, trap density, and fixed charge density. There are a wide variety of processing techniques to deposit dielectric thin films. These include thermal deposition and oxidation, chemical vapor deposition (CVD, MOCVD, and LPCVD), plasma deposition and sputtering, and atomic layer deposition (ALD). Processing parameters and variables involve: chemical species, temperature, pressure, ambient, partial pressure of chemical constituents, and deposition rate. These will be discussed in greater detail with reference to selected specific dielectric films widely used in nonvolatile memory technology.

3.4   GATE MATERIAL FOR NVM DEVICES

In general, gate materials employed for NVM devices are required to be CMOS technology compatible. Gate materials employed in NVM device elements are rarely different from PFET and NFET CMOS gates due to technology scalability considerations and process integration requirements. Typically, CMOS technology had employed either N-doped or P-doped polysilicon gates in the past, which are being progressively changed to metal gates for FET device scaling requirements. In Part II of this book, we have discussed in detail the integration schemes and compatibility requirements for the scaled NVM technology with those of general scaled CMOS technology. Aside from the polysilicon gates, the most common metal gates employed in NVM devices are metal nitrides, namely, TiN and tantalum nitride (TaN). Metal nitride films are compatible to general CMOS technology and provide large work function barriers to prevent undesirable charge injection from the metal gate during writing or erasing. These layers are deposited on top of the blocking dielectric layers followed by the desired thicker metal interconnect levels (e.g., tungsten metallurgy) for the CMOS technology.

3.5 DIELECTRIC CONDUCTIVITY MECHANISMS

3.5.1 Bulk-Controlled Poole-Frenfel Mechanism

This mechanism of conductivity is also known as Internal Schottky Mechanism and often referred to as mechanism of transport due to field enhanced thermal excitation of trapped carriers. Conductivity of a dielectric film exhibiting Poole–Frenkel mechanism is given by:

$$J/E = C \exp[-q\{\Phi e\text{–}tr -(qE/PI \epsilon_0 K_d).5\}/kT]$$

where

    J/E is the conductivity, J being the current density and E being the electric field defined by externally applied, voltage divided by the thickness of the dielectric film that is, V/d
    C is a characteristic material constant
    q is the electronic charge
    $\Phi e\text{–}tr$ is the trap energy depth from the conduction band edge
    PI is the mathematical constant (value 3.14)
    ϵ_0 is the permittivity of the free space
    K_d is the dielectric constant of the dielectric film
    k is the Boltzmann constant (value 1.38046 E-16)
    T is the body temperature of the dielectric film

Many common dielectric films such as silicon nitride (Si_3N_4), wide ranges of silicon oxynitrides (SixOyNz), and alumina (Al_2O_3) exhibit Poole–Frenkel Transport over a major range of voltage stress. As stated earlier, the conductivity is sensitive to temperature and increases with increasing temperature.

Dielectric film parameters such as the trap energy depth $\Phi e\text{–}tr$ and dielectric constant K_d also strongly influence the conductivity of the dielectric film.

Two other bulk-controlled mechanisms are observed in thin dielectric films. The first one is strongly temperature sensitive and is present in dielectric films associated with very shallow traps. This mechanism is called "Hopping" and involves hopping of charge carriers from one shallow trap to another shallow trap due to thermal excitation. The second one exhibits weaker temperature dependence and is called "trap-assisted direct tunneling" or "Resonant Tunneling." This mechanism is displayed at low field and at low current ranges with current density being exponentially proportional to the voltage. Silicon-rich oxides (SROs) and SRNs as well as others exhibit such characteristics at lower field and at room temperature. Schematic representations of all three bulk-controlled mechanisms are shown in Figure 3.4a–c, respectively, for Poole–Frenkel, multiple trap hopping, and trap-assisted direct tunneling.

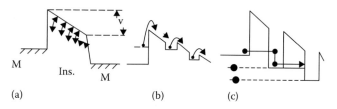

        (a)   (b)   (c)

FIGURE 3.4 Bulk-controlled mechanisms: (a) Poole–Frenkel (Temperature sensitive), (b) Trap Hopping (strong temperature sensitive), and (c) Trap-assisted direct tunneling (weak temperature sensitive).

3.5.2 Electrode-Controlled Quantum Mechanical Tunneling Mechanisms

There are generally three modes of quantum mechanical tunneling, which are all electrode controlled whereby the carriers (electrons or holes) are supplied by the electrode (silicon or metal). These are (a) Direct Tunneling or modified Fowler–Nordheim tunneling; (b) Fowler–Nordheim tunneling; and (c) Enhanced Fowler–Nordheim tunneling.

3.5.3 Direct Tunneling and/or Modified Fowler–Nordheim Tunneling

In this mode of transport, electrons or holes incident at the electrode–dielectric interface facing a potential barrier of trapezoidal type as shown in Figure 3.5a and b and exhibiting a finite probability of movement across the potential barrier, which is exponentially dependent directly on the voltage and inversely on the thickness of the dielectric film or films (case b), when the external voltage is less than the potential barrier. The transport of electrons or holes by this mechanism is characterized by the following equation:

$$Jt \sim De.Pt$$

Where Jt is the tunnel current density and is equal to: I/A, I being the current, and A is the injecting electrode surface area; De is the electron density at the barrier interface where the barrier potential is Φb and Pt is the tunneling probability. The tunneling probability Pt is given by:

$$Pt = \exp(-\Phi b^{0.5}.d)$$

Where d is the dielectric thickness. When two thin dielectric films are involved as could be illustrated for an MNOS device consisting of ultrathin layers of both oxide and nitride as in the case of Figure 3.5b, the transport mode is often referred to as "modified Fowler–Nordheim tunneling." In such examples, direct band-to-band tunneling may take place through both oxide and nitride layers as shown. The tunneling probability function Pt gets modified and incorporates direct probability parameters relating to combined probability through both the layers.

Therefore, direct tunneling current is extremely sensitive to thickness of the dielectric and exponentially reduced as thickness of the dielectric increases. Additionally, Jt is dependent on voltage (not on field), and is electrode controlled through De from the above equation on tunnel current density Jt. It should also be noted that in both variations of direct tunneling, charge transport is not sensitive to temperature.

It may be noted in the above examples of both single and dual insulators associated with MIS structures that when the insulator is very thin, and a positive potential Vi is applied on the metal

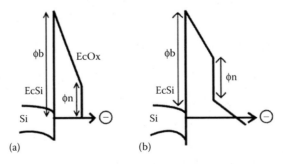

(a) (b)

FIGURE 3.5 Quantum mechanical direct tunneling mechanism (electrode controlled): Two related variations: (a) Direct band-to-band tunneling (single insulator) and (b) Modified Fowler–Nordheim tunneling (dual insulator: oxide and nitride).

electrode with $\Phi b > Vi$, electron flows from the silicon conduction band (cathode) across the trapezoidal energy barrier of the insulator or insulators. Correspondingly, current flows from the metal electrode (anode) to the silicon substrate. For example, for oxide (SiO_2) as in case 1.3.5 (a), Φb for electrons is 3.2 eV, direct tunneling mode would be operative in ultrathin oxide in the thickness range of 1–3.2 nm when an external potential is imposed of <3.2 V for a MOS structure.

It should be noted that direct tunneling is ballistic since no scattering of electrons takes place within the insulating film until after the electrons enter the anode. Additionally, current Jt is not temperature dependent unlike Poole–Frenkel transport. However, as stated earlier, a weak temperature dependence could be observed at low field for silicon-rich insulators where charges are transported from trap to trap by direct tunneling (trap-assisted direct tunneling or resonant tunneling). Another example of such tunneling could be observed in MIS structures when the insulator consists of embedded silicon nanodots sandwiched between two ultrathin oxide films. Electrons tunnel by direct quantum mechanical tunneling first between silicon electrode to the Si-nano dot induced trapping energy states and then from such energy states to the metal electrode. This book will discuss NVM devices based on such transport in section on Direct Tunnel Memories and in section on advanced devices. Many CT devices are also based on direct tunneling of electrons and holes.

3.5.3.1 Fowler–Nordheim Tunneling

When the external voltage across the MIS structure is greater than the energy barriers associated with electrons and holes and the insulator is relatively thicker, charge transport by tunneling takes place through the triangular energy barrier created by the external potential. This is shown in Figure 3.6a; the charged carrier transport mode is known as Fowler–Nordheim tunneling. In this mode of tunneling, the current density J is proportional to the square of the field Ei and is more strongly sensitive to the barrier height Φ_b. The current density Ji is given by:

$$Ji = C\ Ei^2 \exp(-B/Ei)$$

Where C is a Fowler–Nordheim constant related to: (a) electronic charge, q; (b) probability parameters, Φ_b; and (c) effective mass, m*; of electrons or holes for the insulator. The parameter, B is proportional to: $m^{*0.5}$. $\Phi_b^{1.5}$. It should be noted that the transport of carriers and associated current flow is strongly sensitive to the field across the insulator, thereby modulating the characteristics of the triangular energy barrier. In recent FG and CT devices, Fowler–Nordheim tunneling mode or modified Fowler–Nordheim tunneling modes had been the primary mode of charge transport through the tunneling layers of the devices.

Fowler–Nordheim tunneling could be ballistic FN when Vd is in the range of 3–9 V and steady-state FN when Vd is greater than 9 V. This has significance in energy transfer and trap-generation within the insulating layer [2], covered in the next section.

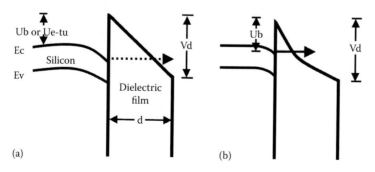

FIGURE 3.6 Fowler–Nordheim tunneling exhibiting triangular energy barrier: (a) Normal or modified Fowler–Nordheim tunneling and (b) Enhanced Fowler–Nordheim tunneling.

3.5.3.2 Enhanced Fowler–Nordheim Tunneling

An enhanced Fowler–Nordheim mode of current transport is created for thicker dielectric films when the electrode interface provides local high field by effectively modifying and lowering the barrier height Φb (or Ub). Such conditions are created by providing textured polysilicon interface and SRI interfaces with high density of silicon nanodots (injector SRI). The modification of the Fowler–Nordheim triangular barrier is schematically shown in Figure 3.6b. Due to such internal field enhancement, both the effective barrier height as well as the effective tunnel distance could be reduced resulting in enhanced charge transport and current flow for the same insulator thickness. The enhancement current density Ji (enh) could be expressed similar to the above Fowler–Nordheim expression by modifying the term B to B (enh) as follows:

$$Ji(enh) = C\ Ei^2\ exp[-B(enh)/Ei]$$

Where B(enh) is a function of Φb(enh) is the reduced barrier height as shown in Figure 3.6b. Current density versus gate voltage characteristics for thin oxide MOS devices covering the thickness ranges of 2.5 (direct tunneling) to 7.5 nm (Fowler–Nordheim tunneling) is shown in Figure 3.7 [2,3]. It should be noted that current density through oxide is enhanced by several orders of magnitude at relatively low voltage when charge transport is operating in direct tunnel mode compared to that of Fowler–Nordheim mode.

Examples of enhanced Fowler–Nordheim mode of charge transport in oxide films had been demonstrated by D. DeMaria et al. [3] and shown in Figure 3.8a–c. A thin film of SRO with silicon nanodots was deposited on top of the SiO$_2$ film in the MIS structure (Figure 3.8a); corresponding modification of the oxide conduction energy band is depicted in Figure 3.8b. Figure 3.8c compares the current–voltage plots with and without interface modification due to the SRO film. Current enhancement of nearly three orders of magnitude due to enhanced Fowler–Nordheim transport mode was demonstrated.

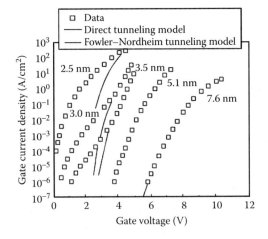

FIGURE 3.7 Current–voltage characteristics in SiO$_2$ films with both direct tunneling and Fowler–Nordheim tunneling. (From Hu, C., Gate oxide scaling limits and projection, *IEDM*, p. 319 © 1996 IEEE.)

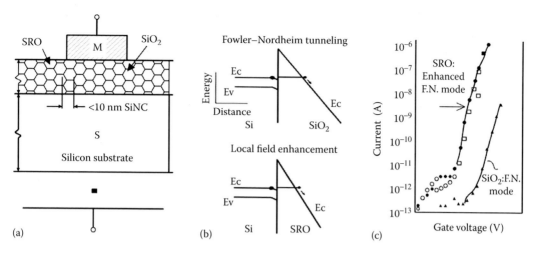

FIGURE 3.8 (a) Enhanced charge injection into SiO_2 due to silicon nanocrystals of silicon-rich oxide (SRO) over layer, (b) Conduction energy band modification associated with silicon nanodots, and (c) Enhanced F.N. current due to the silicon-rich-oxide. (From DeMaria, D. J. et al., *J. Appl. Phys.*, 51, 4830, © 1980 IEEE.)

3.6 CARRIER TRANSPORT MECHANISMS FOR MULTILAYER DIELECTRICS

Advanced NVM devices employ multilayers of dielectric films for tunneling layers and blocking layers. Multilayered trapping dielectric structures have also been suggested to enhance charge retention. Forward and reverse charge transport characteristics could be significantly modified by band-engineered multidielectric gate stack structures for NVM devices. If the dielectric layers are relatively trap-free, charge transport could be characterized in terms of multiple direct tunneling mode or by modified Fowler–Nordheim tunneling mode assuming trapping and detrapping are not involved. Multilayered dielectric structures and related characteristics will be discussed in chapter on band engineering in considerable detail in Part II. In general, transport characteristics in multilayered dielectric structures are modeled by a combination of the different modes outlined above.

REFERENCES

1. A. Bhattacharyya, Consideration of high K dielectric capacitor technology for micro and nano electronics, Micron Technology Technical report, Unpublished, June 2006.
2. C. Hu, Gate oxide scaling limits and projection, *IEEE, IEDM*, 1996, pp. 319–322.
3. D. J. DeMaria, R. Ghez, and D. W. Dong, Charge trapping studies in SiO_2 using high current injection from Si-rich SiO_2 films, APS, *Journal of Applied Physics*, 51, 4830, 1980.
4. P. E. Nicollian, W. R. Hunter, and J. C. Hu, Experimental evidence for voltage driven breakdown models in ultrathin gate oxides, *IEEE International Reliability Physics Symposium*, San Jose, CA, 2000, pp. 7–15.

4 Dielectric Films for NVM Devices

CHAPTER OUTLINE

It should be apparent now that thin dielectric films play a critical role in silicon based NVM devices. We will review some of the most important properties of the dielectric films investigated for NVM applications in greater detail. We will group several selected dielectric films into three groups: (1) conventional dielectric films, (2) common high K dielectric films, and (3) future high K dielectric films. This chapter provides an in-depth discourse on key properties of thin conventional dielectric films for NVM device applications. These include thin films of silicon-di-oxide (SiO_2), silicon nitride (Si_3N_4), different compositions of silicon oxynitrides (SiONs), and silicon-rich insulators such as silicon-rich oxides (SROs) and silicon-rich nitrides (SRNs) employed in gate stack designs for NVM devices prior to the incorporation of higher K dielectric films. Higher K dielectric thin films for NVM devices are discussed in Parts II and III of this book.

Conventional dielectric films are those dielectric films with long history of process development and integration in silicon based microelectronics, including defect control, reproduce ability, high yield production history, and reliability. Top among these films are thermal oxide (SiO_2), CVD, or LPCVD nitride (Si_3N_4), deposited oxide (TEOS-SiO_2), silicon oxynitrides (SiON), deposited, either by CVD or LPCVD processes similar to nitride, and silicon-rich insulators, namely, SROs and SRNs, also deposited by LPCVD processes similar to that of nitride.

Aside from the above dielectric films, several high K dielectric films have been extensively investigated and characterized over the previous decades and employed in NVM devices. Most important among these are: Al_2O_3 (alumina), HfO_2 (Hafnia, or hafnium oxide), ZrO_2 (zirconia or zirconium oxide), Ta_2O_5 (tantalum oxide), TiO_2 (titanic or titanium oxide), and HfSiON (Hafnium silicon oxynitride). This group of high K dielectric films will be labeled in this book as "common high K" dielectric films. It should be noted that Micron Technology in USA has already put higher K interpoly dielectric film into production for their Flash devices during the past decade. Characteristics of these films are discussed in Part II of this book.

We will also discuss other potential high K dielectric films in Part II with the possibility of future applications as well as the possibility of playing significant roles in NVM devices. Some of these are rare-earth metal oxides of lanthanide family called Lanthanides (see also appendix), bimetal and mixed oxides, namely, HfAlO, HfLaO, HfSiO, ZrSiO, TaAlO, $LaAlO_3$; metal nitrides, namely, GaN, AlN, and AlGaN; bimetal oxynitrides, namely, HfLaON, HfTaON, and HfAlON. Although there may be other possibilities, the above list provides sufficient insight into the future high K selection requirements and associated application challenges. This is further discussed in Part III, Chapter 20.

4.0 CONVENTIONAL DIELECTRIC FILMS FOR NVM DEVICES

4.1 THERMAL OXIDE: SiO_2

Thermal oxide produced by oxidation of silicon substrate at elevated temperature followed by an appropriate anneal has been the basic fabric of gate insulator for FET technology since inception to the turn of the century. It is only approximately the last one and a half decade, as the FET devices were vertically scaled, oxide had to be replaced by other dielectric thin films since oxide <2.5 nm had unacceptably high device leakage [1]. Thermal oxide had the most desirable property of reducing the interface density of states << 1E11/cm² when only 2–3 monolayers of silicon is oxidized to form stoichiometric SiO_2. This property cannot be matched by any metal oxides formed on silicon by any known technique.

Additionally, SiO_2 is one of the largest band gap insulators known, readily providing an ideal MIS structure with very low fixed charge density and trap density. Oxide had been the reference dielectric film with the required attributes of reproducibility, manufacturing ability, defect control, and reliability mentioned earlier. Unless otherwise stated, the term oxide in this book refers to thermally grown amorphous SiO_2 film implying the desired density and stoichiometry. The parameter EOT defined in the earlier section will be used throughout this book when the gate insulator stacks of the FET (or NVM element) incorporate one or multiple layers of other insulator films with or without SiO_2. Since gate oxide capacitance is an important FET device design parameter for device scaling, the EOT defines the equivalency in capacitance of any dielectric insulator stack design with corresponding oxide-only equivalent thickness. Therefore, the EOT-i for any insulator "i" of thickness "D-i" and dielectric constant Ki can be expressed as:

$$EOT_i = K_{SiO_2} \cdot D_i / K_i$$

Where K_{SiO_2} is the relative dielectric constant of SiO_2 which is equal to 3.9 times the relative dielectric constant of vacuum or free space of K = 1.

Since all relevant properties of oxide films are well known, dielectric properties of other films for NVM applications will often be compared with similar oxide properties throughout many sections of this book. Separate review of oxide films as such will not be elaborated except the oxide reliability applicable to NVM devices and characteristics of oxide films stressed at high fields of writing and erasing for NVM device charge storage. Reliability of oxide employed as critical elements of the conventional NVM gate stack design is intimately tied with the key NVM properties such as endurance and charge retention. Consequently, considerable attention was paid to investigate high field and high current effects on oxide integrity and reliability.

Carrier transport and energy dissipation: As stated earlier, thermally grown SiO_2 films exhibit extremely low-interface state density (≤1E10/cm²) and intrinsic trap density (≤1E10/cm²) in the thickness range of 1.5–10 nm. Carrier transport mechanisms in such SiO_2 films have been well established. Depending on the film thickness and the voltage drop across the SiO_2 film, carrier transport can be characterized by three types of tunneling mechanisms: (a) Direct tunneling mechanism with or without trap assistance, (b) Ballistic Fowler–Nordheim tunneling, and (c) Steady-state Fowler–Nordheim tunneling [2]. For electron transport, mechanisms (a), (b), and (c) are operative at voltages across the oxide, Vox, respectively, at <3, 3–9, and >9 V for the thickness range mentioned above. When mechanisms (a) and (b) are operative, electrons do not dissipate any significant energy into the oxide during transport, instead, delivers most or all the energy into the anode. When mechanism (c) is operative, electrons may dissipate energy into the oxide by impact ionization creating electron–hole pairs inside the oxide. The associated electron energy is q. Vox for the three mechanisms as above are, respectively, <3, 3–9, and >9 eV. The average kinetic energy electrons delivered to the anode for the corresponding transport mechanisms are, respectively, 0, 06, and >6 eV for mechanisms of direct tunneling, Ballistic Fowler–Nordheim tunneling, and steady-state Fowler–Nordheim tunneling. The maximum energy delivered to the anode is q. Vg (eV) where Vg is the gate voltage. When sufficient energy is dissipated into the oxide, local bonds get broken creating defects or traps. The nature and extent of energy dissipation by the carriers are critical to the creation of traps in the oxide film, thus initiating the degradation process and affecting oxide reliability. It will be discussed in Section 6.3, that oxide degradation by the above process adversely affects NVM device properties such as endurance and retention. It should be noted that for electron transport by direct tunneling with q . Vox = <3 eV, average energy dissipation to the oxide is negligible. Therefore, defect generation and oxide degradation should be expected to be minimal under such stress conditions.

4.1.1 Defect Generation and Oxide Degradation

Traps are defects in oxide and are associated with local variation in potential due to breaking of local (short range) chemical bonds. It was mentioned earlier that intrinsic defects (trap density) in SiO_2 films are very low when SiO_2 is grown under controlled condition. However, when SiO_2 is subjected to significant stress, additional traps are generated. Three different trap generation mechanisms have been identified dependent on film thickness and stress voltage levels. These are called: (1) Anode Hydrogen Release: AHR, (2) Anode Hole Injection: AHI, and (3) Oxide Impact Ionization: OII. Threshold energies for the above three mechanisms are listed in Table 4.1.

Although the generation rate of traps in oxide is low, when the oxide film is stressed below 5 V, trap generation has been observed in oxide even when the gate voltage is as low as 2 V. Below Vox = <2 V, trap generation is negligible.

The presence of traps in thin oxide results in higher leakage current due to trap-assisted tunneling. This is demonstrated in Figure 4.1, showing current density Jox versus gate voltage Vg characteristics for the three cases of oxide: without intrinsic trap (case 1), unstressed oxide with intrinsic traps (case 2), and stressed oxide with stress-generated traps.

Curve 3 above shows significant enhancement in leakage current due to stress-induced generation of traps in oxide film. This is known as "stress-induced leakage current" or SILC. SILC is due to trap-assisted tunneling through bulk trap states.

When oxide is stressed with increasing voltages and consequently at higher current level or fluence, SILC increases with increase in rate of generation of traps. Eventually filamentary conduction paths are created leading to thermal runaway and dielectric breakdown. This is shown schematically in the I–V characteristics in Figure 4.2.

TABLE 4.1

Threshold Energies and Mechanisms for SiO_2 Trap Generation

Mechanisms	Average Anode Kinetic Energy Dissipation qVox (eV)	Average O_x Kinetic Energy KEox (eV)
AHR	>5V	>2V
AHI	>6V	>3V
OII	>>12V	>9V

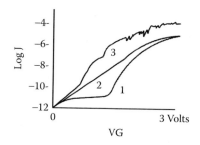

FIGURE 4.1 Trapping effect on leakage in oxide: Curve 1 for unstressed oxide, curve 2 for unstressed oxide containing intrinsic traps while curve 3 for stressed oxide with stress-generated traps.

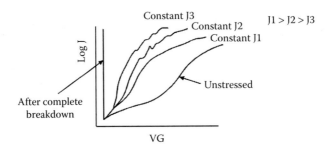

FIGURE 4.2   I–V characteristics in oxide at high current level (fluence).

4.1.1.1   Oxide Reliability

Trap generation at high stress level causes SILC and eventual dielectric breakdown. SILC data are utilized for reliability projection for oxide films. Such information is collected by stressing oxide films at different constant current levels (fluence) and measuring the time to dielectric breakdown known as TDDB. Log TDDB is plotted as a function of Vox and/or 1/Vox as shown in Figure 4.3a and b, respectively, to account for the voltage (or stress) enhancement factor.

TDDB is also often expressed in terms of total charge to breakdown called QBD. QBD is modeled in terms of hole injection and associated trap creation through the mechanisms of AHR or AHI or OII or a combination of these mechanisms and correlates well with experimental data. The breakdown QBD may also be expressed in terms of critical hole fluence expressed as:

$$QBD = C. \exp{-\left[(B + H) / E_{ox} \text{ or } V_{ox}\right]}$$

where:

   C is a constant
   B is the SILC coefficient
   H is the hole-generation coefficient

The charge to breakdown in oxide, QBD versus Vox is shown in Figure 4.4 identifying QBD dependence on different tunneling modes.

It is important to note from the above figure that QBD is enhanced by over six orders of magnitude when the transport modes of electrons and holes in oxide is by direct tunneling as compared to steady-state Fowler–Nordheim tunneling. Therefore, oxide reliability is strongly tied with trap generation and in turn with the mode of transport of charges within oxide. Reliability of ultrathin oxide (tox < 3 nm) is expected to be 6–8 orders of magnitude greater than thicker oxide films such as oxide thickness of 6–10 nm commonly employed for FG Flash devices. Oxide thickness and stress level has great implication in application in NVM devices and key NVM device properties such as endurance and retention. These topics are discussed in Chapter 5.

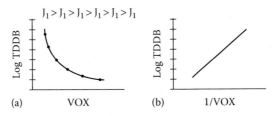

FIGURE 4.3   Time to breakdown (TDDB) in oxide films: (a) log TDDB versus Vox showing exponential reduction in TDDB stressed at high fluence and (b) log TDDB versus 1/Vox.

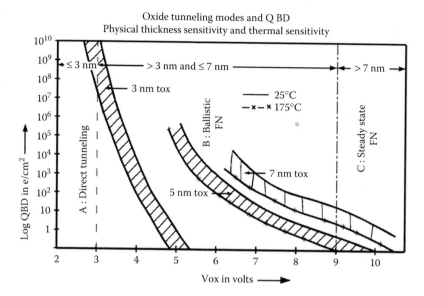

FIGURE 4.4   QBD versus Vox at different tunnel modes.

4.2   CVD OR LPCVD NITRIDE: Si_3N_4

Aside from oxide, historically, the next most investigated and applied thin dielectric film in silicon based technology had been nitride or Si_3N_4. Nitride could be readily deposited by CVD or LPCVD techniques in vapor phase using silane (SiH_4) or dichlorosilane (SiH_2Cl_2) and ammonia (NH_3) with nitrogen as the carrier gas at elevated temperature (typically 900°C). Other methods of obtaining nitride films include PECVD (plasma-enhanced CVD) and ALD techniques at relatively lower temperatures. Nitride films are smooth, amorphous, and adherent to silicon and oxide and contain high density of bulk traps typically in the order of $5E18/cm^3$ for both electrons and holes with large capture cross section. Nitride obtained by high temperature processing exhibit strong Si-N bonds, while lower temperature nitride contains weak Si-H bonds and exhibit process sensitivity. Trap density and trap depth in nitride are process sensitive and dependent on temperature, pressure, ratio and flow rates of reactants (e.g., ammonia to silane ratio) as well as post deposition annealing conditions. Nitride deposited onto silicon exerts tensile stress in silicon, while oxide in silicon exerts compressive stress. Nitride provides diffusion barrier against oxygen and other contaminants and had been widely used for selective oxidation, shallow trench isolation (STI), and surface passivation. Nitride can also be used as hard mask against oxide due to highly differential etch rate between nitride and oxide (e.g., in 7: 1 BHF, oxide etch rate nearly 100× over that of nitride). Nitride is significantly more conductive with lower band gap compared to oxide (Table 4.2). Carrier transport in nitride, as mentioned earlier is bulk controlled, governed by field-enhanced thermal excitation of trapped charges (electrons or holes): Poole–Frenkel mode as mentioned before. Conventional CT devices (e.g., SONOS or MONOS) employ nitride layer for charge trapping and storage to provide nonvolatile characteristics.

4.2.1   Nitride Traps, Trap Creation: Process and Stress Sensitivity

Trap density and trap depth in nitride are strongly process sensitive. Nitride films applied in NVM devices may exhibit varying device characteristics due to variation in nitride process parameters. Electrical and electronic properties of nitride films due to variation in process parameters for CVD and LPCVD nitride films have been extensively studied by A. Bhattacharyya et al., D.W.Dong et al., D. J. Maria et al. and others [3–6]. In general, properties of nitride films are sensitive to ratios of nitrogen, silicon, and hydrogen partial pressures in vapor phase, temperature, and deposition rates. High-quality reproducible and stoichiometric nitride films could be achieved at elevated temperature, high ammonia to dichlorosilane (SiH_2Cl_2) ratio, low partial pressure of hydrogen and moisture, and low deposition rate. Higher density of shallow traps in nitride were identified with the presence of higher hydrogen and silicon content in films exhibiting weak Si-H and N-H bonds as opposed to the desired Si-N bonds in stoichiometric nitride (Si_3N_4) films. Details of process sensitivity of nitride on electronic properties will not be discussed here. However, nitrogen-rich silicon oxynitride films (NR-SiON) [7] and SRN films [4] will be discussed in greater detail and reference of nitride properties will be presented in the context of SiON and SRN films.

Nitride films contain high density of bulk traps. Nonstoichiometric nitride films containing significant amount of Si-H and N-H bonds may exhibit higher trap density with shallow traps. When nitride films are stressed at high field, weaker bonds are first broken such as Si-H bonds and N-H bonds creating defects in nitride and consequently enhanced leakage through nitride. Since nitride is used in conventional FG devices as a part of blocking dielectric and as trapping layer for CT devices (e.g., SONOS), NVM device properties are strongly sensitive to nitride processing and stoichiometry. This will be discussed in more detail in Chapter 6.

4.3 SILICON OXYNITRIDES: SiONs

Silicon oxynitrides, hereafter would be termed as oxynitride or SiON, are a generic group of insulators, which can be best described as amorphous polymers of silicon, oxygen, and nitrogen. Oxynitrides span the entire range between SiO_2 and Si_3N_4 in terms of the oxygen–nitrogen ratio, the refractive indices, and the chemical etch rate. Detailed physical and compositional studies were first reported by Chu et al. [8], and subsequently by M. J. Rand et al. [7]. Rand and Roberts identified two unique compositions of oxynitride: a stress-free insulating film when the composition is Si O1.75 N 0.25 (O/N ratio of 7:1) and another, a radiation resistant composition: Si_2ON_2 (O/N ratio of 0.5:1). Due to the deposition conditions, the above films were of limited value for electronic applications since such films exhibited very high density of interface fixed charge density in excess of 1E12/cm². Bhattacharyya et al. [3] developed for the first time technology for charge-free oxynitride films covering the entire range of O/N ratios as shown in the triple element phase diagram (Figure 4.5) appropriate for FET applications. Critical processing steps to achieve high-quality charge-free nitride and oxynitride films had been explained elsewhere [3]. It is important to note that processing steps were optimized to favor and maximize formation of stronger Si-N bonds and

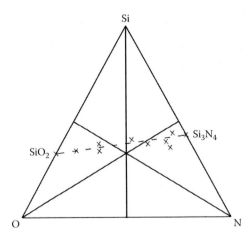

FIGURE 4.5 Si, O, N phase diagram and compositions of Silicon Oxynitrides investigated and characterized. (From Bhattacharyya, A. et al., IBM Internal Technical Reports, 1973 and 1974, *ECS Proceedings*, Los Vegas, CA, 1976.)

to disfavor and minimize formation of weaker and inadvertent formation of Si-H bonds in nitride and oxynitride films. Such processing was essential to achieve charge-free films with reproducible electronic properties relevant for NVM applications. Furthermore, detailed CVD processes were developed to cover the entire spectrum of oxynitride compositions from nearly 5 atomic percent oxygen in SiON film to nearly 60 atomic percent oxygen incorporation in SiON film. Electronic, optical, and chemical properties were investigated and correlated with composition and processing variables. NVM CT devices were fabricated demonstrating superior device characteristics over MNOS and MAOS devices.

Since the development of above SiON films for electronic applications, several other processes were developed for SiON films for electronic applications. For SNOS device application, nitride was replaced by LP CVD oxynitride deposited over oxide to reduce charge injection from the silicon gate [9]. In the early 1990s, considerable development efforts were initiated to replace ultrathin gate oxide for reducing EOT for MOSFETs in order to improve leakage and reliability. Oxynitride films were developed by thermally incorporating nitrogen in thin oxide films by either N_2O or NO RTP (rapid thermal processing) treatment at high temperature. Such processing techniques were employed in NVM devices to replace thin tunneling oxide for SONOS devices [10,11]. Unlike CVD or LPCVD technology, thermal nitridation was diffusion limited and limited in incorporation of wide range of concentration of either nitrogen or oxygen in SiON films, although charge-free high-quality SiON films were readily achieved. In general, based on their electrical properties and electronic applications, the family of oxynitride insulators can be grouped into two subgroups: "oxygen-rich-SiON": OR-SiON and "nitrogen-rich SiON": NR-SiON. The oxygen content of OR-SiON subgroup ranges from approximately 25 atomic percent oxygen to nearly 60 atomic percent oxygen characterized by band gap ranging from approximately 6.5–8.0 eV. On the other hand, the NR-SiON subgroup contains approximately 5 atomic percent oxygen with >50% nitrogen ranging up to 20 atomic percent oxygen with 40% nitrogen (Si_2ON_2) and correspondingly, band gap ranging from 5.7 to 5.2 eV. It is worth noting that within the OR-SiON subgroup, the band gap rapidly increases with increasing oxygen content, whereas, for the NR-SiON subgroup, band gap is slowly raised with oxygen content between 5% and 20%.

Two specific oxynitride compositions have been frequently cited in this book: (1) Si_2ON_2 with 40 atomic percent nitrogen for simultaneously exhibiting high trap density as well as deeper trap compared to Si_3N_4 for trapping applications; and (2) OR-SiON with ~12.5 atomic percent nitrogen and ~54 atomic percent oxygen for replacing SiO_2 as a tunneling layer application interfacing with silicon substrate. The later composition provides interface and transport characteristics similar to SiO_2 with a higher K value (~5.5) and is relatively trap free. The band gap for the above OR-SiON is ~7.5 eV (SiO_2 = 9 eV), compared to Si_2ON_2 of 5.7 eV with K value being the same as that of Si_3N_4. These properties will be further discussed below.

Transport properties of several SiON films were compared with SiO_2 and Si_3N_4 [3] belonging to both subgroups. This is shown in Figures 4.6 and 4.7 where respectively, current I, versus square root of E (Poole–Frenkel plot) and J/E^2 versus $1/E$ (Fowler–Nordheim plot) for the measured MIS capacitors are illustrated. It can be seen that carrier transport for the OR-SiON films are dominated by tunneling mode whereas for the NR-SiON films transport behaviors are dominated by carrier emissions from the trap centers.

4.3.1 Trapping Characteristics of SiON

Trapping characteristics for electrons and holes for the entire range of SiON films were investigated [3]. It was observed that peak trap density for both electrons and holes for all processing conditions existed in films when the atomic nitrogen concentration in films is in the range of 40%–50%; trap density being strongly dependent on the SiH_4/NH_3 ratio during film deposition, the higher the ratio, the higher the trap density. It was also observed that optimized process could yield as much as 2× higher density compared to Si_3N_4 processed at the same SiH_4/NH_3 ratio (no oxygen). Trap density

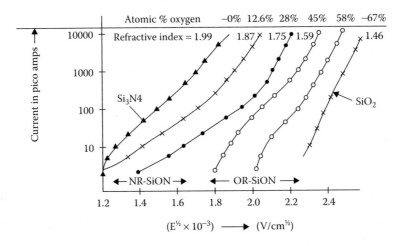

FIGURE 4.6 I versus $E^{0.5}$ plots for nitride and oxynitrides.

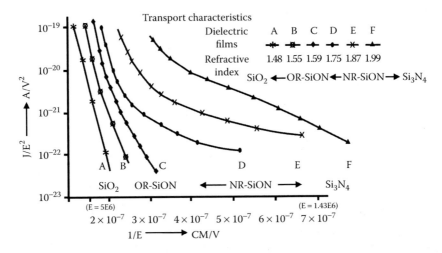

FIGURE 4.7 J/E^2 versus $1/E$ plots for oxynitrides with nitride and oxide as references.

progressively decreased with increased oxygen concentration beyond 20 atomic percent of oxygen in films and dropped sharply above 40 atomic percent of oxygen. This qualitatively explained why transport behavior for OR-SiON films displayed tunneling mode with increased oxygen content in the films.

Trap depths of oxynitride films were obtained from the Poole–Frenkel equation, measured values of maximum dielectric strengths, extrapolated values of conductance at breakdown and previously measured trap depth Utr of nitride of 1.1 eV [12] under similar processing conditions. Computed values of trap depth and band gap for several oxynitride films along with the nitride film are shown along with the measured values of refractive index and dielectric strength in Table 4.2, while trap depth is plotted as a percentage of nitrogen concentration in Figure 4.8. It should be noted that trap depth Utr increases with oxygen concentration in the films and reduces with nitrogen concentration and could be as high as 1.8 eV for approximately 55–60 atomic percent oxygen in the film. Furthermore, oxynitride films exhibited close qualitative relationship between measured refractive indices and electronic properties, namely, permittivity, conductivity, band gaps, and trapping characteristics.

TABLE 4.2

Measured Refractive Index and Oxynitride Films: Compositional Dependence and Dielectric Strength

Film Type	Ref. Index	Atomic Conc.[a]			Diel. Str.[a]	Formula	N/N+O
		O	N	Si	MV/cm		
Nitride	2.0	0	57	43	>11	$SiN_{1.33}$	1.0
NR-SiON	1.88	12	49	39	10–11	$SiO_{0.32}N_{1.26}$	0.8
NR-SiON (Si_2ON_2)	1.84	20.5	41	38	11	$SiO_{0.5}N$	0.67
OR-SiON	1.74	28	32.5	39.5	10	$SiO_{0.71}N_{0.82}$	0.536
OR-SiON	1.65	43.7	17.6	38.7	10	$SiO1.11N_{0.455}$	0.29
OR-SiON	1.55	57.2	9.5	33.3	10	$SiO1.72N_{0.285}$	0.14
Oxide	1.46	66.7	0	33.3	10	SiO_2	0

[a] Measured.

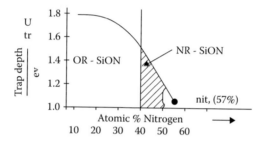

FIGURE 4.8 Trapping depth Utr versus nitrogen percentage in SiON films.

Trap depth in nitride and oxynitride was computed based on the assumption that charge transport is bulk controlled in these films. Except for films with very low nitrogen concentration (≤15 atomic percent nitrogen), such assumption could be justified since the transport characteristics for most SiON films could be described by the Poole–Frenkel mechanism (Figure 4.6) over most of the field range. The calculated trap depth obtained from the Poole–Frenkel equation, measured values of maximum dielectric strengths, extrapolated values of conductance at breakdown, and measured trap depth of nitride of 1.1 eV [12], is plotted against atomic percent of nitrogen in SiON films in Figure 4.8. Trap depth rapidly increases with decreasing nitrogen content for SiON films in the range of 35 atomic percent nitrogen to 50 atomic percent nitrogen. Trap density mentioned earlier in SiON films is also higher in this range and peaked between 40% and 50% atomic nitrogen in SiON films.

4.3.2 MIOS Trapping Comparison between Nitride and Oxynitride

Flat band shift of MIOS capacitors for an oxynitride film of 35 atomic percent nitrogen was compared with that of nitride (stoichiometric Si_3N_4 ~57 atomic percent nitrogen) on long duration stressing. The tunnel oxide thickness for both films was 2.2 nm. Both SiON and Si_3N_4 film thicknesses were 640 nm. Both films had dielectric constant value ~7.0. Flat band versus gate insulator stress characteristics for both films are shown in Figure 4.9. The following should be noted:

1. Electron and hole trapping in nitride are asymmetric. Hole trapping peaks at lower negative gate stress compared to electron trapping for corresponding positive gate bias. This is expected in nitride due to the fact that: (a) when metal gate is biased negative, electrons

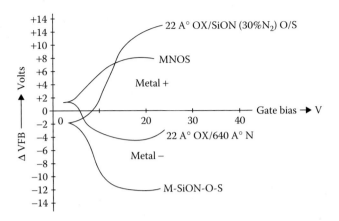

FIGURE 4.9 Characteristics of shift in flat band for M-SiON-O-S structure compared to MNOS.

injected from the gate into nitride compensates the holes injected from silicon, reducing net flat band shift; and (b) the injected holes get trapped deeper into nitride from the oxide/nitride interface due to hole trapping length being nearly 2× (15–20 nm) of electron trapping length (5–10 nm) in nitride. For thinner nitride films, significant part of the injected holes from the substrate during erase (during negative gate stress) gets absorbed in the gate.

2. Electron and hole trapping in oxynitrides are relatively more symmetric. This is due to: (a) reduced electron injection from the metal gate into the SiON layer due to higher energy barrier at the metal/SiON interface; and (b) reduced trapping length of holes in SiON. As a result, net flat band shift for the oxynitride device is significantly larger (thus the window W) compared to the nitride device. Furthermore, due to the difference in trapping lengths, the charge centroids for trapped electrons and holes in device with nitride are further apart when compared to that with oxynitride.

The difference in charge-trapping characteristics between nitride and oxynitride as explained above has important ramification in NVM device properties such as the memory window and write–erase cyclic endurance capability to be discussed in later sections.

As mentioned earlier, the trap density in SiON films are also higher in this range and peaked between 40% and 50% atomic nitrogen in SiON films.

4.3.3 Properties of Oxynitride Films: Co-Relationship with Refractive Index and Composition

Multiple properties of oxynitride films could be correlated with readily measured refractive index, which in turn is dependent on composition of the film. Chemical properties such as etch rate and transport properties will be illustrated here.

4.3.3.1 Refractive Index and SiON Films

Composition: In general, several physical, chemical, and electronic properties of SiON films could be correlated with the refractive index of SiON films. Since SiON films span the entire range of oxygen content in films from Si_3N_4 (zero atomic percent oxygen) to SiO_2 (66 atomic percent oxygen), the nitrogen to oxygen ratio determine many important properties of SiON films. Refractive index has been a simple way to characterize the deposited films and correlates well with the composition of such films. Figure 4.10 plots the entire range of composition of oxynitride films deposited under a wide range of processing variables against the measured refractive index of such films. SiO_2 and

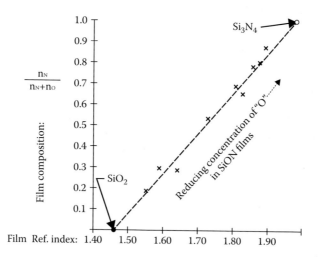

FIGURE 4.10 Relationship between refractive index and composition of SiON films: N/(N+O).

Si_3N_4 films are also included. As can be seen, the refractive index correlates well with the nitrogen to oxygen ratio (or oxygen content) of such films. Therefore, it has become customary to represent a particular oxynitride film by its refractive index. Ude

1. *Etch rate*: Etch rate of SiON films in 7:1 BHF at 35°C is plotted against refractive index of SiON films along with SiO_2 and Si_3N_4 in Figure 4.11. As oxygen concentration increases in films, so does the etch rate. The etch rate varies by nearly two orders of magnitude between nitride and oxide. The differential etch rate between NR-SiON and SiO_2 aids in process integration in ways similar to nitride when nitride films are replaced by oxynitride films.

2. *Conductivity*: As discussed earlier, conductivity in SiON films is dependent on composition. Current (I) through various M-SiON-S structures are plotted at fixed field of 4E6 V/cm against the measured refractive index of SiON films deposited at different silane to ammonia ratios. Variation in conductivity is more pronounced for films with lower

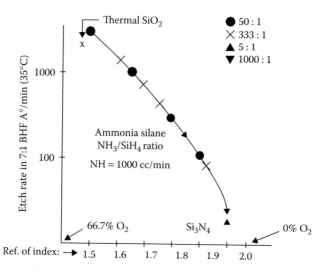

FIGURE 4.11 Etch rate of SiON films in 7:1 BHF versus refractive index with oxide and nitride as references.

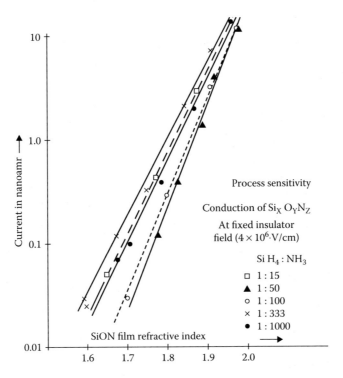

FIGURE 4.12 Conduction through SiON films versus refractive indices at different processing history.

refractive indices (higher oxygen concentration films), and in particular, in films deposited at processing conditions with higher concentration of silane in vapor phase. It had been generally observed that at processing conditions of higher concentration of silane (lower NH3/SiH$_4$ ratio) the deposited films were more conductive due to higher silicon incorporation in films. At the same time, a 10% increase in oxygen concentration in films lower the conductivity of the film by nearly an order of magnitude at the above field. This is shown in Figure 4.12.

4.3.3.2 Dielectric Constant and SiON Films

Similar to the refractive index, the low-frequency dielectric permittivity (K) is sensitive to the composition of the oxynitride film. As can be expected, the dielectric constant of oxynitride with very high nitrogen content (>45 atomic percent of nitrogen) approaches that of nitride whereas oxynitride with very high oxygen content (>55 atomic percent of oxygen) approaches close to that of oxide. Therefore, for SiONs, K value could vary readily in the range between 5 with higher oxygen content to 7 with lower oxygen content. It has also been observed that the K value is also to a smaller degree process sensitive and dependent on atomic percent silicon in SiON film. Atomic concentration of silicon could vary nearly 5% in SiON films by varying silane to ammonia ratio in the vapor phase. With higher silicon content, the measured low-frequency dielectric constant could be higher by as much as 10%.

4.3.3.3 Breakdown Strength and SiON Films

Oxynitride films of all composition range exhibit high breakdown strength >1E7 V/cm comparable to nitride and oxide.

Table 4.3 summarizes the relationship between refractive index and dielectric constant of oxynitride films with composition and other electronic properties such as trap depth and band gap of films.

TABLE 4.3

Electronic and Optical Properties of SiONs: Compositional Dependence

Film Type	Ref. Index	Atomic Conc. O	Low Freq.[a] N	[a]K (1 MHz)	Trap[a] Depth eV	Band Gap[b] eV
Nitride	2.0	0	57	7–8	1.14	5.1
NR-SiON	1.88	12	49	7–8	1.4	5.4
NR-SiON (Si_2ON_2)	1.84	20.5	41	7–8	1.53	5.7
OR-SiON	1.74	28	32.5	6.5–7	1.7	6.7
OR-SiON	1.65	43.7	17.6	6.5	1.75	7.1
OR-SiON	1.55	57.2	9.5	5.4	1.8	8.0
Oxide	1.46	66.7	0	4.0	–	9.0

[a] Measured, [b] Calculated.

4.3.4 Advantage of Oxynitride Films

Since physical, chemical, and electronic properties of oxynitride family of insulators are dependent on composition, these insulators can be tailored to provide many advantages for the silicon based electronic applications. The advantages can be summarized as follows:
NR-SiONs:

1. Selective etchability for process integration
2. Passivation against process contaminants
3. Higher charge trap density and higher trap depth than nitride
4. Higher or comparable dielectric constant to that of nitride
5. Lower conductivity than nitride
6. High breakdown strength

OR-SiONs:

1. Replacement of oxide with lower EOT, leakage and higher charge to breakdown, and consequently higher reliability. This was demonstrated in 1997 by L.K. Han and co-workers by implanting nitrogen into silicon substrate prior to oxidation forming a thin layer of OR-SiON at the interface [13].
2. Tunneling application (Fowler–Nordheim mode at moderate temperature and field)
3. Integration advantages over oxide by preventing boron penetration

4.3.5 Applications of SiONs in NVM Devices

Oxynitride films were applied in early IBM CT NVM devices to overcome limitations of MNOS devices in memory window, charge retention, and cyclic endurance. These devices were called MXOS devices to differentiate them from traditional MNOS devices.

Two general types of SiON films were employed in MXOS devices: one OR-SiON film with >45 atomic percent oxygen (ref. Index 1.6, K = 5.4, 20% nitrogen) and the other NR-SiON film with a composition near Si_2ON_2 with >40 atomic percent nitrogen (ref. Index 1.84, K = 7.0, 20% oxygen). The OR-SiON film was used in combination with ultrathin SiO_2 (1.5 nm) for multilayered tunneling structure. The same oxynitride film was also used as blocking layer to prevent electron injection from the metal gate during erasing due to higher band gap. The NR-SiON film was used to provide higher charge-trapping density (~1E13/cm^2) with deeper energy traps (~1.7 eV) replacing nitride trapping. This replacement of nitride by NR-SiON provided larger window and aided in greater charge retention while minimizing charge centroid displacement during cyclic stressing of writing and erasing. The MXOS device demonstrated vastly superior NVM device characteristics when compared with MNOS devices and maintained superior device characteristics even when compared with subsequently developed SNOS and SONOS NVM devices using nitride as a trapping layer by the general NVM industry. H. E. Maes et al. [14,15] developed and applied LP CVD NR-SiON film replacing nitride in SNOS devices. They demonstrated significantly improved retention and endurance properties of devices compared to standard SNOS devices. H. Fukuda et al. applied N_2O-based oxynitride film replacing SiO_2 as tunnel dielectric [11] for their EEPROM device. They demonstrated significant improvement in cell endurance and reliability. In recent years, T. Ishimaru et al. [16] employed LPCVD SiON of both NR-SiON and OR-SiON types replacing nitride as a charge-trapping layer. Their "MNOS" memory cell used source-side CHE for programming and a combination of electron transport to the gate and hole injection from the gate for erasing by using a thick tunnel oxide (4 nm) MNOS type "device structure." Using SiON as charge-trapping layer, they demonstrated two orders of magnitude enhancement in erase speed as well as charge retention in the memory cell with improved reliability.

The advantage of using oxynitride films in early MXOS devices was illustrated by Bhattacharyya et al. [3]. Enhanced attributes of charge storage (Window), charge retention, and cyclic stress stability (endurance) were demonstrated using multilayered oxynitride films for the device stack structure. Next section on NVM device properties will elaborate on applications of such films in device design. Implication on application of SiON in multilayered films for advanced devices will be covered in Part II of this book on device extendibility.

4.4   SILICON-RICH INSULATORS: (SROs AND SRNs)

Silicon-rich insulators (SRIs), in particular, SROs and SRNs, are unique class of dielectric films by themselves. Considerable research was carried out by IBM researchers during the 1970s and in early 1980s to characterize material and electrical properties of these films and to explore applications in electron devices and in microelectronics technology. The uniqueness of these class of insulators is their existence of stable two-phase structures consisting of stable silicon nanocrystals embedded in the body of a stable stoichiometric bulk amorphous insulator.

High-quality reproducible films were first deposited by Dave Dong of IBM by CVD technique [16], which was soon followed by LPCVD technology developed by A. Bhattacharyya et al. [4]. Physical properties of SROs and SRNs were characterized by the above authors as well as other IBM researchers [17,18]. CVD technique [16] of deposition of thin films of SRIs over silicon substrate or over oxide or nitride involved varying the gas phase ratios of N_2O/SiH_4 and NH_3/SiH_4 mixtures at 700°C to achieve desired Si/O_2 and Si/N_2 compositions, respectively, for films of SROs and SRNs. Whereas, LPCVD technique [4] involved varying N_2O/SiH_2Cl_2 for SROs and NH_3/SiH_2Cl_2 for SRNs at 770°C at a pressure of 0.25 torr.

Solubility of silicon in SiO_2 and Si_3N_4 is limited at any given temperature and pressure. When a large amount of excess silicon is present beyond the solubility limit, the excess silicon precipitates out in the form of silicon nanocrystals ranging in size from <2 to 20 nm. It was found that the concentration, size, and distribution of nanoparticles of silicon depend on the processing conditions and postdeposition annealing and additional thermal budgets and could be altered by the processing variables. It was also observed by Bhattacharyya et al. [4] that silicon nanocrystal size and distribution for the films of SRNs were considerably more stable compared to those of SROs under identical postprocessing thermal budget. This was explained by higher diffusivity of silicon in oxide compared to that in nitride. The unique material and electronic properties of SROs and SRNs originate from the resulting two-phase structures consisting of silicon nanocrystals in oxide and nitride, respectively.

Some of the unique and distinguishing electrical properties of SRIs, and in particular SRNs, will be elaborated here due to their relevance and applications to NVM devices and due to greater thermal and structural stability compared to SROs. Both SROs [17,19] and SRNs [4] have been extensively studied by D. DiMaria et al. and A. Bhattacharyya et al., respectively. Both family of films exhibit some similar general attributes which can be summarized as follows:

1. Both exhibit two-phase structures as explained above and beyond the solubility limits of silicon, which is greater in oxide than in nitride. In the latter case, it is estimated to be less than 2% of excess silicon. Excess silicon content in both family of films could cover the range of 2%–50%. Homogeneous high-quality films are achievable with high excess silicon composition although reproducibility proved to be difficult when the excess silicon concentration exceeds 25%.
2. Optical, electrical, and chemical (e.g., etch rate) could be controlled and modulated by excess silicon concentration, size, and distribution of nanocrystals. SROs and SRNs could be uniquely characterized by their refractive index detailed characterization of LPCVD SRN films.

Table 4.4 lists some of the characteristics of the LPCVD SRN films after post deposition anneal at 1075 for 48 minutes.

4.4.1 Single-Phase SRN

Table 4.3 indicates that the solubility limit of silicon in single-phase SRN is limited to ~2% excess silicon beyond the stoichiometric nitride (Si_3N_4). Single-phase SRN is characteristically similar to nitride from chemical and optical point of view and has a slightly higher refractive index. However, single-phase SRN exhibits higher conductivity compared to the stoichiometric nitride and conductivity increases with increasing concentration of excess silicon in the single phase. Since single-phase SRN could be obtained by increasing vapor phase concentration of SiH_4 or SiH_2Cl_2 in an otherwise similar nitride CVD processing and could be considered as "off-stoichiometric nitride," single-phase SRN is often identified as SiN (or silicon nitride with excess silicon). Transport characteristics follow Poole–Frenkel trapping and detrapping characteristics similar to nitride with comparatively higher trap density and shallower traps dependent on the excess silicon concentration. The dielectric constant of single-phase SRN is found to be somewhat larger that nitride and oxynitrides. However, due to higher conductivity and shallower traps, such films have limited application in NVM devices.

4.4.2 Two-Phase SRN

It should be noted from the above table that stable two-phase structure persists with a wide range of excess silicon content ranging from 2.5% to over 40%. Below the solubility limit, estimated to be <2%, SRN is single phase and should be considered as "Off-stoichiometric" nitride containing excess silicon dangling bonds. Within that range, nitride is more conductive due to the presence of

TABLE 4.4

Characteristics of Annealed SRN Films (1050°C, 45′)

Deposition Ratio R = SiH_2Cl_2/NH_3	Ref. Index	Excess Si %	Dielectric Constant	S/N	Phases
0.01	2.0	0	7.0	0.75	Single
0.25	2.02	<1.5	7.0	0.76	Single
		Type A: Charge trapping			
1.0	2.05	2.5	7.0	0.77	Two
3.0	2.13	6.7	7.0	0.80	Two
5.0	2.17	10.7	7.0	0.83	Two
		Type B: Charge injecting			
10.0	2.4	22.7	7.7		Two
15.0	2.5	34.7	9.0		Two
20.0	2.72	47.0	...		Two

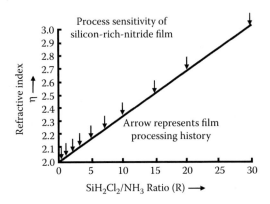

FIGURE 4.13   SRN FILMS: Relationship between refractive index of deposited films and gas phase ratios of dichlorosilane and ammonia during LPCVD deposition.

excess silicon, provides higher trapping density as silicon content increases, and exhibits slightly higher refractive index from 2.0. The two-phase SRNs are grouped into two types: Type A: termed as "Charge Trapping" or CT-SRNs; and Type B: termed as "charge injecting" or CI-SRNs. The material CI-SRN has been labeled in this book as "injector SRN" and abbreviated in this book as IN-SRN or interchangeably i-SRN. The above two terms will be used throughout this book to distinguish the two groups of SRIs due to their distinguishing electrical characteristics to be explained below. The term "injector" SRO was introduced by D. Di Maria (5.15) who developed the first nanocrystal NVM device [19] incorporating Type B of SRO films.

A relationship between the deposition conditions in terms of the gas phase ratio of dichlorosilane (SiH_2Cl_2) and ammonia (NH_3) for the LPCVD process and the resulting composition of the two-phase SRN films defined in terms of the refractive index is shown in Figure 4.13 [4].

4.4.2.1   Electrical Characteristics

Unique and distinguishing electrical characteristics were measured within the two-phase insulator families of SROs and SRNs. Such characteristics were also correlated with the microstructure in the films. Charge trapping could only be observed in SRO films below an excess silicon concentration of 10%–13%, while such limit has been around 7%–10% for SRN films. Within the range of group, A films (CT-SRO or CT-SRN), charge storing capacity progressively reduced with increasing concentration of excess silicon in such films. This is shown in Figures 4.14 and 4.15 for flat band shifts in MIS capacitors due to electron trapping and hole trapping, respectively.

For SRN films (thickness 90 nm) with excess silicon concentration varying from 2.5% to 47%. It should be noted that for both electrons and holes charge storage rapidly diminishes when excess electron concentration of around 10% is reached. No effective charge storage is possible when excess

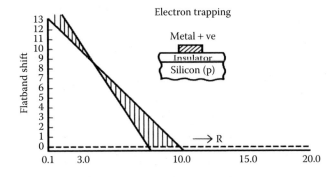

FIGURE 4.14   Electron-trapping characteristics in SRN films.

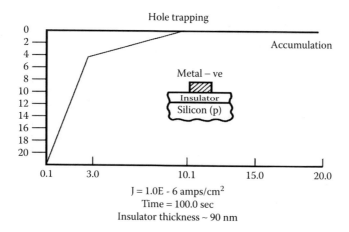

J = 1.0E - 6 amps/cm²
Time = 100.0 sec
Insulator thickness ~ 90 nm

FIGURE 4.15 Hole-trapping characteristics in SRN films.

silicon concentration exceeds such limit. It is also interesting to note that the apparent K values while remaining constant up to such limit kept moving up with higher silicon concentration. The nanocrystal microstructure of annealed films also showed qualitatively that as excess silicon concentration increased, the average size of the nanocrystals was increased. With high excess silicon concentration (ref. Index $\geq$2.4, excess silicon concentration $\geq$20%), the spacing between the nanocrystals was further reduced. Current transport characteristics of SRN films within group A (CT-SRNs) and also within group B (CI-SRNs) are shown in Figure 4.16 where log J (current density) is plotted against E (field) across the MIS capacitors. The same data are replotted in Figure 4.17 where conductivity (LN J/E) is plotted against square root of field (Poole–Frenkel plot).

It should be noted that for CT-SRNs, especially at lower excess silicon concentration, the charge transport mode approximates to that of modified Poole–Frenkel characteristics over a wide field range with a steeper log J versus E slope than the nitride film (Figure 4.16). It is postulated that while the transport is bulk controlled, the mechanism is a combination of temperature/field induced detrapping with tunneling from trapping center to adjacent trapping center ("tunnel-hopping"). The transport characteristics are significantly altered for the CI-SRNs exhibiting many orders of magnitude increase in conductivity and also exponential increase in current density with voltage. These films exhibit "semi-insulating" characteristics with the apparent dielectric constant exceeding that of silicon. It is postulated that transport within the film takes place from one silicon nanocrystal center to the nearest one by direct tunneling, due to local fluctuation of internal potent. It is also noted above that the two-phase SRN films within group A with excess silicon concentration in the range

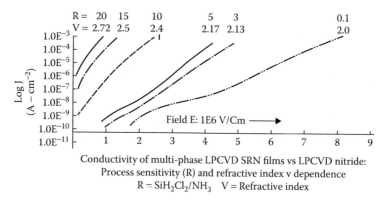

Conductivity of multi-phase LPCVD SRN films vs LPCVD nitride:
Process sensitivity (R) and refractive index v dependence
R = SiH₂Cl₂/NH₃   V = Refractive index

FIGURE 4.16 Current versus field characteristics of SRN films as a function of deposition condition/refractive indices compared to nitride.

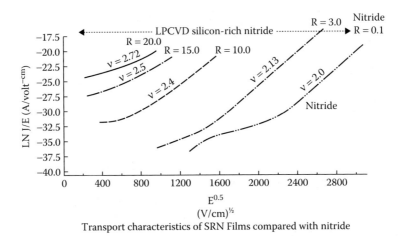

Transport characteristics of SRN Films compared with nitride

FIGURE 4.17   LN J/E versus E plots of SRN films and nitride: Deposition and refractive indices dependence of SRN films.

of 2.5%–10% exhibit the capacity to store charges and, therefore, the films are classified as charge-trapping SRNs or CT-SRNs. However, charge storage capacity decreases with increasing concentration of silicon in such films. It was suggested that with increasing silicon concentration, both the density of trapping centers as well as the depth of the potential well was reduced, thereby reducing the capacity of storing charges. When the excess silicon concentration was further increased, not only the charge storing capacity was lost, but the conductivity of films rose dramatically and the charge transport mode was altered. Additionally, with further increase in silicon concentration, both size and density of silicon crystals increased and the spacing between nanocrystals was significantly reduced. These films (group B) also exhibited very high current density without breakdown. D. Di Maria et al. demonstrated that when group B SRO films were deposited over SiO_2, conductivity of SiO_2 increased by several orders of magnitude and followed enhanced Fowler–Nordheim tunneling with high density of silicon nanocrystals at the SRO/SiO_2 interface acting as charge injecting centers with local field enhancement. Thus the group B films were called "charge injecting" or CI-SRO films. Accordingly, group B SRNs are termed IN-SRNs. IN-SRN films have added advantage over IN-SROs in that their chemical inertness and thermal stability as well as selective etchability made them attractive in passivating local defects in gate oxide and improving oxide reliability [5].

4.4.3   Applications in NVM Devices: IN-SRN Films

The first nanocrystal NVM device was introduced in 1982 by IBM researchers. IN-SRO films were applied to develop an electrically alterable read-only-memory by D. Di Maria et al. who (5.17) called the device as "DEIS" (dual-electron injector structure) EAROM device. The NVM element was a floating-gate device whereby the interpoly oxide was sandwiched between two layers of IN-SRO films serving as enhanced electron injectors for injecting electrons from the control gate to the floating gate for writing and removing electrons from the floating gate to the control gate during erasing. Because of the thermal stability and integration advantages outlined earlier, IN-SRN is a preferred option for application in microelectronics technology. IN-SRN was used to improve gate oxide reliability in early IBM technology [5]. Application of IN-SRN had also been proposed for alterable logic applications [20].

IN-SRN is an effective option to replace floating polysilicon gate element for NVM devices for the following reasons:

(a) IN-SRN provides equal-potential interface to the tunnel oxide underneath, as well as to the interpoly dielectric on top; (b) IN-SRN over-layers provide improved tunnel oxide reliability by lowering the field required for programming and erasing and by enhancing charge to breakdown of the tunnel dielectric, thereby enhancing endurance and also reducing local defect-induced early breakdown; (c) IN-SRN enhances programing and erasing performance and memory window by significantly enhancing tunneling currents through enhanced Fowler–Nordheim tunneling; (d) IN-SRN will reduce parasitic capacitive coupling between neighboring cells ("disturbs" discussed later on) lowering disturbs; and (e) since IN-SRN thickness can be readily scaled, vertical and horizontal scalability of memory cells and voltage scalability of cells are achievable. The enhancement in endurance had been empirically demonstrated by Bhattacharyya [6].

IN-SRN could have many unique applications in advanced NVM stack designs. Such applications could be summarized as:

1. *For tunneling media*: It could be employed as a part of multilevel band-engineered tunneling layer to not only create resonant tunnel barriers (RTB) NVM stack designs to improve stability of memory states for MLS (multilevel store, multiple bit/cell) designs but also to enhance retention by modulating the internal electrostatic potential to reduce back-tunneling of electrons stored in the storage medium (see Part II). IN-SRN when applied at the top of multiple layer tunneling structures provides passivation and integrity of the silicon/insulator interface and enhances erase speed by enhancing FN tunneling of electrons to the substrate during erasure.

2. *For storage media*: It replaces FG as a scalable option creating FP Flash devices. When used with conventional (e.g., Si_3N_4 or Si_2ON_2) or high K (e.g., GaN or AlN) charge-trapping layer, on top of such dielectric layer acts as charge reservoir, enhances programming efficiency, and enlarges memory window. Application of IN-SRN has also been discussed in advanced direct tunnel memory (DTM) cell designs. Many advanced and embedded CT-MLS devices could in principle incorporate IN-SRN as charge reservoir to enhance device characteristics (see Part II).

3. *For blocking media*: When IN-SRN is used as gate/insulator interface, it prevents impurity and contaminant penetration into the dielectric layers and silicon/insulator interface enhancing reliability promoting uniform enhanced charge injection from the gate and preventing localized carrier transport. For both silicided silicon gate and metal gate processing for work function control, IN-SRN could also be employed to facilitate process integration utilizing its selective etch properties.

Many advanced cell design concepts incorporating IN-SRN layers in combination with high K dielectric layers for NVM gate insulator stack designs will be discussed in appropriate follow-on sections in Part II of this book.

4.4.4 Application of CT-SRN Films in NVM Devices

CT-SRNs provides silicon-nanocrystal-induced trapping centers in addition of bulk traps of nitride for charge storage in charge-trapping NVM devices. The key advantage of employing CT-SRN dielectric films is the thickness scalability in gate insulator stack design, thereby reducing the EOT of the stack and the programing voltage. CT-SRN could be effectively scaled in the range of 1.0–1.5 nm for charge-trapping applications [21,22] for advance SONOS type of device applications as compared to nitride which could effectively be scaled to 5 nm only. However, excess silicon concentration is critical for CT-SRN for providing appropriate design goals of programming window (delta Vth) and charge retention. A range of 2%–10% of excess silicon concentration of CT-SRN films was found to be useful for NVM devices [23], although a narrower range of between 2% and 5% was optimal for size and distribution of nanocrystals and to achieve more desirable memory window and retention. The first nanocrystal NVM charge-trapping device was proposed and developed in IBM in 1984 as was discussed in Chapter 2, Figures 2.7 had nearly 8% of excess silicon concentration, which was not optimized. Even at such higher excess silicon concentration, the device exhibited larger memory window compared to control device with nitride trapping as evidenced by nearly 3.5× long-term flat band shift with CT-SRN, when capacitor stack structures of MONOS versus MO CT-SRN OS were compared (Figure 2.7). It should also be noted that charge storage for the CT-SRN stack for both electrons and holes are nearly symmetric similar to the oxynitride (SiON) trapping layers as discussed earlier and in contrast to nitride trapping characteristics. Since CT-SRN films could be significantly thinner, application could essentially provide a planar charge storage possibility in gate stack design providing the potential for unlimited endurance for NVM devices and lower parasitic capacitive coupling between neighboring cells for future scaling of memory cells. This concept is further discussed in the follow-on sections on endurance and disturb and in Part II of this book.

REFERENCES

1. C. Hu, Gate oxide scaling limits and projection, *IEEE, IEDM*, 1996, pp. 319–322.
2. P. E. Nicollian, W. R. Hunter, and J. C. Hu, Experimental evidence for voltage driven breakdown models in ultrathin gate oxides, *IEEE International Reliability Physics Symposium*, San Jose, CA, 2000, pp. 7–15.
3. A. Bhattacharyya, C. T. Kroll, H. W. Mock, and P. C. Velasquez, IBM Internal Technical Reports, 1973 and 1974, *ECS Proceedings*, Los Vegas, CA, 1976.
4. D. W. Dong, D. J. DiMaria, and D. Young, Chemical vapor deposition, ECS, *ECS Proceedings*, 1977, p. 483.
5. D. J. DiMaria, K. M. DeMeyer, C. M. Serrano et al., Electrically alterable read-only-memory using Si-rich SiO_2 injectors and a polycrystalline silicon storage layer, *J. Appl. Phys.*, 52, 4825–4842.
6. A. Bhattacharyya et al., Antifuse structure and process, U.S. Patent 6344373 B1, February 5, 2002.
7. M. J. Rand and J. F. Roberts, *J. Electrochem. Soc.*, 120, 446, 1973.
8. T. L. Chu, J. R. Szenden, and C. H. Lee, *J. Electrochem. Soc.*, 115, 381, 1968.
9. H. E. Maes, The use of oxynitride layers in Nonvolatile S-OxN-OS (silicon-oxynitride-oxidesilicon) memory devices, Chapter 6, in *LPCVD Silicon Nitride and oxynitride Films*, edited by F. H. Habraken, pp. 127–146, Springer-Verlag, Berlin, Germany, 1991.
10. H. Fukuda, M. Yasuda, T. Iwabuchi et al., Novel N_2O-oxynitridation technology for forming highly reliable EEPROM tunnel oxide films, *IEEE Elec. Dev. Lett.*, 12, 587–589, 1991.
11. M. Bhat, D. J. Wristers, L.-K. Han et al., Electrical properties and reliability of MOSFETs with rapid thermal "NO-Nitrided" SiO2 gate dielectrics, *IEEE Trans. Elec. Dev.*, 42, 907–914, 1995.
12. P. Chaudhari, J. Franz, and C. Acker, *J. Electrochem. Soc.*, 120, 999, 1973.
13. L. K. Han, S. Crowder, M. Hargrove et al., Electrical characteristics and reliability of sub-3 nm gate oxides grown on nitrogen implanted silicon substrates, *IEEE, IEDM*, Washington DC, 1997, pp. 643–646.
14. B. H. Yun, Electron and hole transport in CVD nitride films, *Appl. Phys. Lett.*, 27, 256–258, 1975.
15. H. E. Maes and E. Vandekerckhove, Nonvolatile memory characteristics of polysilicon-oxynitride-oxide-silicon devices and circuits, ECS, *Proceedings of ECS*, vol. 87–10, edited by V. J. Kapoor and K. T. Hankins, p. 28, 1987.
16. T. Ishimaru, N. Matsuzaki, Y. Okuyama et al., Impact of SiON on embedded nonvolatile MNOS memory, *IEEE, IEDM*, San Francisco CA, 2004, pp. 885–888.
17. A. Bhattacharyya, R. Bass, W. Tice et al., Physical and electrical characteristics of LPCVD silicon-rich-nitride, *ECS Fall meeting*, New Orleans, CA, October 1984.
18. E. A. Irene et al., On the nature of CVD silicon rich SiO_2 and Si_3N_4 films, *J. Electrochem. Soc.*, 127(11), 2518–2521, 1980.
19. D. J. DiMaria, R. Ghez, and D. W. Dong, Charge trapping studies in SiO_2 using high current injection from Si-rich SiO_2 films, *J. Appl. Phys.*, 51(9), 4830–4841, 1980.
20. A. Bhattacharyya, Reliability improvement in gate dielectric using oxide-IN-SRN gate insulator stack, IBM internal Technical Report (Unpublished), 1985.
21. A. Bhattacharyya, Fundamentals and extendibility of NVM device endurance, ADI associate internal docket, 2001, and U.S. Patents: 7012297, 7250338, 7400012, and 7750395 entitled: "Scalable Flash/NV Structures and Devices with Extended Endurance".
22. R. S. Bass, A. Bhattacharrya, and D. Grise, Nonvolatile memory cell having silicon rich silicon nitride charge trapping layer, U.S.Patent No. 4870470, September 26, 1989.
23. A. Bhattacharyya, Technology development, properties and applications of thin films of SiRich oxides and nitrides, IBM Internal Technical Report, 1984 (Unpublished).

5 NVM Unique Device Properties

CHAPTER OUTLINE

In this chapter, we will discuss three unique device properties of NVM devices.

These are (1) memory window: W, (2) memory state retention or simply retention, and (3) memory programming and erasing cyclic stability or endurance. This section will be followed by a related section which covers all significant aspects of NVM device stack designs to meet the targets related to the above properties of memory states, the stability of such memory states providing nonvolatility not only in terms of time and temperature but also from the standpoint of stress-tolerance associated with the process of programming for the binary states of writing and erasing.

5.1 MEMORY WINDOW W

Silicon based NVM devices, as explained in the beginning, are a variation in an FET device, whereby electronic charges (electrons and/or holes) are stored in the gate insulator stack to create multiple relatively stable threshold states called memory states. These memory states are designed to be stable or "permanent" through the end of life of the device, The memory states can be sensed or "read" by detecting whether the current flows from the source line of the memory device to the bit line or not, when the device memory state is addressed. Hence, the name nonvolatile memory or NVM. The memory "Window" or W is the required margin or difference between the memory states to accurately predict the memory state of the device throughout the lifetime of usage of the device. If the memory is a binary memory with one bit of memory storage capacity per memory cell, the window W is defined as the difference in Vth from Vth-high or Vth(1) to Vth-low or Vth(0). In such case, the window can be simply defined as:

$$W = Vth(1) - Vth(0)$$

When charges are stored in the gate insulator stack, silicon surface potential and consequently the flat band potential are affected altering the threshold of the device. Therefore, assuming other variables such as work functions, Fermi potentials, and interface charges remaining the same, the above definition can be expressed in terms of change in flat band potential between the two memory states of "1" (high Vth of N-channel device) and "0" (low Vth of N-channel device) and can be expressed in terms of the "effective" difference in charge stored in the dielectric stack between the above memory states as:

$$W = Qs'(1) / Ci - Qs'(0) / Ci$$

Where Qs(1) and Qs(0) are charges stored for the memory states of (1) and (0), respectively, Qs'(1) and Qs'(0) are corresponding equivalent charges at the silicon dielectric interface, and Ci is the capacitance of the gate insulator stack and is given by $Xi/Ki = EOT/Kox.e_0$ where Xi is the physical thickness of the dielectric stack, Ki is the effective dielectric constant of the stack, EOT is the effective oxide thickness of the dielectric stack, Kox is the dielectric constant of oxide, and e_0 is the permittivity of the free space = 8.852E-14.

When multiple memory bit per cell is involved, for instance, 2 bit/cell with 4 memory states or 3 bit/cell with 8 memory states, the definition of window still remains the same as the difference between the highest "written" Vth state (most positive value) and the "erased" Vth state with the lowest or most negative Vth value. Multilevel cells (MLCs) or multilevel stores (MLSs) require larger window to ensure accurately sensing with sufficient margin between different memory states with corresponding Vth levels. This will be discussed in Part II, Chapter 15.

The Processes of Programming and Erasing:

It is noted earlier that memory states are created by the process of programming (writing) and erasing a memory cell whereby charges are stored within the gate dielectric stack of the memory cell. Figure 5.1a and b schematically represents the electron flow and storage in the floating gate (FG) for a single transistor FG memory cell during writing and erasing, respectively.

The process of "writing" involves raising the control gate (word-line) potential with respect to the substrate (and source and drain) to a high potential (+Vpp) and holding the potential for an appropriate time period for electrons to overcome the silicon-oxide energy barrier and get stored within the FG. As the electrons accumulate within the FG, the FG potential gets increasingly more negative raising the field between the FG and the control gate. This results in the outflow of electrons from the FG to the control gate through the inter-poly dielectric layer. At the same time, the field between the FG and the substrate is reduced due to the negative charge accumulation in the FG. Finally, a steady state is reached when electron inflow and outflow becomes equal in the FG. On completion of the writing process, the control gate potential drops to zero. The net electronic charge stored in the FG raises the threshold of the cell to Vth-high establishing the memory state "1". The potentials of word-line (control gate) and the FG during writing are shown in Figure 5.2.

The process of "erasing" involves either raising the control gate to a sufficiently high negative potential (−Vpp) with substrate grounded or while holding the control gate to ground, raising the

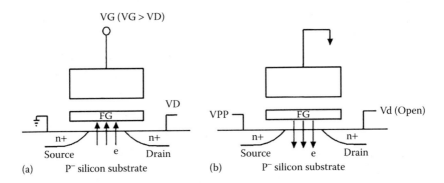

FIGURE 5.1   Charge transfer and storage in FG: (a) writing and (b) erasing.

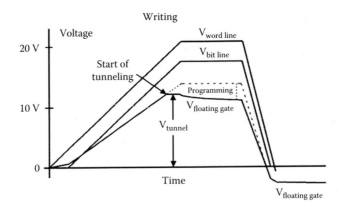

FIGURE 5.2   Word-line and floating gate potentials during writing.

potential of substrate to a high positive potential (+Vpp) during the period of erasing. The electrons stored in the FG return to the substrate as shown in Figure 5.1b by reverse tunneling. The process of erasing results in the progressive rise in potential of the FG as the electrons return to the substrate. This is shown in Figure 5.3. Consequently, the field between the control gate and the FG is raised, while the field between the FG and the substrate is reduced. A steady state is reached again when the electrons inflow from the control gate to the FG equals to the outflow of electrons from the FG to the substrate. On completion of the erasing process, the Vth of the cell is lowered, establishing the memory state "0".

For charge-trapping (CT) devices including nanocrystal (NC) devices, the process of writing and erasing is similar. NC devices provide planar equipotential storage node even though charges are stored in discrete trapping centers. In case of SONOS type of CT devices, charges are stored within the bulk nitride film in discrete trapping centers. When the tunnel dielectric layers in these devices are very thin, for example, ≤3.0 nm of SiO_2, direct tunneling of both electrons and holes is feasible. Writing involves direct tunneling of electrons to the trapping sites, while erasing involves direct tunneling of holes into the hole trapping centers. Therefore, both electrons and holes are stored in these types of devices unlike FG devices where only electrons are involved. Raising of Vth of the device during writing to achieve memory state "1" involves steady-state storage of electrons in the electron trapping centers, whereas lowering of Vth of the device during erasing involves steady-state storage of holes in the hole trapping centers of the trapping dielectric medium of the device, thereby compensating the electrostatic potentials associated with the stored electrons.

In practice, the window of an NVM device and product and the associated margin between the written memory state and the erased memory state is established by taking into account all device, process, and application parameters affecting the memory states. This is schematically shown either in the statistical distribution of memory states shown in Figure 5.4 with the achievable net window or in Figure 5.5 where the Vth spread for the memory states is plotted against the number of write–erase stress cycles with the superimposed Vcc margin depicting the associated program and erase read margins. The net achievable window at the end of life is illustrated in both figures.

5.1.1 Memory Window Generation: Charge Storage and Trap Density

It should be noted that the above discussion on memory window generation and associated charge injection and charge storage and consequently change in potential of the FG or trapping medium is time dependent. This aspect has been more rigorously discussed by G. Groeseneken et al. [1]. A relatively simple assessment is being made here to highlight the physical significance of feature size scaling of devices, the order of magnitude of participation of electronic charges, and the trap

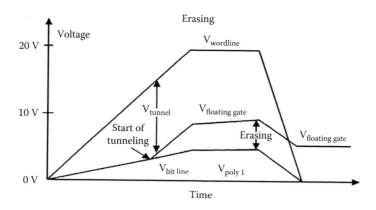

FIGURE 5.3 Word-line and floating gate potentials during erasing

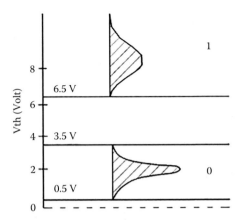

FIGURE 5.4 Memory window definition showing threshold distribution for memory states.

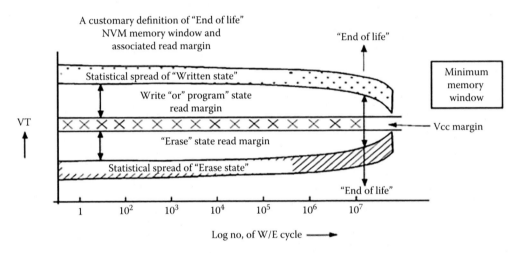

FIGURE 5.5 Alternate definition of memory window with read margins and Vth distribution of written and erased states post EOL W/E cycling.

density, Qs, requirements (for CT devices) for memory window generation. The following simple assumption will be made for such assessment:

1. NVM active cell area is $4F^2$.
2. EOT of the gate insulator stack is 10 nm.
3. Cell writing and erasing time duration is 100 μs each for ±Vpp.
4. Window to be achieved is 4 volts.

Table 5.1 shows the approximate values of electronic current density needed, required trap density for CT devices, total charge in coulomb required to be stored and number of electronic charge associated in the process of generating the memory window.

Several important observations could be made from the above table: These are as follows:

1. Required current transport is quite high to meet the write–erase speed of 100 μs each. For oxide as the tunneling medium, this implies very high field across the tunnel dielectric of ~>10E6 V/cm.

TABLE 5.1

Relationship between Charge Transport, Trapping, and Memory Window

Feature size (nm)	50	35	20
Active device area for Qs (cm^2)	1E-10	4.9E-11	1.6E-11
Electron/hole current (A/cm^2)	6.9E-3	6.9E-3	6.9E-3
Required trap density (N/cm^2)	8.63E12	8.63E12	8.63E12
No. of charge for W = 4V	860	106	138
Qs: Charge stored (C)	1.378E-16	6.8E-17	2.21E-17

2. Required trap density is quite high: for stoichiometric Si_3N_4, typical trap density is around ~5E12/cm^2; for NC devices, the effective trap density could be further reduced due to coulomb blockade and quantum confinement (Chapter 2).

3. Number of electrons per cell is reduced significantly with scaling of the feature size.

It will be discussed later on that this has strong impact on retention and endurance.

5.1.1.1 Vpp and EOT: Impact on Memory Window

In the above example, EOT for the cell is assumed to be 10 nm. However, the stack EOT could easily be twice as large for FG devices since the EOT required for tunnel dielectric alone is around 7–8 nm to meet the nonvolatility requirements of the memory states (see Section 5.2). To meet the voltage coupling requirements between the control gate and the FG, the EOT for the inter-poly dielectric could be in the order of 10 nm with a total stack EOT of ≥18 nm. To meet the electron or hole current requirements during programming, a field of >10E6 V/cm would be required implying a Vpp level required to be >18 V for the FG device. For CT devices such as SONOS or NC devices, the corresponding tunnel oxide thickness is approximately half and would be ~3.5–4.0 nm. For similar blocking layer EOT for all CT device types, the stack EOT would be <14 V. Therefore, the FG device type and the CT device type stack EOTs could be approximately 18 and 14 nm, respectively. This assumes that the same field is required across the tunnel dielectric for either the FG or for the CT devices. Correspondingly, Vpps for programming and erasing will be: >±18 and <±14 V, respectively. From the above discussions related to memory window, the following observations could be made:

1. Achievable memory window in NVM devices is dependent on charge storage density or (trap density for the CT devices), the stack EOT, and the writing and erasing voltages (Vpps) and duration of Vpps. However, these are not independent variables and required to be optimized to achieve the targeted memory window.

2. Memory window is expected to be lower in CT devices due to the limited number of trapping centers available compared to the FG devices.

3. The EOT for the FG devices is expected to be larger than the CT devices, and correspondingly, the Vpp for the FG devices would be higher for programming and erasing.

5.2 MEMORY RETENTION

Memory retention and simply called "Retention" is perhaps the most important device parameter distinguishing NVM devices from other types of memory devices, for example, DRAM. Retention is defined as the stability of the memory threshold states as a function of time, temperature, and application conditions with or without power. NVM devices are expected to retain their memory states throughout the end of life regardless of the application conditions which is customarily defined to be 10 years at an appropriate elevated temperature (typically 85°C or 125°C) after subjecting the memory cells for a range of write–erase cycling (typically 1E5 to 1E6 cycles).

Stability of the memory Vth states for the NVM devices is affected by many factors. These are (a) device type and mechanisms of writing and erasing; (b) time, temperature, application conditions including presence or absence of power; (c) gate insulator and interface material selection (gate stack design); (d) process sensitivity and processing variables including defects, traps, and contaminants present in critical charge storage regions of the device; (e) programming characteristics and programming algorithms; and (f) cell and array design variables. This also induces stresses due to parasitic cross-coupling effects from active cells to neighboring cells in the form of "disturbs" (e.g., drain disturb, gate disturb, and read disturb to be explained later). Memory retention will be discussed here in the context of tunnel-based FG and CT devices employing conventional gate stack designs as explained in earlier sections. In Part II of this book, retention will be discussed in the context of advanced high K stack designs and associated devices. In this section, we will explore charge retention strictly in the context of charge transport within the dielectric layers of the NVM device stack. For such discussion, we will invoke dielectric conductivity and band diagrams for such device stack to explain different charge loss mechanisms affecting retention.

5.2.1   Retention, Charge Transport, and Dielectric Conductivity

Electronic charges stored within the gate stack of NVM devices are surrounded by dielectric medium.

Electrostatic potential created by stored charges establish significant field across either the silicon substrate or across the control gate of the device. Charge transport characteristics and associated conductivity of dielectric films surrounding the stored charge provide important qualitative insight into the charge loss mechanisms affecting charge retention. Figure 5.6 is a plot of current density J versus field E of all conventional dielectric films discussed in Chapter 4. The rationale for the conventional FG and CT NVM stack designs and associated limitations could be

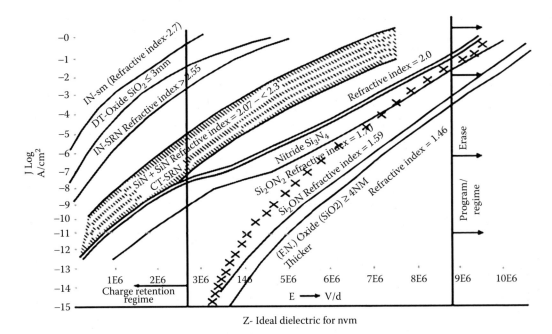

FIGURE 5.6   J versus E plots of conventional dielectric films compared with that of an ideal (hypothetical) single layer dielectric film for tunneling.

appreciated by addressing simultaneously the requirements of programming current and retention current when conventional single layer dielectrics are employed for each of tunneling, trapping/ storage, and charge blocking functions. Two field regions are identified: the program-erase field region which for all conventional NVM devices exceeds 8.5E6 V/cm and the charge retention field region which is typically below 3.0E6 V/cm. It is worth noting that the leakage current (1 and associated conductivity) at charge retention field range can vary by nearly 15 orders of magnitude depending on the films involved!! It is also worth noting that approximately the same orders of magnitude difference in conductivity are required for an NVM device if the device is designed to program-erase at one micro-sec (1E-6s) and hold charges for greater than 10 years (~1E9s).

J versus E plot for an ideal hypothetical dielectric film is superimposed in the above figure which could satisfy programming and retention requirements of conventional NVM device. The plot suggests that while for fields in the range of ~ 9E6V/cm, the current density desired for programming could be similar to that of nitride or NR-SiON of ~1A/cm², the J versus E slope need to be significantly steeper at fields in the range of ~2-4E6V/cm for charge retention. For example, $J = \sim 1E-15A/cm^2$ at E = ~3E6V/cm would be required to provide a current ratio of 1E15 X between the retention field and the program-erase field to meet simultaneously the programming and the retention requirements. Although such a single layer dielectric film is not readily available, multilayered SiON films when appropriately employed for tunneling, trapping, and blocking functions yielded such characteristics [2]. For SiO_2 films, conductivity characteristics show that for films ≥4.0 nm, when employed for tunneling and blocking functions could satisfy retention requirements due to low leakage. However, for very thin oxide films of thickness <3 nm, leakage at low field is many orders of magnitude higher than that required to meet retention requirements. This is due to the fact that carrier transport in such films takes place by direct tunneling, and carriers present at interface of such oxide film would readily tunnel back at modest field. The IN-SRNs have similar conductivity as direct-tunneling oxide and, therefore, are not suitable for charge retention. It has been explained in Chapter 4 that such films could replace FG due to their high charge storage capacity and high conductivity. The CT-SRN dielectric films exhibit almost two orders of magnitude higher conductivity and would yield poor charge retention compared to nitride films due to large concentration of silicon-induced shallow traps. However, in combination with other films, such films could be employed to achieve larger window and higher write–erase performance due to their intrinsic higher conductivity at high field range as long as retention requirements could be achieved through appropriate combination of films of lower conductivity. Compared to oxide, nitride conductivity at retention range is higher by at least six orders of magnitude and due to carrier transport mechanism (Poole–Frenkel) is strongly sensitive to temperature. Consequently, nitride by itself cannot satisfy the retention requirements. Thus, the rationale is a combination of oxide and nitride being employed in SONOS-CT devices both as blocking layer and as CT medium for such devices as well as ONO for IPD combined with thicker SiO_2 for tunneling in FG devices to achieve the retention requirements.

Band diagram is also a convenient tool to recognize various modes of potential charge loss mechanisms affecting charge retention. This is addressed for both FG and CT devices as outlined below.

5.2.2 Charge Loss and Leakage Paths for Floating-Gate Devices

Band diagram and charge loss modes for FG-NVM devices are discussed below.

Current components for FG devices: Figure 5.7 illustrates band diagram for a typical FG device with no external potential either at substrate or at the control gate (standby state). The modulation of the diagram due to electron storage at the FG (Vth "1" state) is also shown by dotted line. The various potential charge leakage modes are identified in the diagram as current components and will be discussed here. A simple assessment will be made to qualitatively identify the major current component affecting charge retention. Several current components in the standby Vth "1" state may

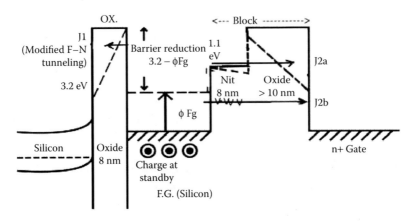

FIGURE 5.7 Band diagram for floating-gate devices at standby state: Charge loss considerations.

be considered. These are J1 (FG/Si): Modified Fowler–Nordheim band to band tunnel current from FG to silicon, whereby the energy barrier is lowered by the band offset reduced by the FG potential: $3.2 - \Phi Fg$; where ΦFg is the FG potential due to the electrons stored by the "writing process." In practice, the energy barrier could be further lowered due either to polysilicon doping or due to the roughness of polysilicon grains of the FG, as qualitatively shown in the diagram at both the FG/tunnel oxide and the FG/nitride "IPD" (inter-poly dielectric) interfaces. J1 could be further modified at the FG/diffusion overlap regions due to the work function differences between the FG and N+ regions of source and drain.

J2a (FG/Gate): This current path may consist of several sub-components: A direct tunneling of charge to nitride trap followed by field enhanced thermally excited detrapping to the nitride conduction band and subsequently transported by modified Fowler–Nordheim tunneling through the blocking oxide. This mode would exhibit greater temperature sensitivity since the current would be dependent on nitride detrapping.

J2b (FG/Gate): The second current path from the FG to the control gate may consist of modified multiple band to band tunneling first from FG to the conduction band of nitride and then from the conduction band of nitride to the conduction band of the blocking oxide. This component would be less sensitive to temperature compared to the J2a component. If the blocking oxide contains significant component of shallow traps, internal trap hopping may be associated in oxide which may or may not have a thermal component. This sub-component is not illustrated in the diagram.

5.2.3   Band Diagram-Based Assessment of FG Current Components

A simple assessment can be made to address the relative values of J1 versus J2a + J2b. For computational purposes, the physical thicknesses of 8.0 nm of tunnel oxide, and inter-poly dielectrics of 8.0 nm of nitride plus 10.0 nm of oxide as shown in Figure 5.7 is used with an EOT of 14.6 nm. Approximate initial field across the tunnel oxide due to electron storage in the FG is assumed to be 3E6 V/cm and corresponding fields across the nitride and blocking oxide is computed and the approximate current values of J1, J2a, and J2b are computed from their transport characteristics. This is shown in Table 5.2.

From the above assessment, one would conclude that the major standby retention charge loss is due to leakage from the FG to the control gate through the nitride and oxide IPD layers.

Leakage characteristics of ONO IPD: Several authors have modeled the leakage characteristics through IPD dielectric layers [2–5]. It was shown that implementing an optimum ONO stack would yield lower leakage and would meet retention requirements. The ONO IPD approach had since been

TABLE 5.2

Assessment of Current Components (Electron Storage Potential across Tunnel Oxide: ΦFg)

Parameter	Tunnel Oxide	IPD Nitride	IPD Oxide
Thickness (nm)	8.0	8.0	10.0
k	3.9	7.0	3.9
Barrier energy (eV)	3.2 − ΦFg	2.1 − ΦFg	1.1 − ΦFg
Field (V/cm)	3E6	0.5E6	1.656E6
Mode	Band to band	FG to nit (conduction band)	Nit to ox. B ->B
Current (A/cm²)	J1 = ~1E-16	J2a = ~1E12	J2b = ~1E12

followed for FG products for more than 15 years [5]. This is shown in Figure 5.8. Data retention affected by electron detrapping and associated Vth shift is shown in Figure 5.9.

5.2.4 Tunnel Oxide Charge Loss Characteristics

For FG flash devices, thermally grown SiO_2 had been successfully employed as the tunnel dielectric to meet the retention requirements. Charge loss mechanisms and reliability issues of tunnel oxide had been well established [5–8]. It has been well known that defects are generated within the oxide film during high field stress cycling associated with writing and erasing. As a result, traps are created within oxide and charge trapping of electrons and holes affect the oxide integrity adversely affecting device reliability [9]. Two mechanisms were associated toward charge loss affecting retention. The first one is electron detrapping [6,8] mentioned earlier, and the second one, "stress-induced leakage current" or "SILC" which was associated with hole trapping in SiO_2. SILC had been studied by Aritome et al. [9,10] who observed that low field leakage was most enhanced when the bias mode was such that the gate was grounded and positive bias was imposed on substrate and diffusions. The implication was that hole fluence from substrate and diffusions get trapped in oxide causing most damage to the oxide film. At lower field, an order of magnitude lower leakage was observed when the oxide was stressed in bipolar mode of writing and erasing to the same high field as compared to unipolar mode of erasing. This is shown in Figure 5.10 [9]. Aside from hole injection, hydrogen release process in SiO_2 was also found to be responsible for SILC [11] causing degradation in oxide. Several activation mechanisms were identified by R. N. Mielke et al. [6] associated with charge loss in tunnel oxide. Charge loss associated with electron detrapping in oxide showed a higher activation

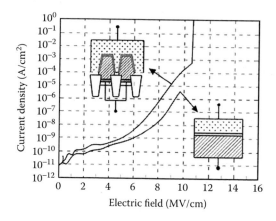

FIGURE 5.8 ONO IPD leakage characteristics for planar and nonplanar test structures in NAND flash design. (From Choi, J.-D. et al., A 0.15 μm NAND flash technology with 0.11μm² cell size for 1Gbit flash memory, *IEDM*, p. 767 © 2000 IEEE.)

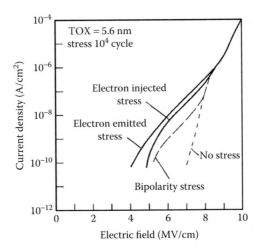

FIGURE 5.9 Stress mode dependence of leakage in 5.6 nm thick tunnel oxide after bipolarity stress, electron-emitted stress, and electron-injected stress. (From Aritome, S. et al., A reliable bipolarity write/erase technology in Flash EEPROMs, *IEDM*, p. 111 © 1990 IEEE.)

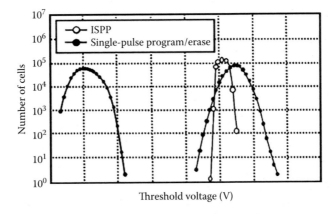

FIGURE 5.10 Cell Vth distribution of 1Mb cells as a function of single pulse versus ISPP demonstrating ISPP advantage. (From Choi, J.-D. et al., Highly manufacturable 1 Gb NAND flash using 0.12 μm process technology, *IEDM*, p. 25 © 2001 IEEE.)

energy of 1.4 eV. A second mechanism is related to contamination in oxide with a somewhat lower activation energy of 1.2 eV, while a third mechanism related to defects in oxide had a significantly lower activation energy of 0.6 eV.

Oxide degradation due to program-erase stressing is minimized by incremental step-pulse programming (ISPP) mode as compared to direct single pulsing to high program-erase voltage levels. The ISPP mode was subsequently widely adopted by the industry. Figure 5.10 demonstrates the ISPP advantage for the 1 Gb flash technology product introduced by Samsung [12]. Oxide degradation also impacts retention during the reading process after the device is subjected to program-erase stressing. ISPP mode also improves read retention. This is demonstrated in Figure 5.11 [5].

5.2.5   Tunnel Oxide Thickness Limit for FG Device

Even with advanced process and defect control for thermally grown SiO_2, the tunnel oxide scalability for NAND flash device is limited to around 8.0 nm for practical product application both

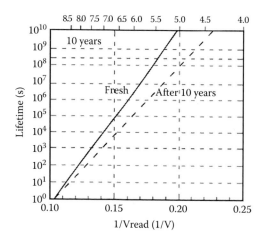

FIGURE 5.11 Read retention before and after 1E5 program-erase cycling of memory cells ensuring 10 years' life time objective of FG memory. (From Choi, J.-D. et al., A 0.15 μm NAND flash technology with 0.11 μm² cell size for 1Gbit flash memory, *IEDM*, p. 767 © 2000 IEEE.)

for retention and reliability reasons. Samsung's 1 Gb NAND chip [5] originally employed 8.5 nm thermally grown SiO_2 to meet the retention and reliability objectives. As the 1 Gb flash chip was made more productive from 217 mm² chip (150 nm technology node) to 132 mm² chip (120 nm technology node) [12], the tunnel dielectric was changed to 7.0 nm of oxynitride (OR-SiON) to meet retention and reliability objectives.

5.2.6 Enhancement of Window and Retention in FG Devices

It has been mentioned earlier that meeting simultaneously large window and retention for FG devices using only oxide as tunnel dielectric and ONO as IPD is limited due to retention and reliability reasons. ONO can be replaced by a combination of OR-SiON and oxide or by OR-SiON alone to meet leakage/retention requirements, reduce stack EOT, lower Vpp, and enhance reliability. Simultaneously, the tunnel oxide could be replaced by OR-SiON alone or by a combination of OR-SiON and IN-SRN replacing SiO_2 and silicon FG, respectively, thereby further reducing EOT and enhancing reliability. The window and program-erase speed could be traded off to achieve large window without compromising reliability. Retention, window, reliability, and endurance enhancements by applying SiON dielectric layers replacing SiO_2 for CT devices had been demonstrated and will be discussed in this section. Additional enhancements in device properties will be addressed in Part II of this book.

5.2.7 Charge Loss and Leakage Paths for Charge-Trapping Devices

5.2.7.1 Band Diagram and Charge Loss Modes for Floating-Gate NVM Devices

Conventional CT NVM devices such as SONOS or MONOS employ oxide-nitride-oxide gate insulator stack as nonvolatile element between the substrate and the gate while charges are stored in the nitride trapping layer. Figure 5.12a and b illustrates band diagrams for SONOS and MONOS devices, respectively. The shift in band edges due to electron storage (post written standby state) and hole storage (post erased standby state) are shown dotted for both cases. The gates for SONOS and MONOS are assumed to be N+ doped poly silicon and TaN, respectively.

The transport mode and current components could be identified in terms of electron flow to the substrate from nitride when the memory state is Vth ("1") labeled as JeS and in terms of hole flow to the substrate from nitride when the memory state is Vth ("0") labeled as JhS. Similarly, electrons and hole flows from nitride to gate at standby states are identified as JeG and JhG for memory states

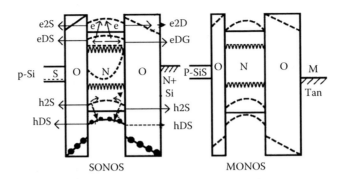

FIGURE 5.12 Band diagrams at standby states: (a) SONOS and (b) MONOS.

Vth ("1") and Vth ("0"), respectively. When multiple steps and modes are involved, the components are identified as sub-components as follows:

JeDS and JeDG or simply eDS and eDG, respectively, for current components define charge loss of electrons to the substrate and gate by direct tunneling; similarly, JhDS and JhDG or simply hDS and hDG define, respectively, current components associated with direct tunneling of holes to the substrate and gate. When dual transport steps are involved, e1S and e1G represent electron transport from nitride traps to the conduction band of nitride as initial steps for eventual transport to substrate and gate, respectively, with e2S and e2G are being direct-tunneling transport steps through oxide toward substrate and gate, respectively. Similarly, for holes, h1S and h1G represent initial hole transport steps from nitride traps to the valence band of nitride associated with movements toward substrate and gate, respectively, while h2S and h2G are direct-tunneling transport components through oxide toward substrate and gate, respectively.

5.2.7.2 Charge Loss from Charged Nitride Floating Node to Substrate: (JeS and JhS)

Electrons: Je1S: Detrapping of electrons from nitride trapping center to the nitride conduction band; followed by Je2S: tunneling of electrons from nitride conduction band to oxide conduction band. Alternately, a single step tunneling process of JeDS: tunneling from nitride interface trap through the oxide preceded by trap hopping within nitride.

Holes: Jh1S: Detrapping of holes from nitride trapping center to the nitride valence band; followed by Jh2S: tunneling of holes from nitride valence band to oxide valence band and then to the substrate. Alternately, a single step tunneling process of JhDS: tunneling from nitride interface trap through oxide preceded by hole trap-hopping within nitride.

Similarly,

5.2.7.3 Charge Loss from Charged Nitride Floating Node to the Gate: (JeG and JhG)

Electrons: Je1G: Detrapping of electrons from nitride trapping center to the nitride conduction band; followed by Je2G: tunneling of electrons from nitride conduction band to oxide conduction band and then to the gate. Alternately, a single step tunneling process of JeDG: tunneling from nitride interface trap through the oxide preceded by trap hopping within nitride.

Holes: Jh1G: Detrapping of holes from nitride trapping center to the nitride valence band; followed by Jh2G: tunneling of holes from nitride valence band to oxide valence band. Alternately, a single step tunneling process of JhDG: tunneling from nitride interface trap through oxide preceded by hole trap-hopping within nitride.

It should be understood that the current flow is in the direction of hole flow and in reverse direction of electron flow through the dielectric layers.

As depicted in the band diagram SONOS/MONOS devices involve transport, trapping, and detrapping of both holes and electrons. Additionally, most typical devices employ very thin tunnel oxide in the thickness range of 1.5–2.5 nm to enable direct tunneling of electrons for writing and for holes for erasing. The charges are trapped at larger tunnel distance deeper in the nitride from silicon/oxide interface to achieve charge retention objective. The blocking oxide film for the ONO stack is designed to be significantly thicker (>4 nm) to prevent stored charge loss to the gate as well as carrier injection (electron injection during erase and hole injection during programming) from the gate to reduce adverse effects on programming and erasing. The evolution of CT devices from MNOS/SNOS types to MONOS/SONOS types enabled the trapping nitride layer for the device to be thinner for stack EOT reduction and scalability. Nitride thickness could be reduced in the range of 5–10 nm and consequently ONO stack EOT to <10 nm and VPP $\leq \pm 10$ V for writing and erasing while meeting device targets of retention and endurance. Since charge transport through tunnel oxide is primarily by direct-tunneling (and secondarily by modified Fowler–Nordheim tunneling for somewhat thicker tunnel oxide), the field required for writing and erasing for the gate insulator stack is lower than that of FG devices. Consequently, stress-induced oxide degradation during writing and erasing is reduced and adverse effect on retention is reduced.

5.2.7.4 Charge Loss Mechanisms in Conventional CT Devices

Charge loss mechanisms of CT devices based on nitride trapping are more complex due to the following reasons:

1. Electron and hole trapping in nitride is asymmetric due to different trapping lengths of electrons (510 nm) and holes (15–20 nm) in nitride (Section 4.3.5). This results in separation of trapped charge centroids of electrons and holes within the nitride film, the electron centroid being closer to the tunnel oxide/nitride interface, whereas the hole centroid is further away. Consequently, charge rearrangement takes place due to write–erase cycling and post cycle baking at elevated temperature. Trapped charge density at both tunnel oxide/ nitride and nitride/blocking oxide interfaces is altered impacting rate of charge loss both to substrate and to gate. For thinner nitride (<10 nm), after write–erase cycling, trapped charge density of holes at the nitride/blocking oxide interface gets significantly higher than trapped charge density of electrons. Thus, retention in such devices becomes sensitive to nitride thickness.

2. Electron and hole injection in nitride are nonuniform due to the difference in band offsets and mobility. Valence band offset is significantly higher and mobility significantly lower for holes compared to those of electrons. Additionally, electron detrapping takes less energy compared to hole detrapping. Consequently, device threshold changes during programming and erasing as well as during standby are dominated by electron movements from energetics point of view. This is further accentuated by incomplete detrapping of holes and net accumulation of positive charge inside nitride after each write–erase cycle. As a result, net charge centroid moves deeper into nitride after repeated write–erase cycling.

The above phenomena of variation in transport, trapping, detrapping, and charge accumulation have been modeled and verified experimentally ([5,14] of Chapter 4, and [13–18] of this chapter). This complicates SONOS/MONOS stability and retention in the following ways:

1. Retention is strongly dependent on stack structure in SONOS/MONOS types of devices. Long-term retention is strongly dependent on nitride stoichiometry, conductivity, processing parameters, and thickness as well as the blocking oxide thickness. Short-term retention after programming is dependent on tunnel oxide in the range of 1.5–2.5 nm due to enhanced detrapping and reverse tunneling of electrons located near the tunnel oxide/ nitride interface. In stack structures consisting of thinner nitride (≤5 nm) and at the same time very thin blocking oxide of thickness range of 2–4 nm, retention of erased state is strongly affected by enhanced hole transport to the gate.

2. Retention is also affected by hole accumulation at the nitride/blocking oxide interface and the resulting electrostatic potential created by the accumulated charge. This further complicates the charge loss mechanism for both MNOS/SNOS and MONOS/SONOS device types. When nitride is sufficiently thick (nitride in SNOS/MNOS >10 nm, nitride in SONOS/MONOS ~10 nm), the effect of built-in potential is to reduce electron detrapping and to increase hole detrapping and loss to the gate. This results in higher rate of charge loss in the erased state compared to the written state as observed.

5.2.8 Comparative Retention Attributes: MNOS/SNOS versus MONOS/SONOS

In spite of the complex nature of charge trapping and detrapping in nitride, MONOS/SONOS devices exhibit generally superior data retention and write–erase cyclic stability compared to MNOS/SNOS devices [13,15,19,20]. This is due to the blocking oxide in SONOS/MONOS devices which (a) reduces charge injection from the gate and (b) charge accumulation during cyclic stressing. A comparative assessment between the two types of device is shown in Table 5.3.

5.2.8.1 Retention Time Modeling for MONOS/SONOS Devices

An approximate estimation of retention time from back tunneling of trapped charge in nitride was provided by F. R. Libsch et al. [16]. It is given by:

$$T = To \exp(Aot.Dot + An\, D)$$

where:

Dot is the tunnel oxide thickness

D is the trapped charge distance in nitride from tunnel oxide/nitride interface

T is the retention time

To is the time constant for charge to move from conduction band to trap or reverse and is in the order of ~1E-14 sec

Aot is the transit coefficient in oxide and is in the order of 1.07E11/m

An is the transit coefficient in nitride related to charge mobility in nitride

The charge loss or change in Vth shows logarithmic-time dependence.

Unlike the FG devices, charge loss in CT devices and associated activation processes are intimately tied to charge detrapping from nitride. Therefore, retention in CT devices is significantly different from those of FG devices. The activation energy varies widely from 0.6 to 1.8 eV without write–erase cycling and was found to be strongly nitride process sensitive. The write–erase cycling was shown to have a very strong degrading effect on retention life time activation energy [21].

TABLE 5.3

Comparative Attributes of SNOS/MNOS versus SONOS/MONOS Devices

Parameter	SONS/MNOS [15]	SONOS/MONOS [13]
Insulator stack	SiO_2: 1.6 nm/Si_3N_4: 28 nm	SiO_2: 2.0 nm /Si_3N_4: 6.8 nm/SiO_2: 7.2 nm
EOT	17.6 nm	13.1 nm
Vpp/Vpe	±15V, 5 ms	±10V, 30 ms, 200 ms
Max P/E field	8.5 MV/cm	7.6 MV/cm
Window	+2V/−4V = 6V	+1.3V/−1.3V = 2.6V
ΔVth(1)	250 mV/decade (electron)	40 mV/decade (electron)
ΔVth(0)	250 mV/decade (hole)	100 mV/decade (hole)

5.2.9 Enhanced Charge Retention for MXOS Devices: Correlation with Band Structures

The complex charge movements within nitride could be significantly altered by confining charge trapping within nitride to a relatively smaller nitride thickness while also reducing charge injection from the metal gate (Reference 5 of Chapter 4). This was achieved by sandwiching a thicker nitride layer of 46 nm in the stack with thin OR-SiON layers (each of 7.5 nm) at both top and bottom of nitride (see Section 5.2.5 on oxynitride). The thinner bottom OR-SiON layer interfacing the ultra-thin tunnel oxide layer (1.7 nm) extended the tunnel distance of stored electrons and holes, thereby virtually eliminating reverse tunneling. The top OR-SiON layer prevented charge injection from the metal gate, thereby enhancing the memory window and reducing charge displacement within the nitride layer. The modified MXOS stack design was compared with two MNOS devices of different tunnel oxide thicknesses of 2.2 and 3.4 nm with same thickness of nitride of 64 nm (Figure 5.13). While both MNOS devices showed accelerated charge loss consistent to their tunnel oxide thickness and poorer retention, virtually no charge loss was observed in the MXOS device fabricated by similar CVD processing of nitride and oxynitride layers. Additionally, the MXOS device showed significantly larger and stable initial window even when compared to the thinner tunnel oxide MNOS device. The band diagram for the stack is shown in Figure 5.14 and the band structures for programming (electron injection and trapping) and erasing (hole injection and trapping) are illustrated in Figure 5.15a and b, respectively. It is to be noted that while the OR-SiON layer over the tunnel oxide significantly increases the reverse tunneling distance preventing charge loss during standby memory states, it does not adversely impact on charge injections and trapping during writing and erasing. This could be understood from the band diagram since once the electrons tunnel through

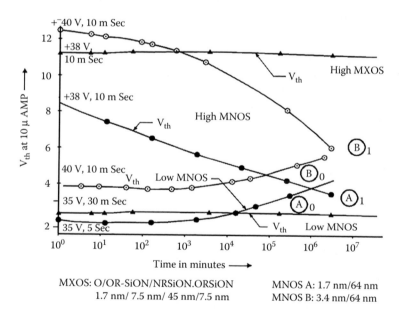

MXOS: O/OR-SiON/NRSiON.ORSiON   MNOS A: 1.7 nm/64 nm
1.7 nm/ 7.5 nm/ 45 nm/7.5 nm   MNOS B: 3.4 nm/64 nm

FIGURE 5.13 Comparison of charge retention characteristics of two MNOS devices of different tunnel oxide thicknesses versus a multilayered oxynitride MXOS device: (a) MXOS device with 17 A tunnel oxide, (b) MNOS device with 22 A tunnel oxide (direct tunneling), and (c) MNOS device with 34 A tunnel oxide (F–N tunneling). The stack design for the devices were: (a) 17 A oxide/75 A OR-SiON/460 A NR-SiON/75 A OR-SiON; (b) 22 A oxide/640 A nitride; and (c) 34 A oxide/640 A nitride. Oxides were thermally grown under similar processing conditions and both nitride and oxynitride films were grown by LPCVD processes under similar processing conditions. The EOT of all stacks were approximately 38.0–39.0 nm. (From Bhattacharyya, A. et al., Properties and applications of silicon oxynitride films, IBM Internal Technical report, October 1976, Unpublished and Declassified.)

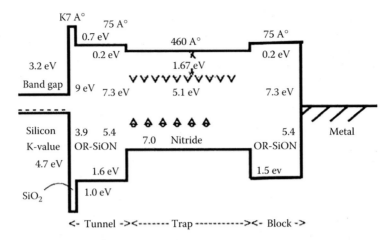

FIGURE 5.14 The band diagram of the multilayered MXOS stack of the device of Figure 6.15. The inset data shows some of the band and trapping parameters of OR-SiON and nitride films used for the device.

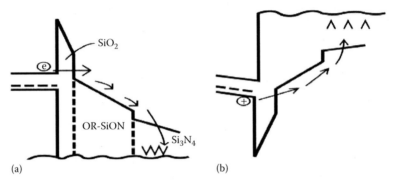

FIGURE 5.15 Band diagrams reflecting injection and trapping: (a) Electrons (writing) and (b) hole (erasing) in MXOS device.

the trapezoidal energy barrier of the ultrathin tunnel oxide, the drift field sweeps the charges away along the conduction band when writing (Figure 5.15a), and the reverse situation for holes along the valence band when erasing (Figure 5.15b). However, during standby, large tunnel distance reduces direct-tunneling probability to be negligible. Multilayered oxynitride films in MXOS structures also vastly improve endurance which will also be covered in this chapter.

5.3 MEMORY ENDURANCE AND ENDURANCE TRAITS OF FG, CT, NC NVMs

Endurance is the capacity of the NVM devices to ensure the stability of the memory states after being subjected to the high field multiple writing and erasing operations. Unlike other silicon based memory devices, for example, SRAM or DRAM with unlimited writing and erasing capability, NVM devices and products such as FG flash and CTNROMS are limited in endurance in the range of 100,000 to 10 million write–erase cycles only, compared to, for example, DRAM memory, which may undergo >1E13 write–erase cycles during the life time of application. For the current multilevel FG flash products (to be discussed in Section 7.4), even achieving 100,000 W/E cycles of endurance has become a serious challenge and sophisticated system level programming/ erasing schemes as well as memory error correction codes (ECC) had to be incorporated to address the challenges associated with devices meeting limited endurance limits. Current devices and products face serious challenges in feature size scalability and expansion of applications due to endurance limitations. Feature size and voltage scalability challenges for NVM devices will be addressed in detail in Part II of this book. A primary objective of this book is also to address the endurance limitations of the current NVM devices and explore potentials to overcome such limitations to develop NVM devices with DRAM-like endurance capability and thereby creating voltage scalable SUM devices with expanded applications with greater compatibility with scaled CMOS technology.

Fundamentally, when the gate stack insulating films are subjected to high field stressing required for high fluence of charge transport through the insulating films during the process of writing and erasing, dielectric films degrade, defects and traps are generated, stress-induced leakage currents are enhanced, and stored charge and associated memory state stability get reduced. As a result, with successive writing and erasing, the discriminating margin between the memory states is reduced to an unacceptable level to read accurately the memory states. This sets the limit of endurance of the NVM device. For conventional FG flash devices employing thermally grown tunnel oxide as the primary medium of charge transport during writing and erasing, the stress-induced degradation of tunnel oxide sets the endurance limit. Typical oxide thickness for such devices had been in the range of 7–9 nm. Endurance for such devices had been <1E6 cycles and typically ~1E5 cycles and had been correlated with oxide QBD in such thickness ranges (see previous section on oxide degradation and QBD).

For SONOS types of CT devices employing significantly thinner oxide, an optimized stack insulator structure could achieve greater endurance by almost two orders of magnitude compared to the FG flash devices as above. Although thinner oxide employed in such devices have higher intrinsic QBD, the endurance process is more complex and the limit could be set by multiple phenomena associated with degradation of both oxide and nitride and charge centroid movement toward nitride/blocking oxide interface during write–erase cycling. During stress cycling, positive charge accumulates due to incomplete erasure and leakage is enhanced due to hole trapping. It has been noted that relatively higher endurance limit has been achieved with optimized NC devices, whereby charge trapping was confined to planar NC regions with stable charge centroid locations. Figure 5.16a–d compares the endurance characteristics of typical optimized FG (a and b), CT SONOS, and metal NC devices with 7.5, 2.0, and 2–3 nm tunnel oxide thicknesses, respectively [22–24]. It is important to note that the gate insulator stack structures are very different in all cases, whereas the tunnel oxide thicknesses for SONOS and gold NC devices is of the same range (<3 nm) for direct tunneling. However, the maximum stress fields during writing/erasing for the FG device and the SONOS devices are significantly high: typically >11E6 V/cm and <10E6 V/cm, respectively, while that for the NC device is in the order of merely 1E6 V/cm. Furthermore, the endurance characteristics are significantly different in FG devices when erasure stress mode changes from uniform erasure to nonuniform erasure [22]. Under nonuniform erasure, the memory window collapses faster (Figure 5.16b). The endurance characteristics show that the FG device (EOT ~18 nm) initiates degradation after less than 100,000 W/E cycles and the SONOS device shows after 10 million cycles (EOT <9 nm) (not shown in Figure 5.16c), while the NC device (EOT ~38 nm) do not exhibit any degradation after 1E9 cycles.

5.3.1 Endurance and Peak Field

Endurance is intimately related to the energy levels of the transporting electrons and holes through the gate dielectric films and the energy dissipation mechanisms of the charged electrons and holes to the local lattice structures of the dielectric films. Bhattacharyya [24] had empirically established that endurance could be approximately correlated to the average "peak programming/erasing stress field" (defined later on in this section) of the gate insulator stack of the NVM device. This relationship is shown in Table 5.4 and Figure 5.17, where number of write–erase cycle limit is plotted against

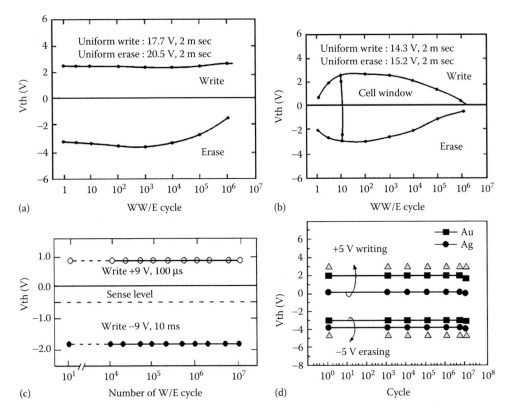

FIGURE 5.16 Endurance characteristics of FG (EOT = 18 nm), (a) uniform erase, (b) nonuniform erase, (c) CT (EOT = <9 nm), and (d) NC (EOT~38 nm) devices. (Reproduced from Paul, A. et al., Comprehensive simulation of programming, erase, and retention in charge trapping flash memories, *IEDM*, pp. 393–396, 2006. With permission of Wiley-IEEE Press.)

TABLE 5.4
Endurance versus Average Stack Peak Field

MV/cm	W/E Cycles
<6	>>1E13[a]
7	~1E13
8	>1E10
9	≥1E4

Source: C. H. J. Wann, and C. Hu, High endurance ultra-thin tunnel oxide for dynamic memory application, *IEDM*, 1995, pp. 867–870; H. C. Wann, and C. Hu, *IEEE Electron Dev. Lett.*, 16(11), 491–493, 1995.

[a] Stack design with ≤6MV/cm average peak field for programming and erasing should exhibit "infinite endurance" or "end-of-life" endurance.

---

the average peak stress field induced in the NVM dielectric stack. Figure 5.17 shows that endurance is strongly sensitive to the "peak field" with an approximate slope of improvement in endurance by ~300X to 1000X for reduction in peak field by 1E6 V/cm. When the peak field exceeds 1E7 V/cm, the maximum endurance could be approximately between 10,000 and 100,000 cycles. However, if the peak field for the device stack could be reduced to <=6E6 V/cm, endurance level could exceed 1E13 write–erase cycles. It should be noted that this is qualitatively consistent to charge to breakdown

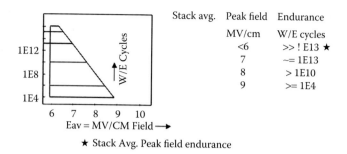

Stack avg. Peak field	Endurance
MV/cm	W/E cycles
<6	>> ! E13 ★
7	~= 1E13
8	> 1E10
9	>= 1E4

★ Stack Avg. Peak field endurance

FIGURE 5.17 Empirical relationship between device endurance and average peak stress field in NVM gate stack design.

model in oxide: QBD (see Chapter 4 on oxide QBD), which shows logarithmic dependence of QBD on voltage or field in oxide and many orders of magnitude improvement in QBD (Figure 4.4) when electron energy is reduced. QBD is dramatically higher for ultrathin oxide (thickness range 1.0–3.0 nm) where electron and hole transport by direct tunneling requires significantly lower energy.

5.3.2 Endurance, Charge Fluence, Dielectric Conductance, and QBD

NVM devices require high electron and hole current density (charge fluence) to achieve appropriate write–erase speeds. However, such memories designed with charge transport through oxide based on Fowler–Nordheim tunneling (e.g., FG flash NVMs) perform write–erase at speeds many orders of magnitude slower than volatile memories such as DRAM or SRAM. As stated earlier, the fundamental reason for this difference in performance is the difference in conductance (mobility) of electrons and holes through oxide compared to those in silicon. Conductance at high field could be compared between thicker oxide employed for FG flash NVMs and ultrathin oxide based on direct tunneling as well as insulators such as CT-SRNs and NR-SiONs. This is shown in Figure 5.18 where charge storage Qt is compared between conventional FG device with thick tunnel oxide and other dielectric media options mentioned above. It could be noted that at higher field ($\geq$1E6 V/cm), conductance could be improved by many orders of magnitude by appropriately replacing thicker tunnel oxide by other options as shown by the above figure. These other dielectric options also provide characteristically enhanced QBD. Both endurance and QBD are high fluence phenomena for the dielectric media and could be cross-correlated. Therefore, it should be anticipated that any dielectric medium with higher

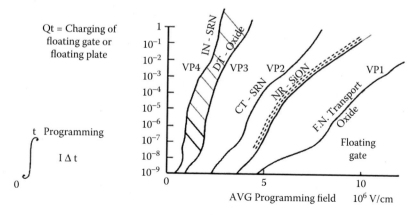

FIGURE 5.18 Qt versus average programming field for thicker oxide, nitrogen-rich (NR) SiONs, charge-trapping (CT) SRNs, thinner direct-tunnel (DT) oxide, and injector (IN) SRNs.

charge conductance and QBD could potentially provide higher endurance. However, it has also been noted in Chapter 4 that IN-SRN has very high conductance, yet lacks the capacity to store charges. IN-SRN when appropriately combined with large band gap insulators such as SiO_2, Al_2O_3 (Alumina), OR-SiON, or others could enhance charge conductance by several orders of magnitude due to local field enhancements. Such application reduces the average peak field for writing–erasing and therefore enhances the device endurance [25].

5.3.3 Endurance and Depend Types: Tunnel Layer Dependency

From the above discussion, it should be apparent that NVM endurance is strongly related to device types associated stack structures and especially associated with the characteristics of tunneling mechanism and medium.

Carrier transport is field dependent for FG flash devices and transport is associated with steady-state Fowler–Nordheim tunneling or modified Fowler–Nordheim tunneling or by Ballistic Fowler–Nordheim tunneling depending on the thickness of the tunnel insulator employed. Tunnel layers are thicker, and stack structures are designed to meet window and retention as well as read-disturb margin requirements for multilevel storage (2 bit/cell or 3 bit/cell). The upper limit for endurance is limited to less than 1 million cycles and more typically 10,000–100,000 cycles limited by tunnel oxide degradation and reliability.

Conventional CT devices, for example, SONOS or MONOS types, on the other hand, typically employ thinner tunneling layer with thickness ranging 2–5 nm of either oxide or a combination of oxide/oxynitride. Such devices are usually single bit per cell memory with write–erase voltage levels lower than that of flash devices and window nearly half of those of FG flash devices designed to provide nearly 10 million cycles of endurance. These devices also demonstrate 10 years of retention and read-disturb immunity. Most products are available in the above two device types for data store and code applications.

Most published literature on NC memories employs direct-tunnel oxide in the thickness range of 2–3 nm with silicon or metal NCs embedded in oxide for charge trapping. The devices exhibit >1E9 cycles of endurance, although no known stand-alone product has been published with such stack structures and typical memory window is smaller than SONOS device types. Although the devices exhibited significantly higher endurance, they showed poorer retention due to reverse tunneling of stored charges. Retention and endurance could be traded by increasing tunnel oxide thickness as discussed in Chapter 2. This will be further discussed in Chapter 13.

Direct-tunnel memories (DTMs) of both FG and CT types employ ultrathin oxide for tunneling exhibiting very high endurance of >>1E11 cycles with limited retention. Table 5.5 summarizes the device types and endurance.

TABLE 5.5
Device Types and Endurance

	FLASH	SONOS/MONOS	NC-CT	DTM
Tunnel oxide (nm)	7.5	3.0–5	2–3	1.5
Endurance cycles	~1E5	~1E7	>1E9	>1E11
Retention	10 years	10 years[a]	Poor	10–1000 s

a Sensitive to oxide thickness and nitride process.

5.3.4 Requirement for High Endurance

Several examples of high endurance devices and associated structures will be cited and some observations on generalized requirements for high endurance device structures will be made. An ultrathin MONOS structure was investigated in 1995 by C. H. J. Wann et al. [25] using 1.2 nm tunnel oxide and 3.5 nm blocking oxide. The write–erase voltage level was ±8 V, and average peak field was ~9E6 V/cm with the voltage drop across the tunnel oxide being 1.1 V. Endurance for the structure exceeded 1E11 cycles and charge retention of 1000 sec was demonstrated at 80 C, and with the CT nitride layer being 7.5 nm. The memory window was shown to be 0.6 V. In 2002, Y.K. Lee and co-workers from Samsung [26] illustrated a FinFET SONOS NVM device with similar stack structure consisting of 1.5 nm tunnel oxide, 5.0 nm trapping nitride, and 4.0 nm of blocking oxide. Window of O.7 V was obtained with similar endurance at write–erase voltage levels of ±8 V, 3 ms. The peak voltage drop across the tunnel oxide was 1.4 V, while the peak field was similar to the one previously mentioned. In 2003, two NC devices were published [27–29] using tunnel oxide thickness of ~2–3 nm, with embedded silicon NC and metal NC and with different thicknesses of blocking oxides. The write–erase voltage levels for both cases were ±5 V with voltage drop across tunnel oxide being <1 V. Both device structures exhibited endurance >1E9 write–erase cycles. In 2004, Tsunoda et al. published a DTM memory with 1.55 nm tunnel oxide [28] and demonstrated greater than 1E12 write–erase cycles of endurance at Vpp of ±4 V, 30–50 ns. From the above examples, we could summarize the requirements for high endurance to be as follows:

1. Tunneling mechanism for charge transport need to be by direct tunneling with voltage drop across oxide to be <2 V and preferred tunnel oxide thickness ≤2.0 nm.
2. Stable trap charge centroid either by means of NC trapping centers or by means of trapping dielectric with deep trap and low conductance.
3. Device structure to minimize hole injection from gate by providing appropriate high energy barrier blocking insulator.

5.3.5 MXOS Device with High Endurance

The earliest known high endurance with simultaneously high retention CT device was developed and applied for an 8K NVM product for internal use in IBM in the mid-1970s [14,30] by A. Bhattacharyya et al. The MXOS device stack structure consisted of 1.7 nm of tunnel oxide with three layers of oxynitrides. The center layer of oxynitride (SiON) of high nitrogen content (NR-SiON: ~50 atomic % nitrogen, ~10 atomic % oxygen) had the characteristics of providing high trap density (~1E13/cm²) and deep traps (1.7 eV) and was intended for the near symmetric charge storage for electrons and holes. The center layer was sandwiched between two relatively higher oxygen content SiON layers (~30 atomic % each of Oxygen and Nitrogen), higher band gap and lower conductivity to enhance charge retention, and to reduce electron injection from the aluminum gate during erasing. All the layers of CVD oxynitrides were deposited simultaneously under similar NH_3/SiH_2Cl_2 ratio with varying partial pressure of oxygen in the vapor phase. MNOS devices were also fabricated under similar conditions and compared with the MXOS devices.

Band diagram of the MXOS stack structure is shown in Figure 5.19. During writing and easing, electrons and holes, respectively, injected from silicon directly tunneled through the ultrathin tunnel oxide and swept by applied field and trapped into the deep traps of the center oxynitride layer. In the standby states (retention state), the charges remained confined to the trapping layer due to higher barrier energy of the higher oxygen containing oxynitride layers and large tunnel distance to substrate and gate. Charge centroids were confined within the trapping layer and nearly symmetric, with minimal net accumulation of holes after each write–erase cycle.

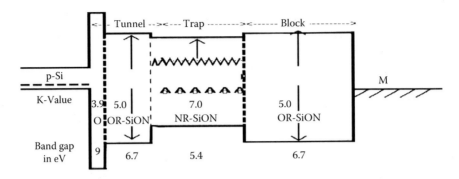

FIGURE 5.19 Band diagram of high endurance multilayered MXOS device using OR-SiON/NR-SiON/ OR-SiON stack for enhanced window and endurance.

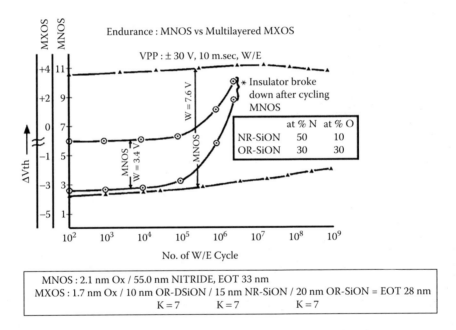

FIGURE 5.20 Comparison of endurance of MNOS and multilayered MXOS devices with OR-SiON/ NR-SiON/OR-SiON gate stack design.

Endurance characteristics of the MXOS device are shown in Figure 5.20 against a 2.1 nm tunnel oxide with 55.0 nm nitride MNOS stack structure. The initial window for the MXOS device was nearly symmetric of 7.6 V compared to the asymmetric window of the MNOS device of 3.4 V when written and erased at ±30 V/10 ms. The MNOS device quickly degraded after 1E5 cycles, while the MXOS device displayed little degradation even after 1E9 write–erase cycles. Post cyclic retention of the MXOS device was similar to that shown in Figure 5.13 with little charge loss satisfying projected 10 years of retention requirements. It should be noted that the peak voltage drop across the tunnel oxide for both MNOS and MXOS devices was of similar magnitude of ~1.8 V during write–erase cycling.

REFERENCES

1. G. Groeseneken, H. E. Maes, J. Van Houdt et al., Basics of nonvolatile semiconductor devices, Chapter 1, in *Nonvolatile Semiconductor Memory technology*, IEEE, edited by W. Brown and J. Brewer, IEEE Press, p. 62, 1997.

2. S. Mori, N. Arai, Y. Kaneko et al., PolyOxide thinning limitation and superior ONO inter-poly dielectric for nonvolatile devices, *IEEE Trans. Electron Dev.*, 38, 270–277, 1991.

3. Bhattacharyya, Modelling of write/erase and charge retention characteristics of floating gate EEPROM devices, *Solid State Electron.*, 27, 899–906, 1984.

4. S. Mori, E. Sakagami, H. Araki et al., ONO inter-poly dielectric scaling, for nonvolatile memory applications, *IEEE Trans. Electron Dev.*, 38, 386–391, 1991.

5. J.-D. Choi, J.-H. Lee, W.-H. Lee et al., A 0.15 um NAND flash technology with 0.11 um² cell size for 1Gbit flash memory, *IEEE, IEDM*, 2000, pp. 767–770.

6. R. N. Mielke, New EPROM data loss mechanism, *Proceedings of the IEEE, IRPS*, 1983, pp. 106–113.

7. V. N. Kynett, J. Anderson, G. Atwood et al., A 90ns 100K erase/program cycle mega bit flash memory, *IEEE J. Solid State Circ.*, SC24(10), 1259–1264, 1989.

8. S. Aritome, R. Shirota, G. Hemink et al., Reliability issues of flash memory cell, *Proc. IEEE*, 81(5), 776–788, 1993.

9. S. Aritome, R. Shirota, R. Kirisawa et al., A reliable bi-polarity write/erase technology in Flash EEPROMs, *IEDM*, 1990, pp. 111–114.

10. S. Aritome, S. Satoh, T. Maruyama et al., A 0.67 um² self-aligned shallow trench isolation cell (SA-STI cell) for 3 V-only 256 Mbit NAND EEPROMs, *IEEE, IEDM*, 1994, pp. 61–64.

11. Y. Mitani, H. Satake, and A. Toriumi, Experimental evidence of hydrogen-related SILC generation, in thin gate oxide, *IEEE, IEDM*, 2001, pp. 129–132.

12. J.-D. Choi, S.-S. Cho, Y.-S. Yim et al., Highly manufacturable 1 Gb NAND flash using 0.12 um process technology, *IEEE, IEDM*, 2001, pp. 25–28.

13. A. K. Banerjee, Y. Hu, M. G. Martin et al., An automated SONOS NVSM dynamic characterization system, *Proc. 5 Nonvolatile Memory Technology Review Conference*, Linthicum Heights, MD, pp. 78–81, 1993.

14. A. Bhattacharyya, C. T. Kroll, H. W. Mock, and P. C. Velasquez, Properties and applications of silicon oxynitride films, IBM Internal Technical report, October 1976 (Unpublished and Declassified).

15. Y. Kamigaki., S. I. Minami, T. Hagiwara et al., Yield and reliability of MNOS EEPROM products, *IEEE J. Solid State Circ.*, 24, 1714–1722, 1989.

16. F. R. Libsch, A. Roy, and M. H. White, Charge transport and storage of low programming voltage SONOS/MONOS memory devices, *Solid State Electron.*, 33(1), 105–126, 1990.

17. A. Paul, C. Sridhar, S. Gedam et al., Comprehensive simulation of programming, erase, and retention in charge trapping flash memories, *IEEE, IEDM*, San Francisco CA, 2006, pp. 393–396.

18. A. Furnemont, M. Rosmeulen, J. VanHoudt et al., Physical modelling of retention in localized trapping nitride memory devices, *IEEE, IEDM*, 2006, pp. 397–400.

19. E. Suzuki, H. Hiraishi, K. Ishii et al., A low voltage alterable EEPROM with metal-oxide-nitride-oxide-semiconductor (MONOS) structure, *IEEE Trans. Electron Dev.*, ED-30, 122–128, 1983.

20. C. C. Chao, and M. H. White, Characterization of charge injection and trapping in scaled SONOS/MONOS memory devices, *Solid State Electron.*, 30(3), 307–319, 1987.

21. T. Ajiki, M. Sugimoto, H. Higuchi et al., Temperature accelerated estimation of MNOS memory reliability, *IEEE, Proceedings of the IRPS*, 1981, pp. 17–22.

22. T. Nozaki, T. Tanaka, Y. Kijiya et al., A 1 Mb EEPROM with MONOS memory cell for semiconductor disk application, *IEEE J. Solid State Circ.*, 26(4), 497–501, 1991.

23. C. Lee, A. Gorur-Seetharam, and E. C. Kan, Operational and reliability comparison of discrete storage nonvolatile memories, *IEEE, IEDM*, Washington DC, 2003, pp. 557–560.

24. A. Bhattacharyya, Scalable flash/NV structures and devices with extended endurance, AdI Associates Internal Docket, August 2001, U.S. Patents: 7012297 (3/14/2006), 7250628 (7/31/2007), 7400012 (7/15/2008), 7750395 (7/6/2010).

25. C. H. J. Wann, and C. Hu, High endurance ultra-thin tunnel oxide for dynamic memory application, *IEDM*, 1995, pp. 867–870; H. C. Wann, and C. Hu, High-endurance ultra-thin tunnel oxide in MONOS device structure for dynamic memory application, *IEEE Electron Dev. Lett.*, 16(11), 491–493, 1995.

26. Y. K. Lee, S. K. Sung, J. S. Sim et al., Multi-level vertical channel SONOS nonvolatile memory on SOI, *IEEE, Symposium on VLSI Technology*, 2002, pp. 208–209.

27. W. Guarini, C. T. Black, Y. Zhang et al., Low voltage scalable nanocrystal flash memory fabricated by templated self-assembly, *IEEE, IEDM*, 2003, pp. 541–544.
28. K. Tusunoda, A. Sato, H. Tashiro et al., Ultra-high speed direct tunneling memory (DTM) for embedded RAM applications, *VLSI Technology Digest*, 2004, pp. 152–153.
29. S. Zirinsky, MXOS n-channel NVM device using multi-layered CVD oxynitride gate insulator.
30. A. Bhattacharyya, FET gate structure (MXOS) for nonvolatile N-channel read mostly memory device, IBM Internal Technical Memo, 1975, Unpublished.

6 NVM Device Stack Design

CHAPTER OUTLINE

This chapter provides a review of the geometric sensitivities for all unique parameters discussed in Chapter 5 for NVM devices in cell design, in feature size scalability, and in addressing parasitic effects. It also addresses various parameters in designing the memory cell features and related gate stack designs for both floating gate and charge trapping flash types of NVM devices to meet the parametric and functional objectives for such devices. The NVM cells and stack designs discussed in this chapter are limited primarily to "normal mode" NVM devices.

6.1 FUNDAMENTALS

Tunnel-based NVM devices require injection of electrons and/or holes either from the silicon substrate (defined as "normal mode NVM") or from the gate electrode (defined as "reverse mode" NVM). The normal mode NVM device stacks are designed to minimize gate charge injections during writing and erasing, while the reverse mode NVM device stacks are designed to minimize charge injections from the silicon substrate. Since virtually all products today are designed for normal mode operation, unless explicitly stated in this book, all design considerations should be assumed to be directed for normal mode devices. Device characteristics discussed earlier in previous chapters assumed normal mode operations. The dielectrics and interfaces as discussed in Chapters 4 and 5 reflect the normal mode considerations for NVM devices.

Silicon substrate (including diffusion regions) interfacing SiO_2 acts as primary source of injecting carriers into the trapping region in the gate insulator stack in conventional NVM devices. Therefore, large energy barriers of 3.2 eV for electrons and 4.7 eV for holes are to be supplied from applied field during writing and erasing for charge injection across the silicon–oxide (SiO_2) interface. Except for oxynitride (MXOS device types), SiO_2 has been the tunnel oxide for all types of NVM devices since inception through 2000. Charge transport through SiO_2 for such conventional devices has been discussed in details in preceding sections in great detail. It was noted that significant charge transport requires high average field more than 10 MV/cm during writing and erasing across the gate insulator stack due to the large barrier energies of SiO_2. It has also been noted earlier that tunnel oxide thickness for conventional FG device and EOT of the gate stack design is typically around 2x compared to those of floating-plate (FP) and CT (SONOS) devices primarily to meet the charge retention requirements. Consequently, the EOT for FG and FP/CT devices are around 20 and 10 nm, respectively, and program-erase voltage levels are typically around ±20 and ±10 V, respectively.

Fundamentally, the NVM stack design involves designing the device stack to generate the end-of-life memory window by injecting and transporting the charges across the tunnel dielectric interface and capturing–storing the charges either in the FG or FP or traps (CT) during the period of the applied potential across the stack such that the end-of-life retention and endurance objectives are met. We shall examine in simple qualitative terms the significance of stack design parameters to meet the above objective.

Assuming the charge transport mechanism to be Fowler–Nordheim tunneling for FG devices or modified/enhanced Fowler–Nordheim tunneling for either FP or CT devices, the charge injected, the charge captured, and the memory window generated could be expressed as follows:

1. Charge injected: Qi = integral of 0 to tpp of Ji.dt
2. Charge captured: QC = Pc.Qi.@b
3. Window: W = QC.EOTi

where $Ji = A.Vpp^2/(EOT)^2 . \exp - \{\Phi_b^{3/2}. (EOT)/Vpp^2\}$; and therefore,

$$W = (B.Vpp^2.Pc.@b.tpp)/(EOT).[\exp - \{\Phi_b^{3/2}.(EOT)/Vpp^2\}]$$

where:
Pc is the capture probability
@b is the blocking efficiency
EOT = (3.9E0.Di/Ki)
tpp is the programming time
Φ_b is the barrier height either for electron or for hole

At steady state, for FG device, the capture probability, Pc, approaches to unity, and by increasing the EOT of the blocking dielectric and Φ_b, @b can approach unity. The design of the blocking dielectric should be such that the leakage current from the FG to the control gate should be many orders of magnitude lower than Ji. Therefore, Vpp is dependent strongly on two parameters: (1) EOT with square-root dependence and (2) Ub, the barrier height with exponential dependence. In Part II of this book, we will examine the importance of high K dielectric and barrier engineering in the gate stack design in scaling Vpp. In this section, we will limit our discussion to the conventional NVM device stack design employing conventional dielectric films described in Section 4.1.

Unlike the CMOS FET stack design which usually deals with a single insulator film with two interfaces, namely, silicon/insulator and insulator/gate (ignoring sidewall), we have noted that the gate stack design for NVM devices is more complex due to multiple dielectric layers and a minimum of four interfaces. This has been explained in Chapter 3 earlier. All unique device properties, window, retention, and endurance, and the process of changing the memory states (writing and erasing), and sensing the memory states (reading) as well as the device reliability are dependent on the characteristics of the dielectric films and interfaces as discussed in the earlier sections. Furthermore, when devices are placed in memory arrays (to be discussed in Section 6.2), additional instability of memory states arises due to parasitic stress-coupling on the neighboring standby memory cells during reading, writing, and erasing of the active cells in the array. Therefore, NVM stack design for the device not only needs to meet the end-of-life targets in terms of window, charge retention, and endurance for worst-case application conditions within the framework of the memory array, but also needs to satisfy the following additional criteria: (a) process compatibility and scalability of the host CMOS technology and associated active and passive device elements; (b) density, yield, and cost as well as reliability and functionality (power/performance) objectives of the overall memory device due to the added complexity of NVM stack requirements; and (c) extendibility of such stack design with fast-changing feature size and voltage scaling of the host CMOS technology. The last item will be discussed in Part II of this book, along with the considerations and complexities associated with multilevel storage of NVM cells. In this section, we will discuss primary considerations of stack designs of conventional FG-NAND flash devices and CT NROM random access NVMs. These devices and stack designs had been successful in not only meeting the unique properties of NVM devices but also have proven to meet the additional criteria of (a) and (b) above over multiple decades of scaled CMOS technology nodes.

6.2 FG-NAND FLASH DEVICES: STACK/CELL DESIGN

The forerunner of current NAND flash stack design was due to R. Kirisawa et al. ([30] of Chapter 2) who employed Fowler–Nordheim tunneling scheme for writing and erasing of an 8-bit string NVM array architecture. Cross sectional schematic view of the stack cell design along the channel and parallel to the bit line was shown in Figure 2.4. Subsequently, as the feature size of CMOS technology was scaled, several process innovations were introduced to achieve not only = ~4F^2 per single level cell (SLC) and = ~2F^2 per multilevel cell (MLC: 2b/cell), but also more planar cell, stack, and isolation designs. Important process innovations included:

1. Incorporation of self-aligned shallow trench isolation scheme [1], for cell density enhancements enabling 32-bit NVM array architecture.
2. Multiple layer floating gate polysilicon to enhance capacitive coupling of the cell and reduce process variation on cell coupling ratio [2].
3. Optimization of tunnel oxide, IPD ONO structures, oxidation processing of polysilicon gate stack, side wall oxidation, etc., to reduce channel resistance of the string [3].
4. STI edge and sidewall profile optimization to reduce cell to cell interference and advanced stack profiling to minimize Vth variations [4].

NAND flash FG devices implemented at the 50 nm technology node and beyond had to deal with multiple parasitic effects due to the storage of very few electrons at the floating gate. Such design required other process innovations for example "air-gap" isolation technology, advanced ECC applications, and others (see Part II of this book). The process innovations as stated above enabled flash NVM technology to take advantage of feature size scalability of each succeeding technology nodes, to advance manufacturability and reliability, to dramatically reduce the cost per bit of flash NVMs, and to establish stability of the stack design and architecture.

Current NAND cell architecture and gate stack design typically consists of a string of 32 flash cells placed between two select gates, associated source line, and bit line at each end as shown in Figure 6.1a and b [5]. Figure 6.1a shows the circuit representation of the NAND array scheme wherein a NAND memory string is shown. Figure 6.1b shows a schematic cross section across the gate stack of the NAND string of Figure 6.1a along the channel parallel to the bit line. The fixed Vth device gates at the two ends of the string: the select gate drain (SGD) for the bit line at the left end and the select gate source (SGS) for the common source line at the right end are shown along with the 32 word lines representation: WL 0 to WL 31 for the NVM cells. The word lines connect to the respective control gates of the individual cells in the string. All 32 bits share the same channel between the defined bit line and source line and the memory bits are accessed serially within the string. The single cell shown in Figure 6.1b shows the self-aligned edges of the floating gate and the control gate parallel to the bit line.

Figure 6.2a and b [5] shows the micrograph and schematic cross section of the single cell in the word-line direction. The shallow trench isolations (STIs) are self-aligned along the width of the devices as shown. The control gate wraps around the floating gate to enhance the capacitive coupling within the cell separated by the IPD dielectric layer which had been until recently an ONO stack. The cell cross section shown is that of an 8 Gb flash memory designed at the 63 nm node showing a <15 nm spacing to be filled by the control gate to reduce the capacitive coupling by field shielding between the adjacent floating gates of the neighboring cells.

6.2.1 Floating Poly Gate, GCR, and STI in Stack Design

We shall now discuss several significant design parameters related to the floating-gate cell design and scalability. This includes both active and parasitic components and geometry effects both horizontally and vertically.

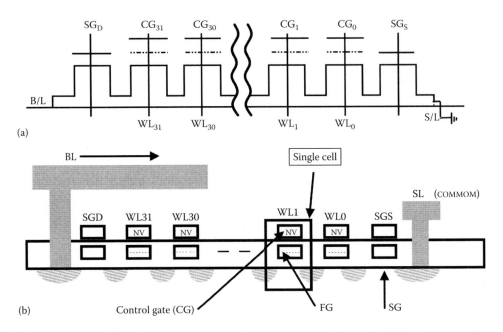

(a)

(b)

FIGURE 6.1 The NAND cell and array: (a) circuit representation of a 32 bit NAND string; and (b) gate stack cross section representation of the same, showing control gate (CG/WL), floating gate (FG), bit-line (BL) at far let, source line (S) at far right; and select gate for bit line (SGD) as well as source line (SGS). (From Watanabe, T., NAND flash memory technology, *2006 IEDM IEEE Short Course on Memory Technology for 45 nm and Beyond*, pp. 1–40, December 2006.)

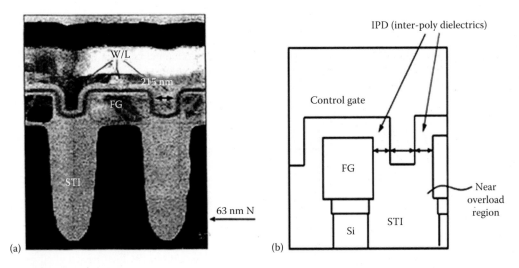

(a)

(b)

FIGURE 6.2 NAND cell stack details: (a) NAND cell stack micrograph. (b) NAND cell schematic along word line showing the non-overlap region of the floating gate by the control gate over the STI region (device width). (From Watanabe, T., NAND flash memory technology, *2006 IEDM IEEE Short Course on Memory Technology for 45 nm and Beyond*, p. 1 © 2006 IEEE.)

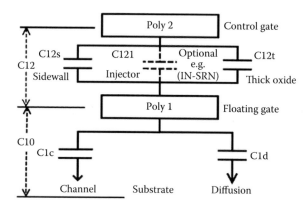

FIGURE 6.3 FG cell: Critical capacitance components and capacitance modeling: And gate coupling-R=SiH$_2$Cl$_2$/NH$_3$ ratio (GCR) of the cell = C12/(C12 + C10).

The floating gate serves as the charge storage equipotential medium capacitively coupling the substrate and the control gate at either end. Ignoring the parasitic coupling of the FG to FG coupling of the neighboring cells, the capacitive components within the cell are shown in Figure 6.3, whereby C 10 represents the net capacitance between the FG and silicon, while C 12 represents the net capacitance between the FG and the CG. Together, C 10 + C12 is considered to be unity (normalized) in terms of capacitive coupling ratio. The gate coupling ratio *GCR* is defined as: C 12/(C10 + C12) or represents the fraction of the potential dropped between the floating gate and the substrate compared to the total potential difference between the gate and the substrate. This is obviously an important design parameter not only for writing-erasing of the device but also for evaluating the memory state stability at application conditions. A design objective for the modern FG cell is to achieve a GCR ~0.6 to optimize, window, retention, write–erase voltage, and speed objectives. Most of the coupling area for the "wrap-around" cell is derived from the two-sidewalls along the word-line direction, where the control gate overlaps the floating gate (Figure 6.5b).

The *floating gate height*, geometry, and placement with reference to STI are important stack design considerations, aside from the insulating layers interfaces and the characteristics of the gate material mentioned earlier. STI is recessed to achieve nearly full overlap of the control gate to the floating gate sidewalls to improve GCR as well as to reduce parasitic FG–FG coupling between adjacent cells by the field shielding effect of the control gate. Profiling of the floating gate sidewalls improves the GCR and control–gate overlap and thereby reduces the coupling ratio sensitivity as a function of floating gate height. Figure 6.4 shows the estimated GCR sensitivity versus floating gate height for a 1 Gb NAND flash product designed at the 150 nm technology node [2]. When field shielding by the control gate is partial as shown in the shaded region of Figure 6.5b, the parasitic coupling component could be significant, both WL to WL and BL to BL capacitive coupling components are dependent on floating gate height and adversely affect Vth. Vth modulation as a function of floating gate height is shown in Figure 6.5c [5]. WL-WL and BL-BL capacitive components are schematically shown in Figure 6.5a and b, respectively. Therefore, aside from the tunnel oxide, IPD insulator stack, the gate characteristics (including gate side wall and WL resistance), and associated interfaces, GCR, FG properties, and STI are important design factors in the FG flash device.

An 8 Gb flash product design with self-aligned floating gate providing fully recessed STI had been implemented and published by Park et al. in 2004 for the first time [4]. Compared to the preceding 1 Gb SLC (1 bit per cell) product at the 120 nm technology node, the succeeding 8 Gb flash provided DLC (2 bit storage per cell) capability and a >4x enhancement in cell density at the 63 nm technology node. Besides the fully recessed STI feature mentioned above, floating gate height was reduced from 170 nm (1 Gb SLC) to 45 nm (8 Gb MLC) with complete overlapping cell feature

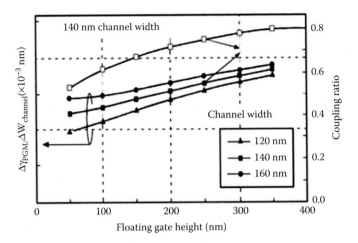

FIGURE 6.4 FG height effects: GCR versus device width sensitivity, GCR increases with increased floating gate height, tunnel oxide and IPD are 8.5 and 15 nm, respectively, FG top width is 170 nm. (From Aritome, S., Advanced flash memory technology and trends for file storage application, *IEDM*, p. 763 © 2000 IEEE.)

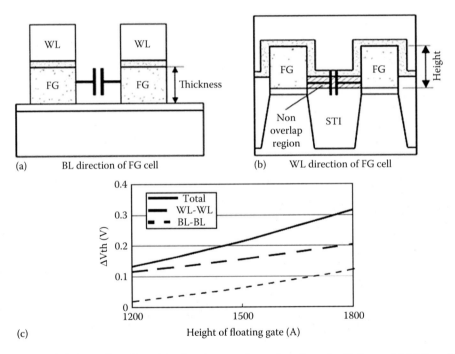

FIGURE 6.5 FG-FG coupling: Capacitive interference due to the adjacent cell charge: (a) Bit line direction, (b) word line direction, (c) Vth modulation versus FG height. (From Watanabe, T., NAND flash memory technology, *2006 IEDM IEEE Short Course on Memory Technology for 45 nm and Beyond*, p. 1 © 2006 IEEE.)

between the floating gate and the control gate, thereby significantly reducing parasitic coupling effects, Vth dispersion, and associated GCR enhancement [4].

The significance of FG height and recessing STI to achieve full overlap of FG by the control gate in the word-line direction are shown in the simulated plots of Vth dispersion versus FG height, and Vth dispersion versus FG sidewall non-overlap are shown in Figure 6.6a and b, respectively.

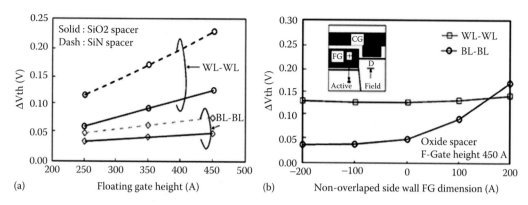

FIGURE 6.6 Vth dispersion: FG height (a) and FG sidewall non-overlap sensitivity (b) using fully recessed STI scheme of self-aligned FG technology of Figure 6.6. (From Park, J.-H. et al., 8 Gb MLC (multi-level cell) NAND flash memory using 63 nm process technology, *IEDM*, p. 873 © 2004 IEEE.)

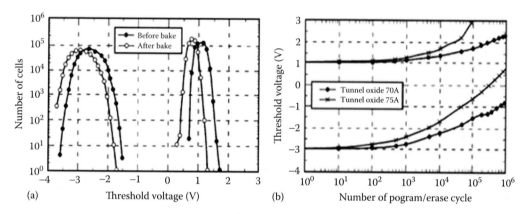

FIGURE 6.7 1 Gb NAND design with 32 bit NAND string [34]. (a) Data retention characteristics post 1E4 W/E cycling, (b) oxide thickness sensitivity on endurance. (From Choi, J.-D. et al., A 0.15 um NAND flash technology with 0.11 um² cell size for 1 Gbit flash memory, *IEDM*, p. 767 © 2000 IEEE.)

It should be noted that while the cell density improved by >4x, the gate stack dielectric layers could not be scaled significantly in order to ensure acceptable leakage, end-of-life window, and retention requirements for 10 years, and additionally to ensuring program disturb requirements [4].

Retention and endurance characteristics of the 1 Gb SLC device are shown in Figure 6.7a and b, respectively. The technology introduced 32 bit NAND string design for the first time to achieve the cell density. The tunnel oxide thickness was optimized to achieve enhanced endurance.

Programming and erasing characteristics of the 8 Gb chip MLC device are shown in Figure 6.8a and b, respectively [2]. As mentioned earlier, fully recessed STI using self-aligned floating polysilicon gate technology was incorporated to reduce neighboring cell capacitive coupling interferences.

The write voltage and time for the 8 Gb FG device was raised to increase charge storage within the floating gate. The device dielectric thickness for both IPD and tunnel layers assures higher level of charge storage with acceptable standby leakage to store significantly higher charge storage for the MLC operation (larger window) without window saturation. The floating-gate flash devices had historically allowed ease of feature size scaling due to the fundamental fact that the program-erase fields and charge transport by Fowler–Nordheim tunneling are vertical and no strong horizontal

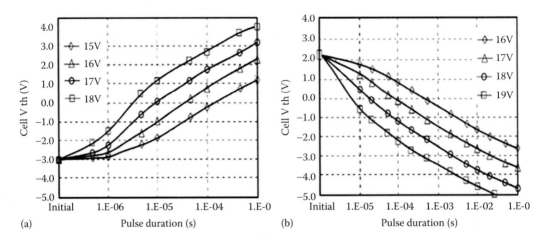

FIGURE 6.8 Programming and erasing characteristics of 8 Gb chip MLC NAND device design: (a) programming characteristics, (b) erasing characteristics. Note no Vth saturation effects for either programming or erasing. (From Choi, J.-D. et al., A 0.15 um NAND flash technology with 0.11 um² cell size for 1 Gbit flash memory, *IEDM*, p. 767 © 2000 IEEE.)

component existed in the device. However, as mentioned earlier, for feature sizes of 50 nm and beyond, the active electronic charge to be stored in the FG is limited and other parasitic effects need to be addressed from the standpoint of the FG stack design. This will be covered in Part II of this book.

6.3 CHARGE TRAPPING DEVICES: CELL/STACK DESIGNS

Cell and stack design considerations for various charge trapping devices are somewhat different from FG devices due to additional limitations of charge storage. While charge storage capacity for the floating-gate devices is limited by the steady-state leakage from the FG to the CG during writing and conversely from the CG to the FG during erasing, the charge storage for the charge trapping devices is limited by either the charge storage density of electrons and holes in the trapping dielectric for the SONOS/MONOS types of CT devices or by the limits associated with the phenomena of quantum confinement and coulomb blockade in nanocrystal (NC) devices. In the case of former type of CT devices, the characteristics of trapping dielectrics in establishing window and window Vth variations, retention, and endurance have been discussed earlier. In the latter type of devices, NC characteristics in terms of size, density, distribution, and work function influencing NVM properties have also been covered. The role of tunneling and blocking layers, gate material, and associated interfaces is basically similar to those of FG devices, although stack and cell designs are required to be optimized on the basis of both electron and hole transport and storage and not based on only electron transport and storage as for the FG type of devices.

Among the two types of CT devices, namely CT-SONOS and CT-NC ("discrete trap" and/or "embedded trap" devices), charges are stored in bulk nitride centers and in nanocrystal associated centers, respectively. These two types of CT devices have been labeled in this book as CT and NC devices, respectively. Random access NOR type of memory array architecture could be built in these types of devices; however, only CT-SONOS types have emerged as the mainstream product to date primarily for embedded code applications. As stated earlier, CT devices typically display faster read and write–erase speed and greater endurance than FG devices. CT-SONOS-based products typically employ contact-less NOR array with virtual ground array (VGA) architecture and are therefore classified as NROM flash devices distinguishing themselves from the slower, denser

NAND flash FG products. However, CT-SONOS stack designs could also be configured in NAND layout to reduce bit density. This will be covered in Part II.

There are two commonly used cell types associated with CT-MNOS/SNOS/MONOS/SONOS types of devices. These are based on the difference in mode of charge transfer from silicon to the nitride trapping centers. The first one is based on hot carrier injection such as hot electron and hot holes. The second one is based on either direct or "modified-Fowler–Nordheim" tunneling (trapezoidal barrier: see Chapter 3) and Fowler–Nordheim tunneling (triangular barrier) of carriers between silicon and nitride trapping centers. Devices are also configured with hybrid mode consisting of either writing by CHE (often by source-side injection) or erasing by Fowler–Nordheim tunneling to reduce program-erase power requirements or reverse, although the reverse approach is usually not common.

Typically, the first type of devices with hot carrier injection mode is employed with a single transistor memory cell which may provide either SLC (1 bit per cell) type of memory in standard cell "read" configuration or an MLC (2 bit per cell) type whereby charges are stored in each localized region above the diffusion edges of the memory FET device and "read" by reversing the source/drain functions of the diffusions (reversing the current flow directions) to achieve 2 bit per cell density. The hot carrier devices provide lower reliability compared to the tunnel-based devices and consequently lower endurance. The advantage of MLC types is higher bit density and lower voltage requirements for write–erase scheme using hot carrier modes. Design examples of this type of devices will be covered here.

The second type of CT devices ("split-gate device") is more common for embedded applications due to higher reliability and endurance. Typically, the NOR cells for these devices employ one-and-a-half transistor cell (the split-gate configuration: see Chapter 2), and a somewhat larger cell size compared to the first type of devices. Stack design and cell configuration of this type of CT devices will also be discussed here. Before illustrating the above two device memory cells and associated characteristics, the sensitivities of various dielectric elements in CT device stack design will be reviewed.

6.3.1 Stack Design Elements and Device Properties: MNOS/SNOS/MXOS/SONOS/MONOS

The conventional CT devices consist of two or three key dielectric elements. For the MNOS/SNOS/MXOS types of subset, the elements consist of a tunnel oxide ranging in thickness of 1.5–4.5 nm with a significantly thicker nitride (SiN) or oxynitride (SiON) based trapping/charge blocking dielectric film. Such film may range in thickness of 10–30 nm depending on the targets to be achieved of device properties and program-erase characteristics ($\pm$Vpp/pulse durations). For the SONOS/MONOS/MX'XOS (X' and X being two different SiON compositions replacing oxide and nitride, respectively) devices, the nitride (or oxynitride) trapping layer and the charge blocking oxide (or oxynitride) layers are considerably thinner and may range between 3 and 10 nm thickness for each layer, while the tunnel oxide thickness may range similar to those of MNOS. Device properties are dependent on the stack design elements and are discussed in Section 6.3.1.1.

6.3.1.1 Tunneling Oxide

Charge transport modes by tunneling through the tunnel oxide layers of CT devices are illustrated in the band pictures: Figure 6.9a–c for electrons and Figure 6.9d and e for holes, respectively. The band diagrams depict the nature of the energy barriers at the silicon–oxide interface and within the dielectric films when subjected to appropriate potentials for writing and erasing. Figure 3.2a and b is band-to-band direct tunneling/modified Fowler–Nordheim tunneling of electrons through the trapezoidal barrier as explained in Chapter 3, whereas Figure 6.9c depicts the triangular energy barrier for the classical Fowler–Nordheim mode for electrons when oxide is thicker (>3.2 nm). The corresponding hole transport modes through the trapezoidal barrier for the thinner oxide are

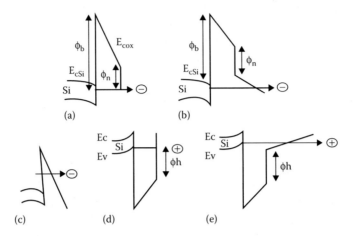

FIGURE 6.9   Band pictures: Various tunneling modes: (a–c) for electrons, (d, e) for holes.

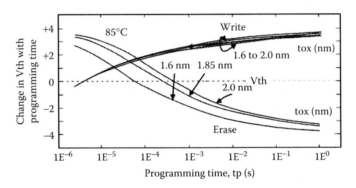

FIGURE 6.10   Programming characteristics for SNOS: Strong tunnel oxide thickness dependence on erase compared to write for SNOS: oxide thickness range of 1.5–2.0 nm. (From Minami, S., and Kamigaki, Y., *IEEE Trans. Electron Dev.*, 38(11), 2519–2526, 1991.)

shown in Figure 6.10d and e. Due to the large band offsets between oxide and nitride (1.1 eV for the conduction band and 2.8 eV for the valence band: see Table 6.1), the electrons and holes once tunneled through the trapezoidal barriers move rapidly by the drift field at the top of the bands until captured within the trapping layer associated with the trapping lengths/trapping cross sections of electrons and holes characteristic of the trapping medium. Transport modes (a), (b), (d), and (e) are strongly sensitive to the tunneling distance, and, therefore, high current density charge transport is feasible when tunnel distance is very small and exponentially decays fast as the tunnel distance is increased (see Chapter 3). Due to significantly higher barrier energy at the interface for holes compared to electrons (4.7 eV for holes vs. 3.2 eV for electrons), and higher effective mass in oxide, hole mobility is lower. Consequently, hole current is significantly lower than electron current for the same tunnel distance and reduces at a faster rate than the electron current. Oxide thickness exhibits a strong dependence on the low Vth (0) state stability due to hole transport sensitivity (retention) as well as the erase speed, as a result.

Tunnel oxide thickness dependence on CT devices was investigated by Y. Yatsuda et al. [6], S. Minami et al. [7], and A. Nughin et al. [9] in the tunnel oxide thickness range of 1.5–2.1 nm regime. In this range of thickness, the initial memory window is larger for higher programming and erase voltage levels and for longer duration. The erase memory state is more strongly sensitive

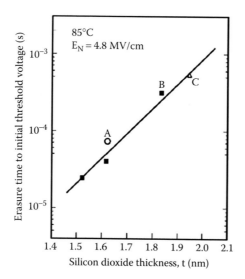

FIGURE 6.11 Strong sensitivity for erasing time on tunnel oxide thickness for CT device. (From Minami, S., and Kamigaki, Y., *IEEE Trans. Electron Dev.*, 38(11), p. xx © 1991 IEEE.)

to programming voltage and stress duration compared to the written memory state showing asymmetry between electron and hole transport characteristics. Tunnel oxide thickness dependence on programming (write or erase) is shown in Figure 6.10 for an SNOS device. Erasing is slower than writing, as should be expected for hole transport. However, charge loss in written state (without cycling) is also at a higher rate due to reverse tunneling of electrons.

Figure 6.11 shows erase time dependence on thickness of the SiO_2 layers. The linear dependence of logarithmic time with the tunnel oxide thickness validates the mechanism of direct tunneling of holes.

Tunnel oxide thickness dependence on charge retention after initial write–erase cycling is shown in Figure 6.12, while similar retention characteristics after 1E5 write–erase cycling are shown in Figure 6.13. Logarithmic time dependence on written and erased memory states can be seen in both cases. After initial write–erase cycling, the charge loss characteristics are similar for both written and erased states. However, the characteristics are different after written state when compared to

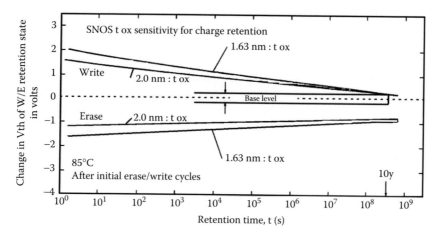

FIGURE 6.12 Data retention sensitivity to tunnel oxide thickness after initial W/E cycling. (From Minami, S., and Kamigaki, Y., *IEEE Trans. Electron Dev.*, 38(11), p. 875 © 1991 IEEE.)

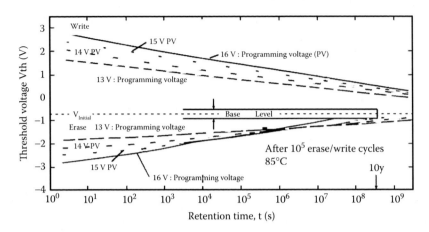

FIGURE 6.13 Data retention sensitivity to programming voltage after 1E5 W/E cycling. Note strong erase sensitivity. (From Park, J.-H. et al., 8 Gb MLC [multi-level cell] NAND flash memory using 63 nm process technology, *IEDM*, p. 873, 2004.)

the erased state. In the written state, the rate of charge loss by direct tunneling increased owing to higher internal potential caused by larger charge storage. The difference in slope between the written state and the erased state is due to the difference in carrier mobility, tunnel distance from the trapping sites (larger for holes), and barrier energy. However, the slopes are altered after the write–erase stress cycling [4]. This is due to increased generation of interface states at the silicon–oxide interface and defects in nitride (shallow traps) as well as possible defect generation in blocking oxide due to high stress field across the dielectric stack. It should be noted that as the erase voltage increased to 16 V, charge loss rate was enhanced. It is postulated that hole-induced damage in blocking oxide is responsible for enhanced leakage. Degradation and oxide reliability at high field had been discussed in Section 4.1.1 earlier. Optimized SONOS stack design employing direct tunneling charge transport had demonstrated 10 years of retention [5]. SONOS device with thicker tunnel oxide–nitride–oxide (2.9 nm tunnel ox/5.0 nm nit/5.0 nm blocking ox) stack design was also published by T. Terano et al. demonstrating >1E5 write–erase endurance and post-endurance retention with tight threshold distribution for a 4 Mb NVM chip design for embedded applications [7].

6.3.1.2 Trapping Nitride (Si_3N_4) and Oxynitride (SiON)

Conventional Ct devices employed nitride or oxynitride layers as charge trapping layer in the stack design. The requirements listed below are satisfied by nitride and NR-oxynitride for the charge trapping layer:

1. Dielectric layer characterized by high trap density (Nt), deep trap energy (Utr), large trap cross section for charge capture for both electrons and holes.
2. Large band offset with respect to both the tunnel dielectric and blocking dielectric.
3. Absence of multiple energy shallow trapping centers.
4. Compatibility with both tunnel dielectric and blocking dielectric to form clean interfaces with minimum fixed interface charges.
5. Thermal, structural, and chemical stability.
6. Process control and reproducible stoichiometry.
7. Selective etch ability.

Band structures for nitride and NR-oxynitride are illustrated in Figure 6.14a and b, respectively. NR–oxynitrides exhibit deeper trap depth and larger band gap compared to nitride.

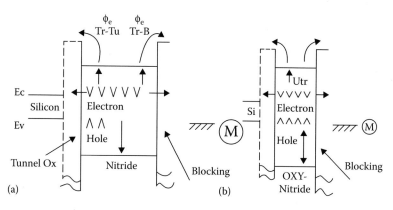

FIGURE 6.14 Band structure of trapping layers: (a) Nitride and (b) NR-oxynitride.

TABLE 6.1
Band Energy and Offsets with Respect to SiO$_2$ for SiN/Si$_3$N$_4$ and SiON Dielectrics

	SiO$_2$	SiN/Si$_3$N$_4$	NR-SiON[a]	OR-SiON[a]
	~43%Si/57%N	5->20%O/50->40%N	25->60%O/36	->8.5%N
Band gap	9.0	5.1	5.3 > 5.8	6.0 > 8.0
Trap depth: Utr eV	–	1.1	1.2- >1.65	>1.65
Band offset:Conduction	0.0	1.1	~1.1- >0.9	<~0.85 - >~0.5
Band offset: Valence	0.0	2.8	~2.6- >2.3	<~2.25- >~0.5

[a] Parameters are composition/process dependent.

Table 6.1 lists the important band parameters for oxide (for tunnel layer and blocking layer), nitride, and NR-oxynitride for trapping layer and OR-oxynitride for blocking layer used for conventional CT devices.

Trapping nitride thickness in recent scaled SONOS devices (both planar and FINFET) typically ranges between 4 and 6 nm [8–12]. Charge trapping in nitride is thickness dependent and hole trapping length is nearly twice as large as electron trapping length which has been discussed in earlier chapters (Chapters 4 and 5). When nitride thickness is equal to or less than 2 nm, no significant charge trapping is observed. For nitride thickness around 5 nm, holes accumulate at and near nitride blocking oxide interface while charge centroid for trapped electrons during writing as well as post write–erase cycling followed by writing exists near that interface due to trapping length, capture cross section, and post write–erase charge displacement consideration for electrons in nitride. As a result, nitride thickness in SONOS is optimal around 5 nm thickness from the standpoint of key device properties such as window, retention, and endurance. When nitride is replaced by NR-SiON of the same thickness (~5 nm), device properties such as window and retention are enhanced due to higher trap density and trap depth of NR-SiON and lower conductivity (Figure 6.6) compared to SiN. Additionally, charge centroid for electrons and holes is symmetric in SiON which aids endurance. General characteristics of NR-SiON are otherwise similar to nitride.

6.3.1.3 Blocking Oxide or Blocking OR-SiON

Band structures of blocking oxide and blocking OR-SiON are shown in Figure 6.15a and b, respectively.

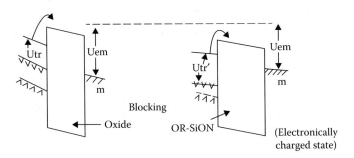

FIGURE 6.15 Band pictures: (a) Blocking oxide and (b) Blocking OR-SiON (electronically charged state).

For SONOS devices, an optimized blocking oxide thickness in the range of 4–5 nm is employed for the ONO stack to minimize retention loss and undesirable charge injection from the gate. The blocking oxide (HTO or TEOX) is deposited by LPCVD process around 800°C. OR-SiON of the same thickness could be effective in replacing blocking oxide and providing lower EOT stack since the dielectric constant of OR-SiON could be 25%–30% larger than that of oxide. Required attributes of blocking dielectric is similar to that of tunneling dielectric and could be summarized as follows:

1. Provide large energy barrier or band offset to minimize stored charge loss.
2. Provide structural integrity and stoichiometry at high field with characteristic low fixed charge, low trap density, and low trapping cross section.
3. High K dielectric with low conductivity to provide low field due to stored charge.
4. Compatibility with gate metal related to (a) preventing undesirable charge injection (high energy barrier at metal–blocking dielectric interface: high work function); (b) low interface stress; and (c) providing chemical passivation at the interface against contaminants during post-metal deposition and processing).
5. Process control, reproducibility, and stoichiometry.
6. Thermal and structural stability.
7. Selective etch ability.

Except item 3, oxide fulfills all the above attributes. OR-SiON provides higher K which is sensitive to oxygen concentration in films while trading off achievable K values with band gap (item 1: see Table 6.1). Higher K dielectric applications will be discussed in Part II of this book.

6.3.2 ONO Stack Design and SONOS Device Properties

Band diagrams for a SONOS stack are shown in Figure 6.16a and b. Figure 6.16a is the band diagram without any applied potential either at the silicon substrate or at the N+ polysilicon gate. Figure 6.16b is the band diagram when a positive potential +Vp is applied at the gate for writing causing electron injection from silicon and trapping in the trapping nitride layer. The stack design corresponds to a SONOS device employing Fowler–Nordheim tunneling with 3.5 nm tunnel oxide to ensure 10 years of charge retention and 1E6 cycles of endurance. The EOT of the ONO stack is 12 nm with 7.0 nm thick trapping nitride and 4.5 nm of blocking top oxide. The device operates at write–erase voltage levels of ±11–12 V of Vp.

Corresponding band diagrams for a similar triple-layer SiON stack replacing ONO with OR-SiON/NRSiON/OR-SiON are shown in Figure 6.17a and b, respectively. Figure 6.17c demonstrates a further enhancement of the basic device by replacing the NR-SiON single trapping layer with a triple trapping layer of NR–SiON/OR–SiON/NR-SiON (e.g., 4 nm of Si_2ON_2/2 nm of OR-SiON/4 nm of Si_2ON_2) while maintaining the total stack EOT constant (total EOT = 12 nm) [8,9]. The same physical thicknesses of tunnel and blocking layers is used as in the SONOS device shown previously in Figure 6.16.

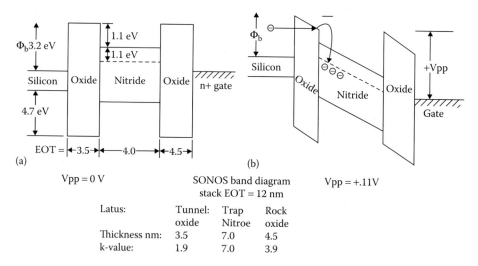

FIGURE 6.16   Band diagram of SONOS stack: (a) Vpp = 0V and (b) Vpp = +11 V.

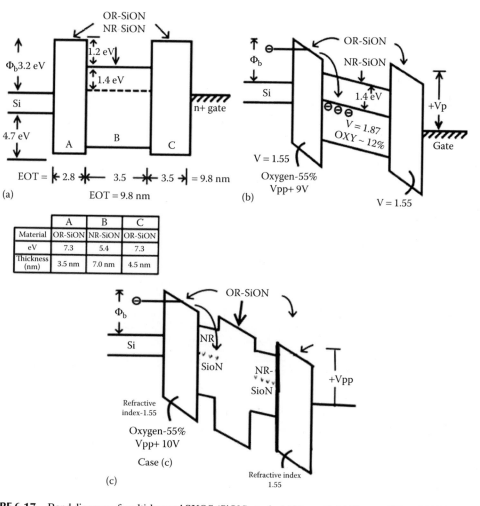

FIGURE 6.17   Band diagram of multi-layered SXOS (SiON) stack: (a) Vpp = 0, (b) Vpp = +9V, and (c) Vpp +10V.

In case of the triple-layer SiON stack design of case (b), whereby oxide is replaced by OR-SiON of refractive index of 1.55 and K = 5.0 (atomic percentage of oxygen and nitrogen ~55% and ~10%, respectively) and nitride is replaced by NR-SiON of refractive index of 1.87 and K = 8.0 (atomic percentage of oxygen and nitrogen ~12% and ~50%, respectively). The band gaps for NR-SiON and OR-SiON are 5.4 and 7.3 ev, respectively. The EOT of the gate insulator stack is 9.9 nm due to higher K values. The device would operate at write–erase voltage levels of ±9V, at improved speed, improved retention, and several orders of magnitude improved endurance aided by an improved memory window compared to the SONOS device cited above. The improved speed comes from lower barrier energy compared to oxide for both electron and hole tunneling. The improved retention comes from deeper traps in the NR-SiON film and greater offset (1.4 and 1.25 eV, respectively, compared to 1.1 and 1.1 eV for nitride). Significantly higher endurance is achieved due to symmetric charge storage in the oxynitride layer and greater charge confinement in SiON compared to nitride. Improved window is due to larger trap density in NR-SiON layer.

For case 6.17c, both window and retention properties are further improved due to the higher energy barrier at the middle OR-SiON layer enabling multilevel storage per unit cell (MLC). This will be further discussed in Chapter 16.

6.3.3   Examples of SONOS Stack Designs

Several examples of SONOS CT device stacks and device characteristics published between 1991 and 2006 will be noted. A split-gate memory cell is shown in Figure 6.17 [12] containing the SONOS NVM stack and associated elements shown in the inset. The device employed 2 nm tunnel oxide (direct tunnel mode), 5 nm nitride, and 4 nm blocking top oxide. The device exhibited nearly 3 V window: Vth(1) = +1 V, Vth(0) = −1.8 V; operated at +9 V, 100 μs write and −9V, 10 ms erase, with >1E7 cycles of endurance as shown in Figure 6.18 [8,9]. The write–erase characteristics are shown in Figure 6.19. The array write–erase–read schemes are discussed in the next section in the NROM array subsection. Due to the ease of integration with CMOS logic technology, lower program-erase voltage requirements and higher endurance, the direct-tunneling-based SONOS CT devices had been widely used in integrated logic embedded applications for code storage and for EEPROM functions. To reduce the cell size, a one transistor nitride NOR cell/array configuration (discussed in Section 7.4.1) was proposed with virtual ground contact-less array known as nitride NOR or NROM [10,11]. The NROM single device cell has been

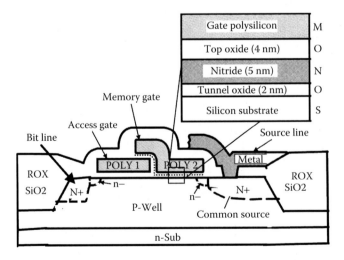

FIGURE 6.18   (a) Programming (W/E) characteristics of the SONOS memory cell of Figure 6.19 and (b) Endurance characteristics of the SONOS memory cell of Figure 6.19. (From Nozaki, T. et al., *IEEE Solid-State Circ. Conf.*, 26(4), 497–501, 1991.)

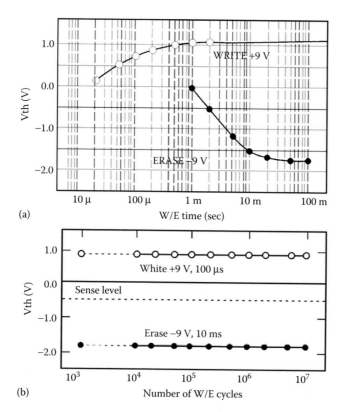

(a)

(b)

FIGURE 6.19 Schematic cross section of a split-channel SONOS memory cell. (From Nozaki, T. et al., *IEEE Solid-State Circ. Conf.*, 26(4), 497–501, 1991.)

widely used since 2001 provided a SLC cell of density 4–6 F × F and employed the ONO stack structures as discussed earlier. To further improve the cell density, localized charge trapping cell structures were proposed and demonstrated by storing charges in the nitride trapping layer locally over each of the diffusion edges of the single device SONOS cell, thereby doubling the bit density yielding 2 bit per cell operation [12–15]. A schematic representation of a planar one device SONOS 2 bit per cell CT device using localized nitride charge trapping is shown in Figure 6.20.

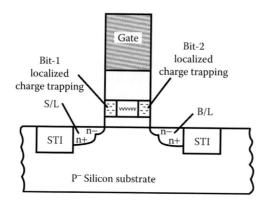

FIGURE 6.20 Example of a 2 bit/cell SNOS device employing localized trapping in nitride. (From Fukuda, M. et al., Scaled 2bit/cell SONOS type nonvolatile memory technology for sub-90 nm embedded application using SiN sidewall trapping structure, *IEDM*, p. 909 © 2003 IEEE.)

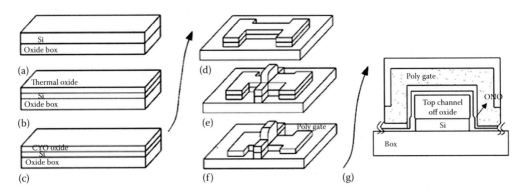

FIGURE 6.21 Fabrication process flow for FinFET SONOS memory cell on SOI. (From Xuan, P. et al., FinFET SONOS flash memory for embedded application, *IEDM*, p. 609 © 2003 IEEE.)

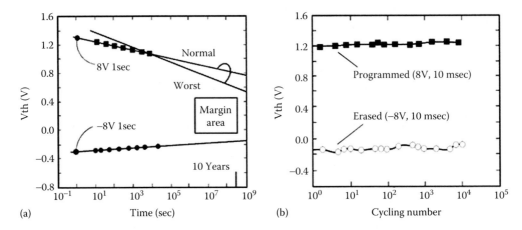

FIGURE 6.22 FinFET SONOS device characteristics of the SONOS memory cell: (a) Retention, (b) Endurance. (From Xuan, P. et al., FinFET SONOS flash memory for embedded application, *IEDM*, p. 609 © 2003 IEEE.)

An NROM device on SOI substrate was also proposed by Y.-H. Shih et al. demonstrating satisfactory retention and endurance by adopting a "Fowler–Nordheim reset scheme" with conventional NROM operation [17].

FinFET vertical channel SONOS device was first published by Y. K. Lee et al. [18] on SOI to overcome feature size scaling limits and short channel effects of planar devices (Figure 6.21). Figure 6.22 shows the cross sectional view of the fabrication process for the FinFET SONOS memory cell using 1.5 nm tunnel oxide/5.0 nm nitride/4.0 nm blocking oxide ONO stack. Figure 6.23a and b shows the retention and endurance characteristics of the FinFET SONOS device, respectively. Subsequently, several authors published scaled SONOS FinFET device characteristics for embedded applications [19,20] both for bulk and SOI technologies.

6.3.4 SONOS Devices with Oxynitride

Nitride trapping layer was replaced by either OR-SiON or NR-SiON layer for MXOS type of devices by T. Ishimaru et al. for embedded NVM applications [21]. The work was published in 2005 and demonstrated enhanced device attributes such as 100X larger retention, 2 orders of

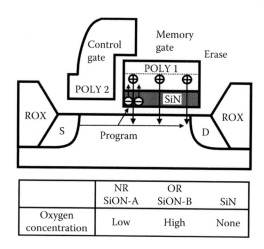

	NR SiON-A	OR SiON-B	SiN
Oxygen concentration	Low	High	None

FIGURE 6.23 Split-channel device comparison: MONOS versus MOXOS devices with NR-SiON (SiON-A) and OR-SiON (SiON-B) replacing Nitride. (From Ishimaru, T. et al., Impact of SiON on embedded nonvolatile MNOS memory, *IEDM*, p. 885, 2004.)

magnitude faster erase, no erase speed degradation up to 1E4 cycles of endurance when compared with an equivalent MNOS control device stack consisting of 4 nm of tunnel oxide and 26 nm of nitride. The memory cell was a split-gate type whereby programming was performed by source-side hot-electron injection and erasing by a combination of hole injection from the gate and electron injection from the trapping layer to the gate. The thick tunnel oxide prevented hole injection from the silicon substrate. The device schematics and the stack attributes of comparing between the trapping layers: nitride as control dielectric case with NR-SiON (SiON-A) and OR-SiON (SiON-B) are shown in Figures 6.23. The bias conditions for the program-erase operations were as follows:

	VAG	VNVMG	VSOURCE	VDRAIN
WRITE: V	12	1.5	6.0	0.8
ERASE: V	15	1.5	0.0	0.0

Program and erase characteristics are shown in Figure 6.24a and b, respectively. Band diagrams under erase conditions for nitride versus oxynitride are illustrated in Figure 6.25a and b, respectively. The erase speed was significantly faster with OR-SiON. Hole current density versus gate potential and electric field plots are shown in Figure 6.26a and b, respectively, for the above cases.

As should be expected, the devices with SiON-A and SiON-B layers exhibit steeper current density versus both gate potential and electric field characteristics compared to the corresponding device with nitride as shown in Figure 6.26a and b while the steepest slope being with SiON-B (OR-SiON layer).

The OR-SiON (SiON-B) device showed significantly enhanced retention of electrons (Figure 6.27a). The endurance characteristics of different device types demonstrate superior endurance of SiON devices (both SiON-A [NR–SiON] and SiON-B [OR-SiON]) as shown in Figure 6.27b, when compared with the endurance characteristics of nitride. The electron trap energy derived from Poole–Frenkel plots for different device types is shown in Figure 6.28. The higher value of trap depth for SiON-B confirms the observed enhanced charge retention characteristics of SiON-B (OR-SiON).

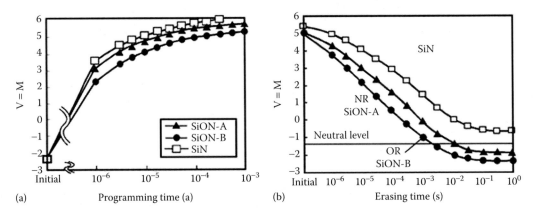

(a) Programming time (a)   (b) Erasing time (s)

FIGURE 6.24 (a) and (b) Write/erase characteristics of devices with Nitride versus SiON-A and SiON-B. (From Ishimaru, T. et al., Impact of SiON on embedded nonvolatile MNOS memory, *IEDM*, p. 885 © 2004 IEEE.)

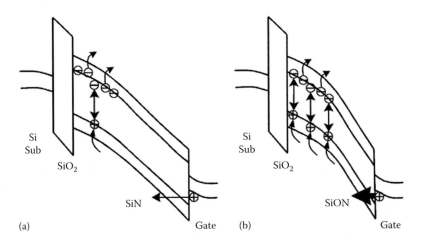

FIGURE 6.25 Band diagrams depicting electron and hole movements under erase conditions for (a) Nitride and (b) SiON. (From Ishimaru, T. et al., Impact of SiON on embedded nonvolatile MNOS memory, *IEDM*, p. 885 © 2004 IEEE.)

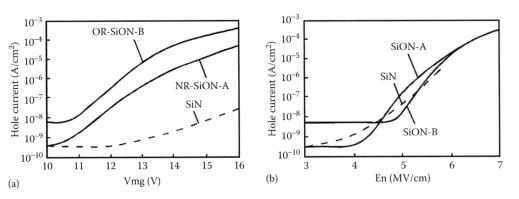

(a) Vmg (V)   (b) En (MV/cm)

FIGURE 6.26 Hole currents for Nitride and SiOn-A and SiOn-B versus (a) Gate Potential and (b) Electric field in trapping dielectric layers. (From Ishimaru, T. et al., Impact of SiON on embedded nonvolatile MNOS memory, *IEDM*, p. 885 © 2004 IEEE.)

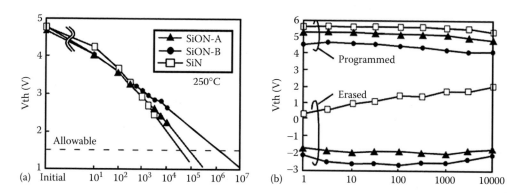

FIGURE 6.27 (a) Retention and (b) Endurance of Nitride, SiON-A and SiON-B. (From Ishimaru, T. et al., Impact of SiON on embedded nonvolatile MNOS memory, *IEDM*, p. 885 © 2004 IEEE.)

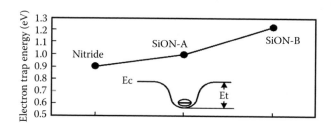

FIGURE 6.28 Electron trap energies Utr for Nitride, SiON-A and SiON-B from P-F plots. (From Ishimaru, T. et al., Impact of SiON on embedded nonvolatile MNOS memory, *IEDM*, p. 885 © 2004 IEEE.)

6.3.5 Nano Crystal Charge Trapping Devices: Stack Design Elements and Device Properties: SiNC/GeNC/MNC

Nanocrystal (NC) CT devices are categorized into either silicon NC or germanium NC or mixed Si-Ge NC and metal NC devices depending on the elements embedded in insulator/insulators to provide trapping centers. Key distinguishing feature in NC-CT devices from the conventional CT devices lies in the NC-based trapping layer from that of the bulk nitride or the SiON trapping layer for the SONOS type of devices. Therefore, the difference in the stack design is due to the trapping centers in the dielectric provided by the NCs embedded in the dielectric. Typically, the NCs are embedded in SiO_2 or other relatively trap-free dielectric medium. However, NCs could also be embedded into trapping dielectrics such as nitride and SiON to provide trapping centers associated with both the NCs and the bulk nitride or SiON films [22]. Otherwise, the role of tunnel layer and charge blocking layer is similar to those of conventional SONOS type of CT devices.

The physics and attributes of NC-CT devices have been discussed in Section 1.5 earlier [22]. Examples of some of the NVM properties characteristic of NC-CT devices will be illustrated here. It should be noted, however, that thus far no commercial product exists based on NC-CT devices.

Silicon NC-CT devices with direct tunnel oxide (1.5–3 nm tunnel oxide thicknesses) typically displayed small memory window (<1 V), limited retention (<1E6 s), fast program-erase speed (<100 μs), and high endurance (>1E9 W/E cycle) [23]. The limitation on retention was due to reverse direct tunneling of trapped charges and was shown to have overcome by introducing multilayer nitride–oxide–nitride (NON) tunnel barrier by S. Baik et al. [24]. High density metal nanodots were embedded in multiple layers within oxide to increase effective NC trap density to overcome window limitations [25,26]. The techniques employed were either sputtering [25] or by pulsed nucleation layer (PNL) method [26]. As much as 5.5 V window was achieved by incorporating two layers WN nanocrystals in oxide separated between ~4 nm deposited oxide. Advanced NC device and stack designs will be discussed in Part II.

REFERENCES

1. S. Aritome, Advanced flash memory technology and trends for file storage application, *IEDM*, San Francisco CA, 2000, pp. 763–766.
2. J.-D. Choi, J.-H. Lee, W.-H. Lee et al., A 0.15 um NAND flash technology with 0.11 um^2 cell size for 1 Gbit flash memory, *IEDM*, 2000, pp. 767–770.
3. J.-D. Choi, S.-S. Cho, Y.-S. Yim et al., Highly manufacturable 1Gb NAND flash using 0.12 μm process technology, *IEDM*, 2001, pp. 25–28.
4. J.-H. Park, S.-H. Hur, J.-H. Leex et al., 8 Gb MLC (multi-level cell) NAND flash memory using 63 nm process technology, *IEDM*, San Francisco CA, 2004, pp. 873–876.
5. T. Watanabe, NAND flash memory technology, *2006 IEDM IEEE Short Course on Memory Technology for 45 nm and Beyond*, pp. 1–40, December 2006.
6. Y. Yatsuda, S. Nabetani, K. Uchida et al., Hi-MNOS II technology for a 64-Kbit byte-erasable 5V only EEPROM, *IEEE Trans. Electron Dev.*, Ed 32(2), 224, 1985.
7. S. Minami, and Y. Kamigaki, New scaling guidelines for MNOS nonvolatile memory devices, *IEEE Trans. Electron Dev.*, 38(11), 2519–2526, 1991.
8. A. Bhattacharyya, Stack design using multi-layered oxynitrides for enhanced NVM attributes, Internal Memo, ADI Associates, November 15, 2001.
9. A. Nughin, A. L. Multsev, and V. A. Miloshevsky, N-channel 256 Kb and 1 Mb EEPROMs, *IEEE Solid-State Circuits Conference*, 1991, pp. 228–229.
10. T. Terano, H. Moriya, A. Nakamura et al., Narrow distribution of threshold voltages in 4Mbit MONOS memory cell arrays and its impact on cell operation, *IEDM*, Washington DC, 2001, pp. 45–48.
11. T. Nozaki, T. Tanaka, Y. Kijiya et al., A 1-Mb EEPROM with MONOS memory cell for semiconductor disk application, *IEEE Solid-State Circ. Conf.*, 26(4), 497–501, 1991.
12. M. White, Y. L. Yang, A. Purwar et al., A low voltage SONOS nonvolatile semiconductor memory technology, *Proceedings of the Sixth Biennial IEEE International Nonvolatile Memory Technical Review*, 1996, pp. 52–57.
13. M. Taguchi, NOR flash memory technology, *2006 IEDM Short Course, Memory Technologies for 45 nm and Beyond*, IEEE 2006.
14. B. Eitan, P. Pavan, and I. Bloom, NROM: A novel localized trapping, 2 bit nonvolatile memory cell, *IEEE Elec. Dev. Lett.*, November, IEEE, 2000, pp. 543–545.
15. Y. K. Lee, J. S. Sim, S. K. Sung, Excellent 2-bit silicon-oxide-nitride-oxide-silicon (SONOS) memory (TSM) with a 90 nm merged triple gate, *IEEE, International Semiconductor Device Research Symposium*, Washington DC, 2003, pp. 489–490.
16. M. Fukuda, T. Nakanishi, Y. Nara, Scaled 2bit/cell SONOS type nonvolatile memory technology for sub-90 nm embedded application using SiN sidewall trapping structure, *IEEE, IEDM*, Washington DC, 2003, pp. 909–912.
17. Y.-H. Shih, H.-T. Lue, K.-Y. Hsieh et al., A novel 2-bit/cell nitride storage flash memory with greater than 1M P/E-cycle endurance, *IEEE, IEDM*, San Francisco CA, 2004, pp. 881–884.
18. Y. K. Lee, S. K. Sung, J. S. Sim et al., Multi-level vertical channel SONOS nonvolatile memory on SOI, *IEEE, VLSI Technology*, Honolulu HI, 2002, pp. 208–209.
19. P. Xuan, M. She, B. Harteneck et al., FINFET SONOS flash memory for embedded application, *IEDM*, 2003, pp. 609–612.
20. J.-R. Hwang, T.-L. Lee, H.-C. Ma, et al., VLSI 2006, #317., 20 nm-gate Bulk Fin-FET SONOS Flash, San Jose CA.
21. T. Ishimaru, N. Matsuzaki, Y. Okuyama et al., Impact of SiON on embedded nonvolatile MNOS memory, *IEDM*, San Francisco CA, 2004, pp. 885–888.
22. A. Bhattacharyya, Si-rich oxides and nitrides, IBM Internal Technical Report, 1984 (Unpublished).
23. W. Guarini, C. T. Black, Y. Zhang et al., Low voltage scalable nanocrystal flash memory fabricated by templated self-assembly, *IEEE, IEDM*, Washington DC, 2003, pp. 541–544.

24. S. J. Baik, S. Choi, U.-I. Chung et al., High speed and nonvolatile Si nanocrystal memory for scaled flash technology using highly field sensitive tunnel barrier, *IEEE, IEDM*, Washington DC, 2003, pp. 545–548.
25. M. Takata, S. Kondoh, T. Sakaguchi et al., New nonvolatile memory with extremely high density metal nano-dots, *IEEE, IEDM*, Washington DC, 2003, pp. 553–556.
26. H. Lim, K. H. Joo, J.-H. Park et al., Nonvolatile MOSFET memory based o high density WN nanocrystal layer fabricated by novel PNL (pulse nucleation layer) method, *Symposium on VLSI Technology*, IEEE, pp. 190–191, 2005.

7 NVM Cells, Arrays, and Disturbs

CHAPTER OUTLINE

This chapter outlines the key evolution of selected NVM cells and arrays with relevance to future NVM designs and programming mechanisms applicable to the present day. The chapter highlights and elaborates some of the relevant materials described in Section 1.2 from historical perspective, leading to the emergence of NVM cells and arrays based primarily on tunneling mechanisms of writing and erasing. After reviewing the appropriate array designs, the various memory state instabilities are described. These are known as "disturbs," required to be addressed for NVM arrays associated with reading, writing, and erasing memory cells within the memory arrays. Mitigation of disturbs is also briefly addressed.

7.1 PRE-SILICON GATE TECHNOLOGY AND NVM CELLS

Silicon based NVM cells are variations of FET cells and have evolved and integrated along with the development of CMOS FET devices, circuits, and technology. The very first floating-gate NVM cell developed by D. Kahng and S. M. Sze of Bell Telephone Systems in 1967 was described in Chapter 2. The cell was a PFET NVM cell with a metal–oxide–metal (floating gate)-thin tunnel oxide–silicon (N-type) substrate (MIMIS) gate stack. This was soon followed by two CT-NVM cells: A PFET MNOS cell and both a PFET MXOS-NVM cell and an NFET MXOS-NVM cell (Chapter 2). The MXOS cells were developed by IBM whereby nitride as a trapping dielectric was replaced by multi-layered silicon–oxynitrides (SiONs) to provide enhanced window, retention, endurance, and reliability [1–3]. The MXOS charge trapping cells operated by modified Fowler–Nordheim tunneling for both writing and erasing. The cell design adopted for the 8K bit PROM (AROS) chip [4] consisted of a two-device cell: a fixed Vth FET device as a select device in series with the variable Vth MXOS-CT device. Schematic of this two-transistor PROM cell (called by IBM as "Alterable Read Only Storage: AROS cell") is shown in Figure 7.1.

7.2 THE FLOATING-GATE FLASH NVM CELLS IN SILICON GATE TECHNOLOGY

With the evolution of silicon gate FET technology, many floating gate planar and nonplanar cells evolved in the 1980s and continued through the 1990s. Multifaceted progress was made in floating-gate cell types and density, array designs, multilevel store capability (multi-bit per cell), and technology integration with existing CMOS technology making the FG-NVM the primary memory product for the NVM industry. Such progress has continued till recently with successful scaling of flash cell designs for each successive generation of feature size scaling. Such progress was aided by the progress made in defect control and yield of the relatively thicker tunnel oxide process technology employed for the FG-NVM cells. Thus, flash FG memory products have fast replaced hard drive memories in applications ranging from large database storage systems to handheld devices, providing superior power performance and reliability.

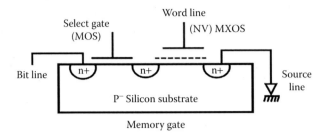

FIGURE 7.1   The two-transistor schematic of the alterable read only storage (AROS) cell: The nonvolatile element employed was of MXOS gate stack design using Silicon Oxy-nitride. (From Bhattacharyya, A. et al., *IBM Tech. Discl. Bull.*, 18(3), 654, 1975.)

7.2.1   Floating-Gate Flash Cells and Arrays

Floating-gate flash memory cell and array designs can be categorized into several groups depending on the specific writing and erasing mechanisms adopted historically in the cell operation.

These are:

1. Channel hot electron (CHE) write and external UV erase
2. CHE write and Fowler–Nordheim tunnel erase through the floating gate (ref. 3.26)
3. CHE write and enhanced Fowler–Nordheim polysilicon to polysilicon erase
4. Enhanced Fowler–Nordheim polysilicon to polysilicon write and erase (nonplanar)
5. Fowler–Nordheim tunneling write and erase from silicon substrate

The Fowler–Nordheim tunneling write and erase mode evolved to provide multiple advantages, namely lower programming power, on-chip single power supply implementation, simpler process integration, higher chip yield with progressive improvement in tunnel oxide defect density reduction, as well as a single transistor cell design with <4F2/bit of memory cell density. This type of operation and cell design became the mainstay for the flash product design during the last decade and will be covered in this section in detail. The cell and array designs for the prior modes of (1) through (4) above had been reviewed by several authors [5–7] and will not be reviewed in this book.

7.2.1.1   The FLOTOX Cell

The first 16 Kb electrically erasable flash memory (EEPROM) product was developed by Intel Corporation in 1980 and authored by W. S. Johnson et al. [8]. The cell known as the FLOTOX cell consisted of two devices, a select transistor and a memory transistor, similar to that shown of AROS cell mentioned in Figure 7.1. FLOTOX was the first NVM cell used in the industrial product which used Fowler–Nordheim tunneling for both writing and erasing. The memory transistor cell had a thinner tunnel oxide region between the silicon substrate and the first polysilicon floating gate, with a tunnel oxide thickness of 8–10 μm in the thinner region. The floating gate was isolated by a thicker oxide layer from the second polysilicon layer which formed the control gate of the nonvolatile memory element. The original FLOTOX cell was a 333 μm^2 cell, which was nearly 40F^2 cell implementation with the feature size F of 3 μm. For writing, a typical voltage of 14 V was applied to the control gate with source, drain, and substrate grounded. For erasing, a large voltage was applied to the drain, while the control gate and substrate are tied to the ground with the source floating.

The schematic cross sections of the cell and the array layout are shown in Figures 7.2 and 7.3, respectively.

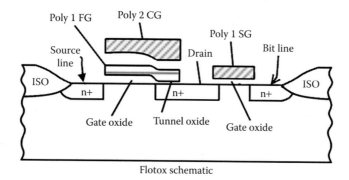

Flotox schematic

FIGURE 7.2 Schematic cross section of FLOTOX cell with Poly 1 select gate. (From Johnson, W. S. et al., 16Kb EEPROM, *IEEE, ISSCC*, p. 271, 1980.)

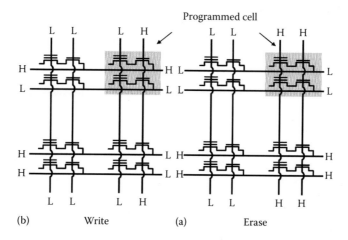

FIGURE 7.3 FLOTOX EEPROM memory array and programming scheme. (From Johnson, W. S. et al., 16Kb EEPROM, *IEEE, ISSCC*, p. 271, 1980.)

7.2.1.2 The ACEE Cell

An improved version of FLOTOX cell called the ACEE cell (array contactless EEPROM cell) was subsequently developed by Texas Instrument Corporation and was published by S. D. Arrigo and coworkers [9]. Their version of FLOTOX was a merged one-and-a half device cell, 5 V only, 256 K bit CMOS EEPROM product with a cell size of 40 μm^2. The cell was implemented in 1.5 μm technology providing a 17.5F^2 cell when F = 1.5 μm. The cell was designed with three thicknesses of thermally grown oxide over silicon, the merged pass-gate element had 50 nm of SiO_2, the floating gate element had 35 nm of SiO_2 while the tunnel oxide region over the extended part of the source had 10 nm of thin SiO_2 with the floating gate overlapping the tunnel oxide region for writing and erasing. The cell featured a self-aligned thick oxide device with buried source and drain. The programming and erasing were accomplished at the 10 nm tunnel oxide node at the source end with the first polysilicon floating gate layer extending into the channel over a 35 nm gate oxide. The second polysilicon layer served as the select/programming gate with a 50.0 nm gate oxide at the drain end of the channel setting the device threshold at the erased state and preventing the device to go into depletion mode (over-erasure) at the erased state.

The device operated in a similar way to that of the FLOTOX cell for writing and erasing. All cells were erased simultaneously prior to selective programming (writing). During parallel erase, all drain nodes were floating while all source nodes were raised to +5 V with the word lines at −11 V. The cell to be written was accomplished by putting +17 V on the word line with 0 V at source node while other cells on the same word line had their source node raised to +5 V, thereby lowering the voltage between the word line and the source nodes of the unselected cells to prevent programming. With all the drain nodes floating, both the unselected word lines and source nodes for the unselected cells were raised to +7 V to prevent any partial writing of the unselected cells. Reading is accomplished by raising the drain node of the select cell to 1.5 V with source at 0 V and the gate (selected word line) at 3.3 V. The word lines of all unselected cells were held at 0 V. During the erased state, the cell current is determined primarily by the select gate, while during the program state by the floating gate potential.

The cell required ±18 V to program and erase, with an erase pulse width of 10 ms. The device window achieved 3.5 V. The memory was organized to be byte (8 bit) programmable and required only 150 µs to program. The device demonstrated >1E5 cycles of endurance and 10 years of data retention. The chip consisted of two 128 Kb of memory arrays, and was organized to be 32 Kb×8.

Unlike earlier cell designs which used CHE programming, the FLOTOX and ACEE chips had the lower programming power due to the Fowler–Nordheim based writing and erasing, enabling on-chip power generation (charge pump circuitry) and single power supply design. The FLOTOX cell, as well as the ACEE cell, was also significant, since these cells demonstrated scalability achieving higher density capability compared to previous nonplanar cells of employing polysilicon to polysilicon erasing schemes.

7.2.1.3   The ETOX Cell

The other significant FG cell design developed by Intel in the late 1980s is known as the ETOX cell [10,11] which also proved to be highly scalable. The ETOX cell combined the tunnel oxide erasing features of FLOTOX with CHE writing into a single transistor flash memory cell design employing a tunnel oxide of 10 nm in thickness. The ETOX cell required only 12 V for programming, thereby reducing field across the tunnel oxide improving yield and reliability. The original ETOX cell was implemented in 1.5 µm lithography, featured a 36 µm² cell (16F × F), and provided a 256 Kb EEPROM chip with in-system reprogrammable. The ETOX cell was scaled to successive lithographic generations to provide 1 Mb memory chip [12] and 16 Mb memory chip [13], the latter using 3.3 V power supply. The ETOX array had the conventional NOR architecture, whereby the cells are connected to a bit line in parallel. The ETOX cell was organized for byte-by-byte writing using CHE injection mode and block erasing (in parallel) by tunnel mode via the positively biased common source line. Each memory cell could be addressed in parallel by accessing individual word line, W/L (control gate, CG), and corresponding bit line, B/L, connected to the drain of the memory cell. The ETOX cell used parallel array (random access) architecture and dual polysilicon technology as explained above. The follow-on to ETOX cells, using similar array architecture and dual polysilicon technology, are DINOR cell [14,15] and HiCR cell [16]. These cells also employed tunneling modes for both writing and erasing. The DINOR cell has been discussed in Section 1.3.

7.3   THE NAND FLASH CELL

Around the same time the ETOX cell was published, Masuoka et al. from Toshiba [17] described a NAND flash cell and array architecture to further reduce the cell size. In the NAND structure, a string of individual one-device cells (currently 32 or 64 cells) are connected in series between a bit line and a source line, thus eliminating contact areas for sources and drains associated with accessing individual memory cells. The first NAND chip was a 4 Mbit EEPROM [18], a 5 V only design and using 1.0 µm design rules. The NAND chip provided a unit cell area per bit of 12.9 μm^2 (12.9 F^2), a ~24% improvement in cell density compared to an ETOX cell with the same feature size (F). The NAND flash readily demonstrated its advantage in cost and scalability over the ETOX-follow-on cells due to its simple cell design and contact-sharing scheme. This advantage continued through the last three decades since the inception of the NAND scheme demonstrating unforeseen growth in volume and applications. The NAND flash memory volume in last decade surpassed the DRAM volume and drove the lithographic scaling in the last several generations of technology nodes. The NAND flash cell design and array architecture will be discussed in greater detail in this section.

7.3.1 Floating-Gate NAND Cell and Array Design

The original NAND cell [17] was conceived to be written by CHE injection mode and to be erased by UV irradiation or by enhanced Fowler–Nordheim tunneling of electrons from the floating gate to the substrate. The 4 Mb NAND chip [18,19] was implemented using conventional double level polysilicon CMOS technology with the addition of a third layer of polysilicon for the floating gate only within the memory array. The technology employed 1.0 μm design rules (F = 1 μm) and 8 bit NAND array strings. The chip organization was 512 KX8 (4Mb) consisting of 512 K of NAND structure cells each with 8 bit NAND strings. Each NAND structured cell consisted of a string of serially accessible 8 bits with two select gates at either end as shown in the SEM micrograph of Figure 7.4a and by the top view and equivalent circuit of the cell in Figure 7.4b [18,20]. A photomicrograph of the 4 Mb EEPROM NAND chip was published by Monodomy and coworkers in 1987 and is shown in Figure 7.5. Some of the key features are described below.

The EEPROM NAND was a 512 K $\times$ 8 memory product and the memory array was divided into four sub-arrays of 1 Mb each consisting of 1024 rows and 1024 columns. The array was programmed in page mode consisting of 1000 pages, each page being of 4 K bits of memory. The memory access time was 1.6 μs. Total programming (writing) time for the chip was 4 s. The chip size was 10.7 $\times$ 15.3 mm.

The memory chip was implemented with three levels of polysilicon and one level of aluminum metal. The CMOS peripheral devices were implemented with 2.0 and 2.5 μm gate lengths for

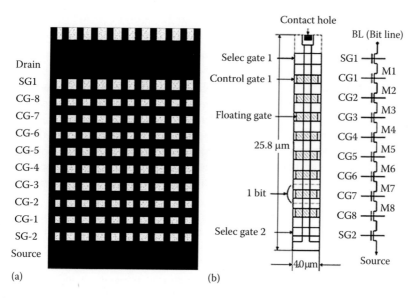

FIGURE 7.4 The NAND cell: (a) SEM micrograph, (b) Top view and equivalent circuit [18,19]. Design and organization of NAND chip. (Momodomi, 412–415, 1988, *IEDM Technology Digest*). Reproduced with permission of IEEE Press.

FIGURE 7.5   Micrograph of 4 Mb EEPROM NAND chip. (From Momodomi, M. et al., New device technologies for 5 V-only 4 Mb EEPROM with NAND structure cell, *IEDM Technology Digest*, p. 552 © 1988 IEEE.) Reproduced with permission of IEEE Press.

NMOS and PMOS FETs, respectively. The gate oxide thickness was 10 nm. The inter-polysilicon dielectric thickness was 25 nm. Each of the nonvolatile memory element for the 8-bit NAND string featured 1 μm gate with 10 nm thick tunnel oxide.

7.3.2   Erasing, Writing, and Reading Operations of 4 Mb NAND EEPROM

Erasing: Erasing was accomplished in blocks of NAND structure cells. Each block consists of 4000 NAND structure cells of 8 bit string (total of 32 Kb). The source of all structure cells was tied. All cells in a block are erased simultaneously. The erase operation consisted of applying potentials to all control gates (word lines) in the associated NAND strings, while the bit lines were floating and applying 20 V to the P-well associated with the array block. At such conditions, electrons stored in the floating gates of each cell tunnel back through the tunnel oxide to the substrate during the erase operation. Since the erasing was tunnel based, the entire block of 4000 NAND bits was erased simultaneously with the minimal energy requirement. The entire memory array was placed in a separate P-well isolated from the P-well containing peripheral circuits. The peripheral circuits also include the high voltage charge pump circuits for writing and erasing. During erasing, select gates at both ends (bit line-end and source line-end) were kept floating. The erase time for the entire block of 32,000 bits (4 K Bytes) was 10 ms (2.5 μs/bit). Therefore, erase operation was quite fast. The erase state had a Vth of −3 V (channel conducting). The cross sectional view of the NAND cells within a bit string has already been illustrated in Figure 2.4 [19].

Writing and programming: Putting the memory cell to the programming or writing state (Hi Vth or "1" state) involved changing the memory state of the cell to an enhancement mode by storing electrons into the floating gate of the cell. Such operation put the cell to a positive value of Vth = +2 V. The specific cell to be written within the string after a block erase was selected by selecting the appropriate bit line associated with the string and by selecting the appropriate word line (control gate) within the string. The selected B/L and the W/L was held at 0 V along with the array P-well. All unselected bit lines were held at a medium voltage level of 7–10 V, reverse biased, with reference to the P-well. At the same time all unselected word lines within the string were raised to the same medium voltage level of 7–10 V to avoid partial writing while the thresholds remaining at 3 V. The select gate at the source end (Figure 7.9) was also held at 0 V turning the FET device off. The select gate at the bit line-end was turned on by imposing 20 V. Since the selected W/L was subjected to +20 V, electrons from the P-well got injected through the tunnel oxide and got stored into the floating gate of the selected cell, thereby raising the Vth of the selected cell from −3 to +2 V. While other cells in the string remained at low state "0" of Vth = −3 V, the selected cell got programmed to state "1". The process then got repeated to any number of cells to be written. The chip also supported a successive programming algorithm where programming could be performed sequentially from the source-side cell to the bit line side cell along the string (from m-8 to m-1 [22]). For high speed programming, a page mode programming was adopted to program a page of 4 Kb. The page programming time was 4 ms.

Reading: For Read operation, bit line is precharged to 5 V and the select transistors adjacent to both the bit line and source line (select gates 1&2) are turned on by applying 5 V to those gates while the source is held at 0 V. Random access "Read" for the select gate is accomplished by applying 0 V to the gate of the selected cell while the gates of all unselected cells in the string were held at +5 V. If the memory of the selected cell is held at "0", that is, Vth = −3 V, the transistor is in depletion mode, and current flows through the entire string since all the other cells act as transfer gates. However, if the memory state of the selected cell is at "1," that is, Vth = +2 V, no current flows through the string since the transistor associated with the cell is in enhancement mode. Reading in NAND string is, therefore, analogous to conventional EEPROM or NAND type of mask ROM. Cell current was designed to be 20 μA. A dynamic sense amplifier circuit was used to detect the current. The typical access time is 1.6 μs. Due to the serial nature of the NAND memory configuration, the NAND array access time is significantly longer compared to a random access NOR EEPROM cell (discussed later on), where the access time could be typically an order of magnitude faster. Erase, write, and read operating conditions are also shown in Figure 7.6 [22].

NAND cell inhibit scheme: When the memory state of a selected cell is programmed or erased or read, a set of appropriate potentials were applied at the selected nodes for memory state change and affirmation as discussed above. All unselected cells in the array must be ensured to remain in their original memory state during such operation of the selected cell. The scheme to ensure the stability of the unselected cells during such operation associated with the selected cell is called the "inhibit" scheme. For the NAND arrays, the inhibit scheme employed was as follows:

Erase inhibit: This was applied to memory blocks deselected from erase operation. For those cells, a high voltage of the order of 20 V was applied to the associated word lines. Under such condition, there would be no effective field between the gate stack of the cells with respect to the P-well being held at 20 V. Therefore, the existing memory states of the deselected blocks would be preserved and any inadvertent erasure would be avoided for cells within those blocks.

Program inhibit: This was applied to all memory cells other than the one selected for programming. Such unselected cells were of two types: (a) those common to the selected word line, and (b) those common to the selected bit line. For the former case, unselected cells belong to different NAND strings having different bit lines, whereas for the latter case cells belong to the same NAND string and share the same bit line. In case

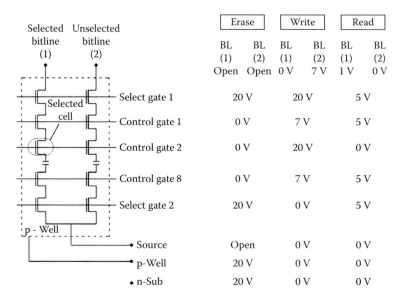

		Erase		Write		Read	
		BL (1)	BL (2)	BL (1)	BL (2)	BL (1)	BL (2)
		Open	Open	0 V	7 V	1 V	0 V
Select gate 1		20 V		20 V		5 V	
Control gate 1		0 V		7 V		5 V	
Control gate 2		0 V		20 V		0 V	
Control gate 8		0 V		7 V		5 V	
Select gate 2		20 V		0 V		5 V	
Source		Open		0 V		0 V	
p-Well		20 V		0 V		0 V	
n-Sub		20 V		0 V		0 V	

FIGURE 7.6 NAND EEPROM operating conditions for erase, write, and read. (From Momodomi, M. et al., New device technologies for 5 V-only 4 Mb EEPROM with NAND structure cell, *IEDM Technology Digest*, p. 412 © 1988 IEEE; Itoh, Y. et al., An experimental 4 Mb CMOS EEPROM with a NAND structure cell, *IEEE ISSCC*, pp. 134–135 © 1989 IEEE; Iwata, Y. and Masuoka, F., A high density NAND EEPROM with block page programming for microcontroller applications, *ISSCC*, Vol. 25, pp. 417–424 © 1990 IEEE.)

of type (a), the program inhibit was achieved by raising the associated bit lines and thereby the channel potentials to a medium voltage level of 7–10 V. This reduced the potential between the unselected channels and floating gates of the cells to a potential level low enough for significant electron injection to the floating gate of the unselected cells preventing partial programming. In case of type (b), for unselected cells belonging to the same bit line, the word lines associated with these cells were kept at a medium potential of 7–10 V so that the potentials between the channel and the control gates of the unselected cells were not sufficient to create any significant electron injection or partial change of the memory state. As stated earlier, cells in each NAND block were programmed sequentially starting from the source side to the bit-line side. This prevents unintentional charging of the unselected cells sharing the same control gate. Such sequence of programming prevented the presence of programmed cells on the drain side of any string, thereby maintaining the required channel potential of the unselected bit lines protecting the stability of the unselected cells along the selected word line from inadvertent memory state change.

7.3.3 Progress in Floating-Gate NAND Cell Design

Historic progress has been made in NAND cell density since its inception nearly two decades back.

NAND flash cells evolved into nearly 5FXF cell for 1 bit per cell (single level cell [SLC] or single level store [SLS]) and since 2004 into 2FXF to 3FXF per bit of memory cell density for 2 or 3 bit per cell (multilevel cell [MLC] or multilevel store [MLS]). Such progress enabled flash memory product availability of 32 Gb per chip in 2008 and recently 64 Gb per chip memory products (Samsung 2008, Micron 2012). The progress in density had been an astounding 16,000x enhancement in two decades. During this period, the feature size scaling progressed nearly 50x linearly from 1000 to 20 nm and 2500x area-wise. The additional factors besides the feature size responsible for density enhancements were: (1) NAND bit string was extended from 8 bit to 64 bit reducing the contact-overhead per bit; (2) reduced cell isolation spacing through advancement in isolation technology; and (3) introduction of one-device planar cell stack design. A cross section of a 32 bit NAND string cell is shown in Figure 7.7 [23] implemented at the 50 nm technology node by Micron Technology Inc.; a cross section of the cell stack across the word line of the center of the string of the same cell design is shown in Figure 7.8 [23]; and the 4 Gb SLC chip developed by Micron in 2005 is shown in Figure 7.9 [23].

Table 7.1 lists key progression of NAND products during the last two decades.

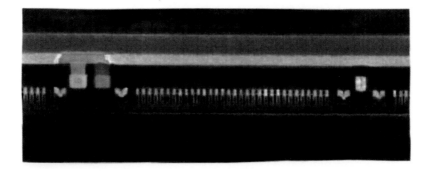

FIGURE 7.7   SEM micrograph of 32 bit NAND string in 50 nm technology. (Courtesy of Micron Technology, Boise, ID, 2005.)

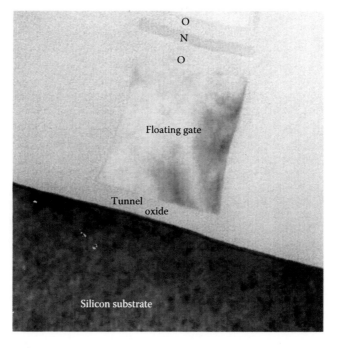

FIGURE 7.8   SEM micrograph of the center of a single bit of the string shown above showing the gate stack structure. (Courtesy of Micron Technology, Boise, ID, 2005.)

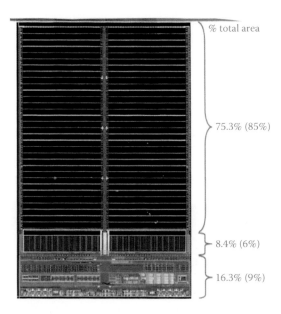

% total area

75.3% (85%)

8.4% (6%)

16.3% (9%)

FIGURE 7.9   4 Gb SLC NAND chip product from Micron Technology in 50 nm lithography where array occupies nearly 75% of the chip. (Courtesy of Micron Technology, Boise, ID, 2005.)

TABLE 7.1
NAND Product Progression

Levels	Year	Bits/Chip	Feature Size	Bits/String
SLC	1990	4 Mb	$>10f^2$	8
	1995	16 Mb	$>10f^2$	NA
	2000/2001	1 Gb	$4.65f^2$	32
	2004/2005	2 Gb/4 Gb	$~5.5f^2$	32
	2008	16 Gb	TBD	32
	2010	64 Gb	TBD	32
MLC	2004	4 Gb	90 nm	
	2007	16 Gb	50 nm	
	2010	128 Gb	30 nm	
	2012	512 Gb	20 nm	

7.4   FLOATING-GATE NOR CELLS AND ARRAYS

The density, scalability, and cost advantages of NAND flash proved to be ideal for data storage and applications where data retrieval speed was not critical such as handheld applications, for example, digital camera and cell phones. However, due to the serial nature of the NAND architecture, accessing the memory had been slower when compared to the parallel random access architecture of the FLOTOX and ETOX type of memories. Therefore, for code storage and embedded system applications, requiring memory access, 10x to 100x faster, the ETOX type of cells were more applicable. The ETOX type of FG devices evolved into a single transistor random access NOR flash memory cell architecture whereby each bit of memory can be addressed and altered independently and in parallel. A follow-on to the ETOX device developed by Mitsubishi and published in 1992 [15,16] was known as DINOR and has been briefly discussed in Chapter 3. The DINOR and other similar NOR devices with the associated contactless virtual ground array (VGA) had become a mainstay for code and embedded applications for the last two decades. The NOR flash memory cells were precursors of the current NROM cells described below and also in further detail in Part II of this book.

7.4.1   The DINOR Memory Cell and the VGA Architecture for NOR Arrays

The DINOR memory cell has been briefly introduced in Chapter 2. Besides being the first NOR-type memory cell employing Fowler–Nordheim tunneling for both writing and erasing, it introduced several technology and architectural features which influenced nonvolatile memory and device technology profoundly. Key elements introduced include, among others, (a) the VGA array architecture in 1992 [14], and (b) asymmetric offset source/drain structure combined with the VGA array architecture and page mode programming enabling lower power consumption and enhanced performance [15]. These elements will be briefly discussed here. An excellent review of DINOR flash memory has been published recently [16].

Figure 7.10 shows a typical circuit representation for a VGA array architecture containing a 3 × 3 memory array. The figure represents a cross-point array configuration using continuous buried n + diffusion lines to form the bit lines and common ground lines (k + 2, k + 1, k, k − 1) and associated control gate word lines (m + 1, m, m − 1). Furthermore, the drain contact for each memory cell is eliminated by providing buried diffusion lines. Metal stitching is employed after multiple word lines or bit lines to reduce series resistances. The technique reduces the cell sizes and relics on the use of asymmetrical floating-gate transistors or proper source and drain for decoding.

When the source and the drain functionality was exchanged, the DINOR array layout density was further enhanced with the VGA architecture with the introduction of contactless array scheme and offset source/drain structure. F–N tunneling for both writing and erasing could be achieved by only selecting the drain overlapped side of the selected cell [15]. This improves device characteristics (Power/Performance). However, the asymmetrical offset device, while significantly reduced power performance compared to the conventional NOR, increased the cell size of the DINOR memory cell. This has been discussed in detail by Ohi and coworkers [15,16]. The DINOR concept of memory array and architecture has been subsequently employed for charge trap NROM devices and arrays.

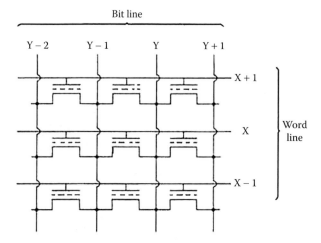

FIGURE 7.10   The virtual ground array (VGA) architecture for FG NOR cells. The architecture provides denser random access memory array by eliminating the common ground line within the array as well as the drain contact for each memory cell. (From Onoda, H. et al., A novel cell structure suitable for a 3V operation, sector erase flash memory, *IEEE IEDM*, p. 599, 1992.)

7.5   CHARGE TRAP NVM CELLS AND ARRAYS

Emergence of NROM devices and arrays:

Around the same time in the early 1990s, simultaneous developments took place in SONOS devices and arrays similar to the offset FG device and arrays. These CT devices were called nitride NOR or NROM devices. These devices had three advantages over the FG DINOR devices. These were:

1. The memory array required only one level of polysilicon simplifying processing and improving yield.
2. The memory array required a smaller number of contacts resulting in higher cell density and improving cost.
3. The cell allowed further scaling.

In addition to above, the SONOS devices demonstrated enhanced endurance and performance.

As a result, code and embedded applications favor the CT-based NROM devices over the FG devices. Discussion on NROM cells and arrays follows.

7.5.1   NROM Cells and Arrays

Conventional NROM cells consist of a split channel FET transistor with a fixed Vth address gate and an overlapping memory gate with drain diffusion connecting the bit line on the one side of the transistor channel and the common source diffusion on the other side of the channel. A schematic cross section with two cells sharing the common source is shown in Figure 7.11. The gate insulator for the address gate is a thicker thermally grown SiO_2. The gate insulator for the memory gate is typically a SONOS stack discussed in earlier sections. Memory gates (W1 and W2) and address gates overlap as shown. The two gates are implemented in polysilicon 2 and polysilicon 1, respectively, and isolated from each other by an oxide layer. The memory operations are explained below:

To Write: Apply +9 V for 100 μs across word line, substrate, and B/L; electrons tunnel into nitride and trapped → Vth gets changed from negative value, for example −2 V to a positive value, for example, +1 V.

To Erase: Apply −9 V for 10 ms across W1, substrate, and B/L; holes tunnel into nitride and trapped -> Vth for cell changed from +1 V to −2 V.

To Read Cell: Address gate turned on by applying Vcc, memory gate is held to 0 V, B/L sense current from the cell.

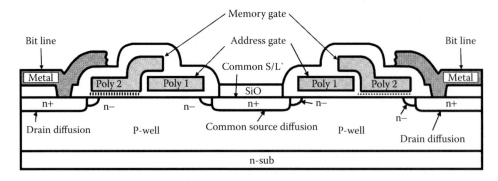

FIGURE 7.11 NROM cross-section: Two split-channel memory cells sharing the common source, the address gates are of fixed Vth with oxide insulator (SOS) while the memory gate in series of SONOS structure splitting the channel between common source and drain (Bit-line). (From Taguchi, M., NOR flash memory technology, 2006 IEDM Short Course; Course Organizer: Rich Liu, IEDM, and IEEE, 2006.)

7.5.2 General Operations of NROM Array

Programming: All cells in the same word line are first erased and then the selected cells are written by electron injection from the substrate to the selected cell by selecting the memory gate and apply +5 V to the gate and simultaneously applying −4 V to the selected drain and the P-well. During this time, all unselected gates were kept at 4 V to keep their Vth unchanged Vth = −2 V ("0") state. Program or write inhibit (so that the unselected cells in the same word line do not get programmed) is applied for unselected cells by applying +5 V to unselected bit.lines.

Erasing and inhibit: While P-well is held at +5 V, the select memory gates/word lines were subjected to −4 V; thus all cells in the same word line received −9 V, and erased simultaneously due to hole injection from the substrate. The unselected word lines were held at +5 V. Therefore, memory states for the unselected word line cells remain unchanged.

Reading: All gates except the select memory gate are held at 0 V, while the select address gate was raised to +5 V. Source is held at 0 V, while drain is limited to <1.5 V. For the unselected cells, reduced drain potential keeps the memory state stability during reading of the selected cell. Electrical characteristics of the SONOS devices were discussed earlier.

7.5.3 Advancement in NROM

Recent NROM device advancement was reviewed by M. Taguchi of Spansion Inc. [25]. NROM cell was implemented in a single transistor contactless implementation in the VGA architecture as shown in Figure 7.12a–c illustrating the cross section micrograph, the layout implementation, and the array circuit, respectively. Such implementation resulted in cell density in the range of 4–8 F^2 compared to the FG NOR implementation.

NROM memory density was advanced by MLC implementation (2 bit per cell design) utilizing localized trapping in nitride over the source and drain regions as discussed in the earlier section on stack design. Reading the 2 bit per cell design involved VGA array implementation to transpose source and drain functionality with localized trapping and sensing the corresponding Vth modulation. This is schematically illustrated in Figure 7.13. Charges could be injected and locally trapped above either end of the diffusion regions [data (1,0) and data (0,1)] as shown in the figure. While the spreading depletion region shields the "drain side" injected charge, the "source-side" charge is reflected in defining the Vth of the memory state.

The challenge associated with accurately sensing memory states for 2 bit per cell design has been addressed.

Reading margin in 2 bit per cell NROM array design was improved by implementing reference cell scheme as shown in Figure 7.14, wherein reference cells were added to word lines to

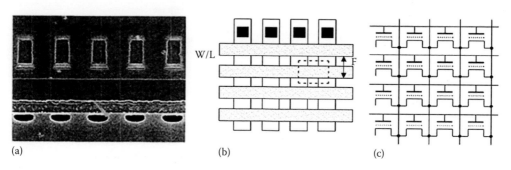

(a) (b) (c)

FIGURE 7.12 Advanced 1 Gb NROM design in VGA: (a) SEM micrograph, (b) Layout, and (c) Circuit representation. (From Taguchi, M., NOR flash memory technology, 2006 IEDM Short Course; Course Organizer: Rich Liu, IEDM, and IEEE © 2006 IEEE.)

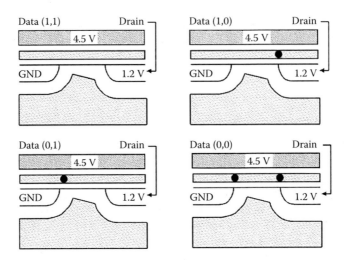

FIGURE 7.13 NROM CELL 2 bit/cell operation and reading by means of localized charge trapping in nitride over the diffusion regions and transposing Source and Drain. (From Taguchi, M., NOR flash memory technology, 2006 IEDM Short Course; Course Organizer: Rich Liu, IEDM, and IEEE © 2006 IEEE.)

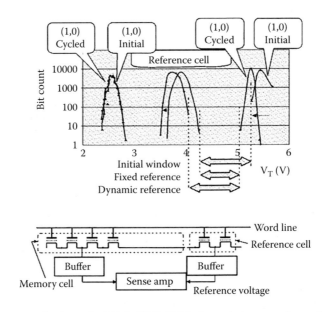

FIGURE 7.14 Reference cell scheme for 2 bit/cell NROM design to improve read margin. (From Taguchi, M., NOR flash memory technology, 2006 IEDM Short Course; Course Organizer: Rich Liu, IEDM, and IEEE © 2006 IEEE.)

experience the same stress as memory cells. During endurance stress cycling the Vth shifts of the memory cells are tracked by the reference cells to provide the dynamic reference of the average of cells with (1,0) and (0,1) memory states as shown, improving reading accuracy for MLC implementation.

Other implementation of NROM array design included improved algorithm of "erase sequencing" [25] to achieve tight Vth distribution.

7.5.4 NROM Scaling and Product Progression

NROM had been successfully scaled from a standalone 1 Mb product at 1200 nm node in 1991 down to 1 Gb standalone product at 90 nm node in 2006 [25]. During the same period, NAND products were scaled from approximately 4 Mb–4 Gb/8 Gb (Table 7.2). It has been noted, however, unlike NAND products, NROM technology had been more compatible with bulk CMOS and NROM devices, and products had been effective in embedded environment requiring and fulfilling a wider range of application conditions and form factors [27]. In recent years, the major application growth of embedded NROM products had been associated with automotive micro-controllers and smartcards exemplifying widely divergent and demanding application conditions and requirements. While cell size and cost per bit are critical driving factors for NAND products and applications, other factors such as higher endurance, faster reading and memory accessibility, broader temperature application range, and lower power consumption (for smartcard) would be more critical for NROM products. At the same time, it is generally recognized that future NVM trend would be driven by larger memory requirements and higher system-level performance.

An NROM 1 Gb product chip implemented in 90 nm lithography has been described by M. Taguchi [25]. The chip featured an initial access speed of 110 ns with page mode access of 25/30 ns. NROM product scalability has been schematically illustrated in Figure 7.15.

Table 7.2 selectively highlights the progress made in NOR/NROM product offerings. Some of these listed in the table have also been briefly discussed and illustrated above.

TABLE 7.2
NOR and NROM Product Progression

Publication Year	1991	2005	2006	2011
Reference	[24]	[25]	[26]	[27]
Technology node (nm)	1200	65	90	30–65
Product	1 Mb	128 Mb	1 Gb	Multiple
Cell size (μm^2)	18.4 ($>13F \times F$)	0.042 ($\sim10F \times F$)	0.042 ($<10F \times F$)	($10–30F \times F$)
Cell feature	Split gate	2 bit/cell	1 and 2 bit/cell	Varied

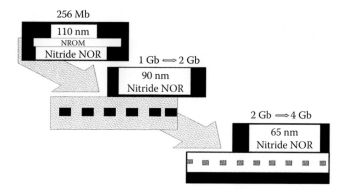

FIGURE 7.15 NROM product scalability road map. (From Taguchi, M., NOR flash memory technology, 2006 IEDM Short Course; Course Organizer: Rich Liu, IEDM, and IEEE © 2006 IEEE.)

7.5.5   NROM Product Potential and Challenges

It had been discussed in earlier sections that CT NROM devices have intrinsic advantages over the FGNAND devices in terms of device characteristics, voltage scalability, as well as CMOS technology compatibility. Furthermore, NROM one-device cell readily provides multilevel storage capability (2 bit per cell design) by means of localization of charge trapping within the trapping dielectric. If the multilevel storage capability could be reliably extended to 3 bit per cell and 4 bit per cell designs, as envisioned by M. Taguchi [25], NROM products could be cost competitive with NAND products and yet deliver enhanced NVM characteristics with applicability for both code and data. While some of the challenges associated with multilevel storage would be similar for both FG and CT devices, NROM would require advanced reading algorithms, sensing schemes, and memory state Vth controls aside from effective schemes for disturb mitigation. Multilevel storage extendibility issues are covered in greater detail in Part II of this book and stack structures for advanced NROM devices are explored.

7.6   DISTURBS AND MITIGATION IN NVM CELLS AND ARRAYS

Disturbs are related to stability of memory states of NVM cells due to operational stresses experienced by the cells when cells associated with the word lines and/or bit lines are read or written or erased. Disturb characteristics differ on cell design, operational conditions, and array architecture and should be evaluated, minimized, and contained within the margins of the memory states of the specific types of nonvolatile device types and arrays. Disturbs associated with the FG devices and associated arrays as well as with the CT devices and associated NROM arrays will be briefly discussed here and the mitigation approaches will be outlined.

7.6.1   Disturbs in FG Devices and Arrays

FG devices are typically implemented in the form of NOR arrays (Figures 7.14, 7.17, and 7.19) and NAND arrays (Figures 7.6 and 7.16) as discussed earlier. There are two types of disturbs associated with the FG memories: one during "Read" operation called "read disturb," and the other during "Writing" or "erasing" operation called "program disturb." These are explained in Section 7.6.1.1.

7.6.1.1   Read Disturb

During the read operation of the cell, the source line is usually held at ground potential, the bit line (drain line) is raised which is connected to a sense amp (Figure 7.16). A potential is applied between the control gate and the substrate via the selected word line to read the selected cell. Beside the

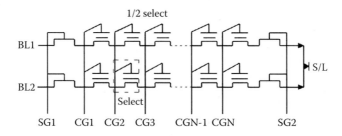

FIGURE 7.16 Schematic circuit representation of NAND arrays identifying the half-select cell: (a) Bit line 2 (BL2) and Control Gate 2 (CG2) are active: selected cell (circled) for reading while other cells on CG2 are exposed to stress: for example, the half-select cell: BL1/CG2 (squared). (b) Select cell (circled) and half-select cell (squared) in a Floating-gate NOR array. Select cell WL2/D2) being erased exposing half-select cell WL1/D2 to stress (drain disturb).

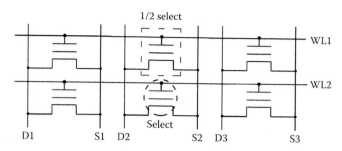

FIGURE 7.17 Select cell (circled) and half-select cell (squared) in a floating-gate NOR array: Select cell Wl2/D2 being erased exposing half-select cell Wl2/D2 to stress (example of drain disturb).

selected cell, all cells in the same word line are thus exposed to the same potential. These cells are called "the half-select cells." Half-select cells are identified for NOR array in Figure 7.17 and for NAND array in Figure 7.18, respectively, as shown below. If some of these half-select cells exist in the low Vth state or the "erased state," the electric field across such cells could be sufficiently high as to cause electron tunneling to the floating gates of those cells raising their thresholds and reducing their sense margins. This is called read disturb. In the worst case, a cell may end up being programmed inadvertently.

Read disturb is strongly sensitive to the leakage characteristic of the tunnel oxide or tunnel insulator due to the stress field across the half-select cells when the cells are at erased state. Therefore, the read disturb is dependent on (a) the integrity of the tunnel oxide or tunnel insulator, (b) the thickness of the tunnel oxide (or tunnel insulator), (c) the coupling coefficient of the stack determining the fraction of the voltage dropped across the tunnel oxide (or tunnel insulator) when the read voltage is applied to the active word line, (d) the number of read operation the cells are subjected to during the lifetime of the memory in the worst-case application conditions, (e) the threshold margin or the sense margin applied to the low threshold state for accurately sensing the erased "0" state (or "00" state), and (f) the leakage characteristics of the tunnel oxide (or tunnel insulator) after being subjected to the targeted end-of-life W/E cycling in the worst-case application conditions to take into consideration any change in the tunnel oxide (or tunnel insulator) integrity [item (a) above] due to factors such as hole trap generation or SILC.

A schematic of read disturb with associated potentials of selected and unselected cells along the word line and potentials of source line and bit line of cells under read condition is illustrated in Figure 7.18.

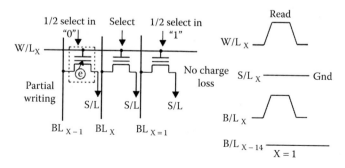

FIGURE 7.18 Example of read disturb in the half-select cell: W/Lx – B/Lx – 1 during reading of the select cell: W/Lx – B/Lx.

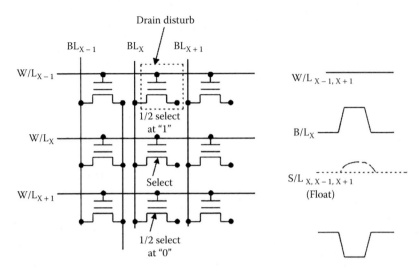

FIGURE 7.19 Illustration of drain disturb in an FG NOR flash array: The select device cell: W/LxB/Lx being erased; the half-select cell: W/Lx – 1 – B/Lx at memory state "1" (high Vth), experiencing drain disturb.

7.6.1.2 Program Disturb

Program disturb typically manifests as "drain disturb." Drain disturb takes place for unselected cells on the same bit line. When a selected cell is being programmed, for example, being erased, floating gate potentials of the unselected cells could be significantly lowered with reference to the bit line (control gate/floating gate turns negative, while bit line is held positive). This causes electron flow from the unselected cells to the drain, lowering the Vth ("1" or the high threshold state) of the unselected memory cells. This is called drain disturb for the affected half-select cell or cells. The situation of drain disturbs and associated potentials of selected and unselected cells are depicted in Figure 7.19.

7.6.2 Disturb Minimization or Mitigation

7.6.2.1 Read-Disturb Mitigation

Read disturb is minimized or mitigated in FG-NVM design by a combination of process and design optimization. The following approaches may be considered from the design point of view: (1) The device stack design is optimized by selecting the appropriate tunnel oxide thickness to meet the retention requirements combined with the cell design to achieve the desired coupling coefficient for optimal operational conditions (writing, erasing, and reading). (2) Read disturb steps are estimated by estimating: (i) the end-of-life read-disturb time, (ii) estimating the read stress across the tunnel dielectric for the erase state during the read operation, (iii) estimating the read-disturb current for case (ii) above, and (iv) finally ensuring that the read-disturb current per (iii) above is significantly lower than that required for the erased state sense margin, taking into account the end-of-life endurance target. If required, steps (1) and (2) are iterated until the desired design optimization is achieved. A case example for the read-disturb estimation will be provided later in this section. *Case Examples*:

Read-disturb: (A NAND flash example)
Array/Stack Design Parameters: Tunnel oxide thickness: 9 nm
EOT: 18 nm, Coupling coefficient: 0.5
Cell Area: 1E-10 cm² (50 nm feature size)

Endurance Target: 1E6 W/E cycles

Array: 32 W/L NAND String; 2 Page

Operation: 4 Read per W/E cycle, 10 us Read

"0" state distribution: 3 V (half-width: 1.5 V); guard band: 1 V

V-Read: +6 V

End-of-life read-time: 10E-6 s X 4 × 2 X 31 × 1E6 = ~3000 s

Stress potential: 6 V + 1.5 V (half-width) + 1.0 V (guard band) = 8.5 V

Read-disturb field: 8.5 × 0.5/9 = 4.72 E6 V/cm;

Leakage current through tunnel oxide: ~5E-13/cm^2

Leakage current required for threshold change of 1.0 V guard band:

No. of electron charge: Nq = 215; the current required:

$$J = Q/Time = Nq \ X \ q/t.Area = 3.44 \ E - 10 \ A/cm^2$$

Since the leakage current through tunnel oxide (5E-13/cm^2) is significantly lower than that required to offset the guard band (3.44 E-10/cm^2), the Vth shift due to read disturb is negligible and the design goal will be met.

7.6.2.2   Mitigation or Minimization of Drain Disturb

Minimization approach of drain disturb is similar to those of read disturb as explained earlier through a combination of process and design optimization. The approach to minimize drain disturb is similar to that of read disturb. However, it should be noted that drain disturb is associated with the instability of the high Vth state (memory state "1") of the unselected cells along the bit line during the erasing operation. Furthermore, the end-of-life erase duration is significantly lower for NVM devices compared to the corresponding read duration.

Drain-disturb steps are estimated by estimating: (i) the end-of-life drain-disturb time, (ii) estimating the stress potentials across the tunnel dielectric for the written state for the half-select cells during the erase operation, (iii) estimating the disturb current (loss of stored electrons from the floating gates of the affected half-select cell or cells) for case (ii) above, and (iv) finally ensuring that the drain disturb current per (iii) is significantly lower than that required for the written state ("1" state) sense margin, taken into account the end-of-life endurance target.

Case Example: Drain disturb: (A NOR flash example):

Stack design parameters: Same as above; Array Organization: No. of cells per B/L = 8; Max. Vth "1" = +5.5 V; Guard Band for memory state "1" = 1.5 V.

Operational condition: (a) 100 μs erase, 1 E6 erase, 7 half-select cells, B/L potential +8.5 V for 100 μs, S/L and Substrate at Ground.

Operational condition: (b) same as (a) except B/L potential floating resulting in peak field for B/L of +4.5 V for less than 1 μsec, while S/L and substrate ramped up simultaneously to the potential of +8.5 V.

For case (a): EOL drain-disturb time = 7x 1 E-6 × 1E6 = ~7 s; stress potential: +7 V;

$$Field = 7.8E6V/cm, \ and \ associated \ leakage \ current \sim 5E - 7A/cm^2;$$

For the guard band of 1.5 V of Vth for state "1", J = 7.3E-8 A/cm^2 which is lower than the associated leakage due to the stress. Therefore, drain disturb is excessive, and cannot be avoided.

For case (b): EOL drain-disturb time is reduced to 7E-2 s; stress potential reduced to +5 V; field reduced to 5.6E6 V/cm and associated leakage current to 5E-11/cm^2. Therefore, no significant drain disturb should be anticipated.

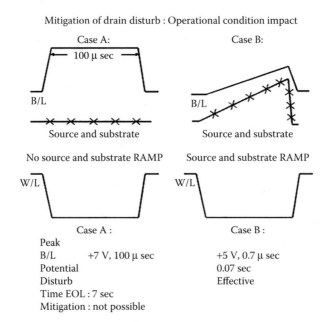

Mitigation of drain disturb : Operational condition impact

FIGURE 7.20   Node potential difference in erase operations for case examples (a) and (b).

The examples illustrate the significance of operational condition to address and mitigate drain disturb. Figure 7.20 illustrates the difference in the various node potentials for the two operational conditions of cases (a) and (b) above.

7.6.3   Disturbs in CT (NROM) Devices and Arrays

Since electrons and holes are trapped in nitride for CT/NROM devices, detrapping of electrons and holes is required for loss of electrons and/or holes to impact the stability of erased state (read disturb) and the written state (program disturb or drain disturb). Another important distinction of CT/NROM devices and arrays from the standpoint of disturb is that typically the stress potentials related to disturbs are significantly smaller (nearly half of those of FG devices) and consequently the impact related to disturb is less.

7.6.3.1   Read Disturb

The applied read voltage for the select word line is low, typically in the range of 0–2.5 V for most CT/NROM designs. Additionally, unselected word lines are usually held at ground. In general, for the half-select cells in the select word line during read, the low Vth state defined by hole trapping exhibits stability due to deeper trap depth, lower effective mass, and higher energy barrier of the valence band. Consequently, read disturb in CT/NROM devices and arrays is of little concern.

7.6.3.2   Program Disturb or Drain Disturb

Program disturb or drain disturb may exist in CT/NROM designs but usually contained by optimizing the stack design and operational conditions of inhibit schemes to minimize electron loss associated with the half-select cells in the select bit line by limiting the rise of bit line potential in the select bit line. Sophisticated programming algorithms are also employed to minimize the disturbs as well as to overcome adverse interference between adjacent memory cells.

REFERENCES

1. A. Bhattacharyya, M. L. Joshi, and C. T. Kroll, FET gate structure for nonvolatile N-channel read-mostly-memory device, *IBM Tech. Discl. Bull.*, 18(6), 1756, 1975.
2. A. Bhattacharyya, M. L. Joshi, D. B. Redding, and R. Silverman, High density single device memory with minimum PL, *IBM Tech. Discl. Bull.*, 18(6), 1751, 1975.
3. A. Bhattacharyya, C. T. Kroll, H. W. Mock, P. C. Velasquez, Properties and applications of SiON films, IBM Technical Report, TR.19.0319, October 20, 1970.
4. A. Bhattacharyya, T. V. Haroun, R. Kanyon, Small system organization for a nonvolatile mass memory, *IBM Tech. Discl. Bull.*, 18(3), 654, 1975.
5. R. Ramswamy, and H. C. Lin, Floating gate planar devices, Chapter 2, in *Non Volatile Semiconductor Memory Technology*, edited by W. D. Brown and J. E. Brewer, IEEE Press, 1977.
6. H. A. R.Wegener, and W. Owen, Floating gate non-planar devices, Chapter 3, in *Non Volatile Semiconductor Memory Technology*, edited by W. D. Brown and J. E. Brewer, IEEE Press, 1977.
7. M. Gill, and S. Lai, Floating gate flash memories, Chapter 4, in *Non Volatile Semiconductor Memory Technology*, edited by W. D. Brown and J. E. Brewer, IEEE Press, 1977.
8. W. S. Johnson, G. Perlegos, A. Reninger et al., 16Kb EEPROM, *IEEE, ISSCC*, 1980, pp. 152–153, 271.
9. S. D'Arigo, G. Imondi, G. Santin et al., A 5V only 256Kb CMOS flash EEPROM, *IEEE, ISSCC*, New York, 1989, pp. 132–133, 313.
10. K. Robinson, Endurance brightens the future of flash, *Electron Component News, "Technology Horizons"*, November 1988, copyright 1988, Chilton Co.
11. V. N. Kunett, A. Baker, M. Fandrich et al., An in-system reprogrammable 32KX8 CMOS flash memory, *IEEE J. Solid State Circ.*, 83, 1137–1162, 1988.
12. K. Seki, H. Kume, Y. Ohji et al., An 80-ns 1 Mb flash memory with on-chip erase/write verify controller, *IEEE J. solid State Circ.*, . 25(5), 1147–1152, 1990.
13. A. Baker, R. Alexis, S. Bell et al., A 3.3V 16 Mb flash memory with advanced write automation, *IEEE, ISSCC*, San Francisco CA, 1994, pp. 146–147.
14. H. Onoda, Y. Kunori, S. Kobayashi et al., A novel cell structure suitable for a 3V operation, sector erase Flash Memory, *IEEE IEDM*, San Francisco CA, 1992, pp. 599–602.
15. M. Ohi, A. Fukumoto, Y. Kunori et al., An asymmetrical offset source/drain structure for virtual ground array flash memory with DINOR operation, *VLSI Technology*, 1993, pp. 57–58.
16. M. Nakashima, and N. Ajika, DINOR flash memory technology, Chapter 7, in *Nonvolatile Memory Technologies with Empkasis on Flash*, edited by J. E. Brewer and M. Gill, pp. 313–336, IEEE Press, 2008.
17. Y. S. Hisamune, K. Kanamori, T. Kubota et al., A high capacitive coupling ratio (Hi CR) cell, for 3 V only 64 Mb and future flash memory, *IEEE, IEDM*, 1993, pp. 19–22.
18. F. Masuoka, M. Momodomi, Y. Iwata, and R. Shirota, New ultra high density EPROM and flash EEPROM cell with NAND structure cell, *IEDM Technology Digest*, 1987, pp. 552–555.
19. M. Momodomi, R. Kirisawa, R. Nakayama et al., New device technologies for 5 V-only 4 Mb EEPROM with NAND structure cell, *IEEE, IEDM Technology Digest*, 1988, pp. 412–415.
20. M. Momodomi, Y. Itoh, R. Shirota et al., An experimental 4 Mb CMOS EEPROM with a NAND-structure cell, *IEEE, IEDM Technology Digest*, 1988, pp. 412–415.
21. Y. Itoh, M. Momodomi, R. Shirota et al., An experimental 4 Mb CMOS EEPROM with a NAND structure cell, *IEEE ISSCC*, 1989, pp. 134–135.
22. Y. Iwata, and F. Masuoka, A high density NAND EEPROM with block page programming for microcontroller applications, *ISSCC*, 1990, Vol. 25, pp. 417–424.
23. K. Prall, 50 nm micron FG flash technology, August 2005, Courtesy Micron Technology (unpublished).
24. T. Nozaki, T. Tanaka, Y. Kijiya et al., A 1 Mb EEPROM with MONOS memory cell for semiconductor disk application, *IEEE J. Solid State Circ.*, 26(4), 497–501, 1991.
25. M. Taguchi, NOR flash memory technology, 2006 IEDM Short Course; Course Organizer: Rich Liu, IEDM, and IEEE.
26. G. Servalli, D. Brazzelli, E. Camerlenghi et al., A 65nm NOR flash technology with 0.042um² cell size for high performance multilevel application, *IEDM*, 2005, pp. 849–852.
27. R. Strenz, Embedded flash technologies and their applications: Status & outlook, *IEDM*, Washington DC, 2011, pp. 211–214.

8 NVM Process Technology and Integration Scheme

CHAPTER OUTLINE

This chapter briefly outlines the evolution of the NVM process technology from its inception to the turn of the century. The NVM technology is an extension of the silicon based FET microelectronics process and integration technology, and closely followed its evolution and transition. Consequently, the NVM process technology started as a metal-gate technology in the 1970s, changed into CMOS silicon gate in the late 1980s, and into CMOS metal gate during the past decade following the ITRS roadmap of CMOS feature size scaling. The chapter emphasizes the unique elements to fabricate the NVM devices, within the framework of the basic CMOS process technology, for logic and peripheral circuits.

8.1 SILICON-GATE CMOS PROCESS TECHNOLOGY HISTORY

The NVM technology is based on silicon-gate CMOS technology and can be considered as an extension of basic CMOS technology. Although the early CMOS technology was an evolution of integration of early metal-gate PFET and NFET technologies, polysilicon-gate CMOS technology evolved in the early 1980s and became an industry standard for both logic and memory processes for large-scale integration. IBM pioneered the deep sub-micron silicon-gate CMOS device design and process technology during the mid-1980s by introducing multiple technology and device design innovations [1–3]. The technology was applied for high-performance process or design [4]. Process and integration features like self-aligned silicon gate, variable-with gate sidewall, silicided gates and diffusion, and "drain-engineered" FETs to contain short-channel effects for deep sub-micron channel length PFET and NFET devices had been widely adopted for successive CMOS technology nodes by the industry. The scaled polysilicon-gate CMOS devices and related integration schemes addressed scaling issues such as short-channel effects, device series resistances, thermal budget management for shallow junction technology, and impurity profile control and reliability. Details of the basic CMOS integration scheme outlined above will not be covered in this section since considerable literature exists on the subject. It should be mentioned, however, that similar schemes were readily adopted into NVM product technology for peripheral devices and circuits, and continued even in recent times.

Basic platform CMOS technology has undergone a transition during the last decade for sub-100 nm technology nodes. During this period, polysilicon-gate technology is being replaced by metal-gate technology to control the PFET and NFET thresholds, gate SiO_2 has been replaced by higher K insulator films to scale EOT as well as control leakage, and "diffusion-early-gate-later" integration schemes have been established for further scaling FET devices. Additionally, planar FET devices are being replaced by 3D FINFET devices to improve short-channel immunity and provide enhanced device transconductance. Most NVM technology has yet to adopt such scheme, especially for NAND Flash products. Advanced integration scheme for NVM technology is covered in greater detail in Part II of this book. This section covers unique technology features associated with NVM device requirements and assumes scaled conventional silicon-gate CMOS technology base for the current technology and products.

8.2 NVM-UNIQUE TECHNOLOGY INTEGRATION FEATURES

8.2.1 High-Voltage Generation, Transmission, and Circuit Requirements

NVM technology uniquely requires high-voltage generation and transmission within a scaled low voltage CMOS base. High voltages are required within the memory array for programming (writing) and erasing the memory bits.

Consequently, starting with a p-silicon substrate, a deep n-well is formed followed by a p-well to allow high-voltage support within the n-well. For a triple-well-CMOS technology, besides the standard n- and p-well formations for PFET and NFET devices, respectively, an additional deep n-well is employed for negative thick oxide high-voltage circuitry. This is shown in the schematic cross-section of Figure 8.1 [5].

An alternate approach had been to implement NAND technology over n-silicon substrate. In such a case, multiple deep wells are formed for peripheral circuits and memory arrays when negative voltage generation was not required. The cross-sectional view of such a scheme is shown in Figure 8.2 [6], which is a reproduction of Figure 7.8 discussed in Chapter 7 while discussing NAND NVM cells and arrays.

High-voltage generation and transmission requires not only deep well formations and, therefore, deep trench isolation in silicon substrate, but also thicker oxide FET devices for the required circuits. These steps are integrated into the silicon substrate at the beginning of the process fabrication since these steps require higher *thermal budget.*

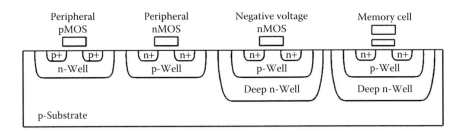

FIGURE 8.1   A schematic cross-section of the triple-well CMOS NVM structure. (From Jinbo, T. et al., A 5V only 16 Mb flash memory with sector erase mode, *IEEE, Solid-State Circuits Conference*, p. 1547 © 1992 IEEE.)

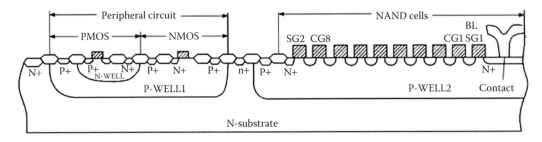

FIGURE 8.2   A cross-sectional view of a NAND EEPROM device. (From Momodomi, M. et al., *IEEE J. Solid-State Circ.*, 24 (5), p. 1238, 1989.)

8.2.2   Self-Aligned Shallow Trench Isolation: SASTI

Self-aligned shallow trench isolation (SASTI) scheme has been widely adopted. It has replaced LOCOS isolation for the current NVM technology. SASTI scheme significantly improved cell density for both NAND and NROM products. NAND cell density enhancement is schematically illustrated in Figure 8.3 by replacing LOCOS isolation by SASTI [7]. More than 40% improvement in cell density for NAND design was accounted employing SASTI. For multi-level cell scheme (covered in Part II), additional density enhancement has been achieved.

Process flow for SASTI in NAND consists of the following steps:

1. Shallow trench patterning followed by trench etching into silicon.
2. CVD oxide deposition filling in the trenches followed by chemical-mechanical polish (CMP) planarization.
3. Deposition of second level of polysilicon for floating-gate formation followed by floating-gate formation employing silicon nitride spacer technique [2].
4. Inter-poly ONO dielectric deposition followed by control gate deposition and patterning. These steps are illustrated in Figure 8.4a–d, respectively [7], as shown below.

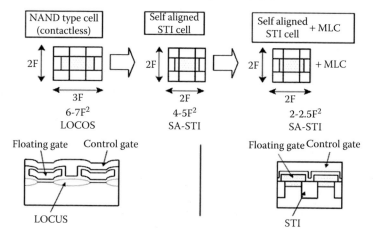

FIGURE 8.3   NAND cell density enhancement SASTI processing scheme. (From Aritome, S., Advanced flash memory technology and trends for file storage application, *IEDM*, p. 763 © 2000 IEEE.)

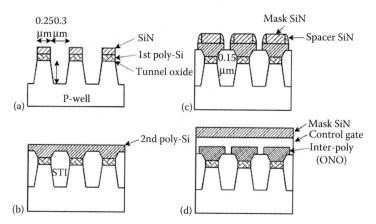

FIGURE 8.4   Process flow and integration scheme for self-aligned STI-cell [7]: (a) STI patterning and etching; (b) oxide deposition, CMP, and second polysilicon deposition; (c) floating-gate formation and SiN spacer; and (d) inter-poly deposition and control gate formation. (From Aritome, S., Advanced flash memory technology and trends for file storage application, *IEDM*, p. 763 © 2000 IEEE.)

8.2.3   NVM Process Flow and Integration Scheme

A typical NVM floating-gate technology may consist of three levels of polysilicon and two levels of metal processes. The first polysilicon gate is employed for peripheral CMOS FET devices, the second level of polysilicon is used for the floating gate, while the third level of polysilicon silicided with tungsten (WSix) defines control gates for the memory arrays. While tungsten metal lines are employed to reduce resistance of bit lines and common source line, a second level of metal which is typically of aluminum serves as a peripheral interconnection.

A conventional process flow for the floating-gate NVM technology is outlined below:

1. Pad structure formation (oxide/nitride/CVD oxide)
2. Deep trench isolation
3. High-voltage gate oxidation plus peripheral FET gate oxidation
4. First polysilicon deposition
5. STI formations
6. Well plus channel implantations
7. Memory tunnel oxide formation
8. Floating-gate polysilicon (second polysilicon) deposition
9. Floating-gate formation
10. Inter-poly ONO dielectric deposition
11. Control gate (third polysilicon) plus tungsten deposition
12. Gate patterning
13. Source/drain implantation and activation
14. WSix diffusion barrier formation
15. Local interconnect formation
16. Gap fill and planarization
17. BPSG passivation
18. Tungsten deposition for bit Line (damascene) (M1)
19. Inter-metal dielectric passivation
20. Aluminum metal deposition (M2)

The basic process flow as outlined above has been further enhanced to improve density and device characteristics beyond the 1 Gb NAND generation. Some such examples are (1) super shallow channel profile to provide superior sub-threshold characteristics of array and select devices, thereby improving program disturb characteristics of the array [8]; and (2) high-voltage recess scheme [9] to improve device breakdown and inverse narrow width effect. Some other critical enhancements are covered in Part II.

8.3   NROM PROCESS FLOW AND INTEGRATION SCHEME

CT-NROM SONOS type of devices and arrays with contact less VGA architecture have been discussed in Chapter 7, Section 7.4.1. It has been noted that such devices are based on localized charge trapping in the nitride layer of the ONO stack and the structure readily integrates in the base-line CMOS technology. Additionally, the NROM cell could store two separate bits in one cell, thereby doubling the density potential. Basically, the NROM process flow is similar to that of NAND technology and yet simpler to integrate into CMOS technology base.

Since NROM arrays could be laid out in cross-point architecture, potential for achieving ~3F2/bit density is feasible for 2bit/cell design with SSTI isolation and post-gate self-aligned S/D junction formation with S/D sharing two adjacent transistors. This approach has been demonstrated by J. Willer et al. [10]. They employed local interconnect scheme and a T-shape improved VGA array architecture to achieve competitive NROM cell density with those of NAND flash

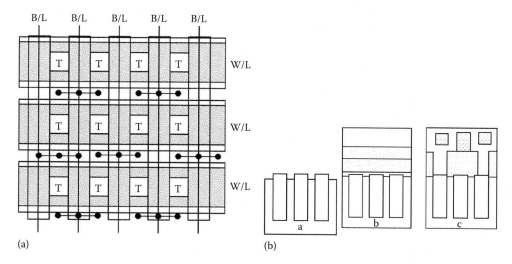

FIGURE 8.5 (a) Virtual ground array architecture: Layout structure of word-lines crossing active area and STI lines and (b) Steps for local interconnect bit line formation scheme: a,b, and c. (From Willer, J. et al., 110 nm NROM technology for code and data flash products, *2004 Symposium on VLSI Technology*, Digest of Technical Papers, p. 121 © 2004 IEEE.)

design. Key process attributes of STI, ONO/word line stack, and local interconnect/Bit line formation are shown in Figure 8.5a and b, respectively. NROM process flow steps are described below.

NROM PROCESS FLOW STEPS: [11]

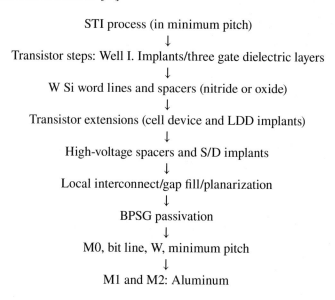

STI process (in minimum pitch)
↓
Transistor steps: Well I. Implants/three gate dielectric layers
↓
W Si word lines and spacers (nitride or oxide)
↓
Transistor extensions (cell device and LDD implants)
↓
High-voltage spacers and S/D implants
↓
Local interconnect/gap fill/planarization
↓
BPSG passivation
↓
M0, bit line, W, minimum pitch
↓
M1 and M2: Aluminum

8.4 NAND PROCESS STEPS

The process flow steps for the NAND device are also illustrated in Figure 8.6. Brief Process Steps outlined as follows:

(a) STI, Tunnel Oxide, FG Poly silicon Formation;
(b) Floating-Gate Formation & Inter-Poly-ONO Dielectric Deposition;

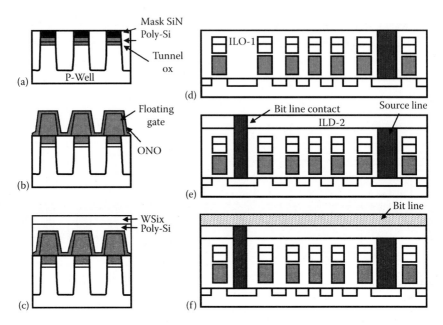

FIGURE 8.6 Schematic representation of key NAND process flow steps. (From Willer, J. et al., 110 nm NROM technology for code and data flash products, *2004 Symposium on VLSI Technology*, Digest of Technical Papers, p. 121 © 2004 IEEE.)

(c) Poly-slicon and Gate WSix Deposition;

(d) Gate, ILD-1, and Source-Line Contact Definition;

(e) ILD-2, and B/L Contact Definition; and finally,

(f) Bit Line formation and Definition

REFERENCES

1. A. Bhattacharyya, R. Mann, E. Nowak et al. A half-micron manufacturable high performance technology applicable for multiple power supply applications, *IEEE & JPS Proceedings, International Symposium on VLSI, Technology, System and Applications*, Taipei, China, 1989, p. 321.

2. A. Bhattacharyya, M. L. Kerbaugh, R. M. Quinn et al., Formation of variable width sidewall structures, U.S. Patent 4776922, October 11, 1988.

3. A. Bhattacharyya, CMOS 2S deep sub-micron device technology unique features, IBM Internal Docket: BU-1-88-031, 1988.

5. H. Schettler, J. Hajdu, K. Getzleff et al., A CMOS mainframe processor with 0.5um channel length, *IEEE, ISSCC*, Washington, DC, 1990, pp. 50–51.

6. T. Jinbo, H. Nakata, K. Hashimoto et al., A 5V only 16 Mb flash memory with sector erase mode, *IEEE, Solid-State Circuits Conference*, 1992, pp. 154–155.

7. M. Momodomi, Y. Itoh, R. Shirota et al., An experimental 4 Mb CMOS EEPROM with a NAND-structure cell, *IEEE, IEEE Journal of Solid-State Circuits*, 24 (5), 1238–1243, 1989.

8. J.-D. Choi, J.-H. Lee, W.-H. Lee et al., A 0.15 um NAND flash technology with 0.11um2 cell size for 1 Gbit flash memory, *IEEE, IEDM*, 2000, pp. 767–770.

9. S. Aritome, Advanced flash memory technology and trends for file storage application, *IEDM*, San Francisco CA, 2000, pp. 763–766.

10. F. Arai, N. Arai, S. Satoh et al., High density ($4.4F^2$) NAND flash technology using super-shallow channel profile (SSCP) engineering, *IEEE, IEDM*, San Francisco CA, 2000, pp. 775–778.

11. J. Willer, C. Ludwig, J. Deppe et al., 110 nm NROM technology for code and data flash products, IEEE & JPS, *2004 Symposium on VLSI Technology*, Digest of Technical Papers, Honolulu HI, 2004, pp. 76–77.

9 NVM Device Reliability

CHAPTER OUTLINE

NVM technology, device, and product reliability are subjects by themselves and are tied with silicon based FET technology and product reliability. The general topic of silicon based FET technology, device, and product reliability is beyond the scope of this book. Since the current silicon based NVM product, device, and technology integration schemes follow those of general CMOS technology for memory, logic, and platform technology products, most of the reliability issues related to FET scaling and device enhancements are common to NVM and will not be discussed in this chapter. This chapter is, therefore, limited in scope by design. Selected reliability topics with respect to NVM stack elements, NVM scalability and extendibility, NVM operational modes, and relevance to SUM device concepts and design are covered in this chapter.

At present, NVM products and associated devices are primarily of floating-gate (FG) NAND, flash and NOR flash, and charge-trapping (CT) SONOS NROM flash types. These products and devices had been successfully scaled over the past three decades along with the technology node scaling of silicon microelectronics. Unique reliability issues related to the products and devices are associated with: (a) process control and integration sensitivities of gate stack design elements; in particular, the dielectric films and the intrinsic characteristics of these films subjected to high-field stressing; (b) specific mode of operation of the memory cells and arrays (both selected and unselected ones) during writing, erasing, and reading; (c) scaling-related parasitic effects of adversely coupling neighboring cells thereby degrading cell characteristics during operational and standby conditions; and (d) understanding and controlling the mechanisms related to localized defect generation and anomalous device or cell leakage issues.

9.1 RELIABILITY ISSUES FOR GATE STACK DIELECTRIC ELEMENTS

Conventional NVM devices had employed oxide (SiO_2) films, interfacing the silicon substrate as the primary tunneling medium for electrons and holes to be injected from silicon, transported through the oxide and to be stored either in the floating silicon gate or in the discrete traps of nitride for the FG and CT devices, respectively. In case of FG devices, the tunneling oxide had been in the thickness range of 6.5–10 nm ("thicker oxide") while in the case of CT devices, the oxide thickness ranged between 1.5 and 4.0 nm ("thinner or ultrathin oxide"). Degradation mechanisms, defect generations, and process/post-processing factors affecting oxide intrinsic characteristics as well as external stress-induced extrinsic characteristics had been extensively investigated over the past three decades [1]. Processing techniques and post-processing treatments to achieve high-quality thermally grown SiO_2 films with extremely low intrinsic defect density covering the entire thickness range as above (1.5–10 nm) have been well established in the industry. Furthermore, post-tunnel oxide integration schemes to minimize mechanical stress in tunnel oxide have been addressed and optimized [2]. Processes for high-quality stoichiometric nitride (Si_3N_4) films in the thickness range of 10–500 nm have also been well established for applications in CT devices as trapping medium or as an element of ONO IP dielectric for FG charge blocking medium [3]. Nitride process sensitivity affecting shallow trap generation and field-induced degradation of Si_3N_4 is also well characterized. Oxide- and nitride-intrinsic characteristics and stress-induced degradation characteristics have been highlighted in Section 5.3 of this book. Role of SiO_2 and Si_3N_4 films in NVM device reliability is discussed below.

9.2 SiO_2 TUNNEL DIELECTRIC FILM

NVM reliability issues relating to the tunnel oxide had been investigated and modeled in terms of "as processed" or "intrinsic" defects (or traps) inherent in unstressed oxide films. Additionally, oxide exhibits external "stress-induced" or "extrinsic" defects (traps) generated in oxide films during application conditions. Both intrinsic and extrinsic defects cause enhanced leakage through the tunnel oxide adversely affecting NVM cell properties such as memory window and threshold stability, retention and endurance. In addition these defects enhance various types of disturbs during operational conditions. However, current-voltage characteristics, stress sensitivities, and reliability impacts are different for intrinsic and extrinsic defects in oxide. Current-voltage signatures of tunnel oxide with intrinsic traps and that with stress-induced traps are shown in Figure 4.1, curves 2 and 3, respectively in Chapter 4. This results in higher leakage through the tunnel oxide. Methodology to separate intrinsic and extrinsic defects was developed by C. T. Liu et al. [1], and was used to characterize tunnel oxide quality and associated reliability. The significance of stress-induced extrinsic defects and related traps and the intrinsic defects and associated traps in tunnel oxide affecting NVM reliability is elaborated here.

9.2.1   Stress-Induced Traps/Defects in Tunnel Oxide: Floating-Gate Cell Issues

9.2.1.1   Erratic Bit Phenomenon

When a "defect-free" or "trap-free" tunnel oxide is stressed above 2V, interface states (with associated traps) as well as bulk traps are generated in the oxide. At lower voltage stress (2–5V), the traps are generated at the anode/insulator interface. Whereas, at higher voltage level of > 6V), anode hole injection occurs creating bulk traps in the tunnel oxide film [4] as previously discussed in Chapter 4. A critical reliability issue known as "erratic bits" [5] was identified in the earlier decades. This was caused by hot hole injection at the anode during program-erase operations generating bulk traps in oxide. The erratic bits issue resulted in a long tail distributions of the "erase" memory state (low Vth state) for the affected memory cells. Examples of erased Vth distribution due to erratic bits and cell Vth characteristics of the erased state as a function of number of cycles are shown below in Figure 9.1a and b, respectively. This phenomenon was largely overcome by modifying the erase scheme from source erase to negative-gate erase, and subsequently to the current method of channel erase [5]. The evolution of the erase scheme also improved the memory endurance.

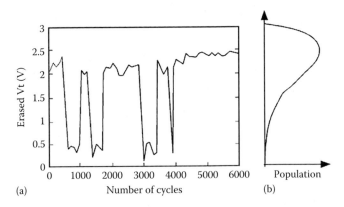

(a)   Number of cycles   (b)

FIGURE 9.1   Erase Vth distribution due to "erratic bit" phenomenon. (a) Erased Vt vs Distribution of memory cells; (b) Erased Vt vs no. of Write–Erase cycling. [5]. [Cappelletti] [2004] p. (489). Reproduced with permission of IEEE Press.

9.2.1.2 Transconductance Degradation

With write–erase cycling, flash cells exhibit shifts in Vth for both written and erased memory states due to charge trapping in the tunnel oxide (Figure 4.9). Increased write–erase cycling leads to: (i) interface states generation and consequently increases in intrinsic Vth of the device, and (ii) electron trapping in the tunnel oxide film generating negatively charged oxide defects or traps resulting in the lowering of the electric field during writing, thereby reducing the tunneling current. The combination of (i) and (ii) above increases the erase Vth with write–erase cycling limiting the read margin to accurately predict the erased memory state. This limitation sets the intrinsic reliability endurance limit of the FG memory cell. The interface generation due to extended write–erase cycling also degrades the transconductance of the cell and reduces the read current adversely affecting the cell performance. The observed degradation of the cells I-V characteristics as a function of write–erase cycling (not shown) validated transconductance degradation [5]. This aspect of reliability issue is addressed through the optimization of the cell/array design and setting an appropriate endurance reliability target which usually range 1E5–1E6 P/E cycles.

9.2.1.3 Data Retention Degradation: Stress-Induced Leakage Current, Anomalous Stress-Induced Leakage Current, or "Tail Cell"

Another important reliability issue is associated with the characteristic enhancement of leakage through tunnel oxide at low field due to the high-field P/E cycling. This is known as stress-induced leakage current (SILC) and had been associated with hot-hole-induced bulk acceptor traps in tunnel oxide, and is covered in Chapter 4.

The reliability impact of SILC results in enhanced Vth variations in memory states during standby due to enhanced data loss in the written state (high Vth memory state retention degradation) as well as enhanced "read disturb" in the erased state (low Vth memory state retention degradation during reading). SILC had been accounted for as a "single-bit fail" associated with trap creation and modeled in thicker oxides (6–10 nm thickness range). It was associated with a single trap generation in oxide and enhanced trap-assisted tunneling through the single trap. However, it had been observed that a certain distribution of cells exhibits significantly higher leakage than what had been related to SILC. The group exhibited strong endurance and thickness sensitivity and could be modeled and correlated with multiple trap-assisted tunneling model (Figure 4.1) of multiple trap-assisted tunneling. These are known as "tail cells". Vth distribution (written state) of tall cells as a function of storage time (retention) at room temperature after 10 P/E cycle and after 1E4 P/E cycle are shown in Figure 9.2a and b, respectively. Oxide thickness sensitivity of tail cells is shown in Figure 9.3a. The tail cells can be baked out and deactivated through a high-temperature

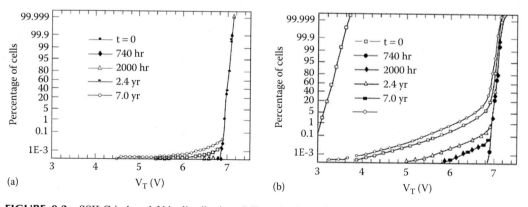

FIGURE 9.2 SSILC-induced Vth distribution: P/E cycle dependence at Room Temperature. (a) 10 P/E cycling and (b) after 10,000 P/E cycling. (From Nicollian, P. E. et al., Experimental evidence for voltage driven breakdown models in ultra-thin gate oxides, *IEEE, 38th Annual International Reliability Physics Symposium*, San Jose, CA, p. 7 © 2000 IEEE.)

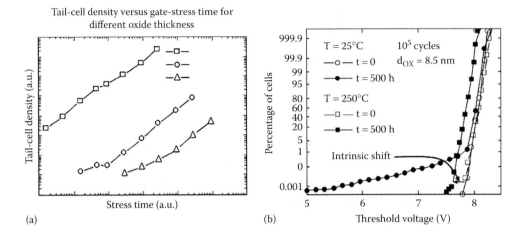

FIGURE 9.3 (a) Tail CELL density: Tox dependence and (b) Vth versus bake characteristics. (From Nicollian, P. E. et al., Experimental evidence for voltage driven breakdown models in ultra-thin gate oxides, *IEEE, 38th Annual International Reliability Physics Symposium*, San Jose, CA, p. 7 © 2000 IEEE.)

bake. This is shown in Figure 9.3b, where the Vth distribution returned to the normal level from that observed at room temperature following the 250°C baking. Additional studies showed that the tail cell characteristics are due to microscopic local defects in tunnel oxide, which could be deactivated at a temperature above 100°C of baking [4].

9.2.1.4 Fast Erase Cells

Another reliability issue which results in single-bit failures similar to SILC and yet have normal current-voltage characteristics is associated with fast erasing feature of FG cell. The enhanced erasing speed of such cells deviates from the normal Vth distribution of the erased cells. Yet, the characteristics of change in Vth (erasure) versus time of fast erasing cells are similar to the normal cells and significantly different from the leaky SILC cell types. The fast erasure was identified to be due to enhanced tunneling caused by interface states and positive charges located at or near the silicon-tunnel oxide interface causing enhanced back-tunneling during erasing. The common elements in reliability between fast erase cells and leaky SILC cells are the occurrence of single-bit failure caused by localized defects in the tunnel oxide although the location and energetics of the electronic levels associated with the defect generation could be different.

9.2.1.5 Drain Disturb

Drain disturb is also a reliability issue in FG devices when a significant drain bias is applied during programming or during erasing, creating channel hot electron or hot-hole injection for the half select cells. Drain disturb manifests in the form of enhanced retention loss for both the written (high Vth) state as well as for the erased (low Vth) state, and has to be accounted and designed out to meet the reliability objectives. Bharat Kumar et al. compared the drain disturb phenomenon for the FG cell with conventional SONOS and nanocrystal cells for the same channel lengths of L = 0.3 μm for all types of cells [10]. This is shown in Figure 9.12 later on in this chapter. Contrary to the CT cells, drain disturb strongly impacts the erased state of the FG cell due to channel hot electron injection produced by the large source to drain voltage. Drain disturb also adversely affects the high Vth half select cells. The reduction in high Vth is due to hole injection at the drain edge; holes originating from band-to-band tunneling inside the drain junctions acquire high energy and gets injected toward the floating gate due to the vertical field between the drain (positive potential) and the floating gate/control gate (negative potential). Drain disturb related reliability issues of CT cells will be discussed below.

9.2.2 CT Cell (SONOS and Nanocrystal) Issues

Oxide-related reliability issues in CT devices could be more complicated due to the thickness variation of tunnel oxide employed in the stack design and the mode of charge transport employed for writing and erasing. Based on the above criteria, the cell characteristics could be subdivided into three broad subgroups: (a) cells based on ultrathin tunnel oxide design with thickness ranging 1.5–2.5 nm whereby transport of electrons and holes occurs by direct tunneling through the oxide layer; (b) cells operating by Fowler-Nordheim or modified Fowler–Nordheim tunneling with thicker tunnel oxide ranging 3.0–4.5 nm; and (c) cells designed typically with thicker oxide with charge transport being accomplished by hot carrier effects (hot electrons and/or hot holes) or by a combination of hot carriers and Fowler–Nordheim tunneling. Since oxide-related reliability issues are strongly sensitive to not only the presence of intrinsic defects in oxide but also to the mechanism of energy release of the electrons and holes to the tunnel oxide due to high-field stressing during writing and erasing, reliability issues vary with the types of CT cell stack design and operating modes. A distinctive feature of type (c) cells is due to localized charge injection and trapping in nitride above the source/drain edges in contrast to cell types (a) and (b), where charges are stored relatively uniformly across the nitride over the entire channel region of the device. In such devices, charge injection, trapping, and charge redistribution, both vertically as well as laterally, play important role in reliability characteristics.

9.2.2.1 Ultrathin Tunnel Oxide Cells: (Type a): Transconductance Degradation

In general, most CT devices of SONOS types employ ultrathin oxide for tunneling. Oxide-related cell reliability and degradation of cell properties of such devices and arrays are driven by voltage drop across the ultrathin oxide and the blocking oxide during writing and erasing (Figures 9.4 and 9.5). In this type of stack design, voltage drop across the tunnel oxide is usually less than 2V, and therefore, stress-related damage in tunnel oxide is limited to interface states generation (Anode Hydrogen release: AHI, (see Chapter 4) and limited hole trap generation near silicon/oxide interface postulated to be aided by the presence of random intrinsic defects in oxide. Consequently, such devices demonstrate transconductance degradation without increase in sub-threshold leakage [5], extended endurance compared to the FG and other types of CT devices (exception: the NC devices, see later), absence of tail cells (anomalous SILC) characteristics of FG devices and arrays. A physical model for oxide degradation in these devices was proposed by J.-H. Yi et al. [6] based on high-energy hole-induced damage in: (a) tunnel oxide during erasing (transconductance degradation) and (b) blocking oxide during writing. A schematic energy band diagram to explain the different effects on ONO stack for

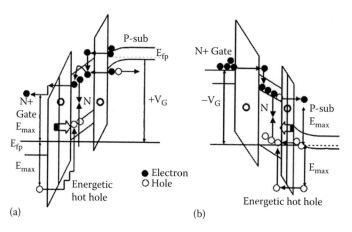

FIGURE 9.4 (a) and (b) Energy band diagrams energetic hole effects in oxide layers. (From Yi, J.-H. et al., Device degradation model for stacked-ONO gate structure with using SONOS and MOS transistor, *IEEE 43rd Annual International Reliability Physics Symposium*, San Jose, CA, p. 604 © 2005 IEEE.)

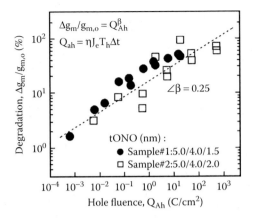

FIGURE 9.5 Transconductance degradations due to anode holes for SONOS transistors. (From Yi, J.-H. et al., Device degradation model for stacked-ONO gate structure with using SONOS and MOS transistor, *IEEE 43rd Annual International Reliability Physics Symposium*, San Jose, CA, p. 604 © 2005 IEEE.)

energetic holes during writing (positive gate bias) and during erasing (negative-gate bias) is shown in Figure 9.6a and b, respectively, indicating blocking oxide and tunnel oxide damage regions induced by high-energy holes. Measured device transconductance degradations for 1.5 and 2.0 nm of tunnel oxide SONOS are plotted in Figure 9.7 as a function of hole fluence. It should be noted that charge to breakdown is significantly higher in these devices (see Section 1.4) compared to FG devices.

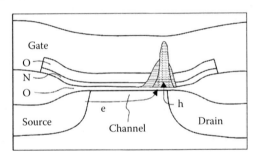

FIGURE 9.6 Injected charge distribution due to programming and erasing in an NROM cell: A schematic cross-sectional illustration. (From Yi, J.-H. et al., Device degradation model for stacked-ONO gate structure with using SONOS and MOS transistor, *IEEE 43rd Annual International Reliability Physics Symposium*, San Jose, CA, p. 604 © 2005 IEEE.)

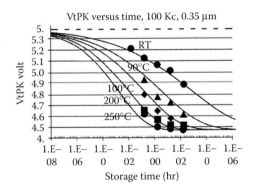

FIGURE 9.7 Retention loss "VtPK" versus storage time: A stretched exponential model fitting. (From Yi, J.-H. et al., Device degradation model for stacked-ONO gate structure with using SONOS and MOS transistor, *IEEE 43rd Annual International Reliability Physics Symposium*, San Jose, CA, p. 604 © 2005 IEEE.)

9.2.2.2   Thicker Tunnel Oxide Cells: (Type b): Hot-Hole Effects

Some SONOS types of cells were designed using thicker tunnel oxide (3–5 nm) to improve retention. The tunneling modes had been by Fowler-Nordheim or by modified Fowler-Nordheim for both writing and erasing. While the reliability issues discussed for type (a) cells and type (b) cells are similar, higher energy holes are involved in such devices, and anode-hole-injection-related reliability issues and lower charge to breakdown should be expected in such cells. Additionally, fast erase-cell possibility may exist due to higher hole trapping at the silicon/oxide interface.

9.2.2.3   Hot-Carrier-Induced CT Cells: (Type c)

Hot-carrier-induced localized effects in oxide and nitride: Several unique reliability issues have been observed for type C cells due to the localized charge injection and trapping of higher energy electrons and holes associated with programming and erasing of these CT cells. These are: (1) post-cycle retention loss; (2) sub-threshold slope (STS) degradation; (3) drain disturb; and (4) drain turn on. These phenomena are briefly discussed below. Channel hot-electron mode for writing and hot-hole mode for erasing have often been employed for embedded applications requiring lower voltages for programming and erasing. Such devices typically employ thicker tunnel oxide (5–7 nm) to meet retention requirements as well as for ease of integration with general CMOS FET devices. Charges are injected locally near the source and/or drain edges and trapped locally in the discrete nitride trapping centers to achieve either SLC or 2bit/cell MLC memory device. Hot-carrier-induced oxide degradations are concentrated over the diffusion regions and localized around nitride discrete traps above the diffusion regions. Such phenomena have been complicated from reliability standpoint, and have been active subjects of investigation consisting of both vertical and lateral effects in oxide and nitride and will be briefly reviewed below.

9.2.2.4   Data Retention Loss

Data retention degradation kinetics had been investigated by J.-H. Yi et al. [6] and M. Janai et al. [7] for NROM products over three different technology nodes of 350, 250, and 170 nm. Cells were subjected to room temperature P/E cycling of ONO stacks employing channel hot electron (CHE) writing and hot-hole-injection (HHI) erasing up to 1E5 P/E cycles followed by baking at temperatures ranging from room temperature up to 250°C, and postcycled mean Vth changes at the written state were analyzed for retention degradation. A schematic representation of the charge distribution due to writing (electron trapping) and erasing (hole trapping) in nitride of an NROM cell is shown in Figure 9.6. Retention loss plotted in terms of the postcycled mean Vth, called "VtPK" as a function of storage time is shown in Figure 9.7 ranging in time from 1E-2 hours to 1E6 hours. The retention loss characteristics (loss of stored electrons) saturates for all cells relative to the initial written state ("1" or high Vth state) around 1.1 ± 0.3V. The data loss kinetics fit stretched exponential model is shown by the solid lines drawn in Figure 9.7. The retention loss is given by:

$$\text{Delta Vth} = \text{Delta Vth-sat.} \left\{ 1 - \exp\left[-(t^{b}/\text{tow})\right] \right\} \text{ and}$$

$$b = T/T_0; \quad \text{tow} = t_0 \exp(U_t/kT),$$

where:

   Delta Vth-sat is the saturation value

   U_t is the activation energy

   T is the temperature, t is the storage time, tow is the time constant, t_0 and T_0 are empirical coefficient

The above saturation characteristics are independent of product, process, and feature-size variations used in the study. The characteristic energy Ut is the activation energy ranging from 0.40 to 1.17 eV and was dependent on design, process, and quality variations of ONO films for the cells investigated. It should be noted that such an activation energy is significantly higher than those obtained from retention degradation studies (0.17–0.26 eV) [9] of FG devices where the electron loss mechanisms from the floating gate has been associated via defects in the tunnel oxide alone. The observed higher activation process in NROM devices has been associated with mechanisms of lateral charge migration in nitride as discussed by M. Janai et al. [7] as opposed to the vertical charge loss through defects in tunnel oxide exhibited in FG NVMs. The suggested mechanisms for NROM NVMs of type (c) cells, proposed by these investigators, were based on the local electron and hole accumulation in nitride over the junction edges (Figure 9.7) [6]. During retention bake, the excess holes laterally migrate in the nitride layer by thermal excitation and are aided by local electrostatic field parallel to the channel to recombine with the trapped electrons injected during the writing cycle. Such mechanism of trap-to-trap lateral hopping was schematically shown in Figure 9.8 and lateral movement to deeper trap energy levels in nitride was postulated to account for varying and higher activation energy as measured [6].

L. Breuil et al. [8] also investigated post-cycle retention loss mechanisms. This was done with split-gate SONOS devices of varying channel lengths. Injection points of electrons and holes were physically separated by the control gate (fixed Vth) at the center of the channel. The loss characteristics of the conventional SONOS devices with overlapping injection points of electrons and holes were compared as discussed above [6]. The split gate device schematic employed for comparison is shown in Figure 9.9. Vth variations of both written and erased states after 1E4 P/E cycles were measured at room temperature as well as 85C for 1E4 secs for both types of devices. The observed post-cycle loss of electrons (written state retention) was interpreted in terms of vertical movement of charges to traps in oxide created by P/E cycling for both the devices. For the split-gate devices, loss of holes for the erased state was explained in terms of hole detrapping.

Charge localization and post-cycle charge movement in SONOS and Nanocrystal devices were also measured and modeled by C.M. Compagnoni et al. [9]. These investigators employed bake-accelerated retention tests on both conventional SONOS and nano-crystal memory cells. Vth shifts before and after bake with respect to drain potential was measured and simulated to identify the nature of electronic charge redistribution in nitride. This was then compared with those for the nano-crystal devices where charges are stored uniformly over the entire channel. The investigation showed that charge profiles after P/E cycling are sensitive to channel lengths of the device, and lateral distribution of electrons is noted for shorter channel SONOS devices. These investigators suggested a combination of both lateral and vertical movement of electrons for charge redistribution in nitride to explain the charge loss mechanisms in SONOS type c devices. Comparison of SONOS charge

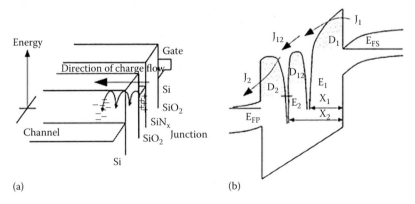

(a)   (b)

FIGURE 9.8 (a) Trapped charge transport; (b) Energy band diagram showing "two-trap" trap-assisted tunneling. (From Yi, J.-H. et al., Device degradation model for stacked-ONO gate structure with using SONOS and MOS transistor, *IEEE 43rd Annual International Reliability Physics Symposium*, San Jose, CA, p. 604 © 2005 IEEE.)

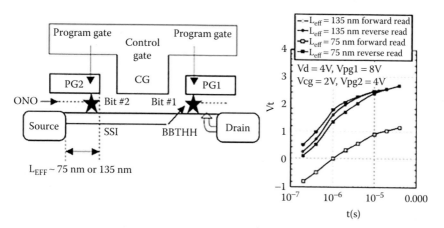

FIGURE 9.9 Split gate device cell schematic and programming characteristics during forward and reverse read. (From Breuil, L. et al., Comparative reliability investigation of different nitride based local charge trapping memory devices, *IEEE 43rd Annual International Reliability Physics Symposium*, San Jose, CA, p. 181 © 2005 IEEE.)

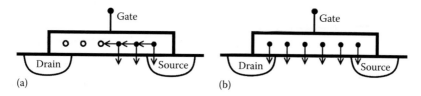

FIGURE 9.10 Retention charge-loss mechanisms compared between SONOS and nanocrystal devices. (From Compagnoni, C. M. et al., Reliability assessment of discrete-trap memories for NOR application, *IEEE 43rd Annual International Reliability Physics Symposium*, San Jose, CA © 2005 IEEE.)

movement with nanocrystal device, where charge trapping is uniform over the channel with vertical loss mechanism, is depicted in Figure 9.10.

9.2.2.5 Sub-Threshold Slope Degradation (STS)

An important reliability issue in hot carrier-injected NROM devices is the sub-threshold degradations due to write–erase cycling and associated degradation in cell trans-characteristics similar to the FG devices [8,9]. STS degradations for both erase-state and program state for conventional SNOS and split-gate SNOS are shown in Figure 9.11a and b for erase-state and 9.11c and d, respectively, for program state, respectively, for each type of devices. The STS degradation has been associated with the interface states generation at the silicon-tunnel oxide interface and stress-induced trap generation in tunnel oxide. STS degradations could be recovered by a 250°C bake for both types of devices. Such recovery through baking validates the mechanism of annealing out the surface states and de-trapping of holes from the tunnel oxide created by hot-hole injection during erasing.

Charge localization after CHE writing has been modeled by CM Compagnonoi et al. for conventional SONOS cells [9]. The degradation of STS as well as the low-current shift of the Id-Vg characteristics of the cell had been explained in terms of the narrow localization of trapped electrons in nitride near the drain edge, providing a narrow high Vth region in series with the lower Vth region of the channel. Their modeling of conduction band edge profile along the source/drain direction as well as the width effect due to a fixed negative charge localization (local electron trapping in nitride) near the drain junction edge affecting the Vth-Vd characteristics of the cell explained the Vth loss of the program state by a combination of lateral electron movement in nitride and vertical electron movement through the tunnel oxide.

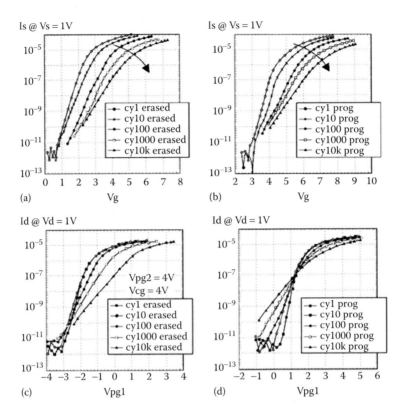

FIGURE 9.11 (a)/(b) Id-Vg characteristics of post-cycled SONOS: Erase state (left), program state (right). (From Breuil, L. et al., Comparative reliability investigation of different nitride based local charge trapping memory devices, *IEEE 43rd Annual International Reliability Physics Symposium*, San Jose, CA, p. 181 © 2005 IEEE.). (c)/(d): Id-Vg characteristics of post-cycled split-gate device: Erase state (left), program state (right). (From Compagnoni, C. M. et al., Reliability assessment of discrete-trap memories for NOR application, *IEEE 43rd Annual International Reliability Physics Symposium*, San Jose, CA, p. 240 © 2005 IEEE.)

9.2.2.6 Drain Disturb

The drain disturb characteristics of SONOS and nanocrystal CT NROM cells are shown along with the FG cells in Figure 9.12. For CT devices, the written-state charge retentions are adversely affected by drain disturb. The characteristics are similar to those of FG devices and the mechanisms behind the characteristics are also due to hot carrier injections as outlined for enhanced electron loss for the FG cells. Charge localization at the drain edge does not seem to impact significantly the enhanced loss mechanism related to drain disturb. However, P. Bharat Kumar et al. demonstrated that for SONOS cells, retention loss for the written state could be significantly enhanced by drain disturb, when stress conditions induce hot-hole injection and short channel device leakage level is significant [10].

HP Belgal et al. had proposed new reliability modeling for post-cycling charge retention of flash memories [11]. An extensive review of basic reliability and related testing and reliability modeling has been provided by Y. Hsia and V.C. Tyree [12].

9.3 RADIATION-INDUCED INSTABILITY AND RADIATION IMMUNITY

Radiation responses of FG and CT devices had been reviewed by G. Messenger [13] and should be consulted for details. For defense applications of NVMs, the radiation sensitivity of FG and SONOS types of CT cells had been measured. Typical characteristics of FG cells exposed to ionizing radiation dose is shown in Figure 9.13a. The data correlates well with the model showing strong Vth

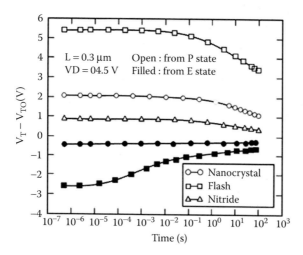

FIGURE 9.12 Drain disturb transient characteristics of FG and CT devices. [10]. [Bharat Kumar] [2005] p. (186). Reproduced with permission of IEEE Press.

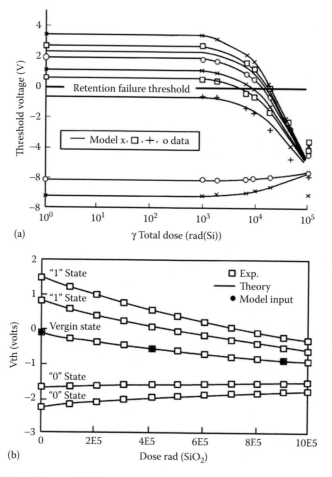

FIGURE 9.13 (a) Shift in Vth versus ionizing dose for FG devices. (b) Model and experimental correlation in Vth for an SNOS NVM. (From Messenger, G., Radiation tolerance, Chapter 7, *Non Volatile Semiconductor Memory Technology*, edited by W. D. Brown and J. E. Brewer, pp. 438–465, IEEE Press, New York © 1997 IEEE.)

instability of high Vth state beyond a total dose of 1E3 dose of gamma radiation in the cell due to electron emission from the floating gate as well as radiation-induced hole creation in oxide. Similar exposure to SONOS types of CT devices of Vth versus dose-dependence is shown in Figure 9.13b. The SONOS devices have shown stronger radiation immunity compared to the FG device stack. By improving the interface characteristics of oxide/nitride interfaces of the SONOS stack, radiation immunity in excess of 1M Rad had been achieved.

REFERENCES

1. C. T. Liu, A. Ghetti, Y. Ma et al., Intrinsic and stress-induced traps in the direct tunneling current of 2.3-3.8nm oxides and unified characterization methodologies of sub-3nm oxides, *IEDM*, 1997, pp. 85–88.
2. J. Om, E. Choi, S. Kim et al., The effect of mechanical stress from stripping nitride to the reliability of tunnel oxide and data retention characteristics of NAND flash memory, *IEEE 43rd Annual Reliability Physics Symposium*, San Jose, CA, 2005, pp. 257–259.
3. A. Bhattacharyya, C. T. Kroll, H. W. Mock, and P. C. Velasquez, IBM Internal Technical Reports, 1973 and 1974, *ECS Proceedings*, Los Vegas, CA, 1976.
4. P. E. Nicollian, W. R. Hunter, J. C. Hu et al., Experimental evidence for voltage driven breakdown models in ultra-thin gate oxides, *IEEE, 38th Annual International Reliability Physics Symposium*, San Jose, CA, 2000, pp. 7–15.
5. P. Cappelletti, R. Bez, A. Modelli, and A. Visconti., What we have learned on flash memory reliability in the last ten years, *IEDM*, 2004, pp. 489–492.
6. J.-H. Yi, J.-H. Ahn, H. Shin et al., Device degradation model for stacked-ONO gate structure with using SONOS and MOS transistor, *IEEE 43rd Annual International Reliability Physics Symposium*, San Jose, CA, 2005, pp. 604–605.
7. M. Janai, and B. Eitan, The kinetics of degradation of data retention of post-cycle NROM nonvolatile memory products, *IEEE 43rd Annual International Reliability Physics Symposium*, San Jose, CA, 2005, pp. 175–180.
8. L. Breuil, L. Haspeslagh, P. Blomme et al., Comparative reliability investigation of different nitride based local charge trapping memory devices, *IEEE 43 rd Annual International Reliability Physics Symposium*, San Jose, CA, 2005, pp. 181–185.
9. C. M. Compagnoni, D. Ielmini, A. S. Spinelli et al., Reliability assessment of discrete-trap memories for NOR application, *IEEE 43 rd Annual International Reliability Physics Symposium*, San Jose, CA, 2005, pp. 240–245.
10. P. Bharat Kumar, R. Sharma, P. R. Nair et al., Mechanism of drain disturb in SONOS flash EEPROMs, *IEEE 43 rd Annual International Reliability Physics Symposium*, San Jose, CA, 2005, pp. 186–190.
11. H. P. Belgal, N. Righos, J. Kalastirsky et al., A new reliability model for post-cycling charge retention of flash memories, *Proceedings of the IEEE international Reliability Physics Symposium*, 40, 2002, pp. 7–20.
12. Y. Hsia, and V. C. Tyree, Reliability and NVSM reliability, Chapter 6, *Non Volatile Semiconductor Memory Technology*, edited by W. D. Brown and J. E. Brewer, pp. 358–437, IEEE Press, New York, 1997.
13. G. Messenger, Radiation tolerance, Chapter 7, *Non Volatile Semiconductor Memory Technology*, edited by W. D. Brown and J. E. Brewer, pp. 438–465, IEEE Press, New York, 1997.

10 Conventional NVM Challenges

CHAPTER OUTLINE

Conventional silicon based NVM devices, technology, and products had an impressive history of evolution over the decades since their inception. These devices followed the evolution of silicon based FET technology and in particular the CMOS technology. As the feature size of silicon technology nodes scaled, the NAND FG NVMs as well as the NROM NVMs were effectively scaled. The success of these devices and related products broadened the application base for the nonvolatile memories. Such a scaling of conventional NVMs continued since inception, from 3 to 5 μm technology node down to 120–150 nm technology nodes until early twenty-first century [1]. In addition, significant advancement in overall technology integration, channel profiling, and STI scaling was achieved by this time frame [2,3]. This resulted in NAND-SLC cell implementations in the range of nearly 5 F^2 cells [4–6] and NOR/NROM cell implementations of nearly 10F×F cells for embedded applications [7,8]. During years 2000 and 2001, NAND products achieved 1 Gb density (Chapter 8: [4,5]) while 32 Mb embedded NOR arrays were reported [8]. In spite of such an outstanding progress, many scaling challenges remained to be addressed. These challenges will be briefly mentioned here and reviewed in detail in Part II of this book.

10.1 SCALING CHALLENGES OF CONVENTIONAL NVM

Primary scaling challenges associated with conventional planar NVMs could be attributed to three factors. These are short channel effects, few-electron effects, and limited vertical scalability with its related write–erase voltage scalability effects. A fourth factor which is dependent on all the above three factors is parasitic device scalability effects and associated cell-to-cell interferences.

10.1.1 Short-Channel Effects (SCEs)

SCE had been fundamental to CMOS FET scalability. SCE had been successfully contained for every technology node by improved CMOS FET device design and integration schemes. The general learning had also been incorporated into the stand-alone NVM devices and technology, and will not be elaborated here. Tri-gate FinFET nonplanar CMOS devices and technology evolved around the same time and provided enhanced FET devices with SCE relief. FinFET NVM devices with conventional ONO stack design was first published by I. H. Cho et al. in 2003 [9]. FinFET NVMs are discussed in Part II. Other nonplanar NVM devices providing SCE relief are also discussed in Part II.

10.1.2 Few-Electron Effects (FEE)

As the feature size is reduced, the number of electrons stored in FG or CT devices gets progressively reduced. This has been illustrated earlier in Table 6.1. The number of electronic charges required to provide memory state separation of 500 mV for an MLC floating-gate cell designed with 25 nm feature size for the state-of-the-art 64 Gb MLC NAND product would be around 50 electrons [10]. FEE fundamentally impacts all NVM parameters and limits scalability and reliability of planar FG and nanocrystal NVMs. Impact on FG memory reliability had been modeled by G. Molas et al. [10,11]. An integrated two transistors "flash-like" memory cell was proposed by Nakazato et al. [12,13] who named the cell as PLEDM cell (phase-state low electron-number drive random access memory cell) to mitigate the FEE effect. The cell was successfully implemented in a compatible CMOS technology by J. H. Yi et al. who named the memory cell as STTM cell (scalable two-transistor memory cell) [14]. Further discussion of FEE is covered in Part II of this book.

10.1.3   Vertical Scalability

Although feature-size scalability of conventional planar NVM devices and products had been impressive since inception, and continues till to date, the voltage scalability of the NVM devices and stack design had been limited due to the employment of oxide and/or oxynitride as tunnel dielectric and oxide or ONO as charge blocking dielectric layers. Consequently, the FG devices required >= to ± 15 V for programming and erasing, and CT devices approximately half that voltage while the power supply voltage level for the CMOS technology got scaled to ≤ 1.8 V. Since the turn of the century, the limitations of oxide and oxynitride were recognized for scaling the CMOS FET devices and high K MIS structures were successfully introduced for vertically scaled CMOS technologies. This provided the impetus for introducing high K dielectric layers into the stack design of NVM devices, thus enabling extended NVM devices with enhanced attributes with voltage scalability. Reduced programming and erasing voltages are required to enhance power/performance trade-offs, enhanced retention, and endurance as well as enhanced reliability through reduction in parasitic effects and cell-to-cell interferences. Part II will cover all aspects of NVM device extendibility for the last decade and beyond.

10.1.4   Limitations and Challenges

Conventional NVM devices and products are of two broad categories: (a) the floating-gate NOR and NAND flashes and (b) the charge-trapping NROM of the SONOS type. Both FG NOR and CT NROM memory products provide random access of the memory bits, while the FG-NAND products provide serial access of memory bits within the memory string consisting of either 32 or 64 or 128 and lately 256 serial bits of memory. NROM products are used primarily for code and BIOS applications requiring faster access and greater endurance (typically, greater than 1E6 W/E cycles). NOR flash products are often used as cache for large storage systems with hard disk drives (HDD) to improve data rate of the storage systems.

NAND flash products are increasingly employed as 2 bits/cell ("MLC or DLC") as well as 3 bits/cell ("TLC") to provide low-cost nonvolatile solid-state storage ("SSD") toward increasing range of applications. These applications include handheld devices such as cell phones, digital cameras, tablets, laptops, as well as toward a variety of consumer electronics and automotive parts and smart cards. NAND flash 1 bit/cell high-performance products are being used in enterprise systems as part of memory hierarchy along with DRAM to satisfy system requirements and back-ups.

During the last decade, NAND flash memories have displaced DRAMs as the major technology driver and also surpassed DRAM in memory revenue. However, the success of NAND flash products through the turn of the 20th century, to displace HDDs, has been limited. The SSDs have been successful in handheld applications requiring unique form factors and lower power. While the cost per bit of NAND flash products has come down dramatically, it has been argued that NAND flash has limited capacity even in TLC mode with advanced packaging when compared with HDD for applications in large storage environments from the standpoints of cost and capacity.

Additionally, TLC-NAND flash products face serious scalability and cost problems beyond implementation in 20 nm technology node. Solutions using the proposed 3D cell designs may add increasing cost in manufacturing. Additionally, limitations in endurance and severe ECC requirements for such memory products may significantly increase system cost overhead and system

TABLE 10.1

Current NVM Device and Product Attributes

Device	NROM	NOR	NAND-SLC	NAND-DLC	NAND-TLC
Cell size	$6F^2$	$<6F^2$	$<5F^2$	$4F^2$	$<4F^2$
VPP	<10 V	<15 V	<15 V	<15 V	<15 V
Power	Low	Medium	Medium	High	Highest
Access time	Fastest	Fast	Medium	Slow	Slow
Cycle time	Slow	Slow	Slow	Slow	Slow
Retention	10 yrs	10 yrs	10 yrs	1 yrs	<1 yrs
Endurance W/E cycles	$>1E6$	$>1E4$	$>1E4$	$>1E4$	$\sim1E4$
ECC	Low	Low	High	High	Highest
Reliability requirements	High	High	High	Medium	Low/medium

complexity. While it had been an impressive growth of flash technology and devices during the last decade, it remains to be seen in what form the silicon based NVM technology and product would evolve during the next decade. Qualitatively, the limitations of conventional NVM devices are summarized in Table 10.1.

Therefore, the limitations of the above conventional NVM devices/products could be summarized to be the following:

1. Limited voltage scaling and power reduction
2. Cycle speed significantly slower than DRAM
3. Very limited endurance and associated reliability requiring increasing ECC requirements
4. Cost reduction at the expense of memory retention for DLC and TLC-NAND flash NVMs

Item (1) requires voltage scalability or vertical scalability of NVM device stack design which in turn requires application of high K dielectric films in NVM stack design replacing conventional dielectric films such as oxide and ONO. Item (2) requires faster electron and hole transport at reduced field across the stack, which requires a combination of band engineering of the stack and high K dielectric films with proper trade-off between programming voltage levels (item 1) and program/erase speed and reliability. Item (3) is related to items (1) and (2) as well as selection of device types and programming/erasing methodology. Item (4) requires an innovative sensing scheme in addition to items (1), (2), and (3). Additionally, it also requires a scalable low-cost device viability, for example, 4 bits of storage per memory cell, and associated technology integration scheme to achieve a multilevel memory array scheme with high-manufacturing yield. In Part II, we will address how to overcome these limitations.

REFERENCES

1. S. Aritome, Advanced flash memory technology and trends for file storage application, *IEDM*, San Francisco CA, 2000, pp. 763–766.
2. A. Goda, W. Moriyama, H. Hazama et al., A novel surface-oxidized barrier-SiN cell technology to improve endurance and read-disturb characteristics for gigabit NAND flash memories, *IEDM*, San Francisco CA, 2000, pp. 771–774.
3. C.-N. B. Li, D. Farenc, R. Singh et al., A novel uniform-channel-program erase (UCPE) flash EEPROM using an isolated P-well structure, *IEDM*, San Francisco CA, 2000, pp. 779–782.
4. J.-D. Choi, J.-H. Lee, W.-H. Lee et al., A 0.15 um NAND flash technology with 0.11um2 cell size for 1 Gbit flash memory, *IEDM*, 2000, pp. 767–770.
5. J.-D. Choi, S.-S. Cho, Y.-S. Yim et al., Highly manufacturable 1Gb NAND flash using 0.12 μm process technology, *IEDM*, 2001, pp. 25–28.

6. F. Arai, N. Arai, S. Satoh et al., High density (4.4F²) NAND flash technology using super-shallow channel profile (SSCP) engineering, *IEDM*, San Francisco CA, 2000, pp. 775–778.

7. Y. H. Song, J. I. Han, J. W. Kim et al., A high-density and low-cost self-aligned shallow trench isolation NOR flash technology with O.14 um² Cell Size, *IEDM*, 2001, pp. 37–40.

8. S. N. Keeny, A 130nm generation high density Etox™ flash memory technology, *IEDM*, Washington DC, 2001, pp. 41–44.

9. I. I. Hwan Cho, T.-S. Park, S. Y. Choi et al., Body-tied double-gate SONOS flash (omega flash) memory device built on bulk silicon wafer, *DRC*, Salt Lake UT, 2003, pp. 133–134.

10. K. Prall, and K. Parat., 25nm 64 GB NAND technology and scaling challenges, *IEDM*, 2010, pp. 102–105.

11. G. Molas, D. Deleruyelle, B. De Salvo et al., Impact of few electron phenomena on floating-gate memory reliability, *IEDM*, San Francisco CA, 2004, pp. 877–880.

12. K. Nakazato, P. J. A. Piotrowicz, D. G. Hasko et al., PLED-planar localized Electron devices, *IEDM*, Washington DC, 1997, pp. 179–182.

13. J. H. Yi, W. S. Kim, S. Song et al., Scalable two-transistor memory (STTM), *IEDM*, New York, 2001, pp. 787–790.

Part II

Advanced NVM Devices and Technology

INTRODUCTION

Part II of this book reviews advancements in silicon based NVM devices and technology since the year 2000 to the present day emphasizing technology and device scalability and extendibility in device concepts and arrays, both planar and 3D devices, and integrating high K films and band engineering and advanced metal gate technology.

In Part I of this book, we have reviewed the impressive progress of the "conventional" NVM devices and products since inception through the year 2000. The progress had been continuing for the past ten-plus years resulting in expanded applications of such devices. Such progress has been achieved by a combination of successful feature-size scaling of devices and arrays and by partially addressing some of the major scaling issues briefly mentioned in Chapter 10 in Part I. During the last decade, while the basic FET device got scaled at an average of 10,000x per decade, the NAND memory cell had been scaled by an average of 8000x per decade. This impressive progress is continuing at the time of publication and is expected to continue for some time. In this part of this book, we will review what has thus far been achieved and what other possibilities may exist to further advance NVM devices and technology.

Major limitations of conventional Silicon gate CMOS technology and device scalability have been overcome during the last decade by the successful introduction of replacing oxide by ultra-thin high K dielectric films and replacing the silicon gate by the metal gate technology with the associated integration scheme of "gate later" approach. As a result, the basic CMOS technology is expected to continue to scale down from the current 20–30 nm node (at the time of publication of this book) to less than 10 nm node within this decade. NVM device scaling is more complex both in gate stack structure and in integration scheme. The basic principles of FET device scalability and associated technology evolution are being successively adopted into the NVM devices. However, NVM devices are already facing serious challenges in memory state stability, data retention, disturb, and endurance. Considerable progress has been made in addressing the above challenges and paving the path toward future extendibility. These items will be addressed and reviewed in this part of this book.

EXTENDIBILITY AND SCALING OF NVM DEVICE AND TECHNOLOGY

Extending NVM device properties and scaling of device and technology features are intimately related. The essential elements following those of CMOS FET devices include horizontal and vertical scaling of device features, maintaining and enhancing essential device properties, reducing energy requirements for device functions, containing parasitic effects, ensuring structural and functional reliability, and integrating with baseline scaled CMOS FET technology cost-effectively. These elements, in turn, entail the following:

1. Reduction in horizontal features of active and parasitic FG and CT device elements including isolation through effective lithographic scaling and integration for every succeeding technology nodes in concert to the baseline CMOS scaling. This will be discussed in "Enhanced Technology Integration" sections in Chapters 14 and 15. The section will also address compatibility considerations with the scaled baseline CMOS technology.
2. Replacement of conventional NVM gate stack elements by high K dielectric films and replacing silicon gate by metal gate technology compatible to CMOS baseline technology. This reduces active gate stack EOT providing vertical (and thereby voltage / power supply) scaling and energy reduction. This will be discussed in "Voltage Scalability" in Chapter 11 and "High K Dielectrics" in Chapter 12.
3. Combination of band engineering and high K dielectrics in stack design with "interface engineering" (to be explained in the Chapters 12 and 13) provides multiple options to extend NVM device properties such as memory window, retention, and endurance. Combining voltage scaling and band engineering along with appropriate high K implementation ensures functional reliability. These aspects will be covered in the Chapter 13 on "Band Engineering" and in follow-on chapters on planar and non-planar devices arrays and (Chapters 15 and 16) and additionally on the last chapter on "advanced device concepts" in Chapter 18.
4. Multilevel NVM memory device stack designs in the framework of scaled NVM technology as mentioned in (1), (2), and (3) will be covered in the chapter on "Planar Multilevel Storage" (Chapter 15).
5. Challenges and potentials of 3D NVM memory devices and arrays to provide density extendibility and product progression will be discussed in the chapter on "Non-Planar and 3D Devices and Arrays" (Chapter 16).
6. The chapter on Part II, "Future Limitations of current NVM Devices and emerging NVMs", Chapter 17, is a short chapter. This chapter highlights the limitations of current and future NROM and NAND devices and limitations on product functionality especially the need to incorporate multifunctionality to gain wider market share in the world of Internet and hand-held devices. The chapter briefly introduces the emerging "non-FET-based" NVM devices with potential of replacing all conventional silicon FET-based memories both volatile and nonvolatile ones.
7. The last chapter, Chapter 18, will provide the rationale for Part III of this book exploring the scope of future extendibility of advanced silicon FET-based device concepts toward unified memory solutions for digital electronics systems primarily based on direct tunneling (DT) transport mode. The section provides some insight on merging memory hierarchy and device concepts of multifunctionality which will lead to Part III of this book on multifunctional devices and "Silicon Based-Unified-Memories" labeled as "SUM."

DEVICE SCOPE AND DEVICE EXTENDIBILITY

It should be emphasized that the focus on NVM device extendibility in this part of this book is based primarily on charge storage within the gate stack either by means of introducing (i) FG or FP: an "uni-potential" medium such as silicon floating gate or injector SRN floating plate or by means of introducing, (ii) CT or NC: a discrete charge residing dielectric medium such as nitride and NR-SiON or nanocrystal embedded dielectric medium. The focus secondarily is also based on the various tunneling modes of charge transport and storage due to the energy efficiency of such modes of operation. Such focus will prevail (with limited exception) throughout this part of this book in the context of device extendibility. Within that context, the chapters will address among other elements, the following device extendibility attributes:

1. Lowering the programming voltage levels and power consumption
2. Reducing cell feature size, controlling device parameters for reduced cell size, containing parasitic effects and disturbs associated with reduced cell size
3. Enhancing device parametric such as Window, Programming speed, Retention, and Endurance
4. Incorporating multilevel storage capability without significant cost overhead

11 Voltage Scalability

CHAPTER OUTLINE

Silicon based NVM devices are FET based. FET devices are being progressively scaled both horizontally, known as "feature-size" scaling as lithography is scaled, as well as vertically, with reduction in gate stack thickness which is intimately tied with power-supply voltage scaling and associated scaling of power requirements. Similar to the FET devices, NVM devices could not only be scaled in feature size, but also in gate stack EOT to enhance operability at reduced power. The reduction in gate stack EOT for NVM device is referred to as "voltage scaling," with the implication of programming (writing and erasing) at lower voltage level and consequently at reduced power. For NVM devices, voltage scalability is, therefore, an integral element in NVM device extendibility. This chapter outlines the voltage scalability challenges for the silicon based NVM devices.

11.1 WHY VOLTAGE SCALING?

Voltage scalability (or vertical EOT scalability) is intimately tied with extendibility of NVM device properties along with feature-size scalability. The feature size scalability in silicon based NVM technology thus far has been primarily horizontal. NVM devices being FET based, have been successful in taking advantage of the feature-size scaling. Consequently, the NVM technology successfully delivered ~4F2/bit of NAND (SLC) products, and ~6F2/bit of NROM products, thereby, establishing NVM technology as primary driver of silicon technology. Yet, the voltage scalability for the NVM products has been limited. As a comparison, it may be noted that the basic CMOS FET devices and associated technology have been scaled during the last three decades at an average of approximately 10,000× per decade from 100 nm to 30 nm technology nodes, while the NAND-NVM and NROM-NVM during the last two decades demonstrated scalability of approximately 8,000× and 4,000×, respectively. In the similar time frame, while the basic CMOS technology was voltage scaled to >4×, the NVM devices were voltage scaled to <<2× only. There exist two primary reasons for limited voltage scaling or "vertical scaling" of the NVM devices. The first one is the high vertical field requirement for the charge transport of electrons and holes, especially for those types of NVM devices where transport mode is based on some form of Fowler–Nordheim tunneling. The second one is the continued employment of SiO_2 as the primary tunnel dielectric and ONO layer structures as the primary (inter-polysilicon-dielectric [IPD]) for charge blocking dielectric design. Such gate insulator stack design results in higher EOT stack design for the memory device. Gate stack design characterized by the thicker SiO_2 for tunneling and thicker ONO layer for blocking has been employed to reduce charge leakage in order to achieve the charge retention objective. For example, even in recent past, the conventional Floating-Gate NAND (FGNAND) stack has had EOT in the order of approximately 15 nm requiring ±17 V to ±19 V for programming and erasing [1,2]. Requirements of such high voltage generation and on-chip routing for related circuitry complicates the technology integration within the framework of current scaled-down low-voltage "scaled CMOS" technology. This results in complexity and cost for the NVM implementation of the silicon technology… The consequence of high voltage requirements not only adversely affects NVM device and application parameters such as retention, endurance and disturbs, but also power requirement and product reliability. This sets the limit of device and technology extendibility. It is worth noting that the operating voltage requirements for the scaled NROM-NVM devices and products of SONOS type are approximately half of those of FG-NAND-NVMs. However, even for the NROMs the voltage levels are still quite high, and adversely affects device parameters similar to those of FGNAND. Therefore, voltage scaling is considered an integral part of NVM device extendibility.

11.2 EARLY HISTORY OF VOLTAGE SCALING

The success of thermally oxidizing silicon to form defect-free thin oxide (SiO_2) film had been the cornerstone of the FET technology. This has been previously discussed in Chapter 3 in Part I of this book. However, the need to replace SiO_2 for the tunnel dielectric as well as for the blocking dielectric in CT MNOS/SNOS/MONOS/SONOS types of devices were realized in early years for voltage scalability and for enhancements in retention and endurance [3–6]. This has been discussed in Chapter 4, Part I by incorporating multilayered SiON (oxynitride) layers replacing MONOS type of stacks with higher K SiON layers lowering stack EOT, thereby (partially) reducing programming and erasing voltage levels and enhancing both retention and endurance. CVD Alumina (Al_2O_3, K = 10; [3]) and ALD Alumina [7] were also attempted in earlier years to replace tunnel oxide for MNOS and replace MOS structure, respectively, to reduce gate stack EOT for NVM and MOS devices. Such earlier attempts were not successful with alumina due to several reasons. These were: (1) high degree of negative fixed charge generation at the Si/Insulator interface; (2) transconductance degradation of MAOS/MAS devices due to interface state generation; and (3) phosphorus diffusion from the gate through alumina to the silicon interface causing mobility degradation of the devices for N+ doped polysilicon gate FETs.

In 2001, an improved process was proposed for incorporating alumina as a gate dielectric to overcome the above constraints [8]. NVM gate stack designs incorporating alumina as tunnel dielectric as well as charge blocking dielectric were also proposed at the same time to reduce the EOT and the programming voltages significantly as well as to improve device endurance by many orders of magnitude [6]. High K dielectric films for NVM device design will be discussed in Section 11.3 in greater detail for NVM device extendibility and voltage scaling.

11.3 RECENT DEVELOPMENTS IN VOLTAGE SCALING

11.3.1 Floating-Gate Devices

Programming voltages have been reduced in floating-gate devices by replacing the conventional ONO IPD with HfO_2 [9] or with alumina adding an over layer of either SiON or SRN [6] or TaN [10] being an interface to the polysilicon gate. A combination of aluminum hafnium oxide with yet higher K value was also reported in 2006 [11]. M. van Durren et al. [12] and R. van Schaijk et al. [13] replaced the ONO IPD with HfSiON dielectric, and achieved voltage scaling and superior device characteristics for FG and CT devices, respectively. Y. Zhang et al. [14] demonstrated reduction of programming voltage by a factor of 2 (Vpp <10 V) by replacing the conventional tunnel oxide by 8 nm of lanthanum aluminate ($LaAlO_3$, K = 27.5). Reduction in programming voltages were reported by Bhattacharyya [15,16] and P. Blomme [17] independently, by replacing both tunnel oxide and IPD ONO with high K dielectric films. A combination of 1.5 nm oxide with 4.0 nm of OR-SiON for tunneling (tunnel insulator EOT = 4.5 nm) and ≤15 nm of Al_2O_3 with 4.0 nm of SRN over layer as IPD interfacing the N+ polysilicon gate (total stack EOT < ~12 nm) was proposed with programming voltage and duration of ±12 to ±14 V, 10 ms. The initial memory window target was >5 V [15]. A second stack design was also proposed employing a novel "reverse-mode" band-engineered stack design [16], whereby the coupling coefficient meant to be ≤0.5 and programming/erasing voltage to be ±12 V, 10 ms for a memory window of ~2 V. The stack design had an EOT of 10 nm and consisted of a tunnel dielectric of SiO_2 + Ta_2O_5 (EOT = 5.0 nm)/FG/30 nm ZrO_2 (EOT = 5 nm)/control gate. The carrier source for the reverse-mode device is the control

TABLE 11.1

Voltage Scaling for Floating-Gate Flash Stacks

Stack Elements and Thickness	EOT (nm)	$\pm$Vpp[a] V
7 nm SiO$_2$/FG/ONO (EOT = 10 nm)	~17	20
10 nm Al$_2$O$_3$/FG/15 nm Al$_2$O$_3$	~10	9–10
4.5 nm SiO$_2$/IN-SRN/FG[b]/IN-SRN/10 nm HfSiON[b]	<7.5	7.5
1 nm SiO$_2$/8 nm HfSiON/IN-SRN/FG[b]/IN-SRN/10 nm HfSiON[b]	<6.5	<6.5
1 nm SiO$_2$/8 nm LaAlO$_3$/IN-SRN/FG[b]/IN-SRN/10 nm La AlO$_3$[b]	<5.0	<4.0

Source: Bhattacharyya, A., and Darderian, G., An enhanced floating gate flash device and process, U.S. Patent Application Publication No. 0067256 A1, March 12, 2009.

[a] Programming Voltage.

[b] FG: P+ doped, K = 14 for HfSiON, K = 27.5 for LaAlO$_3$.

gate unlike the normal mode NVM devices, wherein the carrier source is the silicon substrate. P. Blomme employed a stack design [17] consisting of 2 nm of SiO$_2$ plus 8 nm of ALCVD Al$_2$O$_3$ as tunnel layer (EOT = 5.2 nm). The IPD was 1 nm of oxide plus 8 nm of Al$_2$O$_3$ (EOT = 4.2). The total stack EOT was 9.5 nm and yielded an initial memory window of ~3 V at $\pm$9 V, 3 ms of programming and erasing.

Approach for additional stack EOT reduction and corresponding programming voltage reduction was also exemplified by Bhattacharyya and co-workers [18] by introducing P+ doped floating gate polysilicon sandwiched between two thin layers of IN-SRN with 4.0–4.5 nm of tunnel oxide with progressively introducing higher K IPD as well as tunnel dielectric to lower the stack EOT to achieve $\pm$5V programming voltage target for floating-gate devices. The incorporation of IN-SRN layers improves the reliability of the tunnel dielectric, enhances charge transport while preventing impurity contamination and interface states generation at the silicon/tunnel insulator interface. Table 11.1 illustrates the approach of lowering the programming voltage by almost a factor of 4 from the conventional FG Flash stack design [19].

11.4 CT DEVICE: SONOS VOLTAGE SCALING

TANOS (Tantalum nitride/Alumina/Nitride/Tunnel oxide/Silicon) CT stack design option has been proposed and demonstrated [19,20] as a follow-on multibit per cell design for replacing conventional FGNAND NVM. The stack employed thinner 4 nm tunnel oxide and 15 nm alumina blocking layer and used 7 nm nitride layer for charge trapping.

The programming voltage was scaled down to 17 V with the stack EOT being 14 nm.

The conventional NROM SONOS stack, in comparison to the above TANOS stack as previously illustrated in Chapter 6 in Part I, typically employs 2.0 nm of tunnel oxide, ~7 nm of trapping nitride, and ~5 nm of blocking oxide with a stack EOT of ~11 nm and programming voltage in the order of $\pm$10 V. The NROM is a single bit/cell design with memory window approximately half of the above TANOS cell. During the last decade, several investigators have proposed stack designs replacing oxide in SONOS layers with high K insulators for voltage scaling and enhancement of device characteristics [21–27]. With the exception of [21], which created a novel charge trapping dielectric layer by plasma nitridation of oxide to contain 22% nitrogen to provide high density of deep traps, other proposals consisted of high K dielectric films to replace either (a) charge blocking layer, or (b) both tunnel layer and charge blocking oxide layer, or even (c) all layers of SONOS including the nitride trapping layer. In all cases, the EOT for the stack was reduced to lower the programming voltages. Table 11.2 lists the proposed stack options and the resulting voltage scaling for the stack design with the initial memory window and programming parameters achieved thereby.

TABLE 11.2

Voltage Scaling for CT SONOS Stack Designs

Stack Elements and Thicknesses (nm)				Vpp	Window	
Tunnel	Trap	Block	EOT nm	W (ms)/E (ms)	V	Ref./Yr
7.0-x*	x* (~1 nm)	5	11.5	$Vd = 4V, Vg = 6$&/ $Vg = 4, Vd = -6$	>1 V	[20] (2006)
2.0 oxide	14 nm HfO_2	7.5 oxide	12	+10V, 10 ms/−10, 10 ms	1.5 V	[21] (2003)
1.5 or-SiON+1.7 Al_2O_3	7 NR-SiON	15 HfO_2	7	+5V, 1 ms/+8, 1 ms	>2 V	[22] (2004)
1.5 SiO_2+OR-SiON	4.0 NR-SiON	8 Al_2O_3	7	+6V, 1 ms/−6, 1 ms	>2 V	[23] (2004)
1.5 SiO_2+OR-SiON	4.0 NR-SiON	6 HFO_2	5	+5V, 1 ms/−5, 1 ms	>2 V	[24] (2004)
1.5 SiO_2 +OR-SiON	5 HfO_2	8 Pr_2O_3	3.5	+3V, 1 ms/−3, 1 ms	~2 V	[25] (2004)
2.5 SiO_2	5 Si_3N_4	7.5 Hf Al O	7.5	+6V, 50 ms/−10, 50 ms	>1 V	[26] (2006)
3.0 Al_2O_3	10 HfO_2	10 Al_2O_3	7	+ 6V, 2 ms/−6, 2 ms	>1.4 V	[27] (2006)

It should be noted that with the exception of the first SONOS stack, all other proposals involved charge transfer (for writing/erasing) by direct tunneling and voltage scaling by means of EOT reduction using high K films to replace oxide and nitride charge blocking dielectric layers. Subsequently, high K dielectric films were employed for both tunneling and charge blocking functions. Eventually, all high K layers were used for CT device functions of tunneling, trapping, and charge blocking to reduce stack EOT, thereby reducing the programming voltage levels. In several examples, the tunnel layer consisted of one or two monolayers of SiO_2 at the silicon interface followed by ultra-thin or-SiON (1 nm film) serving as the tunnel layer. Maximum voltage scaling could be achieved by appropriate selection of high K dielectric films in the stack design as well as band engineering the stack structure to optimize the NVM device characteristics which are discussed in subsequent chapters.

Chapter 12 is devoted to the topics of high K dielectric films, while Chapter 13 provides an in-depth discussion on band engineering as requirements for voltage scaling and device extendibility and optimization for all types of NVM devices.

REFERENCES

1. D. Kang, S. Jang, K. Lee et al., Improving the cell characteristics using low-k gate spacer in 1Gb NAND flash memory, *IEDM*, San Francisco CA, 2006, pp. 1001–1004.
2. M. Noguchi, T. Yaegashi, H. Koyama et al., A high-performance multi-level NAND flash memory with 43nm-node gloating-gate technology, *IEDM*, 2007, pp. 445–448.
3. A. Bhattacharyya, FET gate structure (MXOS) for non-volatile N-channel read mostly memory device, IBM Internal Technical Memo, 1975 (Unpublished).
4. A. Bhattacharyya, C. T. Kroll, H. W. Mock et al., Properties and applications of silicon oxynitride films, IBM internal Technical Report, TR. 19.0399, October, 1976 (Unpublished).
5. S. Zirinsky, MXOS n-channel NVM device using multi-layered CVD oxy-nitride Gate insulator, *ECS Fall meeting Proceeding*, Los Vegas, CA, 1977.
6. A. Bhattacharyya, Scalable flash/ NV structures and devices with extended endurance, Micron Technology Internal Docket, August, 2001, U.S. Patents: 7012297 (3/14/2006), 7250628 (7/31/2007), 7400012 (7/15/2008), 7750395 (7/6/2010).
7. J. H. Lee, K. Koh, N. Lee et al., Effect of polysilicon gate on the flatband voltage shift and mobility degradation for ALD-Al2O3 gate dielectric, *IEDM*, 2000, pp. 645–648.
8. A. Bhattacharyya, Scalable gate and storage dielectric, U.S. Patents: 6743682 (6/1/2004), 6998667 (2/14/06), 7528043 (5/5/09).
9. P. Blomme, A. Akheyar, J. Van Houdt, and K. De Meyer, Data retention of floating gate memory with SiO2/high-K tunnel or interpoly dielectric stack, *IEEE Proceeding, IEEE DRC*, 2004.
10. K. Kim, Technology for sub-50nm DRAM and NAND Flash Manufacturing, *IEDM*, New York, 2005, pp. 333–336.

11. H. M. Choi, K. Y. Park, S. H. Lee et al., Novel high-k inter-poly dielectric for sub 50 nm flash memories, *Proceedings, IEEE, NVSMW*, 2006.

12. M. van Duuren, R. van Schaijk, M. Slotbloom et al., Performance and reliability of 2- transistor FN/FN flash arrays with hafnium base high-K inter-poly dielectrics for embedded NVM, *NVSMW*, Monterey CA, 2006, pp. 48–49.

13. R. van Schaijk, M. van Duuren, N. Akil et al., A novel SONOS memory with Hf SiON/Si3N4/Hf SiON stack for improve retention, *NVSMW*, 2006, pp. 50–51.

14. Y. Zhang, S. Hong, J. Wan et al., Flash memory cell with LaAlO3 (k=27.5) as tunnel dielectric for beyond sub-50 nm technology, *Proceedings IEEE NVSMW*, 2005.

15. A. Bhattacharyya, NVM product/technology migration strategy, ADI Associates Internal Memo, December, 2004 (Unpublished).

16. A. Bhattacharyya, Asymmetric band-gap engineered flash with enhanced dtat retention, U.S. Patents: 6784480 (8/3/2004), 6950340 (9/27/05), 7072223 (9/4/206).

17. P. Blomme, J. D. Vos, L. Haspeslagh et al., Scalable floating gate flash memory cell with engineered tunnel dielectric and high-K (Al2O3) interpoly dielectric, *Proceedings, NVSMW*, Monterey CA, 2006, p. 52.

18. A. Bhattacharyya, and G. Darderian. An enhanced floating gate flash device and process, U.S. Patent Application Publication No. 0067256 A1, March 12, 2009.

19. C.-H. Lee, C. Kang, J. Sim et al., Charge trapping memory cell of TANOS (Si-Oxide-SiN-Al2O3-TaN) structure compatible to conventional NAND flash memory, *Proceedings NVSMW*, Monterey CA, 2006, p. 54.

20. Y. Park, J. Choi, C. Kang et al., Highly manufacturable 32Gb multi-level NAND flash memory with 0.0098um2 cell size using TANOS (Si-Oxide- Al2O3-TaN) cell technology, *IEDM*, 2006, pp. 29–32.

21. H. Sunamura, K. Masuzaki, M. Terai et al., Ultra uniform threshold voltage in SONOS type non-volatile memory with novel charge trap layer formed by plasma nitridation, *IEEE NVSM Workshop*, Monterey CA, 2006, pp. 70–71.

22. M. She, H. Takuchi, T.-J. King, SONNS memory: Improvement over SONOS flash memory, *IEEE Device Research Conference*, Salt Lake UT, 2003, pp. 55–56.

23. A. Bhattacharyya, K. Prall, and N. C. Tran, Band-engineered multi-gated channel non-volatile memory device with enhanced attributes, ADI Associates Internal Docket: July 2004, U.S. Patent: 7279740, 10/9/2007.

24. A. Bhattacharyya, A very low-power memory sub-system with progressive non-volatility, ADI Associates Internal Docket, August, 2004, U.S. Patent: 7276760, 9/25/2007.

25. A. Bhattacharyya, Defining high speed, high endurance scalable NVM device using multi-mechanism carrier transport and high density structures for contactless arrays, ADI Associates Internal Docket, September, 2004, U.S. Patent: 7244981, 7/17/2007.

26. A. Bhattacharyya, A novel low power battery operated non-voltile memory device and associated gate stack, ADI Associates Internal Docket, October 2004, U.S. Patent: 7612403, 11/03/2009.

27. Y. N. Tan, W. K. Chim, W. K. Choi et al., Hafnium aluminum oxide as charge storage and blocking-oxide layers in SONOS-type nonvolatile memory for high speed operation, *IEEE Transaction on Electron Devices*, 53, 2006, pp. 654–662.

12 High K Dielectric Films for NVM

CHAPTER OUTLINE

The importance of high K dielectric films in NVM stack designs to derive enhanced NVM device properties has been discussed in Chapters 4 and 5 in Part I. General properties of select high K films of interest in NVM applications have also been discussed in Chapter 3 in Part I. In this chapter, we will discuss in greater detail of the characteristics of (A) common high K dielectric films and (B) potential high K dielectric films. We will review current and future application potentials of dielectric films of groups (A) and (B) in scaled CMOS technology and NVM device stack designs. After the introductory Sections of 12.1 and 12.2, this chapter is essentially divided into two inter-related segments. The first segment from Sections 12.3 through 12.14 reviews published characteristics of select high K films from the standpoint of replacing oxide and oxynitride gate dielectrics for scaled CMOS FET devices. The second segment from Sections 12.15 through 12.24 focuses on applicability of such high K film characteristics for the NVM device stack designs.

12.1 HISTORICAL PERSPECTIVE

The evolution in digital microelectronics technology is intimately tied to the successful miniaturization of CMOS FET transistors both in horizontal dimension through feature size scaling and in vertical dimension through oxide scaling. This process of miniaturization not only reduced power (lower power supply requirement) and enhanced performance (enhanced transistor drive capability and transconductance) but also enabled digital functions at reduced cost. In familiar technical term, this is known as "technology scaling" and "FET device scaling." As such technology scaling continued to around 100 nm technology node, it was apparent that oxide (SiO_2) could not be scaled beyond ~4 nm thickness due to excessive leakage. Consequently, oxygen-rich-oxynitride (OR-SiON) was introduced as the gate dielectric for CMOS FET devices to contain leakage while minimizing the required silicon-insulator interface state density equivalent to SiO_2 at the same time. Additional quest for appropriate higher K gate dielectric continued to enable future technology scaling. It was not until the turn of the century that viable ultra-thin high K dielectric film technology was firmly established and integrated to enable replacement of oxynitride gate dielectric for scaled CMOS FET devices.

The application of high K dielectric film preceded that of the FET gate insulator for nearly a decade in DRAM capacitor structures, replacing conventional oxide/nitride films, enabling capacitor scalability. This reduced the DRAM cell size, providing higher charge density per unit area with lower leakage. Alumina (Al_2O_3) films replaced the conventional capacitor structures, more than doubling the charge storage density. However, the requirement of high K capacitors for DRAM has been less stringent than those for the FET devices.

Alumina film was also successfully applied as the blocking medium, replacing the nitride/oxide structure for the SONOS device for flash memory applications by C. H. Lee et al. [1]. The device and the memory cell were subsequently called the TANOS charge-trapping memory cell [2]. The TANOS (tantalum-nitride/alumina/nitride/oxide/silicon) stack design not only demonstrated programming voltage reduction and leakage but also prevented erase-saturation of conventional SONOS devices by introducing a high-energy TaN metallic barrier between polysilicon gate and alumina, thereby significantly reducing unwanted electron injection from the gate during erasing. This has been discussed earlier in Chapter 11, Section 11.4.

Before discussing the characteristics of specific thin high K dielectric films of interest, it would be appropriate to briefly highlight the general requirements of any high K films for integrated silicon device technology for logic (CMOS FET), memory (DRAM), and NVM applications as well as requirements unique to the device types. This helps establish the baseline for application of high K dielectric films not only to any silicon based integrated technology but also to the common CMOS platform technology as well as the integration requirements toward future SOC (system-on-chip) applications.

12.2   GENERAL REQUIREMENTS

All applications require high K films to be amorphous, thermally stable, non-reactive at the interfaces, chemically inert, thickness (EOT) scalable, selectively etchable, reproducible, large band gap, high K value, low leakage, field-stress-stable, high breakdown strength, and stable properties for the end-of-life application conditions (both thermal and electrical). Additionally, the films should exhibit both low level of intrinsic defects (material defined) and low level of extrinsic defects (process and stress induced). Intrinsic defects imply presence of intrinsic centers of non-stoichiometry at the elemental/molecular level, generating short-range disorder in local bonding, thereby creating defects (or "traps") and fixed charges which cannot be removed by postdeposition thermal treatment. Common extrinsic defects imply "traps" generated by postprocessing conditions and by thermal/electrical stresses. A common example of the latter type of defects responsible for SILC affecting NVM device reliability has been highlighted for SiO_2 films [3,4].

It should be noted, however, that while the general requirements of high K dielectric films as stated earlier for the silicon devices could be similar, functional requirements for device types could be significantly different. Therefore, selection and application of specific dielectric film for the specific device type would be dictated by not only the technology integration requirements for the device type but also the specification of electronic characteristics that fit the device functional requirements. For example, the primary requirement for a DRAM capacitor would be to exhibit charge storage capacity of 20–30 fF/cell at the leakage level 1E-3A/cm² and at a potential across the capacitor of 1.25 V. However, for a scaled CMOS FET device, at EOT 1 nm, the functional leakage requirement required to be satisfied at 1 Volt across the gate insulator after being subjected to a stress fluence of 1E3 C/cm² would be <1E2 A/cm². Obviously, these functional requirements are quite different. Similarly, the leakage requirements of a blocking dielectric film at 1 Volt equivalent of stored charge in the floating gate or in the charge-trapping layer need to be many orders of magnitude lower to meet the retention specification of the NVM device (although at significantly reduced field) after being subjected to 1E6 W/E cycles. Figure 12.1 illustrates the

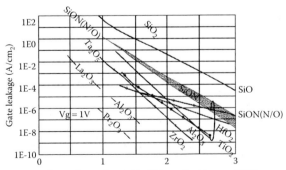

EOT : Electrical equivalent oxide thickness (nm) (source ITRS)

FIGURE 12.1   Leakage characteristics of common Hi K films. (From Skotnicki, T., Short course on applications of HK dielectrics, *Symposium VLSI Technology* © 2004 IEEE.)

wide variation in gate leakage for common single metal high K films and HfSiON films compared to SiO_2 films [3,4].

12.3 COMMON HIGH K DIELECTRIC FILMS FOR NVM GATE

In Chapter 4 of Part I, single metal oxide films, namely, Al_2O_3 (Alumina), ZrO_2 (Zirconia), HfO_2 (Hafnia), Ta_2O_5 (Tantalum Pentoxide), and TiO_2 (Titania) were selectively defined along with HfSiON as "The common Hi K Dielectric films for NVM devices." Key electronic properties of the above metal oxide films have been provided in Tables 3.1 through 3.3.

The initial phase of high K film development was motivated for DRAM applications. Extensive investigation was subsequently undertaken to develop and characterize potential ultra-thin high K films mentioned above for gate-insulator applications for CMOS technology. The scaling limitations of SiO_2 and oxynitride films for gate insulator applications were recognized before the turn of the 20th century. The focus for development and characterization of the Hi K films mentioned above was to investigate the scaling advantages of associated FET devices over those of oxide and oxynitrides in terms of parameters such as transconductance enhancements, drive current enhancements, voltage scalability, and associated improvements in power and performance of devices, circuits, and functions. The link between such characterization and applications between FET gate dielectric and NVM stack dielectrics should be quite apparent in addition to their obvious direct applications for all support circuits for NVM memory designs. The FET gate-insulator interfaces silicon substrate at one end and gate material at the other end. Similarly, the tunnel layer interfaces the silicon substrate, while the blocking dielectric interfaces the gate electrode in the other end. For the tunnel layer of an NVM device, all considerations related to interface states, defects, fixed charges, and interface layer formation relevant to FET device equally apply to the NVM gate stack designs. This also includes all considerations of process/physical/chemical/thermal/field-related stability of electronic parameters as well as integration considerations. Similarly, for the blocking layer, all considerations related to gate-induced charge injection, work-function requirements, leakage through the dielectric with positive potential to the gate will be relevant from NVM stack design and integration standpoint. Since an NVM device is, in principle, a variation of an FET device, FET device scaling principles are also applicable to the NVM memory device stack design in terms of EOT reduction and voltage scaling. Some of the key device properties such as electron and hole carrier mobility are sensitive to process parameters and interface characteristics in both types of devices. Furthermore, dielectric reliability considerations especially under thermal- and application-induced field stress would be of great significance aside from basic electronic and transport fundamentals in the selection of dielectric films for NVM stack designs.

It has been mentioned in Chapter 11 on voltage scaling that using higher K films for tunnel and blocking layers reduces programming voltages, thereby, enhancing not only power-performance tradeoffs, but also improving endurance and reliability by reducing programming stress fields across the gate insulator stack. An additional advantage should also be worth mentioning. Voltage reduction in NVM device programming reduces various disturbs associated with nearest neighbor cell-to-cell parasitic capacitive coupling. This should be particularly relevant for floating-gate NVM devices as the feature size is scaled.

Another aspect of consideration is that, from the scaled NVM technology integration standpoint, it should be highly desirable that appropriately selected high K films should simultaneously be applicable not only for the NVM gate-insulator stack design but also as the gate insulator for both the PFET and the NFET devices for the CMOS peripheral device design requirements. Such possibility facilitates integration of different device sets and reduces integration costs for the NVM device. Additionally, such approach aids in providing multifunctional platform CMOS technology solutions and SOC solutions, enriching functionality, reducing product cost, and improving reliability.

It should be mentioned that in general, integrating high K dielectric films in baseline CMOS technology as well as in NVM technology requires many challenges to overcome. Some of these challenges are process sensitivity of critical parameters, thermal instability at high temperatures, lower band gap, higher leakage, stronger reactivity with silicon, limited process control, and reproducibility and higher sensitivity to specific ambient. As a result, many years of R&D activities precede successful applications of high K dielectric films in devices.

12.4 FET GATE-INSULATOR REQUIREMENTS

The critical requirements to replace SiO_2 for FET gate dielectric applications while addressing device scalability requirements are as follows:

1. Stable amorphous and adherent films in the temperature range of room temperature to 1000°C in the physical thickness range of 1–100 nm. Such films should provide EOT scalability down to ~1 nm.
2. Relatively defect-free with uniform and reproducible composition characterized by fixed charge density and trap density in such films of <1E11/cm².
3. Stable non-reactive interface with silicon substrate yielding interface state density <<1E11/cm². The implication is that the Si bonds at the interface are required to be saturated with high-strength covalent bonds formed with the molecular elements of the dielectric. However, for many high K dielectric materials, such bonds tend to be ionic.
4. The dielectric material should have a band gap >5 eV, and electron and hole barrier height each greater than 2 eV at the silicon interface. Current transport through the dielectric should be governed by tunneling which is independent of temperature and not by trap-assisted thermally dependent transport. The conductivity of the dielectric should be several orders of magnitude lower than either SiO_2 or SiON films at a field strength of ~5E6 V/cm. It should be noted that films of oxide, nitrogen incorporated oxide, and SiON films, reached scalability limit at EOT≤2.5 nm due to unacceptable device leakage through the gate dielectric at such field strength.
5. The destructive breakdown strength of the dielectric should be >>6E6 V/cm.
6. Integration requirements for FET gate application consists of compatibility with the gate material, selective etchability, chemical inertness to contaminants, dopant and postprocessing environments (temperature, pressure, ambient), and intrinsic property of annealing of defects/damages caused by postprocessing requirements such as ion-implantation, plasma-radiation, and gate/back-end processing, and so on.

12.5 THIN AMORPHOUS FILMS OF Al_2O_3, ZrO_2, HfO_2, Ta_2O_5, AND TiO_2 FOR GATE-INSULATOR APPLICATIONS

Table 12.1 provides key attributes of the above thin metal oxide films when compared to SiO_2 films. It also highlights some of the challenges for their applicability as gate dielectric for FET devices or for tunnel/blocking layers for NVM devices. For example, TiO_2 is strongly reactive with silicon substrate/SiO_2 forming silicate (K~10) in a modest processing environment and readily forms conductive silicide with N+ polysilicon gate with very low band offset. Such silicide formation with low barrier energy creates undesirable excessive charge injection (electron) at very low stress (negative gate potential). Additionally, the leakage through TiO_2 film at 1.5 nm EOT (physical thickness of 30 nm) is unacceptably high. Furthermore, dopants such as boron (e.g., boron doping required for PMOS-FET device) and oxygen diffuse readily at a modest processing temperature through TiO_2 films. Therefore, in spite of the intrinsic high K value of TiO_2 films, such film cannot be effectively applied for FET/NVM devices without passivating both silicon and gate interfaces, thereby compromising the effective K value of the device stack design.

Amorphous metal oxide thin films as specified above can be deposited over silicon substrate and over conventional thin dielectric films by a variety of processing techniques both at room

TABLE 12.1

Single Metal Oxide Thin Film Attributes for SiO$_2$ Film Replacement

Film Type	SiO$_2$	Al$_2$O$_3$	HfO$_2$	ZrO$_2$	Ta$_2$O$_5$	TiO$_2$
Dielectric constant	3.9	9.0	25	25	26	80
Thermal/structural stability of amorphous state at >500°C	Hi	Hi	Lo	Lo	Hi	Lo
Oxygen diffusion through film	–	Lo	Hi	Hi	Lo	Hi
Oxygen vacancy (local nonstoichiometry/shallow trap	Lo	Yes	Yes	Yes	Yes	Yes
Boron/contaminants diffusion	Hi	Hi	Hi	Hi	Hi	Hi
Silicate formation at Si interface (IL: Int. layer) due to processing	–	Lo[a]	Med	Med	Lo	Very Hi
Electronic band offset eV: Φb-e	3.2	2.2	1.55	1.7	0	0.1
Leakage @ EOT = 1.5 nm	Hi	Lo	Lo to Med	Lo to Med	Very Hi	Very Hi
N+ polysilicon gate compatibility[c]	Yes	No[b]	No[b]	No[b]	No[b]	No[b]

[a] Requires very low moisture to prevent negative fixed charge generation at interface and prevent silicate formation ([7] of Chapter 11).

[b] Forms silicide, requires metallic interface layer such as TaN or TiN between N+ gate and insulator interface to reduce injection.

[c] All high K gate processes today use metal gates.

temperature and at elevated temperatures. The resulting properties of these films are sensitive to processing techniques and processing parameters. All these thin films can be formed (a) by reactive sputtering either by initially depositing a very thin metal film followed by oxidation in oxygen plasma or by simultaneous sputtering of metal targets in an oxygen plasma; or (b) by an appropriate CVD technique such as PECVD (plasma enhanced chemical vapor deposition), MOCVD (metal oxide CVD), or ALD (atomic layer deposition); or (c) by thermal oxidation in the presence of oxygen and ozone or by electrochemical oxidation. Some oxide films may also preferably be formed by chemical or electrochemical anodization at room temperature especially for reactive metals such as TiO$_2$. Other options include metal oxidation at relatively high temperatures due to thermodynamic and kinetic considerations. For thermal and structural stability and stoichiometry, thin films are annealed at higher temperatures in an oxygen-bearing environment or appropriately annealed in an inert ambient such as nitrogen or argon or in forming gas or NH$_3$ after deposition on an appropriate electrode or substrate. The postdeposition anneal is also intended to reduce intrinsic defects, improve film quality, and reduce leakage. Reactive metal oxides such as films of TiO$_2$ or Al$_2$O$_3$ readily react at relatively low temperatures and at low partial pressure of moisture with silicon substrate to form stable thin film silicates, effectively lowering the dielectric constant and adversely affecting the interface characteristics. Common interface challenges and technology integration involve reduction of fixed negative-charge density at the interface as well as containing interface states generation, maintaining local compositional uniformity and stoichiometry, and preventing contaminants and dopant penetration into the thin film and silicon insulator interface. Single metal oxide films are prone to non-uniform oxygen concentration at the molecular level (oxygen vacancy or defect formation). This results in undesirable shallow trap generation, higher leakage through the dielectric film, mobility degradation, and Vth instability of the FET device. Such challenges are often addressed by introducing chemically stable and passivating interface layers at both top and bottom interfaces of the metal oxide films at the expense of compromising the EOT objective of the gate-insulator stack design. It is worth noting that metal silicates formed at the interface of silicon and the metal oxide have typically K values in the order of 10–11, often significantly lower than that of the metal oxide. However, metal aluminates typically provide higher K values and greater thermal stability. Mixed-metal aluminates may be preferred in some cases, which will be discussed later in this section. It is also worth mentioning that doping a small amount of rare earth metal such as lanthanides (La, Y, Dy, Er, etc.) or appropriate transition metal (e.g., Zr) may significantly improve dielectric characteristics and stability of reactive metal oxides. Mono-layers of such oxides can be deposited by molecular beam deposition or by ALD techniques over silicon and under the electrode. Additionally, such metal may be incorporated by implantation during thin-film processing of metal oxide by previously mentioned techniques.

It has been noted in the above table and in Part I earlier that the K values of the above metal oxide high K films range in K values from Al_2O_3 in the range of 9–12 [5] to TiO_2 being 60–80, while the band gap ranges from Al_2O_3 being the highest of 8.7 eV to TiO_2 being the lowest of 3.5 eV. It should also be noted that for FET applications, since leakage or conductivity in these films varies by many orders of magnitude (at an appropriate application condition), both EOT and leakage characteristics are to be considered simultaneously for applications of these films for FET and NVM devices. Table 3.3 and Table 12.1 in this chapter illustrated such consideration. It was noted in the DRAM application that the figures of merit for Al_2O_3, ZrO_2 (K ~ 20), and HfO_2 (K ~ 25) are approximately the same and around ~ 2X of that of SiO_2, whereas the figures of merits for Ta_2O_5 (K ~ 26) and TiO_2 (K ~ 80) are not significantly better than that of SiO_2 in spite of higher K values due to the higher conductivity and lower band gap of these films.

The limitations of TiO_2 and Ta_2O_5 for gate-insulator applications will be further discussed later on in Section 12.9.

Key characteristics of specific high K metal oxide films will be briefly highlighted in sections below followed by two selective high K films—the first one is oxynitride of Hafnium: HfSiON, while the second one is a rare earth aluminate: $LaAlO_3$. Both HfSiON and $LaAlO_3$ are stable high K films and exhibit promising characteristics for future NVM device applications.

12.6   ALUMINA (Al_2O_3)

Thin Al_2O_3 films have been the subject of most investigations after the conventional dielectric films discussed in Part I for device applications. Early application of CVD Al_2O_3 films to develop MAOS NVM devices was abandoned due to the formation of high density of negative fixed charge at the silicon/insulator interface and transconductance degradation on write/erase stressing. J. H. Lee et al. [5] confirmed high density of fixed negative-charge creation and mobility degradation in NFET cobalt-silicide-polysilicon gate (phosphorus doped n+) MIS devices fabricated using ALD process, which employed ozone as an oxidizing agent at 450°C to oxidize ultra-thin aluminum films. Postformation annealing was carried out at 850°C in N^2 ambient. For the dielectric Al_2O_3

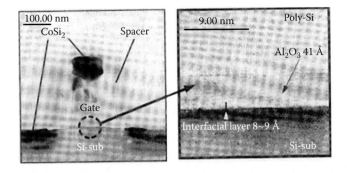

FIGURE 12.2   Cross-sectional TEM of gate stack with Al_2O_3 dielectric, $CoSi_2$-Si-gate, silicate IL. (From Lee, J. H. et al., Effect of polysilicon gate on the flatband voltage shift and mobility degradation for ALD-Al_2O_3 gate dielectric, *IEDM*, p. 645 © 2000 IEEE.)

film of 4.1 nm K ~ 10, the interface layer was ~0.9 nm of Alumina-phospho-silicate (K ~ 5.5), resulting in the effective K value of the stack in the range of 7–8. Cross-sectional TEM of the FET device and the interfacial layer formed is shown in Figure 12.2. While the leakage characteristics improved by three orders of magnitude compared to SiO_2 (EOT ~2.8 nm), the transconductance degradation ranged from 22% to 41% for nMOS and 36% to 74% for pMOS compared to SiO_2-MOSFETs. The lower values were associated with in-situ doped phosphorus polysilicon gates, while the higher values were associated with implanted phosphorus and boron into polysilicon gate for NFET and PFET devices, respectively [5]. Similar characteristics of high negative fixed charge density (~1E13/cm^2) and mobility degradation were also confirmed by D. A. Buchanan et al. [6] with 1.3 nm EOT Al_2O_3 film produced by atomic layer CVD process using NFET device with conventional n+ polysilicon gate.

Nonstoichiometric silicate ($Al_x Si_y O_z$) formation in the interfaces results in unquenched Al-O bond and consequently high value of fixed negative charge and lower effective K values. It was observed that such silicate formation is owing to significant partial pressure of moisture which is inadvertently present in all common processing schemes of forming ultra-thin Al_2O_3 films on silicon substrate and associated surface preparation [7]. Improved silicon surface preparation was proposed and ultra-thin aluminum film deposition and subsequent ozone oxidation scheme was introduced to eliminate silicate formation at the interface [7]. In the proposed processing scheme, a thin layer (0.5–1.0 nm) of IN-SRN was deposited in nitrogen ambient on top of the Al_2O_3 film, passivating the surface from subsequent phosphorus or boron diffusion during polysilicon gate doping for CMOS FET devices. The scheme improved scalability, device characteristics, and reliability of Metal-Alumina-Silicon FET devices.

Leakage characteristics of the polysilicon-gated Al_2O_3 gate stack is shown in Figure 12.3, whereby ultra-thin Al_2O_3 films were formed by three different deposition techniques: Atomic layer CVD (ALD), Metal Oxide CVD (MOCVD), and UHV-PVD (MBE). It should be noted that the leakage characteristics for any given EOT is relatively insensitive to the deposition technique [8]. However, as should be expected, leakage should be strongly dependent on interface structures of the specific film stack, and permittivity and barrier height of the high K film.

Stress stability of Al_2O_3 films were compared with SiO_2 films of comparable EOT of ~2.3 nm. This is shown in Figure 12.4 in terms of 10 years of lifetime extrapolation. The mean time to fail is significantly longer for Al_2O_3 films compared to that of SiO_2 films.

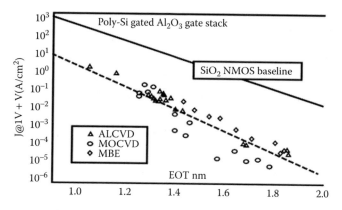

FIGURE 12.3 Leakage characteristics of ultra-thin Al_2O_3 films deposited by different processing techniques. (From Lee, J. H. et al., Effect of polysilicon gate on the flatband voltage shift and mobility degradation for ALD-Al_2O_3 gate dielectric, *IEDM*, p. 645 © 2000 IEEE.)

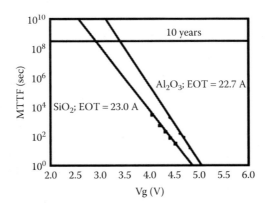

FIGURE 12.4   Mean time to failure comparison of Al_2O_3 versus SiO_2 films of EOT ~2.3 nm. (From Lee, J. H. et al., Effect of polysilicon gate on the flatband voltage shift and mobility degradation for ALD-Al2O3 gate dielectric, *IEDM*, p. 645 © 2000 IEEE.)

12.7   HAFNIA (HfO$_2$)

Among the IV A group of metals in the periodic table of elements Ti, Zr, and Hf; Hafnium (Hf) forms the most stable oxides with the highest heat of formation of 271 Kcal/mol. Ultra-thin films of HfO_2 have been successfully deposited over silicon substrate for MOS FET gate insulator by reactive dc sputtering [9,10], by CVD process at 500°C using O_2 and $C_6H_3HfO_4$ with N_2 [11], by ALD process [12], as well as by Jet Vapor Deposition (JVD) process at room temperature [13]. The sputtering and the CVD processing consisted of an initial process of interface passivation either by depositing an ultra-thin hafnium metal film first [9] or by treating the interface with NH_3 at 700°C to form an ultra-thin SiON layer. Such a scheme with optimum oxygen modulation reduces the interface silicate layer down to ~0.5 nm of physical thickness, enabling gate stack EOT in the range of 1–1.5 nm. This is shown in TEM micrographs of Figure 12.5a and b.

It should be also worth noting that the HfO_2/gate interface was protected by the scheme of in-situ deposition of amorphous Si at 540°C [11] directly on HfO_2 and postdeposition in-situ annealing in an inert nitrogen ambient at higher temperatures. The objective is to prevent interface silicide formation. In the ALD and JVD process schemes, HfO_2 films were directly deposited on silicon substrate after appropriate cleaning of the silicon surface without a separate passivation scheme. Such a scheme results in a somewhat thicker interface, silicate formation of nearly 1 nm as well as silicide formation at the HfO_2/ Gate interface. This has been reported by Kim and co-workers in 2001 [12]. Stack EOT in the range of 1.5–2 nm was achieved using a conventional non-in-situ processing scheme. Due to the interface layers, the effective stack dielectric constant of the stack could be in the range of 10–12 (K for HfO_2 ~ 24).

Thin amorphous HfO_2 films were known to be stable up to 700°C [9]. Stability of ultra-thin HfO_2 films after RTA anneal and postdiffusion impurity activation anneal up to 1000°C in nitrogen have been demonstrated by several investigations [10–12]. HfO_2 films exhibit significantly lower leakage by three to five orders of magnitude when compared to equivalent EOT ultra-thin SiO_2 films regardless of the processing schemes discussed above. This is shown for gate stacks in Figure 12.6 for sputtered HfO_2 and CVD HfO_2 films, respectively.

Current-voltage characteristics for MIS capacitors under substrate injection and under gate injection of Pt/JVD HfO_2/n-Si structures are shown in Figure 12.7, respectively to investigate transport mechanisms through HfO_2 films of EOT = 1.63 nm. While substrate injection fits Schottky emission (tunneling mechanism), the gate injection fits Poole–Frenkel conduction from traps in HfO_2 films with trap depth of 1.5 eV. It should be noted that the barrier height of silicon/HfO_2 interface is 1.13 eV, which is significantly lower than the trap depth of HfO_2. Therefore, tunneling becomes the dominant transport mechanism in thin HfO_2 films.

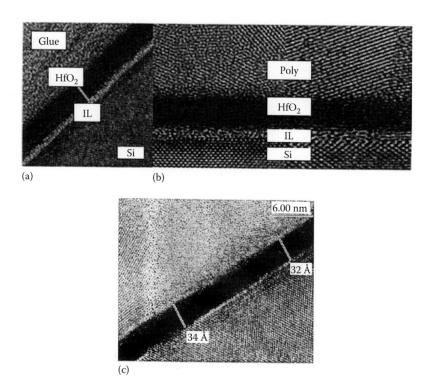

FIGURE 12.5 (a) TEM micrographs of HfO_2/n+ doped polysilicon gate stack: (a) sputtered HfO_2 film after post deposition anneal ([6], of Chapter 11) and (b) CVD HfO_2 film after p-gate implant plus gate activation anneal (900°C 60 s, plus 950°C 60 s). (From Lee, S. J. et al., High quality ultra-thin CVD HfO_2 gate stack with Poly-Si gate electrode, *IEDM*, p. 31 © 2000 IEEE.)

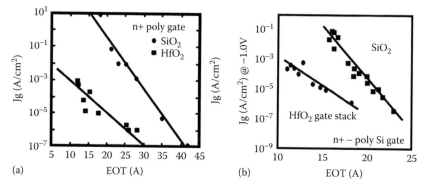

FIGURE 12.6 Leakage characteristics of HfO_2 films in comparison with SiO_2 obtained by two different deposition schemes: (a) Sputtering. (From Lee, B. H. et al., Ultrathin hafnium oxide with low leakage and excellent reliability for alternative gate dielectric application, *IEDM*, p. 133 © 1999 IEEE.) and (b) CVD. (From Lee, S. J. et al., High quality ultra-thin CVD HfO_2 gate stack with Poly-Si gate electrode, *IEDM*, p. 31 © 2000 IEEE.)

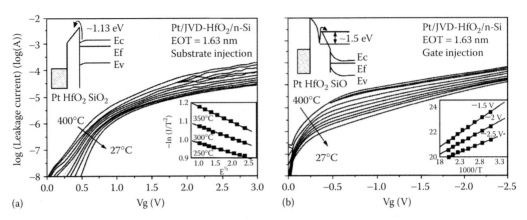

FIGURE 12.7 Ig-Vg temperature dependence for HfO$_2$ films with Pt gate, under (a) substrate injection, and (b) gate injection. (From Zhu, W. et al., HfO$_2$ and HfAlO for CMOS: Thermal stability and current transport, *IEDM*, p. 463 © 2001 IEEE.)

High-frequency C-V plots exhibit negligible hysteresis in HfO$_2$ gate stacks with either TiN/Al gate or n+ doped polysilicon gate by doping POCl$_3$ or by phosphorus implant. This is shown in Figure 12.8, which confirms negligible trapping under such conditions.

HfO$_2$ films exhibit excellent stress stability independent of processing techniques of deposited films, approaching those of Al$_2$O$_3$ films. Figure 12.9 shows stress-induced leakage current (SILC) characteristics of HfO$_2$ stack of 1.35 nm of EOT deposited by sputtering [9]. Negligible change in current density versus gate bias could be observed for films at stress current density of 10 mA/cm^2 up to 50 C/cm^2 of fluence. Stress stability for CVD HfO$_2$ films of EOT 1.09 nm was also characterized by S. J. Lee et al. [11]. This is shown in Figure 12.10a: Jg versus Vg plots and 12.10b: Jg versus Stress-time plots for CVD deposited ultra-thin films. The film stacks exhibit no shift in characteristics before and after stressing.

HfO$_2$ films have several advantages over other comparable high K single metal oxide films. These are relatively lower leakage compared to Al$_2$O$_3$ [8], relatively greater thermal stability compared to ZrO$_2$, TiO$_2$ and so on (see later), relatively higher mobility compared to Al$_2$O$_3$ [8,12], and

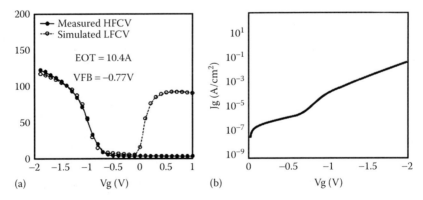

FIGURE 12.8 (a) Measured high frequency CV and simulated low frequency CV, and (b) Ig-Vg of HfO$_2$ gate stack with a-Si gate doped with phosphorus-implantation. (From Zhu, W. et al., HfO$_2$ and HfAlO for CMOS: Thermal stability and current transport, *IEDM*, p. 463 © 2001 IEEE.)

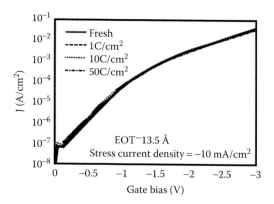

FIGURE 12.9 SILC characteristics of thin HfO$_2$ films. (From Lee, B. H. et al., Ultrathin hafnium oxide with low leakage and excellent reliability for alternative gate dielectric application, *IEDM*, p. 133 © 1999 IEEE.)

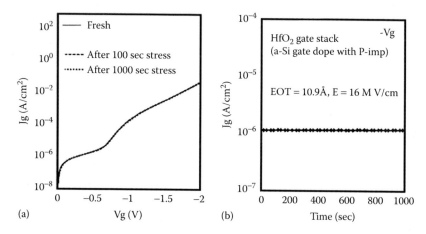

FIGURE 12.10 (a) Ig-Vg after constant voltage stress and (b) Jg-time during constant voltage stress of HfO$_2$ gate stack with N+ doped a-Si gate. (From Lee, S. J. et al., High quality ultra-thin CVD HfO$_2$ gate stack with Poly-Si gate electrode, *IEDM*, p. 31 © 2000 IEEE.)

scalability down to EOT ~1 nm [11]. These attributes make HfO$_2$ ultra-thin gate stack a viable high K option for FET gate application. However, application of HfO$_2$ for EOT around 1 nm and sub 1 nm regime faces several challenges (2.19) [14]. These challenges include (a) mobility degradation with polysilicon gate; (b) limited thermal budget for S/D activation after gate formation (conventional integration scheme of "gate-first" process); (c) high negative fixed charge density of ≥2 E12/cm^2 in spite of high-temperature RTA activation (RTA) anneal of 1000°C [12] when HfO$_2$ is formed directly over silicon; and (d) relatively high trap density due to injected charges [8]. High fixed charge and interface state density adversely impact device reliability, causing Vth shift and Vth instability. Such adverse effects could be minimized, resulting in improved mobility and reliability, by incorporating an ultra-thin (0.51.0 nm) interface layer of OR-SiON (K ~ 5) at the silicon interface coupled with 1.0–2.0 nm of NR-SiON (K ~ 7) or thicker layer of IN-SRN (K > 10) at the polysilicon gate/HfO$_2$ interface for process integration and passivation against dopant penetration and oxygen diffusion. An example of SiON incorporation at the silicon interface has been provided by Gusev et al. [8] who demonstrated significant improvement in NFET device channel mobility by such "interface engineering." However, interface engineering is done at the expense of EOT while improving overall device characteristics and reliability. It should be noted that mobility degradation could be minimized by applying a metallic interface such as TaN [15] or Ta Silicide [16]. Consideration of interface engineering for high K film applications and integration requirements are equally applicable for tunnel layer and blocking layer for advanced NVM device. This is discussed in Chapter 14.

12.8 ZIRCONIA (ZrO_2)

Similar to HfO_2 (K ~ 25, band gap ~5.65 eV), ultra-thin ZrO_2 has been widely investigated as possible replacement of SiO_2 due to its comparable dielectric constant (K ~25, [17]) and band gap (5.7 eV, [18]). Unlike more reactive metal oxides, namely, TiO_2 and Ta_2O_5, ZrO_2 like HfO_2 could be directly deposited over silicon substrate without requiring a passivating barrier layer to prevent reaction with silicon and inter-diffusion at the silicon/insulator interface. Thermal stability of amorphous ZrO_2 films is limited to temperature <400°C, but similar to HfO_2 films, amorphous phase could be stabilized to significantly higher temperatures by reacting with silicon to form silicate and/or by reacting with aluminum to form aluminate [19]. While formation of hafnium silicate significantly lowers the dielectric constant of the film, zirconium silicate film has the advantage of a relatively higher K value of 10–12 [20]. It was also shown that zirconium silicate films exhibit low interface state density of <2E11/cm² as well as lower fixed charge density of 5E11/cm² compared to that of HfO_2 films of >1e12/cm². Additional advantages of ZrO_2 films are (a) selective etchability in HF similar to SiO_2 which aids process integration and (b) higher K value of Si-Doped Zr- aluminates with K > 15 [19] compared with Hf-aluminates of K = 12 (at ~40 atomic % Al). Multiple processing techniques were employed to deposit ultra-thin films of ZrO_2 films over silicon substrate. These include:

1. Sputtering a Zr target in a mixture of Ar and O_2 at room temperature [17] followed by either forming gas or O_2 anneal at 400°C–500°C, or at room temperature to 400°C [21]; Y. Ma et al. [17] used the nitride gate replacement process, whereby S/D processing and high-temperature impurity activation were carried out prior to ZrO_2/polysilicon gate formation.
2. Rapid thermal CVD on silicon substrate at 500°C with or without pre-deposition NH_3 anneal of silicon surface at 700°C, followed by postdeposition N_2 anneal at 700°C–900°C [20].
3. ALD deposition of ZrO_2 directly on silicon substrate, followed by LPCVD process of amorphous silicon deposition and subsequently conventional gate patterning and S/D processing for device fabrication including high-temperature RTA for impurity activation [12].
4. Pulsed-laser-ablation deposition (PLAD) of ZrO_2 film in O_2 ambient on HF-treated silicon substrate without subsequent annealing [18]. The S/D for the FET was formed prior to the gate formation.

Properties of ZrO_2 films as well as the characteristics of the interface silicate formation depended on processing history. Sputtered and CVD films [17,20,21] yielded lower EOT devices of ~1 nm, while ALD [12] and PLAD [18] yielded EOTs in the range of 1.71.8 nm. Interface silicate formation was the least (0.5 nm) for the CVD process, while in the sputtering technique, lower EOT and less interface silicate formation was observed at higher power during deposition. The ALD and PLAD processes resulted in thicker interface silicate formation of ~1 nm. TEM micrographs of ZrO_2 films with interface silicate are shown in Figure 12.11a–d below for films processed by CVD, sputtering, ALD, and PLAD processing, respectively.

Regardless of the processing techniques of film deposition, leakage through ZrO_2 films was in the orders of magnitude lower than that of SiO_2 films of equivalent EOT. Leakage levels as low as 20 mA/cm² for CVD films (EOT ~0.9 nm) and 30 mA/cm² for sputtered films (EOT ~ 1.13 nm) at +1 V were demonstrated. Leakage current was also characteristically low for ALD and PLAD films. Leakage versus EOT of sputtered ZrO_2 films is compared with corresponding SiO_2 films in Figure 12.12a. Such characteristics are similar to those of HfO_2 films. J-V and C-V characteristics

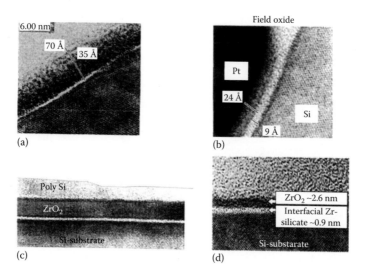

FIGURE 12.11 Cross-sectional TEM micrographs of ZrO_2 films: (a) CVD, (b) Sputtered, (c) ALD, and (d) PLAD processing techniques showing different degree of interface silicate formation. (From Kang, L. et al., MOSFET devices with polysilicon on single-layer HfO_2 high K dielectrics, *IEDM*, pp. 35–38 © 2000 IEEE; Koyama, M. et al., Thermally stable ultra-thin nitrogen incorporated ZrO_2 gate dielectric prepared by low temperature oxidation of ZrN, *IEDM*, pp. 459–462 © 2001 IEEE; Ragnarsson, L.-A. et al., High performance 8A EOT HfO_2/TaN low thermal budget n-channel FETs with solid phase epitaxy regrown (SPER) junctions, *VLSI Symposium*, p. 234 © 2005 IEEE; Yamaguchi, T. et al., Band diagram and carrier conduction mechanism in ZrO_2/Zr-silicate/ Si MIS structures fabricated by pulsed-laser-ablation deposition, *IEDM*, pp. 19–22 © 2000 IEEE; Manchanda, L. et al., Si-doped aluminates for high temperature metal-gate CMOS: Zr-AlSi-O, a novel gate dielectric for low power applications, *IEDM*, pp. 23–26 © 2000 IEEE.)

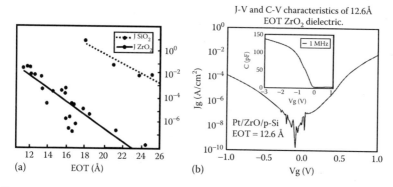

FIGURE 12.12 (a) Leakage characteristics of ZrO_2 films compared to SiO_2 films and (b) J-V and C-V characteristics of sputtered ZrO_2 MOS-CAP. (From Qi, W.-J. et al., MOSCAP and MOSFET characteristics using ZrO_2 gate dielectric deposited directly on Si, *IEDM*, p. 145 © 1999 IEEE.)

of Pt/ZrO_2/silicon MOS capacitors containing sputtered 1.26 nm EOT ZrO_2 films are also shown in Figure 12.12b, exhibiting the low leakage level at 1V.

SILC and time-dependent dielectric breakdown (TDDB) in 1.26 nm EOT ZrO_2 MOS-CAPs were investigated by W.-J. Qi et al. [21]. SILC characteristics under substrate and gate injection are shown in Figure 12.13a and b, respectively, at fluence up to 100 C/cm². The time-dependent characteristics at stress current of 50 mA/cm² under both stressing polarities were also investigated [21]. No significant SILC was observed. The ZrO_2 films displayed high Ibd and large QBD (>100 C/cm²), as well as very small charge trapping in the order of 10–20 MV. It should also be noted that thin ZrO_2 films exhibit high breakdown strength in excess of 25 MV/cm.

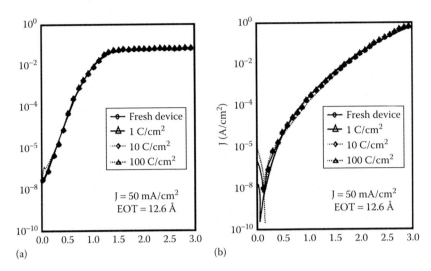

(a)   (b)

FIGURE 12.13 SILC characteristics at stress current of 50 mA/cm²: (a) substrate injection, (b) gate injection with platinum gate electrode. (From Qi, W.-J. et al., MOSCAP and MOSFET characteristics using ZrO₂ gate dielectric deposited directly on Si, *IEDM*, p.145 © 1999 IEEE.)

Transport mechanisms in ultra-thin ZrO_2 films have been investigated by T. Yamaguchi et al. [18] for both sputtered and PLAD deposited films. Unlike thin HfO_2 films, electronic charge transport was found to follow Frenkel–Poole mechanism with weak temperature dependence for both sputtered and PLAD films, with the trap energy depths being process sensitive, and 1.0 eV and 0.8 eV, respectively. However, hole transport was found to follow Fowler–Nordheim tunneling, similar to that of HfO_2 films.

Figure 12.14 shows the weak temperature dependence of the leakage current characteristics of $Pt/ZrO_2/p\text{-}Si$ and the band diagram illustrating electron transport [21].

Both PFET (EOT = 1.5 nm, [21] EOT = 1.26 nm, [21] with metal gate and NFET with polysilicon gate (EOT = estimated in the range of 1—2 nm, [12]) and TiN gate (EOT = 1.2–1.5 nm, [19]) were fabricated and characterized. PFET devices demonstrated comparable mobility to that of SiO_2 within 15%. NFET devices with polysilicon gate demonstrated gate-dimension-dependent

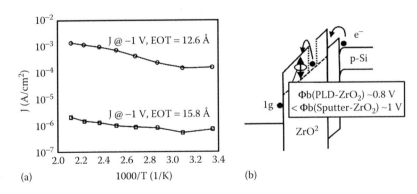

(a)   (b)

FIGURE 12.14 $Pt/ZrO_2/p\text{-}Si$ structure (a) temperature dependence of electron transport (leakage) mechanisms of electrons showing weak temperature dependence in ZrO_2 films and (b) energy band diagram illustrating electron transport of metal-ZrO_2IL-silicon structure. (From Qi, W.-J. et al., MOSCAP and MOSFET characteristics using ZrO_2 gate dielectric deposited directly on Si, *IEDM*, p. 145 © 1999 IEEE.)

leakage possibly due to interface/edge silicide formation, especially for long-channel devices. However, NFET devices with TiN gate showed excellent device characteristics. The implication is that appropriate passivation of polysilicon/ZrO_2 interface is required for compatibility and work-function tuning.

12.9 TANTALUM OXIDE (Ta_2O_5) AND TITANIUM OXIDE (TiO_2)

There has been considerable early investigation into and interest in Ta_2O_5 and TiO_2 films due to higher K values for DRAM capacitor applications. Subsequently, applicability of these films for scaled FET device gate-insulator films has been explored in recent years. It has been noted earlier that conductivity (leakage) of thin films of both Ta_2O_5 and TiO_2 films are unacceptably high and TiO_2 films are strongly reactive to silicon to form silicates and have relatively lower thermal stability as amorphous film. Additionally, while oxygen diffuses through thin films of TiO_2 and reacts with the silicon substrate to form oxide at relatively low temperatures, oxygen non-stoichiometry in Ta_2O_5 films creates oxygen-atom vacancy, inducing defects and trapping centers in such films. To achieve lower EOT, both film types require interface engineering between the silicon substrate and the insulator film as well as process integration schemes whereby S/D activation precedes gate formation ("gate later") and/or low thermal budget metal gate technology.

12.9.1 TiO_2

Feasibility of TiO_2/Si_3N_4 gate dielectric with EOT = 1.5 nm has been demonstrated using low temperature processing, and aluminum-gated FET devices were characterized by X. Guo et al. [22]. The gate stack consisted of 12 nm of TiO_2/1.5 nm of Si_3N_4 (EOT 1.55 nm), with leakage current density lower by two orders of magnitude compared to SiO_2 films. TEM Micrograph of the gate stack layers and the associated leakage characteristics are shown in Figure 12.15. While the FET drive currents and transconductances are comparable to those of direct tunneling gate oxide transistors, their characteristics were inferior to those of HfO_2 transistors of equivalent EOT, possibly due to lower carrier mobility.

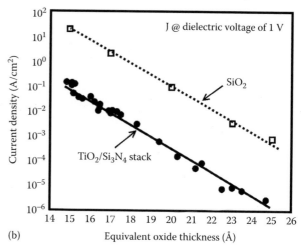

(a)   (b)

FIGURE 12.15 Al/TiO_2/Si_3N_4/Si gate stack (EOT = 1.5 nm) characteristics (a) TEM micrograph with nitride interface layer and (b) leakage comparison with SiO_2 films. (From Guo, X. et al., High quality ultra-thin (1.5nm) TiO_2/Si_3N_4 gate dielectric for deep submicron CMOS technology, *IEDM* © 1999 IEEE.)

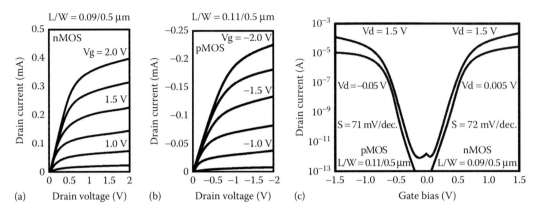

FIGURE 12.16 (A) Ta_2O_5 FET Id-Vg plot and leakage characteristics; (B) FET Id-Vd plots (a) NMOS and (b) PMOS. (From Inumiya, S. et al., Conformable formation of high quality ultra-thin amorphous Ta_2O_5 gate dielectrics utilizing water assisted deposition (WAD) for sub 50 nm damascene metal gate MOSFETs, *IEDM*, p. 649 © 2000 IEEE.)

12.9.1.1 Ta_2O_5

Ultra-thin Ta_2O_5 films of EOT < 1.0 nm has been fabricated and MIS structures have been characterized [23], with interface engineering demonstrating very low leakage compared to SiO_2 films, similar to TiO_2 with interface nitride. The bottom passivation layer was grown for Ta_2O_5 in NO ambient to form an SiON layer, while the aluminum/TiN was used as electrode for the MIS capacitors.

Device feasibility for Ta_2O_5 gate dielectrics was demonstrated by S. Inumiya et al. [3] at EOT = 1.6 nm. The gate stack consisted of an ultra-thin SiON interface layer using NO gas for thermal oxynitride formation, LPCVD Ta_2O_5 deposition followed by in-situ H_2O annealing at temperatures 200°C–300°C, with Al/TiN gate over layer low temperature processing. The source/drain process and activation preceded the gate formation of the FET devices. The leakage characteristics and the Id-Vg plots of the nMOS and pMOS devices are shown in Figure 12.16a, while the Id-Vd plots are shown in Figure 12.16b. The device characteristics were modest when compared to those of HfO_2 and ZrO_2 gate dielectrics.

12.10 BIMETAL OXIDES AND ALUMINATES OF HAFNIUM AND ZIRCONIUM

When aluminum is doped with hafnium or zirconium to form oxides, bimetal oxides are formed, often termed as "aluminates." The properties of such aluminate films depend on the atomic concentration of aluminum in bimetal oxides. In general, aluminum incorporation increases the amorphous state thermal stability and lowers leakage at the expense of lowering the K value, thereby increasing the EOT of the film. NFET devices fabricated with zirconium aluminate gate dielectric and TiN gate exhibited excellent device characteristics with an effective K ⤳ 15 [17,19]. Devices with hafnium aluminate have shown to improve leakage over HfO_2 by two orders of magnitude [8], however, at the expense of an effective K value ~10. Using ultra-thin nano laminates of alumina (Al_2O_3) and hafnia (HfO_2) with composite composition of Al_2O_3x-$HfO_2$1-x, Gusev et al. demonstrated a reduction of six orders of magnitude in leakage current compared to the SiO_2 films of comparable EOT [8]. This is shown in Figure 12.17.

12.11 ALUMINATES OF LANTHANIDES

In recent years, rare earth metal oxide films have drawn considerable interest for high K films for applications in scaled CMOS technology. The rare earth metals belong to the lanthanum (La, atomic no. 57) family (La, Pr, Nd, Sm, Gd, Dy, and Er), and their single metal oxides are called

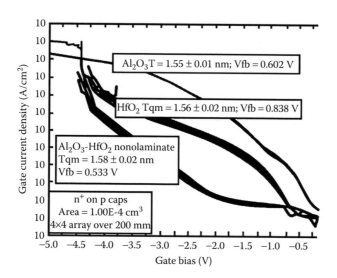

FIGURE 12.17 Ig-Vg plots of 1.6 nm ultra-thin alumina-hafnia-laminate versus same EOT Al_2O_3 and HfO_2 films. (From Gusev, E. P. et al., Ultrathin high-K gate stacks for advanced CMOS devices, *IEDM*, p. 451 © 2001 IEEE.)

lanthanides. These oxides possess larger band gap and higher temperature stability, with their lattice parameter matching that of silicon. The aluminate of lanthanum oxide, $LaAlO_3$, is particularly attractive for NVM device applications due to its relatively higher thermal stability as a film as well as its ability to form a non-reactive interface with silicon substrate [24]. $LaAlO_3$ is *discussed in greater detail in 12.11.1 below due to its possible applications in NVM devices.*

12.11.1 Lanthanum Aluminate ($LaAlO_3$)

$LaAlO_3$ films were deposited by sputtering of $LaAlO_3$ pellets at room temperature under high vacuum by Li et al. [25], demonstrating good thermal stability and low leakage. However, they observed silicate formation at the interface as deposited as well as after postdeposition anneal in vacuum [24,25]. By sputter depositing $LaAlO_3$ films, at 700°C, at 4E-7, Torr followed by postdeposition anneal in nitrogen. Suzuki et al. showed that interface layer formation could be eliminated and superior film properties with higher dielectric constant (K = 25) and extremely low leakage could be achieved in these films even at 0.3 nm EOT [26]. Their investigation confirmed that high-temperature deposition is required to eliminate interface layer formation after postdeposition anneal, lower leakage, fewer intrinsic defects, and higher K in such films. The TEM image of MIS structure for high temperature processed 2 nm thick film with molybdenum (Mo) gate is shown in Figure 12.18a. Unlike La_2O_3 and other lanthanides and also other high K metal oxide films, interface layer formation was confirmed to be absent even after all postdeposition high-temperature processing.

Leakage current characterization of high-temperature processed films showed that leakage current films of EOT ≤1 nm could be five orders of magnitude or lower than corresponding oxide films. This is shown in Figure 12.18b, where the leakage level of optimized HfSiON films as reported by Koike et al. [27] is also plotted for comparison. Current-voltage plots of thin films exhibit no significant temperature dependence indicative of charge transport mode by tunneling. This is shown in Figure 12.19a, while Figure 12.19b displays the energy band diagram of Mo/$LaAlO_3$/Si structure extracted from XPS studies [25]. $LaAlO_3$ films also displayed very high breakdown strength and high stress stability after very high current injection. This is shown in Figure 12.20.

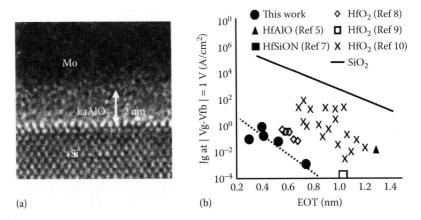

(a)   (b)

FIGURE 12.18 (a) TEM image of Mo/LaAlO$_3$/Si structure exhibiting no interface layer formation; (b) Current density versus EOT plot of LaAlO$_3$ films at Vg-Vt-fb = 1V compared with SiO$_2$ and HfSiON. (From Suzuki, M. et al., Ultra-thin [EOT = 3A] and low leakage dielectrics of La-aluimate directly on Si substrate fabricated by high temperature deposition, *IEDM*, p. 445 © 2005 IEEE.)

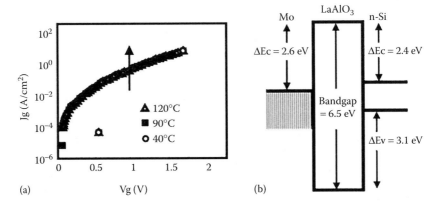

(a)   (b)

FIGURE 12.19 LaAlO$_3$ film (a) temperature dependence of current density, (b) energy band diagram determined from XPS. (From Suzuki, M. et al., Ultra-thin [EOT = 3A] and low leakage dielectrics of La-aluminate directly on Si substrate fabricated by high temperature deposition, *IEDM*, p. 445 © 2005 IEEE.)

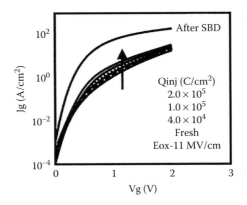

FIGURE 12.20 Stress dependence of leakage current of Mo/LaAlO$_3$/Si capacitors. (From Suzuki, M. et al., Ultra-thin [EOT = 3A] and low leakage dielectrics of La-aluminate directly on Si substrate fabricated by high temperature deposition, *IEDM*, p. 445 © 2005 IEEE.)

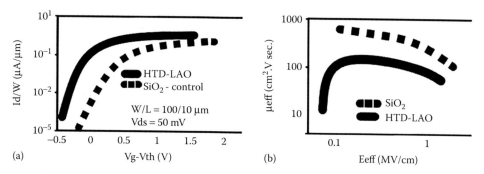

FIGURE 12.21   LaAlO$_3$ and SiO$_2$ FET comparison (a) NFET device current and (b) mobility. (From Suzuki, M. et al., Ultra-thin [EOT = 3A] and low leakage dielectrics of La-aluminate directly on Si substrate fabricated by high temperature deposition, *IEDM*, p. 445 © 2005 IEEE.)

NFET device characteristics and electron mobility of devices with EOT = 0.8 nm were compared with corresponding oxide FET devices of EOT = 3.2 nm. This is shown in Figure 12.21a and b. While drive current was higher than that of SiO$_2$ films (much thicker), mobility was about 50% of that of SiO$_2$. Mobility degradation was postulated to be due to interface defects and high density of measured fixed charge at the interface. This is suggestive of more process development work to further improve the interface characteristics for applications in CMOS FET devices.

12.12   NITRIDES AND OXYNITRIDES OF HAFNIUM AND ZIRCONIUM

Silicon oxynitrides played very significant roles in the evolution of microelectronics and nano electronics technology and devices for decades. This has been discussed in Part I. Nitrides and oxynitrides provide unique and distinctive physical and chemical properties enabling technology and devices. The unique properties are derived fundamentally from strong O-N and Si-N bonding. These insulators are amorphous and stable at high temperatures, passive with very low diffusivity of all elements including hydrogen, provide selective etchability against silicon and SiO$_2$, and exhibit wide range of reproducible and desirable electronic properties with breakdown strength >10E6V/cm. Consequently, these insulators have been widely used, ranging from passivation, patterning (block mask), and isolation to direct device applications such as FG and CT-NVM devices. OR-SiON has been the bridge in gate-insulator scalability between SiO$_2$ and high K insulators below 3.5 nm SiO$_2$ and high K to contain scaled FET device leakage in the range of EOT of ~2.0 to ~4.0 nm.

Further scaling of FET gate required application of high K ultra-thin dielectric films. It has been noted earlier that even in the case of most promising single metal high K oxides such as HfO$_2$, the conventional "gate-first" process integration scheme with polysilicon gate would be challenging due to limited thermal stability at high temperatures. However, nitrides and oxynitrides of such metals (Hf or Zr) exhibit significantly higher thermal stability without significantly lowering the K value. Of all the nitrides and oxynitrides of Hf and Zr, the most investigated and perhaps the most promising has been HfSiON [28–34]. La-doped HfSiON to form HfLaSiON was also investigated to further improve thermal stability and scalability [35]. HfLaSiON films were formed by a molecular beam deposition (MBD) of ultra-thin La$_2$O$_3$ capping layer over ALD HfSiON film followed by deposition of TaN by either ALD or PVD. Subsequently, a 1070C spike anneal was used in a "gate-first" process to drive La into HfSiON. HfLaSiON film yielded <1 nm EOT FET gate with low leakage, high mobility, and high thermal stability and reliability comparable to HfSiON with a somewhat higher K value (K ~ 16). Other high-temperature stable thin films of hafnium-based oxynitrides applicable to scaled (gate-first) metal-gated (e.g., TaN/Ir2Si) CMOS FET technology included HfLaON (K = 20, Eb ~ 6 eV), which was developed by plasma nitridation of PVD HfLaO [36]; HfTaON (K > 15, Eb ~ 6 eV) [37], deposited by reactive sputtering followed by postdeposition anneal [38]; and HfAlON (K ~ 15, Eb = 6.8 eV), fabricated by PVD deposition of HfAlO followed by plasma nitridation and PDA at 800C [38]. All the above Hf-based oxynitrides are compatible with TaN metal gate.

Zirconium nitride or more appropriately, ZrSiON, could also be promising [39], but thus far investigation in such films has been limited. Therefore, discussion in this section will be limited to HfSiON films.

12.13   HAFNIUM SILICON OXYNITRIDE (HfSiON)

HfSiON films are discussed here in detail because of many desirable attributes [28,29]. It has been shown by R. van Schaijk et al. that when oxide layers are replaced by HfSiON layers in SONOS NVM devices, superior performance with enhanced data retention could be achieved [29]. HfSiON has a K value which is weakly composition dependent but typically ≥14. The ultra-thin films have a band gap of 6.9 eV, and with electron and hole band offset (with reference to silicon substrate) being of 3.0 and 2.8 eV respectively. Ultra-thin HfSiON films can be produced by several techniques. To avoid an interface layer and Hfsilicate formation in the final film, a sub-1 nm oxide (SiO_2) is initially incorporated by treating the silicon surface either by ozone or by chemical oxidation prior to an MOCVD deposition of HfSiON. The HfSiON film is deposited using hafnium tetra-t-butoxide (HTB) and disilane (Si_2H_6) at a relatively low temperature ~260C, [28,31,34,35]. Other deposition techniques include ALD [30] and plasma nitridation of HfSiO deposited by MOCVD [31]. The films are thermally stable and remain amorphous after conventional high-temperature S/D anneal and after postdeposition anneal (PNA) in low-pressure N_2 or an N_2-diluted O_2 ambient.

Leakage current of HfSiON films is typically three orders of magnitude or lower than comparable ultra-thin SiO_2 films [3]. The electric and structural properties of hafnium silicon oxynitride (HfSiON) with high Hf/(Hf+Si) ratios were investigated, focusing on the role of Hf-N bonds inside the material. The results show that the existence of Hf-N bonds in the films is responsible in producing a higher dielectric constant film with high thermal stability. Thermally stable amorphous high K stacks with EOT of 0.6 nm and with a factor of 10E 5 X reduction in gate leakage was demonstrated, (when compared to that of SiO_2 films,) using ultra-thin HfSiON with high Hf and high N concentrations, Current density versus Voltage characteristics of a 1.28 nm HfSiON films are compared with corresponding SiO_2 films, as shown in Figure 12.22 [28]. Leakage current transport mechanism shows weak temperature dependence, and fits Frenkel–Poole mechanism for electrons injected from the substrate. The mechanism is indicative of electron movement through trap hopping and the field dependence of the current is due to lowering of the potential barriers of traps.

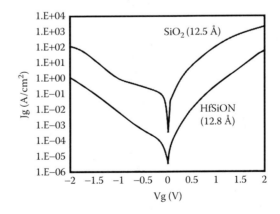

FIGURE 12.22   I-V characteristics of HfSiON. (From Shanware, A. et al., Reliability evaluation of HfSiON gate dielectric film with 12.8A SiO_2 equivalent thickness, *IEDM*, p. 137 © 2001 IEEE.)

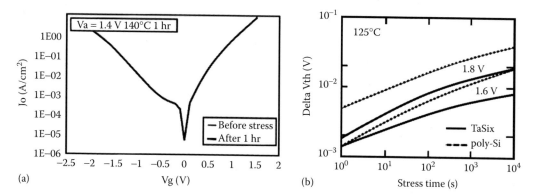

FIGURE 12.23 (a) I-V stress stability of HfSiON NFET device and (b) PBTI characteristics of a TaSix/ HfSiON MOSFET compared with Poly-Si/HfSiON. Constant voltage stress was applied at 125°C. Higher immunity was seen for TaSix. (From Shanware, A. et al., Reliability evaluation of HfSiON gate dielectric film with 12.8A SiO$_2$ equivalent thickness, *IEDM*, p. 137 © 2001 IEEE.)

HfSiON films exhibit high breakdown strength (>20 E6 V/cm), no significant trapping (no significant change in Flatband), and stress stability. The I-V characteristics, as shown in Figure 12.23a, and C-V (not shown) showed no detectable change in characteristics before and after stress, confirming such attributes [28]. The Ta-silicided device showed superior postbias-temperature instability (PBTI) and higher drive current. PTBI was shown in Figure 12.23b. HfSiON films also exhibited negligible SILC. Extensive reliability studies were conducted on ultra-thin HfSiON films ranging in thickness of 0.9 nm EOT [31] to 2.0 nm EOT [30,33,34]. Properly processed HfSiON films with an Hf to Si ratio of 3 to 1 demonstrated excellent reliability.

FET device characteristics for HfSiON gate dielectrics were investigated for polysilicon-gated device versus Tasilicided gate devices with 0.9 nm EOT. Both showed good carrier mobility and S factor (64 MV/decade, see Figure 12.24a). Figure 12.24b compares Id-Vd characteristics of an nMOS device with a TaSix gate versus a polysilicon gate. The authors demonstrated excellent S factor for FET devices (W/L=10/1 μm) [28].

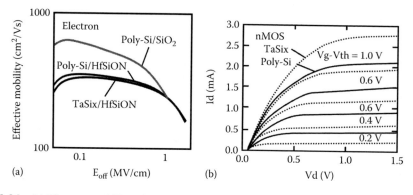

FIGURE 12.24 (a) Electron mobility of TaSix/HfSiON and n+ Poly-Si/HfSiON. Universal curve for Poly Si/SiO$_2$ is also shown for reference; (b) TaSix versus Poly-Si Gate Id-Vd comparison versus gate electrode for nMOS device. (From Shanware, A. et al., Reliability evaluation of HfSiON gate dielectric film with 12.8A SiO$_2$ equivalent thickness, *IEDM*, p. 137 © 2001 IEEE.)

12.14   COMPARISON OF BAND DIAGRAMS OF HfO₂, HfSiON, ZrO₂, AND LaAlO₃

Published band diagrams for HfO₂, HfSiON, and ZrO₂ films against silicon substrate and aluminum or platinum electrodes are shown in Figure 12.25a–d, respectively, for ultra-thin films. Multiple techniques are often employed to obtain the band gap and assumption on effective mass through the dielectric film is made to derive the barrier energy from current density versus field plots at different temperatures. Photo electron emission yield versus photon energy methodology is also often used to determine the band diagram. Additionally, processing techniques and history, film thickness, interface characteristics, etc. all influence the values obtained. It should, therefore, be assumed that the energy values are only approximate. A wide variation in the band gap ranging from 5.7 eV to 7.8 eV for ZrO₂ films has been reported by investigators [18,19].

Transport mechanism studies have shown that in all films, hole transport follows the Fowler–Nordheim tunneling mechanism and has no temperature dependency. However, while electron transport in HfO₂ films is governed by tunneling, electron transport in HfSiON and ZrO₂ films exhibits weak temperature dependence and fits the Poole–Frenkel mechanism of thermally aided field emission from traps to the conduction band once electrons get injected from either the silicon substrate (gate positive) or the gate electrode (gate negative). The implication is that leakage through HfSiON and ZrO₂ films could be relatively more sensitive to non-stoichiometry and processing history, and may vary depending on the trap density and trap energy levels in these films. It should be noted that transport characteristics in silicon oxynitrides have been sensitive to the oxygen/nitrogen ratio in such films and transport mechanisms change from tunneling (OR-SiON) to Poole–Frenkel mechanism (NR-SiON) depending on the composition of such films. HfSiON films should, therefore, be no exception.

It should be apparent from the band diagrams that electron injection from the silicon substrate should be expected to be the dominant transport mode due to significantly lower barrier energy (1.13 eV for HfO₂ and ~1.5 eV for ZrO₂). In contrast, HfSiON and LaAlO₃ have much higher barrier energy of 3.0 eV and 2.4 eV, respectively. From a hole injection standpoint, all films exhibit high barrier energy approaching >~3 eV. It should be further noted that using an aluminum gate or n+ polysilicon gate would result in high electron injection for both HfO₂ and ZrO₂ films when the gate

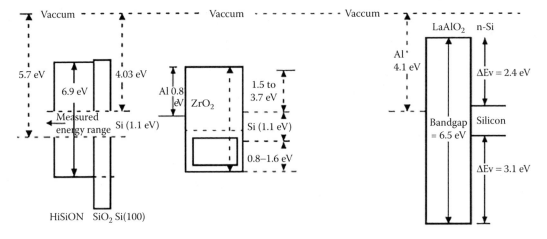

FIGURE 12.25   Band diagram of HfO₂, HfSiON, ZrO₂ and LaAlO₃ thin films with aluminum gate for comparison. (From Yamaguchi, T. et al., Band diagram and carrier conduction mechanism in ZrO₂/Zr-silicate/ Si MIS structures fabricated by pulsed-laser-ablation deposition, *IEDM*, p. 19 © 2000 IEEE; Manchanda, L. et al., Si-doped aluminates for high temperature metal-gate CMOS: Zr-AlSi-O, a novel gate dielectric for low power applications, *IEDM*, p. 23 © 2000 IEEE.)

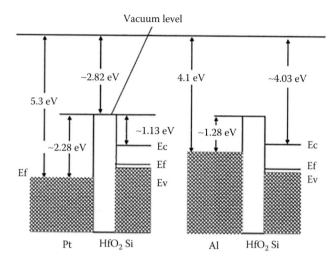

FIGURE 12.26 Band diagram of HfO$_2$ capacitors with aluminum gate on right and platinum gate on left. (From Zhu, W. et al., HfO$_2$ and HfAlO for CMOS: Thermal stability and current transport, *IEDM*, p. 463 © 2001 IEEE.)

is stressed with negative potential due to relatively low barrier energy with Al electrode (1.28 eV for HfO$_2$, and 1.6 eV for ZrO$_2$). This suggests that high work-function metal electrode (e.g., Pt) should be considered to minimize leakage. The difference between a band diagram with a platinum gate and that with an aluminum gate for HfO$_2$ films is exemplified in Figure 12.26 [13]. High-energy barriers for both electrons and holes and nearly symmetric barrier heights for both electrons and holes (~3 eV) and large band gap (6.9 eV) in HfSiON make such high K dielectric films an attractive choice not only for scaled CMOS FET devices but also for tunnel dielectric and blocking dielectric for NVM devices. A similar case could be made of LaAlO$_3$ films except appropriate metal electrode (Mo or Pt) would be required or metallic interface like TaN would be required to reduce gate injection for such films. Applications of these films will be discussed further in this section.

12.15 COMMON HIGH K FILMS FOR NVM APPLICATIONS

Thus far, we have discussed select common high K films for scaled CMOS FET device application from the point of view of replacing the conventional dielectric films used during the early decades. We have reviewed all the basic characteristics of the dielectric films of interest and common characteristics relevant to FET device gate insulators for device enhancements as well as dielectric characteristics relevant to NVM stack designs especially for tunnel layer and blocking layer applications. We will first identify some unique functional requirements of high K films for NVM stack design followed by considerations of applicability of high K dielectric films for NVM devices.

12.15.1 Unique Functional Requirements of High K Films for NVM

Unique functional requirements for NVM devices are memory window, writing and erasing voltage levels and speed, retention, and endurance. These requirements necessitate unique sets of characteristics for high K ultra-thin films over and above those required for scaled FET device applications and interface characteristics common to both FET's and NVM's tunnel and blocking layers discussed earlier. These requirements could be briefly summarized below and will be explained in greater detail through application examples later in this section.

12.15.2 Tunnel Dielectric Film

Memory window requirements and charge loss through the tunnel layer are of primary consideration for selecting the tunnel layer in addition to the silicon/insulator interface requirements similar to that of FET gate insulators. Yielding a large memory window at a modest field across the tunnel layer would be the ideal goal which would require high electron fluence from the silicon substrate into the tunnel dielectric. Such high current density at lower voltage drop implies not only a higher K dielectric film with low EOT but also low barrier energy Ube at the silicon/insulator interface. However, such requirement would only enhance charge loss by back tunneling unless the dielectric conductivity rises by many orders of magnitude as the field across the dielectric is raised. However, to the author's understanding, such ideal dielectric film characteristics have not yet been found. To provide an order of magnitude understanding of such requirements, in an ideal environment, a current flow of 1E-3 A/cm^2 at an average field of 6 MV/cm across the dielectric reducing to 1E-15 A/cm^2 at a field of 2 MV/cm, and consequently with a conductivity reduction ranging between 1E-18/4E6 and 1E-15/4E6A/MV-cm [J/E] would be required if we assume charging time between 1E-6 to 1E-3 s range and 1E9 s of end-of-life (EOL) retention requirements. This means conductivity through the tunnel layer needs to rise by ~3.5–4.3 orders of magnitude per MV/cm of field across the film. In reality, for most known dielectric films investigated, conductivity rises between 1 and 2 orders of magnitude in this field range. Therefore, no single tunnel dielectric would fulfill such unique requirements. The answer lies in "tunnel engineering" or "band engineering" to provide an asymmetric barrier for the electron flow, which will be discussed in Chapter 13. Therefore, from the standpoint of selecting the tunnel layer film, a high K dielectric film with barrier energy in the range of 1.5–2.5 eV, fulfills all other requirements, and is, therefore, usually sought for single tunnel layer stack designs.

12.15.3 Trapping Dielectric Film

Amorphous dielectric films exhibit a short-range order of elements linked with either covalent or ionic bonding. When the local composition of elements varies from overall composition and local linkages between elements are disturbed, "defects" and/or "traps" are created in such films. Most dielectric films contain a certain level of intrinsic or characteristic traps in thermodynamic equilibrium. These regions may reflect local fluctuations in lattice potentials to attract or repel electrons or holes or may remain overall electrically neutral. Intrinsic trap levels at room temperature may vary widely from one type of dielectric to another. For example, the intrinsic trap density in amorphous SiO_2 films is nearly three orders of magnitude lower (~1E 10/cm^2) than that in amorphous Si_3N_4 films (5-10E12/cm^2). Additionally, "charge capture probability" or "trapping probability," which depends on capture cross-section of the dielectric films, could vary significantly from film to film depending on their intrinsic characteristics. This characteristic determines the charge fluence requirement to trap injected charges. Figure 12.27a shows trapped electron density against injected charges for oxides, HfO_2 and Al_2O_3, while Figure 12.27b illustrates capture probability for Al_2O_3 and HfO_2. Both Al_2O_3 and HfO_2 exhibit many orders of magnitude higher capture probability than that of SiO_2 films and are characteristically independent of stress bias, For oxide films, the capture probability is bias dependent and requires significantly higher bias field. It should be worth noting that in nitride films and in nitrogen-rich oxynitride films, capture probability approaches unity (not shown here).

Dielectric films may exhibit significantly higher concentration of defects and traps beyond the intrinsic trap level dependent on process/ambient-induced and thermally induced stresses which may only be partially removed by appropriate annealing. These are called extrinsic traps. Traps are also induced by subjecting the dielectric film to a high electrical field. These are called field-induced traps and often lumped into the "extrinsic trap" group. There could also be multiple energy levels characteristically associated with certain dielectric films. For example, thin films of Al_2O_3, HfO_2, and AlN are known to exhibit multilevel traps. For most high K single metal

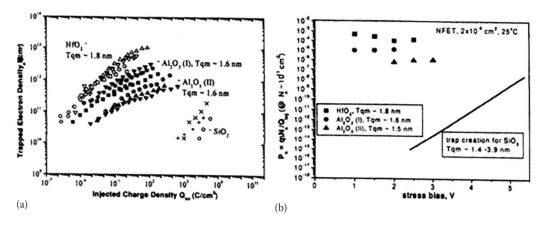

(a)   (b)

FIGURE 12.27 (a) Trap density as a function of injected charge for Al_2O_3 and HfO_2 and (b) trapping probability for high K gate stacks (symbols) and rate of trap generation for SiO_2 (solid line). (From Lee, J. H. et al., Effect of polysilicon gate on the flatband voltage shift and mobility degradation for ALD-Al_2O_3 gate dielectric, *IEDM* © 2000 IEEE; Gusev, E. P. et al., Ultrathin high-K gate stacks for advanced CMOS devices, *IEDM*, p. 451 © 2001 IEEE.)

oxide films, oxygen vacancy (local loss of oxygen) is known to create shallow level traps over and beyond their characteristic intrinsic traps.

For charge-trapping devices, the trapping dielectric layer is required to have high density of intrinsic reproducible traps with deep energy levels such that electrons and holes could "permanently" get trapped in the potential well. While several high K films exhibit high density of deep energy level intrinsic traps, some of these films may not be considered for NVM devices because of integration challenges, thermal and chemical stability at elevated temperature, and reactivity with the processing environment. Typically high K charge-trapping films should exhibit single level "stable" traps of >5E12/cm² of density of intrinsic traps uniformly distributed with energy level >1 eV without being prone to generating shallow energy level traps during processing or during external field stressing. As should be seen later in this section, the number of high K dielectric films fulfilling such requirement is rather limited.

12.15.4 Blocking Dielectric Film

The primary functional role of the blocking layer for the normal mode NVM device is twofold: (a) to prevent electron injection from the gate during erasing and writing, and (b) to prevent the trapped electron (during standby state of retention) loss to the gate through the blocking layer. A higher K blocking layer allows a larger voltage drop across the tunnel layer, thereby ensuring that the silicon substrate acts as the primary source of electrons and holes during writing and erasing, respectively. At the same time, a lower field across the blocking layer ensures that the gate remains passive during both writing and erasing from the standpoint of charged carrier injection. The blocking dielectric and the gate interface need to be compatible to prevent charge injection. Depending on the characteristics of the blocking dielectric film, either a passivating layer is required between the n+ polysilicon gate and the blocking layer or a thermally stable metallic interface layer with high work function such as TaN, TiN, or WN would be required. Additionally, to ensure no trapped charge loss during the standby state, the high K film should exhibit very low leakage. This implies that higher K films with larger Ube and Ubh would be preferred. In principle, many of the requirements for the blocking dielectric high K films are similar to those of the FET gate dielectric including relatively low density of intrinsic and extrinsic traps, and insensitivity in the formation of shallow traps.

12.16 REVIEW OF HIGH K DIELECTRIC APPLICATIONS FOR CURRENT NVM DEVICES

During the last decade, high K dielectrics have been widely explored for NVM devices to replace conventional oxide for tunneling, conventional nitride and oxynitride for trapping, and conventional oxide (for SONOS) or ONO-IPD (for FG) for charge blocking. We will briefly review such select applications here. Follow-on sections will discuss other relevant investigations.

12.16.1 High K Dielectric for Charge Tunneling

Alumina was introduced by S. Jeon et al. [39], HfO_2 was investigated by X. Wang et al. [40], and R. van Schaijk et al. incorporated Si-rich HfSiON [29] as tunnel dielectric layers for their various stack designs replacing tunnel oxide. The triple high K stacks of S. Jeon et al. consisting of 3 nm Al_2O_3 for tunneling with 10 nm HfO_2 for trapping and another 10 nm of Al_2O_3 for blocking (EOT~7 nm) demonstrated low-voltage F-N operation for programming at $\pm6V$, 1–2 ms, providing a memory window of 1.4 V. Post1E5 W/E cycle retention was improved to ~54% of the original window baked at 85°C by D2 post-metal annealing at 400°C, 30 min of the stack, thereby improving interfacial quality and reliability of the stack. The Al_2O_3 and HfO_2 films were deposited by sputtering in argon. X. Wang et al. compared two stack designs: Si/HfO_2 (4.8 nm)/Ta_2O_5 (6.4 nm)/HfO_2(10 nm)/ TaN called "MHTHS" and Si/HfO_2/Ta_2O_5/Al_2O_3(4.0 nm)/TaN called "MATHS" with the same stack EOT of ~5.0 nm for both designs and same thicknesses for tunnel layers of HfO_2 and trapping layers of Ta_2O_5. They found Al_2O_3 as blocking dielectric was more effective in reducing leakage to the gate during retention compared to HfO_2 as a blocking dielectric [40]. The lower barrier energy of HfO_2 as tunnel dielectric compared to SiO_2 demonstrated higher programming speed for both stack designs. However, the MATHS stack design proved superior in achieving a larger initial window of ~2.5 V and EOL charge retention of ~56% at 85°C compared with the MHTHS stack of initial window of ~0.9 V with charge retention ~48%. For the MATHS design, programming and erasing could be achieved at ±6 V, 1 ms as well as very fast speed of 1 us at ±10 V. Excellent device endurance was demonstrated for the stack design even at W/E cycling at ±10 V, 1 ms. It should be noted that the EOT for the tunnel dielectric of HfO_2 was ~1 nm, which enhanced charge tunneling, possibly by direct tunneling, and provided very high dielectric strength for such ultra-thin HfO_2 films. The HfO_2 were deposited on NH_3-treated silicon interface using MOCVD processing technique followed by postdeposition anneal. All other films in the stack design used PVD processing for the metal deposition followed by oxidation treatments for Ta_2O_5 and Al_2O_3 films at low temperatures of 550°C and 400°C, respectively. PVD TaN gate formation was carried out at 700°C. Instead of HfO_2, R. van Schaijk et al. employed a nonstoichiometric silicon-rich HfSiON as tunnel dielectric and a near-stoichiometric HfSiON as blocking dielectric, with a conventional nitride trapping dielectric for their stack design [29]. The silicon-rich HfSiON with Si/Si+Hf = 0.77 provided an Ube = 2.5 eV compared to HfO_2 of 1.5 eV and Ubh = 3.4 eV (compared to oxide of 4.6 eV and HfO_2 of 3.1 eV). However, the silicon-rich HfSiON reduced the K value to ~6 compared to the stoichiometric HfSiON K value of ~12–14. The silicon-rich HfSiON provided a high-enough energy barrier for electrons to minimize reverse tunneling of electrons during standby and yet low enough compared to oxides to achieve high electron injection during programming with nearly similar Ubh for holes compared to HfO_2 for faster erase and significantly lower than that of SiO_2. The near-stoichiometric HfSiON-blocking layer provided the characteristic lower leakage of HfSiON (compared to HfO_2) for reducing charge loss to the gate. The stack consisted of 3–4 nm of HfSiON (EOT ~ 2–2.6 nm)/6 nm of nitride/10 nm of HfSiON with a total stack EOT of ~9–10 nm. The stack demonstrated nearly 80% charge retention at room temperature and nearly 60% charge retention at 85°C at the EOL of 10 years with an initial memory window of >2.0 V when written and erased at ±12 V, 0.5–1.0 ms. The device exhibited minimal endurance degradation up to 1E7 cycles at ±12 V, 1 ms Vpp.

12.16.2 High K Dielectric for Charge Trapping

Nitride replacement with higher K dielectric layers was motivated by several considerations. These include larger memory window possibility by seeking higher trap density dielectric with lower bulk conductivity (deeper trap) with consequently lower leakage and improved retention. Additionally, the band structure of such material would be preferred with lower conduction band offset and higher valence band offset with reference to silicon in comparison to nitride to establish primarily the electron transport during writing and erasing, and reduce or eliminate hole transport during erasing. Such a band structure would enhance both programming speed (writing and erasing both being solely determined by electron mobility through the tunneling dielectric) and reliability (reducing hole-induced interface states generation as well as trap generation in dielectric layers). Additionally, a higher K would aid in reducing programming voltage levels and field within the dielectrics aiding endurance.

Nitride exhibits a trap density of 5E12/cm^2, trap depth of 1.1 eV, and band offsets (with reference to silicon) of ~2.1 and 2.0 eV for electrons and holes, respectively. Nitride was replaced in early years with Si_2ON_2 (see Chapter 4), with higher valence band offset demonstrating improved window and retention for MXOS devices. Several dielectric films have been investigated to replace nitride. These include in descending K values: Ta_2O_5 (K = 26), HfO_2 (K = 24–25), HfAlO (K = 14–17, 10% Al), AlN (K = 10) [41], and GaN (K = 10) [42]. All of these films exhibit higher bulk trap density ($\geq$1E13/cm^2), deeper traps $\geq$1.5 eV with Ta_2O_5 reported to be the deepest of 2.7 eV [40]. All films exhibit lower conduction band offsets, with Ta_2O_5 and AlN (both ~0.1 eV) and GaN (−0.65 eV) being very low. All the above films also exhibit significantly higher valence band offset being ~3 eV or greater compared to nitride, AlN being the highest of 4.4 eV. Within the framework of published data, stack structures with the above trapping layers displayed faster programming and improved retention. The best retention results thus far have been demonstrated by the work of K.-H. Joo et al. using GaN dielectric as the trapping layer using oxide tunneling [43]. More systematic investigation would be required to establish relative merits of the above trapping dielectrics.

12.16.3 High K Dielectric for Charge Blocking

It has been discussed before that high K Al_2O_3 was first introduced in stack designs for charge blocking for both FG and CT types of devices to enhance NVM device attributes. For gate electrode compatibility and to prevent undesirable charge injection from the gate, thin films of TaN, TiN, or WN had been successfully employed. Further improvements in device characteristics were observed by optimizing the composition of hafnium aluminate (HfO_2) × (Al_2O_3)1-x to achieve lower leakage and higher K value for the blocking dielectric. Application of stoichiometric HfSiON as a blocking layer was also shown to achieve desirable NVM characteristics.

12.17 APPLICABILITY OF HIGH K DIELECTRIC FILMS FOR NVM GATE STACK DESIGN

We shall now discuss the applicability of the common high K films as well as potentially other high K films for replacing conventional FG-NVM and SONOS types of CT-NVM devices. As discussed in Part I, NVM devices operating in the normal mode have requirements of dielectric films to satisfy three distinct functions: the tunneling medium, the charge storage medium, and the charge blocking medium.

In order to address the applicability of high K films, one could examine the effective functional requirements for the tunneling medium, the charge storage medium, and the charge blocking medium from the point of view of general characteristics of such films matching the device requirements. However, we will take a non-conventional approach to discuss the merits and limitations of high K films discussed above by actually considering the design of an NVM device stack with single layer high K films for each of the three above functional requirements and analyze a plausible stack design, and its merit and shortfalls.

12.17.1 Application Case Exercise

In this exercise, we consider the high K film applicability for a CT-NVM device stack design for CMOS platform compatibility reasons. These reasons are (a) the tunnel layer for NVM and gate dielectric layer for CMOS FET could be the same; and (b) there exists greater compatibility of CT-NVM devices with scaled CMOS platform technology from the standpoint of power supply requirements, since such devices require nearly half the programming voltage compared to the FG-NVM devices and consequently nearly half the EOT requirement for the stack compared to that of the FG-NVM stack.

12.17.1.1 Case Example of the CT-NVM Device Stack Design and High K Film Applicability

Let us consider a set of flexible stack design objectives of a scaled CT-NVM device with all high K films discussed earlier with the intent of satisfying each functional element of the NVM stack with the best fit high K dielectric film. Since the functional requirements of tunnel and charge blocking layers are essentially similar for both CT-NVM and FG-NVM devices, such considerations of dielectric films should be applicable to both device types of devices with the exception of the element for charge storage. The design objective for such a device is as follows:

Vpp: ± 5 to ± 7 V, 1–10 ms (lower Vpp, faster speed preferred)
End-of-life (EOL) Window: **Weol** = 1 V (SLC); **Retention:** = 1–10 years. (1E8 ~ 1E9 sec.);
Endurance: >>1E7 W/E cycles

Vpp requirement broadly sets the stack EOT to be approximately 5.5 ± 0.5 nm to 8.5 ± 1.0 nm, assuming <10 MV/cm of average field across the dielectric stack for writing and erasing (see Part I, Chapter 6 on NVM device properties). This should also satisfy the above endurance objective. If one makes an additional assumption that the charge centroid should be at the middle of the charge-trapping layer of the stack to maximize retention and minimize leakage (a somewhat symmetric stack design concept of nearly similar EOT for each of the stack elements from charge centroid standpoint), the resulting stack design could be simple and would require a lower value of Vpp. Therefore, a best case and simplified design with minimum iterations could be the following for considerations of applicability of the appropriate high K elements:

Best case:
Tunneling element: <~1.5–2 nm EOT
Trapping/Storage element: <~2 nm EOT
Blocking element*: ≥2 nm EOT

* Design of blocking element is not thickness constrained and customarily thicker to reduce leakage to the gate at the expense of higher EOT for the stack and higher Vpp.

We shall now examine the relative merits of high K dielectric films to fulfill the above stack design objective items 2 and 3, that is, (a) the required initial window, (b) the standby leakage, (c) the EOL window, and (d) the retention requirements. We will assume that electrons and not holes play major roles in generating the memory window, charge loss/leakage during standby due to the electrostatic field of the stored charge and define the retention characteristics for the high K films in the device stack. Such assumption is reasonable since in most high K films, hole mobility is low and barrier energy with reference to silicon is higher. Electronic characteristics of most common dielectric films investigated appear to confirm such assumption.

12.18   HIGH K FILMS FOR TUNNELING

12.18.1   Electron Transport and Current Requirements for Window Generation

Electron transport mechanism should be expected to follow F–N Tunneling or modified F–N Tunneling since the K value for the high K tunnel dielectric film of interest should be ≥ 10, which implies that the physical thickness of such film would be >3.5 nm for 1.5 nm EOT or thicker for thicker EOT, and therefore, direct band-to-band tunneling would not be the operative mechanism. It has been noted in Chapters 4 and 5 in Part I that F–N tunneling current is strongly sensitive to the energy barrier offset, Ub [reducing proportional to exp (-Ub powered 3/2)] while enhancing with the field, E, proportional to the square of the field. For SiO_2 as the tunnel dielectric, with a high barrier energy offset from silicon conduction band of 3.2 eV, the required current transport of ~1E-2A/cm², for example, imposes a very high E field of +~11 MV/cm across the tunnel SiO_2 for 10 ms of writing speed. Such high field not only limited the programming voltage scalability but also impacted the endurance and reliability as discussed in Part I. In addition to the requirements associated with the scaling of gate dielectric for FET devices such as low leakage at gate voltage, high breakdown strength, high stress reliability, and so on, the tunnel dielectrics for NVM devices require high electron transport (high fluence) at modest field for higher speed of writing and erasing to achieve desired memory window while at the same time extremely low leakage through the tunnel dielectrics at standby conditions. While high fluence requires low band offset, low leakage requires high band offset. This makes selection of a single high K dielectric film for tunnel layer a challenging problem. Nevertheless, let us examine the attributes of single metal oxide films discussed earlier and HfSiON along with two other metal oxide films of lanthanide family La_2O_3 and Pr_2O_3 (see appendix for lanthanides) for application for tunneling layer for NVM devices. Table 12.3 lists the EOT limit (maximum EOT) and corresponding physical thickness limit for current transport (electron fluence) of ~ 1E-3 A/cm² required at voltage drop across the tunnel layer of 1.5V to achieve the required memory window for writing (or erasing), for programming time of 10 ms assuming the objectives of stack design set earlier. Other parameters of interest, namely, band gap (Eb), band offset with reference to silicon substrate (Ub), average stress field across the tunnel layer during programming, and breakdown strength of the dielectric films, are also provided. Relevant electronic properties of SiO_2 tunnel dielectric and OR-SiON tunnel dielectrics are presented as references for comparison with similar parameters of high K dielectric films in Tables 12.2 and 12.3. respectively.

TABLE 12.2
Reference Dielectric Films: Tunneling Characteristics[a]

Ref. Films	K	Eb ev	Ub/e ev	EOT[b] nm	Actual[b] nm	Programming MV/cm	Breakdown MV/cm
SiO_2	3.9	9	3.2	3.5	3	4.3	>11
OR-SiON	5	>7.2	3	3.75	4	4	>11

[a] For tunnel current of 1E −3 A/cm² at 1.5 V drop across the tunnel layer.

[b] Maximum value.

TABLE 12.3

High K Metal Oxide and HfSiON Dielectric Films: Tunneling Characteristics[a^]

Film	Al_2O_3	HfO_2	ZrO_2	La_2O_3	Pr_2O_3	TiO_2	HfSiON	$LaAlO_3$
K	10	25	25	30	31	80	14	25
Eb	8.7	5.7	5.7–7.8	4.3	5.1	3.5	6.9	6.5
Ub-e (ev)	2.75	1.5	1.5–3.5	2.3	2	0	3	2.4
EOT (nm)[a]	2.3	2	2.3	<1.5	<1.0	3.9	5	<1.0
Thickness (nm)[b]	5.75	12.5	<10	<10	<7.0	78	17.5	<7.0
Program Field (MV/cm)	2.6	1.2	<1.5	<2.0	<2.0	<0.2	<0.8	<2.0
Breakdown field (MV/cm)	>10	>10	>10	>10	>10	>>10	>>10	>10

[a] For tunnel current of $1E-3$ A/cm^2 at 1.5 V drop across the tunnel layer.

[b] Maximum value.

From the 12.2 and 12.3, the following observations could be made:

1. Unlike SiO_2, high K films provide greater margin between programming field and breakdown field, thereby enhancing the reliability of the NVM device. Programming field is lowest for TiO_2 with highest K value with very little barrier energy (offset Ube nearly 0), followed by Ta_2O_5 for similar reasons. This should be anticipated since the energy required for tunneling should be small and electron injection from the substrate is enhanced by the field. It is interesting to note that, however, HfSiON provides similar results in spite of higher Ube compared to other metal oxide films. Among the other high K films, HfO_2 has the greater margin. It has been discussed before that the high K values of TiO_2 and Ta_2O_5 are not realizable over silicon substrate due to their reactivity with silicon to form significantly lower K silicates. From implementation point of view, the effective K values are significantly lower when a lower K passivation layer is introduced between the high K film and silicon. In addition, integrating these films into base CMOS technology for gate dielectric had been challenging. Therefore, for practical considerations, HfSiON and HfO_2 would be preferred from current transport and stress field considerations. $LaAlO_3$ may also be very attractive since it exhibits better thermal stability and high-temperature process feasibility directly on silicon coupled with leakage comparable to Pr_2O_3 (see later).

2. The maximum EOT values (and correspondingly the physical thickness numbers) to achieve the desired current transport suggest that electron transport could be significantly enhanced in films with higher values compared to the others on a relative basis. Here again, similar considerations suggested in observation 1 holds with the implication that from implementation point of view, HfSiON film should perform the best. It should also be noted that the lowest EOT (max) value was exhibited by Pr_2O_3 and $LaAlO_3$ films, followed by the La_2O_3 film, despite their relatively lower Eb values compared to Al_2O_3. These lanthanide films exhibit significantly lower conductivity at relatively high stress field and correspondingly lower leakage. Such attributes could be of advantage in NVM stack design, which will be further discussed.

3. From the evaluation in Table 12.2, one would surmise that the possibility of voltage scalability of ±5V to ±7V for Vpp could be feasible by replacing SiO_2 with high K films from the stand point of required charge transport during programming.

12.19 LEAKAGE AND RETENTION FOR HIGH K TUNNEL DIELECTRIC FILMS

Leakage and retention requirements for tunnel dielectrics for meeting the EOL window objective is discussed here. We shall now examine the characteristics of high K metal oxide films and HfSiON from the standpoint of charge leakage during standby (once the charge is stored) to meet the retention objective stated earlier. In this estimation, we shall assume the charge loss by electron transport alone and only by leakage to the silicon substrate due to the internal electrostatic field created by the charges stored in the trapping layer. We shall first estimate (a) the charge stored during writing (high Vth state), (b) the internal potential thus created (and thereby the memory window at the gate), and (c) the leakage requirement to meet the EOL window/retention objective. Then we will evaluate/estimate how the dielectric films of interest stack up to meet such objective. It should be noted that the evaluation will be qualitative, comparative, estimate only, and non-rigorous based on the first principle.

12.19.1 Charge Stored during Writing and Internal Potential

We will make the following assumption to get an estimate of charge stored Q during writing for the above stack:

1. The charge centroid is at the center of the stack: 2.5 nm from substrate interface.
2. The peak field during writing is 50% of the pulse duration and 50% of all fluence (electron flow) is trapped to arrive at the steady state.
3. The current density at the peak field is 1E-3 A/cm².

Using the simple electrostatic equation of:

Change in Vth (at the gate): Q/C = (J. t. EOT)/ Kox. Eo where Kox = 3.9 and Eo = 8.853E-14 and t = 2.5 ms, J = 1E-3, EOT = 2.5 nm, we obtain, change in Vth = 1.8 V (see Chapter 5).

Thus we obtain the initial internal potential to be 0.9 V and associated average EOT field to be approximately ~0.9/2.5E-7 = 3.6 MV/cm.

The leakage requirement:

Assuming that the internal potential could only drop to (0.9–0.5) V = 0.4 V (so that EOL window at the gate ≥1 V), during the life time of ~1E9 sec (≥10 years). The allowed leakage current can be estimated to be: J (leakage) = [0.4 (potential drop). Dielectric Capacitance]/[1E9 (life time). Area], this results in leakage current required to be ≤5.52 E-16 A/cm²!!!

Note that this is a very low leakage requirement. If we cannot achieve such low leakage, we may have to compromise retention objective.

12.19.2 Leakage Characteristics for High K Films for the Tunnel Applications

If the stored charge potential drops from the initial state to the end-of-life from 0.9 V to 0.5 V, the field across the tunnel layer drops from 3.6 MV/cm to 2.0 MV/cm. Table 12.4 provides the approximate leakage characteristics of the high K films of interest in these field ranges [44].

It should be apparent from the above table that none of the single dielectric films would satisfy the leakage requirements; however, $LaAlO_3$ and Pr_2O_3 may come close. Although some significant R&D effort is required to appropriately integrate ultra-thin Pr_2O_3 films in NVM technology, $LaAlO_3$ is a real possibility (see appendix on lanthanides). Alternately one could consider Al_2O_3, or a combination of HfO_2/Al_2O_3 [8] or ZrO_2 or La_2O_3 films or even HfSiON or HfO_2 with compromised stack design objectives.

TABLE 12.4

Leakage Characteristics of High K Films (Estimate from [44])

Film[a]	Leakage at 3.6 MV/cm A/cm^2	Leakage at 2.0 MV/cm A/cm^2
Al_2O_3	1.00E-09	1.00E-11
HfO_2	1.00E-07	1.00E-09
ZrO_2	3.00E-10	1.00E-11
La_2O_3	1.00E-09	1.00E-11
Pr_2O_3	1.00E-12	1.00E-14
TiO_2	1.00E-08	1.00E-10
HfSiON	<<1E-10	<1E-13
$LaAlO_3$	1.00E-12	1.00E-14

[a] All Film EOT = 2.5 nm, Approximate values.

The exercise above highlights the challenges to incorporate single Hi K films for tunnel dielectric applications. The solution lies in "Band Engineering" using multi-layer high K dielectric films such that an "asymmetric" energy barrier is established for electrons (or holes) in the stack design [45] (see Chapter 13). While the energy barrier for electrons at the silicon/high K interface could be relatively small for high carrier injection at relatively low field for writing or erasing, the trapped charges could face a significantly larger energy barrier to return to the silicon substrate (or to leak to the gate, see later), and thereby reduce the leakage current to the desired level to meet the retention objective. This will be discussed in detail in the section on band engineering.

12.20 HIGH K FILMS FOR CHARGE TRAPPING

For SONOS types of CT devices, nitride had been widely used as a charge-trapping medium and continues to enjoy the same role even today in spite of its limited K value (K = 7). Nitride had a long development history, extensively characterized, easy to integrate in silicon based technology, and highly reliable and reproducible. Nitrogen-rich-silicon-oxy-nitride (NR-SiON) films have very similar characteristics to those of nitride and, with optimum composition, have the advantage of higher trap density and deeper trap depth compared to nitride without lowering the K value and also had been employed effectively for the charge-trapping layer in CT-NVM devices. Transport characteristics of these films have been discussed in Part I of this book. Trapping and detrapping characteristics of charges in nitride and NR-SiONs have been associated with field-enhanced thermal excitation of trapped carriers and are characterized by the Poole–Frenkel parameters discussed in Part I Chapters 3 and 4. Field and temperature dependence of trapping and detrapping in these films have been investigated in earlier years by Bhattacharyya et al. [46] and in recent years by H. T. Lue et al. [47]. These will be used as references when other higher K films are compared and the characteristics of these films are listed in Table 12.5.

There had been limited investigation in the past on higher K films for charge-trapping replacements of nitride and NR-SiONs. Aside from nitride and oxynitrides, Al_2O_3 was also investigated and characterized. In recent years, there has been renewed interest to investigate trapping and detrapping characteristics of several other lower band gap high K films for potential replacement of nitride for the charge-trapping layer in NVM devices. These include AlN (K = 10), GaN (K = 10), ZnO, HfO_2 (K = 24), and others. These high K films exhibit

TABLE 12.5

Charge-Trapping Characteristics of Si_3N_4, Si_2ON_2, and Al_2O_3 (Reference Dielectric Films)

Film	Process	Ub ev	K	Capture x Sec	Depth ev	Density #/cm²	Transport Mechanism
				Trapping Parameters			
Si_3N_4	CVD/LPCVD	5.1	7	1E13–1E14	1.1	5.00E+12	Poole–Frenkel
Si_2ON_2	CVD/LPCVD	6.5	7	1E13–1E14	1.67	1.00E+13	Poole–Frenkel
Al_2O_3	CVD/ALD	8.7	10-Sep	1.00E17	1.5–2.0*	2.00E+12	Tunneling

characteristically high density of intrinsic (material) and extrinsic (process-induced defects) traps. Additional traps are also generated during high field stressing [8]. Some of the promising high K dielectric films for trapping are AlN (Aluminum Nitride), GaN, ZnO, HfAlO (Hafnium Aluminum Oxide), HfO_2, Ta_2O_5, and TiO_2 in order of their higher K values. In general, most of these films exhibit transport, trapping, and detrapping through Poole–Frenkel mechanism. Additionally, the trapping parameters are process and stoichiometry sensitive as seen in nitrides and oxynitrides. The characteristics of some of these films from charge-trapping points of view are shown in Table 12.6.

Film Stability:

a) Thermal		?	?	High	OK	Low	Low	Ok
b) Chemical		OK	OK	?	OK	OK	OK	Low
c) Interface with dielectric:		OK	OK	OK	OK	OK	OK	Low
Postprocessing thermal:		NA	NA	950C	800C	700Cª	700Cª	Poor
		RTA	RTA	RTA	RTA	RTA	RTA	RTA

ª Limited thermal budget.

TABLE 12.6

Charge-Trapping Characteristics of Selected High K Films for Nitride and NR-SiON Replacements

Parameter	ZnO	GaN	AlN	HfAlO	HfO_2ᵇ	Ta_2O_5	TiO_2	
				Films				
Dielectric constant K	NA	10	10	14–17	24	26	80	
Band gap Eb, evª	3.37	3.39	6.0	6.2	5.7	4.5	3.5	
Band offset@, Ub-e/h, ev	0.27/3.64	0.95/3.24	3.6/0.3	1.7/3.4	1.5/3.1	0.1/3.3	0.0/3.4	@with respect to Si
Band offset#, Ub-e/h, ev	4.4/1.28	4.15/1.46	2.5/0.5	1.5/1.3	1.65/1.65	3.1/1.4	3.2/2.3	# with respect to SiO_2
Trap density, No/cm²	>>1E13	>>1E13	>1E13	8E12	1E13	>1E13	>1E13	
Trap depth, Utr-e, ev	NA#	NA#	Multiple	Multiple 0.3/1.5	Multiple 0.3/1.5	2.7	NA	
Capture cross-section	NA	NA	NA	NA	~1E116	NA	NA	

ª Exhibit superior retention.

ᵇ Shallow traps due to oxygen vacancy (0.3–0.5 ev) could also be present (process sensitive).

OBSERVATIONS:

1. Band energy offsets for both electrons and holes with references to silicon interface as well as oxide interface were noted for all dielectric films in the table. The former provides the barrier energy electrons and holes face with reference to silicon, while the latter provides the energy barrier for conventional cases when the tunneling dielectric is SiO_2 interfacing the trapping dielectric. Also noted was the measured electron trap depth obtained from Poole–Frenkel plots of the trapping dielectric films from temperature dependence of conductivity of dielectric films. For devices where charge trapping and charge loss mechanisms are dominated by electron motion, the higher the barrier energy and the higher the trap depth, the lower would be the stored charge loss with consequent enhancement in retention. It would therefore be desirable in such stack designs to select the trapping dielectric that would provide both higher Φb-e and Utr-e. From such consideration, AlN and Si_2ON_2 should be considered for options to replace nitride, even though the K values are relatively low for such dielectric films. The advantage of Si_2ON_2 over nitride as a charge-trapping dielectric has been discussed in Part I, since it has the integration advantages of nitride coupled with higher density deeper energy traps and lower conductivity by an order of magnitude compared to nitride. GaN appears to be an ideal trapping layer due to its large conduction band energy offset, deep traps, and passivating characteristics. K.-H. Joo et al. demonstrated excellent charge retention characteristics at room temperature with <40 MV/decade charge loss and 10 years of retention using a gate stack consisting of 3.5 nm of tunnel oxide and 20 nm of blocking HfO dielectric and Al gate [48]. This suggests deeper trap depth in GaN although no direct measurement of trap depth was reported. However, very little additional NVM device data exists employing GaN as a trapping dielectric. In case of AlN, more investigation is needed to address the integration and interface challenges. The presence of shallow traps in HfO_2 and HfAlO [49] (which is process sensitive) could be challenging for general applicability, but if successfully eliminated, could be attractive replacements for nitride due to their significantly higher K values lending the stack to be more scalable. HfAlO is more thermally stable and therefore would be preferred over HfO_2. In spite of the higher K values, of Ta_2O_5 and TiO_2, and in the former case, deep energy traps, these films exhibit difficult integration challenges,

2. Due to their strong chemical reactivity. But if the challenges are solved, these could have future application possibilities. Both GaN and ZNO could be exciting future options due to their negative conduction band offset (NCBO), very high density deep traps [42] whereby larger window and longer retention could be simultaneously achieved. However, further characterization and integration challenges need to be addressed.

3. Band energy offsets for both electrons and holes with references to silicon interface as well as oxide interface were noted for all dielectric films in the table. The former provides the barrier energy electrons and holes face with reference to silicon, while the latter provides the energy barrier for conventional cases when the tunneling dielectric is SiO_2 interfacing the trapping dielectric. Also noted was the measured electron trap depth obtained from Poole–Frenkel plots of the trapping dielectric films from temperature dependence of conductivity of dielectric films. For devices where charge trapping and charge loss mechanisms are dominated by electron motion, the higher the barrier energy and the higher the trap depth, the lower would be the stored charge loss with consequent enhancement in retention. It would therefore be desirable in such stack design to select the trapping dielectric that would provide both higher Ub-e and Utr-e. From such consideration, AlN and Si_2ON_2 should be considered for options to replace nitride, even though the K values are relatively low for such dielectric films. The advantage of Si_2ON_2 over nitride as charge-trapping dielectric has been discussed in Part I, since it has the integration advantages of nitride coupled with higher density deeper energy traps and lower conductivity by

an order of magnitude compared to nitride. In case of AlN, more investigation is needed to address the integration and interface challenges. The presence of shallow traps in HfO_2 and HfAlO (which is process sensitive) could be challenging for general applicability, but if successfully eliminated, could be attractive replacements for nitride due to their significantly higher K values lending the stack to be more scalable. HfAlO is more thermally stable and therefore would be preferred over HfO_2. In spite of the higher K values of Ta_2O_5 and TiO_2 and in the former case deep energy traps, these films exhibit difficult integration challenges but if solved could have future application possibilities. Both GaN and ZNO could be exciting future options due to their NCBO, very high density deep traps [42] whereby larger window and longer retention could be simultaneously achieved. However, further characterization and integration challenges need to be addressed.

4. All dielectric films noted above exhibit high value of Ub-h compared to nitride films. In general, hole mobility in dielectric films is significantly lower than electron mobility with few exceptions. Stack designs are favored for electron motion compared to hole motion, if charge loss could be contained and retention objectives could be met. However, to improve low Vth (erase) state stability of the device, a higher K film with relatively higher Ub-h (with reference to silicon) would be desirable. From this point of view, Hf AlO would be most desirable due to simultaneously providing higher K (14–17) and significantly higher Ub-h when compared with nitride, provided shallow trap and integration challenges could be successfully addressed.

5. It should be noted that all films listed in the table exhibit higher trap density compared to nitride. This is obviously a key requirement for selecting such high K film to be classified as "trapping dielectric film."

6. Another parameter of significance is the barrier energy symmetry between Ub-e and Ub-h when referred against silicon substrate especially when the counter gate electrode is also silicon gate. It should be noted that both Si_3N_4 and Si_2ON_2 films are relatively more symmetric in terms of Ub-e and Ub-h.

From the above observations, we could draw the following conclusion about applicability of the trapping dielectric for our case example of stack design:

Preferred selection: (a) HfAlO film (EOT = 1.5–2.5 nm, physical thickness ~5–8 nm, assuming other above-mentioned challenges are resolved or alternately; (b) Si_2ON_2 film with EOT = 2–2.5 nm and physical thickness ~3–4 nm.

12.21 HIGH K FILMS FOR CHARGE BLOCKING

As the name suggests, the purpose of the high K charge blocking layer is to prevent charge leakage to the gate of the stored charges during standby state to achieve retention objective as well as to block charge injection from the gate (of opposite polarity) during programming to minimize charge compensation at the storage node for effective programming. Therefore, the blocking high K film should have the characteristics of providing high barrier energy not only at the trapping layer/blocking layer interface, but also at the blocking layer/gate interface. The latter condition requires gate material compatibility with the blocking dielectric. This way the blocking layer and the gate selection need to be together taking into consideration technology integration and postprocessing requirements. The blocking layer is designed to be thicker to reduce charge transport to the gate during standby, thereby reducing the electrostatic field across the blocking layer. Since by design, charge loss to the gate should be avoided, the blocking layer should be one of large band gap and trap-free dielectric film with leakage requirements similar to that of the tunneling film. Therefore, the referred tunneling layer may also be considered for the blocking layer as long as the gate interface is compatible. With the above consideration in mind, we will make an additional assumption that our stack design will allow metal nitride gate such as TiN or TaN or WN for compatibility with

TABLE 12.7

High K Films for Charge Blocking Layer Applicability

Film	OR-SiON	Al_2O_3	HfO_2/Al_2O_3 (4)	HfSiON	$LaAlO_3$ (230)
K	5.0	9–10	12–14	14	27.5
Band gap ev	7.2	8.7	NA	6.9	6.5
Ub-e ev/Si	3.0	2.75	NA	3.0	2.4
Ub-h ev:Si	3.1	4.85	NA	2.8	3.1
Leakage @ 2 nm	1E-8/1E-10	1E-10/1E-13	1E-12/1E-15	1E-8/1E-11	1E-13/1E-16
Minm EOT, nm: to avoid leakage	~5 nm	~4 nm	~3.5 nm	~5 nm	~3 nm
Charge-trapping probability	1E-14/1E-10	1E-5/1E-6	1E-3/1E-5	<1E-10	NA
Stability					
Thermal	Good	Good	Medium	Good	Good
Chemical	Good	OK	OK	Good	OK
Electrode	n+Poly	TaN	TaN	Poly	TaN

Note: Leakages are estimate only.

n+ polysilicon gate at the blocking dielectric interface so that we will not be otherwise restricted in the selection of high K blocking dielectric film. Table 12.7 provides the key factors to consider in selecting the high K blocking dielectric among the high K films discussed earlier.

OBSERVATIONS:

The leakage levels and minimum EOT requirements to avoid charge loss at standby through the blocking layer are approximate estimates only, but these reflect the characteristics of the dielectric films. The best blocking layer would be $LaAlO_3$, which would require the metallic interface (TaN) for compatibility. The option HfSiON would require a thicker blocking film to reduce the field and consequently the leakage level but have the advantage of conventional polysilicon gate technology without requiring a buffer TaN layer.

From the process integration point of view, all options except SiON and HfSiON would require limited thermal budget after the stack formation.

12.22 DEVICE DESIGN OBJECTIVES AND DIELECTRIC SELECTION OPTIONS

It is clear from the above exercise that simple selection of a single layer of high K film to meet functional requirements of tunneling, charge storage, and blocking to fulfill design objective of the NVM device is difficult to attain. When integration issues are taken into consideration, there could be limited choice and design objective would be compromised. However, application of appropriate high K films provides the following possibilities:

1. Vpp reduction and/or write/erase speed enhancement
2. Endurance enhancements over conventional devices
3. Enhancement of device reliability
4. Greater tradeoff between memory window and retention

While there could be several options and compromise toward the stack design objective in the application exercise undertaken above, two general approaches could be considered in selecting the high K elements for our stack design: (a) conventional CMOS integration approach with higher thermal budget and conservative, lower cost (R&D), highly compromised stack design approach; and (2) more aggressive, higher R&D stack design approach with least relaxed design objective.

For Case (A): Potential stack design options are:

A1: Si/HfO$_2$-Al$_2$O$_3$ (tunneling) [2 nm]//Si$_2$ON$_2$ (trapping)[2.5 nm]//HfSiON (blocking)[4.5 nm]/poly; total stack EOT = 9 nm or

A2: Si/HfSiON (tunneling) [3 nm]//HfO$_2$ (trapping)[1 nm]//HfSiON (blocking)[5 nm]/poly; total stack EOT = 9 nm

Both of the above options of stack design would use conventional CMOS processing, the tunnel layer as EOT for CMOS FETs, compatible processing schemes for high K layers, and meet the following NVM characteristics:

Vpp <±8V, for programming duration of 1–10 ms
EOL window >1 V
Retention >1 month
Endurance >1E7 cycles

For Case (B): Potential stack design options are:

B1: Si/HfO$_2$/Al$_2$O$_3$ (tunneling) [1.5 nm]//HfAlO (trapping) [1.5 nm]//HfO$_2$/Al$_2$O$_3$ (blocking) 3 nm/# Or

B2: Si/LaAlO$_3$ (tunneling) [1.5 nm]//HfAlO (trapping) [1.5 nm]//LaAlO$_3$ (blocking)3 nm/ ## the gate electrode requires TaN as buffer metallic interface in both options still using standard CMOS processing. TIF stack EOT for both cases would be 6.0 nm. The design is expected to meet the following NVM objective:

Vpp <±5.5 V, for programming duration of 1–10 ms
EOL window >1 V
Retention >1 Year
Endurance >1E7 cycles

Scaled NVM device design will not only require high K films, but also will require considerations of integration issues including scaled CMOS FET devices for peripherals and band engineering. These topics will be covered in subsequent sections.

12.23 OTHER POTENTIAL FUTURE HIGH K DIELECTRIC FILMS FOR NVM DEVICES

We have discussed in greater details selected single metal oxide films for their historical significance—a specific hafnium aluminate (HaAlO), a specific lanthanum aluminate (LaAlO$_3$), and a specific hafnium oxynitride (HfSiON)—because of their favorable characteristics to replace oxide and oxynitride for CMOS gate dielectric as well as for potential applications as tunnel layer and blocking layer for NVM devices. In Appendix A, we have included, respectively, a historical summary of high film for DRAM application, potentials of bimetal oxynitrides and lanthanides. In recent years, considerable investigations have been undertaken to explore potential high K ultra-thin films for microelectronics and nanoelectronics applications. These include multi-element oxides, silicates, oxynitrides, aluminates, and carbides to seek out unique characteristics for extended silicon device applications. The list is exhaustive and will be out of scope of this book. Some of the films which stand out for potential NVM device applications are:

Complex oxides: Hf AlO$_3$ and HfLaO
Complex silicates: HfSiO and ZrSiO
Complex oxynitrides: Hf LaON, HfTaON, and HfAlON

Although we will not be specifically describing these films in detail in this section, we may introduce such films and other high K films in other sections pointing out the unique characteristics of advantage in applications of such high K films from device perspective.

The potential for simultaneous application of high K films for charge trapping and charge blocking with compatible gate electrode while using conventional SiO_2 for tunneling was demonstrated by C. H. Lai et al. [41]. Their application of a stack consisting of SiO_2/AlN/HfAlO/IrO_2 demonstrated fast erase (Vpp = ±13 V, 100 μs), large window (W = 3.7 V, EOL W = 1.9 V), and good retention at 85°C post-1E4 W/E cycling. For erasing at −13 V, 1 ms, an initial window of 5.5 V and EOL W = 3.4 V at 85°C was achieved.

The potential for applications of all three functions of tunneling, trapping, and blocking with high K layers to achieve lower programming voltages, and improve retention and reliability was demonstrated by S. Jeon et al. [39,50]. The authors characterized stacks consisting of 3.0 nm of Al_2O_3 film for tunneling, 10 nm of HfO_2 film for trapping, and 10 nm of Al_2O_3 film for blocking (stack EOT ≤9 nm) and demonstrated ±6V programming, in 1–2 ms, with 1.4 V of initial memory window with a 10-year EOL window of 0.53 V after 1E5 W/E cycles. To improve interfacial quality, high pressure H_2 or D_2 annealing was performed for the capacitor stacks at 400°C for 30 minutes.

12.24   INTEGRATION OF HIGH K FILMS IN SILICON BASED CMOS NVM TECHNOLOGY

Incorporation of high K thin amorphous dielectric films in NVM technology, devices, and products requires, in principle, similar attributes discussed for conventional dielectric films in Part I. These are charge control; silicon substrate and gate interface compatibility; chemical, physical, structural, and thermal stability at processing and postprocessing conditions; chemical passivity during processing and postprocessing environments; selective etchability for device patterning; low leakage requirements; high breakdown strength and application-induced stress stability and reliability. Such requirements impose considerably greater challenges for successful incorporation and application of high K dielectric films for MOS FET devices as well as for NVM devices. In terms of electronic properties, the requirements of high K dielectric films for NVM tunnel dielectric film interfacing the silicon substrate at one end and the blocking dielectric film interfacing the gate electrode at the other end are similar in many respects to those of the CMOS FET gate insulators interfacing silicon substrate at one end and gate electrode at the other end.

These are further elaborated in this section.

In recent years, silicon technology has undergone a major shift in the approach of technology integration prompted by device-scaling requirements, use of high K as a device gate insulator replacing oxide or oxygen-rich oxynitrides and containing channel doping, short channel effects, and mobility enhancements for the ultra-short-channel FET devices. "Gate-first/diffusion-later" approach of silicon gate technology scheme has been replaced by shallow diffusion, low thermal budget and "diffusion first/gate later" integration scheme with metal gate and optimized work-function control uniquely for both PFET and NFET devices coupled with strained channel has been incorporated. This shift has an important implication for NVM technology as well due to the high K stack structures required for advanced NVM devices and high-performance peripheral circuitry requirements for the sub-hundred nano-meter technology nodes. This topic will be discussed in detail in Chapter 14.

REFERENCES

1. C. H. Lee, K. I. Choi, M. K. Cho et al., A novel SONOS structure of SiO_2/SiN/Al_2O_3/Al_2O_3 with TaN metal gate for multi-giga bit flash memories, *IEDM*, Washington DC, 2003, pp. 613–616.
2. C.-H. Lee, C. Kang, J. Sim et al., Charge trapping memory cell of TANOS (Si-Oxide-SiN-Al2O3-TaN) structure compatible to conventional NAND flash memory, *NVSMW*, Monterey CA, 2006, pp. 54–55.

3. S. Inumiya, Y. Morozumi, A. Yagishita et al., Conformable formation of high quality ultra-thin amorphous Ta_2O_5 gate dielectrics utilizing water assisted deposition (WAD) for sub 50 nm damascene metal gate MOSFETs, *IEDM*, 2000, pp. 649–652.

4. H. Sesaki, M. Ono, T. Yoshitomi et al., 1.5 nm direct-tunneling gate oxide Si MOSFET s, *IEEE Trans. Electron Dev.*, 43, 1233–1242, 1996.

5. J. H. Lee, K. Koh, N. I. Lee et al., Effect of polysilicon gate on the flatband voltage shift and mobility degradation for ALD-Al2O3 gate dielectric, *IEDM*, 2000, pp. 645–649.

6. D. A. Buchanan, E. Gusev, E. Cartier et al., 80nm polysilicon gated n-FETs with ultra-thin Al_2O_3 gate dielectric for ULSI applications, *IEDM*, 2000, pp. 223–226.

7. A. Bhattacharyya, Scalable gate and storage dielectric, ADI Associates Internal Publication, 2001, U.S. Patents: 6743681 (6/1/2004) and 6998667 (2/14/2006).

8. E. P. Gusev, D. A. Buchanan, E. Carrier et al., Ultrathin high-K gate stacks for advanced CMOS devices, *IEDM*, New York, 2001, pp. 451–454.

9. B. H. Lee, L. Kang, W.-J. Qi et al., Ultrathin hafnium oxide with low leakage and excellent reliability for alternative gate dielectric application, *IEDM*, Washington DC, 1999, pp. 133–136.

10. L. Kang, K. Onishi, Y. Jeon et al., MOSFET devices with polysilicon on single-layer HfO_2 high K dielectrics, *IEDM*, San Francisco CA, 2000, pp. 35–38.

11. S. J. Lee, H. F. Luan, W. P. Bai et al., High quality ultra-thin CVD HfO_2 gate stack with Poly-Si gate electrode, *IEDM*, San Francisco CA, 2000, pp. 31–34.

12. Y. Kim, G. Gebara, M. Freilier et al., Conventional n-channel MOSFET devices using single layer HfO_2 and ZrO_2 as high-k gate dielectric with poly-silicon gate electrode, *IEDM*, Washington DC, 2001, pp. 455–458.

13. W. Zhu, T. P. Ma, T. Tamagawa et al., HfO_2 and HfAlO for CMOS: Thermal stability and current transport, *IEDM*, Washington DC, 2001, pp. 463–466.

14. Y. Akasaka, K. Miyagawa, T. Sasaki et al., Impact of electrode-side chemical structures on electron mobility in metal/HfO_2 MISFETs with sub-1nm EOT, *VLSI Symposium*, Kyoto, Japan, 2005, p. 228.

15. M. Koyama, K. Suguro, M. Yoshiki et al., Thermally stable ultra-thin nitrogen incorporated ZrO_2 gate dielectric prepared by low temperature oxidation of ZrN, *IEDM*, 2001, pp. 459–462.

16. L.-A. Ragnarsson, S. Seven, L. Trojmanm et al., High performance 8A EOT HfO2/TaN low thermal budget n-channel FETs with solid phase epitaxy regrown (SPER) junctions, *VLSI Symposium*, 2005, p. 234.

17. Y. Ma, Y. Ono, L. Stecker et al., Zirconium oxide based gate dielectrics with equivalent oxide thickness of less than 1.0nm and performance of submicron MOSFET using a nitride gate replacement process, *IEDM*, New York, 1999, pp. 149–152.

18. T. Yamaguchi, H. Satake, N. Fukushima et al., Band diagram and carrier conduction mechanism in ZrO2/Zr-silicate/ Si MIS structures fabricated by pulsed-laser-ablation deposition, *IEDM*, 2000, pp. 19–22.

19. L. Manchanda, M. Green, R. van Dover et al., Si-doped aluminates for high temperature metal-gate CMOS: Zr-AlSi-O, a novel gate dielectric for low power applications, *IEDM*, 2000, pp. 23–26.

20. C. H. Lee, H. F. Luan, W. P. Bai et al., MOS characteristics of ultra-thin rapid thermal CVD ZrO_2 and Ze silicate gate dielectrics, *IEDM*, 2000, pp. 27–30.

21. W.-J. Qi, R. Nieh, B. H. Lee et al., MOSCAP and MOSFET characteristics using ZrO_2 gate dielectric deposited directly on Si, *IEDM*, Washington DC, 1999, pp. 145–148.

22. X. Guo, X. Wang, Z. Luo et al., High quality ultra-thin (1.5nm) TiO_2/Si_3N_4 gate dielectric for deep submicron CMOS technology, *IEDM*, 1999, pp. 137–140.

23. H. F. Luan, S. J. Lee, C. H. Lee et al., High quality Ta_2O_5 gate dielectrics with tox, eq <10A, *IEDM*, 1999, pp. 141–144.

24. X.-B. Lu, Z.-G. Liu, Y.-P. Wang et al., Structure and dielectric properties of amorphous $LaAlO_3$ and $LaAlO_xN_y$ films as alternative gate dielectric materials, *J. Appl. Phys.*, 94, 1229–1234, 2003.

25. A.-D. Li, Q.-Y. Shao, H.-Q. Ling et al., Characteristics of $LaAlO_3$ gate dielectrics on Si grown by metal-organic chemical vapor deposition, *Appl. Phys. Lett.*, 83, 3540–3542, 2003.

26. M. Suzuki, M. Tomita, T. Yamaguchi et al., Ultra-thin (EOT=3A) and low leakage dielectrics of La-aluinate directly on Si substrate fabricated by high temperature deposition, *IEDM*, 2005, pp. 445–448.

27. M. Koike, T. Ino, Y. Kamimuta et al., Effect of Hf-N bond on properties of thermally stable amorphous HfSiON and applicability of this material to sub-50nm technology node LSIs, *IEDM*, Washington DC, 2003, pp. 107–110.

28. A. Shanware, J. McPherson, M. R. Visokay et al. Reliability evaluation of HfSiON gate dielectric film with 12.8A SiO_2 equivalent thickness, *IEDM*, 2001, pp. 137–140.

29. R. van Schaijk, M. van Duuren, N. Akil et al., A novel SONOS memory with Hf SiON/Si3N4/Hf SiON stack for improve retention, *NVSMW*, 2006, pp. 50–51.

30. G. Lucovsky, Asymmetries in the electrical activity of intrinsic grain-boundary and O-atom vacancy defects in HfO_2/ZrO_2, at their interfaces with SiO_2: A possible show-stopper for CMOS high K devices, *IEEE SISC*, December 1–3, 2005, Arlington, VA.

31. J. P. Kim, Y. S. Kim, H. J. Lim et al., HCI and BTI characteristics of ALD HfSiO(N) gate dielectrics as the compositions and the post treatment conditions, *IEDM*, San Francisco CA, 2004, p. 125.

32. S. Inumiya, Y. Akasaka, T. Matsuki et al., A thermally stable sub-0..9nm EOT TaSix/HfSiON gate stack with high electron mobility, suitable for gate-first fabrication of hp45 LOP devices, *IEDM*, 2005, p. 27.

33. M. A. Quevedo-Lopez, S. A. Krishnan, D. Kirsch et al., High performance gate-first HfSiON dielectric satisfying 45 nm node requirement, *IEDM*, Washington DC, 2005, p. 428.

34. K. Torii, K. Shirishi, S. Miyazaki, Physical model of BTI, TDDB and SILC in HfO_2-based high-k dielectrics, *IEDM*, 2004, pp. 129–132.

35. M. Houssa, M. Aoulaiche, S. Van Eishocht et al., Negative bias temperature instabilities in HfSiOn/TaN-based pMOSFETs, *IEDM*, San Francisco CA, 2004, pp. 121–124.

36. P. D. Kirsch, M. A. Quevedo-Lopez, S. A. Krishnan et al., Band edge n-MOSFETs with high-k/metal gate stacks scaled to EOT=0.9nm with excellent carrier mobility and high temperature stability, *IEDM*, 2006, pp. 629–632.

37. C. H. Wu, B. F. Hung, A. Chin et al., High temperature stable [Ir3Si-TaN]/HfLaON CMOS with large WorkFunction difference, *IEDM*, San Francisco CA, 2006, pp. 617–620.

38. X. Yu, C. Zhu, M. Yu et al., Advanced MOSFETs using HfTaON/SiO_2 gate dielectric and TaN metal gate with excellent performances for low standby power applications, *IEDM*, Washington DC, 2005, pp. 31–34.

39. Y. Park, J. Choi, C. Kang et al., Highly manufacturable 32Gb multi-level NAND flash memory with 0.0098um² cell size using TANOS (Si-Oxide- Al2O3-TaN) cell technology, *IEDM*, 2006, pp. 29–32.

40. S. Jeon, S. Choi, H. Park et al., Triple high-k stacks ($Al_2O_3/HfO_2/Al_2O_3$) with high pressure (10atm) H2 and D2 annealing for SONOS type flash memory device applications, *4th IEEE Conference on Nanotechnology*, Munich, Germany, August 2004, pp. 53–55.

41. Y. N. Tan, W. K. Chim, W. K. Choi et al., High-K charge trapping layer in SONOS-type nonvolatile memory device for high speed operation, *IEDM*, New York, 2004, pp. 889–892.

42. C. H. Lai, C. C. Huang, K. C. Chiang et al., Fast high-k AlN MONOS memory with memory window and good retention, *IEEE, 63rd DRC*, Santa Barbara CA, 2005, pp. 99–100.

43. K.-H. Joo, C.-R. Moon, S.-N. Lee et al., Novel charge trap devices with NCBO trap layers for NVM or image sensors, *IEDM*, San Francisco CA, 2006, pp. 979–982.

44. D. S. Yu, A. Chin, C. H. Wu et al., Lanthanide and Ir-based dual metal-gate/HfAlON CMOS with large WorkFunction difference, *IEDM*, San Francisco CA, 2005, pp. 649–652.

45. H.-T. Lue, S.-Y. Wang, Y.-H. Hsiao et al., Reliability model of bandgap engineered SONOS (BE-SONOS), *IEDM*, San Francisco CA, 2006, pp. 495–498.

46. T. Skotnicki, Short course on applications of HK dielectrics, *Symposium VLSI Technology*, 2004.

47. A. Bhattacharyya, C. T. Kroll, H. W. Mock et al., Processing and characterization of CVD thin silicon oxynitride films, *ECS*, Las Vegas, CA, 1976.

48. S. Seki, T. Unagami, and B. Tsujiyama, Electron trapping levels in rf-sputtered Ta2O5 films, *J. Vac. Sci. Technol. A*, 1, 1825–1830, 1983.

49. Y. N. Tan, W. K. Chim, W. K. Choi et al., Hafnium aluminum oxide as charge storage and blocking-oxide layers in SONOS type nonvolatile memory for high-speed operation, *IEEE Trans. Electron Dev.*, 53, 654–662, 2006.

50. X. Wang, and D.-L. Kwong, A novel high-K SONOS memory using TaN/Al_2O_3/Ta_2O_5/HfO_2/Si structure for fast speed and long retention operation beyond the 45 nm generation, *IEEE Trans. Electron Dev.*, 53, 78–82, 2006.

13 Band Engineering for NVM Devices

CHAPTER OUTLINE

High K dielectric films and gate–insulator work function control had been successfully solving the scaling issues of FET devices since the turn of the century. Gate stack for a silicon based NVM device is significantly more complex when compared to the gate stack design of an FET device and consists of multiple dielectric layers and interfaces. These interfaces and dielectric layers have unique roles in regulating carrier transport and storage in order to achieve the unique NVM properties of memory window, charge retention, and endurance as well as material integrity and device reliability. Electronic transport and storage is fundamentally quantum mechanical and band properties of solids play a pivotal role. This chapter highlights the role of band engineering to enhance extendibility and unique device properties of scaled NVM devices in the context of multilayered gate stack designs for the NVM devices.

Historically, oxide and nitride thin films had been successfully incorporated into the different types of NVM devices until the turn of the century. This has been discussed in Part I in considerable detail. The large band gap of oxide and high energy offsets for electrons and holes at the silicon/oxide interface provided the charge retention requirements for the floating-gate devices with optimized ONO inter-poly dielectrics (IPDs) for the FG-NVM device types. Aided by the energy efficient Fowler–Nordheim charge transfer mode for writing and erasing coupled with contactless NAND array architecture and one-device cell configuration and multiple bit per cell capability provided a highly impressive low-cost NAND flash nonvolatile memory solution to the industry. This memory product drove the silicon technology scaling for the last decade. The faster NROM product with higher endurance and contactless NOR/RAM architecture based on the energy efficient modified F–N tunneling transport mode (band-to-band tunneling through ultrathin oxide and trapping in nitride) and SONOS stack design displaced other options for BIOS and code applications for the entire industry.

However, this dramatic growth of silicon based NVMs is being challenged by other NVM possibilities due to the scalability problems of conventional NVM devices and ever growing demand on enhancement of memory characteristics such as speed, density, power, and cost. The limitations on base-line CMOS technology has been overcome by successful replacement of oxide gate dielectrics by high K dieletric films and metal gate integration replacing silicon gate. The NVM device and technology has been slower to introduce such changes in the gate stack design since multiple layers of thin dielectrics and interfaces are required for such design and the integration is more complex. The limitations of oxide, nitride, and conventional NVM designs have been elaborated in Part I of this book. In the last section of Part I, we have pointed out that conventional stack design using single-layer oxide as tunnel dielectric is limited in not only scaling the programming voltage level but also in achieving simultaneously faster programming (writing and erasing), enhanced retention and endurance.

The significance of voltage scaling and high K applications has been discussed in the previous chapters. To achieve simultaneously scaled and enhanced NVM device characteristics (programming speed, memory window, retention, and endurance), such concepts have to be effectively integrated with band-engineered NVM stack designs. This will be discussed in considerable detail in this chapter.

This chapter is divided into several sections. Initially, we will explore the basic concepts of electronic transport (electrons and holes) through the various types of energy barriers presented by single and multiple dielectric films interfacing silicon and each other and the gate. During such discussion, our focus will be related to the dielectric films assuming that electrons (or holes) are available at the silicon–insulator (or gate–insulator) interface, and for simplicity, we will ignore the band bending associated with such interfaces. We will discuss the role of the dielectric films in modulating the energy barriers when subjected to internal and external fields ("band bending"). We will discuss the concepts of "barrier energy lowering" and "barrier thinning" from the point of view of electronic (both electrons and holes) transport. We will classify different types of energy barriers into their key attributes. We will then investigate the desired attributes of tunneling, storage, and blocking mediums required to achieve device requirements and to complement/address trade-offs through energy barrier engineering or "band engineering." We will explore the potential of single-layer versus multiple-layer dielectrics in applications of: (a) writing and erasing (e.g., programming voltage levels and programming speed), (b) in charge storage, (c) in charge blocking (e.g., retention), (d) in cyclic stress stability (endurance), and (e) reliability. We will illustrate multiple applications of band engineering for specific enhancements of device properties, namely, memory window, retention and endurance, and limitations of achieving such enhancements simultaneously. We will, thereafter, combine all elements in the applications of stack designs for FG and charge-trapping (CT) devices to simultaneously achieve optimal device properties to overcome limitations of conventional stack designs mentioned in Part I of this book.

Our focus on stack design will be oriented primarily for normal mode devices where silicon substrate plays the active role for charge source and sink and the gate electrode to remain passive during device operations. However, limited illustrations of application of band engineering to optimize reverse- mode NVM device characteristics as well as multimode NVM device characteristics will also be presented.

13.1 REVISITING BAND DIAGRAM AND BAND ENGINEERING FOR CONVENTIONAL NVMs

13.1.1 FG Devices

Let us revisit the conventional band diagrams and associated energy barriers and briefly highlight the limitations faced and modifications attempted using oxide and nitride films. This sets the back ground for other higher K film applications for NVM stack designs. We illustrate in Figure 13.1 the band features and associated band bending at four different states for a typical conventional FG device with an SiO_2 tunnel dielectric layer and an ONO stack for the inter-polysilicon gate charge blocking layer. Figure 13.1a displays the band features and energy barriers at the initial state of the device at time zero when gate potential Vg = 0 V. Energy barriers at all interfaces without band bending and thicknesses of all dielectric films are shown. Figure 13.1b displays conduction band bending for all dielectric layers during the "writing" process (Vg ≤ ~+Vpp) assuming primarily electron transport from silicon substrate by the F–N tunneling. Figure 13.1c similarly displays conduction band bending for all dielectric layers during the "erasing" process (Vg ≤ ~–Vpp). Figure 13.1d displays the band features at the "standby" state after writing following the electronic charge stored in the floating gate due to the writing process when the gate potential is reduced to zero. This state is also called "written standby state" or "high Vth memory retention state" for NFET-based NVM devices.

All gate stacking layers are identified for a typical case of EOT ~ 20 nm with Vpp ≥ ±20 V across the gate for programming. The tunnel oxide thickness is assumed to be 6.0 nm. The ONO stack EOT ≤ 14 nm consisting of ~5.4 nm oxide, 6.5 nm nitride, and ~5.0 nm oxide (EOT ~14.0 nm).

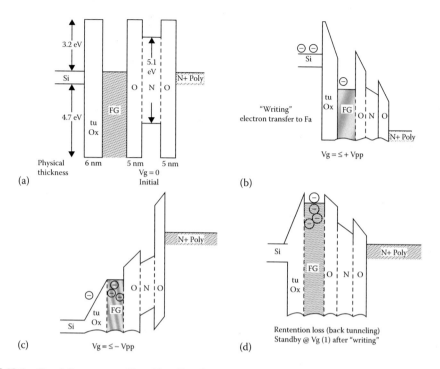

FIGURE 13.1   Band diagrams and band bending for the conventional FG device at different application conditions: (a) At initial state: VG = 0 V, (b) During writing with +Vpp at the gate, (c) During erasing with −Vpp at the gate, and (d) At the standby state during the "written state," when Vth is high showing charge loss by back tunneling to Si.

We assume that at the beginning of the post-written standby state, the floating gate potential is raised to 3 V due to electron accumulation reflecting into a memory window of change in Vth = 6 V at the gate (coupling coefficient of 0.5).

Figure 13.1a shows the band diagram of the gate stack for the FG device without any external potential applied either to the silicon substrate or to the gate. The barrier offsets for electrons and holes at the silicon/oxide interface are shown as well as the barrier energies interfacing the FG and the n+ polysilicon gate. When a +Vpp is applied at the n+ poly gate, the potential energy of the gate is lowered while the potential energy for the silicon bands is raised. Electrons get preferentially injected from silicon substrate into the tunnel oxide across the triangular energy barrier formed by band bending in response to the Vpp imposed at the gate. The field across the tunnel oxide is the highest by design, the coupling coefficient being 0.5 (half of the Vpp is dropped across the tunnel oxide). The electrons tunnel through the oxide and gets stored at the floating gate as shown in Figure 13.1b. This raises the potential energy of the floating gate. Consequently, the device threshold is raised to the memory state Vth (1) [or Vth high] and "writing" is accomplished. When −Vpp is applied at the gate for erasing, the field is reversed and silicon substrate is at lower energy state with respect to the potential at the floating gate. Therefore, electrons readily leave the floating gate and return to the silicon substrate. This is shown in Figure 13.1c. As electrons leave the floating gate, the floating-gate energy is lowered until electrons are removed from the floating gate and steady state is reached. The device threshold is lowered to the memory state Vth (0) and "erasing" is accomplished.

When erasing is not followed after writing, electronic charges remain at the floating gate raising the electrostatic potential of the floating gate and memory state Vth (1) remained. Electrons stored in the floating gate alters the band structure as shown in Figure 13.1d. At standby, both silicon substrate and n+ poly gate are at lower energy state compared to the floating gate. However, the field is stronger at the silicon substrate end as shown. Therefore, electrons preferentially tunnel back to the silicon substrate. This accounts for the charge loss during retention lowering the threshold level of the Vth (1) state and is known as the retention loss for the memory state.

13.1.2   Band Diagram for CT-SONOS Devices

Conventional CT-SONOS band diagram was reviewed in Part I in several places. SONOS band diagram has been discussed in Part I Chapter 5 (Figure 5.12a). For thicker oxide SONOS devices, electron transport provides the primary modes of current flow during writing, erasing, and standby similar to the FG devices as shown in Figure 13.1b–d and also in Figure 6.14a and b of Part I, where charges are stored in nitride traps. For thinner oxide SONOS devices, where direct tunneling of electrons and holes take place, band diagram illustrating transport of electrons is illustrated 6.16a and b in Part I. Similar considerations of charge loss at standby apply for CT devices as well. However, the mechanisms of charge loss could be different depending on the stack design and the thickness of the tunnel oxide as well as the blocking oxide and the relative roles the electrons and the holes play in device characteristics. It is important to point out that when ultrathin oxide is involved (oxide thickness << 3.2 nm), the tunneling probability is very high and transport is most energy efficient and sensitive to the tunneling distance regardless of the triangular (modified F–N tunneling) or trapezoidal (direct band-to-band) tunneling modes. In such stack designs, the NVM device properties such as programming speed and charge retention are strongly affected by the tunneling dielectric thickness.

13.1.3   Limitations of Conventional NVM Stack Designs

Conventional stack designs for all silicon based NVM device types employ a single dielectric layer (typically SiO_2) for tunneling and a single charge storage medium. With the exception of direct tunnel memory (DTM) devices, all others are distinguished in terms of the storage medium employed in the gate stack. The charge storage medium for FG and FP devices being one of high conductivity necessitates thicker film of tunneling SiO_2 to meet charge retention objectives. Charges are stored in dielectric potential well in CT and NC devices and, therefore, the tunnel SiO_2 layer could be thinner to meet charge retention objective with associated reduction in programming voltage.

For all the above device types consisting of single SiO_2 layer for tunneling and single storage medium, the trade-offs between simultaneously achieving high programming current (high charge fluence for large memory window) and charge retention at standby is limited. This is due to the fact that the quantum mechanical tunneling mechanisms are strongly sensitive to the high interface barrier energy, the effective tunneling distance associated with the thickness of the tunnel layer, and the permittivity of the film. Additionally, even for CT and NC devices, typically employing nitride for bulk trapping and silicon NC in oxide, the charge storage capacity, the energy offset, and the depth of the potential well for stable charge storage are also limited.

In general, the above observation is valid for conventional NVM devices regardless of the specifics of tunneling mechanisms operative in such device structures and the geometry of the tunnel barrier, even though the tunneling probability and tunneling current could be significantly different depending on the specifics of the tunnel layer. For the NVM device stack designs with ultrathin tunnel layers (tunnel oxide thickness <3.2 nm) with characteristic trapezoidal energy barrier (Figure 3.5, Part I), the quantum mechanical tunneling probability is very sensitive to the thickness of the oxide film, exponentially reducing with the thickness to zero at a distance of 3.2 nm. The tunneling current is directly proportional to the voltage drop across the tunnel layer and at the same time proportional to the inverse exponential of the square root of the interface barrier energy Ub–e or Ub–h for electrons and holes, respectively. Therefore, high density of tunneling current is achievable as long as the tunneling distance is short and typically, ≤2 nm. For the stack designs with thicker tunnel layers, with characteristic triangular energy barrier (Figure 3.5, Part I), the tunneling current is strongly dependent directly on the square of the field across the tunnel layer and inverse exponentially to the 1.5th power of [Ub–e] or [Ub–h]. Therefore, with triangular barrier, high current density could be achieved only for cases when both high field and lower barrier energy are present. For oxide with large barrier offsets (Ub–e = 3.2 eV and Ub–h = 4.7 eV with respect to silicon), such current density for thicker films are achievable only at extremely high field close to the breakdown strength of SiO_2. Therefore, conventional FG devices with thicker oxide would be limited in terms of programming speed (milliseconds or longer) even when the peak programming field is greater than 10 MV/cm. This high field requirement limits the reduction in scaling of the programming voltage as well as the endurance of the device as discussed earlier in Part I. Additionally, to meet the retention objective, the reduction in tunnel oxide thickness for FG device is limited affecting scalability. For the CT and NC devices, while the programming field across the tunnel oxide could be lower, charge storage capacity is limited thereby limiting the memory window and retention is compromised when the tunnel distance is shorter due to enhanced back tunneling of charges.

Conventional DTM devices with ultrathin SiO_2 layer for tunneling yields limited nonvolatility and memory window while providing significantly higher endurance, lower programming voltage operability, and higher programming speed compared to the devices mentioned above. DTM devices provide the framework for several SUM devices covered in Part III of this book. Band-engineered multilayered DTM devices are treated separately in subsections 13.4 and 13.6 due to their potential applicability and extendibility.

13.2 BAND ENGINEERING USING SINGLE AND MULTILAYER DIELECTRICS

In recent years, SiO_2 as gate insulator film has been replaced in scaled CMOS FET devices by higher K dielectric films to lower EOT and to aid device scalability. For FET-based NVM devices replacing SiO_2 film for tunneling layer by higher K metal oxide or other appropriate higher K dielectric film allows EOT reduction for voltage scalability discussed in Chapter 12 as well as integration compatibility with the base-line CMOS technology. In Section 12.15, we have discussed in detail the characteristics of higher K dielectric films and their applicability in various elements of NVM stack designs for NVM device extendibility. In this chapter, we will discuss the significance of single and multilayer higher K dielectric films for band-engineered NVM device stack designs to achieve enhanced device attributes.

13.2.1 Single-Layer Tunnel Dielectric Films: Conventional SiO_2 versus High K Films

NAND flash stack design successfully employed thicker (>6 nm) SiO_2 films for the tunnel dielectric. Although SiO_2 film had the disadvantage from charge injection standpoint due to high barrier energy for electrons and holes at the silicon/SiO_2 interface, the high barrier energy prevents back tunneling of charges during standby enhancing the retention characteristics required for nonvolatility. In contrast, when a high K film with lower band offset replaces oxide with physically equivalent or even thicker film with lower EOT, Fowler–Nordheim tunnel current (with triangular barrier) could be significantly enhanced even at a modest field (e.g., 6–7 MV/cm). This is due to a combination of lower barrier energy and effectively lower EOT (implying higher field across the tunnel layer even though average field across the insulator stack remains modest). However, when a high K film as mentioned above is applied as a tunnel dielectric, while the programming speed is improved, the standby charge loss is also increased due to the increased back tunneling through the film. This adversely affects retention. It has also been noted and exemplified in the previous subsection that a single-layer high K tunnel dielectric film would be limited in providing simultaneously high current transport in one direction (e.g., during writing) and very low leakage in the reverse direction (e.g., during standby, even though at reduced field), due to the symmetric nature of the barrier energy and the rate of charge loss requirements to meet retention target of ~1E9 Seconds.

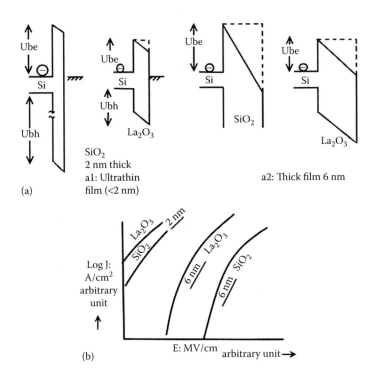

FIGURE 13.2 a1/a2 single-layer dielectric films of SiO$_2$ and La$_2$O$_3$: (a1) Ultra-thin film <~2 nm: Direct tunneling; (a2) >>2 nm: Modified FN tunneling. b1/b2 Log J versus E characteristics of ultrathin: b1 (~2 nm) and thicker: b2 (~6 nm) films of SiO$_2$ and La$_2$O$_3$: All single-layer films.

The conceptual picture as discussed above could be illustrated through band diagrams both for physically ultrathin layers of 2.0 nm (a1) each of SiO$_2$ (Ub = 9.0 eV, K = 3.9) and of La$_2$O$_3$ (Ub = 4.3 eV, K = 30 and thicker layers of 6.0 nm (a2) each of SiO$_2$ and of La$_2$O$_3$ as shown in Figure 13.2 (a1/a2). Corresponding J versus E characteristics for film thicknesses of 2.0 nm (b1) and 6.0 nm (b2) for single dielectric layers each are shown in Figure 13.2 (b1/b2) respectively.

It should be noted that for cases as shown in Figure 13.2a for both SiO$_2$ and La$_2$O$_3$ films, if the charges tunneled through the 2.0 nm of the films are stored in a floating gate or in a trapping dielectric at the interface, leakage back to the silicon substrate would be significant, even at the low internal field generated by the stored charge due to reverse direct tunneling. This is demonstrated in the J versus E characteristics for the 2 nm films of SiO$_2$ and La$_2$O$_3$ as shown in Figure 13.2 (b1). However, for the thicker 6 nm films, leakage would be significantly lower as shown by the curves on the right of the above figure due to the change in the tunneling mode [Figure 13.2 (b2)]. The thicker films would require significantly higher field for the required programming current and consequently slower programming speed to achieve the desired memory window. Therefore, simultaneous achievement of high programming speed and large memory window with memory retention would be difficult for cases of single layer of dielectric film employed as the tunneling medium. The symmetric nature of band configuration could be overcome by incorporating multiple dielectric films with equivalent EOT with different barrier energies. This is discussed in Section 13.2.2 below.

13.2.2 Multilayer Tunnel Dielectric and Associated Concepts of "Band Engineering"

Let us now consider that instead of a single layer of 6 nm of SiO$_2$ tunnel dielectric, we consider tunneling with three layers of 2 nm each consisting of SiO$_2$/La$_2$O$_3$/SiO$_2$ as tunnel layer. The band diagram for such a dielectric stack is shown in Figure 13.3a where the dotted curve

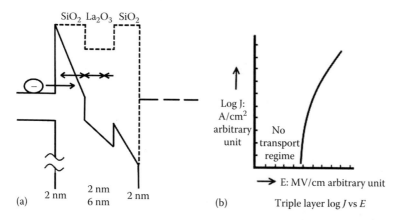

FIGURE 13.3 (a) Illustration of band-engineering concept of multiple-layer tunneling stack consisting of SiO₂/La₂O₃/SiO₂ layers of 2 nm thickness each (b) J versus E characteristics of the stack showing "no transport regime" and onset of current transport.

represents the conduction band diagram of the triple layer when there is no or little positive potential at the floating gate at the right. When a strong positive potential is applied as is the case for writing the device, the floating gate potential will be positive and the conduction bands would be bent as shown in the solid curves. Electrons will be injected from silicon substrate and tunnel through the triangular barrier of the first 2 nm layer of SiO_2 film and subsequently would be swept away along the conduction band of the other layers due to their higher energy state aided by the external field. This mode is commonly known as "enhanced F–N tunneling" due to the nature of the barrier formed, what may also be termed as "modified direct tunneling" from conceptual point of view. The effective tunnel distance in such mode is only 2 nm. Consequently, large fluence is achieved at relatively lower average field across the stack. This facilitates faster programming. After the charges gets stored in the floating gate, the reverse band bending is small and would be similar to the initial state as shown dotted in Figure 13.3a. The stored charges would face high barrier energy and tunnel distance of 6 nm eliminating the possibility of reverse direct tunneling. This is shown by the arrow with the X sign. Therefore, the stored charges would be retained. The triple-layer tunnel structure as shown could, therefore, provide simultaneously higher programming speed, memory window, and enhanced retention. The J versus E characteristics for the triple stack is shown in Figure 13.3b. Once the onset of conduction is achieved at the higher field as shown (field required for the triangular barrier to form in the first SiO_2 layer) the current will rise faster with the field as shown. At lower field below the onset, no current flows as indicated in Figure 13.3b showing the "no transport regime." This is the regime providing retention enhancement. By employing multiple layers of dielectric films of appropriate band structures, the forward and reverse transport of electronic charges could be vastly modulated. Such approach of stack design is called "band engineering" and is employed to achieve desired attributes of NVM devices such as programming speed and memory state retention.

13.2.3 Band Engineering Objectives through Multilayered Dielectric Films

We have illustrated in the above example how by replacing single layer of dielectric film by multilayered dielectric films and thereby creating barrier energy asymmetry called band engineering, NVM device parameters could be enhanced and limitations of single tunneling layer device structure could be overcome. Such concept of band engineering will be extensively covered in this section. The motivation behind band engineering for gate stack design may be summarized as follows:

1. To enhance tunneling current for a range of thickness of tunnel dielectrics for both writing and erasing at moderate to high field to enhance programming speed without adversely affecting retention.
2. To enhance charge storage and charge retention during standby state by providing appropriate asymmetric tunnel barrier for stored charges to minimize charge loss to either silicon substrate or to the gate throughout the operational life of the device. This enhances nonvolatility.
3. To enhance endurance by reducing stress field and trap generation within the dielectric layers of the device stack, thereby improving functionality and reliability.
4. To facilitate scalability and extendibility of NVM devices for future technology nodes.

13.2.4   Band-Engineering Design Options: Distinction and Classification Based on Single and Multilayered Tunneling Modes and Dielectric Thickness

It has been explained earlier that for single-layer dielectric films as tunneling layer, transport of electronic charges could be explained in terms of two basic energy barrier types as follows:

1. The triangular energy barrier applicable to thicker dielectric films with Fowler–Nordheim tunneling mode with associated variations such as "enhanced Fowler–Nordheim tunneling mode" and others (Chapter 3, Part I).
2. The trapezoidal barrier applicable to thinner dielectric films with direct band-to-band tunneling.

For electronic charge transport in multilayered tunnel dielectric gate stack structures, it is necessary to distinguish the dielectric film layers of "thicker" tunneling dielectric films (each layer with EOT ≥ 3.2 nm) from those of ultrathin or "thinner" dielectric films (EOT ≤ 3.0 nm). The key rationale for such distinction lies in the extent of energy transfer of electronic charges to the surrounding dielectric lattice during transport through the specific dielectric film layer or layers. This distinction is essential in designing extendible NVM device designs due to the recognition that significant energy of electronic charges get transferred to the dielectric film lattice in thicker dielectric films during Fowler–Nordheim tunneling, whereas insignificant energy gets transferred during transport through ultrathin single or multilayered dielectric film structures when transport mode is either by "Ballistic" tunneling or by single or multiple "Direct" tunneling. Such distinction is relevant for multilayer ultrathin films even when the total physical thickness of the composite layers is significantly thicker than 3.2 nm. Electronic transport through such multilayered structures is governed by band-to-band tunneling or by direct tunneling through each individual layer with minimal energy exchange to the surrounding lattice. In such cases, we would broaden the definition of device types as "direct tunnel device," even though the precise modes of electronic transport may constitute either (a) multiple trapezoidal barrier (multiple direct tunneling) or (b) one or more ultrathin triangular barrier (traditionally termed as "enhanced Fowler–Nordheim") or a combination. In such cases, we would emphasize the physical nature of the individual dielectric layer and will not be concerned in the barrier geometry distinction of triangular or trapezoidal configurations as applicable to single dielectric-layer structures. Henceforth, all multilayered "DTM" types of devices will be grouped as "multilayered thin dielectric" devices regardless of the combined total physical thickness of the tunnel layers in the NVM stack design.

Appropriately, band-engineered DTM devices are discussed in detail in Part III of this book to achieve multifunctional SUM devices (MSUMs). We will therefore, address band-engineering options into two separate subgroups: First for the thicker multilayer tunnel dielectric films (Section 13.3) and thereafter, the second subgroup dealing with thinner multilayer tunnel dielectric films (Section 13.4). During such discussions, we will introduce the concepts of "barrier energy lowering" and "barrier thinning" and different types of barrier designs to achieve NVM device extendibility. Section 13.5 will discuss application of band engineering to achieve specific NVM device attributes as well as device enhancements with multiple examples currently being adopted in products. This will be followed by a separate Section 13.6, dedicated for direct tunnel NVM devices for future potential applications in SUM devices. Finally, in Section 13.7, we have examined how multilevel and multifunctional NVMs could be designed using band engineering.

13.3 BAND-ENGINEERING OPTIONS FOR THICKER MULTILAYER TUNNEL DIELECTRICS

This section will explore multiple options for band engineering to address NVM device parameter for thicker multilayer tunnel dielectrics. First we will review earlier work on band engineering with conventional dielectric films. Thereafter, we will address other options and examples of band engineering employing combination of conventional dielectric films and high K films as well as examples with all-high-K films in the gate stacks.

Multilayered dielectric films are often considered with varying barrier energies to aid asymmetric carrier transport in order to achieve simultaneously higher programming current desired to achieve programming speed and memory window and simultaneously reduced reverse current flow to the substrate to enhance retention. We will initially examine electronic transport through double-layer dielectrics in subsections 13.3.1 and 13.3.2 followed by triple-layer dielectrics in subsection 13.3.3. We will first discuss the general concepts and thereafter will provide specific illustrations.

13.3.1 Double-Layer Dielectrics

There could be four potential options in general terms for any two-layer dielectric system: the first layer could be relatively low K with high barrier energy Ube or Ubh ("D01") followed by the second layer with relatively high K with low barrier energy ("D10"). This combination is labeled as D01/D10. The reverse combination is D10/D01 implying the first dielectric to be relatively high K with low barrier energy and the second dielectric with low K with high barrier energy, with combination labeled as D10/D01. The other two combinations would be high K with high barrier energy interfacing with low K with low barrier energy: D11/D00, and the reverse D00/D11 consisting of low K with low barrier energy interfacing with dielectric layer of high K with high barrier energy. The four combinations are schematically shown in Figure 13.4a–d, respectively, with silicon substrate at one interface and silicon FG/alternately n+ poly gate at the other interface.

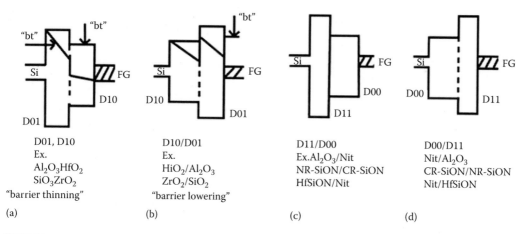

FIGURE 13.4 Double-layer dielectric band options, (a) D01/D10, (b) D10/D01, (c) D11/D00, and (d) D00/D11 D01 (LO k-HI Eb; D10: HI k-LO Eb; D11: HI k-Hi Eb; D00: LO k-LO Eb).

We will apply a certain potential both positive and negative simulating conditions of "writing" and "erasing" assuming charges incident get stored at the floating gate and silicon substrate being the active source and sink for electronic charges. Additionally, we will discuss the barrier characteristics of each options from the point of view of charge loss from the FG to silicon substrate (reverse tunneling) when the FG assumes certain electrostatic potential due to charge accumulation qualitatively simulating the "standby" conditions of NVM memory. To qualitatively explain the effects of such combinations on band bending and carrier transport, we will make the following simple assumptions:

Total EOT for the two layers is equal to or greater than 7 nm, equally divided between the two dielectrics so that the potential drop in each of the layers is the same regardless of the difference in physical thickness (due to the difference in K values). Each layer is thick enough that the relevant energy barriers are triangular and electron transport follows Fowler–Nordheim tunneling mode. Furthermore, we will, for this simple explanation, disregard the difference in electron effective masses between the two dielectric media, the interface and fixed charge effects, and assume no trapping during charge movement through the dielectrics.

13.3.1.1   Option A: Low K High Barrier/High K Low Barrier (D01/D10)

This choice was illustrated in Figure 13.4a and exhibits so called "barrier thinning" or "K effect." This option when subjected to an external field sufficiently high to be equivalent to +Vpp for writing on the gate and correspondingly a coupled positive potential at the floating gate will bend the conduction band more steeply for the low K dielectric interfacing the silicon substrate compared to the high K dielectric film (for the same voltage drop, the lower K film with same EOT will be thinner). As a result, electrons need to travel shorter distance to reach to the conduction band of the first layer and the drift field sweeps the charges through the conduction band of the second layer to be collected at the floating gate. This is shown in Figure 13.5a. This is called "barrier thinning" which effectively reduces the tunnel distance at high field to the thickness of the first layer and consequently the current flow increases strongly with field, and electron flow through the second layer is reduced or eliminated. Thus, writing speed will be enhanced with this type band option when compared with a single-layer low K high barrier film of the same EOT

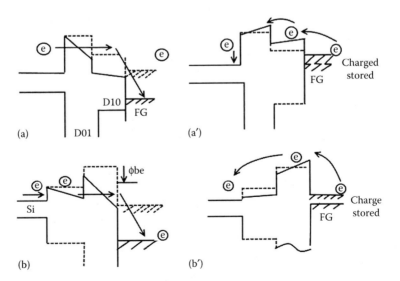

FIGURE 13.5   Barrier thinning (a) for writing; (a′) during retention; barrier lowering (b) for writing (b′) during retention.

(e.g., for the case of a single layer SiO$_2$ film) where the band bending will be uniformly less steep across the thickness and electrons need to travel through the entire thickness of the dielectric film across the triangular barrier.

When a sufficiently high erase field is applied equivalent to −Vpp at the gate and correspondingly coupled high negative potential at the floating gate, once electrons assume large enough potential energy above the Ub–e (low barrier energy) of the second dielectric film interfacing the floating gate, the steep reverse field with the barrier thinning will apply (not shown). Consequently, the erase speed of electrons returning to the silicon substrate will also be faster.

For electrons stored in the floating gate (after being written), the electrostatic potential will raise the energy of the floating gate. Since the standby field is considerably lower compared to the field required for writing and erasing, even if barrier thinning applies, electrons have to travel through the high K film as well as most of the low K film to reach the silicon substrate overcoming both barriers and longer tunnel distance. Consequently, electron loss to the substrate will be reduced. This is schematically shown in Figure 13.5a′ to be reflected in retention characteristics.

13.3.1.2 Option B: High K Low Barrier/Low K High Barrier (D10/D01)

This choice was illustrated in Figure 13.4b, and it exhibits so called "barrier lowering" or "barrier offset effect." This option when subjected to an external field sufficiently high to be equivalent to +Vpp for writing on the gate and correspondingly a coupled positive potential at the floating gate will bend the conduction band more steeply of the D01 layer (the low K dielectric) interfacing D10, the high K layer, the later interfacing the silicon substrate. Additionally, since D10 has lower barrier energy electrons will be injected at relatively lower +Vpp value and tunnel through the conduction band of D10 more readily to be incident at the D10/D01 interface. Subsequently, electrons will tunnel through a shorter distance of D01 due to higher field and steeper banding of the D01 layer. Finally, the drift field sweeps the charges to be collected at the floating gate. This is shown in Figure 13.5b. In this situation, as shown in the figure the effective energy barrier for the electrons at the D10/D01 interface is lowered since electrons have already assumed higher energy from the external field. This is called "barrier lowering." Consequently, electron flow is enhanced compared to the same EOT for a single-layer of D01 dielectric film. As a result, writing speed will also be enhanced with this type of band option. When a sufficiently high erase field is applied equivalent to −Vpp at the gate and correspondingly coupled high negative potential at the floating gate, electrons need to assume larger potential energy above the Ube (low barrier energy) of D01, the second dielectric film interfacing the floating gate. The steep reverse field with the barrier thinning will apply (not shown) similar to the writing case for the previous option. Consequently, the erase speed of electrons returning to the silicon substrate will also be faster.

For electrons stored in the floating gate (after being written), the electrostatic potential will raise the energy of the floating gate. Since the standby field is considerably lower compared to the field required for writing and erasing, even if barrier thinning applies, electrons have to travel through the high barrier of the low K film, which determines the rate of loss of electrons to the substrate. Subsequently, electrons face no effective barrier through the D10 layer to reach the silicon substrate. The nature of the energy barrier of D01 layer will determine the retention characteristics of this option. This is schematically shown in Figure 13.5b′.

13.3.1.3 Option C: High K High Barrier/Low K Low Barrier (D11/D00)

This choice was illustrated in Figure 13.4c. In this option, the band features appear similar to the first option of D01/D10 except that the K values of the films are reversed. At the conditions of writing, barrier thinning will be less pronounced since band bending will be less steep due to higher K value of D11. Consequently, the writing current will be lower and thereby the speed of writing. However, when the external field is reversed during erasing, the consequence of lower barrier energy of D00 and barrier lowering of DII will aid in faster erasing when compared to the other two cases discussed above. Retention conditions will be similar to erase and therefore will result in faster charge loss during high threshold state standby condition. For most applications, this may not be a preferred option.

13.3.1.4 Option D: Low K Low Barrier/High K High Barrier (D00/D11)

This choice was illustrated in Figure 13.4d. Here the band features appear similar to the second option of D10/D01 except that the K values of the films are reversed. Although barrier lowering during writing will prevail similar to the second option, it will be less pronounced due to the "K" effect of band bending of D11 layer. Less current flow will be expected compared to the second case and therefore less speed of writing. Erasing will also be slower due to the "K" effect of less band bending. However, for similar reasons, charges stored will retain longer compared to the second option.

It is evident, therefore, that replacing thicker oxide (5–7 nm EOT) with two-layer dielectric combination could significantly alter the charge injection characteristics and thereby the programming characteristics as well as the charge retention characteristics by the mechanisms of barrier thinning and barrier lowering. The specific combination of the characteristics determines the degree of enhancements in programming characteristics and retention enhancements. In general, however, Low K High barrier/High K Low barrier (D01/D10) and High K low barrier/Low K High barrier (D10/D01) options yield the best characteristics from NVM device point of view. The band-engineering options and their expected characteristics discussed above can be summarized in Table 13.1. Specific examples of combinations are also listed.

13.3.2 Current–Voltage Characteristics of Double-Layer Dielectrics

A specific example of a two-layer combination of dielectrics could be illustrated and could be compared with an equivalent single-layer SiO_2 film [1,10]. The band diagrams for a single-layer SiO_2 film (EOT ~5 nm) is compared with equivalent two-layer combination of the first type: D01/D10, consisting of 3.5 nm of SiO_2/8 nm of ZrO_2 films, are shown in Figure 13.6a and b, respectively.

When a voltage Vpp is imposed on the silicon gate of the MIS structures, the band bending for the SiO_2/ZrO_2 two-layer structure is shown in green in Figure 13.6b. Corresponding J versus V

TABLE 13.1

Band Options and Characteristics for Double-Layer Dielectrics

Layer Combinations	Writing	Erasing	Vth High Retention	Remarks	Examples
D01/D10: Lo-HiK Hi-Lo barrier	Fast	Fast	Good	Barrier thinning	SiO_2/Si_3N_4, SiO_2/ZrO_2, Al_2O_3/HfO_2, Al_2O_3/TiO_2, SiO_2/SiON, SiO_2/HfO_2, Al_2O_3/Ta_2O_5
D10/D01: Hi-Lo K, Lo-Hi barrier	Fast	Fast	Good	Barrier lowering	SiON/SiO_2, N/O, HfO_2/SiO_2, ZrO_2/SiO_2, TiO_2/SiO_2, TiO_2/Al_2O_3, TiO_2/HfSiON, TiO_2/OR-SiON
D11/D00: Hi-Lo K, Hi-Lo barrier	Slow	Very fast	Poor	High leakage	Al_2O_3/Si_3N_4, ZrO_2/Si_3N_4, Al_2O_3/SiON
D00/D11: Lo-Hi K, Lo-Hi barrier	Moderate	Moderate	Very good	Less leakage	NR-SiON/Al_2O_3, Si_3N_4/Al_2O_3, Si_3N_4/LaAlO_3, Si_3N_4/HfSiON, Si_3N

Examples: SiON/SiO_2 or N/O; HfO_2 or ZrO_2 or TiO_2/SiO_2 or Al_2O_3 or HfSiON or OR-SiON. D11/D00: Hi-Lo K, Hi-Lo barrier: Slow, V. Fast, Poor, High Leakage.

Examples: Al_2O_3/Si_3N_4 or SiON; ZrO_2/Si_3N_4, etc.

D00/D11: Lo-Hi K, Lo-Hi barrier: Moderate, Moderate, V. Good

Less Leakage Examples: NR-SiON or Si_3N_4/Al_2O_3 or HfSiON or LaAlO_3.

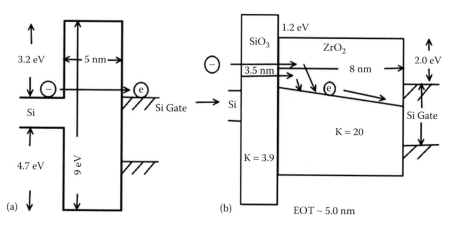

FIGURE 13.6 Band diagram of MIS structures: (a) Single-layer SiO₂ and (b) Dual-layer SiO₂/ZrO₂ with equivalent EOT. Note: (J vs. V p-lots for above structures are shown in Figure 13.9 (b) and (z) respectively from references [1,10]).

characteristics for the single-layer SiO_2 and equivalent SiO_2/ZrO_2 two-layer are shown in Figure 13.9b and z, respectively. The effective tunnel distance for the two-layer film was reduced to the SiO_2 thickness due to steeper band bending of SiO_2 in the dual-layer film (as previously explained) resulting in internal field enhanced Fowler–Nordheim tunneling for the dual-layer film as illustrated in Figure 13.6b. The current density (J) versus the voltage plot shown in Figure 13.9. Figure 13.9 illustrates the following: (i) Steeper J–V slope at lower voltage and (ii) Enhanced charge flow at higher voltage for the two-layer film when compared with that of single-layer SiO_2 film. The current density level at retention Jr (Jr-sl for single-layer SiO_2, Jr-dl for two-layer; see Figure 13.9) corresponds to the potential across the layers Vr as shown in Figure 13.9. The current density ratio for the dual layer (dl) between programming and retention, that is, J (dl)/Jr (dl) is significantly greater than the corresponding ratio: J-sl/Jr-sl for single-layer SiO_2 illustrating both retention and programming enhancements in dual-dielectric band-engineered structure. Therefore, the dual-dielectric structure by providing steeper voltage (field) dependence on current density is not only desirable for retention enhancements with associated lower charge loss at reduced field during standby, but also enhancements in programming speed. Additionally, the thicknesses of the individual double-layer films could be optimized to reduce the EOT of the gate insulator stack and thereby the Vpp while maintaining such enhancements.

13.3.2.1 Other Examples of Double-Layer Dielectrics

The above example illustrates the enhancements due to barrier thinning. As pointed out in the table, many other possibilities exist for barrier thinning. Notable other combinations with oxide at the silicon interface (providing stable interface with low density of fixed charge and interface states) could be SiO_2/HfO_2 (5.7 eV, K = 24), $SiO_2/HfSiON$ (6.9 eV, K = 14), and SiO_2/La AlO_3 (6.6 eV, K = 27.5). Early implementation of double layer was developed by Bhattacharyya et al. and had been oxide/OR-SiON and Ref. [2] for details, providing enhanced device properties as discussed in Part I. Oxide-nitride double layer for tunneling combined with HfO_2 layer for trapping has also been investigated in recent times by Wang et al. [3]. The advantage of double layer in enhancing program, erase, retention, and endurance was demonstrated by the above authors.

Several double-layer dielectric structures based on barrier lowering (D10/D01 type) had been proposed in recent years by Bhattacharyya et al. [4,5]. The double-layer tunneling structure for one of the devices [4], consisted of OR-SiON (7.3 eV, K = 5) interfacing silicon substrate with Al_2O_3 (8.8 eV, K = 10). The other device [5] consisted of HfO_2 (5.7 eV, K = 24) interfacing silicon substrate with La Al O_3 (6.5 eV, K = 27.5). Characteristics of both of the above devices will be discussed later in this section.

13.3.3 Crested Barrier and Variot Barrier: Triple-Layer Dielectrics

13.3.3.1 Crested Barrier

Triple dielectric layers had been of considerable interest in NVM stack design for blocking and tunneling functions to achieve desired electronic transport characteristics. ONO inter-poly dielectrics had been widely used for decades to control charge loss to the gate for the conventional NAND flash stack design. K. K. Likharev [6] first proposed a triple-layer dielectric tunneling structure consisting of low barrier-high barrier-low barrier symmetric tunneling structure to demonstrate "barrier lowering" of the high barrier dielectric and enhanced current transport compared to a single high barrier dielectric for tunneling. This is also known as "barrier offset effect." The concept demonstrated higher programming speed at lower power without impacting retention. Likharev called this type of band engineering as "crested tunnel barrier." Likharev employed a triple layer consisting of nitride/AlN/nitride between the floating gate and the n+ poly gate to exemplify the barrier lowering for a reverse mode NVM. The barrier lowering is shown in Figure 13.7. Following the above concept, S. J. Baik et al. built a FG normal mode memory device with noncrested tunnel barrier [7], and demonstrated that such device is highly field sensitive and could achieve an acceptable memory window at high speed programming of 10 µs with significantly reduced programming voltage of +8 V.

13.3.3.2 Variot Barrier

"Barrier thinning" concept of barrier engineering was first conceived in double-layer and triple-layer dielectrics by P. Blomme et al. [8,9]. They termed the concept as a variable oxide thickness dielectric barrier or "Variot" dielectric barrier [10] consisting of two and three dielectric layers with different K values. The two-layer low K/high K barrier (D01/D10) as shown earlier (Figure 13.5b) is asymmetric, whereas a triple dielectric combination of low K/high K/low K (as the previous example of $SiO_2/ZrO_2/SiO_2$) would be a symmetric Variot barrier. While a symmetric three-layer Variot provides uniformly enhanced current flow in both directions during writing and erasing (from silicon substrate to the storage node and from the storage node to the silicon substrate), a two-layer asymmetric Variot barrier enhances the current flow during writing while relatively reducing the current flow during erasing. Similar to the crested barrier concept, the Variot barrier provides the capability of highly field sensitive stack design enabling lower voltage programming as well as Vpp-speed trade-offs while meeting retention objective. Barrier thinning concept for two-layer and three-layer dielectrics compared to the single layer and corresponding J–V characteristics is illustrated in Figures 13.8 and 13.9, respectively.

Bhattacharyya [11] had proposed several triple dielectric-layer tunneling structures for FG and CT stack designs with both Variot and crested barriers for enhanced field sensitivity. A Variot stack design for FG stack design with EOT ~ 9 nm is shown in Figure 13.10, consisting of 2 nm SiO_2/8 nm

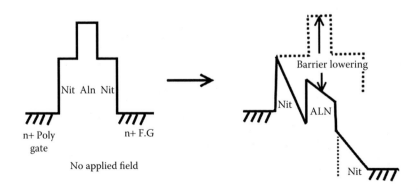

FIGURE 13.7 Triple-layer crested barrier showing barrier lowering. (From Likharev, K. K., Riding the crest of a new wave in Memory, *IEEE Circuits and Devices*, pp. 17–21, July 2000.)

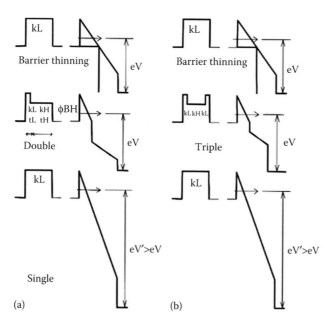

FIGURE 13.8 Band diagram and associated band bending for double dielectric (asymmetric) and triple dielectric (symmetric) demonstrating barrier thinning for Variot barriers compared to single dielectric low K barrier. (From Govoreanu, B. et al., *IEEE Electron Device Lett.*, 24(2), p. 99 © 2003 IEEE.)

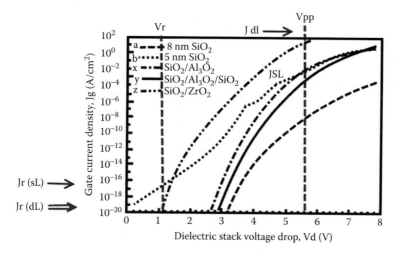

FIGURE 13.9 Current density versus voltage drop across the single layers of SiO_2 (a and b), double-layer Variot and triple-layer Variot showing enhanced field sensitivity of charge transport with such schemes for SiO_2/Al_2O_3(x), SiO_2/ZrO_2(z), and $SiO_2/Al_2O_3/SiO_2$(y): Variot barriers. (From Govoreanu, B. et al., *IEEE Electron Dev. Lett.*, 24, 99, 2003.)

HfSiON/2.0 nm SiO_2 Variot barrier with 10 nm HfSiON blocking layer. The stack provides 4 V of memory window at ±8V Vpp, 100 μs write, and 1 ms erase. Similar results were reported by P. Bloom et al. [9,12] who used Al_2O_3 instead of HfSiON with equivalent EOT and provided memory window at Vpp +9 V. Their device demonstrated 10 years of retention and 1E5 W/E cycle endurance.

For SONOS stack design, lower EOT (<~6 nm) and programming voltage level of ±5 V Vpp was cited [11], using crested barrier consisting of 1.5 nm Si_3N_4/2.5 nm HfSiON/1.5 nm Si_3N_4

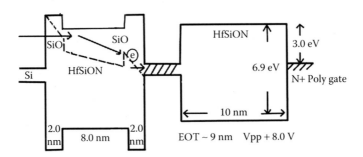

FIGURE 13.10 Stack design example of Variot triple-layer oxide/HfSiON/oxide barrier for FG gate stack design. (From Bhattacharyya, A., Discrete trap non-volatile device for Universal memory chip application, *ADI Associates internal publication*, U.S. Patents: 7436018 [10/14/2008], 7786516 [10/14/2010], 8143657 [03/27/2012], 2005.)

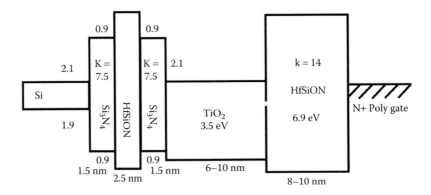

FIGURE 13.11 Stack design example of crested barrier SONOS NVM with nitride/HfSiON/nitride films for tunneling and TiO_2 film for trapping and HfSiON film for blocking. (From Bhattacharyya, A., Discrete trap non-volatile device for Universal memory chip application, *ADI Associates internal publication*, U.S. Patents: 7436018 [10/14/2008], 7786516 [10/14/2010], 8143657 [03/27/2012], 2005.)

triple-layer symmetric tunneling stack with TiO_2 trapping layer and HfSiON blocking layer. This stack design is shown in Figure 13.11.

13.3.3.3 Variot versus Crested Barrier for Charge Tunneling

As noted above the triple-layer low K/high K/low K Variot barrier provides barrier thinning due to what is called the "K" effect, whereas the triple layer low barrier/high barrier/low barrier crested barrier provides barrier energy lowering of the high barrier due to what is known as the "barrier offset effect." Since field dependence of charge transport by tunneling is more sensitive to tunnel distance compared to barrier height, Variot barrier comparatively provides greater enhancements on charge transport compared to the crested barrier. This was explained theoretically and demonstrated experimentally by J. Buckley et al. [13]. These authors simulated and compared the J–V characteristics of triple-layer $HfO_2/SiO_2/HfO_2$ (crested barrier) with that of triple-layer $SiO_2/HfO_2/SiO_2$ (Variot barrier) for the same EOT (~5 nm) stack design.

This is shown in Figure 13.12. Their results and conclusion favored the Variot barrier due to relative thickness of the triple layer selected for the same overall EOT. However, it is believed that optimum combinations of Variot and Crested barriers with appropriate selection of high K dielectric films could yield superior results over either of the barrier types.

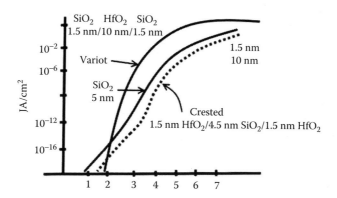

FIGURE 13.12 J versus V characteristics of Crested ($HfO_2/SiO_2/HfO_2$) versus Variot ($SiO_2/HfO_2/SiO_2$) barrier stacks with overall EOT~5 nm compared with single-layer SiO_2 film. (From Buckley, J. et al., *Experimental and theoretical study of layered tunnel barriers for nonvolatile memories, Solid State Device Research Conference, Proceedings ESSDERC*, p. 55, 2006.)

13.3.3.4 Combining Crested Barrier and High K Dielectrics

Crested barrier band engineering could be combined with higher K dielectric layers to achieve desired combination of higher programming speed as well as superior charge retention. Examples of such double-layer and triple-layer tunneling structures are shown in Figure 13.13a and a′ for FG-NVM and CT-NVM applications, respectively, and Figure 13.13b for CT-NVM applications [3,4]. In all stack design, the SiO_2 layer interfacing silicon substrate is replaced by a higher K (K = 5–7) SiON dielectric film layer, which forms stable interface with silicon substrate similar to SiO_2. In the FG stack example of Figure 13.13a, the triple barrier is symmetric with SiON/Al_2O_3/SiON design (crested barrier for barrier lowering); however, the higher K alumina layer at the center also promotes the "K" effect (barrier thinning). In case of the triple layer of Figure 13.13b, the middle alumina layer is followed by a still higher K layer of HfO_2 with higher band offset. This combines

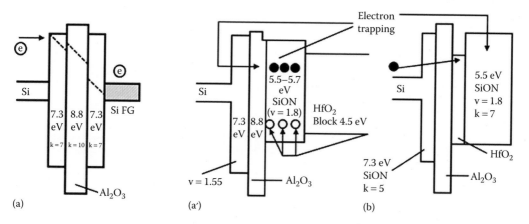

FIGURE 13.13 Examples of multilayered tunneling structures combining crested barrier with high K: (a) three-layer crested barrier with FG, (a′) two-layer crested barrier with CT SiON, and (b) three-layer crested barrier with band offsets for higher performance and retention. (From Wang, Y. Q. et al., Fast erasing and highly reliable MONOS type memory with HfO_2 high-k trapping layer and Si_3N_4/SiO_2 tunnel stack, *IEEE IEDM Technical Digest*, pp. 971–974, 2006; Bhattacharyya, A. et al., Band- engineered multi-gated-channel non-volatile memory device with enhanced attributes, *ADI Associates Internal Publication*, U.S. Patents:7279740, 10/9/2007; 7749848, 7/6/2010; 8063436, 11/22/11, July 2004.)

both the "K" effect and the "band offset effect" (barrier lowering) when current flow is considered from either direction combined with a deeper energy CT dielectric NR-SiON (K = 7) layer (higher band offset compared to nitride) for enhanced charge storage as shown.

13.3.4   Progressive Band Offset (Barrier)

Another type of multilayered barrier structure was proposed by Bhattacharyya [5], which is called "progressive band offset" barrier or simply PBO. PBO may be considered as an extension of the two-layer band offset structures discussed earlier. PBO may consist of either progressive conduction band offset or progressive valence band offset multilayered dielectrics or a combination of both. An example of a triple-layer conduction band offset could be: SiO_2 (3.2 eV Ube, K = 3.9)/ HfO_2 (1.5 eV Ube, K = 24)/Ta_2O_5 (0.1 eV Ube, K = 26) interfacing silicon substrate. Similarly, a triple-layer progressive valence band offset interfacing n+ polysilicon gate, for example, may consist of: La_2O_3 (0.9 eV Ubh, K = 30)/Y_2O_3 (2.2 eV Ubh, K = 15)/Al_2O_3(4.85 eV Ubh, K = 10) dielectric layers. The two-triple layers exemplified above could be combined in the gate stack design with either a silicon FG device or an IN-SRN FP device to provide a complete gate stack design as shown in Figure 13.14.

When only one of the triple PBO scheme is employed, either for writing or for erasing, the design would provide asymmetric barrier. Transport of carriers will take place much faster in one direction and slower in the other direction. Depending on the stack design, either a very fast write or a very fast erase could be achieved based on the direction of the PBO. At the expense of process complexity, such devices could be very high performance for both write and erase as exemplified in Figure 13.14. Such stack design could be applicable for multilevel storage to be discussed later on. Charge storage medium could be either FG or FP mentioned earlier or optionally, in discrete trapping medium such as NR-SiON or AlN for CT devices. For nanocrystals (NC) devices, platinum (Pt) nanocrystals embedded into a high K dielectric, for example, Al_2O_3 or Ta_2O_5 may be considered for charge trapping.

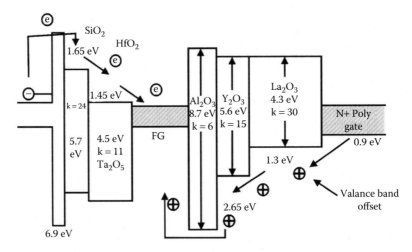

FIGURE 13.14   FG-NVM stack design with progressive band offset barriers, triple-layer conduction band offset for writing and triple-layer valence band offset for erasing. (From Bhattacharyya, A., High performance multi-level band–Engineered (ET/DT) NVM device, *ADI Associates internal publication*, U.S. Patents: 7429767, 09/30/2008; 7553735, 06/30/2009; 7579242, 08/25/2009; 8159875, 04/17/2012, December 2004.)

13.4 BAND-ENGINEERING OPTIONS FOR DIRECT TUNNEL MULTILAYER TUNNEL DIELECTRICS

It has been discussed earlier that the primary advantages of charge transport by direct tunneling is high current density at low voltage across the tunnel layer and little or no energy transfer by the electronic charges to the surrounding dielectric lattice, thereby ensuring film integrity and reliability. For NVM devices with appropriate stack designs employing ultrathin multiple layers of tunnel dielectric films to ensure charge transport by direct tunneling, higher programming speed, larger window, and enhanced endurance could be achieved simultaneously. However, such designs may exhibit enhanced charge loss by reverse tunneling due to the built-up of internal potential associated with stored charges adversely affecting NVM retention. Direct tunneling current depends linearly on the potential across the tunnel barrier and exponentially reduced with tunnel distance coupled with the square root of the energy barrier at the interface. Therefore, multiple direct tunneling layers could be introduced in the stack design with optimized tunnel distances and barrier energies such as to create asymmetric barriers to reduce reverse direct tunneling while promoting forward direct tunneling. This implies also that while injecting energy barriers should be low, the trapped charge barriers should be high and back tunneling distance should be kept large. We will explore different options of two-layer and three-layer dielectric films to achieve such objectives.

13.4.1 Double-Layer Direct Tunnel Dielectrics

For a double-layer direct tunnel structure, it would be desirable to consider the first layer interfacing the silicon substrate should be of lower bandgap with lower electron and hole barrier energies in order to promote charge injection at lower applied potential. The second layer in such cases should be of higher bandgap and higher barrier energies for electrons and holes. The band diagrams for such layers are schematically shown in Figure 13.15 for the three types of charge storage medium, namely, (a) the FG, (b) the Discrete trap, and (c) the NC-induced trapping. Examples of the first layer (layer A) could be HfO_2, Si_3N_4, or NR-SiON with bandgap in the range of 5–5.5 eV, while the

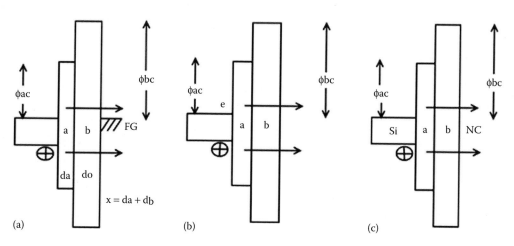

FIGURE 13.15 Lo/Hi barrier band diagram schematic for two-layer direct tunnel dielectric structure, (a) FG, (b) discrete trap, and (c) NC embedded trap for charge storage options.

second layer (layer B) could be SiO_2 (9.0 eV), Al_2O_3 (8.7 eV), GeC (7.1 eV), HfSiON (6.9 eV), and $LaAlO_3$ (6.6 eV). Higher K second layer with higher barrier, for example, Al_2O_3 would be preferred for enhancing retention as explained in C1.1 earlier.

13.4.2 Modified Double-Layer Direct Tunnel Barrier

Two modified double-layer direct tunnel barriers which will be discussed in greater detail later on in sections on device design should be worth mentioning in this section. Both operate on the mechanisms of successive direct tunneling within each dielectric layer. Both incorporate a unique layer between the two direct tunnel dielectric films to modulate charge transport to achieve unique enhanced properties. The first one was published by S. Kim et al. [14] and is called "resonant tunnel barrier" (RTB) whereby the tunnel barriers were modulated by sandwiching a thin film of either amorphous silicon or amorphous germanium of ~1 nm of thickness. This created unique sets of energy states for quantum mechanical tunneling reducing threshold dispersions of programmed states as well as enhancing programming speed, retention, endurance, and disturbs. The second one was published by R. Ohba et al. [15] who incorporated 1.5 nm silicon nanocrystal layer between two tunnel oxide layers of 1 nm in thickness. The nanocrystal induced charge centers provided repulsive electrostatic field for charges stored within the silicon nitride trapping layer, thereby reducing charge loss and enhancing retention. This modified double-layer direct tunnel barrier is called by the above authors as "double tunnel junction" (DTJ). Variations of modified DTJ have been employed in developing several CT SUM devices. This is discussed in Part III of this book.

13.4.3 Triple-Layer PBO Direct Tunnel Barrier ("DT-PBO")

Similar in concept to the previously discussed PBO barrier for thicker dielectric layers, triple-layer direct tunneling structures could be envisioned for simultaneously achieving: (a) very fast programming, (b) significant saving in programming energy, (c) Significantly enhanced endurance, and (d) good retention (>1E6 sec). Charge transport involves multistep direct tunneling being aided by barrier thinning as well as internal field-aided charge movement. Examples of triple-layer structures with ultrathin SiO_2 (<2 nm) interfacing silicon substrate may consist of several options as listed below:

Option 1: SiO_2 (Ub = 9 eV, K = 3.9)/SiON (Ub = 5.5 eV, K = 7.0)/HfO_2 (Ub = 5.7, K = 24)/trapping layer

Option 2: SiO_2 (Ub = 9 eV, K = 3.9)/Si_3N_4 (Ub = 5.1 eV, K = 7)/Pr_2O_3 (Ub = 5.1, K = 31)/trapping layer

Option 3: SiO_2 (Ub = 9 eV, K = 3.9/HfO_2 (Ub = 5.7, K = 24)/Pr_2O_3 (Ub = 5.1, K = 31)/trapping layer

Option 4: SiO_2 (Ub = 9 eV, K = 3.9)/HfO_2 (Ub = 5.7, K = 24)/TiO_2 (Ub = 3.5, K = 60)/trapping layer

The triple-layer stack options for tunnel layers as exemplified above could be designed to obtain EOT of ~3 nm to achieve superior end-of-life retention by nearly five orders of magnitude and even greater when compared to any single-layer direct tunnel structure. The option 4 example for the triple-direct-tunnel layer PBO stack design is illustrated in the form of band characteristics during initial state, writing state, and erasing state in Figure 13.16a–c respectively.

Such a structure when employed for a charge-trap NVM device (Direct-Tunnel PBO or DT-PBO NVM), is expected to exhibit not only end-of-life (EOL) endurance, enhancements in power reduction, programming speed, reliability, but also significantly enhanced memory retention.

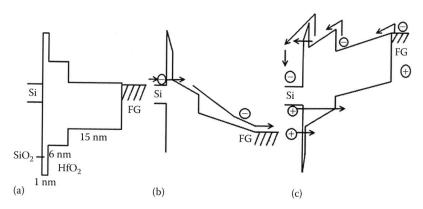

FIGURE 13.16 Band diagram for $SiO_2/HfO_2/TiO_2$ triple-layer direct tunnel structure with EOT = 3 nm, for a DT-PBO NVM design: (a) initial state, (b) writing state, and (c) erasing state. (From Bhattacharyya, A., High performance multi-level band –Engineered (ET/DT) NVM device, *ADI Associates internal publication*, December 2004, U.S. Patents: 7429767, 09/30/2008; 7553735, 06/30/2009; 7579242, 08/25/2009; 8159875, 04/17/2012; Bhattacharyya, A., Advanced scalable nano-crystal device and process, *ADI Associates internal publication*, November 19, 2004, Retitled: Scalable multi-functional and multi-level nano crystal NVM, U.S. Patents 7476927, 01/13, 09; 7759715, 07/30/2010; 7898022, 2011; 8242554, 08/14/2012; Bhattacharyya, A., A novel low power nonvolatile memory and gate stack, *ADI Associates internal publication*, October 21, 2004, U.S. Patent 7612403, 12/22/2009.)

13.5 APPLICATIONS OF BAND ENGINEERING FOR SPECIFIC NVM DEVICE ATTRIBUTES

Band engineering addresses all the significant elements in NVM device stack design. In Part I, all significant elements and their role in NVM stack design was addressed (Chapters 5 through 7). These include the silicon/insulator interface, the tunneling elements, the trapping or charge storage elements, the blocking elements, and the insulator/gate interface. In the earlier sections, we have discussed the importance of voltage scalability and incorporation of high K dielectrics in NVM stack designs. In this chapter, we have thus far reviewed the significance of multilayered dielectrics in charge transport. In this subsection, we will explore enhancements of various specific NVM device attributes by appropriate combination of these concepts. We will apply such concepts to all the subgroups of NVM devices discussed in Chapter 4, namely: (a) the FG/FP type, (b) the SONOS discrete trap type, (c) the NC embedded trap type, and the (d) direct tunnel (DT) memory type to illustrate the significance of band engineering. The specific NVM enhancement attributes that will be addressed are: (A) Vpp lowering and associated programming power reduction; (B) Programming speed enhancements; (C) Window enhancements; (D) Retention enhancements; (E) Endurance enhancements; and (F) Applications in multilevel memory devices.

13.5.1 Vpp Lowering and Associated Reduction in Programming Time and Power

13.5.1.1 Floating-Gate and Floating Plate Devices

For FG or FP devices, two-layer or three-layer dielectrics with band engineering could be effectively employed for Vpp lowering and achieving enhanced speed/power trade-offs while meeting the end-of-life retention targets.

There are at least two approaches for multilayered tunneling structures to address the above objectives. Both of these could be based on the concept of "barrier thinning" involving either a two-layer tunneling structure (D01/D10 i.e. High energy barrier/Low energy barrier) or a three-layer tunneling structure (Variot). A simple yet evolutionary approach would be to replace the thicker SiO_2 single dielectric layer with a two-layer tunneling structure where the first layer would be a thinner SiO_2

layer to preserve the desired Si/SiO2 interface characteristics. The second layer interfacing the SiO_2 layer could be a "trap-free" conventional dielectric of higher K value and lower bandgap (D10) such as SiON or SRN discussed in Part I of this book. Alternately, the second layer could be a higher K dielectric such as ZrO_2. In either case, such double-layer tunneling design may also be identified as "Variot" or "PBO" types of barrier.

The second approach is to replace SiO_2 at the silicon interface with an appropriate higher K dielectric and combine with additional single or multilayers dielectrics to achieve the desired barrier characteristics. While the second approach yields lower EOT and consequently lower Vpp, the integration is relatively more complex as noted previously. The second approach depending on the selection of the high K interface dielectric layer and corresponding barrier energy with reference to silicon may be elements of any of the multi-layered barrier types discussed before. We will briefly illustrate several examples of each approach as follows:

13.5.1.2   Conventional NAND Flash Devices

Two-layer tunneling media with oxide or OR-SiON interfacing silicon:

Conventional NAND flash devices with single-layer oxide for tunneling were limited in scaling oxide thickness to ~6.57 nm to ensure retention objectives when amorphous silicon or n+ doped silicon for the floating gate was employed. Purely from leakage point of view a 4.5 nm defect-free oxide would be sufficient if (a) preferred leakage path to n+ source/drain diffusion region could be eliminated by p+ doping of the floating gate, thereby adding the work function barrier by ~1 eV (without adversely affecting boron diffusion to the silicon interface); and (b) eliminating the interface roughness of the conventional n+ silicon FG/oxide to eliminate enhanced FN injection due to interface barrier lowering [16]. Alternately, the second layer could be an SRN layer with lower silicon content followed by an IN-SRN layer. The IN-SRN layer serves as the charge storage layer for the floating plate device replacing the floating gate. The dual SRN layers could be deposited at the same time (see Part I). Both requirements of (a) and (b) are served with this approach and boron penetration issue at the silicon/oxide interface in integrating silicon gate technology (PFET device) is eliminated since boron diffusion is prevented by the SRN layer. The double-layer barrier option of D01/D10 type discussed above could be readily employed with 4.5 nm SiO_2 (alternately OR-SiON) interfacing silicon with consequent reduction in EOT and enhanced current transport by barrier thinning (Variot barrier) during writing and barrier lowering (crested barrier) during erasing. Such scheme simultaneously lowers Vpp, enhances electron currents in both directions speeding up programming/erasing without adversely affecting retention. Table 13.2 lists some possible stack options with consequent EOT/Vpp reduction. The reference stack is a conventional floating gate stack with 7 nm SiO_2 for tunneling and 13 nm EOT for ONO interpoly dielectric (IPD) between the floating gate and polysilicon control gate for comparative Vpp reduction.

It should be noted that although not specifically mentioned in Table 13.2, oxide/nitride could also form as a conventional double layer for tunneling of the above Variot barrier type [17]. It should be further noted that oxide could be replaced by OR-SiON (~8 eV) of the same physical thickness with an overall improvement of EOT, with other attributes much the same. Furthermore, when the high K dielectric interfacing oxide has large conduction band offset with respect to oxide Figure 13.9, the electronic current is further enhanced due to the enhanced drift field imposed on the electronic charges tunneled through the oxide layer.

13.5.1.3   Two-Layer Tunneling Media with Oxide Replaced by High K Dielectric Interfacing Silicon

The double-layer barrier of D01/D10 type as above could also be conceived with oxide being replaced by an appropriate high K dielectric film with or without "interface engineering" as explained in the previous section on high K dielectrics. Both HfSiON (K = 14, Ub = 6.9 eV), and La AlO₃ (K = 27.5, Ub = 6.5 eV) could be considered for replacing oxide since these films could be readily deposited with clean interface with silicon substrate and with high temperature stability as discussed in the previous section. Table 13.3 illustrates examples of high K D01/D10 type where the D01 would be HfSiON or $LaAlO_3$ replacing SiO_2.

TABLE 13.2

EOT and Vpp Reduction (Rvpp) with Two-Layer Tunneling Dielectrics for FG/FP NVM NAND Flash Design [58,63]

Dielectric Layers[b]	EOT (nm)	UbD10 (eV)	Storage Gate/Plate	Blocking Layers	EOT (nm)	Vpp[a] (V)	Rvpp
4.5 nmOx/SiON	7	7.0	P+FG or FP	SiON/Al$_2$O$_3$	12.5–14	12–14	<0.7×
4.5 nmOx/SRN	6.5	5.1	P+FG or FP	SiON/Al$_2$O$_3$	11	10	~0.5×
4.5 nmOx/SRN	6.5	5.1	P+FG or FP	SRN/HfSiON	10	9	0.45×
4.5 nmOx/SRN	6.5	5.1	P+FG or FP	SRN/LaAlO$_3$	8.5	8	0.4×
4.0 nmOx/ZrO$_2$	4.75	6.5	P+FG or FP	SRN/LaAlO$_3$	6.75	6	0.3×

Source: Bhattacharyya, A., (a) Enhanced PBO devices and tunneling structures for NVM designs, ADI Associates Internal Publications: December 2004 (Unpublished), (b) Enhanced RTB-based Device Designs, ADI Associates Internal Publications, August 2005 (Unpublished), (c) Further advancement in Band-engineered NVM stack designs and implementation concepts, October through December, 2007 (Unpublished) and Bhattacharyya, A. and Darderian G. U.S. Patent Application, 0067256 A1, March 12, 2009.

[a] Reference stack 7 nm oxide/FG/ONO, stack EOT 20 nm, Vpp ≥ ±20 V, 10 ms.

[b] Physical thickness of first and second layers is equal.

The below Table 13.3 illustrates that appropriate barrier engineering with double-layer tunnel dielectrics (asymmetric barrier) could lower the Vpp as much as one-fourth of the conventional Vpp and consequently the programming power could be reduced by nearly a tenth!

13.5.1.4 Three-Layer Tunneling with Barrier Engineering with Oxide or with High K Interfacing Silicon

Two-layer barrier engineering for tunneling is intrinsically asymmetric from the standpoint of carrier transport during writing and erasing while lowering Vpp as illustrated in the Tables 13.2 and 13.3. The three-layer barrier engineering based on Fowler–Nordheim tunneling could provide symmetric barriers for both writing and erasing as well as capable of providing greater retention while moderately trading off stack EOT and Vpp due primarily to greater tunneling distance. This is exemplified for the three-layer Variot barrier of SiON/La AlO$_3$/SiON in Table 13.3. Multiple examples of symmetric and asymmetric triple-layer tunneling dielectrics for FG and CT devices with crested barriers have been provided demonstrating Vpp reduction. These include HfO$_2$/SiO$_2$/HfO$_2$ [5], SiON/Al$_2$O$_3$/SiON [4], and Si$_3$N$_4$/HfSiON/Si$_3$N$_4$ [11] among others as examples of symmetric crested barriers and HfO$_2$/SiO$_2$/LaAlO$_3$ as example of asymmetric crested barrier [11]. Some of these stack designs will be discussed in appropriate sections of this part of this book in greater detail.

TABLE 13.3

EOT and Vpp Reduction Using Double- and Triple-Layer Tunneling Dielectrics with High K Film Replacing Oxide (Variot Barrier)

Dielectric Layers	EOT (nm)	Storage Gate/Plate	Blocking Layers	EOT (nm)/Vpp (V)	Rvpp
LaAlO$_3$/Si$_3$N$_4$	2.7	P+FG or FP	Si$_3$N$_4$/LaAlO	5.5 nm/5.0 V	0.25×
HfSiON/Si$_3$N$_4$	2.7	P+FG or FP	HfSiON	5.5 nm/5.0 V	0.25×
SiON/LaAlO$_3$/SiON	5	P+FG or FP	HfSiON	8.0 nm/<7.0 V	<0.35×

Reference stack: 7 nm oxide/FG/ONO: STACK EOT 20 nm, *Vpp* ≥ ±20 V, 10 ms.

13.5.2   CT/NC/DT NVMs

Conventional devices of discrete trap (CT) and embedded trap (NC) NVMs primarily employed thinner oxide for direct band-to-band tunneling through the oxide layer for both electrons and holes with partial charge trapping at the oxide/nitride interface (CT) and additional charge transport through the triangular barrier (modified Fowler–Nordheim tunneling) of nitride and subsequent trapping in the bulk nitride (CT) or in the nanocrystal traps (NC), respectively. In DT devices, transport mode is by direct tunneling through thin oxide only. Since the oxide is thinner in these types of devices and high current density is achieved at relatively lower Vpp, the programming voltages are significantly lower compared to FG or FP devices. However, additional Vpp reduction and associated programming power reduction could be achieved by appropriate incorporation of band engineering in these types of devices as well. Many examples of double-layer and triple-layer band-engineered tunneling structures with and without high K dielectrics exist in publications during the last decade to demonstrate scalability of these devices. Some of such examples will be selectively illustrated here. Several others will be discussed in subsequent sections. Stack structures of CT/NC/DT NVMs illustrating Vpp reduction and speed enhancements are shown in Table 13.3, above. Conventional typical SONOS device is shown as reference for comparison.

In Table 13.4, shown below, the crested barriers are abbreviated as "C" and the Variot barriers as "V" respectively in the "Remarks" column. As a thumb rule, if the EOT of the tunneling layers is in

TABLE 13.4
Examples of Vpp Reduction and Speed Enhancements for CT/DTM Devices[b]

Stack Type	Tunnel Layers	EOT nm	Storage Layers	Blocking Layers	STACK EOT (nm)	Programming ±Vpp/tpp V/µs	Remarks
			A: Double-Layer Tunneling				
SONOS	SiO_2/ZrO_2	<~2.5	Si_3N_4 or NR-SiON	HfSiON ~7 nm	~6 nm	5 V/<100 µs	[4] V
SONOS	$SiON/Al_2O_3$	<~2.2	SiON	HfO_2 ~10 nm	~7 nm	5 V/1 µs W–9 V/ 1 µs E	[4] C
SONOS	$SiON/HfO_2$	<2.2	SiON	HfO_2 ~10 nm	~7 nm	5 V/1 µs W–9 V/1 µs E	[4] V
NC	$HfO_2/LaAlO_3$	~2.5	Ge NC in HfSiON	$SRN/LaAlO_3$	~7–8 nm	±6 V/<100 µs	[19] MLC/ MF[a] C
			B: Triple-Layer Tunneling				
SONOS	$Si_3N_4/SiO_2/Si_3N4$	4.3	Si NC in SiO_2 SiO_2		~10 nm	11.2 V/100 µs	[7] C
SONOS	$HfO_2/SiO_2/LaAlO_3$	2	Ta_2O_5		<6 nm	±5 V/1 µs W/E	[5] MLC[a] C
NC	$HfO_2/LaAlO_3/$ La_2O_3	2	Pt NC in AlN		<6 nm	±5 V/1 µs W/E	[5] MLC[a] C
SONOS	$HfO_2/SiO_2/LaAlO_3$	2	Si_3N_4		<6 nm	±5 V/1 µs W/E	[11] C
SONOS	$Si_3N_4/HfSiON/$ Si_3N_4	2	TiO_2		<6 nm	±5 V/1 µs W/E	[11] C
			C: Multiple Direct Tunneling with Progressive Band Offset (PBO)				
SONOS	$SiO_2/Si_3N_4/HfO_2$ or Pr_3O_3 or TiO_2	<2.0	SiON		2.5[b]	1.5 V, <1 µs	[22][c]

[a]   See Section 13.7: MLC, Multilevel Cell; MF, Multifunctional Cell.

[b]   Reference stack: 1.5 nm SiO_2/5 nm Si_3N_4/4.5 nm SiO_2; EOT~9 nm, Vpp ± 9 V, tpp~1 ms.

[c]   Fast write, small window ~0.5 V (Battery Operable).

the range of 0.3–0.4× of the stack EOT and the stack EOT $\leq$ 8.0 nm, then one could assume that an increase in Vpp by 1 V would result in approximately an order of magnitude enhancement in writing or erasing speed.

Example of progressive band offset (PBO) multiple DT structure is also exemplified in the table. PBO provides faster write and/or Vpp scaling capability to very low voltage levels. Such applications are illustrated in appropriate sections of this book.

In Part III, we will discuss multiple SUM stack designs with very low Vpp.

13.5.3 Applications of Band Engineering in Other NVM Property Enhancements

13.5.3.1 Programming Speed Enhancement through Band Engineering

We have reviewed how band engineering enhances tunneling currents for writing and erasing and Vpp scaling. We have also noted that leakage level by back tunneling needs to be as low as 1E-16 A/cm^2 to meet the retention requirements for NVM devices for 10 years. We have also shown that tunneling current and programming speed are interrelated; higher the tunneling current at a given Vpp, faster would be the programming. Additionally, for a given stack design, higher the Vpp, tunneling current would increase thereby increasing the programming speed assuming the retention requirement is fixed. We have shown that Vpp can be modulated by band engineering to meet the tunneling current enhancement. Therefore, programming speed could also be enhanced by band engineering. For FG devices, by means of both barrier lowering (crested barrier, [14]) and barrier thinning (Variot barrier, [10]), thereby, significantly enhancing the tunneling current and consequently the programming speed. For example, it has been shown in Figure 13.9 that for the same voltage drop across the same EOT tunnel-layer options, electronic current enhancement of SiO$_2$/ZrO$_2$ double tunnel layer could be >1E4× compared to single-layer SiO$_2$ tunnel dielectric.

This implies that programming speed could be enhanced by >1E3× by appropriate band engineering of tunneling layers. Since overall stack EOT could also be reduced using high K dielectrics for trapping and blocking layers, lower Vpp and enhanced programming speed could be achieved simultaneously.

Therefore, compared to a conventional FG Oxide-ONO stack design with Vpp at ±20 V, 10 m sec for writing or erasing, an appropriate band-engineered high K stack could provide less than, ±8V, <1 μs programming capability. Several authors have proposed such enhancements as exemplified in the Tables 13.4a–c discussed above [4,5,7,11,19,22,58,64].

13.5.3.2 Memory Window Enhancements through Band Engineering

To enhance memory window NVM stack design simultaneously requires large tunneling current at programming field and extremely low charge loss to both the silicon substrate and the gate electrode during the standby field. The standby field would be higher due to higher charge storage associated with the larger memory window. As a result, large memory window and NVM retention are intimately coupled.

Memory window and retention trade-offs:

For all NVM device types discussed earlier, certain basic considerations are common in establishing appropriate memory window and retention trade-offs. Larger memory window and stable memory states require high charge fluence during programming field and very low charge fluence (leakage) at standby (internal) field generated by higher density of charges stored. It was discussed earlier that the stack design objectives should be to provide a programming current to leakage current ratio at Vpp and Vr (internal potential during standby) of ~ >1E13 (current densities of 1E-3 A/cm^2 and 1E-16 A/cm^2, respectively, at Vpp and Vr) to meet window and retention objectives. Even larger memory window and stable memory states could be achieved with greater enhancements in programming current and lowering of leakage current during standby. Band engineering employing multilayered tunneling and trapping with appropriate low leakage blocking dielectric could effectively modulate the J–V slopes of the gate insulator stack to enable the desired window-retention trade-offs. Band engineering to achieve steeper J–V slope has been demonstrated by Govoreanu et al. [1,8,10] (see also Figure 13.9). Several investigators have demonstrated that by appropriate band engineering with employing a combination of low leakage high K dielectrics or a combination of oxide and high K dielectrics as exemplified in Tables 13.1 through 13.3, programming current could be enhanced by >1E4× resulting in larger memory window as well as lower leakage [10,12,20–22]. The advantage of simultaneously lowering Vpp is also noted.

Additionally, optimization of band-engineered blocking dielectrics and gate electrode work functions are required to reduce leakage to the gate and charge injection from the gate–insulator interface. The significance of introducing high work function metal gate such as tungsten (W, 4.6 eV), platinum or iridium (Pt/Ir, 5.3 eV), or palladium (Pd, 5.0 eV), or nitrides of tungsten, titanium, or tantalum (WN or TiN or TaN, >4 eV) has already been discussed in Part I to reduce gate injection from the control gate during operational conditions. Choi et al. have demonstrated that application of W/WN top electrode interfacing Al_2O_3 blocking layer reduced the interpoly dielectric leakage of the NVM stack by 1E4× at a field of 6 MV/cm [23].

By combining the above design aspects of band engineering of NVM stack, J. Buckley et al. [13] have demonstrated that using a Variot tunnel barrier consisting of $ZrSiO_4/ZrO_2/ZrSiO_4$ (K = 10), and silicon gate passivated with metal nitride at the stack insulator interface, not only low Vpp of ±6V, along with excellent memory retention could be achieved but also simultaneously high programming speed with desired memory window.

For devices with larger tunnel distance of FG and CT types, high fluence from silicon substrate during programming conditions and low leakage to the substrate and the gate at standby internal field with higher density of charge storage (higher electrostatic field at standby) should be common requirements to achieve a larger window. However, in case of CT devices (SONOS and NC types), larger window requires not only high density of traps with high capture probability, but also deep energy traps such that once the charge is captured, it could be retained during operational and standby conditions throughout the life of the device. For DT devices, while high charge fluence is easier to achieve due to the tunneling mode, back tunneling could only be prevented by creating a combination of (a) deep potential well of the trapped charge and (b) a repulsive potential for the trapped charge to return to the substrate since the tunneling distance is short, and (c) asymmetric barriers for reverse tunneling creating longer reverse direct tunneling distance. These approaches with illustrations for specific device types will be discussed below.

13.5.3.2.1 FG Devices

FG NVMs could store large density of electronic charge at the floating gate, thereby enhancing memory window provided charge loss could be controlled to achieve retention objective.

It should be noted that for FG stack design, in addition to the band engineering of tunneling layers and blocking layers with high work function passivation for the gate/insulator interface, the coupling coefficient should be optimized to achieve larger window. This is customarily done by optimizing the blocking high K dielectric thickness to be of higher EOT compared to the tunnel-layer EOT so that for a given stored potential at the floating gate, the high state would have a higher Vth ("1" state) and consequently larger memory window.

13.5.3.2.2 CT Devices

Compared to the FG devices, the memory window for CT/NC devices would be limited to the available trap density of discrete traps at the interface and bulk of the trapping dielectrics, and/or the NC-induced embedded traps within the constraints of quantum confinement and Coulomb blockade. Trapping characteristics could however be improved by embedding nanocrystals within a discrete trapping dielectric layer. J. Fu et al. showed that when Si nanocrystals are embedded into nitride, the SONOS memory showed large window, and enhanced programming speed without compromising retention [24]. Trapping characteristics are dependent on material and process parameters, which has been explained in Part I for the conventional devices and in the previous section for the high K trapping dielectrics. Fundamentally the requirements of large fluence during programming and reduced charge loss to the substrate and the gate during standby are equally applicable for these devices to achieve enhanced window and retention. Band engineering could also be employed to achieve such objectives. Historical examples of multilayered SiON band-engineered structures were discussed and illustrated previously in Part I to simultaneously provide window and retention enhancements for thicker CT devices as early as in 1976 (Part I, Chapter 5); and also recently in 2005 (Part I, Chapter 6).

High K band-engineered structures could be considered for enhancement of window and retention fulfilling the premise of high fluence and deep level high density trapping at low Vpp. Ta_2O_5 films exhibit deep trapping levels of 2.7 eV [25]. X. Wang et al. demonstrated a single tunnel layer stack structure with a Ta_2O_5 trapping layer and with 4.8 nm HfO_2 tunnel layer combined with Al_2O_3/TaN blocking layer yielding retention loss of <75 MV/decade, with memory window of >2 V and, <1 ms programming speed at ±8 V of writing and erasing [26]. A second example of SONOS equivalent single-layer dielectric stack structure is shown in Figure 13.17a and b based on high K dielectric characteristics discussed in the preceding section. The stack consists of a GaN trapping layer ($\Phi b = 3.9$ eV, K = 10) sandwiched between a tunnel layer and a blocking layer of different thicknesses of La AlO_3 ($\Phi b = 6.6$ eV, K ~ 25) as shown. The conduction band offset of GaN is over 3 eV with respect to that of $LaAlO_3$ coupled with deep level trapping in GaN. This ensures extremely low retention loss, which has been confirmed by K.-H. Joo et al. [27]. Additionally, leakage through $LaAlO_3$ dielectric is extremely low at low field, ensuring retention of large memory window post writing. $LaAlO_3$ exhibits high fluence at moderate programming field enabling high programming speed and large memory window schematically shown in Figure 13.17a in the band diagram. La AlO_3 has a respectable barrier energy with respect to the conduction band of silicon substrate to minimize back tunneling (Ube = 2.4 eV), yet not as high as SiO_2 (Φb–e = 3.2 eV), thereby providing higher fluence at moderate field across the tunnel dielectric. A typical stack design consisting of 10 nm $LaAlO_3$ for tunneling, 10 nm GaN for trapping, and 20 nm La AlO_3 for blocking (EOT ~ 8.7 nm) is estimated to provide at programming condition of ±7.5 V Vpp, 10–100 μs, an initial memory window of >6 V with end-of-life window of ~5 V suitable for MLC NVM memory functionality. Lower programming field also ensures higher device endurance and reliability. Similar device characteristics could be achieved using nanocrystals of ZnO embedded in La AlO_3 (NC device) instead of GaN, although memory window would be limited by Coulomb blockade and quantum confinement. However, excellent retention for the stack would be expected [27]. Both of the above examples exhibit common features of: (i) High fluence characteristics of electron injection during programming due to relatively lower interface energy barrier compared to Si/SiO_2, (ii) deep energy level of trapping, and (iii) high electron barrier for charge loss either to gate or to the substrate with high work function gate interface.

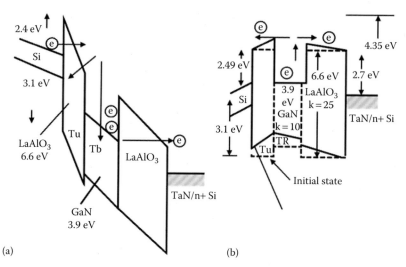

FIGURE 13.17 Example of band-engineered (single layer) high performance, high K CT device with enhanced memory window and retention band schematic during (a) writing and (b) retention.

Therefore, the CT version of the device stack as discussed above consists of $Si/LaAlO_3$ (tunneling)/GaN (trapping)/$LaAlO_3$ (blocking)/TaN with electronic parameters identified for each layer in Figure 13.17a and b during writing and during retention respectively as shown below.

Several other examples could be provided with enhanced window, high speed, and good retention for CT devices employing thinner SiO_2 or SiON for tunneling ($\leq$3 nm) but employing higher K deep trap dielectrics such as AlN or AlGaN or with low leakage blocking dielectric such as HfAlO or $LaAlO_3$ with compatible gate electrode, for example, IrO_2 or TaN [26–29].

13.5.3.3 "Trap Engineering"

Enhanced memory window and retention using band-engineered multilayer-trapping dielectrics:

Analogous to the tunnel engineering using multilayered dielectrics considered earlier, multiple trapping layers could be considered to maximize charge trapping and minimize retention loss. Instead of a single-layer trapping dielectric, a three-layer trapping dielectrics consisting of crested barrier (D00/D11/D00: low barrier/high barrier/low barrier) could be considered to provide multiple energy level higher density trapping with reduced charge loss to the substrate and gate. Integrated with double-layer or triple-layer Variot tunnel layers for high fluence, such structures of "trap engineering" provide simultaneously enhanced window and retention at the expense of processing complexity of stack design.

For conventional dielectric films, such trap-engineered structures could be: nitride/oxide/nitride (NON) or better yet Si_2ON_2 ($\Phi b = 6.5$)/OR-SiON ($\Phi b = 7.5$)/ Si_2ON_2 ($\Phi b = 6.5$) or $Si_2ON_2/Al_2O_3/$ Si_2ON_2. The combination $Si_2ON_2/SiON/Si_2ON_2$ (Part I, Figure 6.17c) can be easily deposited by CVD or LPCVD process as a single step by appropriately modulating silicon and nitrogen sources (e.g., SiH_2Cl_2 for Si and NH_3 for N) during deposition. However, better results could be achieved by the triple layer: $Si_2ON_2/Al_2O_3/Si_2ON_2$ since alumina could also provide high density deep traps and lower conductivity with proper processing. Such structures will be further discussed in the next subsection on enhanced retention. Following the above concept, Z. Huo et al. demonstrated enhanced memory window and superior retention in TANOS device (discussed later) by replacing nitride trapping layer with nitride/alumina/nitride (NAN) trapping layer [30].

For higher K dielectric films, there could be multiple combinations by combining higher bandgap trapping layers such as HfAlO ($\Phi b = 6.2$, K = 14–17) sandwiched between two layers of HfO_2 or GaN or Ta_2O_5 or TiO_2. The simpler processing solution could be $HfO_2/HfAlO/HfO_2$, which could all be produced in a single-step CVD or sputtered process or alternately by appropriate ALD process.

The approach of combining tunnel engineering and multilayer trapping could be considered for yielding multibit per cell memory design (MLC) aiding bit density and reducing cost per bit. Additionally, functionally enriched (multifunctional) NVM device could be envisioned to enhance applications. In a later subsection, we will provide some such examples.

Again, as explained earlier, high work function nanocrystals embedded in the trapping layers in multiple planes would provide larger window and enhanced retention similar to above CT devices provided integration challenges could be overcome using such structures. Examples of such structures are provided in section on advanced devices and in Part III.

13.5.4 Memory Retention Enhancement through Band Engineering

Memory retention properties of silicon based NVMs differentiate these devices from other silicon based memories. Therefore, much attention is focused to enhance retention of memory states through band engineering of NVM stack design. Memory retention limitations of conventional oxide and nitride based FG and CT devices have been discussed in detail in Part I of this book and enhancements provided by multilayered oxynitrides have been illustrated. Recently Gilmer et al. published "best in class" performance for MANOS-type charge trap flash nonvolatile memory devices [31] through improved program/erase (P/E), endurance and retention" by combining band-engineered tunnel oxide, band-engineered SiNx for trapping, and band-engineered alumina for blocking with effective work function gate electrode achieving enhanced memory window and retention with >1E5 endurance with nearly zero charge loss over 24 hours at 150°C. The above work was a follow-on of the TANOS (TaN/Al$_2$O$_3$/nitride/oxide/silicon) CT cell [32–38]. The TANOS device first published by C. H. Lee et al. [32] had since become a basic reference of comparison for other subsequently evolved SONOS follow-on devices encompassing metal-gated band-engineered CT (MG-BE-CT) stack designs.

We shall briefly review the characteristics of TANOS device especially from the standpoint of stability of the memory states and associated retention properties. Further discussions on TANOS cells will be covered in the multilevel cell design section.

13.5.4.1 The TANOS Device

The TANOS (TaN/Al$_2$O$_3$/nitride/oxide/silicon) device deviated from the classical SONOS stack through the introduction of high K alumina as the blocking layer and an over layer of tantalum nitride to provide a higher work function metal interface between alumina and the n+ polysilicon gate. The TANOS device, therefore, could be considered to be the first in introducing "metal-gated" SONOS type CT-NVM device for NOR and NAND applications [32]. This CT-NVM device introduction was motivated to address the feature size scalability issues of silicon-FG devices related to the FG to adjacent FG isolation and adjacent cell to cell parasitic coupling problems (FG–FG capacitive coupling and associated disturb) [35]. The higher work function TaN metal gate also reduced the electron injection from the gate during erase preventing erase saturation, thereby providing larger memory window. Additionally, the high barrier energy of Al$_2$O$_3$ coupled with the high negative fixed charge density in Al$_2$O$_3$ reduced stored charge leakage to the gate enhancing retention [33]. The original TANOS device employed a 2.5 nm tunnel oxide [32]; consequently charge loss during standby due to reverse electron tunneling to the substrate was significant reducing the end-of-life window to approximately one-third of the original value. This is shown in Figure 13.18.

The TANOS stack was subsequently optimized to thicker tunnel oxide of ~4.0 nm to prevent retention mode reverse direct tunneling of stored electrons in nitride and also optimized to thinner trapping nitride (6.5 nm) and thicker Al$_2$O$_3$ blocking layer (15 nm). Such stack optimization also suppressed hole injection during writing and erasing while enhancing electron transport during writing and erasing. This significantly improved both programming speed as well as retention [34]. Further optimization in stack design involved introduction of 5 nm tunnel oxide and improved process integration

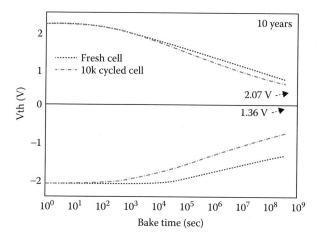

FIGURE 13.18 Data retention characteristics of the original TANOS device consisting of 2.5 nm oxide/12 nm of nitride/and 10 nm of Al$_2$O$_3$/TaN/Poly-Si gate design. (From Lee, C. H. et al., A novel SONOS structure of SiO$_2$/SiN/Al$_2$O$_3$ with TaN metal gate for multi-giga bit flash memories, *IEEE IEDM Technical Digest*, p. 613 © 2003 IEEE.)

employing W/WN/TaN gate. This resulted in improved retention and larger EOL memory window enabling MLC (multibit/cell) TANOS design [35,36]. Key process parameter and characteristics of the 32 Gb TANOS flash memory chip fabricated using 40 nm CMOS technology node are listed below:

Technology: 40 nm CMOS with triple well and triple metal with PVD W bit line and flowable oxide ILD for bit line.
Chip features and design characteristics: 32 Gb memory chip; chip size: 230 mm²; page mode programming (800 μs); block erasing (20 ms), read acces: 25 μs transfer/25 ns burst cycle.
Memory cell features: TANOS: Unit cell size: 0.0098 μm²; gate stack: M/WN/TaN/Al₂O₃ (15 nm)/SiN (6.5 nm)/SiO₂ (4.0 nm).

TANOS device was further improved by replacing the nitride trapping layer with a band-engineered multilayer CT structure ("trap-engineered") consisting of nitride/alumina/nitride triple layer.

The improvement provided not only a very large memory window of ~10 V but also significantly enhanced retention and reliability while reducing post cycling operational charge loss to <0.5 V [37]. Advanced integration technique and lithography were adopted to provide a 3 bit/cell MLC memory design at 27 nm technology node. Highly manufactural and reliable 64 Gb NVM memory product was achieved yielding a unit cell size (including overhead) of 0.00375 μm² [38]. It is believed that the recently announced 3D 128 Gb NAND chip from Samsung Corporation incorporates the above CT NAND cell stack design.

13.5.4.2 Other TANOS-Follow-on Band-Engineered Devices with Enhanced Retention

Y. Q. Wang et al. investigated the device characteristics of the TANOS follow-on device with a two-layer oxide nitride (Variot) tunnel barrier combined with a higher K HfO₂ trapping dielectric [39]. The stack was identified as "DTL2" stack. A separate stack option identified as "DTL1" stack consisted of SiO₂/HfO₂/Al2O₃/TaN where only the trapping layer nitride of the TANOS stack was replaced by HfO₂. The DTL1 stack demonstrated higher writing speed compared to the TANOS stack due to more favorable conduction and valence band offsets. However, the DLT2 double tunnel layer stack with oxide (2.9 nm) plus nitride (1.6 nm) demonstrated simultaneously higher writing and erasing speed, enhanced retention and endurance. The characteristics of DTL2 device retention are shown in Figure 13.19a, while the enhanced endurance of the DTL2 device compared to the TANOS device are shown in Figure 13.19b.

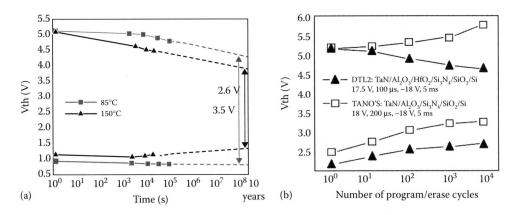

FIGURE 13.19 Device characteristics comparison between TANOS and DTL devices: (a) Retention and (b) Endurance. (From Wang, Y. Q. et al., Fast erasing and highly reliable MONOS type memory with HfO₂ high-k trapping layer and Si₃N₄/SiO₂ tunneling stack, *Electron Devices Meeting, 2006, IEEE IEDM Technical Digest*, p. 971 © 2006 IEEE.)

When oxide and nitride are replaced by higher K dielectrics and multilayer band-engineered stack structures are envisioned, many options are feasible to enhance memory retention and simultaneously other NVM device attributes for all types of NVM devices. In the next subsection, we will examine the memory retention enhancement in greater detail employing band-engineered multilayered structures of high K dielectrics in the stack design. The objective would be to achieve optimal tradeoffs of retention with other NVM device parameters.

13.5.4.3 Band-Engineered Stack Requirements to Achieve Enhanced Retention: Devices Designed to Operate in Fowler–Nordheim Tunneling Mode

This subsection will briefly discuss some of the critical requirements for retention enhancements focused on CT/NC devices. Selected examples of band engineering will be illustrated toward retention enhancement for thicker tunnel devices operating in F–N mode or modified F–N mode.

Retention enhancement for stack design needs to meet the following criteria:

1. Transport through high K dielectric films with characteristic lower conductivity for equivalent EOT. Examples of such high K films provided in Section 2.3.4 are: Pr_2O_3, Al_2O_3, $LaAlO_3$, and HfSiON.
2. Charge trapping in dielectrics with stable deep level traps and less sensitive to process-induced shallow trap generation. Examples of such trapping dielectrics are: Si_2ON_2, GaN, AlN, Ta_2O_5 and TiO_2. As discussed in the previous section, all above dielectric layers provide lower conduction band energy offset compared to silicon nitride and higher valence band energy offset aiding electron transport and preventing hole transport during programming and erasing enabling simultaneously enhanced retention and programming performance. Alternately, for NC devices, trapping charges in nanocrystals with deep quantum well is the key requirement. Generally, metal nanocrystals with high work functions such as Pt and Ir (5.3 eV), Pd (5.0 eV), W (4.6 eV), and semiconductor such as Ge (4.0 eV) exhibit deep quantum well.
3. Trapped charge facing asymmetric high conduction band or valence band barriers preferably greater than 2.5 eV to minimize leakage. This can be accomplished by band engineering.
4. Reverse tunneling distance from the trapping plane to both the silicon substrate and to the gate is significantly large to eliminate the possibility of direct tunneling mode of charge loss (direct tunneling probability to be zero).

An example of a band-engineered three-layer stack replacing oxide with higher K HfSiON for tunneling and blocking fulfilling conditions (1), (3), and (4) is provided by R. van Schaijk et al. [40]. The tunnel-layer barrier energy was optimized and reduced to 2.5 eV from stoichiometric HfSiON of 3.0 eV by increasing the ratio of Si/Si+Hf from ~0.5 to 0.77 ("HfSiON-b"). This improved the writing speed while lowered the k-value. The blocking layer was near-stoichiometric with higher barrier energy and lower electron leakage ("HfSiON-a"). The stack consisted of 4.0 nm HfSiON-a/6 nm of Si_3N_4/10 nm HfSiON-b (stack EOT of ~9.4 nm), exhibiting programming at ~±12 V, 5 ms providing enhanced window and programming speed compared to the stack option with 2.2 nm tunnel oxide and either HTO or HfSiON-b for the blocking layer with similar stack EOT [40]. The retention was significantly enhanced as shown in Figure 13.20a. The data retention at different temperatures is also shown in Figure 13.20b with characteristic thermal sensitivity of the detrapping of bulk charges from nitride trapping layer.

For further enhancement of device retention and performance, a modified stack was proposed by replacing nitride trapping layer with either a GaN trapping layer or an AlN trapping layer or an appropriate NC-based trapping layer satisfying the conditions of item (4) above [41]. The band diagram of the modified device is shown in Figure 13.21b and compared with that of R. van Schaijk et al. as in Figure 13.21a. It is estimated that the proposed all-high-k replacement will lower the writing and erasing voltage levels to <±11 V, at 1 ms with larger memory window and enhanced

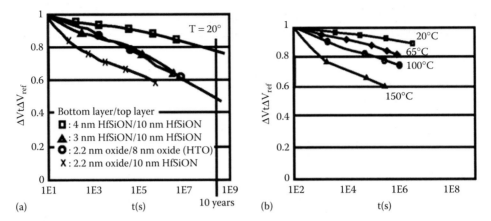

FIGURE 13.20 Retention enhancement with band-engineered high K tunnel layer and blocking layer using off stoichiometric HfSiON for tunneling (Si/Si+Hf = 0.77) optimizing Ube and using near-stoichiometric HfSiON (S/Si+Hf = 0.47) with conventional Si_3N_4 trapping layer. (From van Schaijk, R. et al., A novel SONOS memory with HfSiON/Si_3N_4/HfSiON stack for improved retention, *Proceeding IEEE NVSMW*, p. 50 © 2006 IEEE.)

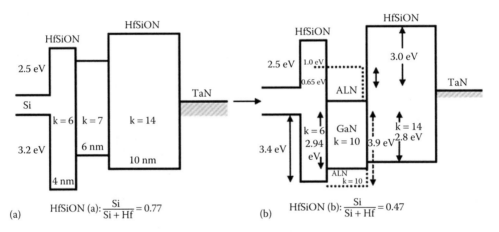

FIGURE 13.21 (a) Band diagram of the device of R. van Schaijk et al. [40] and (b) Band diagram of modified all high-k stack design replacing nitride trapping with either GaN or AlN trapping. Further enhancement in retention and device characteristics is suggested thru option (b). (From Bhattacharyya, A., Novel band-engineered nano-crystal NVM device utilizing enhanced gate injection, *ADI Associates internal publication*; U.S. Patent 7629641, 12/22/2009, December 4, 2004.)

retention. It should also be noted that such stack structure would be relatively easier to integrate similar to the referenced stack [40] since both GaN and AlN could exhibit similar differential etch characteristics as nitride as well as providing barrier to diffusion of contaminants similar to nitride. These integration issues will be further discussed in the next section.

13.5.4.4 Multiple Examples of Band-Engineered Stacks with Enhanced Retention

Several other examples of stack designs fulfilling all above requirements and enhancing retention properties will be exemplified. These stack design options are illustrated in Figures 13.22 and 13.23 and are described below [5,18,23,39,58].

The stack design of Figure 13.22a: PBO barrier with deep-offset trapping: This fulfills all the above requirements of retention enhancement with a triple-layer tunnel barrier of PBO type discussed previously in Section 13.5.2 (Table 13.4C). The design provides enhanced retention, large

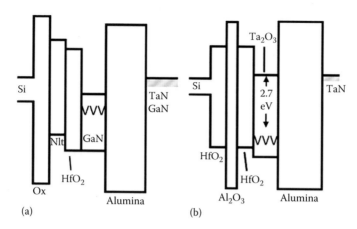

(a)   (b)

FIGURE 13.22 Illustration of first 2 of 4 schemes of band-engineered stack designs fulfilling retention requirements: (a) PBO barrier with deep-offset trapping and (b) crested barrier with similar trapping. (From Bhattacharyya, A., (a) Enhanced PBO devices and tunneling structures for NVM designs, ADI Associates Internal Publications: December 2004 (Unpublished), (b) Enhanced RTB-based Device Designs, ADI Associates Internal Publications, August 2005 (Unpublished), (c) Further advancement in Band-engineered NVM stack designs and implementation concepts, October through December, 2007 (Unpublished).)

window, and high programming speed characteristic of PBO tunnel barriers. The tunnel barrier consists of a triple layer of oxide/nitride/hafnia with progressive conduction band offset which is designed to fulfill requirements (1) and (4) above. The trapping layer is either a GaN discrete trap layer or a TiO_2 layer with embedded Pt nanocrystal to provide deep quantum well trapping [23], thus fulfilling requirement (2). The blocking layer of alumina with tantalum nitride or titanium nitride barrier satisfies the requirement (3) for low standby leakage and negligible charge injection from the gate preventing erase saturation. It should be noted that the design could be simplified by eliminating the nitride having a two-layer oxide/hafnia tunnel barrier of type D01/D10 (double-layer Variot, Section 13.3.3.2) to achieve retention objectives as well as the programming speed and memory window.

The stack design of Figure 13.22b: Crested barrier with deep-offset trapping: The design illustrates a triple-layer crested tunnel barrier consisting of hafnia/alumina/hafnia or hafnia/oxide/hafnia or $HfO_2/LaAlO_3/HfO_2$, alternately a two-layer version of type D00/D11 ensuring conditions (a) and (d) with deep (2.7 eV) discrete trapping in Ta_2O_5 and/or optionally Pt nanocrystal trapping embedded in Ta_2O_5 [5,42] or Ge nanocrystal embedded in the HfO_2 layer interfacing TaA_2O_5 [18,19,35,42–45]. Blocking layer could be similar to the first stack design satisfying conditions (2) and (3). Such device would exhibit superior low Vth retention and could be tailored for multilevel multifunctional NVM design to be discussed later on.

The stack design of Figure 13.23a: 2-layer crested barrier with single or dual-energy level trapping: The stack design option, similar to the first and second provides single-layer trapping at dual-energy levels while the fourth stack option provides double-layer trapping with different distance (and thereby different internal field during standby and different rate of charge loss) from the gate electrode and the silicon substrate. Both the third and the fourth options are shown to have $HfO_2/LaAlO_3$ double-layer tunneling (D00/D11: Crested barrier as the second one) satisfying conditions (a) and (d). One of the trapping layer for the third design is the same as the first trapping layer of the forth stack design. This layer may consist of either one energy level of GaN discrete trap or Ge nanocrystals embedded in Al_2O_3 [5] or La_2O_3 or OR-SiON or SiO_2 [33,41,45] (depending on the process integration simplicity) or combining dual-energy levels of trapping with GaN and Ge nanocrystals. As mentioned above, the dual-energy level trapping could provide multilevel storage and/or multifunctional storage. The blocking layer for both the third and fourth designs consists of a thicker HfSiON layer with titanium nitride or tantalum nitride barrier satisfying conditions of (b) and (c).

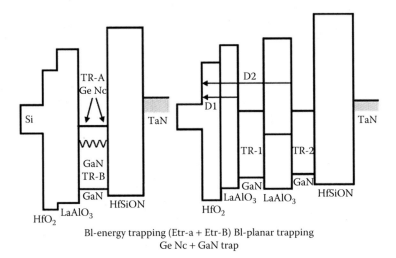

Bl-energy trapping (Etr-a + Etr-B) Bl-planar trapping
Ge Nc + GaN trap

FIGURE 13.23 Illustration of last 2 of 4 schemes of band-engineered stack designs fulfilling retention requirements (a) single or dual-energy single-layer trapping and (b) bi-planar/double-layer trapping. (From Bhattacharyya, A. (a) Enhanced PBO devices and tunneling structures for NVM designs, ADI Associates Internal Publications: December 2004 (Unpublished), (b) Enhanced RTB-based Device Designs, ADI Associates Internal Publications, August 2005 (Unpublished), (c) Further advancement in Band-engineered NVM stack designs and implementation concepts, October through December, 2007 (Unpublished).)

The stack design of Figure 13.23b: 2-layer crested barrier with bi-planar level trapping: However, in case of the fourth design option, a second trapping layer of discrete or embedded trapping (or a combination consisting of Ta_2O_5 deep traps and Pt nanocrystal deep well) is incorporated and sandwiched between the blocking layer of HfSiON and a thinner layer of the same dielectrics for enhancements of multifunctional and multilevel trapping [5]. The charge trapping takes place at distinctly different locations separated by high-energy barriers yielding multilevel and/or multifunctional characteristics. The illustrations are intended to demonstrate multiple possibilities in band-engineered CT devices to achieve retention enhancements with enhanced functionality at the cost of process and integration complexity.

It is worth noting as illustrated above that in CT devices charge loss kinetics during charge storage (standby) could be modulated by altering transport mechanisms by means of storing charges either at the same storage plane with distinctly different energy depths by altering "detrapping kinetics," or at the same energy depth at distinctly different location by altering "band bending kinetics" (trapezoidal to triangular barrier). Therefore, such schemes could be employed to achieve bimodal or multimodal retention and could yield multilevel storage and/or multifunctional memory. Follow-on sections will discuss such NVM devices.

13.5.4.5 Endurance Enhancement through Band Engineering

It has been discussed in Part I that fundamentally endurance (in thicker dielectrics NVMs) is limited by stress-induced trap generation within the tunneling layers (and also the blocking layers) enhancing SILC and eventual collapse of the memory window during W/E cycling. It has also been shown that endurance could be enhanced if the average W/E field for the stack design could be lowered especially in relationship with the breakdown strength of the composite stacking layers [5]. It has been seen in the previous sections that incorporating high K layers replacing SiO_2 lowers Vpp, thereby lowering average W/E field and consequently enhancing endurance. Combining high K layers with band engineering as discussed in this section earlier further lowers the W/E field for the same fluence required for memory window and therefore provides superior trade-offs between memory window and endurance while meeting retention objective. It has also been shown in the earlier sections that several ultrathin high K dielectric films exhibit significantly high breakdown strength (>15 MV/cm) and high speed programming could be achieved at relatively lower field through band engineering of such multilayered dielectrics. Therefore, appropriate combination of these attributes in stack design could enhance endurance by several orders of magnitude.

In order to achieve unlimited endurance, however, the transport mechanism has to be by DT mode to ensure virtually no transfer of energy from transporting charges to the surrounding dielectric lattice. This implies that if the NVM device could be designed whereby programming (W/E) takes place only by direct tunneling and charges can be retained to achieve simultaneously memory window and retention requirements, then the NVM design could achieve window, retention, and infinite endurance at the same time. The thicker dielectric stack designs for FG, CT, and NC devices thus far do not meet such requirements. We shall now explore the DT class of NVM designs, which has the best potential of achieving infinite endurance and at the same time may optimize the appropriate trade-offs in speed, memory, window, and retention to overcome existing limitations of NVM devices through applications of band engineering in DT devices. This approach has been employed in several proposed SUM devices contained in Part III of this book.

13.6 BAND ENGINEERING FOR DIRECT TUNNEL NVM DEVICES

DT NVMs (or DTMs) were introduced in Chapter 4 (Part I) and enhancements of NVM device properties employing multilayered band-engineering options for the direct tunneling dielectric structures have been outlined in Section 13.4 earlier. The term "direct tunneling" is broadened in this book to include tunneling through the thinner (ultra-shallow) dielectrics across triangular barrier, which is customarily called enhanced F–N tunneling. The effective tunnel distance in such layers (at high field) is within the physical thickness of such dielectric layers and typically <2.5 nm, characterized by band-to-band tunneling regardless of the barrier geometry. All such devices are considered as DTM devices in this book. Interest in such DTM devices and structures have grown in recent past due to the inherent energy efficiency of charge transport and ease of integration of such transport mode in other types of NVM devices. Most investigations for the DTM devices were directed to the enhancement in retention in such devices, which has been the critical challenge. Band-engineering concepts discussed thus far have been applied to the DTMs to overcome the retention limitations.

All the retention enhancement concepts discussed in subsections of 13.5.4 above were applied to improve retention in DTM devices. In order to increase the reverse tunneling distance to reduce standby charge loss, multiple direct tunneling layers were incorporated using ultra-shallow multiple dielectric layers for tunneling.

13.6.1 The Band-Engineered Multiple Direct-Tunnel High K Device

Historically such concept was first proposed in the context of a DT FG-NVM design in 2001 [45]. The stack structure contained a first DT dielectric layer of oxide with an over layer of IN-SRN containing silicon nanocrystal in an ultrathin nitride layer followed by another DT layer of Ta_2O_5 with an over layer of floating gate or floating plate (charge storage, "CS- SRN" or "IN-SRN" layer). The blocking layer is a high K layer of ZrO_2 with a passivation layer of IN-SRN for an aluminum metal gate device. The stack structure is shown in Figure 13.24. The device was called, "asymmetric band gap-engineered" (ABE) DTM nonvolatile memory. The device solved the retention problem, provided unlimited endurance, and at the same time provided very fast low Vpp programming. The entire stack EOT could be designed to be in the range of 4–6 nm and would operate at programming voltage ±<5 V, <1 μs. The stack structure may consist of 1.5–2.0 nm of oxide/1.5–2.0 nm of IN-SRN/1.5–2.0 nm of Ta_2O_5/5 nm of CS-SRN (refractive index~2.3)/5–10 nm of ZrO_2/2 nm of IN-SRN with either silicon or aluminum control gate. The device could functionally replace DRAM for most applications with higher density and greatly reduced operating power. The device functions as CT device when the FG or FP is replaced by nonstoichiometric silicon nitride with higher silicon concentration or by Si_2ON_2 as explained earlier. In principle, similar approach as

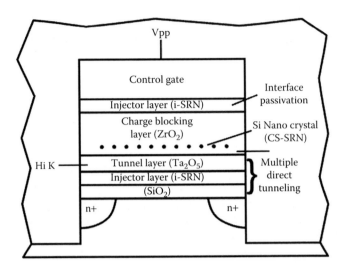

FIGURE 13.24 Band-engineered multiple-layer direct tunnel NVM stack structure to provide infinite endurance and improved retention. (From Bhattacharyya, A., Asymmetric band gap-engineered nonvolatile memory device, *ADI Associates internal publication*, U.S. Patents: 6943065 [9/13/15], 7072223 B2 [7/6/2006].)

above has been incorporated in several of the proposed SUM device stack designs employing appropriate high K multilayer films. This is discussed in Part III of this book.

13.6.2 Double Tunnel Junction DTM

A similar device concept was published by R. Ohba et al., first in 2003 [46] and subsequently in 2004 [47], 2005 [48], 2006 [49], 2007 [50], and 2008 [51]. The authors called the device "double tunnel junction" or "DTJ" device, wherein a silicon nanocrystal layer was embedded between two layers of ultrathin tunnel SiO_2, each layer of 1 nm thickness with a nitride trapping layer and a top oxide blocking layer as shown in Figure 13.25 [48]. Compared to the conventional single tunnel junction, DTJ significantly improved charge retention in spite of very thin tunnel oxide by the nano-crystal induced Coulomb blockade and quantum confinement discussed in Chapter 2.

Both the above approaches of either the asymmetric band gap-engineered ABE device DTM [45] or the DTJ DTM [46–51] provided reverse repulsive potential for the charges stored in FG/FP [45] and nitride [46,48,50] due to the charges present in silicon nanocrystal quantum wells while both simultaneously provided large energy barrier of oxide to reduce leakage of the stored charges to the substrate. As a result, retention was vastly improved. Retention characteristics of the DTJ DTM devices are shown in Figure 13.26a and b, the former showing excellent temperature stability

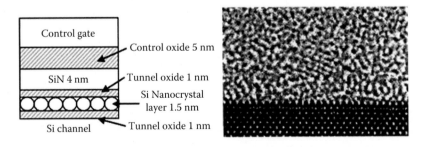

FIGURE 13.25 "Double tunnel junction" DTM stack structure. (Ohba, R. et al., 35 nm floating gate planar MOSFET memory using double tunnel junction tunneling, *IEEE IEDM Technical Digest*, p. 35 © 2005 IEEE.)

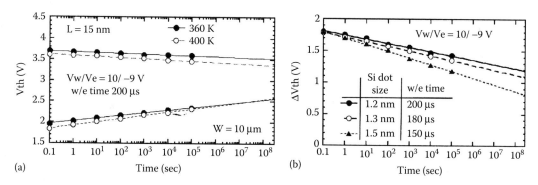

FIGURE 13.26 Retention characteristics of double tunnel junction DTM: (a) Retention versus temperature and (b) retention versus silicon dot size and impact on programming time. (From Ohba, R. et al., 15 nm planar bulk MONOS-type memory with double tunnel junction layers using sub-threshold slope control, *Electron Devices Meeting, 2007, IEEE IEDM Technical Digest*, p. 77 © 2007 IEEE.)

while the latter showing retention sensitivity due to silicon dot size. As should be anticipated, due to Coulomb blockade and quantum confinement, smaller silicon dot size provides higher reverse potential due to higher density of electrons trapped during programming with consequently lower rate of charge loss during standby. The write/erase time for programming is also longer to overcome the higher electrostatic potential of the charges stored in the nanocrystal due to the above quantum effects.

13.6.3 BE-SONOS DTM

The "BE-SONOS" device was published first in 2005 by H. T. Lue et al. [52] and subsequently optimized in 2006 [53,54] and further optimized in 2010 [55]. BE-SONOS device employs a classical three-layer ultrathin O-N-O layers (Variot barrier) for tunneling combined with 7.0 nm of nitride layer for trapping and 9.0 nm of blocking oxide by partially thermal conversion of the trapping nitride layer to generate interfacial trap at the interface between the nitride trapping layer and the blocking oxide layer, thereby improving the leakage to the gate. The O-N-O tunnel layer was initially 1.5 nm of thermal oxide with 2.0 nm of LPCVD nitride and 1.8 nm of top HTO oxide [52]. The stack was successively optimized using thinner bottom oxide of 1.3 nm SiO_2/2.0 nm Si_3N_4/and 3.5 nm SiON [54]. The composition of the SiON was tuned to further enhance the hole injection during erase by lowering the valence band barrier and at the same time increasing the tunnel distance of SiON to improve retention.

As explained earlier, the Variot barrier induces barrier thinning at high field modifying the direct tunneling characteristics of electrons (during writing) to effectively tunneling through the first barrier only (triangular barrier) and field-aided drift through the conduction bands of successive barriers due to high field band bending of the successive barriers, thereby minimizing the effects of successive barriers for the electron transport due to field-aided offsets of the conduction bands of the successive layers. This is usually categorized as enhanced F–N tunneling (Figure 3.6b, Part I) or "modified direct tunneling." From band-engineering standpoint, the triple-tunnel layer of O-N-O in the BE-SONOS device is analogous to the Variot barrier [11] exemplified earlier in Figure 13.10. For, ultrathin oxide during erase, similarly, enhanced hole injection takes place primarily through the first layer when the field is reversed. However, during retention when the internal field is low, band bending is negligible, the tunnel distance is too long and the barriers are effective for all the tunnel layers. This way reverse direct tunneling could be eliminated providing long retention. The O-N-O stack of the BESONOS device exhibited the above effects for the triple-layer tunneling and further improved the erase performance by tuning the valence band barrier with the SiON layer while providing excellent retention. Such attributes as above common to the earlier example and the BESONOS device is illustrated in the band diagrams of the generalized band diagram below to explain the erasing and retention properties of such band-engineered stack designs in Figure 13.27a and b, respectively.

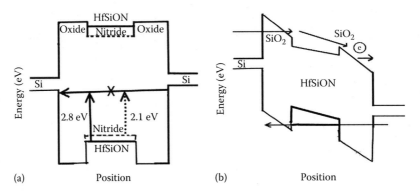

FIGURE 13.27 BE-SONOS band diagrams for erasing (a) and retention (b). (From Bhattacharyya, A., Discrete trap non-volatile device for Universal memory chip application, *ADI Associates internal publication*, U.S. Patents: 7436018 [10/14/2008], 7786516 [10/14/2010], 8143657 [03/27/2012], 2005 [11]; Lue, H.-T. et al., BE-SONOS: A bandgap engineered SONOS with excellent performance and reliability, *IEEE IEDM Technical Digest, IEDM*, p. 555, 2005.) [52].

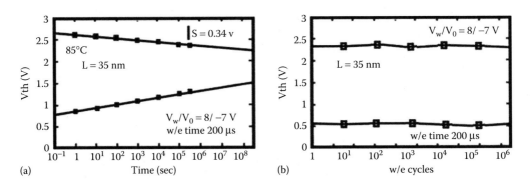

FIGURE 13.28 BE-SONOS device characteristics: (a) Retention and (b) endurance. (From Lue, H.-T. et al., BE-SONOS: A bandgap engineered SONOS with excellent performance and reliability, *IEEE IEDM Technical Digest, IEDM*, p. 555 © 2005 IEEE.)

The BE-SONOS devices as originally conceived demonstrated large window, excellent retention, and programming at an average field ~10 MV/cm with 1–10 ms W/E capability (erase was slower than write). Erase performance was significantly improved by the SiON top-layer design [52]. The retention characteristics of the original design is shown in Figure 13.28a and the endurance in Figure 13.28b. Long-term retention for BE-SONOS was demonstrated to be insensitive to the variation of the tunnel-layer thicknesses and P-F parameters and temperature dependence was found to be solely dependent on trapping nitride properties [52]. Improved results were achieved using noncut trapping layer demonstrating no lateral movement of charges in nitride as previously postulated [55]. The improved retention characteristics of noncut ONO stack is shown in Figure 13.29.

13.6.4 Resonant Tunnel Barrier DTM Device

A novel DTM was first introduced by S. Kim et al. in 2005 based on multiple direct tunneling whereby electronic charge injection and transport take place only during preferred ranges of applied potentials across the tunnel layers [56]. The device was aptly called as "RTB" NVM device. Due to its unique capability of limiting threshold dispersions of memory states, such device could be preferred in multilevel storage, a topic which we will discuss separately in a follow-on section. The RTB device was created with a thin layer of amorphous silicon (a-Si) of 1.5 nm sandwiched

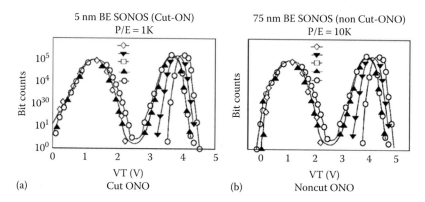

FIGURE 13.29 BE-SONOS retention enhancement using noncut nitride trapping. (From Hsieh, C.-C. et al., A novel BE-SONOS NAND flash using non-cut trapping layer with superb reliability, *IEEE IEDM Technical Digest, IEDM*, p. 114 © 2010 IEEE.)

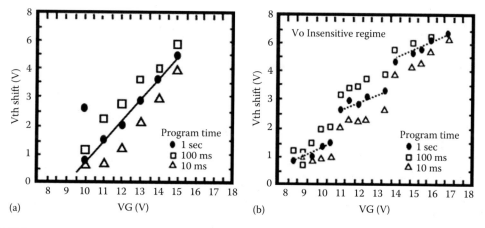

FIGURE 13.30 Vth-shift versus programming voltage (a) SONOS and (b) RTB. (From Kim, S. et al., Robust multi-bit programmable flash memory using a resonant tunnel barrier, *Electron Devices Meeting, 2005, IEDM Technical Digest*, p. 861 © 2005 IEEE.)

between two thin layers of SiO_2 each of thickness of 1.5 nm. Unlike SONOS devices where Vth-shift linearly increases with applied voltage as shown in Figure 13.30a, the Vth-shift abruptly jumps at specific applied voltage levels in the RTB devices as shown in Figure 13.30b. This provides a natural memory state Vth control. In addition, the RTB devices demonstrated superior retention and endurance when compared with SONOS devices. Figures 13.31a and b demonstrate the retention comparison while Figure 13.32 exhibits the endurance characteristics. The devices were fabricated by conventional CMOS processing.

RTB devices could be further scaled and improved by taking advantage of replacing oxide with higher K dielectric layers and incorporating deep level trapping and higher K blocking. Several such options to achieve lower programming voltage levels, faster speed, enhanced retention and endurance were proposed by Bhattacharyya [57,58]. Some such options are listed in Table 13.5.

13.6.5 Progressive Band Offset DTM Device

The PBO-DTM device was proposed and developed in 2004 [5,19]. The PBO concept has been briefly discussed earlier in Section 13.3.4. The PBO-DTM provides very low EOT stacks and

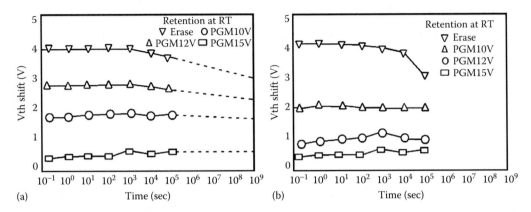

(a)   (b)

FIGURE 13.31 Retention characteristics: (a) SONOS and (b) RTB. (From Kim, S. et al., Robust multi-bit programmable flash memory using a resonant tunnel barrier, *Electron Devices Meeting, 2005, IEDM Technical Digest*, p. 861 © 2005 IEEE.)

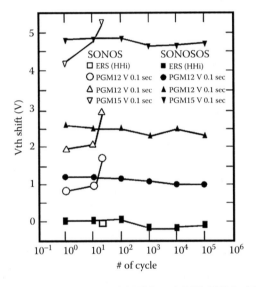

FIGURE 13.32 Endurance comparison between SONOS and RTB-DTM. (From Kim, S. et al., Robust multi-bit programmable flash memory using a resonant tunnel barrier, *Electron Devices Meeting, 2005, IEDM Technical Digest*, p. 881 © 2005 IEEE.)

consequently can be written at very low Vpp making such device very energy efficient. In addition, The PBO-DTMs could achieve very fast writing speed in less than microseconds. Like other DTMs such devices will exhibit EOL endurance. The PBO triple-layer tunneling stack may range in EOT between 1 nm and 3 nm. With higher K deep trapping layers such as Ta_2O_5 (K = 26) or GaN (k = 10) or Pt NC in TiO_2 (k = 60) combined with high K blocking dielectric with very low conductivity such as Pr_2O_3 (k = 31) or $LaAlO_3$ (k = 25), the entire PBO-EOT stack could be in the range of 2–5 nm with Vpp levels ranging from 1.5 V to 3 V. As mentioned earlier, while the writing speed could be very fast the erase speed could be slower due to the barrier asymmetry. However, with block erase memory architecture, PBO-DTM could be appealing for low power multifunctional memory applications in future. Several examples of PBO-DTM stack designs are provided in Table 13.6 [19].

Charge trapping could be an either of deep energy trapping or of deep quantum well trapping by embedding metal nanocrystals into Ta_2O_5 or GaN or TiO_2 or a combination.

TABLE 13.5

Stack Structure and Attributes of Advanced RTB-DTMs:[a] [58] All Stacks Designed for EOL Endurance

Tunnel Layer	Trapping Layer	Blocking Layer	EOT	Vpp/Time	Remarks
1.2 nm SiON 1 nm Si 1.2 nm SiON	5 nm TaO_5	10 nm HfSiON	<6 nm	±5 V/1 ms	Enhanced retention
1.3 nm SiON 1 nm Si 1.2 nm SiON	5 nm TaO_5	7.5 nm Al_2O_3	6 nm	±5 V/1 ms	Enhanced retention
1.4 nm HfO_2 1 nm Si 1.5 nmnHfO_2	5 nm TaO_5	10 nm HfSiON	4.5 nm	±3.5 V/100 μs	Speed
1.5 nm ZrO_2 1 nm Si 1.5 nm ZrO_2	Pt NC in ZrO2	10 nm HfSiON	4.5 nm	±3.5 V/100 μs	Speed
1.5 nm HfSiON 1 nm Si 1.5 nm HfSiON	Pt NC in HfSiON	7.5 nm Al_2O_3	<5 nm	±4.0 V/100 μs	Speed, window, retention, endurance
1.2 nm SiON + 1 nm Si+ 1.2 nm SiON	5 nm TaO_5	10 nm HfSiON	<6 nm	±5 V/1 ms	Enhanced retention
1.3 nm SiON + 1 nm Si+ 1.2 nm SiON	5 nm TaO_5	7.5 nm Al_2O_3	6 nm	±5 V/1 ms	Enhanced retention
1.4 nm HfO_2 1 nm Si + 1.5 nm HfO_2	5 nm TaO_5	10 nm HfSiON	4.5 nm	±3.5 V/100 μs	Speed
1.5 nm ZrO_2+ Si+ 1.5 nm ZrO_2	Pt NC in ZrO_2	10 nm HfSiON	4.5 nm	±3.5 V/100 μs	Speed
1.5 nm HfSiON +1 nm Si + 1.5 nm HfSiON	Pt NC in HfSiON	7.5 nm Al_2O_3	<5 nm	±4.0 V/100 μs	Speed, window, retention, endurance

Source: Kim, S. et al., Robust multi-bit programmable flash memory using a resonant tunnel barrier, Electron Devices Meeting, 2005, IEDM Technical Digest, pp. 861–864, 2005.

[a] All stacks designed for EOL endurance.

RTB will be further discussed in the subsection on multilevel NVMs and in Part III.

TABLE 13.6
PBO-DTM Stack Options

Tunnel Layer	EOT	Trapping Layer	Blocking Layer	Stack EOT	Vpp/Time
SiO_2/SiN/HfO_2	1.5 nm	Ta_2O_5	Pr_2O_3	<4 nm	3 V/100 μs
SiO_2/HfO_2/Pr_2O_3	1.3 nm	Pt in TiO_2	Pr_2O_3 or $LaAlO_3$	<3 nm	< 2.5 V/<10 μs
SiO_2/HfO_2/TiO_2	1.0 nm	Ta_2O_5	$LaAlO_3$	<2.5 nm	<2.0 V/<1 μs

13.7 BAND ENGINEERING FOR MULTILEVEL (MLC) AND MULTIFUNCTIONAL (MF) NVMs

By providing simultaneously large memory window, programming speed and retention enhancement, and appropriate retention/endurance trade-offs, band engineering provides the flexibility of multibit per cell storage (MLC devices) as well as multifunctional capability within a memory cell [18]. Some stack design examples were mentioned in Table 13.4. The context of window, speed and retention enhancements, and TANOS devices were exemplified for MLC applications. Due to their significance in driving future memory cost, power and performance, and future system architecture, these topics are further discussed in Section 13.7.3. In this subsection, we will provide several additional band-engineered stack examples to illustrate multilevel cell potentials and potentials of enriched functionality within an NVM cell through band *engineering*.

13.7.1 Single Polarity Vpp for Both Writing and Erasing

The schematic cross section of the 2-bit per cell NROM-NVM split-channel cell is shown in Figure 13.33 and the band diagram for the stack design is shown in Figure 13.34 [4], respectively. The stack design employs a double-layer tunneling of D00/D11 type (crested barrier), which enhances electron injection at modest write field (direct tunneling) for programming while providing low leakage (high retention) at low field. During erase hole injection primarily takes place from the gate electrode by F–N tunneling due to the larger hole energy barrier (valence band offsets) of the ORSiON/Al$_2$O$_3$ tunnel dielectric. The device operates at the "dual" mode with silicon substrate for electron and polysilicon gate (optionally IN-SRN or TaN metal gate) on top of HfO$_2$ for

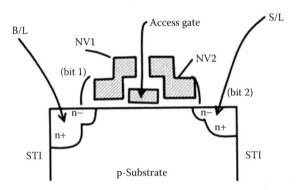

FIGURE 13.33 Split-channel 2-bit cell with single polarity programming and enhanced attributes. (From Bhattacharyya, A. et al., Band-engineered multi-gated-channel nonvolatile memory device with enhanced attributes, *ADI Associates Internal Publication*, U.S. Patents:7279740, 10/9/2007; 7749848, 7/6/2010; 8063436, 11/22/11, July 2004.)

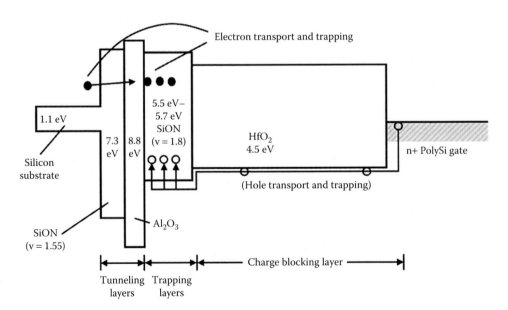

FIGURE 13.34 Band diagram for the 2-bit/cell NROM split-channel device with single polarity programming. (From Bhattacharyya, A. et al., Band-engineered multi-gated-channel nonvolatile memory device with enhanced attributes, *ADI Associates Internal Publication*, U.S. Patents:7279740, 10/9/2007; 7749848, 7/6/2010; 8063436, 11/22/11, July 2004.)

enhancing erase speed] for hole injection. The stack consists of 1.5 nm of OR-SiON (K = 5)/1.5 nm Al_2O_3/4.5 nm of Si_2ON_2 (K = 7)/15 nm of HfO_2 with a total EOT of ~7 nm. The tunnel SiON layer interfacing silicon has a band gap of 7.3 eV, whereas the trapping Si_2ON_2 with deep energy traps of 1.7 eV and higher trap density of ~1E13/cm² has a bandgap of 5.5 eV. The Vpp at the gate is held at 6 V, 1 µs with bit line potential at 0.25 V and source line held at Vdd (current flowing, access gate at Vdd), whereas during erase, bit line remains open (no current flow) and a larger Vpp of +9 V, 1–10 µs is applied at the respective program gate (NV1 or NV2). The substrate is held at ground for both writing and erasing. The SLC cell uses one program gate (not shown) but operates in similar ways. The device provides memory window of >3 V, retention >1E8 sec, and endurance >>1E10 W/E cycles. Aside from enhancements of device attributes compared to SONOSNROM, single polarity for both writing and erasing provides ease of process integration and peripheral circuitry for high voltage generation and routing adding functionality. The device will be further discussed in subsequent section.

13.7.2 NVMs with Progressive Nonvolatility

NVM memory subsystems based on NVM devices built on conventional CMOS silicon gate technology have been proposed for either of SOC (system-on-chip) architecture or for a chip-set combining processor containing SRAM caches and memory chips containing NVM arrays. Such NVM memory subsystems feature progressive nonvolatility without the requirements of conventional DRAMs [59,60]. The basic CMOS PFET and NFET devices employed ORSiON dielectric films (ref. index = 1.55, K = 5.0) for the polysilicon gate, which is also the tunneling layer for the NVM device sets. The OR-SiON layer thicknesses were varied for different NVM arrays within the same chip to achieve different degree of nonvolatility for the NVM arrays. Except for the thickness of the tunneling layers, the stack structures for the NVM arrays were identical and consisted of OR-SiON/Si_2ON_2/in-SRN/Al_2O_3 with or without TaN layer interfacing the polysilicon gate. Three different OR-SiON thicknesses were considered for the three different NVM arrays: the fastest requiring refresh every 10 s had 1.5 nm OR-SiON, the intermediate nonvolatile array required refresh every day with 2.5 nm thick OR-SiON and the longest retention EOL NVM memory array had 4.5 nm thick ORSiON requiring no refresh (retention >10 years). The 2.5–3.0 nm Si_2ON_2 in combination with the 5–7 nm IN-SRN layer was the trapping and storage medium for the stack while the 10 nm Al_2O_3 layer was the charge blocking dielectric layer. The EOT of the stack was ~8 nm and Vpp was ±6.5 to ±7.5 V. The programming pulse durations for the three arrays were, respectively, 0.1 µs, 30 µs, and 1 ms (1000 µs), respectively. The charge transport modes were primarily by direct tunneling through the thinner OR-SiON layers and subsequent trapping and storage in the NR-SiON/IN-SRN layers. For the 4.5 nm thick OR-SiON, transport was by FN tunneling, all charges were stored in the IN-SRN floating plate layer and little or no hole tunneling during erase was involved.

For the intermediate thickness OR-SiON array (2.5–3 nm OR-SiON), tunneling and erase was primarily by electron tunneling while some hole tunneling was associated with erasure operation. In case of the thinnest OR-SiON array, both electron and hole tunneling were operative by direct tunneling. All arrays displayed enhanced endurance while the array with the thinnest tunneling layer showed EOL endurance due to direct tunneling mode of charge transport. This faster array functionally replaced the need for DRAM in the system design. The programming part of the power requirements for the memory system was significantly lower than that of the conventional DRAM-based memory subsystems since the refreshing power requirement is significantly reduced. Additionally, cost per bit is significantly lower due to improved memory density compared to DRAM or embedding more complex DRAM memory array within the SOC architecture. The band-engineered stack structure for the arrays is shown in Figure 13.35. The memory subsystem architecture is schematically shown in Figure 13.36.

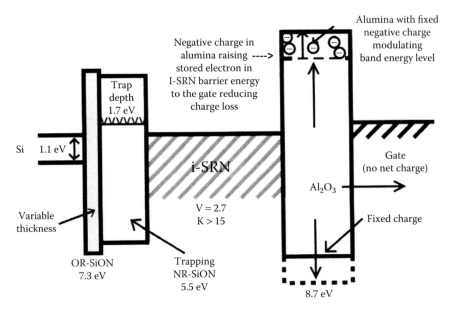

FIGURE 13.35 Band diagram for stacks with progressive nonvolatility by varying thickness of OR-SiON tunnel dielectric layer in each memory array. (From Bhattacharyya, A., Low power memory subsystem with progressive non-volatility, *ADI Associates internal publication*, (MIDD 39); U.S. Patent: 7385245, 06/10/2008, August 2, 2004.)

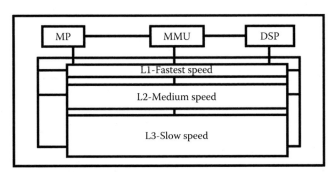

FIGURE 13.36 Schematic memory subsystem concept containing progressive NVM nonvolatility, thickness of OR-SiON tunnel layer in NV cells for L1 is the thinnest (direct tunneling) and L3 is the thickest (FN tunneling). Nonvolatility of L1, L2, and L3 are respectively: >10 sec, >!E5 sec and >1E9 sec; while corresponding cycle times are <300 ns, <100 μs, and <3 ms, respectively. (From Bhattacharyya, A., Low power memory subsystem with progressive non-volatility, *ADI Associates internal publication*, (MIDD 39); U.S. Patent: 7385245, 06/10/2008, August 2, 2004.)

13.7.3 Multibit per Cell (MLC) NVM

In the previous two subsections, we have provided multiple examples of stack designs for simultaneously achieving large memory window and enhanced retention through band engineering: the key requirements for multibit per cell designs (MLCs). We have noted that both TANOS and BE-SONOS stacks had provided MLC capability by providing large window and enhanced retention. Additionally, we have illustrated examples and concepts of multiplanar trapping stack designs (Section 13.5.4) and RTB (Section 13.6.4) to further advance multibit storage capabilities by enabling discrete memory states in the former case and by reducing Vth dispersions of the memory states in the latter case. A separate dedicated section will cover advance MLC designs. In this section, two novel band-engineered designs will be illustrated to exemplify multibit per cell devices.

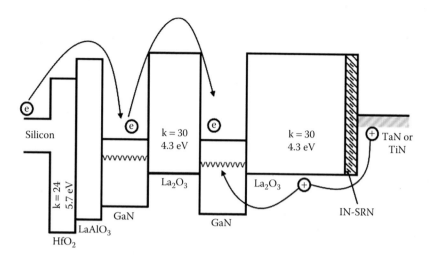

FIGURE 13.37 Band diagram for a stack design with two different trapping planes for a novel duel-bit cell design. The stack consists of HfO$_2$/LaAlO$_3$/GaN/La$_2$O$_3$/GaN/La$_2$O$_3$/TiN layers. (From Bhattacharyya, A., High performance multi-level band–Engineered (ET/DT) NVM device, *ADI Associates internal publication*, U.S. Patents: 7429767, 09/30/2008; 7553735, 06/30/2009; 7579242, 08/25/2009; 8159875, 04/17/2012, December 2004.)

The first one is based on charge trapping and compensation at two different trapping planes with the associated tunnel distances from silicon substrate and unique trapping energy levels and unique compensation (erasing) schemes. This is illustrated in the band diagram in Figure 13.37 [5].

The stack design contains two trapping layers at distances d1 and d2 from silicon substrate/ insulator interface, d2 > d1 as shown. Both the trapping layers could be 5 nm of either GaN or AlN and alternately and preferably the first layer closer to silicon substrate to be GaN and the second one to be AlN, respectively. The first trapping layer could also be embedded Pt metal dots placed between the layers of LaAlO$_3$ and La$_2$O$_3$ (the second and third tunneling layers; this option creates more process complexity but improves endurance and programming speed). The second trapping layer is placed sandwiched between two layers of La$_2$O$_3$, the latter being the blocking layer with an over layer of IN-SRN for passivation against process-induced oxygen vacancy and to achieve enhanced hole injection from the polysilicon or metal gate (TaN or WN/W).

As the band diagram depicts, the device operates on two tunneling modes. The first tunneling mode Tu1 is a direct tunneling mode through a crested barrier consisting of 1.5 nm each of HfO$_2$ and LaAlO$_3$. The second tunneling mode is a modified F–N tunneling through the layers of HfO$_2$/ LaAlO$_3$/GaN/La$_2$O$_3$, Tu2, for electron trapping into the second trapping layer further away from the silicon substrate. A single-step erase is followed by a three step programming to achieve 2-bit (duel-bit) cell storage. The stack design consists of 1.5 nm HfO$_2$/1.5 nm of LaAlO$_3$/5 nm of GaN/5 nm of La$_2$O$_3$/5 nm of GaN/10 nm of La$_2$O$_3$/5 nm of IN-SRN/ TaN-W structure. Even though many layers of stack structure are involved, multiple layers could be processed in one step (e.g., the last four layers of GaN/La$_2$O$_3$/GaN/La$_2$O$_3$ could be fabricated by sputtering in one step). The total EOT for the stack as above is ~6.5 nm (physical thickness ~33 nm). The erase consists of a negative 6 V pulse for 1 ms to put the device at the Vth (00) state of −2 V. The write consists of a successive pulse of +4 V, 0.1 μs [Vth (01)], +5 V, 10 μs [Vth (11)], and +6 V, 1 ms [Vth (10)] to provide the four memory states as summarized in Table 13.7.

It should be noted that at the highest writing voltage of +6 V, 1 ms the Vth of the memory state (10) is lower (Vth = +2 V) due to hole injection from the gate to the second trapping layer compensating the electron storage of the preceding writing condition of 5 V, 10 μs when the second trapping-layer stores additional electrons injected from the silicon substrate. During the first lower

TABLE 13.7

MLC States for Duel-Bit Cell

(2-Bit of Storage per Cell)

Memory State	Vth
1	+1 V
10	+2 V
11	+3 V
00 (erase state)	−2 V

voltage short pulse of +4 V, 0.1 µs only the electrons injected by direct tunneling gets trapped in the first trapping layer raising the Vth to +1 V (memory state 01). The memory simultaneously exhibits large window, long retention, and high endurance since the average peak field across the stack <10 MV/cm.

Another novel band-engineered design could be considered for achieving 2-bits per cell design. This involves band-engineered tunneling of electrons from silicon substrate during writing, and band-engineered tunneling of holes from the gate electrode during erasing while trapping charges (electrons and holes) within different unique energy levels (intrinsic level for the trapping layer and extrinsic level of deep quantum well induced by nanocrystals) created in Ta_2O_5 film by embedding Pt nanocrystals.

The band diagram for such concept is shown in Figure 13.38 with a triple-layer tunneling of $HfO_2/Al_2O_3/HfO_2$ forming an effective crested barrier interfacing silicon for electron injection while a triple-tunnel layer of $La_2O_3/Y_2O_3/Al_2O_3$ with a PBO interfacing TaN gate for hole injection and tunneling. The trapping consists of two deep levels of trapping within Ta_2O_5 of intrinsic electron trapping at 2.7 eV deep and Pt-induced hole trapping at 1.4 eV deep [5]. While both electrons and holes are stored within deep trap levels thus created in Ta_2O_5 film, the high energy barriers on both directions of substrate and gate ensure very low leakage at standby to provide excellent retention. The structure can be produced by multiple target sputtering technique or by ALD processing

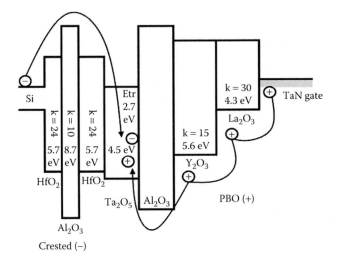

FIGURE 13.38 Band diagram for gate stack demonstrating electron tunneling from silicon for writing and hole tunneling from gate for erasing: Electron tunneling using crested barrier while hole tunneling using PBO barrier. (From Bhattacharyya, A., High performance multi-level band–Engineered (ET/DT) NVM device, *ADI Associates internal publication*, U.S. Patents: 7429767, 09/30/2008; 7553735, 06/30/2009; 7579242, 08/25/2009; 8159875, 04/17/2012, December 2004.)

with appropriate post-deposition anneal. The stack structure could simultaneously yield very high programming speed (write/erase) in 10–100 ns, low Vpp of ±4–5 V with large window (Vth-11: >+5 V), excellent endurance and >10 years of retention. Multiple memory states are created within the large memory window of >8 V between the highest written state commonly designated as Vth (11) and the erased state designated as Vth (00) by appropriate writing pulses as commonly used in MLC programming scheme. The MLC programming scheme has been discussed in detail in a later section.

13.7.4 Multifunctional Cell (MFC) NVM

Multifunctionality is an essential attribute of SUM-NVM devices. This topic will be covered extensively in Part III of this book. Band engineering is one of the key elements in NVM stack design to achieve multifunctionality providing functionality of both DRAM and conventional NROM or FLASH NVMs to replace both in system designs and memory hierarchy. Multiple approaches of band-engineered stack designs to simultaneously achieve multifunctionality of DRAM and conventional NVM devices in the same chip have been proposed [5,11,18,19,61,62]. One illustration of band engineering for a relatively simple implementation in conventional CMOS technology will be illustrated here demonstrating simultaneous DRAM and NVM functionality [62].

The stack design is based on the DTM device explained earlier. It consists of 1.5–2.0 nm of OR-SiON/4.0 nm of trapping layer of Si_2ON_2 with a 5–10 nm of IN-SRN over layer acting as an extended storage plate as well as an enhanced coupling medium between the floating node and the control gate. A thin oxide with a thicker layer of Al_2O_3 with a TaN interface acts as the blocking dielectric. The oxide/Al_2O_3 interface provides high density of fixed negative charge to significantly improve memory retention and provide larger window. The short pulse lower voltage writing provides DRAM functionality through charge trapping closer to the silicon substrate with retention >>10 s and infinite endurance with operating speed 10–100 ns. The long pulse larger voltage programming provides charge storage both in the Si_2ON_2 film and I-SRN floating plate providing larger window and 10 years of nonvolatility. Band diagram for the stack is shown in Figure 13.39a. The device could be further improved in speed and retention by replacing the single-layer SiON

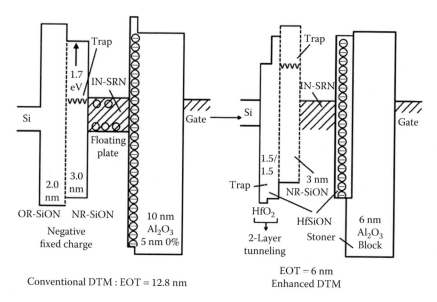

FIGURE 13.39 (a) and (b) DTM based multifunctional NVMs [58,62]; (a) Single-layer SiON tunnel dielectric implementation and (b) Dual tunnel dielectric crested barrier implementation containing HfO₂/HfSiON tunnel layer. Fixed negative charge at Oxide/Al₂O₃ blocking to improve retention.

tunnel dielectric with a double tunnel-layer crested barrier consisting of 1.5 nm of HfO_2/1.5 nm of HfSiON with the remaining structure same as earlier with thickness optimization. This is shown in 13.39b. The EOT for either stack could be designed to be ~6 nm with Vpp ±4 V.

13.7.5 Reverse Mode and MultiMode NVM Devices

Conventional FG and CT devices operate on the normal mode whereby charges are injected from silicon substrate and by design charge injection from the gate is suppressed or reduced. The device design approach is similar in concept to that of an FET device where the gate serves as a field plate to the silicon channel and ideally without a current flow from the gate to the substrate. During FET operation, conventional NVM devices operating by the normal mode had to overcome several problems, the primary one being hole injection from silicon during erase operation and subsequent defect creation in oxide, SILC, and eventual leakage and breakdown affecting reliability.

This has been discussed in Part I. Likharev first proposed what is called the "crested barrier" to enhance NVM device characteristics through band engineering and applied the concept to a "reverse mode" device whereby the silicon substrate remained passive during writing (electron injection) and erasing (hole injection) and all charge injection took place from the gate electrode [6]. Subsequently, several band-engineered reverse mode and dual-mode device concepts were proposed by Bhattacharyya [5,11,41,45,58]. While this book is primarily focused on normal mode devices, two examples would be of interest here: (a) A reverse mode Hi K device compatible to current FET design and (b) a dual-mode device based on PBO to provide very fast writing and erasing.

13.7.5.1 High K Reverse Mode Device

The stack design consists of the first layer of HfSiON, which could be identical to the gate dielectric for peripheral FET devices followed by a TiO_2 film for trapping and a crested barrier of nitride/HfSiON/nitride interfacing the gate (polysilicon or metal gate) [11]. The band diagram for the device is shown in Figure 13.40. The device provides lower programming voltages (lower EOT stack design: low Vpp), good retention, high endurance, and reliability.

13.7.5.2 High-Performance Dual-Mode Device

An example of a very high-performance device could be provided whereby electron injection for writing is accomplished by electron injection from silicon substrate and hole injection for erasing is accomplished by hole injection from the gate [5]. The tunneling structures for interfacing silicon substrate and gate are triple layers of PBOs type for very high speed performance. The triple-layer

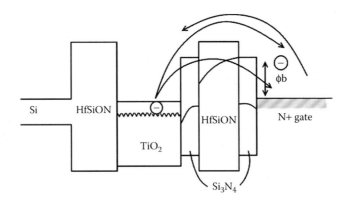

FIGURE 13.40 Band diagram for HfSiON/TiO_2/Nit-Hafnia-Nit/gate, reverse mode device with enhanced reliability. (From Bhattacharyya, A., Discrete trap non-volatile device for Universal memory chip application, *ADI Associates internal publication*, U.S. Patents: 7436018 [10/14/2008], 7786516 [10/14/2010], 8143657 [03/27/2012], 2005.)

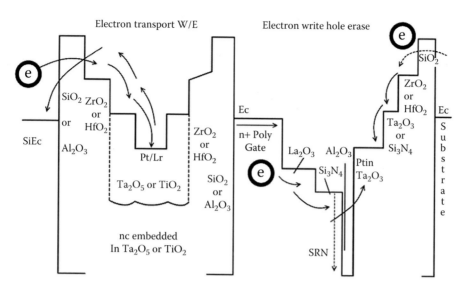

FIGURE 13.41 A stack design band diagram for a high-performance dual-mode device. (From Bhattacharyya, A., High performance multi-level band-engineered (ET/DT) NVM device, *ADI Associates internal publication*, U.S. Patents: 7429767, 09/30/2008; 7553735, 06/30/2009; 7579242, 08/25/2009; 8159875, 04/17/2012, December 2004.)

interfacing silicon is Si-SiO$_2$/ZrO$_2$/Ta$_2$O$_5$ whereas the triple-layer interfacing gate is also the same in reverse order that is: GateSiO$_2$/ZrO$_2$/Ta$_2$O$_5$. The trapping layer could be AlN or GaN as a separate layer or the thicker Ta$_2$O$_5$ central film could be embedded with a layer of Pt nanocrystals. SiO$_2$ could be replaced by Al$_2$O$_3$ and ZrO$_2$ could be replaced by HfO$_2$ to achieve similar results. The band diagram for such structure is shown in Figure 13.41. Due to the nature of PBO and ultrathin films of SiO$_2$ and ZrO$_2$ (typically of thickness 1.5–2 nm), very high speed programming and erase could be accomplished. Multiple modes of operation could be feasible in such stack design.

13.7.6 Multimechanism NVM Devices

Band engineering aids multimechanism charge transport [58,61]. This will be covered in Enhanced Memory section.

REFERENCES

1. B. Govoreanu, P. Blomme, M. Rosmeulen et al., A figure of merit for flash memory multi-layer tunnel dielectrics, Springer Wien, New York, 2001, pp. 270–273. And also P. Blomme, B. Govereanu, M. Rosmeulen, J. Van Houdt and K. De Meyer, Multilayer tunneling barriers for nonvolatile memory applications, *DRC Conference Digest*, June, 2002, pp. 153–154.
2. A. Bhattacharyya, C. T. Kroll, and P. C. Velasquez, MXOS n-channel NVM device using multi-layer CVD oxynitride gate insulator, *ECS Fall Meeting Proceeding*, Los Vegas, CA, 1977.
3. Y. Q. Wang, D. Y. Gao, W. S. Hwang et al., Fast erasing and highly reliable MONOS type memory with HfO$_2$ high-k trapping layer and Si$_3$N$_4$/SiO$_2$ tunnel stack, *IEEE IEDM Technical Digest*, San Francisco CA, 2006, pp. 971–974.
4. A. Bhattacharyya, K. Prall, and L. C. Tran, Band- engineered multi-gated-channel non-volatile memory device with enhanced attributes, *ADI Associates Internal Publication*, July 2004, U.S. Patents:7279740, 10/9/2007; 7749848, 7/6/2010; 8063436, 11/22/11.
5. A. Bhattacharyya, High performance multi-level band –Engineered (ET/DT) NVM device, *ADI Associates internal publication*, December 2004, U.S. Patents: 7429767, 09/30/2008; 7553735, 06/30/2009; 7579242, 08/25/2009; 8159875, 04/17/2012.

6. K. K. Likharev, Riding the crest of a new wave in Memory, *IEEE Circuits and Devices*, July, 2000, pp. 17–21.
7. S. J. Baik, S. Choi, U.-I. Chung, and J. T. Moon, High speed and nonvolatile Si nanocrystal memory for scaled flash technology using highly field-sensitive (NON) tunnel barrier, *IEEE IEDM Technical Digest*, Washington DC, 2003, pp. 545–548.
8. P. Blomme, B. Govoreanu, and M. Rosmeulen, Insulating barrier, EUR. Patent Application 1253646 A1, October 30, 2002.
9. P. Blomme, A. Akheyar, J. Van Houdt et al., Data retention of floating gate memory with SiO_2/high-K tunnel or interpoly dielectric stack, *Proceeding, IEEE DRC*, 2004.
10. B. Govoreanu, P. Blomme, M. Rosmeulen et al., VARIOT: A novel multilayer tunnel barrier concept for low-voltage nonvolatile memory devices, *IEEE Electron Device Lett.*, 24(2), 99–101, 2003.
11. A. Bhattacharyya, Discrete trap non-volatile device for Universal memory chip application, *ADI Associates internal publication*, 2005, U.S. Patents: 7436018 (10/14/2008), 7786516 (10/14/2010), 8143657 (03/27/2012).
12. P. Blomme, J. D. Vos, L. Haspeslagh et al., Scalable floating gate flash memory cell with engineered tunnel dielectric and high-K (Al2O3) interpoly dielectric, *Proceeding IEEE NVSMW*, Vol. 179, 2006.
13. J. Buckley, B. De Salvo, G. Molas et al., Experimental and theoretical study of layered tunnel barriers for nonvolatile memories, *Solid State Device Research Conference, Proceedings ESSDERC*, 2006.
14. S. Kim, S. J. Baik, Z. Huo et al., Robust multi-bit programmable flash memory using a resonant tunnel barrier, *Proceeding IEEE NVSMW IEEE NVSMW*, Washington DC, 2004.
15. R. Ohba, Y. Mitani, N. Sugiyama et al., 35 nm floating gate planar MOSFET memory using double tunneling junction, *Electron Devices Meeting, 2005, IEEE. IEDM Technical Digest*, New York, 2005, pp. 853–856.
16. A. Bhattacharyya, Thin gate stack structure for NVM cells and method for forming the same, *ADI Associates internal publication*, April, 2006, U.S. Patents: 7662693, 7956426.
17. Y. Q. Wang, D. Y. Gao, W. S. Hwang et al., Fast erasing and highly reliable MONOS type memory with HfO_2 high-k trapping layer and Si_3N_4/SiO_2 tunneling stack, *Electron Devices Meeting, 2006, IEEE IEDM Technical Digest*, 2006, pp. 971–974.
18. A. Bhattacharyya, Advanced scalable nano-crystal device and process, *ADI Associates internal publication*, November 19, 2004, Retitled: Scalable multi-functional and multi-level nano crystal NVM, U.S. Patents 7476927, 01/13, 09; 7759715, 07/30/2010; 7898022, 2011; 8242554, 08/14/2012.
19. A. Bhattacharyya, A novel low power nonvolatile memory and gate stack, *ADI Associates internal publication*, October 21, 2004, U.S. Patent 7612403, 12/22/2009.
20. Y. Zhang, S. Hong, J. Wan et al., Flash memory cell with LaAlO3 (k = 27.5) as tunnel dielectric for beyond sub-50nm technology, *Proceeding IEEE NVSMW*, 2005.
21. A. Bhattacharyya, K. Ahn, and L. Forbes, Metal-insulator-metal capacitors using Pr_2O_3 dielectrics (Capacitors and methods with Praseodymium oxide insulator) *ADI Associates internal publication*, September, 2006; U.S. Patents:796389 and 946620.
22. A. Bhattacharyya, Scalable logic and DRAM FET technology using Pr_2O_3 and other lanthanides as gate insulator and storage capacitor dielectric films, *ADI Associates internal publication*, March 2007, Retitled, Lanthanide dielectrics with controlled interfaces, U.S. Patents:8153497, 04/10/2012.
23. H. M. Choi, K.-Y. Park, S.-H. Lee et al., Novel Hi-K inter poly dielectric for sub 50 nm flash memories, *Proceeding IEEE NVSMW NVSMW*, 2006.
24. J. Fu, K. D. Buddharaju, S. H. G. Teo et al., Trap layer engineered gate-all-around vertically stacked twin Si-nanowire nonvolatile memory, *Electron Devices Meeting, 2007, IEEE IEDM Technical Digest*, Washington DC, 2007, pp. 79–82.
25. S. Seki, T. Unagami, and B. Taujiyama, Elctron trapping in rf-sputtered Ta_2O_5 films, *J. Vac. Sci. Technol.*, A1, 1825–1830, 1983.
26. X. Wang, and D.-L. Kwong, A novel high K SONOS memory using TaN/Al2O3/Ta2O5/HfO2/Si for fast speed and long retention operation beyond the 45nm generation, *IEEE Trans. Electron Dev.*, 53, 78–82, 2006.
27. K.-H. Joo, C.-R. Moon, S.-N. Lee et al., Novel charge trap devices with NCBO trap layers for NVM or image sensors, *IEEE IEDM Technical Digest, IEDM*, San Francisco CA, 2006, pp. 979–982.
28. C. H. Lai, C. C. Huang, K. C. Chiang et al., Fast high K AlN MONOS memory with large memory window and good retention, *Device Research Conference Digest, 2005, Proceeding, IEEE DRC*, 2005, pp. 99–100.
29. A. Chin, C. C. Laio, C. Chen et al., Low voltage high speed SiO_2/AlGaN/Al LaO_3/TaN memory with Good retention, *Electron Devices Meeting, 2005, IEEE IEDM Technical Digest*, 2005, pp. 158–161.

30. Z. L. Huo, J. Yang, G.-H. Lim, and S. J. Baik, Band engineered charge trap layer for highly reliable MLC flash memory, *VLSI Technology*, June, Kyoto, Japan, 2007, pp. 138–139.
31. D. C. Gilmer, N. Goel, H. Park et al., Engineering the complete MANOS-type NVM stack for best in class retention performance, *IEEE IEDM Technical Digest*, Baltimore MD, 2009, pp. 439–442.
32. C. H. Lee, K. I. Choi, M. K. Cho et al., A novel SONOS structure of $SiO_2/SiN/Al_2O_3$ with TaN metal gate for multi-giga bit flash memories, *IEEE IEDM Technical Digest*, Washington DC, 2003, pp. 613–616.
33. S. Jeon, J. H. Han, J. Lee et al., The impact of work-function of metal gate and fixed oxide charge of high-K blocking dielectric on memory properties of NAND type charge trap flash memory devices, *IEEE Electron Device Lett.*, 39–40, 2005.
34. Y. Shin, J. Choi, C. Kang et al., A novel NAND-type MONOS memory using 63nm process technology for multi-gigabit flash EEPROMs, *IEEE IEDM Technical Digest*, Washington DC, 2005, pp. 337–340.
35. C.-H. Lee, C. Kang, J. Sim et al., Charge trapping memory cell of TANOS (Si-Oxide-SiN-Al2O3-TaN) structure compatible to conventional NAND memory, *IEEE Electron Device Lett.*, 54–55, 2006; also Multi-level NAND flash memory with 63nm-node TANOS (Si-Oxide-SiN-Al2O3-TaN) cell structure, *VLSI Symposium*, Monterey CA, 2006, pp. 26–27.
36. Y. Park, J. Choi, C. Kang et al. Highly manufacturable 32Gb multi-level NAND flash memory with $0.0098um^2$ cell size using TANOS (Si-Oxide-Al203-TaN) cell technology, *Electron Devices Meeting, 2006, IEEE IEDM Technical Digest*, 2006, pp. 29–32.
37. Z. Huo, J. K. Yang, S. H. Lim et al., Band engineered charge trapping layer for highly reliable MLC flash memory, *VLSI Technology symposium*, Kyoto, Japan, 2007, pp. 138–139.
38. C.-H. Lee, S.-K. Sung, D. Jang et al., A highly manufacturable integration technology for 27nm 2 and 3bit/cell NAND flash memory, 5.1.1–5.1.4 *IEEE IEDM Technical Digest*, San Francisco CA, 2010.
39. Y. Q. Wang, D. Y. Gao, W. S. Hwang et al., Fast erasing and highly reliable MONOS type memory with HfO_2 high-k trapping layer and Si_3N_4/SiO_2 tunneling stack, *Electron Devices Meeting, 2006, IEEE IEDM Technical Digest*, San Francisco CA, 2006, pp. 971–974.
40. R. van Schaijk, M. van Duuren, N. Akil et al., A novel SONOS memory with $HfSiON/Si_3N_4/HfSiON$ stack for improved retention, *Proceeding IEEE NVSMW*, 2006, pp. 50–51.
41. A. Bhattacharyya, Novel band-engineered nano-crystal NVM device utilizing enhanced gate injection, *ADI Associates internal publication*, December 4, 2004; U.S. Patent 7629641, 12/22/2009.
42. A. Bhattacharyya, Embedded trap direct tunnel non volatile memory, *ADI Associates internal Publication*, July 15, 2004; U.S. Patent 7365388, 04/29/2008.
43. R. Gupta, W. J. Yoo, Y. Wang et al., Formation of Si-Ge nanocrystal in HfO_2 using in-situ CVD for memory application, *Appl. Phys. Lett.*, 84, 4331–4335, 2004.
44. M. Kanoun, A. Souifi, T. Baron, and F. Mazen, Electrical study of Ge nanocrystal-basedMOS structures for p-type NVM application, *Appl. Phys. Lett.*, 84, 5079–5081, 2004.
45. A. Bhattacharyya, Asymmetric band-gap engineered nonvolatile memory device, *ADI Associates internal publication*, U.S. Patents: 6943065 (9/13/15), 7072223 B2 (7/6/2006).
46. R. Ohba, N. Sugiyama, J. Koga et al., Silicon nitride trap memory with double tunnel junction, *VLSI Technology, 2003, Digest of Technical Papers*, Kyoto, Japan, 2003, pp. 35–36.
47. R. Ohba, Y. Mitani, N. Sugiyama et al., Impact of stoichiometry control in double junction memory on future scaling, *Electron Devices Meeting, 2004, IEEE IEDM Technical Digest*, San Francisco CA, 2004, pp. 897–900.
48. R. Ohba, Y. Mitani, N. Sugiyama, and S. Fujita, 35 nm floating gate planar MOSFET memory using double tunnel junction tunneling, *IEEE IEDM Technical Digest*, Washington DC, 2005, pp. 873–876.
49. R. Ohba, Y. Mitani, N. Sugiyama et al., 25nm planar bulk SONOS-type memory with double tunnel junction, *Electron Devices Meeting, IEEE IEDM Technical Digest*, San Francisco CA, 2006, pp. 959–962.
50. R. Ohba, Y. Mitani, N. Sugiyama et al., 15nm planar bulk MONOS-type memory with double junction tunnel layers using sub-threshold slope control, *Electron Devices Meeting, 2007, IEEE IEDM Technical Digest*, 2007, pp. 75–78.
51. R. Ohba, Y. Mitani, N. Sugiyama et al., 10nm bulk-planar SONOS-type memory with double tunnel junction and sub 10nm scaling utilizing source to drain direct tunnel sub-threshold, *Electron Devices Meeting, 2008, IEEE IEDM Technical*, San Francisco CA, 2008 pp. 1–4.
52. H.-T. Lue, S.-Y. Wang, E.-K. Lai et al., BE-SONOS: A bandgap engineered SONOS with excellent performance and reliability, *IEEE IEDM Technical Digest, IEDM*, Washington DC, 2005, pp. 555–558.
53. E.-K. Lai, H.-T. Lue, Y.-H. Hsiao et al., A multi-layer stackable thin-film transistor (TFT) NAND-type flash memory, *Electron Devices Meeting, 2006, IEEE IEDM Technical Digest*, San Francisco CA, 2006, pp. 41–44.

54. A. Bhattacharyya, Scalable flash/NV structures and devices with extended endurance, *Micron Technology Internal Docket*, August, 2001, U.S. Patents: 7012297 (3/14/2006), 7250628 (7/31/2007), 7400012 (7/15/2008), 7750395 (7/6/2010).

55. H. T. Lue, S.-Y. Wang, Y.-H. Hsiao et al., Reliability model of bandgap engineered SONOS (BE_ SONOS), *IEEE IEDM Technical Digest, IEDM*, San Francisco CA, 2006, pp. 495–498.

56. C.-C. Hsieh, H.-T. Lue, K.-P. Chang et al., A novel BE-SONOS NAND flash using non-cut trapping layer with superb reliability, *IEEE IEDM Technical Digest, IEDM*, San Francisco CA, 2010, pp. 114–117.

57. S. Kim, S. J. Baik, Z. Huo et al., Robust multi-bit programmable flash memory using a resonant tunnel barrier, *Electron Devices Meeting, 2005, IEDM Technical Digest*, Washington DC, 2005, pp. 861–864.

58. A. Bhattacharyya, (a) Enhanced PBO devices and tunneling structures for NVM designs, ADI Associates Internal Publications: December 2004 (Unpublished), (b) Enhanced RTB-based Device Designs, ADI Associates Internal Publications, August 2005 (Unpublished), (c) Further advancement in Band-engineered NVM stack designs and implementation concepts, October through December, 2007 (Unpublished).

59. A. Bhattacharyya, Enhanced multi-bit non-volatile memory using resonant tunnel barrier, (MIDD53), *ADI Associates internal publication*, August 18, 2005, U.S. Patent No. 7482561 (01/27/2009); U.S. Patent N0. 7867850 (1/11/2011).

60. A. Bhattacharyya, Scalable integrated logic/NV memory technology with associated gate stack structure and process, (MIDD 34) ADI Associates internal publication, May 23, 2004; U.S. Patent: 7544990, 06/09/2009.

61. A. Bhattacharyya, Low power memory subsystem with progressive non-volatility, *ADI Associates internal publication*, August 2, 2004, (MIDD 39); U.S. Patent: 7385245, 06/10/2008.

62. A. Bhattacharyya, Defining high speed, high endurance scalable NVM device using multimechanism carrier transport and high density structure for contact-less arrays, *ADI Associates internal publication*, September 7, 2004, U.S. Patent: 7244981, 07/17/2007.

63. A. Bhattacharyya and G. Darderian, An enhanced Floating Gate Flash Device and Process. U.S. Patent Application Publication No. 0067256 A1, March 12, 2009.

14 Enhanced Technology Integration for NVM

CHAPTER OUTLINE

With the ever-increasing application base, silicon based microelectronics and nanoelectronics technology and products have assumed an indispensable role in virtually every human activity. As a result, continued evolution is being sought in cost reduction, power reduction and enhancements in functionality, and performance and reliability of memory and logic products as well as integrated solutions such as SOCs. Toward such goals, small and large system solutions are being addressed with appropriate architecture in considering memory hierarchies limited to SRAM for performance and endurance, and NVMs for cost, functions, system speed, and reliability, ideally displacing other memory options and integrating both logic and memory in chip-level (SOC) or in package level (e.g., System-in-Package or SIP). This approach is likely to continue especially with the potential evolution of advanced planar and 3D NVM multifunctional devices and SUMs. In this chapter, we will first provide a brief overview of system-level integration of memory and logic technology followed by enhanced NVM device-technology integration schemes in greater detail.

14.1 FUNCTIONAL INTEGRATION AT INTERCONNECT AND PACKAGING LEVELS

NVMs have been increasingly employed in electronic devices and systems, large and small, to improve power/performance/cost/reliability trade-offs. From very-large-end distributed-memory systems employed for ever-expanding Internet activities to smart controllers for automotive and handheld devices such as cell phones, NVMs improve memory throughput, power reduction, code protection, and update, as well as system reliability. This is achieved either in combination of HDD to form hybrid large-volume storage for large database systems and enterprise computing systems as well as storage for high-performance PCs or as stand-alone storage subsystems for numerous devices for consumer applications. For portable and hand-held and especially "form-factor" sensitive appliances such as cell phones, NVMs are increasingly decreasing the role of both DRAM and HDDs.

At the processor-end, a combination of DRAM-NVDIMM is being increasingly employed to improve power/performance/cost trade-offs. The specific memory/logic/I-O architecture and integration scheme is application dependent and optimized to meet functional specifications of any system or sub-system. The functional integration falls into two broad categories: (a) system-in-package or "SIP" approach and (b) system-on-chip or "SOC" approach. These approaches are briefly outlined here.

14.1.1 System-In-Package: SIP

This integration scheme is widely applied for a broad range of memory requirements ranging from large-end to laptop PC/cell phone/other portable applications. Processors and logic chips such as DSPs and memory management units (MMUs) are interconnected with diverse stand-alone memory chips such as SRAMs, DRAMs, and Flash memories in the memory hierarchy of the associated system along with the input/output devices (I/Os) within the appropriate package or package sets. A schematic representation of the SIP integration scheme is shown in Figure 14.1. The figure shows the concept of the main memory bus communicating with the memory hierarchy at different levels (SRAM, DRAM, FLASH, HDD, etc., the last one not shown) with different latencies managed by the MMU control logic system. The MMU, in turn, is interconnected with the processors, DSPs, and I/Os with appropriate bus architecture (not shown). A simpler version of SIP may contain a single logic chip consisting of a microprocessor, MMU, and I/O logic (an SOC chip) interconnecting with NROM, DRAM, and FLASH chips as shown in Figure 14.2. Such an SIP could be applicable for portable applications. A third SIP integration scheme is proposed for a hybrid memory containing NOR Flash for permanent code and data, and SRAM connecting with an SOC chip to provide endurance (SRAM). The code and data are transferred from Flash to SRAM at the beginning of the processor cycle only and returned to the nonvolatile storage at the end of the processor cycle (not shown). The SIP integration scheme is shown in Figure 14.3 [1–3].

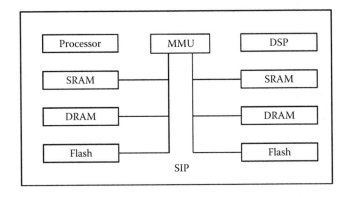

FIGURE 14.1 A general SIP scheme integrating processor, DSP and MMU with memory hierarchy of SRAM, DRAM, and NVMs. (From Bhattacharyya, A., A very low power memory sub-system with progressive non-volatility, ADI Associates Internal Docket, 8/2/2004; U.S. Patent: 7276760 [10/02/2007], 7385245 [06/10/2008]; Bhattacharyya, A., SRAM devices and electronic system comprising SRAM devices, ADI Associates Internal Docket, U.S. Patent 7898022 [1/2011]; Bhattacharyya, A., High performance capacitor-less one transistor memory device in bulk and SOI technology, ADI Associates Internal Docket, 4/2003, U.S. Patent 7432562 [10/07/2008], 7625803 [12/01/2009], 7968402 [06/28/2011], 81255003 [02/28/2012].)

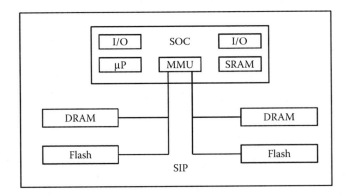

FIGURE 14.2 A simpler SIP scheme for portable system containing an SOC logic chip with memory chips.

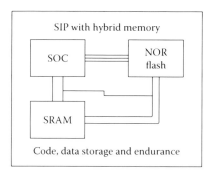

FIGURE 14.3 A SIP integration scheme with SOC combining SOC, NROM, and SRAM to enhance endurance through SRAM while providing nonvolatile code and data storage in the NROM Flash. (From Bhattacharyya, A., Integrated two-device multiple configuration alterable ROM/SRAM/General purpose nonvolatile memory, ADI Associates Internal Docket, 6/2004; U.S. Patent 7728350 [6/1/2010].)

14.1.2 System-On-Chip: SOC

With the progressive scaling of CMOS technology and associated FET (pMOS and nMOS transistors) devices, microelectronics and nanoelectronics have witnessed revolutionary enhancements in integrating memory and logic functions. An SOC has been the natural outgrowth of such integration capability. SOC has been accomplished by integrating high-density logic, memory, and I/O functions within a single chip of silicon substrate by incorporating many levels of wiring interconnects (as many as 10 or more vertical levels of wiring) in the back-end-of-line (BEOL) of silicon microelectronics or nanoelectronics technology. An SOC to date may integrate logic functions such as microprocessors, registers and I/Os, and MMU along with functional memory modules of DRAMs and NROM flash devices. Such a conceptual integration scheme is shown in Figure 14.4a. In near future, such an SOC could evolve to replace DRAM modules to "NVDRAM" modules [4], enhancing SOC nonvolatility and SOC power reduction. This is shown in Figure 14.4b.

In near future, SOC schemes could replace many functional integration schemes of SIPs as shown in Figures 14.1 through 14.3, providing single-chip solutions and speed-power-nonvolatility

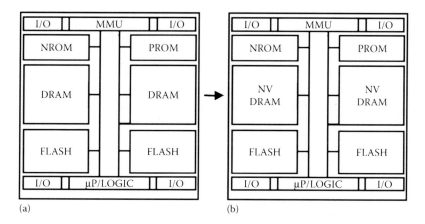

(a)   (b)

FIGURE 14.4 (a) and (b) currently available SOC integration scheme of logic and memory modules. (From Bhattacharyya, A., High performance capacitor-less one transistor memory device in bulk and SOI technology, ADI Associates Internal Docket, 4/2003, U.S. Patent 7432562 [10/07/2008], 7625803 [12/01/2009], 7968402 [06/28/2011], 81255003 [02/28/2012].)

enhancements by replacing DRAM and HD functions within a package into a piece of silicon. This will be further discussed in Part III in the context of SUM applications.

14.1.3 Other Examples of Functional Integration

Two other examples of functional integration will be mentioned here as examples of current and future integration possibilities. One is known as Hybrid Memory Cube (HMC) jointly developed by Micron Technology and Open Silicon [5]. The second one was proposed in 2006 by M. Taguchi of Spansion Inc. [6] called high-density SIM (HDS). The HMC scheme is shown in Figure 14.5, where customized ASIC logic functions are integrated with DRAM and other logic chips interconnected over a silicon wafer. Such a scheme significantly improves performance and power over conventional DIMM applications. The HDS provides SOC solutions of integrating CPU, encryption logic, and sensors with RAM and NROM functions into HDS as shown in Figure 14.6 [6].

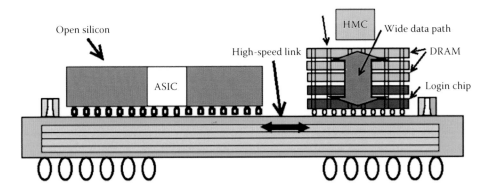

FIGURE 14.5 The HMC integration scheme. (From Dillon, J., Patel, H., Hybrid memory combining SRAM and NOR flash for code and data storage, Flash Memory Summit, © 2012 IEEE.) Reproduced with permission of IEEE Press.

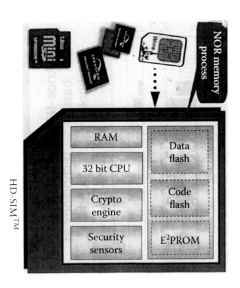

FIGURE 14.6 The high density SIM (HDS) Integration scheme. (From Taguchi, M., NOR flash memory technology, Spansion Inc., 2006 IEDM Short Course, IEEE organizer Rich Liu, Memory Technologies for 45 nm and Beyond, December 2006.) Reproduced with permission of Wiley-IEEE Press.

14.2 NVM INTEGRATION AT MEMORY LEVEL

As discussed in the previous section (Section 14.1), the integration schemes of NVMs could also be divided into two subsets: (1) embedded NVMs and (2) stand-alone NVMs.

14.2.1 Embedded NVMs

For embedded NVMs, several integration assumptions are made. These include that a platform technology in a given technology node provides not only the scaled NFET and PFET transistors, but also NVM memory arrays with appropriate isolation schemes, and high-voltage generation and distribution circuitry required for programming and sensing. Additionally, contents of memory arrays in the embedded technology reflect proper trade-offs in terms of application requirements, cost, and process complexities. For example, a platform technology may or may not incorporate embedded DRAMs since it requires more complex processing steps adversely affecting density, yield and cost, power, and performance. Instead, the memory requirements could be met by a combination of on-board SRAM, PROM, and NVM [4]. System requirements for overall memory performance (latency, storage, power, etc.) in such cases are met by appropriate memory hierarchy and architecture, and related I/Os and logic design through optimization of densities of SRAM, PROM, and NVMs, eliminating the need for DRAMs. Design considerations of eliminating DRAMs, which also affects integration, may include the following:

1. Intermittently storing appropriate data and instruction into the NVM arrays, thereby minimizing the number of write/erase cycles required for on-board NVM arrays
2. Parallel operations in writing and erasing NVM arrays to improve programming performance
3. Multiple NVM arrays with different attributes in terms of storage density (SLC, DLC, TLC, etc.), access time (NROM, NAND, etc.), retention and endurance (stack design) [2]

It should be noted that in the above discussion, it is assumed that SOC and "embedded" applications for NVMs are similar and a platform technology integration scheme for a given technology node contains both logic and NVM memory elements.

14.2.2 Stand-Alone NVMs

Stand-alone NVMs could be based on both FG and CT types. NAND FG NVMs have the largest market share and application base due to the availability of SLC/DLC/TLC capabilities, highest bit density, and lowest cost/bit. With DLC and TLC designs, increasing logic overhead is required for programming and sensing as well as ECC to ensure memory-state reliability and to meet performance specification and endurance targets for NAND FG NVMs. NOR-FG NVMs are usually applied for code storage and special applications requiring higher performance. CT-NROM stand-alone is often used for radiation-sensitive applications and other applications requiring higher endurance, higher performance, and lower programming power. However, such applications for Rad Hard usages may be limited due to buried N+ source and drain in the cell design. While CT-NROM is widely used in embedded applications due to simpler integration requirements with base CMOS technology, stand-alone in both NOR and NAND configuration and multiple-bit per cell configurations are gaining increasing acceptance due to scalability and reliability as well as integration compatibility with scaled CMOS technology and device design.

14.3 INTEGRATION AT FRONT-END-OF-LINE (FEOL) LEVEL

Rest of this section is devoted to a discussion on integration of various current and future NVM devices with CMOS technology in the sub-100 nm technology node. It should be assumed that interconnectivity of elements in memory cell design is limited to two levels of polysilicon and two levels of metals.

All additional metal levels in silicon technology are assumed to be driven by functional integration of logic elements, I/Os, and peripheral logic requirements for other functions associated to the memory arrays (such as MMU and ECC functions) and will not be discussed here. The discussion here is focused on integration topics in FEOL related to NVM devices and arrays, and related progression, manufacturability, and scalability within the framework of basic CMOS technology and scaled FET devices.

Table 14.1 provides an approximate guide for the basic CMOS platform technology and FET devices in the timeframe of 2010 to 2020 in the context of sub-50 nm technology nodes. Contrary to the previous technology generations of planar polysilicon gate CMOS devices, the present and future nodes assume both planar and nonplanar (FinFET) FETs, and metal gate, both bulk and SOI technology. Additionally, as should be expected, higher K gate dielectrics for FET devices incorporating compatible work function-tuned gate/insulator interface become essential requirements for FET device design to achieve FET Vth control and transconductance/mobility targets. Therefore, scaled NVM technology requires compatibility with the changes in basic CMOS technology to accommodate and integrate feature size scaling.

Table 14.2 lines the anticipated NVM device types and characteristics expected to emerge during the sub-50 nm timeframe compatible to the CMOS technology for both stand-alone and embedded

TABLE 14.1
Platform CMOS Technology Features for Sub-50 nm Generations

Time Frame:	2010–2014	2014–2017	2017–2020
Volume Manufacture Nodes	28–32 nm	22–25 nm	14–20 nm
CMOS Technology:	Planar Bulk/PD SOI	PD/FD SOI	FD SOI
FET Device:	Planar FETs	Planar/Fin-FETs	Fin-FETs
Vdd:	1.2–1.5 V	1.0–1.2 V	0.8–1.0 V
Vth; N/P	±0.5 V	±0.3 V	±0.2 V
Id-sat @ Vdd; N/P:	~750/450	~900/600	~1000/750
Gate Ins. Stack:	SiON/Hi K	Hi K (K>10)	Hi K (K>15)
Gate Ins. EOT:	1.5 nm	1.2 nm	1.0 nm
Gate Ins. Leakage:	< 1E-2A/cm^2	< 1E-2A/cm^2	< 1E-2A/cm^2
Gate Process:	Gate First	Gate First	Gate First
Gate:	Metal	Metal	Metal
For example	FuSi-Ni/TiN/W	TaN/TaC	TBD

TABLE 14.2
Anticipated NVM Embedded and Stand-Alone Device Features for Sub-50 nm Generations

Time Frame:	2010–2014	2014–2017	2017–2020
Volume Manufacture Nodes:	28–32 nm	22–25 nm	14–20 nm
Technology Base:	Planar CMOS	Planar/FinFET CMOS	FinFET/3D CMOS
Peripheral FETs: N/P: Vth:	±0.5 V	±0.4 V	±0.3 V
NVMs: TYPE A: FG/FP	NAND/NOR	NAND/NOR	NAND/NOR
TYPE B: CT	NROM	NAND/NROM	NAND/NROM
SLC & MLC	YES	YES	YES
MULTIFUNCTIONAL:	NO	YES	YES
DEVICE CHARACTERISTICS:	TYPE A/TYPE B	TYPE A/TYPE B	TYPE A/TYPE B
WINDOW	>4 V/> 2 V	>3 V/> 1.5 V	>2V/> 1 V
RETENTION	>1E 8 sec	>1E 8 sec	>1E 8 sec
ENDURANCE: W/E Cycles	>1E4/>1E6	>1E4/>1E6	>1E4/1E6

NVMs. It is anticipated that the NAND arrays will evolve into 3D products by 2017 and CT devices with high K stack design will progressively displace FG/FP devices during the coming decade. It is also assumed that while FinFET peripheral CMOS devices will assume greater significance in logic applications, the FinFET NVMs (primarily CT types) will have a shorter life-time and will transition to nonplanar 3D devices and arrays during 14–20 nm manufacturing technology nodes. It is further anticipated that multifunctional NVMs and SUMs will emerge during the above timeframe displacing DRAMs in the system architecture, providing enhanced functionality and endurance and minimizing current ECC overheads. Many of the relevant topics are discussed in other sections to follow.

14.4 NVM TECHNOLOGY/DEVICE INTEGRATION SCHEMES

It has been broadly outlined earlier that the NVM integration schemes are driven by cost of implementations and application requirements. These criteria, in turn, are associated with (a) compatibility with scaled CMOS technology at any given technology node which drives cost of mass-scale production and applicability, and (b) appropriate memory array and stack design requirements to fulfill memory specifications for application requirements. Compatibility includes enablement of logic functions and support circuitry in the context of stand-alone as well as embedded memory options for FG/FP, CT, NC, and DT memory devices, both planar and nonplanar types. Compatibility in stack design not only addresses the above options but also ensures appropriate trade-offs of basic CMOS device and NVM device parametrics in meeting application requirements with minimal process complexity. The criteria (a) and (b), therefore, are interdependent. Present and future technology/device integration schemes are addressed here, keeping in view the above criteria and associated options.

14.4.1 Compatibility with CMOS Platform (Embedded) Technology

It has been mentioned earlier in this book that basic CMOS FET technology had successfully addressed and altered the previously established silicon-gate integration schemes to meet device challenges associated with scaling below sub-100 nm nodes as outlined in Table 14.1. These challenges include (a) Vdd scaling; (b) Vth scaling, short channel effects (SCEs), and drain-induced barrier lowering (DIBL) containments incorporating metal gate technology integration; (c) leakage containment for incorporating high K gate dielectric replacing SiO_2; and (d) overall device reliability issues related to high K gate stack design and interface control. Such challenges were met by adopting what is known as the "Metal Gate First" integration scheme whereby the gate formation is followed by shallow diffusion technology and low thermal budget activation scheme for dopants activation such as spike annealing or laser-pulsed annealing including post-activation forming gas anneal to achieve low-interface state density and targeted work functions for FET devices. NVM technology needs to be compatible with the basic CMOS technology especially for embedded applications. Recently, a dual-channel (Si and Si–Ge) high K metal gate (first) low-cost high-performance low-power CMOS technology integration scheme was published by S. Krishnan et al. [7].

14.4.2 Compatibility with Gate Stack Design

It should be highly desirable that any integration scheme yield the desired device attributes with minimal process complexity and cost. Device attributes for NVM technology not only need to attain all NVM device attributes such as memory window, retention, endurance programming speed, and so on, but also scaled device attributes such as sub-threshold characteristics, SCE and DIBL control, Leakage (Ioff), Speed (Id), Transconductance (Gm), and so on. Therefore, the integration scheme should be flexible yet as much as possible compatible in terms of interface characteristics and selection of high K dielectric for different devices. Common elements for consideration are:

1. Substrate (Si or Si–Ge with appropriate doping) and associated substrate/dielectric interface from the standpoint of interface state density and fixed charge density
2. High K/Gate interface from the standpoint of work-function targets and stability including charge injection control from the gate to the insulator during operational conditions
3. Common S/D technology with reduced device series resistance

Certain integration challenges associated with high K gate stack design and metal gate are worth emphasizing:

Many properties of high K amorphous films are process and stoichiometry sensitive. Additionally, adverse characteristics include (a) structural instability of transition from amorphous to crystalline state during process integration requiring postdeposition temperature, ambient, and stress control; (b) local variation in composition affecting bond strength and trap generation; (c) lower band gap and associated higher leakage and lower breakdown strength; (d) reactive interface with silicon substrate resulting in high interface state density and fixed charges; and (e) difficulty in process control and reproducibility.

Incorporating metal gate technology requires addressing thermal stability, electro migration immunity, and interface stability involving incorporation of passivation under layer, over layer, and high work-function requirements.

These challenges are similar to both FET gate stack integration and NVM gate stack integration, the latter being more complex due to the incorporation of multiple layers of high K dielectrics in the stack design.

Therefore, taking into account the above considerations, NVM technology integration schemes for both planar and nonplanar options should be expected to incorporate the following common elements:

1. STI and Deep Trench Isolation Scheme including appropriate well (p-well and n-well) formations
2. Selective multi-layer gate insulator stack depositions and patterning for PFET, NFET, and NVM devices employing selective etch masking (e.g., silicon or silicon-nitride hard mask wherever appropriate) and patterning techniques
3. Selective multi-layered interface and gate formations for work function and unwanted charge injection control
4. Source/Drain shallow junction engineering including LDD/Spacer/selective S/D epitaxy
5. "Stress or strain engineering" techniques for mobility enhancements such as SMT, CESL, DCL, or DSL techniques [8–10]
6. Final low temperature anneals
7. ILD, contact schemes, metal levels, postmetal anneals, and BEOL technology

For NVM device and technology, elements as per items (a) through (c) are of particular interest. For planar technology, items (b) and (c) define devices, NVM cells, and arrays whereas for nonplanar device structures such as FIN-FETs or 3D devices, items associated with element (a) could be uniquely different.

Planar and nonplanar implementations are discussed separately.

14.4.3 Compatible Planar Implementation: Illustration of Integration Schemes ("SMLK" Concept)

14.4.3.1 Well and Isolation Schemes

NVM devices and arrays typically require programming voltage level of 5X-10x Vdd. Therefore, NVM arrays are placed into dedicated p-wells using deep trench isolation and N+ isolation guard rings.

The process sequences are similar to those of platform logic technology where multiple voltage sources are generated on-chip to carry out a diverse set of functions. All required well formations and isolation schemes (including STI for FET device isolation and local isolation within the NV memory array for cell-to-cell isolation) involve high-temperature processing and are carried out at the initial stage of process integration prior to device fabrication.

14.4.3.2 Compatible CMOS FET Gate Stack and NVM Gate Stack

Since device parametrics are unique to NFET, PFET, and NVMs, the integration scheme should be flexible yet as much as possible compatible, as mentioned earlier, in stack design to reduce process complexity. The three elements identified as the substrate/insulator interface, insulator/gate interface, and S/D technology could be addressed as follows:

The first element is typically addressed by growing an ultra-thin layer of SiO_2 film of 0.7 to 1.0 nm thickness using ozone and subsequently converting the layer into oxygen-rich SiON by treating with nitrogen plasma [11]. Optionally, an ultra-thin OR-SiON layer could be deposited by other techniques mentioned earlier.

The second element is relatively more complex since it involves the three unique gate stack structures, namely, NFET, PFET, and NVM gates. One approach has been to employ Ni-Si "FUSI-Gate" with customized alloying for NFET and PFET addressing work-function tuning [12–14]. A second approach could be to incorporate a dedicated metal gate for both PFET and NFET types of devices including NVMs (n-type or p-type) with an appropriate thin under layer dielectric to control the interface dipole and consequently the initial Vth and work functions for each type of devices [15]. Since charge injection control is also critical for NVM gate stack design and EOT for such device is significantly larger than that for CMOS FET devices, it is necessary to customize high K/metal gate interface to optimize individual device characteristics.

The third element is commonly addressed by forming metal silicide using limited thermal budget processing, namely, laser annealing or spiked annealing for shallow junction technology. The optional approach employs raised source-drain junctions by selective epitaxy. For device performance enhancement, additional process steps called "strain engineering" are often employed especially for PFET devices. For example, additional processing steps may consist of incorporating compressively stressed nitride liner ("CSL" technique, [3]). Strain engineering may be optional for NVM technology. One potential gate stack integration scheme for 22–25 nm technology node is schematically illustrated in Figure 14.7. The approach may be called "single metal multiple high K" (SMLK) integration scheme following the processing steps suggested by S. Kubicek et al. [15]. The three device stacks are shown in Figures 14.7a–c, respectively, for PFET (Vth = −0.3 V), NFET (Vth = +0.3 V), and the NVM devices. Such a scheme should be expected to meet device characteristic targets at the 22–25 nm node as suggested in Tables 14.1 and 14.2.

The common integration scheme and compatibility in the above stack design for all devices consist of the following:

1. Common silicon insulator interface of 1 nm of OR-SiON providing low-interface state density
2. Common TaN interface and aluminum metal gate
3. Common Ni-Silicided diffusion contacts to lower device series resistance

Additionally, the high K layer interfacing OR-SiON could all be of the same thickness of 2.5 nm as shown in Figure 14.7 consisting of either HfSiON (k = 14), HfTaON (k = 20), or HfLaON (k = 20). The leakage characteristics of 2.5 nm HfSiON, HfTaON, and HfLaON, are, respectively, 1E-3 A/cm^2, 1.5E-4A/cm^2, and 1E-5A/cm^2 compared to 1A/cm^2 for SiO_2 [16–18]. The multi-layered gate stack for CMOS PFET and NFET could be in the EOT range of 1.6–1.8 nm depending on the specific high K dielectric layers selected. Top layers of Al_2O_3 for PFET and La_2O_3 for NFET

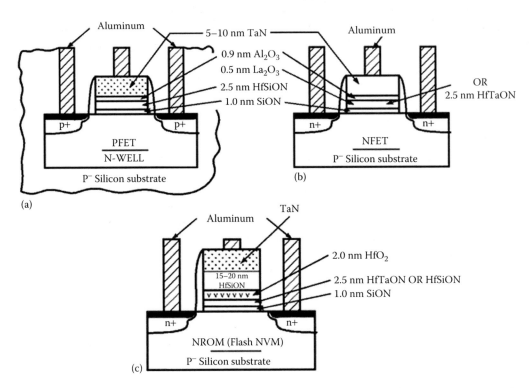

FIGURE 14.7 Compatible gate stacks for (a) PFET, (b) NFET, and (c) NVM with common integration scheme for 22–25 nm node.

aids in work-function control and thereby the thresholds of the devices. In case of the NVM device, the charge storage layer may either consist of 2 to 5 nm of SRN for an FP device or 2 to 5 nm of HfO_2 for a CT device or 2 to 5 nm of W NC embedded in HfSiON for an NC device. The charge blocking layer could be 15 to 20 nm of HfSiON or HfTaON, with the stack EOT ranging from 5 to 7 nm depending on the thickness and high K dielectric layer selection enabling Vpp ~ ±5 V for writing and erasing.

The stack processing steps and the associated integration schemes for the above three devices are illustrated in Figure 14.8 following the SMLK integration approach whereby each device stacking layer could be optimized separately. In such schemes, only the interface layer with silicon and the metal gate are common, while each device stack is uniquely customized. The PFET, NFET, and NVM layers are listed below:

PFET:Al/TaN/Al_2O_3/HfSiON/OR-SiON: layers a, b, and c being, respectively, OR-SiON, HfSiON, Al_2O_3

NFET:Al/TaN/La_2O_3/HfTaON/OR-SiON: layers d and e being, respectively, La_2O_3, Hf TaON

NVM:Al/TaN/HfSiON/HfO_2/HfTaON/OR-SiON: layers f, g, and h being, respectively, HfTaON, HfO_2, HfSiON

The overall process integration sequence is outlined in Table 14.3:

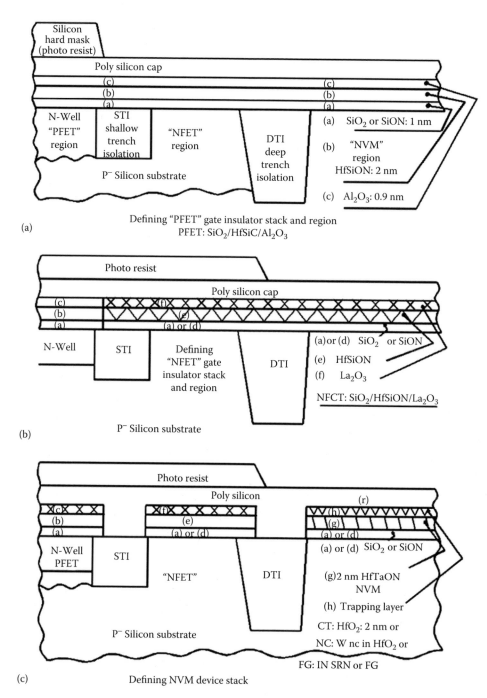

FIGURE 14.8 Customized stack processing and integration steps following SMLK scheme: (a) Defining "PFET" gate insulator stack and region; (b) Defining "NFET" gate stack; and (c) Defining NVM device stack. (From Aoyama, T., Maeda, T., Torii, K. et al., Proposal of new HfSiON CMOS fabrication process (HAMDAMA) for low standby power device, *IEDM*, pp. 95–98, 2004; Kim, Y.H., Cabral, Jr., C., Gousev, E.P., et al., Systematic study of work function engineering and scavenging effect using NiSi alloy FUSI metal gates with advanced gate stacks, *IEDM*, 2005, pp. 657–660; Lauwers, A., Veloso, A., Hoffmann, T. et al., CMOS integration of dual work-function phase controlled NI FUSI with simultaneous silicidation of NMOS (NiSi) and PMOS (Ni-rich silicide) gates on HfSiON', *IEDM*, pp. 661–664, 2005; Kubicek, S., Schram, T., Paraschiv, V. et al., Low Vt CMOS using doped Hf-based oxides, TaC-based metals and laser only anneal, *IEDM*, pp. 49–52, 2007.)

TABLE 14.3

Process Integration Sequence

Deep Trench isolation

↓

STI PROCESS

↓

ANNEAL

↓

BORON Ion Implant for PFET N-WELL

↓

Interface Layer Formation /Plasma

Nitridation/Anneal forming SiON

↓

PFET/NFET/NVM DEVICE STACK PROCESSING (Figure 4.8)

↓

GATE DEFINITIONS /ANNEALS

↓

Gate Stack Removal over Isolation and Removal

↓

Offset Spacer Formation

↓

Halo/Extension Ion Implant

↓

Side Wall Spacer Formation

↓

Source/ Drain Ion Implantation and Activation

Anneal (spike/laser anneal)

↓

Ni-Si Formation over Diffusion Regions/Anneal

↓

(Liner Formation/Optional)

↓

Nitride/Oxide Deposition and CMP

↓

Contact Definitions

↓

Aluminum or Tungsten Metal Deposition (M1)

↓

Post Metal Forming Gas Anneal

↓

Inter-Metallic-Dielectric

↓

Second Level Metal (M2)

14.5　NVM DEVICE TRANSITION AND INTEGRATION CHALLENGES

It has been stated earlier that until recently, progress in NVM device and technology was limited to successful feature size scaling of conventional devices and associated NAND, NOR, and NROM Flash NVM products. Therefore, the integration schemes continued along the planar CMOS silicon-gate technology base. The pace of adoption of planar and nonplanar scaled CMOS FET device

TABLE 14.4

SUB-50 nm Flash Products/Devices: Scaling Challenges

	FG-NAND	FG-NOR	CT-NOR/NROM
Products	SLC	SLC	SLC
MLC	2 bit/cell/3 bit/cell	2 bit/cell	2 bit/cell-> potentially 4 bit/cell
No. of Memory States:	4/8	4	2 for 2 bit->4 for 4 bit
Programming mechanism:	F–N Tunneling	CHE Write/F-N Erase	CHE Write/Tunnel Erase
Scaling challenges:			
Tunnel Oxide Scaling Limit:	~5.0 nm	~5 nm	Interface oxide: ~0.7 nm
Gate Coupling Ratio Scaling	Same as NAND	None	
Voltage Scalability		Same as NAND	Reduced Voltage Scalability
"Too Few Electron" for MLC	Same as NAND	Containable	
Retention/Endurance		Same as NAND	Retention
FG to FG Cross Talk		Same as NAND	"Second-bit-effect" (SBE)
OTHER ISSUES:			
Short Channel Effects (SCEs)	SCE		
Junction Breakdown (JB)	JB		
Short-Channel Effects (SCEs)	Read Disturb (RD)		

technology with high K gate stack design and metal gate has been slower and challenging. This is particularly true for integration when gate first integration scheme is considered as outlined in Table 14.1.

Table 14.4 highlights the challenges associated with the various NVM-FLASH devices in transitioning to sub-50 nm scaling.

14.6 ADDRESSING NVM DEVICE/ARRAYS CHALLENGES AND INTEGRATION

The challenges outlined in the Table 14.4 for specific device/product types are being explained below:

14.6.1 NAND Flash Devices and Arrays

During the last decade, FG-NAND flash NVM has driven the feature size scaling per ITRS roadmap due to the ever-expanding application base for such memory. The driving force has been the progressive reduction in cost per bit through DLS (2 bit/cell) and TLS (3 bit/cell) product availability. As the feature size gets reduced further, continued momentum in application would require solving the challenges outlined not only at the cell level as above but also at the programming and sensing levels combined with reduced cost of ECC overhead at the memory system level to ensure overall cost reduction. In this sub-section, we will limit our discussion to the cell/array framework. Subsequent sections will address other factors.

The NAND array architecture need not be limited to FG/FP types of devices and could extend to other types of NVMs such as CT, NC, and DTMs as well. In this discussion, we would assume NAND architecture to be cost driven and not performance driven (serial access of memory bits) whereas for higher performance code applications, we would assume random access NOR architecture.

Future cost-driven NVM applications may be addressed within two broad categories: (a) embedded NAND memory with rich logic functions within a single chip of SOC type, and (b) stand-alone NAND memory with appropriate MMU or flash controller with high-performance I/O capability for communication with the host devices. While NAND (a) can be tailored for a wide range of applications dependent on storage requirements, NAND (b) could be employed for enterprise and cloud computing applications for large database storage systems. For embedded NAND applications, a relatively simpler logic-compatible technology integration scheme would be desirable to reduce cost and complexity overhead. However, for large database storage, more complex NVM cell and associated integration scheme could be justified to provide power saving, enhanced performance, and reliability when compared against HDD options. For higher performance, higher endurance code applications and tunnel-based NROM architecture would be appropriate and should be discussed later on.

The FG/FP NAND scaling challenges associated with fundamental items of (1) "too few electron" and FG to FG cross talk at minimum lithographic pitch could be overcome by configuring the array in 3D mode with relaxed geometry. Additionally, FG's height and thickness limitations could be overcome with FP approaches. GCR scaling problems could be obviated by replacing FG/FP charge storage approach with CT approach (SONOS-type). Tunnel oxide limits of 5.0 nm could be overcome by either "double tunnel junction" [7] or by "double tunnel layer structures" [19] described in detail in Chapter 13, with an appropriate trapping layer. As explained in previous sections, using high-temperature stable metal gate (e.g., TaN or WN with W metal) and high K blocking dielectrics such as HfSiON or HfLaON, the overall EOT of gate stack could be reduced.

Consequently, programming and erasing voltage levels could be scaled while achieving targeted retention and endurance goals.

From general considerations, the scaling challenges for FG-NOR devices are similar to those of NAND with respect to tunnel oxide thickness limit, "too few electrons," GCR, voltage scalability, retention, and endurance. Similar solutions apply for FG-NOR devices.

14.6.2   NOR Flash and NROM Devices and Arrays

NOR Flash and/or NROM NVM devices are preferred for code-driven and higher performance applications requiring significant logic-memory mix, SOC, and cost-effective SIP replacements. With proliferation of portable electronics (cell phones, smart phones, tablets, etc.) and other consumer applications (such as automotive, smart appliances, educational toys), NOR/NROM applications are expected to grow during the next decade. Ease of integration with CMOS logic technology, combined with limited memory requirements, drives such applications. While planar technology and associated integration could continue, nonplanar structures such as FinFET and dual gate may get momentum for such applications for sub-22 nm nodes since FinFET CMOS could emerge as a platform technology during the next decade. Between planar FG-NOR and CT-NOR (Table 14.4), the later provides advantages of higher density and scalability without the constraints of tunnel oxide thickness limitations, CGR, and programming voltage level (and consequently power). However, currently both types of cell employ CHE for writing, while FG-NOR uses tunnel erase and CT-NOR employs hot-hole tunneling for erase. CHE generation requirement limits channel length scalability for such devices to ~100 nm. For both FG-NOR and CT-NOR cases, defect generation in the insulator during applications limits endurance, retention, and reliability, although better endurance is achieved in CT devices due to thinner tunnel oxide and lower average field during endurance cycling. FG-NOR exhibits a larger window used for multilevel storage, while CT-NOR could operate by localized trapping over diffusion regions to achieve 2 bit/cell capability. Both types of planar cell suffer from scaling issues such as "too few electrons," SCE, junction breakdown, and hot-carrier-induced retention, endurance, and reliability limits as stated in the table. While FG-NOR exhibits FG-FG cross talk and Program-disturb issues, the CT-NOR faces lateral charge transfer related "second-bit" and read disturb issues related to scalability. Scalability of NOR devices based on CHE requirements would be limited to ~0.1 nm in channel length. However, such adverse effects could be eliminated by adopting tunneling mechanism for charge transport for the CT-NOR devices. CT-NOR FinFET devices could be considered to solve "too few electron" and SCE issues. Therefore, it is likely that NOR applications may evolve into the following transitions for 22 nm and beyond:

FG-NOR (CHE-erase)

↓

CT-NOR /NROM (CHE-erase)

↓

Tunnel-Based CT-NOR/NROM

REFERENCES

1. A. Bhattacharyya, A very low power memory sub-system with progressive non-volatility, ADI Associates Internal Docket, 8/2/2004; U.S. Patent: 7276760 (10/02/2007), 7385245 (06/10/2008). [MIDD39].
2. A. Bhattacharyya, Integrated two-device multiple configuration alterable ROM/SRAM/General purpose nonvolatile memory, ADI Associates Internal Docket, 6/2004; U.S. Patent 7728350 (6/1/2010).
3. A. Bhattacharyya, High performance capacitor-less one transistor memory device in bulk and SOI technology, ADI Associates Internal Docket, 4/2003; U.S. Patent 7432562 (10/07/2008), 7625803 (12/01/2009), 7968402 (06/28/2011), 81255003 (02/28/2012).
4. A. Bhattacharyya, SRAM devices and electronic system comprising SRAM devices, ADI Associates Internal Docket, U.S. Patent 7898022 (1/2011). [MIDD 47].
5. J. Dillon, and H. Patel, Hybrid memory combining SRAM and NOR flash for code and data storage, Flash Memory Summit 2012.
6. M. Taguchi, NOR Flash Memory Technology, Spansion Inc., 2006 IEDM Short Course, IEEE organizer Rich Liu, Memory Technologies for 45 nm and Beyond, Dec. 2006.
7. R. Ohba, Y. Mitani, N. Sugiyama et al., 25 nm planar bulk SONOS-type memory with double tunnel junction, *IEDM*, San Francisco CA, 2006, pp. 959–962.
8. H.T. Huang, Y.C. Liu, Y.T. Hou et al., 45 nm High-K metal gate CMOS technology for GPU/NPU applications with highest PFET performance, *IEDM*, Washington DC, 2007, pp. 285–288.
9. H. Ohta, N. Tamura, H. Fukutome et al., High performance Sub-40 nm Bulk CMOS with Dopant Confinement Layer (DCL) techniques as a strain booster, *IEDM*, Washington DC, 2007, pp. 289–292.
10. S. Mayuzumi, J. Wang, S. Yamakawa et al., Extreme high Performance n and p MOSFET boosted by dual metal high K Gate damascene process using top cat DSL on 100 substrate, *IEDM*, 2007, pp. 293–296.
11. T. Aoyama, T. Maeda, K. Torii et al., Proposal of new HfSiON CMOS fabrication process (HAMDAMA) for low standby power device, *IEDM*, San Francisco CA, 2004, pp. 95–98.
12. Y.H. Kim, C. Cabral, Jr., E.P. Gousev et al., Systematic study of work function engineering and scavenging effect using NiSi alloy FUSI metal gates with advanced gate stacks, *IEDM*, Washington DC, 2005, pp. 657–660.
13. A. Lauwers, A. Veloso, T. Hoffmann et al., CMOS integration of dual work function phase controlled NI FUSI with simultaneous silicidation of NMOS (NiSi) and PMOS (Ni-rich silicide) gates on HfSiON', *IEDM*, Washington DC, 2005, pp. 661–664.
14. Z. Luo, Y.F. Chong, J. Kim et al., Design of high performance PFETs with strained Si channel and laser anneal, *IEDM*, Washington DC, 2005, pp. 495–498.
15. S. Kubicek, T. Schram, V. Paraschiv et al., Low Vt CMOS using doped Hf-based oxides, TaC-based metals and laser only anneal, *IEDM*, Washington DC, 2007, pp. 49–52.
16. X. Yu, C. Zhu, M. Yu et al., Advanced MOSFETs using HfTaON/SiO2 gate dielectric and TaN (+Al) metal gate with excellent performance for low standby power applications, *IEDM*, Washington DC, 2005, pp. 31–34.
17. J. Buckley, M. Bocquet, G. Molas et al., In-depth Investigation of Hf-based High-k dielectrics as storage layer of charge trap NVMs, *IEDM*, San Francisco CA, 2006, pp. 251–254.
18. Y.Q. Wang, D.Y. Gao, W.S. Hwang et al., Fast erasing and highly reliable MONOS type memory with HfO2 high-k trapping layer and Si_3N_4/SiO_2 tunneling stack, *IEDM*, San Francisco CA, 2006, pp. 971–974.
19. S. Krishnan, U. Kwon, N. Moumen et al., A manufacturable dual channel (Si and SiGe) high-k metal gate CMOS technology with multiple oxides for high performance and low power applications, *IEDM*, 2011, 28.1.1.

15 Planar Multilevel Storage NVM Devices

> **CHAPTER OUTLINE**
>
> Multilevel storage capability within a single memory cell has been the most distinguishing feature for the silicon based nonvolatile memories and arguably the single-most feature that has been driving the memory market during the last decade and potentially the decades to come. This chapter reviews the current state-of-the-art in planar multilevel storage NVM devices and novel and extendable memory concepts for future applications.

Perhaps the most distinguishing feature of Si-NVMs among various silicon based memories is the capability of storing multiple bits of memory in a single device cell yielding NVM cells to provide highest bit density for any given generation of feature size or technology node. The incredible growth of FG-NAND flash products during the last decade has been attributed to the rapid cost per bit reduction of NAND products providing 2 bit/cell (Double level storage, or, "DLS," alternately, Double level store per memory cell, or "DLC") and 3 bit/cell (Triple level storage, or "TLS," alternately, Triple level store per memory cell, or "TLC") memory devices. This capability had led to the fastest memory growth rate of Flash memory displacing DRAM in volume and revenue and driving feature size scaling for the past several generations. Feasibility of 4 bit/cell CT-NROM ("Quad NROM") has also been published. In this section, we will review various approaches to create multilevel storage cells and associated challenges both for FG-NVM devices and CT-NVM devices. Sections 15.1 through 15.6 describe planar multilevel cell ("MLC") NAND devices while Sections 15.7 through 15.9 cover both uniformly charge trapped and localized charge trapped ("MNSC") MLC devices. Section 15.10 outlines the future of such devices. Section 15.11 outlines recent developments on stackable planar MLC NAND in 3D configurations. Finally, in Section 15.12, we outline limitations of planar NAND devices.

15.1 MULTILEVEL NVM: EARLY DEVELOPMENTAL HISTORY

The proposal for multilevel NVM device emerged in 1995. The concept of subdividing the programmed (written) memory window (defining the maximum Vth shift from the erased state [Vth-low] to the highest written state [Vth-high]) into multiple memory states and appropriately sensing (reading) such memory states had been published by Bauer et al. [1] as well as by G.J. Hemink et al. [2]. T. Jung and coworkers in the following year published the 2 bit/cell reading and programming circuitry and the "pulse and verify" scheme consisting of programming pulse sequence to achieve and predict targeted program levels with self-compensation for device variations [3]. With the pulse and verify scheme, reproducible multilevel FG-NAND devices were demonstrated by the authors for practical applications. Several technologies, device design, and application enhancements were incorporated in early years to enable multilevel FG-NAND NVMs. These include: (a) Self-aligned shallow trench isolation (SASTI) Integration schemes [4–7] discussed in Chapter 14, enabling larger memory window, lower memory state dispersion, and desirable gate coupling ratio (GCR); and (b) Large Initial Memory Window ([Vth-high]-[Vth-low]) and memory state stability (retention). These are obvious requirements to achieve multiple memory states required for multilevel device design. For FG devices, this also implies well-controlled GCR requiring optimal device stack design for an appropriate programming pulse condition to yield the desired levels of initial and end-of-life window.

General criteria associated with obtaining large memory window, initial and end-of-life memory window stability, and GCR has been discussed in detail for FG devices earlier. Scalability limits of tunnel oxide and associated charge fluence limiting the values of memory window for conventional FG devices has already been discussed before. In recent years nonplanar "wrap-around" cell with overlapping control gate in the word-line direction was introduced in the sub-100 nm node regime to enhance GCR for SLC and MLC designs. For MLC operations, the memory window is subdivided into multiple stable memory states. Appropriate error-free sensing of such memory states is, therefore, vital. Factors affecting such stability for MLC requirements will be elaborated in Section 15.4 below, after reviewing related factors such as FG to FG coupling and GCR for MLC memory cell design standpoint. Cell structural scalability issues for planar and nonplanar NAND-FG-cells will be discussed in Section 15.2 and subsequent sections.

This chapter is divided into three subchapters or parts: Part A, consisting of Sections 15.1 through 15.6, describes the planar MLC NVM devices based on floating-gate cell NAND flash designs, which has been the work-horse of SSD products in the industry. Part B, consisting of Sections 15.7 through 15.9, describes the planar MLC cells and attributes based on charge trapping gate stack designs applicable to both NROM and NAND products with greater compatibility with CMOS logic technology. Part C, the final subchapter is in Section 15.10, which discusses the future planar flash technology and current challenges being faced with planar designs.

15.2 PLANAR FLOATING-GATE MLC NAND FLASH DEVICES

15.2.1 FG to FG Capacitive Coupling (Cell-to-Cell Coupling: CCC) for MLC NAND Design

FG to FG parasitic coupling and related memory state modulation has been discussed in Part I of this book (Figure 6.5) for conventional SLC-NAND design. Such interference is of significantly greater importance in MLC operation since such interference is cell geometry sensitive and increases significantly when cell dimensions and cell-to-cell spacing are reduced. From design consideration, Vth variation due to FG to FG parasitic coupling should be reduced to <0.2 V for a given technology node. For example, for 63 nm technology node, this was achieved by J-H Park et al. of Samsung Electronics [8] by reducing the floating gate height and by using low k oxide spacer between the adjacent floating gates. An 8 Gb NAND flash product chip fabricated by the above technology with a chip size of 135.6 mm^2 was demonstrated in 2004, which translated to an overall <4.3 F^2/bit with all support circuitry overhead. Key process parameters for the chip consisted of a gate stack design of 6.0 nm tunnel oxide/4.5 nm polysilicon floating gate with an ONO blocking IPD of 14.5 nm EOT. The control gate was a W/polysilicon gate stack. Other feature included an STI isolation of 200 nm depth. Figure 15.1a and b illustrates the cross-sectional TEM images of the NAND cell array parallel to the word-line direction and bit-line direction, respectively.

The NAND memory cell illustrated above was aligned with SA-STI in the bit-line direction while the control gate "wraps-around" the floating gate to enhance coupling and GCR in the word-line direction. This approach not only enhances the GCR but also provides shielding effect reducing parasitic capacitive coupling between adjacent cells during operational conditions. However, the gap filling becomes a challenge when the wrap-around cell is further scaled. At the 63 nm technology node cell design, the IPD filling for such cell resulted in a gap of <15 nm to be filled by the control gate [8,9].

The above "wrap-around" cell with narrower spacing between floating gates could not be scaled with feature size scaling due to the dramatic increase of parasitic cell-to-cell coupling (CCC). The CCC increases unwanted interference between cells during operational conditions. The increase in CCC ratio with feature size reduction is plotted in Figure 15.2a [10]. Beyond 43 nm node, the normalized CCC ratio increases dramatically as shown. To meet the CCC-related challenge, the advantage of using "air-gap" technology was discussed by D. Kang [13] and was subsequently introduced as shown in Figure 15.2b. It may be noted that for the 19 nm technology node, air-gap CCC

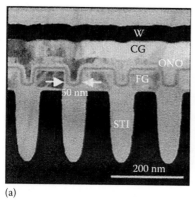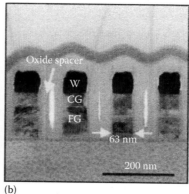

(a) (b)

FIGURE 15.1 NAND string cross-sectional TEM: (a) Parallel to word-line, (b) parallel to bit-line. (From Park, J.-H., Hur, S.-H., Lee, J.-H. et al., 8 Gb MLC [Multi-level cell] NAND flash memory using 63 nm process technology, *IEDM*, pp. 873–876 © 2004 IEEE.)

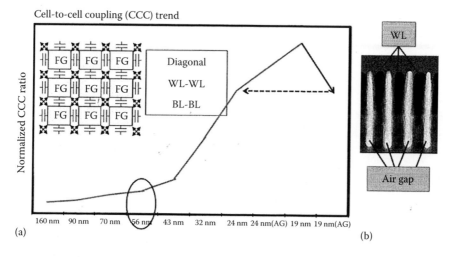

(a) (b)

FIGURE 15.2 (a) Normalized CCC ratio versus NAND cell technology scaling node: CCC reduction with "air-gap", (b) SEM photo of the air-gap. (From Li, Y., 3 bit per cell NAND flash memory on 19 nm technology, Sandisk, *Flash Memory Summit*, Santa Clara, CA, 2012.) Reproduced with permission from SANDISK.

is equivalent to the "Non-airgap" option for the 24 nm node, a reduction of 24% in CCC with "air-gap." The "air-gap" technology was applied for the 19 nm technology node to develop 64 Gb MLC NAND flash product by N. Shibata from Toshiba Corporation [11]. However, CCC control is limited at sub-20 nm regime even with the air-gap technology. For sub-20 nm regime, additional scaling challenges for such Flash cell were also related to geometry-induced high field effects, as well as structural reproducibility and instability of the "wrap-around" gate shapes. These issues adversely affect device reliability at sub-20 nm dimensions. However, an integration technology at 20 nm was demonstrated by K.W. Lee and coworkers from Hynix Semiconductor in 2011 featuring SPT patterning technology employing air-gap with separate gate etch process and optimized polysilicon deposition process for the polysilicon control gate [12].

To control CCC, reduction of K value of FG to FG isolation was required. The effect of "air-gap" on cell Vth variation had been discussed by D. Khang et al. [13] when compared to oxide and nitride. Oxide with a lower K value reduced CCC.

For continued scaling of NAND-FG cell, nonwrap around planar cells were introduced with or without air-gap technology replacing IPD between the adjacent floating gates [10–14]. Cell design considerations of some of these MLC cells will be discussed later on in this section.

15.3   GCR FOR MLC NAND DESIGN

Conventional FG-NAND SLC devices are required to be designed to achieve a GCR > = 0.6 to meet the following requirements: (a) retention of FG charge, (b) reduction/elimination of charge compensation by controlling electron injection during erase, (c) designing programming voltage (Vpp) and programming duration (tpp) at around 20 V and 10 ms, respectively, (d) targeting memory window around ~± 3 V and related electron and hole fluence during writing and erasing, and (e) limiting writing and erasing fields low enough (~10–12 MV/cm) to meet endurance and reliability targets.

Since tunnel oxide thickness and ONO IPD are not significantly scalable, MLC designs up to 43 nm node employed the "Wrap-around" cell design approach as discussed earlier to achieve higher value of GCR with limited extendibility until air-gap technology was introduced with nonwrap around planar cell design for sub-24 nm designs. However, even with such approaches, high GCR could not be achieved. Consequently, device parameters such as memory window, retention, endurance, and reliability are compromised for MLC design compared to SLC design for the NAND NVMs. Conventional ONO IPD charge blocking layer had to be replaced with single or multilayer higher K IPD such as Al_2O_3. IPD options to overcome above limitations for the FG-NAND MLC designs have already been discussed in Chapters 12 and 13 where application of other higher K dielectrics such as HfSiON and $HfAlO_3$ have also been discussed.

15.4   FG-MLC CELL DESIGNS FOR MEMORY LEVELS, STABILITY, SENSE MARGIN, AND RELIABILITY

Schematic Vth (memory) levels and distribution for SLC and MLC 2-bit/cell and MLC 3-bit/cell are shown in Figure 15.3.

Figure 15.3 exhibits not only the Vth distributions for the SLC, DLC, and TLC cells but also the sense margins for the memory states represented by the shaded regions between the memory states.

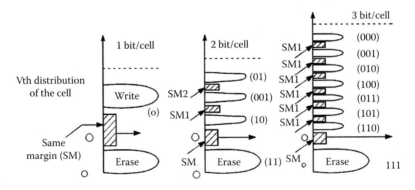

FIGURE 15.3   Schematic representations of memory levels for 1-bit/cell (SLC), 2-bit/cell (DLC), and 3-bit/cell (TLC) and Vth distributions of FG-NAND cells and associated sense margin.

In all cases, the lowest Vth state defined by the memory state of 1 (SLC), or 11 (DLC), or 111 (TLC) represents the lowest Vth state corresponding to the "erased" memory state. The highest memory state is the highest Vth achievable by the programming condition and is limited by the program disturbs. This highest memory state is labeled as "0" for SLC, "01" for DLC, and "001" for TLC memories, respectively. All sub-memory states in MLC design are created in between the "erased" state and the highest written Vth state by multiple programming steps which for TLC design could typically consist of three-step programming. Between each memory states, adequate margins are required to reliably identify and sense sub-memory states for discrimination and error-free reading of such memory states. MLC design involves taking into consideration all factors affecting the stability and variations (Dispersion) of each memory states, thereby establishing budgets associated with Vth margins and memory window requirements.

Factors affecting memory window are: (a) Feature size scaling which results in reduced number of charges (electrons and holes) affecting larger percent of Vth shift per electronic charge; (b) programming step size; (c) adjacent cell interference during writing and erasing due to CCC; and (d) noise.

Application related Vth dispersions associated with each memory states may include: (a) Retention loss; (b) endurance (write–erase cyclic stressing); (c) temperature cross-shift and charge loss due to thermally induced detrapping; and additionally (d) variation in cell geometry.

Both memory window (Vth width of the memory state) and Vth dispersion affect the *sense margin* for each memory state and has to be accounted for ensuring sensing, read performance, and memory reliability.

Sense margin: Factors affecting sense margin are related to the above-mentioned factors affecting Vth width and Vth dispersion. For reliable addressing of each of the MLC memory states, Vth margin budgets need to be established taking into consideration (1) program and read disturb; (2) effects on retention due to program and erase cycling; and all application conditions such as temperature and noise affecting loss of sense margins.

It should be noted from Figure 15.3 that the sense margins are progressively reduced from SLC design to DLC design and from DLC design to TLC design as the number of memory states between the "erased state" and the highest "written state" is increased. Sense margin addresses both the memory state sensing accurately and the reading performance. For DLC, two additional memory states are introduced (a total of four states) while for TLC six additional states are introduced (a total of eight states) between the erased state and the highest written state. Each memory state with its unique dispersion in Vth creates the requirement of sense margin in MLC design a special challenge. Various factors affecting sense margin are: stack design, process reproducibility, sense circuitry, high-voltage circuit performance variations, as well as variations in device characteristics such as retention and endurance. Kim et al. employed a statistical approach of relevant variables in optimizing sense margins for MLC design [14].

As feature size is scaled, sense margin for MLC design is reduced and implementation requires optimization in many areas. These are: (i) Cell coupling, (ii) cell-to-cell interference, (iii) programming algorithms in voltage and time, (iv) pass-word-line voltage margin, (v) word-line and bit-line delay reduction (RC time constant reduction), (vi) sense-amp and high-voltage- regulator-circuit optimization, (vii) establishing enhanced "program-verify-read" algorithm before programming to ensure memory state validity, and (viii) other device design variables impacting the sense margins [14–16]. A detailed discussion of some of these elements is out of scope of this book. Characteristics of program-verify algorithm and cell attributes for 3 bit/cell FG-NAND flash technology are illustrated in Figure 15.4a–g, and are reproduced from the investigations carried out by H. Nitta and coworkers [16] for 30 nm node FG-NAND flash memory design.

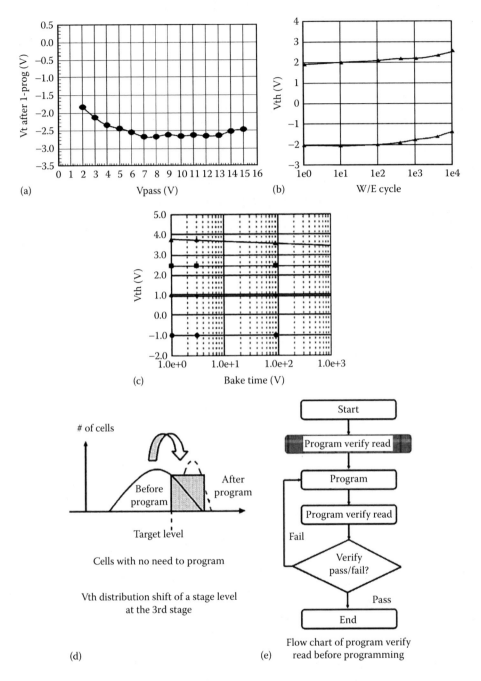

FIGURE 15.4 3 bit/cell program-verify read before programming verification. (a) Vpass characteristics after program "1"; (b) memory cell endurance characteristics; (c) memory state retention after baking; (d), (e) program-verify-read algorithm (right); test-characteristics (left), cells requiring re-program shown to demonstrate memory state stability and sense margin improvement. (*Continued*)

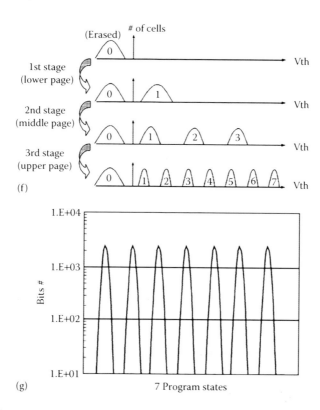

FIGURE 15.4 (Continued) 3 bit/cell program-verify read before programming verification. (f) multi-stage programming using 3 bit/cell programming using program-verify-read algorithm; and (g) final distribution of all seven program states demonstrating state stability and margin. (From Nitta, H., Kamigaichi, T., Arai, F. et al., Three bit per cell floating gate NAND flash memory for 30 nm and beyond, *IEEE 47 Annual International Reliability Physics Symposium*, p. 307, © 2009 IEEE.)

15.5 ADVANCED TECHNOLOGY: FG-MLC EXTENDIBILITY CHALLENGES

In addition to the scaling challenges discussed earlier and process, device, system solutions, and optimization mentioned above, scaling beyond 20 nm node creates yet additional challenges for floating-gate designs not only for conventional "wrap-around" FG cell approach but also for the thin planar polysilicon FG devices. These challenges include: (a) Sub-lithographic patterning to achieve desired word-line and bit-line pitches, (b) nonextendibility of "wraparound" cell design (discussed earlier), (c) high electric field and associated charge-loss problems between word-lines, (d) cell-side-wall/ STI fringing field and device length and width geometry effects, (e) read-current reduction due to higher substrate doping to contain short-channel effects, and (f) structural reproducibility and stability of FG and CG features for conventional cell design at sub-20 nm dimensions.

A combination of advanced technology and system solutions has been addressing the above challenges in recent years with considerable success. These will be briefly outlined here.

15.5.1 Sub-Lithographic Patterning

Sublithographic patterning beyond the photolithographic capability had been successfully developed and implemented in CMOS technology during the 1980s in IBM to develop deep-submicron CMOS transistors [17–19]. These techniques were known as "Sidewall Image Transfer" or SIT technique or "Spacer Patterning technology" or SPT. Such technology was subsequently implemented to reduce cell sizes of DRAM and FLASH devices [20,21]. The techniques were further extended in recent years by C.-H. Lee et al. [22] and by J. Hwang et al. [23] and applied to sub-20 nm bit-line pitch features called ArF immersion SPT and Quad SPT. A schematic diagram for Quad SPT (QSPT) is shown in Figure 15.5. Using such technique, Hynix semiconductors implemented advanced air-gap technology between the word-lines achieving sub-20 nm word-line pitch separating the word-lines [19]. Additionally, FG shapes were optimized called "FG slimming" enabling void-free filling of the control gate for the "wrap-around" memory cell. TEM cross-section of thus achieved sub-20 nm cell pitch is shown in Figure 15.6.

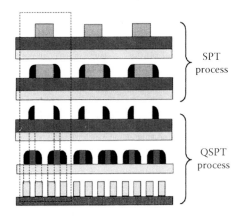

FIGURE 15.5 Schematic diagram of quad SPT (QSPT) technique. (From Hwang, J., Seo, J., Lee, Y. et al., A middle-1× nm NAND flash memory cell [MIX-NAND] with highly manufacturable integration technology, *IEDM*, pp. 199–202. [Hynix] © 2011 IEEE.)

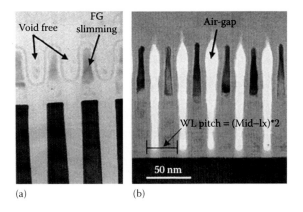

FIGURE 15.6 (a) Cross-section TEM demonstrating FG slimming, (b) well-patterned air-gap and CG patterning implementing void-free word-line "wrap-around" cell feature using QSPT. (From Hwang, J., Seo, J., Lee, Y. et al., A middle-1× nm NAND flash memory cell [MIX-NAND] with highly manufacturable integration technology, *IEDM*, pp. 199–202. [Hynix] © 2011 IEEE.)

15.5.2 Effects of Advanced Technology and Improved Program Algorithm on FG-MLC Device Design

Nitta and coworkers reported several improvements in cell design with the employment of air-gap using QSPT technology [16]. GCR improved nearly 10% at 20 nm and nearly 20% at 15 nm nodes, respectively.

Additionally, when the advanced technology was combined with an improved V pass optimization algorithm for the neighboring cells during programming of the active cell, parasitic lateral field imposed on the neighboring half-select cells was reduced with consequent improvement in retention of the memory state of the neighboring half-select cells. Nitta et al. also noted nearly 60% improvement in programming speed due to optimization of V pass potential during programming through V pass algorithm optimization during programming [16].

15.5.3 STI/Sidewall Fringing Field and Cell Geometry Effects for Planar FG Cell

H-T Lue and coworkers have intensively investigated the STI/side-wall fringing field effects on memory window and turn-on characteristics (sub-threshold slope (SS)) of thin polysilicon planar FG devices when the device lengths and widths are reduced in the range of 20–50 nm dimensions [24]. They studied FG thickness in the range of 3–10 nm with ONO CT-IPD for enhanced charge retention (nitride charge-trapping) and large memory window, and cell structures of either uniformly planar control gate in the WL direction or STI defined T-shaped control gate design in the WL direction. In both cell designs, the control gate was of P+ doped polysilicon. Large area devices (W/L = 120/40 nm) showed large initial P/E window, larger for the uniformly planar CG cell (~9 V) without program saturation. However, both memory window and turn-on characteristics were significantly degraded exhibiting strong L effect at L = 20 nm and also W effect as W was reduced. The effects were more severe for the T-shaped control gate design. Figure 15.7 illustrates device Vth roll-off versus channel length for the uniformly planar CG cell at initial, written and erased conditions demonstrating the memory window sensitivity on device length. Figure 15.8 shows the Id–Vg characteristics showing that the sub-threshold slope (SS) is degraded only at the written state and not in the erased state. The degradation has been explained through 3D TCAD simulations due to fringing field associated sidewall turn-on of the device. It was shown that the E-field reduction at smaller L is the primary cause of degradation and other parameters such as higher P-well doping,

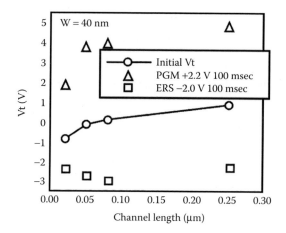

FIGURE 15.7 Vth roll-off and P/E memory window degradation due to channel length scaling. (From Lue, H.T. et al., *IEDM*, pp. 203–206 © 2011, IEEE.)

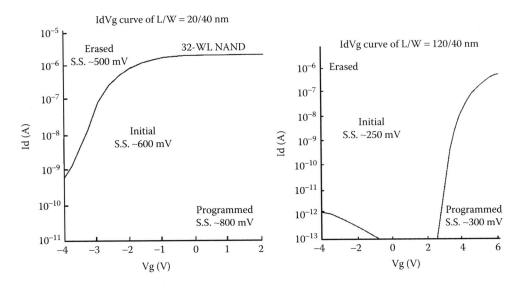

FIGURE 15.8   Id–Vg characteristics of uniformly planar CG flash cell: (a) L/W=120 / 40 nm, (b) L/W=20/40 nm showing L effect on the programmed state. (From Lue, H.T. et al., *IEDM*, pp. 203–206 © 2011, IEEE.)

and thinner FG thickness had minor impact on the memory window roll-off. Optimization of short-channel effect and confinement channel E-field were key factors of consideration for sub-20 nm planar FG cell.

15.5.4   Controlling Read-Current Reduction

Read performance and reliability are major issues in scaling due simultaneously to (a) reduced sense margin and (b) read-current reduction caused by higher substrate doping required to contain short-channel effects. To control read-current reduction, (a) improved sense circuit design, (b) sense algorithm enhancements, and (c) transistor junction optimizations were implemented to improve read performance.

15.6   PLANAR AND WRAP-AROUND FLOATING-GATE FLASH DESIGNS: EXTENDIBILITY ISSUES

The extendibility issues and the progress for conventional "wrap-around" floating gate cell have been discussed above. Concerns were raised over further extendibility beyond 20 nm technology node in planar technology.

This was addressed by P. Bloome and coworkers in 2010. They proposed a novel "dual-layer" planar floating-gate NVM cell structure combining conventional polysilicon with a thin film of conductive TaN as floating gate in planar configuration [28]. A fully planar floating gate without the side-wall capacitive coupling (no-wrap-around) eliminates the cell to cell parasitic interference at the expense of a) significantly reducing GCR (Figure 6.3), and consequently b) highly increased field across the inter-FG dielctric films between the FG to CG as well as between the FG and the substrate. This would lead to unacceptable write and erase saturation and memory window as well as high field associated reliability issues. The authors addressed the issues by not only employing a high work-function metal control gate (minimize electron injection from the gate during erase mitigating erase saturation), but also a dual n-poly/TaN conductive floating gate with high blocking dielctric (Al_2O_3) in place of ONO. The authors demonstrated significant reduction in write satura-tion, >4 V memory window, and desirable NAND parametrics for the planar FG-NVM cell and elimination of cell-to-cell parasitic interference.

Concurrently, S. Jayanti and coresearches demonstrated that TaN conductive floating gate could be scaled down to 1 nm and yet demonstrate large memory window for MLC capability when high K low leakage blocking dielectric such as HfAlO is employed between the FG and the CG [30]. Such approach clearly established extendibility of planar FG-NAND technology beyond 20 nm technology node.

While planar FG-flash devices have the advantage of reduced FG–FG interference and gap-filling limitations, planar polysilicon FG is still, until recently, a 3D structure unlike the floating plate option which could be of <2 nm height. The geometry effects of polysilicon "wrap-around" floating gate have been discussed earlier. The size effects of thin planar polysilicon FG and asso-ciated challenges in 20 nm regime and beyond have been outlined earlier. There may yet exist several extendibility challenges with planar FG designs as mentioned earlier. In case of floating plate option as suggested earlier containing highly conductive I-SRN silicon nanocrystal layers (<1 nm nanocrystal size), floating plates need to be isolated to avoid lateral charge migration to the neighboring cells both in WL and BL directions to avoid undesirable interference. However, Micron technology has demonstrated that conductive FG plates could be scaled down to =<5.0 nm and yet large memory window (>15 V) could be achieved. Very recently, Micron technology has proposed a planar nonwrap around cell with enhanced IPD and conductive FG to overcome some of the limi-tations of the "wrap-around" cell scalability issues [25]. This has resulted an 128 Gb NAND TLC product offering from Micron Technology at the 20 nm technology note to be illustrated later on. Some of the memory cell features and chracteristics are presented below.

A TEM cross-section of such a memory cell is shown in Figure 15.9a and b along BL and WL directions, respectively. Figure15.9c demonstrates the memory window characteristics of the cell [25].

15.6.1 High K IPD and Metal Gate for Planar Conductive FG Cell

Ramaswamy and coworkers demonstrated that optimization of high K blocking layer leakage and high work-function compatible metal gate are key requirements to achieve large memory window, meeting retention and endurance targets, and further scalability for the planar conductive floating-gate cell [25]. The authors compared other planar (nonwrap around) memory cell/stack options including charge trapping nitride with different Si/N ratios with and without band-engineered ONO tunnel oxide and CT-Pt and Ct-Rh nanocrystals for charge trapping with similar high K blocking

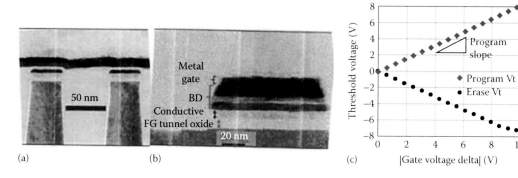

(a)　　　　　　(b)　　　　　　(c)

FIGURE 15.9 TEM pictures of conductive FG mMemory cell: (a) Along BL direction, (b) along WL direc-tion, and (c) proramming characteristics. (From Ramaswamy, N., Graettinger, T., Puzzilli, G. et al., Engineering a planar NAND cell scalable to 20 nm and beyond, *IEDM*, 2013; Courtesy of Micron Technology, Boise, Idaho.)

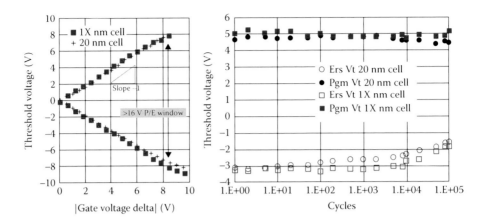

FIGURE 15.10 Device scalability of conductive FG cells at 15 nm technology: (a) Memory window characteristics for 1X nm cell versus 20 nm cell and (b) endurance characteristics compared between 20 nm cell and 1X nm cell. (From Ramaswamy, N., Graettinger, T., Puzzilli, G. et al., Engineering a planar NAND cell scalable to 20 nm and beyond, *IEDM*, 2013; Courtesy of Micron Technology, Boise, Idaho.)

layer and metal gate and concluded that the planar conductive FG option yielded superior memory window and program erase characteristics, Vth stability, retention, and other NVM device characteristics. Comparison of various planar cell characteristics from their investigation for field edge capacitors and memory arrays fabricated both at 50 nm node and 20 nm node are reproduced in Table 15.1. Scalability of such FG stack design for 15 nm node were confirmed as shown in Figure 15.10a and b

TABLE 15.1
50 and 20 nm Planar Floating-Gate Device Options and Attributes

	ETO-SN-CTF	NC-CTF	Conductive FG	Conductive FG
	50 nm	50 nm	50 nm	20 nm
Program voltage for +6 V Vfb (10 ms)	19.6	19.8	19.9	20.4
Erase voltage for −5 V Vfb (5 us)	−21.7	−25.5	−21.5	−19.3
Prog slope	0.78	0.86	0.91	0.96
Erase slope	0.8	0.73	0.78	0.96
Capacitor charge loss in mV (11 V initial window +6/−5) post 150°C, 24 hr bake				
	2030	240	300	210
50 nm array				
	ETO-SN-CTF	**NC-CTF**	**Conductive FG**	
	50 nm	50 nm	50 nm	
Program voltage for +6 V Vfb (10 ms)	24.7	20.9	23.2	
Erase voltage for −5 V Vfb (5 ms)	−21	−23.3	−21.2	
Prog slope	0.65	0.8	0.95	
Erase slope	0.84	0.8	1	
Arry charge loss in mV (150°C 24 hr bake)				
	200	640	60	

Source: N. Ramaswamy, T. Graettinger, G. Puzzilli et al., Engineering a planar NAND cell scalable to 20 nm and beyond, *IEDM*, 2013; Courtesy of Micron Technology, Boise, Idaho.

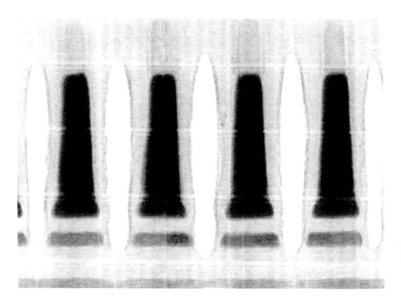

FIGURE 15.11 Conductive FG cells at 15 nm node. (From Ramaswamy, N., Graettinger, T., Puzzilli, G. et al., Engineering a planar NAND cell scalable to 20 nm and beyond, *IEDM*, 2013; Courtesy of Micron Technology, Boise, Idaho.)

demonstrating memory window and endurance, respectively. The optimized metal gate consisted of ALD TaN for thermal stability and high work function compared to other options [25]. A TEM cross-section of FG cells for 15 nm node design along the B/L direction is shown in Figure 15.11.

Other challenges and potential solutions are considered below.

15.6.2 GCR Reduction Yielding Smaller Memory Window

It has been discussed earlier that conventional ONO IPD scaling was limited due to degradation in memory retention. When planar FG cell area is scaled, GCR is reduced and consequently the memory window. Such challenge could be addressed in at least two ways for the planar NAND designs. These are: (a) incorporation of higher K IPD as suggested by Bhattacharyya and Darderian [26] and others [27–30]; and (b) by band-engineered charge trapping IPD for charge-trapping flash (CTF) memory cells as proposed by H.T. Lue et al. [24]. The band-engineered CT-IPD consisted of ONONO stack of 1.3 nm oxide/2 nm of nitride/2.5 nm of oxide/6 nm of nitride and 6 nm of oxide. Excess charges (both electrons and holes) tunnel through the thinner layers of oxide and nitride and get trapped within 6 nm of nitride layer of the IPD stack.

15.6.3 Parameter Degradation: STI Edge-Fringing Field Effect

It has been shown that as the feature size is reduced, strong lateral fringing field at the STI sidewall degrades the vertical field during programming. As a result, the sub-threshold slope in the programmed state is adversely affected. This results in memory window and endurance degradation [24]. Such degradation can be prevented by a modified STI process and a modified STI structure proposed by Bhattacharyya [31]. After the SA-STI region was defined, a thin layer of Al_2O_3 was deposited both along the vertical and horizontal interfaces of silicon substrate by ALD technique followed by an appropriate anneal. Prior to the deposition of Al_2O_3 layer, a thin layer of IN-SRN liner layer could optionally be deposited to provide an unipotential interface between the silicon and Al_2O_3 layers (Figure 15.12). This approach facilitates the termination of the fringing field. Subsequently, the STI oxide is deposited and planarized following the conventional processing steps. The interface

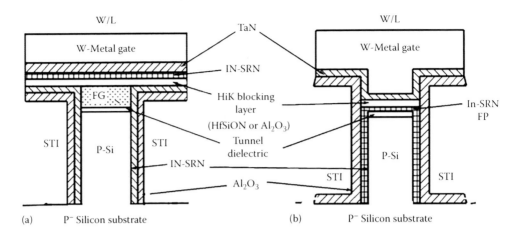

FIGURE 15.12 Modified STI structure and conventional silicon-gate device stack for FG-flash NVM for 20 nm node and beyond for eliminating STI fringing field effect. (a) For FG type of Cell; and (b) For FP or Charge-Trap type of Cell. (From Bhattacharyya, A., Proposed novel STI and gate stack design for FG-NAND flash NVMs for 20 nm node and beyond, A. Bhattacharyya, ADI internal disclosure, December, © 2012 IEEE.)

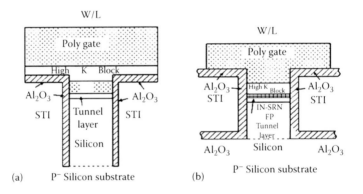

FIGURE 15.13 Modified STI structure integrated for Metal-Gated FP-Flash NVM for 20 nm node and beyond for eliminating STI fringing field effect using Al2O3/IN-SRN interface. (From Bhattacharyya, A., Proposed novel STI and gate stack design for FG-NAND flash NVMs for 20 nm node and beyond, A. Bhattacharyya, ADI internal disclosure, December, © 2012 IEEE.)

alumina layer provides high density of fixed negative charge at the silicon/oxide interfaces, thereby raising the parasitic field-oxide threshold and eliminating lateral fringing field. The corresponding device and STI structures are illustrated in Figures 15.12 and 15.13, respectively, for silicon gate and metal gate technology variations. The enhanced trapping layer for the MLC stack design suggested by Bhattacharyya consists of $Si_2ON_2/Al_2O_3/Si_2ON_2$, which provides large memory window and excellent retention. The tunnel layer is a thick OR-SiON layer, which replaces conventional oxide, the blocking layer being HfSiON or IN-SRN/Al_2O_3 characterized by low leakage [32].

It is worth noting that for MLC applications of FG-NAND-cells, same amount of charge loss results in large amount of Vth shifts adversely affecting both data retention and endurance for the memory cell. Cell endurance, data retention, operational window, sense margin, and program disturb are primary challenges required to be overcome for extendibility of MLC capability of FG-NAND cells. Consequently, parameter targets for FG-MLC NAND flash NVMs had been decreasing with feature size scaling of the FG-flash cells. Simultaneously, the error correction code (ECC) requirements of the memory products had been increasing exponentially to ensure MLC capability and data integrity. Such challenges and mitigation approaches to ensure MLC data integrity will be discussed in Section 15.10 (15°C).

15.7 CURRENT STATE-OF-THE-ART IN PLANAR MLC FG-NAND FLASH DEVICE AND PRODUCTS

Due to the extendibility issues for the planar MLC FG-NAND flash device and application constraints outlined above, there had been a growing concern of extendibility of MLC FG-NAND flash device and technology for the 20 nm nodes and beyond. Concurrently, there had been an increasing R&D efforts during the past years to seek various approaches of band-engineered CT devices both SONOS and nanocrystal types to replace FG-NAND flash memory devices and products. The success of high volume production of FG-NAND flash 64 Gb/chip products from Micron Technology in 2011 in 20 nm node and subsequently 128 Gb/chip in 2013 in 16 nm node followed by other major suppliers namely Samsung, Toshiba, Macromix, and Hynix announcing production of similar products in 20 nm and mid-teen nm technology nodes had put to rest such concern for implementation at least up to 15 nm node [22–25,33–43]. A brief discussion of some of the product features and challenges met will be highlighted below.

It should, however, be noted that the scalability of planar FET-based devices beyond 15 nm node has already been challenged by tri-gate FinFET devices and gate-all-around ("GAA") device structures (e.g., nano-wire FETs). The GAA vertical channel device structures would be ideal for scalability to sub-10 nm nodes. R&D and productization efforts are being strongly persued by the industry for vertical channel 3D device structures. At the time of the final preparation of this book, SAMSUNG had already announced for sampling their 48-mask-level Vertical-NAND ("V-NAND") 256 Gb TLS (3-bit/cell MLC) chip [44]. The V-NAND chip is implemented with 21 nm feature size to compete against previously announced planar 128 Gb NAND chip implemented with 16 nm feature size from Micron Technology, which is discussed in detail in this section. The vertical channel NVM devices are the subject matter of the next chapter of this book.

Micron Technology, jointly with Intel corporation led the way of scaling the MLC FG-flash devices through engineering of several potential device options for 20 nm and sub-20 nm nodes. The sub-20 nm TLC chip (3 bits/cell) with 128 Gb of memory is shown in Figure 15.14 providing the smallest planar-channel triple level cell in the industry. The chip size was 146.5 mm^2 [33,34].

Toshiba Corporation described their 24 nm, 151 mm^2 64 Gb MLC NAND in 2011 [36] followed by 19 nm, 112.8 mm^2 64 Gb MLC FG-flash technology and product in 2012 [37]. The product was further characterized by Y. Li from SanDisk Corporation [38]. Toshiba introduced both 2 and 3 bit/cell designs, the former offering higher performance and reliability while the latter providing higher bit density, implementing air-gap technology to reduce parasitic FG coupling effects and metal silicide gate to reduce control gate RC delays. Toshiba also introduced all-bit line (ABL) chip architecture

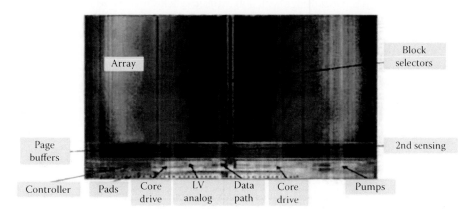

↳ Smallest triple level cell (TLC) die in industry

FIGURE 15.14 20 nm 128 Gb, 146.5 mm 2 TLC chip from Micron Technology. (Courtesy of Micron Technology.)

to reduce chip size and peripheral overhead. Additionally, program algorithm was modified with three-step programming to reduce program disturb, parasitic coupling, Vth dispersion of memory states due to residual cell-to-cell coupling (RCCC), thermal cross-shift (TCS) and noise, improving performance and reliability, and to contain reduced memory state margins due to scaling.

Samsung enabled their high performance 64 Gb MLC (2 and 3 bit/cell) NAND products at 21 nm node [40,41] and sub-20 nm node [42] through a combination of technology enhancements [22,39], peripheral circuit design enhancements, ECC enhancements, and program algorithm enhancements. The key technology enhancements consisted of (1) reduction of initial Vth distribution of memory states of "wrap-around" cell using "Self-Aligned-Double Patterning" lithography for NAND strings, and (2) novel tunnel oxidation to raise initial Vth of the device combined with advanced "dual-pulse" channel doping scheme to reduce cell-to-cell FG coupling and improve Vpass program disturb margin, active-edge peak E-field reduction to improve retention leakage and device reliability. Key circuit innovations included: (a) High speed data path architecture and I/O circuit enhancements to improve double data rate (DDR) interface and I/O bandwidth, and (b) memory operational schemes including enhanced "On-Chip- Randomizer" scheme for programming, soft data readout scheme for improved ECC, and incremental bit-line precharge scheme to reduce power consumption. WL-to-WL FG coupling interference was reduced by introducing enhanced program algorithm called "correction before coupling" (CBC) reprogram and P3-Pattern pre-pulse scheme; the latter reduced BL-to-BL coupling interference. Another technique was incorporated to reduce program disturbance called "Inhibit Channel-Coupling-Reduction" or ICCR. The 64 Gb chip was implemented in wave pipeline architecture and provided 400 Mb/sec DDR at 21 nm node [40] and 533 Mb/sec DDR at sub 20 nm node [41]. The chip was implemented in a triple-well 3-metal technology. The chip provided TLS capability with a chip size of 162.4 mm^2.

Hynix followed a similar approach to that of Toshiba by incorporating advanced air-gap technology to suppress charge loss between the programming control gates and neighboring half-select floating gates to reduce lateral high electric field during programming [23]. Their investigation showed that when their CoSi-based WL air-gap technology was combined by the next neighbor (N+1) WL bias scheme, the electric field between the active CG and next neighboring FG dropped by ~35% and consequently charge loss-induced delta Vth shift improved by ~300 mv. Other technology improvements incorporated were: (a) Quad Spacer Patterning technology (QSPT) to achieve the WL pitch objective, (b) FG slimming for void-free filling of the wrap-around control gate along the word-line direction thereby achieving BL pitch objective, preventing electrical depletion in CG and enhancing cell current. Hynix semiconductors published their technology design rules for sub-20 nm ("Middle-1× nm) technology around similar time frame. Hynix published their very high performance 64 Gb NAND flash product design with 800 MB/sec synchronous DDR interface in 2012 [43].

While Micron Technology took the lead in FG cell design transitioning from the "wrap-around" cell design to the planar conductive FG cell design, other memory producers continued the path of scaling the "wrap-around" cell overcoming the challenges discussed above. The complexity of the "wrap-around" cell geometry and associated STI fringing field effects appear to be relatively more difficult to scale deeper into the 10 nm node compared to the planar approach. 3D vertical channel nonplanar designs are being actively investigated for the 10 nm node NVM FG and CT-flash designs. The current FG-NVM technology is entering an interesting path in the next decade to establish the cost/performance tradeoffs for the future NVM options.

15.7.1 Planar Charge-Trapping MLC/MNSC Flash Devices

Floating-gate devices demonstrated capability to store uniformly large amount of electronic charges within the floating gate medium enabling MLC capability as discussed above. Silicon gate provides large equipotential interface between the gate and silicon channel, thereby resulting in large memory state window at the control gate enabling subdivision of the window into multiple states for MLC operations. However, such memory window is strongly sensitive to floating-gate geometry and GCR and thereby on the feature size scaling and variations thereof. This is especially true for the "wrap-around" cell design employed thus far. Conductive planar floating-gate design as exemplified above in very recent years exhibiting scalability appears to be a viable option in sub-20 nm regime when combined with high K blocking layer of very low leakage and metal gate with high work function. Charge trapping devices, however, stores charges in discrete traps in nitride or in similar trapping dielectric medium of both electrons and holes either uniformly across such medium or locally and selectively at or near the source and/or drain diffusion edges due to high locally induced field. Additionally, there had been a lot of investigations in the past decade to apply not only high K blocking layers combined with high work-function metal gates for the stack designs of charge trapping devices, but also band-engineered designs of tunnel layers and trapping layers to enhance window retention and endurance. Additionally, the integration complexity of a charge-trapping device with base-line CMOS technology is simpler and less expensive to implement (with the exception of nanocrystal charge trapping types). Since most charge trapping devices operate primarily by either in the modified Fowler-Nordheim transport or by direct tunneling transport modes by design, these devices exhibit higher endurance compared to the floating-gate devices, higher SILC immunity, and lower parasitic effects. Charge trapping device design is not GCR constrained and have the advantage of simpler planar design. Therefore, it is generally assumed that charge trapping MLC devices could be more scalable, reliable, and extendable with enhanced device attributes when compared to the floating gate counterpart.

15.7.2 Uniform Charge Storage (CT-UNSC or CT-CMLC) and Local Charge Storage (CT-MNSC) Cells

The effect of uniform charge injection and trapping on device threshold is similar to that of the floating-gate devices in developing multilevel storage (MLC) capability. As such the FET-NVM devices and sensing scheme are similar to those of floating-gate devices. However, when charges are injected and stored locally in the gate dielectric medium, such storage affects the FET-NVM device turn-on characteristics. By storing the charges locally over the trapping dielectric at over each of the source and drain diffusion nodes and sensing the threshold effects by reversing the current flow direction, the two regions of localized charge storage could be distinguished providing dual storage capability within one device. This capability of localized storage can be distinguished from uniform storage type of MLC or MLS (multilevel store) devices of either of the FG and CT types. These types of devices could be differentiated from the conventional types and is proposed to be called as "multiple node storage devices" (MNSD) or "multiple node storage cell" (MNSC). The conventional devices could therefore be termed as "uniform node storage devices" (UNSD) or "uniform node storage cell (UNSC)." Therefore, CT NVMs could provide both conventional MLC and MNSC types of multiple bit per cell capability. We will first discuss planar CT-MLC or UNSC cells in Section 15.7. Planar MNSC cells will be discussed in Section 15.8.

Planar conventional uniform charge trapping devices such as conventional SONOS and MONOS types and silicon nanocrystal devices exhibit limited memory window and retention and therefore limited in MLC capability. Exceptions were demonstrated with multilayered oxynitride devices where all MONOS layers of oxide and nitride are replaced by appropriate compositions of oxynitrides for tunneling, trapping, and blocking layers enhancing window and retention as previously discussed in Part I. Vertical scaling of conventional SONOS/MONOS layers adversely affects both window and retention. Horizontal scaling of SONOS/MONOS devices affect short-channel effect as well as threshold instability due to "too few electrons" whereby loss of unit electron causes larger shift in Vth. These limitations have already been discussed in the context of SLC devices. However, as discussed in earlier sections in this part of this book, high K application and band engineering in stack design provide considerable potential of extendibility of device characteristics of MONOS types of devices enabling MLC capability. In the following section, we will explore potential of enhanced SONOS/MONOS devices for MLC applicability. We will also explore MNSC applicability of similar devices in Section 15.8.

15.8 PLANAR CHARGE-TRAPPING MLC DEVICES: ENHANCED SONOS/MONOS

Multiple concepts of simultaneously enhancing memory window and retention with planar CT devices enabling MLC capabilities have been introduced and illustrated in Chapter 13 previously (Sections 13.5.3 and 13.5.4). Several variations of high K stack designs using conventional poly-silicon gate as well as high work-function metal gates were discussed. In the context of MLC design, the following reference of earlier developments should be noted: Bhattacharyya [47,48,61] for large memory window, H-T. Lue et al. [60] and S. Kim et al. [45] for enhanced retention.

MLC capability in FG devices is established by subdividing the largest charge storage in the floating gate reflected in the memory window established between the erased state and the (highest) written state. In CT devices, MLC capability could be established not only through the above approach of enhancing the memory window but also due to the discrete nature of storage of charges in dielectric media through charge storage at different trap depths and at different tunneling distances (from silicon substrate or gate), thereby trapping charges at varying depths and distances through appropriate programming altering memory window and Vth. Through band engineering of both tunneling layers ("tunnel engineering") and trapping layers ("trap engineering"), multiple options of MLC capability could be established. Multiple stack design structures reflecting such approaches have been covered in the section on band engineering while discussing enhancements in window, retention, and other NVM characteristics. To derive appropriate device parameters for MLC capability, multilayered structures are preferred for tunneling and trapping instead of conventional approach of single layer of oxide for tunneling and single layer of nitride for trapping. This increases process complexity. Therefore, among the many options discussed in Chapter 15, we will illustrate only selected options for MLC design whereby multiple layers of dielectric films could be combined in single step reducing process complexity and processing cost, thereby improving integration possibility from practical point of view.

Several stack design options to provide MLC capability for planar CT NVMs are provided in the Table 15.2. Table 15.2 shows several examples of single trapping layer with tunnel-engineered structure. Other options (not shown) could also be considered using multilayered trap- engineered structure with the same tunnel-engineered structure options shown in the table to achieve larger memory window as discussed in Chapter 13. All options have: (a) ultrathin triple-layer VARIOT asymmetric tunnel layers with multiple direct tunneling modes for programming; (b) single high K blocking layer with low leakage and compatible high work-function gate to prevent gate injection, and (c) dielectric film selection such that multiple films could be processed in single step and limiting the number of unique films in the stack design to minimize process integration complexity and cost. In these cases of single-trapping layer options, the programming voltage level Vpp would be lower in the range of ± 9–11 V for the stack EOT in the regime of 11 to 13 nm for the pulse duration of 1 ms. In case of trap-engineered options, the stack EOT is higher in the regime of 15.5 to 16.5 nm and consequently the programming voltage level would be around ± 14 to 15 V for the same pulse duration of 1 ms. The objective for

TABLE 15.2

Stack Design Options for Planar CT-MLC Devices

	Single Trapping Layer Options				
Tunnel Layers (nm)	**Trapping (nm)**	**Blocking (nm)**	**Gate EOT (nm)**	**Vpp/Tpp**	**Window**
(a) [a]Ox (1.5)/Si (1.5)/HTO (1.5)	$Si_2ON_2(6.0)$	Al_2O_3 (12)	TaN/W: 11.8	±10 V/1 ms	>9 V
(b) #Ox(1.5)/SiON(1.5)/HTO (2.0)	$Si_2ON_2(6.0)$	Al_2O_3 (12)	TaN/W: 12.9	±11 V/1 ms	>9 V
(c) ^SiON(1.5)/Si_2ON_2(1.5)/SiON (2.0)	$Si_2ON_2(6.0)$	Al_2O_3 (12)	TaN/W: 11.9	±11 V/0.1 ms	>9 V

[a] RTB Barrier [13.55]; #Conv VARIOT Barrier [13.9]; ^Special VARIOT Barrier [13.10]

each option to yield large memory window enabling 3 bit/cell operability. Additional objectives include superior endurance and reliability compared to planar FG options.

15.8.1 Stack Design Options for Planar CT-MLC Devices

15.8.1.1 Single Trapping Layer Options

Three options are illustrated here with a single-layer trapping in oxynitride of composition Si_2ON_2 of 6 nm in thickness. The trapping characteristics of Si_2ON_2 is superior to that of nitride and has been discussed in Part I. Common to all options are trapping in deep single energy traps while reverse direct tunneling probability is zero and multiple layer processing in single step. In all three examples below, the differences are in the trapping layers and consequently in associated device characteristics with common elements being the trapping layer, the blocking layer of Al_2O_3, and TaN/W gate electrode. In all cases, the tunneling layers consist of three ultra-shallow layers, which could be produced by appropriate controlled single-step processing within an LPCVD reactor. This is given in Table 15.2.

Option (a) not only provides large enough memory window for MLC capability (2 or 3 bit/cell), but due to RTB barrier provides threshold dispersion for all memory states [45]. Option (b) provides asymmetric reverse tunnel barrier with longer tunnel distance, which enhances retention and stability of all "written-state" memory states. Option (c) is also a Variot barrier whereby oxygen-rich oxynitride replaces oxide with lower band gap and higher K, while higher K ultrathin Si_2ON_2 center layer enables multiple direct tunneling of electronic charges during writing. The lower barrier energy at the silicon interface of SiON compared to SiO_2 coupled with lower tunnel layer EOT enables enhancements in electron injection during writing improving writing speed for the MLC device. The band diagram for option (c) is shown in Figure 15.15.

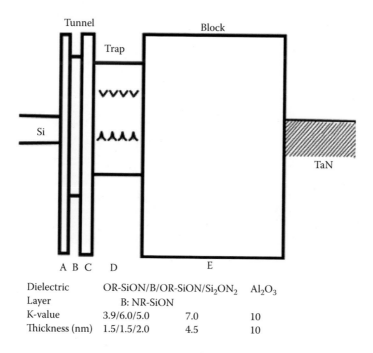

Dielectric	OR-SiON/B/OR-SiON/Si_2ON_2	Al_2O_3
Layer	B: NR-SiON	
K-value	3.9/6.0/5.0	7.0	10
Thickness (nm)	1.5/1.5/2.0	4.5	10

FIGURE 15.15 Band diagram: Single trapping stack MLC structure of EOT = 11.9 nm designed with Variot-tunneling option with oxygen-rich oxynitride replaces oxide for tunneling to achieve ± 11 V, 0.1 ms W/E providing a large initial memory window of >9 V.

15.8.1.2 Multiple Trapping Layer Options

Single trapping layer MLC stack design with tunnel engineering to provide asymmetric tunnel barrier (e.g., Variot barrier) can be replaced by multilayer "trap-engineered" structure discussed in Section 3.1.1 to provide additional window range and memory state stability as well as multiple memory states capability while keeping the attributes of asymmetric tunnel barrier. This could be achieved effectively by adding a triple-layer trapping section in the stack at the expense of additional process steps. Multiple trapping concepts involving dual-trapping dielectric layers (charge centroids separated by a higher bandgap dielectric barrier has been proposed by K.-H. Joo et al. [46] and also by A. Bhattacharyya [47]. This type of trap-engineered structure is analogous to crested barrier stack design from charge flow point of view (as adopted in tunnel-engineered structure). This approach reduces stored charge leakage at lower internal field during retention and has been elaborated in Chapter 13. Additionally, charge centroids are placed in different tunnel distances separated by higher energy barrier in between. Other concepts include placing nanocrystals in dielectric layer or layers at different distances from silicon/insulator interface modulating tunnel distance variations similar to above in trapping and/or placing nanocrystals of significantly different work functions to take advantage of trapping energy variations [48–50]. These later concepts are more complex from process integration point of view and have been previously discussed. Concepts of employing multiple trapping dielectric layer with reduced processing variables will be discussed here since these options will be relatively easier for CT-MLC designs in near future. The stack design options of single-layer trapping as illustrated by options (a), (b), and (c) above could simply be expanded for multilayered trapping case by replacing Si_2ON_2 trapping layer with nearly an equivalent EOT of triple layer such as $Si_2ON_2/Al_2O_3/Si_2ON_2$ whereby 3–4 nm of Al_2O_3 is sandwiched between 4 and 5 nm of Si_2ON_2 layers. Such an option (option d) will increase the stack EOT by 3–4 nm and therefore the Vpp level to up to $\pm$ 14 V, but would significantly improve window margins and MLC capability. The Vpp level could be lowered to nearly the same as single-layer trapping options by replacing Al_2O_3 blocking layer with thinner lower leakage higher K dielectric layers such as ~10 nm of HfSiON (K = 14) with EOT reduction by ~2 nm or by ~8 nm of $LaAlO_3$ (K = 25) with EOT reduction by ~3.5 nm. In the latter case, Vpp could be equivalent to the single trapping layer case. Option (d) characteristics and the corresponding band diagram (Figure 15.16) are illustrated below for the multiple trapping layer option of case (b). Note that by repeating the dielectric layers, process integration complexity is reduced.

Tunnel Layers (nm)	Trapping (nm)	Blocking	EOT	Vpp/Tpp	Window (2.0)
O × (1.5)/SiON (1.5)/HTO	$Si_2ON_2/Al_2O_3/Si_2ON_2$	Al_2O_3	~16 nm	14 V/1 ms	>>9 V

15.8.2 Advanced CT-MLC Stack Design Concepts with Multimode and Multilayer Trapping

Unlike FG-NVM devices, MLC capability in CT-NVM devices could be accomplished by more than one ways. Conventional SONOS follow-on devices with large memory window can be subdivided into multiple memory states similar to the FG-NVM devices. Since CT-NVM devices store charges in discrete traps, transport of charges and trapping/detrapping is dependent on trapping characteristics of the dielectric media, trap energy depth, and location of trap-charge centroids with reference to the silicon substrate and the gate. Therefore, by altering programming conditions with appropriate stack designs, charges can be stored either at (a) multiple charge centroids or at (b) multiple charge trapping energy depths. These two concepts are shown in Figure 15.17a and b, respectively. These characteristics can be utilized in establishing and sensing multiple memory states for MLC capability.

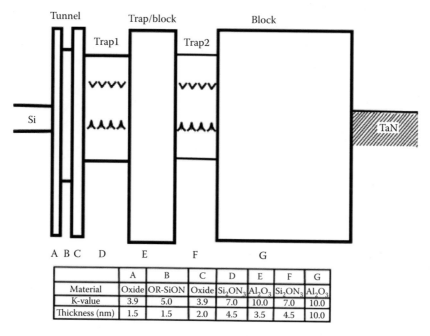

	A	B	C	D	E	F	G
Material	Oxide	OR-SiON	Oxide	Si$_2$ON$_3$	Al$_2$O$_3$	Si$_2$ON$_3$	Al$_2$O$_3$
K-value	3.9	5.0	3.9	7.0	10.0	7.0	10.0
Thickness (nm)	1.5	1.5	2.0	4.5	3.5	4.5	10.0

FIGURE 15.16 Band diagram for the CT-MLC triple trapping layer stack design.

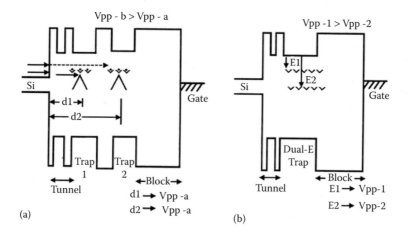

FIGURE 15.17 CT device concepts for MLC: (a) Physically separated charge centroid and (b) energetically separated charge-trapping centers.

Multiple charge centroid stack design can be readily implemented by the stack design approach discussed above in II B wherein two trapping layers, for example, Si$_2$ON$_2$ with associated charge centroids were physically separated by a higher band gap dielectric such as SiO$_2$ or Al$_2$O$_3$. Although trap depths are equivalent, the tunnel distances for the charge centroids will be different and therefore, trap filling during programming will occur at different Vpp for each case. This is illustrated schematically by means of the band diagram shown in Figure 15.17a. The band diagram is schematically similar to that of Figure 15.16. The two trapping layers Tr-a and Tr-b are separated by a higher bandgap dielectric (e.g., Al$_2$O$_3$), which also constitutes the thicker blocking layer. The distances from silicon interface for the charge centroids of Tr-a and Tr-b layers are shown respectively to be d2 and d2. During programming, Tr-a gets filled first at Vpp-a and Tr-b gets filled when Vpp-b is applied, such that Vpp-b > Vpp-a. When both charge-trapping layers are filled, the largest window is programmed and memory state (111) is obtained with the highest value of Vth. At the erased state, all charges are detrapped obtaining the lowest Vth state of (000). Six more memory states can be created by appropriate trap filling and detrapping: of Tr-a ([001] and [100]), trap filling and detrapping of Tr-b ([011], and [110]) and partial filling and emptying of both Tr-a and Tr-b ([101] and [010]). This is illustrated in Figure 15.17a.

Multiple trap energy depth design involves interlayering two trapping dielectric layers of significantly large difference (≥2 eV) in conduction band offset. As an example, the trapping layers could consist of ultrathin layers (~1 nm) of nitride and GaN successively deposited on top of each other to a total thickness of 6–8 nm (3–4 successive layers of each dielectric). Such layers could be deposited by various techniques such as sputtering, ALD, and MOCVD processes. Schematic band diagram for such a stack design is shown in Figure 15.17b. It should be noted that the conduction band offset for nitride and GaN are, respectively, 1.1 and 4.15 eV, with respect to SiO_2, a difference >3 eV. As illustrated in the band diagram below, the traps for the dielectric with lower band offset will be filled at relatively higher programming voltage Vpp1 (e.g., for nitride) during the writing process, only after the trap-filling process is completed for the higher band offset dielectric layers at Vpp2, such that Vpp1 >> Vpp2 (e.g., in the case of GaN). However, detrapping will require higher voltage for GaN during "erasing" process compared to nitride due to higher barrier energy and trap depth of GaN. Analogous to the earlier option of separated charge centroid stack design, MLC levels of different memory states could be established by selected trapping and detrapping in the two trapping layers with different band offset such as nitride and GaN. The interlayered nitride/GaN band diagram is illustrated in Figure 15.18 [61, Chapter 13].

Multiple trapping layer options as described above yield larger memory window enabling 3 bit/cell storage (8 memory states). This is discussed below.

15.8.3 Three-Bit/Cell Storage

Either of the above two concepts could provide 3-bit/cell MLS design (8 memory states) by means of appropriate programming for writing and erasing. This is further illustrated schematically in Figure 15.19 for the case in Figure 15.17b with two trapping energy levels of E1 and E2. The erased state (000) is achieved when all charges are detrapped. The highest written state (111) is achieved when both E1 and E2 are filled. By appropriate writing conditions, each of E1 or E2 is primarily filled generating two sub-memory written states (110) and (100). By detrapping these states, additional two sub-memory states (101) and (001) are created. The two additional memory states reflect partial trap filling of both E1 and E2 (011) and detrapping (010).

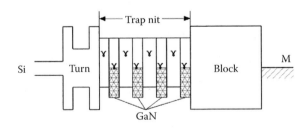

FIGURE 15.18 Interlayered band diagram with multilayered Si_3N_4/GaN trapping. (From Bhattacharyya, A., A novel stack design concept for charge trapping NVM cell with band engineered tunneling and tricking layers providing large memory window and 3 bit/cell MLC capability, ADI associates, internal memo, August 2007, unpublished.)

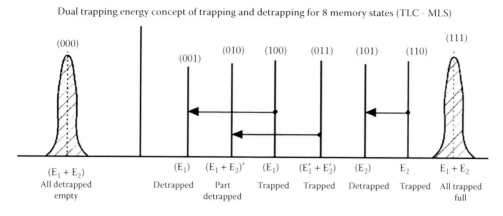

Dual trapping energy concept of trapping and detrapping for 8 memory states (TLC - MLS)

FIGURE 15.19 Sub-memory states for 3 bit/cell storage with dual-trapping energy stack design.

15.9 ADDITIONAL EXAMPLES OF CT-MLC MULTILAYER NVM STACK DESIGNS

15.9.1 DCSL Cell: 2 Bit/Cell MLC Technology

In Chapter 13 earlier, we have provided multiple band engineered planar CT-MLC stack design examples and referenced relevant publications. These include "BE-SONOS" and related device types "TANOS" and TANOS-FOLLOW-ON devices, and "double tunnel junction" and associated device designs (Chapter 13). These will not be further discussed. Several other examples of multilayer and multiplanar trapping involving embedded nanocrystals and a combination of trapping dielectric and nanocrystals have been also covered in Chapter 13. Only selected recent examples of CT-MLC planar charge-trapping device examples will be illustrated below.

G. Zhang and coworkers demonstrated a dual-charge storage layer (DCSL) stack structure containing ZrO_2 and nitride trapping layers fulfilling 2 bit/cell design requirements. Distinctive trapping levels uniquely associated with trap filling in ZrO_2 and Si_3N_4 provided Vth control for the sub-memory program states of (10) and (01) unlike conventional single-layer trapping MLC stack characteristics. The stack designs, the band characteristics multilevel endurance and retention are shown in Figures 15.20 through 15.23. The device characteristics demonstrated were superior when compared with other SONOS devices. The device structure, process flow, and associated band diagram for the DCSL is shown in Figure 15.20a–c respectively. Band diagram during programming, Figure 15.21a and b, and erasing, Figure 15.21c, are exemplified. The introduction of the ZrO_2 layer in the stack created two additional memory states: [10] and [01] as shown in Figure 15.21a and b

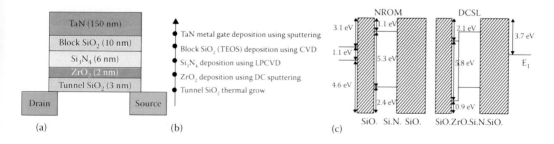

FIGURE 15.20 DCSL memory: (a) Device structure, (b) process flow, and (c) band diagram. (From Zhang, G., Hwang, W. S., Bobade, S. M. et al., Novel ZrO_2/Si_3N_4 dual-charge storage layer to form step-up potential wells for highly reliable multilevel cell applications, *IEDM*, pp. 83–87 © 2007 IEEE.)

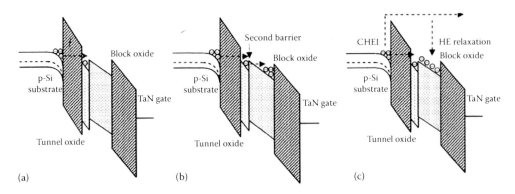

FIGURE 15.21 Energy band diagram of DCSL at programming conditions: (a) [10], (b) [01], (c) [00]. (From Zhang, G., Hwang, W. S., Bobade, S. M. et al., Novel ZrO$_2$/Si$_3$N$_4$ dual charge storage layer to form step-up potential wells for highly reliable multi-level cell applications, *IEDM*, pp. 83–87 © 2007 IEEE.)

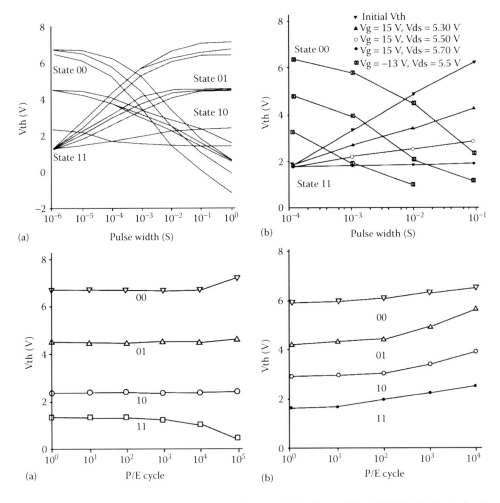

FIGURE 15.22 (a) DCSL memory: Programming characteristics for multi-level P/E (left); Single level conventional NROM (right), (b) DCSL memory: Endurance characteristics for multi-level P/E (left); Single level conventional NROM (right). (From Zhang, G., Hwang, W. S., Bobade, S. M. et al., Novel ZrO$_2$/Si$_3$N$_4$ dual charge storage layer to form step-up potential wells for highly reliable multilevel cell applications, *IEDM*, pp. 83–87 © 2007 IEEE.)

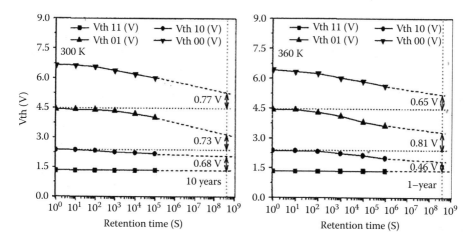

FIGURE 15.23 DCSL memory: Device retention property. (From Zhang, G., Hwang, W. S., Bobade, S. M. et al., Novel ZrO$_2$/Si$_3$N$_4$ dual charge storage layer to form step-up potential wells for highly reliable multilevel cell applications, *IEDM*, pp. 83–87 © 2007 IEEE.)

respectively. The muti-level programming characteristics of DCSL device is shown in Figure 15.21a demonstrating distinct separation of memory states due to the difference in trap-filling energetics between ZrO$_2$ and Si$_3$N$_4$ trapping layers. Endurance (Figure 15.22a for DCSL and b for SONOS) and retention (Figure 15.23a for DCSL and b for SONOS) demonstrated improved device characteristics of DCSL when compared to SONOS device [51].

15.9.2 DSM Cell: 4 Bit/Cell MLC Technology

Another version of planar charge-trapping MLC device was described by C.W. Oh et al. who employed both front-gate and back-gate storage scheme to yield 4-bit/cell device design [52]. Figure 15.24a shows the device cross-section while Figure 15.24b shows the I–V characteristics

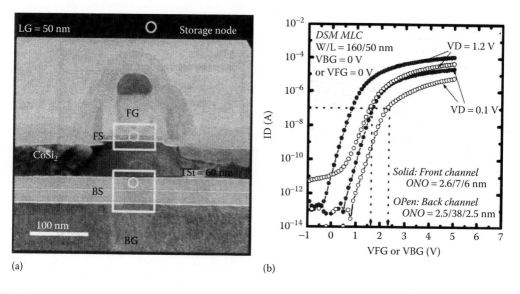

FIGURE 15.24 (a) SEM picture of 50 nm gate length for DSM CELL and (b) I–V characteristics for both front and back channels. (From Oh, C. W., Kim, N. Y., Kim, S. H. et al., 4-bit double SONOS memories (DSMs) using single level and multi-level cell schemes, *IEDM,* pp. 967–970 © 2006.)

of a double SONOS memories (DSM)-MLC front-gate and back-gate channel characteristics. It should be noted that the gate insulator stack used triple ultrathin direct tunnel dielectric layers: $O_1N_1O_2$ and N_2 trapping layer plus O_3 blocking layer similar to those used for BE-SONOS stack design [63]. The cell design provides four memory states at the frontside (FS) for 2 bits and an additional four memory states for the additional 2 bits at the backside (BS) with ONO layers.

Both FS and BS demonstrated similar program/erase characteristics, with Vth shift of ~4 V in each case. Minimal interference when either side is operating, and satisfactory retention at 150°C for 2 bit/cell memory levels (4 memory states) in each side were demonstrated. Programming was accomplished by channel hot electron injection while erase by hot hole injection. Reverse read was demonstrated in one direction [52].

15.10   PLANAR CHARGE-TRAPPING MNSC DEVICES AND EXTENDIBILITY: MNSC-NROM AND NAND

The concept of localized trapping at the diffusion edges of a SONOS FET-NVM device to achieve 2 bit/cell memory capability was first published by B. Eitan et al. [53]. Channel hot electron programming in either "source" or "drain" side and band-to-band hot hole injection erasing were employed to create diffusion-edge injections of charged carriers into localized regions of trapping into the trapping dielectric layer. This modulated the threshold at either ends of the device and sensed by reversing the direction of the current flow. Thus, the cells were sensed for 2 bit/cell NROM design by placing the cells into VGA architecture [54]. Conventional devices used thicker tunnel oxide SONOS stack with 5 to 7 nm thickness coupled with 4 to 6 nm of trapping nitride and 5 to 7 nm of blocking oxide to meet retention requirements of 10 years [55–57]. Conventional MNSC devices (both NANDs and NROMs) are attractive options to FG MLCs since the basic challenges of GCR and FG to FG coupling interference issues are absent in such devices. MNSC-NROM devices and products have already been designed to yield nearly 5 F^2/bit for SLC and 2.5 F^2/bit for DLC, and 1.25 F^2 for 4 bit/cell ("Quad-NROM") memory designs in 130 nm technology node (1 Gb Quad-NROM) by SAIFUN SEMICONDUCTORS Limited of Netanya, Israel. An excellent review on the device, technology, and product reliability had been provided by Eli Lusky and coworkers in the 2008 book publication by IEEE press edited by J. Brewer and M. Gill [76]. Reliability of the Quad-NROM product had been assessed by B. Eitan and coworkers [64] of SAIFAN demonstrating excellent product reliability with no single bit failures and featuring faster programming speed (300 MB/sec write speed), advanced memory state sensing, and ECC schemes [77].

Planar MNSC device and technology may be limited in scalability to approximately 50 nm feature size. This is due to (a) hot carrier-induced programming and erasing, (b) minimum separation requirements of localized trapping nodes (to reduce "second-bit" effect related to stress-temperature-time-induced redistribution of electrons and holes with characteristic variation of energy trapping levels of electrons and holes and mobility within the trapping dielectric medium), and (c) subthreshold leakage due to short-channel effects. However, such device concepts could be extended to sub-50 nm technology nodes using vertical channel configurations, potentially providing extremely high density Quad-NROM-NVMs and QUAD-NAND-NVMs with significantly enhanced performance and reliability compared to FG-NAND flash technology. Furthermore, the device/technology has the integration advantage over the conventional n-channel FG-device from base CMOS silicon gate or metal gate technology compatibility perspective. Investigations on P-channel MNSC-NAND device and technology provide further potential of extendibility of MNSC device concepts for NANDs and NROMs. This is further discussed below.

Investigations by S.H. Gu et al. suggests that the gate length for the above types of device could be scaled to 30 to 50 nm without adversely affecting device properties [59]. Device issues of conventional MNSC cells relates to high current requirement and leakage associated with hot electron programming at shorter gate lengths and Vth instability induced by interface states generation/hole trap generation during (BTBT) erasing. These issues have been successfully addressed by

H.T. Lue et al. [60] and by Y.H. Shih et al. [57]. Thicker tunnel oxide was replaced by ultrathin triple dielectric tunnel layers of oxide/nitride/HfO$_2$ [61] or by ONO [60] to achieve erasure by means of direct tunneling, thereby eliminating hot-hole injection-induced damages of BTBT erasure. Such approach improved device reliability significantly. Additionally, programming was accomplished at relatively modest current level due to band-to-band direct tunneling associated with lower energy channel electrons. Band-to-band direct tunneling reduced programming power requirements by reducing programming current as well as improving device retention, endurance, and reliability characteristics. Additionally, H.T. Lue et al. [60] demonstrated excellent immunity of the P-channel NAND array against program disturb and read disturb as compared to the conventional FG-flash NAND array. A triple-layer tunnel ONO structure was used by the above investigators consisting of ultrathin layer of 1.5 nm oxide/2.0 nm nitride/1.8 nm oxide VARIOT barrier stack design. The trapping nitride layer was of 7.0 nm of nitride and the blocking oxide was 8.0 nm with N+ poly gate. The following briefly outlines the programming and sensing schemes employed by H-T Lue and coworkers for their P-channel MNSC-NAND device and the resulting characteristics achieved of such device.

15.10.1 Erase

The Reset/Erase operations for the NAND array are shown in Figure 15.25a. The self-converging characteristics of erasing were displayed in Figures 15.25b and c for the NVM devices for −18 V

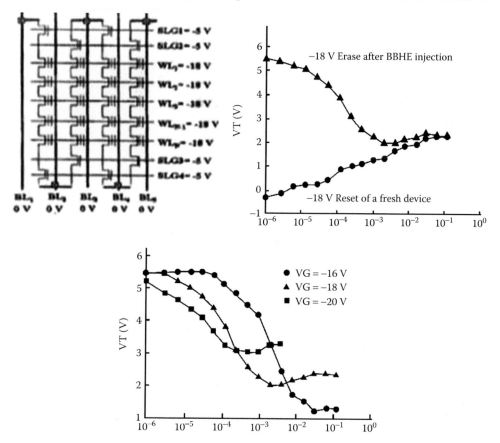

FIGURE. 15.25 Reset/Erase operation and characteristics for the P-channel NVM Device. (From Lue, H. T., Wang, S.-Y., Lai, E.-K. et al., A novel P-channel NAND-type flash memory with 2 bit/cell operation and high programming throughput >20 Mb/sec, *IEDM*, pp. 341–344 © 2005 IEEE.)

erase in 15.25b and as a function of different erase voltage levels (−16 V, −18 V, −20 V) in 15.25c after band-to-band hot electron (BBHE) injection.

15.10.2   Sensing (Reading)

The sensing method follows the reverse read method with suitably switching the Source-line-gate (SLG) functions in selecting Bit-1 and Bit-2 of the cell. Vpass is applied to all the unselected pass gates and Vread is applied to the bit line [60].

15.10.3   Programming (Writing)

As mentioned earlier, programming (writing) involves band-to-band direct tunneling through the triple-layer ONO dielectrics due to the ultra-shallow thickness of each layer. Even though this is conventionally classified as BBHE mode of tunneling, the energetics of electrons are lower due to multistep direct tunneling. For programming, a positive voltage was applied to the selected WL and a negative voltage to the BL, with pass gate and selected SLG activated at the same time. The programming current for such mode of transport was around 10 nA, which is considerably lower than typical FG-NAND device. Operationally, such low current can easily pass through a NAND string with significantly lower programming power. This allows page programming technique at significantly higher programming throughput compared to the current FG-NAND approach. As should be expected, programming current is sensitive to drain potential and increases nearly exponentially with enhanced (−ve) drain bias (enhanced hot electron injection).

Programming (written) Vth for the device (device length = 0.16 um) for the reverse bit (Bit-1) and forward bit (Bit-2) is dependent on Vread as should be expected. This dependency is shown in Figure 15.26a. A larger Vth window between Bit-1 and Bit-2 of the cell could be achieved by applying larger Vread. Figure 15.26b shows the Vth versus programming time characteristics using Vread = −2 V demonstrating discrimination of the memory states of >1 V between the two bit.

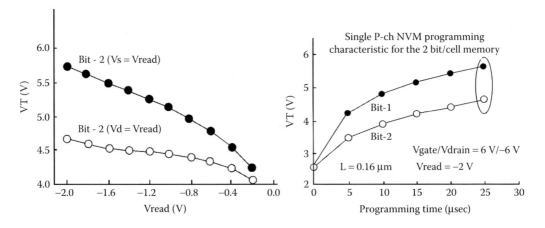

FIGURE 15.26   Programming characteristics of MNSC-SONONOS Device: (a) Vth versus Vread for Bit-1 and Bit-2, (b) Vth versus programming time for Bit-1 and Bit-2. (From Lue, H. T., Wang, S.-Y., Lai, E.-K. et al., A novel P-channel NAND-type flash memory with 2 bit/cell operation and high programming throughput >20 Mb/sec, *IEDM*, pp. 341–344 © 2005 IEEE.)

15.10.4 Disturbs

P-channel MNSC-SONONOS devices exhibit strong immunity toward read disturb and program disturb. Even with a large gate overdrive and Vds = −2 V, read disturb was not observed since channel hot hole is negligible in P-channel MNSC devices (not shown here) [60]. The work of Lue et al. demonstrated excellent program disturb immunity showing that Vth is stable even when the programming time of ~30 μs can be exceeded by 1000× to 30 ms.

15.10.5 Endurance and Retention

The endurance and retention characteristics of the MNSC-SONONOS 16-string NAND arrays are illustrated in Figure 15.27a and b, respectively. As explained earlier since hot hole injection for erasing was eliminated, excellent endurance was achieved and SILC was not observed. Excellent retention capability both at 25°C and 150°C was demonstrated as shown in Figure 15.27b. Due to the asymmetric nature of the ultrathin triple-layer VARIOT barrier, reverse direct tunneling was prevented as explained in Section 3.5.1 and retention characteristics after 1E5 W/E cycles were vastly improved.

15.11 ENHANCED CT-MNSC DEVICES

15.11.1 Band-Engineered MNSC Device

It should be noted that the MNSC-SONONOS device discussed above operated at Vpp in the range of 18–20 V. Following the stack design scheme outlined in Section 15.7.1 for CT-MLC devices, Bhattacharyya proposed in 2004 a modified MNSC-SONONOS type of device design with Vpp in the range of ± 9 V to ± 11 V with significantly enhanced device properties [61]. The design employed an ultrathin triple-layer VARIOT tunnel barrier consisting of OR-SiON (K = 5.5)/Si_2ON_2 (K = 7)/ HTO (K = 4.0) instead of conventional ONO for higher programming speed, yet superior retention with an over layer of Si_2ON_2 for trapping dielectric and Al_2O_3 (K = 10), or HfSiON (K = 14) as blocking dielectric. The gate could be either N+polysilicon or TaN/W metal gate as previously suggested. The stack EOT could either be ~11.5 or ~10 nm with corresponding Vpp ~± 10.5 V, 0.1 ms, or Vpp ~± 9 V, 0.1 ms, respectively, for Al_2O_3 or HfSiON blocking layers. Compared to the SONONOS device, this design reduces the EOT stack by > 50%, provides deeper level higher density trapping further enhancing window, retention, and endurance as well as enhanced performance at reduced

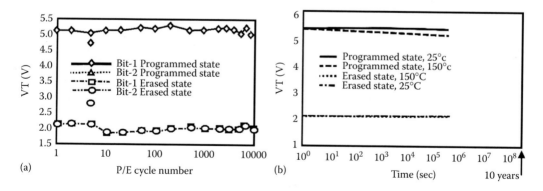

FIGURE 15.27 (a) Endurance and (b) retention characteristics of MNSC-SONONOS P-channel device exhibiting superior characteristics compared to conventional FG-NAND Devices. (From Lue, H. T., Wang, S.-Y., Lai, E.-K. et al., A novel P-channel NAND-type flash memory with 2 bit/cell operation and high programming throughput >20 Mb/sec, *IEDM*, pp. 341–344 © 2005 IEEE.)

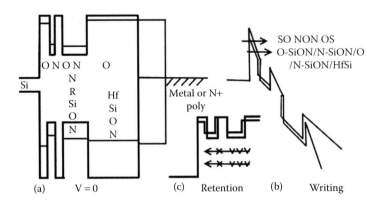

FIGURE 15.28 Band diagram comparison between enhanced MNSC device with high K blocking layer compared with SONONOS device. (a) Band diagram without external potential; (b) Band-bending when the "Write" potential is imposed; and (c) Band diagram during charge retention. (From Bhattacharyya, A., CT-flash memory cells for multi-bit per cell nonvolatile memory utilizing localized charge injection and trapping, ADI Internal Technical Memo, May, 2004.)

power. As previously mentioned, the stack design simplifies technology integration since all the layers consisting of tunneling and trapping could be deposited by a single-step LPCVD processing. The design could achieve voltage scaling by nearly ~ 0.5× compared to the device discussed by H.T. Lue et al. [60]. Band diagrams of the device proposed by Bhattacharyya [61] is compared with the SONONOS device in Figure 15.28a–c, respectively, for prestressed state, programming state, and retention state. The enhanced device is expected to provide significantly larger memory window (>~4 V) compared to the SONONOS design. Therefore, a 3 or a 4 bit/cell design could be feasible.

15.11.2 Dual-Gated Enhanced CT-MNSC Device: 2 Bit/Cell and 4 Bit/Cell Designs

A bulk version of backside CT-NVM device in dual-gated CT-NVM device concept consisting of 2-bit storage was published by R. Ranica and coworkers [63]. The NVM device was based on bulk "Silicon-on-nothing" (SON) processing technology developed by M. Jurczak et al. [62]. The top-gated planar PMOS-FET was a fixed Vth device, which is employed as an access gate, while the back-gated NVM was a 2-bit/cell SONOS-like MNSC device with localized trapping in nitride below the source and drain nodes. The proposed CT-NVM device structure was incorporated into the backside of the thin PFET silicon device and had an O/N/O gate insulator stack with the substrate acting as the NVM gate. A SiGe layer served as an isolation between the top PFET and the bottom NVM. The device is conceptually similar to an MNSC-NVM device disclosed in 2004 to be described following the above device [64].

The 2 bit/cell write and read operations are schematically shown in the Figure 15.29 [63]. The difference of Vth between the source side and drain side read programmed at drain side, is shown in Figure 15.29a and measured Id/Vg characteristics on a cell programmed at drain side (Drain potential = −4 V, Tpp = 1 ms) is shown in Figure 15.29b. By reversing the read voltage, a difference of 0.35 V was achieved for discriminating the memory state as shown. The proposed 2 bit/cell operation involved storing charges in nitride at both source side and drain side and sensing appropriately.

In 2004, bulk and SOI implementation of 2 bit/cell MNSC-NVM back-gated devices were proposed characterized by enhanced window, performance and other enhancements in device attributes [64]. Subsequently, similar concepts to 4 bit/cell device design were also proposed [65]. These will be discussed briefly below.

The 2 bit/cell device structures in bulk and SOI implementations are illustrated in Figure 15.30a and b, respectively, and the stack design elements for the bulk gate is shown in Figure 15.30c [64]. The device, in principle, operates similarly to that of the SONOS-MNSC device of Ranica et al. as

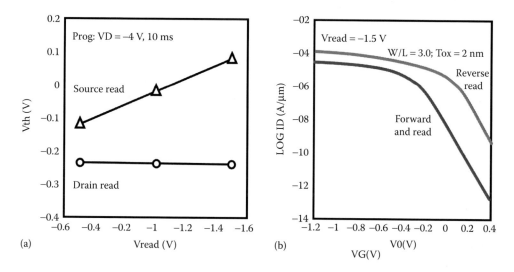

FIGURE 15.29 (a) Vth versus V-read characteristics for source side and drain side read of programmed device at Vd = −4 V, 10 ms programming, (b) Id–Vg Characteristics. (From Ranica, R., Villaret, A., Mazoyer, P. et al., A new 40 nm SONOS structure based on backside trapping for nanoscale memory, *IEEE T. Nanotechno.*, 4, pp. 581–587, © 2005 IEEE.)

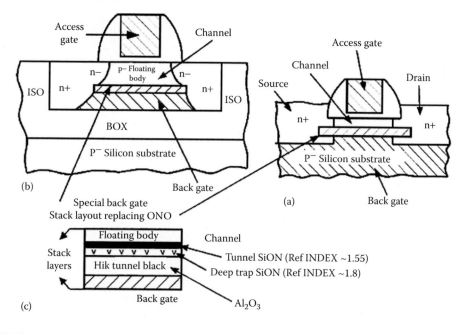

FIGURE 15.30 Enhanced backside MNSC 2 b/cell structure: (a) bulk, (b) SOI, and (c) stack elements. (From Bhattacharyya, A., Back-sided trapped non-volatile device, Micron Technology internal Publication, 2004–0927, July 12, 2004; U.S Patent Publication 257324, December 11, 2008, U.S. Patent: 7749848 [7/6/2010] and 7851827 [12/14/2010].)

described above. However, instead of a SONOS stack with nitride trapping layer, the device employs either a single-layer OR-SiON as a tunnel layer or a triple-layer PBO direct tunneling structure, for example, ultrathin layers of SiO_2/OR-SiON/NR-SiON coupled with deep trapping SiON and higher K blocking layer (e.g., Al_2O_3) to lower the stack EOT and the programming voltage levels, as well as to enhance the programming speed. The device was proposed for both NOR and NAND

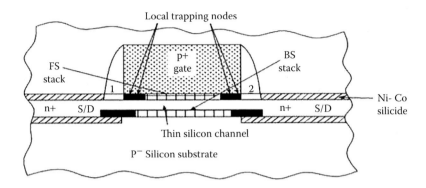

FIGURE 15.31 4 bit/cell-MNSC device schematic cross-section. (From Bhattacharyya, A., A novel 4 bit/cell dual channel MNSC device, Internal Memo, ADI Associates, September 2007, unpublished.)

architecture and for SLC (1 bit/cell) and DLC (2 bit/cell) applications. Similar to the SONOS-MNSC device, the top gate was used to read the memory states. Band representation of the device is essentially similar to that shown in Figure 15.15.

By implementing NVM stack structures both under the top gate and back gate with common thin p-body of silicon to form the channel for both top-gated and back-gated devices, a 4 bit/cell CT-MNSC device could be created. Such a device structure is schematically shown in Figure 15.31 [65]. Storage capacity could thus be doubled. Additionally, advanced band-engineered gate stack design could be incorporated for both the front gated and back-gated devices as discussed for the 2 bit/cell design. The concept has the potential to yield ~ 1F 2/bit NROM density with higher NVM performance and endurance displacing DRAM for many applications as well as < 1F 2/bit NAND cell density to low cost high volume storage applications with scalability beyond the 20 nm technology node.

15.12 FUTURE OF PLANAR MLC NVM DEVICES

15.12.1 Future of Planar Multilevel Storage NVM Technology, Device, and Products

The explosive growth of silicon based NVM during the past decade has been primarily due to the success of planar MLC FG-NAND product scalability from 120 nm node to 22 nm node (> 5×) resulting in the delivery of NAND bytes from 3E17 bytes in 2002 to 3E19 bytes in 2012! This implies a 100-fold increase in application with a product CGR of 70% per year. During the same period of time NAND price per bit dropped 1/50th of that at 2002 while the DRAM price per bit dropped 1/10th. Such outstanding achievement has been associated with the NAND DLC (2 bit/cell) and TLC (3 bit/cell) product offerings. Now that the 20 nm product scalability (64 and 128 Gb/chip) has been successfully demonstrated through high volume production and 15 nm scalability challenges have been addressed, it would be reasonable to assume that the growth of such products will continue at equal or greater CGR up to 2020. The key requirements would be continued expansion in applications through productivity and architectural advancements, cost/price/power reduction, and performance/reliability enhancements not only at the device/chip/package level but at the subsystem/system level.

Historically, DLS NAND product was offered in 2001 and TLS product was introduced in 2009. TLC-NAND accounted for the major expansion of NVM bits and cost/bit reduction of NVM products. To keep the momentum going, one should anticipate a 4 bit/cell design/product introduction in near future. However, MLC capability with increasing bit/cell device design creates multiple challenges beyond the realm of device design, technology integration, and device characteristics. MLC NAND designs with increasing number of memory states (e.g., 8 memory states for TLC-NAND) with shrinking sense margins and increasing Vth dispersions for each memory states, creates unique challenges in memory chip design and memory subsystems. Recently a disturb-free programming scheme was proposed by R. Shirota et al. based on recombining electrons in the surface states with accumulated holes before programming thereby dramatically suppressing program disturb [75]. Future NAND devices would be expected to incorporate such scheme. Increasing bit/cell design adversely impacts endurance, retention, raw bit error rates (RBER), write speed, data input/output throughput (IOP) thereby adversely impacting memory and system performance and reliability, thereby limiting application expansion. Challenges associated with peripheral/sensing circuit design, storage controller multitasking design, ECC capability, throughput enhancement design and architecture, and effective packaging (stackable package such as TSV) are also important challenges toward meeting product/application goals for multilevel designs. Detailed discussions on these subjects are beyond the scope of this book. Some of these will be briefly discussed in the last section of this part of this book. In this section, we will focus on the planar device perspective for 10 nm node and beyond by examining factors that might influence planar cell design viability for 4 bit/cell design from the challenges faced in technology, device scalability, and product development established thus far down to the 15 nm node.

From technology perspective, both NROM in embedded environment and NAND flash in standalone environment had been successful in application and market expansion during the last decade. This is expected to continue through the next decade within the framework of 10 nm node. NROM will continue to be "Advanced CT-based," driven by higher performance and endurance requirements and ease of integration into base-line CMOS platform technology and is at least expected to partially replace DRAM requirements providing power reduction and superior cost/performance trade-offs for a diverse set of portable applications. The form-factor could be SOC or SOP depending on the application requirements. The stand-alone NAND flash technology could be a scalable extension of the 15 nm technology being manufactured by Micron Technology [15], to implement an advanced form of planar "conductive floating plate" (CFG/CFP) cell as discussed earlier or a planar "fusion FP/CT" (FPT) cell conceptually equivalent to that proposed by H.T Lue et al. [66]. The fusion FP/CT cell may consist of a conductive floating plate combined with a trap-engineered multilayered structure or simply a thin layer of nitride or NR-SiON whereby injected charges are mostly stored in the trapping over layer providing enhanced retention while the ultrathin conductive FP layer provide equipotential medium with unlimited electron density of states. Such multi-layer scheme of storage medium extends the capacity of charge storage in planar flash memory design confirmed by the investigations of P. Blomme et al. [28] and S. Jayanti et al. [29]. The tunnel layer/layers could be band-engineered to enhance charge injection and to minimize back tunneling as explained in Section 2.1, while the blocking layer could be of optimized high K to lower charge loss as discussed by Ramaswamy et al. [25]. The 4 bit/cell NAND flash technology will be based on effective cost/bit taken into consideration all the overhead associated with the peripheral circuits, ECC requirements, and throughput capability required for application expansion and cost reduction as mentioned above. It is also conceivable that nonplanar options such as FinFET-Flash or Full 3D Flash (see Section 6.3.1.2) may fulfill the market expansion requirements more effectively compared to the planar options if the overall cost/bit could be reduced for nonplanar options.

From NVM device scalability perspective, we have previously discussed four NVM device options discussed in Part I and Part II of this book and their evolutions. These are

1. FG/FP options based on equipotential storage medium and single thicker tunnel dielectric ($\geq$5 nm EOT) with high K IPD (for ONO replacement) and metal gate (for silicon-gate replacement)
2. The CT-band-engineered multidielectric stack design with discrete trapping planes combined with high K low leakage blocking dielectric and high work-function metal gate (for SONOS/MONOS replacement)
3. The CT-single or multilayered embedded nanocrystal trapping planes with other attributes similar to option 2 (NC device)

4. CT-multilayered direct tunnel memories with or without PBO and trapping/blocking layer designs optimized for long retention and low leakage (DTM follow-on)

It should be noted that as the device design evolved to address scalability and charge-loss issues, many of the distinctions of early stack design features got blurred. For example, the 5 nm thick "conductive FG cell" (Figures 15.9 and 15.11), [15] applicable to 15 nm node cell design consisted of low leakage high K blocking dielectric and TaN metal gate similar in principle to that of CT-TANOS cell discussed in the context of SONOS follow-on CT devices. The above "FG" cell could aptly be termed as "conductive floating plate" cell since the classical polysilicon floating gate height and sidewall sensitivity issues of "wrap-around" cell were eliminated by the design and dimension of the storage medium. Since all the four stack design could, in principle, adopt tunneling and blocking features to enhance charge transport and retention at storage node and minimize stored charge loss to either gate or substrate, the key distinction should therefore be based on (a) storage medium, that is, equipotential storage versus discrete storage and (b) charge transport mechanism and barrier energetics, that is, electron only versus electron and hole transport and low energy transport mode (e.g., direct tunneling) versus higher energy transport mode (e.g., F–N tunneling or hot carrier transport). We will, therefore, re-classify the planar scalable NVM devices into the following categories and combinations applicable to sub-15 nm node and beyond. This is shown in Table 15.3.

In all above possible options, appropriate MLC/MNSC memory window requirements define the required number of subdivided memory states and associated stability, which in turn would determine the options for tunneling layer or layers and the selection of blocking/gate layer or layers. The energetics of programming mode or modes would determine the options for endurance and reliability limits to match the application requirements. Let us now explore the options for 10 nm node 4 bit/cell NROM devices and stand-alone NAND flash devices. The primary objective would be to enhance the cost/performance objectives for embedded NROM for the high end (enterprise-like applications). For the NAND flash, the primary objective would be continued lowering of effective cost/bit (with overhead) not only for portable applications for the low end but also for large database storage systems to expand into the arena of cost/throughput sensitive HDD storage.

For embedded NROM applications, the device option could be relatively simpler. For 4 bit/cell NROM design, either the discrete trap B-L or B-Hy, alternately C-L or C-Hy could be the preferred options in the form of 4-bit enhanced CT-MLC DSM cell type [52] for the B-type and alternately, enhanced 4-bit MNSC cell type [64,65] for the C-type. Larger memory window achievable in the C-type could be of distinct advantage for higher speed reading and reducing memory latency enabling replacement of DRAM bits for many applications and providing cost and power advantages. The specific selection would depend on ease of integration into platform CMOS technology and application requirements. It should be noted that the C-type device consisting of the fusion FG/CT storage elements has the additional advantage of improved retention and scalability to the "few-electron regime" since the number of electrons participating in programming could be as low

---

TABLE 15.3
Reclassification of Planar Scalable NVM Devices

Storage Medium	Storage Characteristics	Energetics
Floating plate: A (L, M,H, Hy)	S ingle polar (electron only)	L: Low energy (DTM-based)
Floating plate: A' (L, M,H, Hy)	Bipolar (electron + Hole)	M: Medium energy (F.N.-based) single polar (electron)
Discrete trap: B (L, M,H, Hy)	(only)	H: High energy (Hot E/H-based)
Discrete trap: B' (L, M,H, Hy)	Bipolar (electron + Hole)	HY: Hybrid energy mode-based
Fusion plate/trap: C (L, M,H, Hy)	Single polar (electron only)	
Fusion plate/trap: C' (L, M,H, Hy)	Bipolar (electron + Hole)	

as 16 electrons [24] at 10 nm node for the requirement of 1 V of memory window! The other advantage would be the "edge effect" (STI fringing field effect) immunity due to the conductive FP [66].

For NAND flash applications, the key challenges would be to overcome (a) "few electron effect," (b) "the fringing field effect," (c) the Vth dispersion effects of memory states, (d) the endurance limitations, (e) the latency reduction, and (f) RBER reduction. All above items are interrelated, limits application expansion and adds cost/bit overhead. The device type: Floating Plate A-M currently adopted by Micron technology [15] shows promise and could be further extended with band-engineered multilayer tunneling structure to improve issues related to items (c), (d), (e), and (f) enabling 4 bit/cell design at 10 nm node. Alternately, the Fusion Plate C-M device type with advanced tunneling and blocking layers incorporating 4-bit MNSC cell type in NAND architecture could be considered for the 10 nm node.

15.12.2 Multiplanar Stackable NAND Devices and Technology

In order to improve memory bit density in a single piece of silicon-chip, Samsung Electronics in recent years proposed several stackable schemes of integrating planar devices. Conventional planar MLC FG-NAND device arrays were integrated in a 3D "double-stacked" configuration while sharing the bit-line structure [67]. This effectively doubled the bit density within a single chip. Another such scheme at device level was based on back-gate NVM devices called DSM developed by C.W. Oh et al. [52] and has been discussed earlier. A third such scheme was also proposed earlier by S.M. Jung et al. of Samsung by integrating multiple planar NAND arrays on top of each other using stackable single-crystal silicon layers on insulating films containing TANOS-NAND devices and arrays [68]. The Stacking scheme is illustrated in Figure 15.32a while vertical SEM pictures of the stacked NAND arrays are shown in Figure 15.32b. From the definition adopted in this book, such scheme would still be considered as planar devices since silicon channels are in X–Y planes as compared to the 3D devices where silicon channels could be formed in X–Y–Z planes.

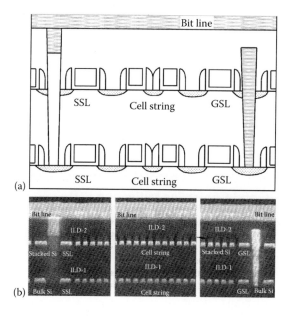

FIGURE 15.32 Integrated Multi-planar NAND arrays: (a) stacking scheme of TANOS-NAND string in multiple planes and (b) SEM photograph of stacked planar NAND cell strings separated by a thick dielectric layer. (From Jung, S.-M., Jang, J., Cho, W. et al., Three-dimensionally stacked NAND flash memory technology using stacking single crystal Si layers on ILD and TANOS structure for beyond 30 nm node, *IEDM*, pp. 37–40 © 2006 IEEE.)

Recently, Samsung has announced a 128 Gb FG-NAND product which employs CT-TANOS vertical channel devices in their "3D VNAND" technology. 3D devices are discussed in the next section. Macronix International also proposed a double-layer stackable TFT NAND flash array using trigated BE-SONOS device on polysilicon over oxide with hybrid (horizontal/vertical) channel [70].

15.13 ADDRESSING CURRENT PLANAR MLC FG-NAND FLASH/SSD LIMITATIONS AND SCALABILITY ISSUES

We have thus far addressed planar MLC NV devices of both FG and CT types primarily from device/stack design and technology integration perspectives and noted how the challenges of device properties due to scaling have been met up to the 20 nm technology node for current NAND products at the chip level. We have noted in Sections 15.4 through 15.6 that as the MLC FG-NAND flash was scaled, NVM parameters such as retention, endurance, and sense margins got degraded due to FEE and other parasitic effects. It is significant to note that MLC NAND chips integrated with the memory controller (memory management unit or MMU) constituted the solid-state drive or SSD. SSD had been the primary driver of NVM bits and flash technology during the last two decades.

In Sections 15.3 through 15.5, we have discussed some of the extendibility issues of MLC FG-NAND flash devices at the cell and array levels adversely impacting NVM parameters such as endurance, data retention, disturb, and affecting data integrity and reliability. In this section, we will discuss the scalability challenges and issues addressed from the SSD and application perspective in order to recognize the limitations and challenges that planar MLC-FGNAND FLASH need to address further down the road. This will lead us to discuss the rationale to address nonplanar and 3D devices in Chapter 16 and application requirements in Chapter 17 to fulfill market requirements of silicon based NVMs in future decades.

In MLC FG-NAND flash technology, both device parameters such as retention and endurance and functional parameters such as sense margin, data integrity, uncorrectable bit error rate (UBER) or RBER, read and write latency, and associated reliability are highly coupled. Extendibility limitations are reflected in creating reliability issues associated with data integrity, displacing and distorting Vth distributions of MLC memory states and activating undesirable mechanisms such as trapping and detrapping of charges, SILC and related disturbs. Consequently, to address data integrity and reliability, technology dependent functions such as ECC and technology independent functions such as block management and drivers as well as advanced programming/reading algorithms (e.g., "read retry"), implementable in software/firmware had to be expanded. Additionally, ECC free novel solutions need to be created. However, such approaches increase overhead expenses and reduce performance. The coupling of device properties impacting MLC functionality and extendibility and mitigation approaches are briefly discussed below.

15.13.1 MLC Endurance (W/E Cycling)-Related Issues on Data Integrity: ECC and Controller Mitigation [71]

It has been discussed earlier that as the feature size of the floating gate is reduced, the number of electrons/cell decreases. Typically, as the technology node is reduced from 100 to 20 nm, the number of available electron for a Vth shift of 100 mv is reduced from nearly 50 to 10. Therefore, loss or gain of one electron at 20 nm node results in a 10% shift in Vth! This severely impacts both retention and endurance for the FG-MLC device. Consequently, functional data integrity is adversely impacted in the form of RBER and UBER.

As feature size is scaled, RBER increases with P/E cycles. ECC has been increasingly used to improve data retention and endurance as the technology node is reduced [72,73] and to restore the desired UBER level. P/E endurance reduction for SLC and MLC (DLC) is compared against feature size in Figure 15.33. Requirement to increase ECC to achieve equivalent UBERs is also shown in Figure 15.33 demonstrating exponential rise in ECC requirements for MLC functionality and data integrity.

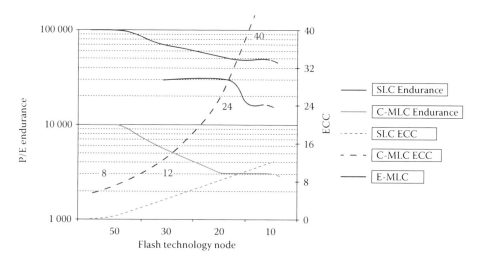

FIGURE 15.33 P/E Endurance degradation comparison of SLC and DLC NAND with feature size. Note exponential increase in ECC requirement for DLC to achieve equivalent UBERs. ECC Management of ECC and "ECC Free Solution." (From Yoon, H., and Tressler, G. A., Advanced flash technology status, scaling trends and implications to enterprise SSD technology enablement, *Flash Memory Summit*, IBM Corporation, Santa Clara, CA, 2012.) [72].

Traditional solution of ECC management had been implemented in hardware within the processor, which also performed block management and driver functions. While ECC is NAND flash technology dependent and therefore cannot be updated in software/firmware, block management and driver functions are updatable since these functions are technology independent.

Traditional solutions were updated to address ever-increasing ECC requirements to an "ECC Free Solution" by rearranging the above functions for SSD into (a) technology dependent memory plus controller units with closely coupled ECC to the NAND technology and dedicated bulk management architecture with managed interface with the processor; and (b) technology independent design of processor and software updatable driver interfacing the host. This approach has been effective in handling increased ECC requirements and is known as "Fully Managed Solution." Schematics for Traditional and Fully Managed Solution are shown in Figure 15.34 [74].

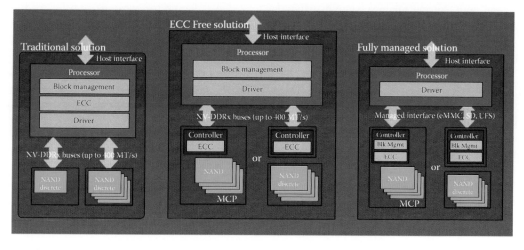

FIGURE 15.34 Schematics of traditional and fully managed ECC solutions for SDD. (From Abraham, M., How to handle increasing ECC?, Micron Technology, *Flash Memory Summit*, Santa Clara, CA, 2012.)

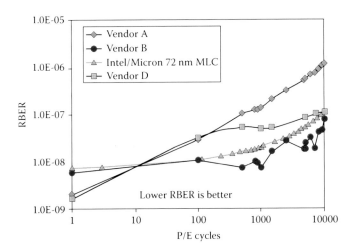

FIGURE 15.35 SCALED MLC NAND RBER impact due to P/E cycling. (From Fricky, R., Data integrity on 20 nm SSDs, *Flash Memory Summit*, Santa Clara, CA, 2012; Park, K.-T. et al., A 45 nm 4 Gb 3-Dimentional double stacked multi-level NAND flash memory with shared bitline structure, *ISSCC*, pp. 510–632, 2008.)

15.13.2 Endurance and Overprovisioning [68,71]

Additionally, the adverse impact of W/E cycling is minimized in the controller by increasing over-provisioning of memory bits within the array so that certain memory bits are not overstressed and also reducing write amplification by trading off cost and performance. It should be noted that with smaller feature size, memory organization trend involved increasing page sizes containing more planes and spare bytes per sector (which increases sequential throughput) and more pages per block (which reduces the chip size).

15.13.3 Other MLC NAND Data Integrity Issues

There could multiple other issues impacting data integrity when MLC NAND is scaled. Some of these are (a) page mapping related BL-BL, WL-WL, diagonal cell-to-cell interferences, and (b) P/E cycling both causing memory state distributions to widen reducing the sense margin for reading the memory state Figure 15.35 [68]. Advanced signal processing algorithms are introduced in the memory controller to optimize programming algorithm to minimize widening of Vth distributions and simultaneously incorporate advanced "read retry" methodology of reading first the neighboring cells to determine their state and subsequently reading the desired cell multiple times with a high and low reference scheme. Such schemes mitigate the effect of cell-to-cell interference at the expense of increased read latency. Cell-to-cell interference is schematically shown in Figure 15.36a and increased interference sensitivity to technology node scaling is depicted in Figure 15.36b. An additional data refresh scheme was also proposed to alleviate concerns about power-on data retention. The scheme known as background data refresh or BDR ensures that static data are refreshed regularly [68].

Advanced system level signal processing algorithm and strong ECC capability will be critical requirements for reliable MLC design for sub-20 nm FG flash technology with compromised P/E capability. Recently Shirota has proposed a new "disturb free" programming scheme for 20 nm node NAND flash memory [75].

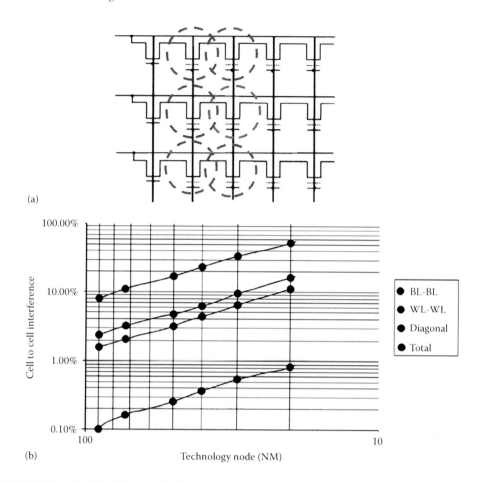

(a)

(b)

FIGURE 15.36 MLC NAND: (a1/a2): Schematic representation of various cell–cell interferences, (b) cell–cell interference dependency on technology node. (From Fricky, R., Data integrity on 20 nm SSDs, *Flash Memory Summit*, Santa Clara, CA, 2012.)

REFERENCES

1. M. Bauer, R. Alexis, G. Atwood et al., A multilevel-cell 32 Mb flash memory, *Cell/ISSCC*, 1995 pp. 132–133,
2. G. J. Hemink, T. Tanaka, T. Endoh et al. Fast and accurate programming method for multi-plevel EEPROMs, *VLSI Technology*, 1995, pp. 129–130.
3. T. Jung, Y.-J. Choi, K.-D. Suh et al., A 3.3 V, 128 Mb Multi-level NAND flash memory for mass storage applications, *ISSCC*, San Francisco CA, 1996, pp. 32–33.
4. S. Aritome, S. Satoh, T. Maruyama et al., A 0.67 u² self-aligned shallow trench isolation cell (SA-STI cell) for 3 V-only 256 Mbit NAND EEPROMs, *IEDM*, 1994, pp. 61–64.
5. K. Shimizu, K. Narita, H. Watanabe et al., A novel high-density 5F² NAND STI cell technology suitable for 256 Mbit and 1 Gbit flash memories, *IEDM*, 1997, pp. 271–274.
6. K. Imamiya, Y. Sugiura, H. Nakamura et al., A 130 mm², 256-Mbit NAND flash with shallow trench isolation technology, *ISSCC*, 1999, pp. 1536–1543.
7. S. Aritome, Advanced flash memory technology and trends for file storage application, *IEDM*, San Francisco CA, 2000, pp. 163–166.

8. J.-H. Park, S.-H. Hur, J.-H. Lee et al., 8 Gb MLC (Multi-level cell) NAND flash memory using 63 nm process technology, *IEDM*, San Francisco CA, 2004, pp. 873–876.

9. K. Kim, and J. Choi, Future outlook of NAND flash technology for 40 nm node and beyond, *21st IEEE NVSMW*, Monterey CA, 2006, pp. 9–11.

10. Y. Li, 3 bit per cell NAND flash memory on 19 nm technology, Sandisk, *Flash Memory Summit*, Santa Clara, CA, 2012.

11. N. Shibata, 19 nm 112.8 mm^2 64 Gb Multi-level flash memory with 400 Mb/s/pin 1.8 V toggle mode interface, *Flash Memory Summit*, Santa Clara, CA, 2012.

12. K. W. Lee, S. K. Choi, S. J. Chung et al., A highly manufacturable integration technology of 30 nm generation 64 Gb multi-levl NAND flash memory, *VLSI Tech*, 2011, pp. 70–71

13. D. Kang, S. Jang, K. Lee et al., Improving the cell characteristics using low-k gate spacer in 1 Gb NAND Flash Memory, *IEDM*, 2006, pp. 1001–1004.

14. Y. G. Kim, S.-H. Lee, D.-H. Kim et al., Sense margin analysis of MLC flash memories using a novel unified statitical model, *7th International Symposium On Quality Electronic Design*, 2006 p. 189.

15. I. Noguchi, T. Yaegashi, H. Koyama et al., A high performance multi-level NAND flash memory with 43 nm node floating gate technology, *IEDM*, Washington DC, 2007, pp. 445–448.

16. H. Nitta, T. Kamigaichi, F. Arai et al., Three bit per cell floating gate NAND flash memory for 30 nm and beyond, *IEEE 47 Annual International Reliability Physics Symposium*, Montreal QC, 2009, p. 307.

17. A. Bhattacharyya, M. L. Karbaugh, R. M. Quinn, J. A. Robinson., Formation of variable width sidewall structures, IBMCorpn, U.S.Patent 4776922, October 11, 1988.

18. A. Bhattacharyya et al., A half-micron manufacturable high performance technology applicable for multiple power supply applications, Proceedings, Intl. VLSI, Technology, System and Applications, 1989, p. 321.

19. H. Shettler, W. Hang, K. J. Getzlaff, C. W. Starke, and A. Bhattacharyya, A. CMOS mainframe processor with 0.5 um channel length, *IEEE SSC*, 25, 1166–1177, 1990.

20. K. Prall et al., 25 nm 64 Gb NAND technology and scaling challenges, *IEDM*, 2010, pp. 102–105.

21. K. W. Lee, S. K. Choi, S. J. Chung et al., A highly manufacturable integration technology of 20 nm generation 64 Gb multi-level NAND flash memory, *VLSI Technology Symposium*, pp. 70–71, 2011.

22. C.-H. Lee, S.-K. Sung, D. Jang et al., A highly manufacturable integration technology for 27 nm 2 and 3 bit/cell NAND flash memory, *IEDM*, San Francisco CA, 2010, pp. 5.1.1–5.1.4 (Samsung).

23. J. Hwang, J. Seo, Y. Lee et al., A middle-1X nm NAND flash memory cell (MIX-NAND) with highly manufacturable integration technology, *IEDM*, 2011, pp. 199–202. (Hynix).

24. H.-T. Lue, Y.-H. Hsiao, K.-Y. Hsieh et al., Scaling feasibility study of planar thin floating gate (FG) NAND flash devices and size effect challenges beyond 20 nm, *IEDM*, Washington DC, 2011, pp. 203–206.

25. N. Ramaswamy, T. Graettinger, G. Puzzilli et al., Engineering a planar NAND cell scalable to 20 nm and beyond, *IEDM*, Monterey CA, 2013.

26. A. Bhattacharyya, and G. Darderian, An enhanced floating gate flash device and process, U S Patent Application Publication No. 0067256 A1, March 12, 2009, ADI internal disclosure, MIDD66, 2006.

27. S. Raghunathan, T. Krishnamohan, K. Parat et al., Investigation of ballistic current in scaled floating-gate NAND FLASH and a solution, *IEDM*, 2009, pp. 819–822.

28. P. Blomme, M. Rosmeulen, A. Cacciato et al., Novel dual layer floating gate structures as enabler of fully planar flash memory, *VLSI*, Honolulu HI, 2010, pp. 129–130.

29. S. Jayanti, X. Yang, R. Sun, and V. Misra, Ultimate scalability of TaN metal floating gate with incorporation of high-K blocking dielectrics for flash memory applications, *IEDM*, San Francisco CA, 2010, pp. 5.3.1–5.3.4.

30. N. D. Spigna, D. Schinke, and S. Jayanti et al., A novel double floating-gate unified memory device, *VLSI and Sustem-On-Chip*, 2012, pp. 53–58.

31. A. Bhattacharyya, Proposed novel STI and gate stack design for FG-NAND flash NVMs for 20 nm node and beyond, A. Bhattacharyya, ADI internal disclosure, December, 2012.

32. A. Bhattacharyya, Trap-engineered stack design for both FG/FP NAND flash as well as CT NAND Flash NVMs compatible with proposed novel STI for sub-20 nm node implementation, A. Bhattacharyya, ADI internal disclosure, December, 2012.

33. Intel Press Release, Intel, micron extend NAND flash technology leadership with introduction of world's first 128 Gb NAND device and mass production of 64 Gb 20 nm NAND, Intel Newsroom publication, December 6, 2011.

34. D. McGrath, Micron sampling 16 nm NAND, EE Times, July 16, 2013.

35. T. Kamigaichi, F. Arai, H. Nitsuta et al., Floating gate super multi level NAND flash memory technology for 30 nm and beyond, *IEDM,* San Francisco CA, 2008, pp. 1–4.

36. K. Fukuda, Y. Watanabe, E. Makino et al., A 151 mm 2 64 Gb MLC NAND flash memory in 24 nm CMOS technology, *ISSCC,* 2011, pp. 198–199.

37. N. Shibata, 19 nm 112.8 mm 2 64 Gb Multi-level flash memory with 400 Mb/s/pin 1.8 V toggle mode interface, *Flash Memory Summit* 2012, Santa Clara, CA. (Toshiba).

38. Y. Li, 3 bit per cell NAND flash memory on 19 nm technology, *Flash Memory Summit* 2012, Santa Clara, CA. (SanDisk).

39. C.-H. Lee, S.-K. Sung, D. Jang et al., A highly manufacturable integration technology for 27 nm 2 and 3 bit/cell NAND flash memory, *IEDM,* 2010, pp. 98–101.(Samsung).

40. C. Kim, J. Ryu, T. Lee et al., A 21 nm high performance 64 Gb MLC NAND flash memory with 400 MB/s asynchoronous toggle DDR interface, *IEEE J. of Solid. St Ckts,* 47(4), pp. 981–989, 2012. (Samsung).

41. K.-T. Park, O. Kwon, S. Yoon et al., A 7 MB/s 64 Gb 3-Bit/Cell DDR NAND flash memory in 20 nm-node technology, *IEEE ISSCC,* San Francisco CA, 2011, pp.212–213. (Samsung).

42. D. Lee, I. J. Chang, S.-Y. Yoon et al., A 64 Gb 533 Mb/s DDR interface MLC NAND flash in Sub-20 nm technology, *IEEE ISSCC,* San Francisco CA, 2012, pp. 430–431.(Samsung).

43. H. Huh, C. W. Jeon, C. W. Yang et al., A 64 Gb NAND flash with 800 MB/sec synchronous DDR interface, *4th IEEE Intl. Memory Workshop (IMW),* 2012. (Hynix).

44. K. Gibb, First look at samsung's 48L 3D V-NAND Flash, EE Times, 4/6/2016.

45. S. Kim, S. J. Baik, Z. Huo et al., Robust multi-bit programmable flash memory using a resonant tunnel barrier, *IEDM,* Washington DC, 2005, p. 881.

46. K.-H. Joo, C.-R. Moon, S.-N. Lee et al., Novel charge trap devices with NCBO trap layers for NVM or image sensors, *IEDM,* Washington DC, 2006, pp. 979–982.

47. A. Bhattacharyya, Stack designs using multi-layered oxynitrides for enhanced NVM attributes, Internal Memo, ADI Associates, November 15, 2001.

48. A. Bhattacharyya, High perfromance multi-level non-volatile memory device, ADI Associates Internal Docket, December 4, 2004: 2004–1350, U. S. Patent: 7429767 (9/30/08), 7579242 (8/25/09), 8129213 (3/6/12).

49. A. Bhattacharyya, Memory cell comprising DRAM nanoparticles and NVM nonoparticles, original docket titled "Advanced scalable Nano-crystal NVM device and Process", ADI Associates Internal Docket, August 2005, US Patent: 7759715 (7/30/2010), 8193568 (6/5/2012).

50. A. Bhattacharyya, A novel stack design concept for charge trapping NVM cell with band engineered tunneling and trapping layers providing large memory window and 3 bit/cell MLC capability, ADI associates, internal memo, August 2007 (unpublished).

51. G. Zhang, W. S. Hwang, S. M. Bobade et al., Novel ZrO_2/Si_3N_4 dual charge storage layer to form step-up potential wells for highly reliable multi-level cell applications, *IEDM,* Washington DC, 2007, pp. 83–87.

52. C. W. Oh, N. Y. Kim, S. H. Kim et al., 4-bit double SONOS memories (DSMs) using single level and multi-level cell schemes, *IEDM,* San Francisco CA, 2006, pp. 967–970.

53. B. Eitan, P. Pavan, I. Bloom et al., NROM: A novel localized trapping, 2 bit nonvolatile cell, *IEEE Elec. Dev. Lett.,* 21, 543, 2000.

54. M. Taguchi, NOR flash memory technology, *IEEE IEDM* short course, 2006.

55. W. J. Tsai, N. K. Zous, C. J. Liu et al., Data retention behavior of a SONOS type two-bit storage flash memory cell, *IEDM,* Washington DC, 2001. pp. 714–717.

56. W. J. Tsai, N. K. Zous, M. H. Chou et al., Cause of erase speed degradation during two-bit per cell operation of a trapping nitride storage flash memory, *Proceedings Reliability Physics Symposium,* 2004, pp. 522–526.

57. Y. H. Shih, H.-T. Lue, K.-Y. Hsieh et al., A novel 2-bit/cell nitride storage flash memory with greater than 1 M P/E-cycle endurance, *IEDM,* San Francisco CA, 2004, pp. 881–884.

58. B. Eitan, G. Cohen, A. Shappir et al., 4-bit per cell NROM reliability, *IEDM,* 2005, pp. 539–542.

59. S. H. Gu, T. Wang, W. P. Lu et al., Characterization of programmed charge lateral distribution in a two-bit storage nitride flash memory cell by using a charge pump technique, *IEEE Trans. On Electron Devices,* 53, pp. 103–108, 2006.

60. H. T. Lue, S.-Y. Wang, E.-K. Lai et al., A novel P-channel NAND-type flash memory with 2 bit/cell operation and high programming throughput > 20 Mb/sec), *IEDM,* Washington DC, 2005 pp. 341–344.

61. A. Bhattacharyya, CT-flash memory cells for multi-bit per cell nonvolatile memory utilizing localized charge injection and trapping, ADI Internal Technical Memo, May, 2004 (unpublished).

62. M. Jurczak, T. Skotnicki, M. Paoli et al., SON (silicon on nothing)-a new device architecture for the ULSI era, *Symposium VLSI, Tech. Digest,* pp. 29, 1999.

63. R. Ranica, A. Villaret, P. Mazoyer et al., A new 40 nm SONOS structure based on backside trapping for nano-scale memory, *IEEE T. Nanotechnology,* 4, pp. 581–587, 2005.

64. A. Bhattacharyya, Back-sided trapped non-volatile device, Micron technology internal Publication, 2004–0927, July 12, 2004; U.S Patent Publication 257324, December 11, 2008, U.S. Patent: 7749848 (7/6/2010) and 7851827 (12/14/2010).

65. A. Bhattacharyya, A novel 4 bit/cell dual channel MNSC device, Internal Memo, ADI Associates, September 2007 (unpublished).

66. H.-T. Lue, P.-Y. Du, T.-H. Hsu et al., A novel planar floating-gate (FG)/charge-trapping (CT) NAND device using BE-SONOS inter-poly dielectric (IPD)., *IEDM,* Baltimore MD, 2009, pp. 827–830.

67. S.-M. Jung, J. Jang, W. Cho et al., Three-dimensionally stacked NAND flash memory technology using stacking single crystal Si layers on ILD and TANOS structure for beyond 30 nm node, *IEDM,* 2006, pp. 37–40.

68. R. Fricky, Data integrity on 20 nm SSDs, *Flash Memory Summit,* Santa Clara, CA, 2012.

69. E.-K. Lai, H.-T. Lue, Y.-H. Hsieh et al., A multi-layer stackable thin-film transistor (TFT) NAND-type flash memory, *IEDM,* San Francisco CA, 2006, pp. 41–44.

70. K.-T. Park, D. Kim, S. Hwang et al., A 45 nm 4 Gb 3-Dimentional double stacked multi-level NAND flash memory with shared bitline structure, *ISSCC,* 2008, pp. 510–632.

71. H. Yoon, and G. A. Tressler, Advanced flash technology status, scaling trends and implications to enterprise SSD technology enablement, *Flash Memory Summit,* IBM Corporation, Santa Clara, CA, 2012.

72. J. H. Yoon, NAND flash scaling challenges, *Flash memory Summit,* Santa Clara, CA, 2012.

73. E. Lusky, A. Shappir, G. Cohen et. al, "NROM Memories", Chapter 13 *Nonvolatile Memory Technologies with Emphasis on Flash,* Edited by J. Brewer and M. Gill, IEEE Press, Published by John Wiley & Sons, 2008, pp. 624–658.

74. M. Abraham, How to handle increasing ECC?, Micron technology, *IEEE Flash Memory Summit,* Santa Clara, CA, 2012.

75. R. Shirota, C.-H. Huang, H. Arakawa et al., A new disturb-free programming scheme in scaled NAND Flash memory, *IEDM,* Washington DC, 2011, pp. 9.3.1–9.3.4.

16 Nonplanar and 3D Devices and Arrays

CHAPTER OUTLINE

It has been recognized for some time that as feature size scalability continues beyond the sub-20 nm node to sub-10 nm node, FET device scalability will be severely constrained by fundamental effects such as short channel effects (SCEs) and doping effects in planar configurations. Added to that, for electron-based NVM devices, would be few-electron effects (FEEs) for NVMs adversely affecting functionality. For the last several years, significant R&D efforts have been directed toward "quasi-3D" and 3D NVM devices and arrays to establish viable cost-effective solution toward meeting ever-increasing demand for nonvolatile memory. The "planar-vertical" channel FINFET-NVMs had been previously established. The vertical channel surround gate NVMs were published before the twenty-first century, although the integration proved to be complicated. This chapter briefly highlights the progress made toward planar-vertical (FINFET) designs, GATE-ALL-AROUND (GAA) vertical channel designs either in the form of "Pillar" configuration or in trench-based vertical column designs. The newly emerging Nanowire MOSFET-based NVMs will also be discussed within the context of GAA structures. In addition, the recently announced full 3D NVM cell structures and NAND devices will be briefly described.

The scaling limitations of all planar NVM cells covering NROM and NAND cells and arrays could potentially be overcome by implementing full 3D devices and arrays. It is generally believed that full 3D NVM technology would mature into products during the period of 2014 through 2020 and would gain increasing acceptance starting at 15 nm node. It is also expected that challenges associated with added process and architectural complexity and associated cost will be met during the above time frame. If financially successful, full 3D NVM devices could be the mainstay at 15 nm technology and beyond. Several full 3D NAND cells have already been announced. At the time of writing of this book, no nonplanar NVM NROM product has been announced, although productization of such devices is being actively highlighted. At the same time, Samsung has announced shipping 3D NAND products in volume adopting a CT SONOS type of memory cell. Both Toshiba and Micron Technology have announced to be in early production phase with their 3D NAND products. This section will briefly cover a noncomprehensive account of such nonplanar and 3D device characteristics and current status of such devices either being investigated or in early production phase.

As the NVM devices are scaled for sub-15 nm node, the planar devices are constrained by SCEs, "FEE" [1] and STI nonuniform fringing field effects [2]. The evolution of planar and nonplanar gated FET structures is schematically shown in Figure 16.1. As should be noted that the gate-all-around (GAA) structure provides the greatest capacitive coupling between the gate and the channel, the latter could be both in x-y plane (defined planar in this text) and in x-z or y-z plane (defined vertical in this text) or a combination of both. By incorporating NVM devices operating in nonplanar vertical channels, the planar size effects could be reduced or eliminated. With CMOS platform technology adopting FinFET devices as scaled base-line CMOS SOC approach to contain SCE, considerable investigations were made during the past decades to demonstrate the viability of CT-FinFET NVMs [3] among other nonplanar NVMs. FinFET SONOS has been introduced in Chapter 5 of this book.

The Gate-All-Around (GAA) structure provides for the greatest
capacitive coupling between the gate and the channel.

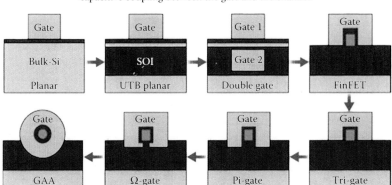

FIGURE 16.1   MOSFET device evolution: Planar and non-planar FET devices. (From Liu, T-J K., FinFET history, fundamentals and future, 2012 Symposium on VLSI Technology Short Course, June 11, 2012, UC Berkeley.)

Other nonplanar devices will be briefly reviewed in this section along with the FinFET SONOS. These include trench-based vertical channel NROMs and NANDs, GAA-nanowire NVM devices, stacked surround-gate NVM devices, and full 3D NVM devices.

16.1   VERTICAL CHANNEL AND NONPLANAR NVM DEVICES: HISTORICAL EVOLUTION

Planar NVM devices operate fundamentally as planar FET devices. As the device geometries, namely gate length/channel length and gate width are scaled for each succeeding technology node, several FET device design challenges have to be overcome. These challenges are: "short channel effect" (SCE), "narrow width effect" (NWE), "substrate doping effect" (SDE), and isolation "fringing field effect" (FFE) and associated device characteristics such as leakage, transconductance degradation, and reliability issues. To obviate such scaling problems, vertical channel [4–7], and 3D FET device structures [27] evolved for application in sub-100 nm featured technology nodes. Two FET device structures drew considerable attention earlier. These were: (a) "FinFET" Transistor [5] published in 2000 and (b) "surrounding gate transistor" (SGT) published in 1988 [4].

The SGT transistor was applied to develop a Floating channel type SGT memory by T. Endoh et.al in 1999 [8]. The authors demonstrated NAND programming and erasing characteristics similar to planar NAND structured cells and proposed such device in 2001 for application toward 16/64 Gb NAND products.

The FinFET transistor evolved as a follow-on to the double gate planar transistor to enhance scalability and improve logic performance [5]. Feasibility of vertical channel FinFET SONOS device for 2 bit/Cell NVM applications using ONO stack was first demonstrated by Y.K Lee et al. in 2002 [6,7] which was soon followed by other investigators providing optimized devices for embedded and NAND applications [3,9,10,11,22,23,26,27,30,31,32].

We will review the progress thus far in nonplanar and 3D NVM device concepts and array proposals in four subsections. In keeping with the continuity of the previous section, we will first describe some nonplanar trench-based device concepts using local injection and storage (MNSD/MNSC type) yielding compact 2 bit/cell designs especially suitable for future NROM applications. We will then briefly describe some similar FG-NAND structures. Subsequently, we will describe the FinFET SONOS and GAA device concepts again primarily for embedded and stand-alone higher performance future NROM/SOC applications. A follow-on to GAA concept is the nanowired device structures which hold considerable promise in sub-10 nm technology time frame and will be briefly covered within this subsection. In the third subsection, we will review the SGT-NVM devices and particularly high-density SGT-NAND device proposals. In the final subsection, we will briefly review the current status on full 3D-NVM devices, cell structures, and arrays for NAND implementation.

16.2 NONPLANAR MULTIBIT/CELL VERTICAL CHANNEL CT DEVICES

The first vertical channel device was published by Y.K. Lee et al. who built a dual-gate "FinFET" like structure on a p-type SOI substrate [6,10]. Silicon over the BOX was etched both in the x-z plane and the y-z plane, while the gate wrapped along the y-z plane and gate-self-aligned diffusions were formed along the x-z plane. The ONO stack formed vertically consisted of 1.5 nm of tunnel oxide/5.0 nm of trapping nitride and 4.0 nm of blocking oxide. On top of silicon, a thicker oxide of 100 nm was formed for top channel current suppression. The SOI silicon thickness was 120 nm over BOX, while the effective device width was 57 nm. The device yielded a memory window of ~1.5 V at a programming condition of ±8 V, 3 ms. The device demonstrated good endurance and retention. The device fabrication steps have been illustrated in Figure 6.21 [10]. The endurance and retention characteristics of the FinFET SOI NVM device were also shown in Figure 6.22 [6].

In principle, local charge injection and nitride trapping could be accomplished over either diffusion edges (not demonstrated) to provide 2 bit/cell operation.

FinFET NVMs will be further discussed in a later section in this chapter.

Several vertical channel trench-based CT-NVM devices were proposed during 2004 to overcome multiple limitations of planar FG and CT devices [12–14]. These limitations were (a) feature size scalability especially the SCE for FG and CT devices; (b) programming and erasing performance limitations of classical FG-ONO-IPD and scaled SONOS devices; (c) endurance limitations of classical FG and SONOS devices; (d) cell density limitations of FG/CT devices; (e) over-erasure protection of CT devices; (f) FG to adjacent FG parasitic coupling interference limitations; (g) design margin limitations of FG and CT MLC devices, and (h) device reliability limitations of planar devices of both types. The proposals provided compact multiple bit/cell device concepts combined with enhanced device characteristics for NROM and NAND flash devices and implementation flexibility in the form of NOR, AND, VGA, and NAND arrays [12,13,15]. We will first briefly review selected NROM device and array schemes in the next subsection and subsequently the FG device and array schemes in a later subsection.

16.2.1 Trench-Based Vertical Channel CT NROM

An example of the trench-based vertical channel device NROM device schematic is shown in Figure 16.2 and the NROM contactless array layout is shown in Figure 16.3. The devices are formed inside a trench with deep buried N+ diffusion acting as common source lines which could be all tied together outside the array for layout efficiency. The bit-line contacts for buried N+ diffusions are also shown in Figure 16.2. N+ Doped polysilicon rails run parallel to the word-line directions (in y-z plane, Figure 16.2) underneath the silicon surface and make self-aligned connect ("SALI") with the buried diffusion regions formed inside the P-substrate. Each buried diffusion regions are shared with two adjacent trench cells acting as a "link-diffusion node" as shown in Figure 16.2.

Each cell could be operated as 2 bit/cell or 4bit/cell by localized trapping (QUAD-NROM) over each diffusion nodes and sensing for memory states by reversing the roles of source and drain as explained in the earlier section (as CT-MNSC devices). With a "Self-aligned local interconnect" (SALI) layout as shown in the figures, bit density for NROM cells inside the array could be as low as < 1.0 F²/bit (4 bit/cell). The cells are linked by surface N+ diffusion regions with SALI bit lines as mentioned earlier. Two buried N+ gates are formed inside each trench and run parallel to the buried N+ doped polysilicon rails in the y-z plane acting as word-lines for the memory cells. The NROM array layout is detailed in Figure 16.3. It illustrates the self-aligned local interconnect (SALI) features for bit-line contacts within the array while all other contacts (source lines and word-lines) are provided outside the array yielding a very dense memory array layout.

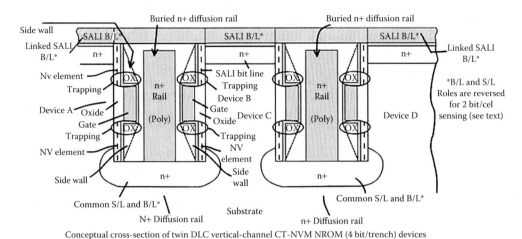

Conceptual cross-section of twin DLC vertical-channel CT-NVM NROM (4 bit/trench) devices

FIGURE 16.2 Twin 2 bit/cell or 4 bit/cell (QUAD-NROM) vertical channel CT-NVM NROM devices in trenched silicon with buried N+ diffusion rails (trench bit capacity as much as 8 bits). (From Bhattacharyya, A., Extremely high density two-bit-per-cell planar-vertical embedded trap, NVM cell and contactless array, Micron Internal Docket: 2004–0943, 2004, U.S. Patents: 7671407 [3/2/2010], 7838362 [11/23/2010].)

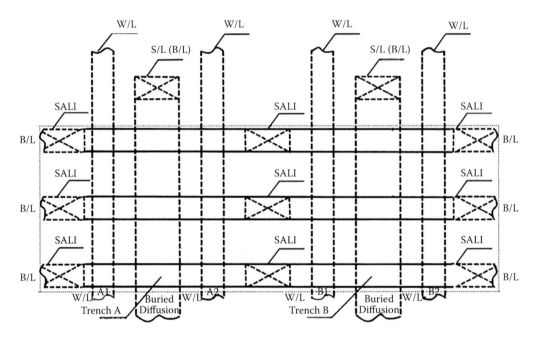

FIGURE 16.3 Contact-less NROM array layout with SALI features for buried contacts. (From Bhattacharyya, A., Extremely high density two-bit-per-cell planar-vertical embedded trap, NVM cell and contactless array, Micron Internal Docket: 2004–0943, 2004, U.S. Patents: 7671407 [3/2/2010], 7838362 [11/23/2010].).)

The stack design schemes to provide higher performance erasing and writing are illustrated in Figure 16.4a and b. The vertical channel CT-NVM device in the proposal was a split channel device with the center part of the channel had a single layer of thick oxide providing a higher fixed Vth defining a stable Vth for the erase state of the NVM device providing over-erasure protection. Two options of the variable stack elements of the device are shown with the buried diffusion subsection of the device in Figure 16.4a and b. In Figure 16.4a, the NV CT stack elements consisting of a tunnel insulator of OR-SiON, a deep trap SiON trapping insulator (replacing nitride) with an

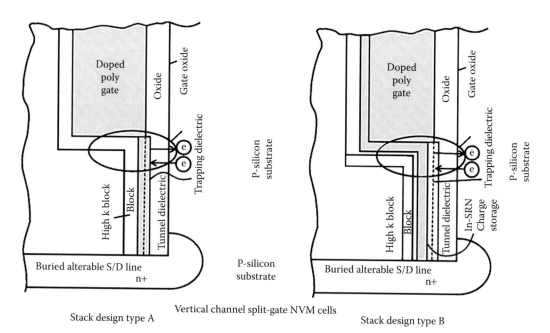

Stack design type A

Vertical channel split-gate NVM cells

Stack design type B

FIGURE 16.4 Gate insulator stack design schemes to enhance programming performances: (a) OR-SiON/ NR-SiON/Al₂O₃, (b) OR-SiON/NR-SiON/in-SRN/Al₂O₃. (From Bhattacharyya, A., Extremely high density two-bit-per-cell planar-vertical embedded trap, NVM cell and contactless array, Micron Internal Docket: 2004–0943, 2004, U.S. Patents: 7671407 [3/2/2010], 7838362 [11/23/2010]; Bhattacharyya, A., Defining high speed high endurance scalable NVM device using multimechanism carrier transport and high density structures for contact-less Arrays, Micron Internal Docket: 2004–1008, 9/1/2004; U.S. Patents: 7244981 [7/17/07], 7553735 [6/30/09].)

over layer of high K blocking layer. In Figure 16.4b, an additional injector SRN (i-SRN) layer was added over the NV stack elements and continued on the top of the polysilicon control gate (W/L) underneath the charge blocking dielectric to enhance hole injection during erase achieving high erase speed.

It should be noted that the cell design is readily amenable to other band engineered stack design schemes discussed in Chapter 13. Some of those schemes were incorporated in split-gate vertical channel devices to enhance NVM device performances [12–15] including DTM and PBO options to enhance device characteristics. The scalability of split-gate charge-trapping devices down to 20 nm technology node has been investigated recently in CEA LETI by L. Masoero and coinvestigators [16].

16.2.2 Vertical Channel DTM-NROM

Planar split channel DTM device was introduced in Chapter 2 in Part I. Band engineered stack designs for DTM and PBO devices have been discussed in detail in Chapter 13. Vertical channel split gate DTM device was proposed in 2004 to achieve scalable NROM arrays in sub-50 nm nodes with enhanced device attributes [14]. The high-performance DTM cells shared common source line and control gate within a trench with dedicated SALI bit lines contacting STI bound surface N+ drain diffusions as shown in the vertical cell design in Figure 16.5.

The band diagram (Figure 16.6) utilizing OR-SiON tunnel layer in place of oxide with deep trapping SiON layer combined with I-SRN equipotential charge storage layer and high K Al₂O₃ blocking layer with high density of fixed negative charge, provided compact NROM density, short

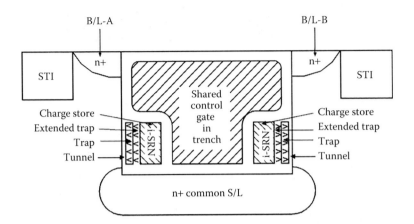

FIGURE 16.5 Schematic cross-section of high-performance vertical channel split-gate NROM DTM with shared source line and control gate. (From Bhattacharyya, A., A vertical cross-point high density split channel direct tunnel non-volatile memory, Micron Internal Docket: 2004–0929, August 2, 2004, U.S. Patents: 7365388 [4/29/2008], 7671407 [3/2/2010].)

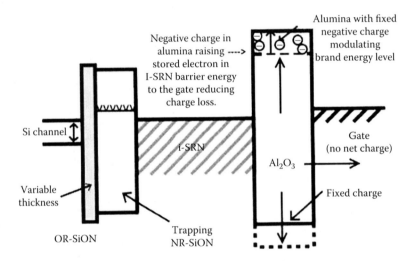

FIGURE 16.6 Band diagram for the above vertical channel DTM-NROM device; presence of fixed negative charge at the i-SRN/Al$_2$O$_3$ interface shifts the Al$_2$O$_3$ band diagram as shown in the solid line and provide enhanced retention and memory window. (From Bhattacharyya, A., A vertical cross-point high density split channel direct tunnel non-volatile memory, Micron Internal Docket: 2004–0929, August 2, 2004, U.S. Patents: 7365388 [4/29/2008], 7671407 [3/2/2010].)

channel immunity, large window, and enhanced retention, endurance and reliability when compared to SONOS stack design.

The cross-point array layout scheme employed SALI bit-line contacts improving the cell density with second level metal (M2) stitching the common source lines for all cells outside the array. In the layout shown below (Figure 16.7), both word-lines and source lines were tied with the M2 level. However, for embedded applications, the third level metal could be employed for the common source line while the cell pitch could be defined by M1 and closely spaced M2 level enhancing array bit density.

Other vertical channel CT-NROM devices and arrays has also been published demonstrating higher performance and disturb immunity [17].

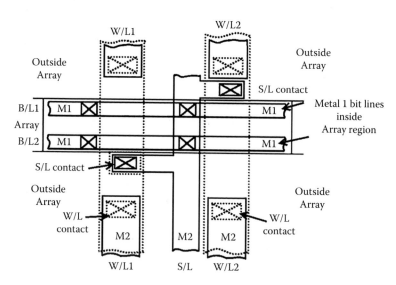

FIGURE 16.7 Array layout for the vertical channel SLC/MLC DTM-NROM. (From Bhattacharyya, A., A vertical cross-point high density split channel direct tunnel non-volatile memory, Micron Internal Docket: 2004–0929, August 2, 2004, U.S. Patents: 7365388 [4/29/2008], 7671407 [3/2/2010].)

16.2.3 Vertical Channel FG-NAND

TRENCH-BASED VERTICAL CHANNEL FG-NAND DEVICES AND ARRAYS BASED on the split-gate vertical channel concept discussed in the previous Section 16.2.2. Several FG-vertical channel devices were proposed in 2004 and subsequent time frame and NAND arrays were published [12,18–20] employing such concept. The cross-section of such a cell concept is schematically shown in Figure 16.8a, and nonvolatile stack elements are shown in Figure 16.8b. Each trench consists of 2-bit/cell whereby the bottom N+ diffusion is shared between the two bits as well as the surface diffusions shares and links two adjacent NAND bits. The insulator stack between the floating node and the control gate consists of a layer of high K dielectric such as Al_2O_3 sandwiched between two thin layers of injector SRN layers to simultaneously enhance erase speed and low leakage to the control gate (larger memory window). The scheme provides positive voltage erase as well as positive voltage write for higher programming performance. The erasure involves enhanced electron removal from the floating gate to the control gate due to enhanced FN tunneling provided by the SRN layers. FG-NAND writing could be accomplished either by the highly efficient source-side injection or by tunneling since the source potential efficiently couples the floating gate while also capacitively boosting the control gate (pulsed high). The NAND array layout is otherwise similar to the planar array shown in Figure 16.8. The circuit schematic for the vertical channel NAND is shown in Figure 16.9.

16.3 FINFET AND GATE-ALL-AROUND (GAA) NV DEVICES

16.3.1 FinFET Technology and Devices

FinFET development was initiated in UC Berkeley through DARPA initiative in mid-1997 to address FET scaling requirements for down to 25 nm gate length and beyond [21]. Since the publication of FinFET transistor in 2000, the device has been scaled for over the past decade through successive technology nodes both in SOI technology and in bulk technology with dualgate vertical channel as well as trigate schemes [3,9,20,22–26]. High-performance 22 nm FinFET CMOS devices with high-k metal gate stack design were reported by C.C. Wu et al. in 2010 [25]. FinFET SONOS flash devices

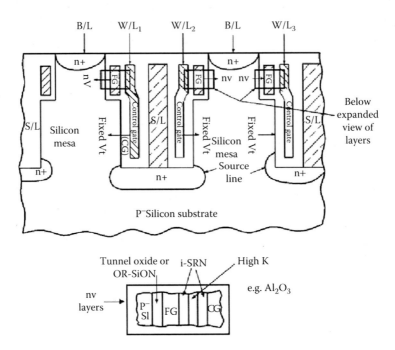

FIGURE 16.8 Trench-based vertical channel 2 bit/ trench FG cell design sharing surface and buried link diffusions (top figure); NV element stack construction (bottom figure). (From Bhattacharyya, A., Scalable high density, high performance hybrid planar-vertical NV memory cells and contactless arrays, Micron Internal Docket: 2004–0942, 2004, U.S. Patents: 7166888 [1/23/07], 7273784 [9/25/07], 7964909 [6/21/2011].)

demonstrated significantly reduced random telegraph noise (RTN) compared to planar SONOS [26]. A 22 nm SoC platform technology featuring 3D trigate was introduced by Intel Corporation in 2012 [27]. It has been assumed that for general CMOS logic and platform technology, Fin-FETs continue to hold promise down to 15 nm node even though the parasitic effects due to tapered fin geometry and source-drain parasitic capacitive coupling to the gate, and higher device resistance may limit the performance extendibility of doped FinFET devices [27,28]. It has been published that major foundries such as Global Foundries and Samsung are in preparation to offer 14 nm FinFET logic technology in volume in 2014–2015 time frame [29]. However, there has been considerable controversy as to the merits of scalability of FinFET devices for 10 nm node and beyond and options such as fully depleted planar SOI, nanowired (GAA, see later) 3D devices, tunnel FETs have been proposed as potential candidates [28–31]. Regardless of the nature of the scalable device beyond 14 nm node, it is assumed here that the manufacturable platform technology could be compatible to embedded NV-NROM and NAND arrays for a wide range of future SOC and SOP applications through 2020. Therefore, we will briefly review the published FinFET NROM and NAND flash devices below.

A 30 to 80 nm gate length trigate-SONOS NVM device on fully depleted SOI (50 nm silicon with 100 nm BOX) was fabricated and characterized for the first time in 2004 by M. Specht et al. [22]. The stack design consisted of 3 nm of tunnel oxide/4 nm of trapping nitride and 4.8 nm of blocking oxide. The memory window of 1.1 V end-of-life was achieved with +11.5 V of 3 ms of writing and −11.5 V of 100 ms of erasing. The device structure, retention, and endurance are shown in Figure 16.10a–e, respectively. The short channel characteristics both experimental and simulated are illustrated in Figure 16.11a and b, respectively, demonstrating excellent transfer characteristics (short channel immunity) up to 30 nm gate length. The devices demonstrated good program and read disturb immunity as well (not shown).

Historically, J-R Hwang and coworkers fabricated and characterized in 2005 for the first time a bulk FinFET SONOS device with 20 nm gate length [23]. The fabrication process steps and the

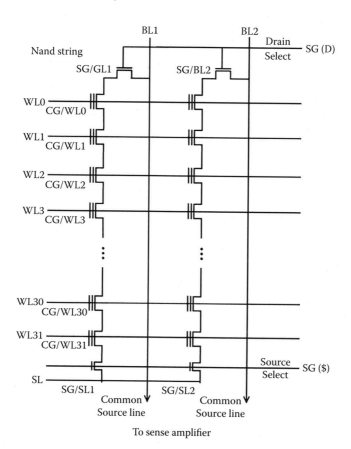

FIGURE 16.9 FG-NAND flash array circuit schematic of vertical channel trench-based design. (From Bhattacharyya, A., Scalable high density, high performance hybrid planar-vertical NV memory cells and contactless arrays, Micron Internal Docket: 2004–0942, 2004, U.S. Patents: 7166888 [1/23/07], 7273784 [9/25/07], 7964909 [6/21/2011].)

device processing structures have already been illustrated previously in Figure 6.21. Their device demonstrated 1E4 W/E cycle retention and >1.5 V of postendurance end-of-life memory retention. Programming characteristics and memory window for their device are shown in Figure 16.12a and b, respectively.

S.H. Lee and coworkers investigated and compared TANOS-NAND FinFET device with planar TANOS device characteristics and observed significant improvement in postcycling characteristics of FinFET device over planar equivalent [21]. Bake retention characteristics of the device at 200 C before cycling and after 1000 W/E cycling are shown in Figure 16.13a and b, respectively. The erase state retention of planar device was inferior to the FinFET device before cycling. After cycling the erase characteristics of the planar device retention got comparatively worse while that of the FinFET device was significantly superior.

The TANOS-NAND strings and SEM image of the cell with the schematic representation of TANOS stack are illustrated in Figures 16.14 [27] and 16.15 [32], respectively.

Friederich et al. investigated optimization of p+ doped trigate stack design for SONOS NAND arrays for MLC operability and potential extendibility for future technology nodes using SOI trigate CMOS technology [24]. The SOI wafers used were of 50 nm silicon thickness and 100 nm of BOX. The ONO stack design consisted of 3.0 nm of tunnel oxide and 9.0 nm each of trapping nitride and blocking oxide to achieve nearly symmetric memory window of 6 V with Vth of nearly−3 V (erase) and +3 V Vth-high (write). The programming conditions were Vg = ±20 V, at write time of 100 us

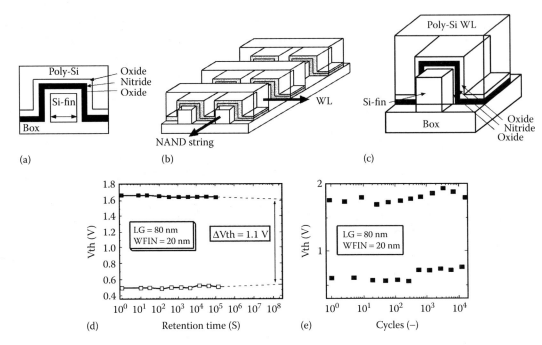

(a)   (b)   (c)

(d)   Retention time (S)   (e)   Cycles (−)

FIGURE 16.10   SONOS-tri-gate device characteristics: (a,b,c) device structure, (d) retention, and (e) endurance. (From Specht, M., Dorda, U., Dreeskornfeld, L. et. al., 20 nm tri-gate SONOS memory cells with multiple level operation, *IEDM*, pp. 1083–1086 © 2004 IEEE.)

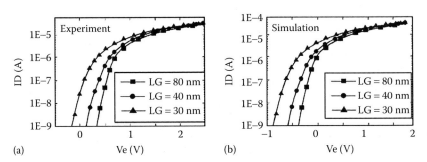

(a)   Ve (V)   (b)   Ve (V)

FIGURE 16.11   Transfer characteristics for several gate length trigate SONOS devices: (a) experimental, and (b) simulated results. (From Specht, M., Dorda, U., Dreeskornfeld, L. et. al., 20 nm tri-gate SONOS memory cells with multiple level operation, *IEDM*, 2004, pp. 1083–1086.) Reproduced with permission of Wiley-IEEE Press.

and erase time of 2 ms. The memory window demonstrated efficient erasing with p+ doped polysilicon compared to n+ doped polysilicon gates yielding symmetric window, tighter Vth distribution of multilevel Vth memory states, and superior disturb immunity due to higher work function of p+ gate providing improved blocking properties of unwanted gate injection of charges. The device structure exhibited excellent retention and endurance for DLS (2bit/cell) operation. The NAND strings were fabricated down to 50 nm gate length at 50 nm feature size (F) with 25 nm Fin width and 35 nm spacing providing word-line feasibility down to 15 nm Fin width in a pitch of 100 nm (2F). The design provided 2F x 2F cell size with 2bit/cell capability. The scalability simulation showed that the design as such is scalable down to F = 32 nm. For sub-30 nm node, inter-string distance between the adjacent Fins has to be reduced and metal gate would be required to offset poly depletion effects. Simulation study demonstrated design feasibility at 15 nm Fin width.

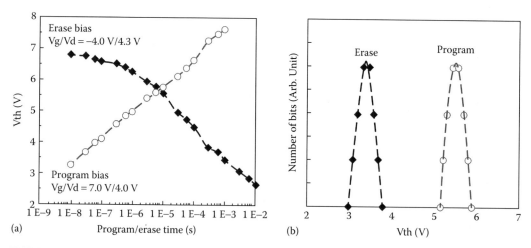

FIGURE 16.12 (a) Programming characteristics, and (b) memory window of bulk FinFET SONOS. (From Hwang, J.-R., Lee, T.-L., Ma, H.-C. et al., 20 nm gate bulk-FinFET SONOS flash, *IEDM*, pp. 161–164. © 2005 IEEE.)

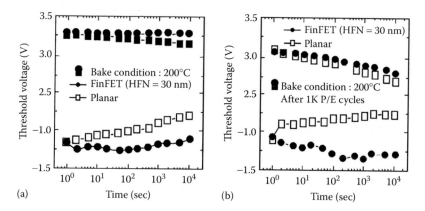

FIGURE 16.13 Bake retention comparison of FinFET TANOS device with planar equivalent at 200°C: (a) before cycling stress, and (b) after 1000 W/E cycling. (From C.-C. Yeh, C.-S. Chang, H.-N. Lin et al., A low operating power FinFET transistor module featuring scaled gate stack and strain engineering for 32/28 nm SoC technology, *IEDM*, pp. 34.1.1–34.1.4 © 2010 IEEE.)

Figure 16.14a illustrates a schematic cross-section of a NAND string on SOI with the silicon Fin and ONO stack with polysilicon trigate across a word-line. Figure 16.14b shows the schematic representation of a double NAND string memory array in 3D [27]. Figure 16.15 shows an SEM image of fabricated double NAND string array with silicon Fins [32]. Figure 16.16 shows a TEM cross-section of double NAND strings with 23 nm Fin width and 40 nm Fin height with Fin half pitch of 60 nm is shown above from the investigation of Jan and coworkers [32]. Extensive device characteristics were investigated by the same authors and highlighted below [32].

Transfer characteristics of a trigate device are shown in Figure 16.17 demonstrating excellent subthreshold characteristics both at erased state and at the maximum written (Vth-high) state. The device has a 20 nm Fin-thickness and in a NAND string with 6 word-lines and 2 select gates. Gate length sensitivity on Vth for both high and low states is shown in the inset for gate lengths of 50 nm and 100 nm.

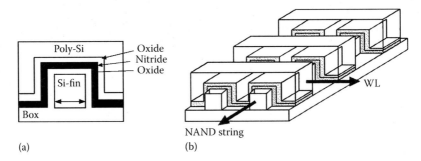

FIGURE 16.14 Schematic cross-sections of SONOS-MLC NAND in SOI: (a) Word-line cross-section of a single cell, (b) 3D view of double NAND tri-gate array. (From Chiu, J.P., Chou, Y.L., Ma, H.C. et al., Program charge effect on random telegraph noise amplitude and its device structural dependence in SONOS flash memory, *IEDM*, pp. ISBN 978-1-4, 244-5639-0 © 2009 IEEE.)

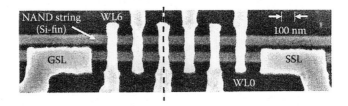

FIGURE 16.15 SEM image top view of fabricated double NAND string. (From Jan, C.H., Agostinelli, M., Deshpande, H. et al., RF CMOS technology scaling in High-k/metal gate, era for RF SoC [system-on-chip] applications, *IEDM*, pp. 647–650. © 2010 IEEE.)

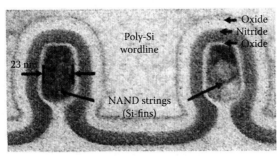

- Fin string pattering in (110) : HSQ resist, E-beam lithography
- Dry etching
- Sacrificial oxide
- Thermal tunnel oxide in diluted O_2
- LPCVD Nitride
- Thermal oxidation of nitride
- In-situ phosphorous or boron - doped 100 nm poly-Si
- Word line patterning : HSQ resist, E-beam lithography, dry etching
- Ultra thin spacers of 12 nm
- As-implantation, RTP anneal

FIGURE 16.16 SEM image top view of fabricated double NAND string. (From Jan, C.H., Agostinelli, M., Deshpande, H. et al., RF CMOS technology scaling in High-k/metal gate, era for RF SoC [system-on-chip] applications, *IEDM*, pp. 647–650. © 2010 IEEE.)

The DLC (2 bit/cell) capability is illustrated in Figure 16.18. The three programmed levels of memory states—01, 10, and 11—with clean separation between the states for sensing for the double NAND string are displayed for the device [32].

Programming and erasing characteristics were compared between n+ gate versus p+ gate. Their data demonstrated the merits of p+ gate over n+ gate in achieving more symmetric memory window and erasure efficiency [32]. P+ gated device exhibited superior endurance and greater window stability compared to N+ gated device. Additionally, excellent retention was observed for the p+ gated device after 10000 W/E cycling compared to that of n+ gate device. W/E cycling was performed at $V_{pp} = 20$ V, 100 us for write and Ve = 19 V, 2 ms erase conditions for both p+ gate and n+ gate devices. Retention data for the erased state as well as for the three program level states are

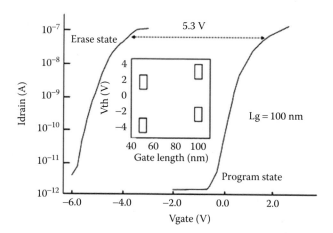

FIGURE 16.17 Transfer characteristics of a tri-gate transistor in SoC technology. (From Jan, C.H., Agostinelli, M., Deshpande, H. et al., RF CMOS technology scaling in High-k/metal gate, era for RF SoC [system-on-chip] applications, *IEDM*, pp. 647–650. © 2010 IEEE.)

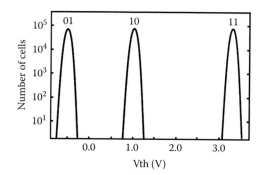

FIGURE 16.18 Measured DLC memory programmed states for the tri-gate NAND device. (From Jan, C.H., Agostinelli, M., Deshpande, H. et al., RF CMOS technology scaling in High-k/metal gate, era for RF SoC [system-on-chip] applications, *IEDM*, pp. 647–650. © 2010 IEEE.)

presented in Figure 16.19 after 10000 W/E cycling. Good retention characteristics for all memory states were observed with extrapolated 1.2 V Vth level separation at the end-of-life of 10 years at 25°C. High temperature bake stability of the memory states postendurance cycles confirmed device Vth stability [32].

Read disturb of the erased state was also characterized and is shown in Figure 16.20. The measurement was made for device with gate length of 100 nm at V-read stress level of 4.5 V. The read disturb was minor at Vth = −2 V up to at least 1 E7 read operations [32].

The scalability of such devices has been mentioned previously. It should be noted that for the above design, for 32 nm node, the inter-Fin spacing was reduced to 8 nm.

CT-FinFET Advantages:

CT-FinFET NAND appears to have the following advantages over planar FG-NAND:

1. FG-FG parasitic coupling is minimized.
2. SCE minimization due to strong electrostatic gate control of the channel.
3. Larger effective width at any technology node providing reduction of FEE in terms of memory window and Vth stability, and easier integration issues with baseline scaled CMOS (FinFET) Platform technology.

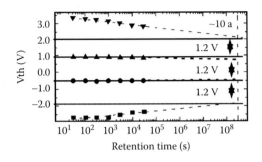

FIGURE 16.19 Retention characteristics of p+ gated tri-gate SONOS device. (From Jan, C.H., Agostinelli, M., Deshpande, H. et al., RF CMOS technology scaling in High-k/metal gate, era for RF SoC [system-on-chip] applications, *IEDM*, pp. 647–650. © 2010 IEEE.)

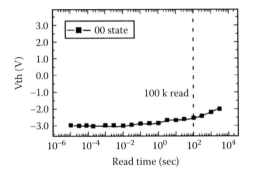

FIGURE 16.20 Read disturb characteristics of p+ gated tri-gate SONOS device. (From Jan, C.H., Agostinelli, M., Deshpande, H. et al., RF CMOS technology scaling in High-k/metal gate, era for RF SoC [system-on-chip] applications, *IEDM*, pp. 647–650. © 2010 IEEE.)

16.3.2 Gate-All-Around (GAA) and Nano-Wired (NW) FET (TSNWFET) and NVM Devices

It was mentioned in the beginning that among all nonplanar FET devices, the GAA structure provides the greatest capacitive coupling between the gates and the channel (Figure 16.21). The evolutionary "Twin Silicon Nanowire MOSFETs" (TSNWFET) with GAA structure was fabricated and characterized by several authors of Samsung Electronics R&D center [32–35] using bulk CMOS technology. The TSNWFET holds considerable promise for silicon based technology extendibility even beyond the ITRS roadmap of 7 nm feature size. The TSNWFET showed no Vth-roll-off, excellent subthreshold swing, very low DIBL and high saturation current capability. In 2006, as low as 4 nm radius with gate length of 15 nm showed excellent PFET and NFET device characteristics as published by the same group of researchers [35]. During the same year another group from Institute of Microelectronics, Singapore Science Park reported nanowire diameter sensitivity, channel orientation, and temperature effects of NWFET devices as well as the fabrication schemes for the 3D stacked SiGe Nanowire array and GAA p-MOSFETs [36,37]. Subsequently J. Fu et al. from the above institute published device characteristics of 5 nm diameter twin Si-Nanowire SONOS device with silicon nanocrystal/nitride trapping layer demonstrating 1 us programming speed, wider memory window, and excellent retention and endurance [38]. The first GAA-SONOS nanowire device with 3D memory array architecture was published by A. Hubert et al. who demonstrated the feasibility of the SONOS-GAA-NV memory consisting of up to four levels with 6 nm diameter of crystalline nanowire and associated memory architecture [39]. The process flow was used to demonstrate a variety of highly coupled gate architectures including SONOS FinFET, a variety of SONOS GAA

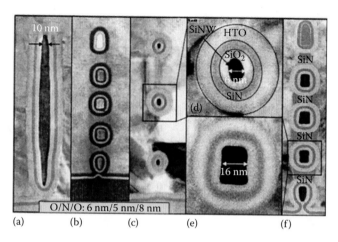

FIGURE 16.21 TEM cross-sections: (a) SONOS-FinFET, (b),(c),(d),(e) 3D stacked nanowires, and (f) O-Flash. (From Hubert, A., Nowak, E., Tachi, K. et al., A stacked SONOS technology, up to 4 levels and 6 nm crystalline nanowires, with gate-all-around or independent gates [O-Flash], suitable for full 3 d integration, *IEDM*, pp. 637–640 © 2009 IEEE.)

Nanowires as well as a GAA-O-flash architecture. A cross-sectional TEM picture of several variations in GAA-SONOS devices is shown in Figure 16.21.

The process flow for fabrication of the various structures and associated schematics is illustrated in Figure 16.22 showing that the initial Si/Si-0.8Ge-0.2 super lattices need to be tailored for each device configurations: for example, for FinFET: Si 195 nm/ SiGe 20 nm; for 4 levels of 3D nanowire and O-flash: [Si 30 nm/SiGe 25 nm] × 4, etc. The hydrogen annealing and dry sacrificial oxidation were implemented to achieve 6 nm nanowires and 10 nm wide FINFETs. The SONOS stack consisted of 6 nm tunnel oxide grown thermally followed by LPCVD deposition of 5 nm trapping nitride and 8 nm of blocking oxide. O-flash required a nitride insulator layer to fill the gaps between the nanowires.

The Id-Vg characteristics of 3D stacked nanowire devices with 80 nm gate length showed excellent subthreshold characteristics with 70 MV/decade of sub-threshold slope (SS) and 41 mV/V of DIBL, both being significantly better compared to the FinFET device. The Id-Vg characteristics are shown in Figure 16.23a and the fast programming characteristics with large programming window are shown in Figure 16.23b for the three-channel stacked SONOS-nanowire device. Programming characteristics are shown in Figure 16.23c [39].

Endurance and retention characteristics of SONOS-nanowire devices are shown in Figure 16.24a and b, respectively. Postendurance (after 10000 W/E cycling) window was reduced from 4.6 V (uncycled) to 4.0 V window. Figure 16.24c showed the temperature dependence of charge retention. The memory window reduction at 200°C after 10 years was less than 50% as shown below.

Program/erase performance was studied as a function of nanowire diameter. This is shown in Figure 16.25. Performance could be improved by more than two orders of magnitude from ms to <10 us for a memory window = 4 V, by reducing the nanowire diameter from 26 to 8 nm [39].

The independent "double gate" 3D nanowires as proposed in the O-flash device were investigated to address the feasibility of localized charge trapping under each gate separately, thereby providing the capability of 2 bit/cell design (double gate providing 4 bits of memory). Such feasibility was successfully demonstrated. Additionally, by selectively oxidizing the Si-Ge layer between the stacked nanowire gates, the channels of each device were isolated from each other in the stack. 3D nanowired MLC O-flash devices hold considerable promise in extending silicon based nonvolatile memory beyond the sub-10 nm feature size. This will be further mentioned in the context of SUM devices in Part III of this book [40]. Hubert and coworkers proposed a nanowire-based full 3D flash memory consisting of different levels accessed by bit lines laid horizontally with vertical columns of nanowires consisting of access gates and word-lines for each levels integrated to form the 3D flash

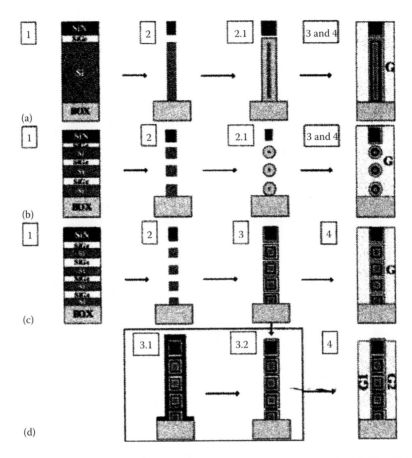

FIGURE 16.22 Process flow and schematics for different structures: (a) SONOS-FinFET, (b) 3-level, (c) 4-level 3D stacked nanowires, and (d) 3D stacked O-Flash. (From Hubert, A., Nowak, E., Tachi, K. et al., A stacked SONOS technology, up to 4 levels and 6 nm crystalline nanowires, with gate-all-around or independent gates [O-Flash], suitable for full 3 d integration, *IEDM*, pp. 637–640 © 2009 IEEE.)

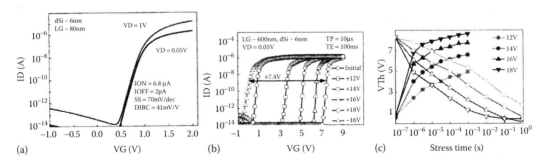

FIGURE 16.23 (a) Id-Vg characteristics of stacked SONOS-nanowire device, (b) programming characteristics (Id-Vg), and (c) programming and erasing characteristics. (From Hubert, A., Nowak, E., Tachi, K. et al., A stacked SONOS technology, up to 4 levels and 6 nm crystalline nanowires, with gate-all-around or independent gates [O-Flash], suitable for full 3 d integration, *IEDM*, pp. 637–640, 2009 © IEEE.)

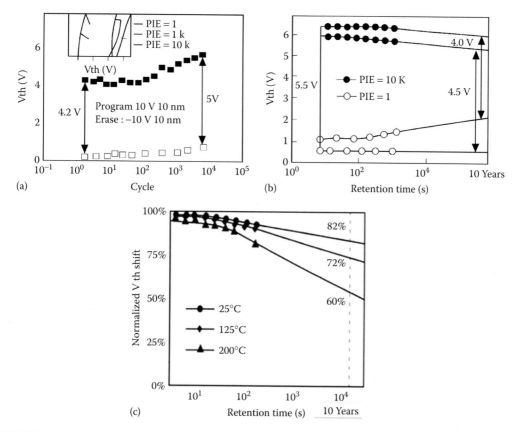

(a) Cycle

(b) Retention time (s)

(c) Retention time (s)

FIGURE 16.24 Stacked SONOS-nanowire device characteristics: (a) endurance, (b) retention before and after 10000 W/E cycling, and (c) temperature dependence on retention. (From Hubert, A., Nowak, E., Tachi, K. et al., A stacked SONOS technology, up to 4 levels and 6 nm crystalline nanowires, with gate-all-around or independent gates [O-Flash], suitable for full 3 d integration, *IEDM*, pp. 637–640, © 2009 IEEE.)

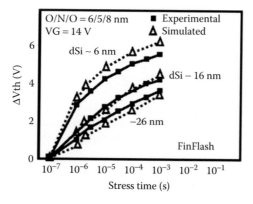

Stress time (s)

FIGURE 16.25 P/E performance versus nanowire diameter. (From Hubert, A., Nowak, E., Tachi, K. et al., A stacked SONOS technology, up to 4 levels and 6 nm crystalline nanowires, with gate-all-around or independent gates [O-Flash], suitable for full 3 d integration, *IEDM*, pp. 637–640, © 2009 IEEE.)

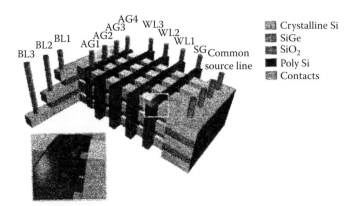

FIGURE 16.26 A full 3D SONOS flash memory proposal based on GAA nanowires (a) close up views of full 3D memory, and (b) equivalent circuit schematic. (From Hubert, A., Nowak, E., Tachi, K. et al., A stacked SONOS technology, up to 4 levels and 6 nm crystalline nanowires, with gate-all-around or independent gates [O-Flash], suitable for full 3 d integration, *IEDM*, pp. 637–640, © 2009 IEEE.)

memory. The proposed architecture for such a 3D flash memory is shown in Figure 16.26 [38]. Figure 16.26a shows the 3D features identifying various layers at the inset, whereas the equivalent circuit for the memory array is shown in Figure 16.26b.

During 2012 (at the time of initial preparation of this book) the baseline CMOS technology in production has been in the 22 to 24 nm node technology range as per ITRS road map for semiconductors. From 2014 through 2024, ITRS road map predicts progression in technology scaling of 15 nm node at 2016, 11 nm node at 2020, and at the end of road map at 7 nm node around 2024 time frame. Full 3D stacked silicon based nanowire NVMs could be a viable option at around 10 nm technology and beyond and could be embraced in the form of SOC and SOP platform technology. It is postulated that the stacked GAA FETs or equivalent concepts (e.g., nanowire FETs) may provide the highest layout efficiency for the silicon based nano electronics. CT-SUM device concepts as formulated in Part III of this book could be appropriately configured into 3D-GAA-Nanowire arrays providing high-performance and high-capacity solutions for 10 nm technology node and beyond. Additionally, multifunctional and multilevel SUM-nanowire arrays have the potential of eliminating the current memory system hierarchy by appropriately replacing volatile SRAM and DRAM arrays as well as less reliable and slower performing HDDs vastly improving system power/performance/ cost characteristics. This will be further explained in Part III of this book.

16.4 SURROUND GATE NV DEVICES (SGT)

SGT has been the precursor for GAA structures and assumed historical significance inspiring evolution of trench-based DRAM cell designs. As mentioned earlier, the concept was readily applied in DRAM and NAND-NVM cell designs. Even though the SGT-based NAND-NVM may not have been cost-effective, the concept should be worth reviewing due to its historical significance and the possibility of deriving potential other more cost-effective solutions applicable in sub-50 nm technology nodes, derived from SGT-GAA concepts.

T. Endoh et al. published the first stacked-SGT (S-SGT) NAND flash memory array containing vertically stacked two select gates and two NAND memory cells within one silicon pillar demonstrating 2 F2/bit SLC flash density feasibility [41,42]. The processing steps to form the SGT vertical pillars were described and device characteristics were presented [41]. A common source line connecting the buried N+ region was fabricated at the bottom of each pillar, while the top N+ diffusion at the surface contacted the bit line for each NAND string. An SEM photograph of the S-SGT structured cell is shown in Figure 16.27.

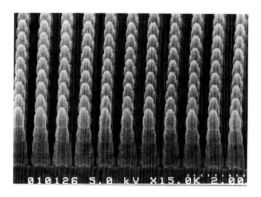

FIGURE 16.27 SEM photograph of the first S-SGT structured cell. (From Endoh, T., Kinoshita, K., Tanigami, T. et al., Novel ultra high density flash memory with a stacked-surrounding gate transistor [S-SGT] structured cell, *IEDM*, pp. 33–36, © 2001 IEEE.)

The stacked SGT silicon pillar was formed by successively repeating a process of etching silicon substrate and masking the silicon sidewall and subsequently oxidizing exposed silicon to form tunnel oxide for NV element and gate oxide for select gate, forming select and floating gates, selectively forming nitride and oxide over the FG, followed by S/D and CG processing. Subsequently, word-line and bit-line formation (E) were accomplished. The processing steps were discussed in detail by Endoh et al. [41]. The structure is of historical significance but was limited by large cell size.

The stack design for the SGT-FG-NAND consisted of 9.8 nm of tunnel oxide and on top of the FG the ONO IPD stack had an EOT of 16 nm. All cells were initially erased by F–N tunneling of holes at −30 V, 10 ms and selected cells were programmed by electron injection from the substrate at Vpp = 20 V, 1 ms of programming. Figure 16.28a shows the programming and erasing characteristics of S-SGT-NAND cells. Figure 16.28b shows the Vth shift of the selected and unselected cells as a function of programming time when the unselected state was held at Vpp/2. The unselected cell shows no program disturb.

16.4.1 Perspective of S-SGT and GAA Device Structures

Due to the GAA characteristics of these structures and vertical channel implementation capability, these structures need not suffer from scalability constraints of planar structures due to SCE and FEE. However, the implementation and integration of these nonplanar structures in base CMOS technology are relatively more complex and thereby face cost constraints. In recent years, the

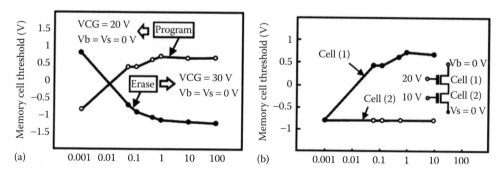

FIGURE 16.28 Programming and disturb characteristics of S-SGT-NAND cells: (a) programming and erasing, (b) selected and unselected cell characteristics during programming. (From Endoh, T., Kinoshita, K., Tanigami, T. et al., Novel ultra high density flash memory with a stacked-surrounding gate transistor [S-SGT] structured cell, *IEDM*, pp. 33–36, © 2001 IEEE.)

processing schemes for the pillar for S-SGT and nanowires for NWFET were provided with additional options such as "punch-and-plug" processing [43] and "FGSG" processing [40]. These will be discussed in greater detail later on in the context of full 3D device schemes. Some of the above versions could be implemented in vertical trenched configuration as opposed to pillar configuration with buried source line. The number of cells in a given vertical string in series could be traded off as a function of yield and process complexity and the bit-line contact on the surface could be shared to optimize and enhance both speed and density. The cells could be envisioned to be SLC or DLC or higher bit count per cell with appropriate stack design for cost, complexity, speed, endurance, retention trade-offs. The S-SGT devices, for example, could embrace multiple layered band engineered stack designs as discussed in earlier sections to provide larger window and enhanced attributes for MLS operations overcoming limitations discussed earlier for both NROM and NAND products.

16.4.2 Extendable S-SGT NV Devices and Arrays

SGT and S-SGT device concepts were extended in 2004 whereby a range of scalable devices, namely, PROM, capacitor-less DRAM (SGCL-DRAM), NVMs, and multifunctional memory devices and arrays were proposed within the same silicon substrate following compatible processing schemes [44]. Schematic cross-sections of SGT PROM, NROM/flash, and capacitor-less DRAM devices, as proposed are shown, respectively, in Figure 16.29a–c, respectively.

The polysilicon surround gate and the stack designs for NROM and SGCL-DRAM are illustrated in Figure 16.29a and b, respectively.

The PROM and NROM devices and arrays could be identical in the gate insulator stack design and with respect to (i) tunneling distance large enough and (ii) the charge centroid within the trapping dielectric being far enough from the silicon (pillar) interface to ensure life-time memory retention (>> 10 years). In case of SGCL-DRAM cell (Figure 16.29c), the effective tunneling distance for the charge transport was made thinner (<2 nm EOT) either with single tunneling layer or with multilayered tunneling dielectric and the trapping distance from silicon interface to the charge centroid was optimized for faster programming and erasing (in ~10–100 ns) with retention traded off to achieve memory refresh frequency requirement in the range of 1000–10000 sec to ensure memory integrity. Device stack and gate cross-sections for NROM and SGCL-DRAM are shown in Figure 16.30a and b, respectively.

The SGCL-DRAM devices not only achieve significantly higher bit density but also consume significantly lower standby power due to short channel immunity several orders of magnitude lower refresh requirements. Examples of gate stack design options are shown in Table 16.1, and expected device attributes are shown in Table 16.2.

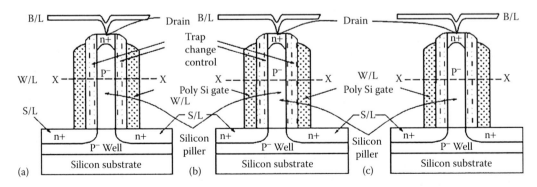

FIGURE 16.29   SGT devices: (a) PROM, (b) NROM/FLASH, and (c) SGCL-DRAM. (From Bhattacharyya, A., Integrated surround-gate multifunctional memory device, Micron Internal Docket: 2004–1348, 5/26/2004, Patent Application abandoned.)

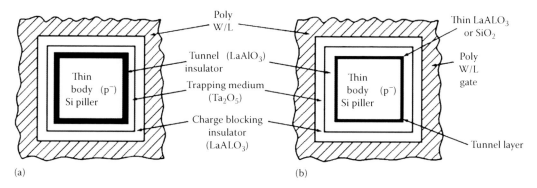

(a)   (b)

FIGURE 16.30 SGT device stack/gate cross-section: (a) NROM, (b) SGCL-DRAM. (From Bhattacharyya, A., Integrated surround-gate multifunctional memory device, Micron Internal Docket: 2004–1348, 5/26/2004, Patent Application abandoned.)

TABLE 16.1
Stack Design Options for SGT-NVM Devices

Stack Design Options for SGT-NVM Devices			
Device Types	**Tunnel Layer/Layers**	**Trapping Layer/Layers**	**Blocking Layer/Layers**
A: PROM/NROM/FLASH: EOT	>2.5 ~ 2.5 nm	~2.5 nm	~3.0 nm
B: SGCL-DRAM: EOT	~1.0 ~ 2.0 nm	~2.0 nm	~2.0 nm
Options:	OR-SiON or $LaAlO_3$	Si_2ON_2 or GaN	$LaAlO_3$
	SiO_2/SiON/SiO_2	AlN or TiO_2 or Ta_2O_5 or HfO_2	HfSiON or Al_2O_3

TABLE 16.2
Device Characteristics of SGT-NVM Devices

B: Device Characteristics of SGT-NVM Devices			
Device Types EOT	**Vpp/ Tpp Initial W EOL W**	**Retention**	**Endurance**
PROM/NROM FLASH ~8 nm	±8v/1ms >2v >1v	>1E8 s	>1E6 W/E
SGCL-DRAM <5 nm	±5v/0.1 µs >1v >0.5v	>1E3 s	>1E14 W/E

16.4.3 Multibit S-SGT Devices and Arrays

16.4.3.1 Surround Gate Floating Gate

S.J. Whang and coworkers from Hynix Semiconductor Group published in 2010 a multibit 3D "SSGT" FG-NAND flash cell for 1 Tb file storage application [43]. The four-string NAND array consisted of two top select gates with shared bit line at the top surface and common buried source line at the bottom of the trench. The device has a polysilicon channel for the surrounding floating gates featuring high coupling ratio and lower voltage cell operation. With +15 V writing and −13 V erasing a large memory window of 9.2 V was achieved suitable for multibit per cell capability. FG to FG interference of as small as 12 mV/V was demonstrated. The 3D S-SGT-FG-NAND structure is shown in Figure 16.31.

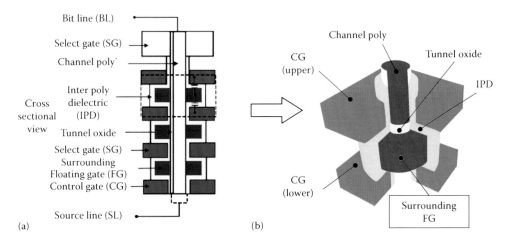

FIGURE 16.31   A surround gate floating-gate 3-D NAND structure: (a) 4-String NAND array; (b) 3D Cell structure showing surrounding Floating gate and Control gate. (From Whang, S. J., Lee, K. H., Shin, D. G. et al., Novel 3D dual control gate with surrounding floating gate [DC-SF] NAND flash cell for 1 Tb file storage application, *IEDM*, pp. 668–671, © 2010 IEEE.)

16.4.3.2   Stacked Gate-All-Around or Surround Gate CT NROM (SGAA-NROM OR SGT-NROM)

An SGT variation in a split channel design has been previously discussed, whereby the NROM cell with multibit/cell localized trapping could share the same surround gate. In that cell design, the N+ diffusion at the top of the silicon pillar acts as common diffusion node for the bit line of two adjacent NROM cells located in adjacent trenches, while the bottom N+ diffusion linked the two devices placed within a trench. Another variation in stacked GAA approach is discussed below for MLC-NROM and extended to MLC-NAND design concepts to achieve very high bit density. In this concept, both the top N+ diffusion and the top N+ diffusion can interchangeably act as bit line and source line as appropriate for multilevel sensing. The fixed Vth element at the middle of the vertical channel sets the low Vth memory state for the multibit cell. The bottom N+ ring serves as the "conventional" source line node. However, the role can be reversed. Different amount of charges could be stored locally at the near surface edge (edge A) and the near bottom edge (edge B) of the cell by programming the diffusion edges differently and independently. As stated above, the role of bit line and source line could be reversed to achieve multilevel sensing. By storing different amount of charges at different edges, 2 bit/cell (or 4 levels of memory state or even 3 bit/cell) (8 levels of memory state) could be achieved. The cross-section of the SGT-NROM cell is shown in Figure 16.32. A relatively simple DTM stack design could be considered to achieve large enough memory window for DLC operation. For example, a stack shown in the Figure 16.34, consisting of 2.5 nm SiO_2–1.6 nm Si_3N_4 (dual-layer tunneling) combined with 18 nm HfO_2 for trapping and 10.5 nm of Al_2O_3 for blocking with TaN interface (EOT = 11 nm) is estimated to provide nearly 8 V memory window programming-erasing at less than ±15 V, 1 msec conditions. The NROM design concept of features of Figures 16.32 and 16.33 could be vertically extended with SGT capability for high-density NAND flash designs shown in Figures 16.34 and 16.35, and is further discussed in this section [40].

16.4.3.3   SGT Multifunctional NROM

Another variation in SGT single cell NROM could be considered which could function simultaneously as capacitor-less DRAM as well as NVM. This family of devices will be discussed in greater detail in Chapter 8 in this part and also in Part III on SUM. These will be classified as "Multifunctional

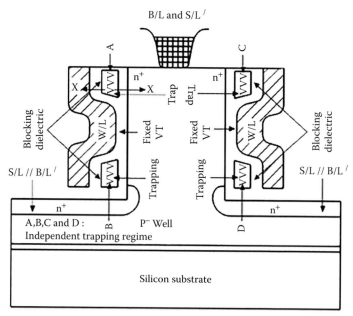

Layers through X-X

Metal or poly silicon / TaN / Al_2O_3 / HFO_2 / Si_3N_4 / SiO_2 / P^-Si

FIGURE 16.32 The split channel SGT-NROM device. (From Bhattacharyya, A., Extendable multi-level SGT-CT NROM and NAND flash devices with enhanced characteristic for multifunctional NVM technology applicable to 20 nm technology node and beyond, Internal Memo, ADI Associates, April 12, 2013.).

NVMs." Such multifunctionality could be achieved by appropriate gate stack designs in at least two different ways, based on band engineered gate stack designs discussed in Chapter 13:

- Multienergy single trapping plane design: Trapping plane with two well defined yet considerably different trapping energy depths for electrons and holes with asymmetric tunnel barrier. Such structures would provide bimodal retention characteristics of the device.

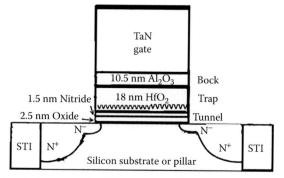

Typical stack structure for 8 V memory window at $Vpp^- +-18$ V, 10 mS

FIGURE 16.33 Illustration of a typical stack structure for high density MLC 3D NAND flash design. (From Bhattacharyya, A., Extendable multi-level SGT-CT NROM and NAND flash devices with enhanced characteristic for multifunctional NVM technology applicable to 20 nm technology node and beyond, Internal Memo, ADI Associates, April 12, 2013.)

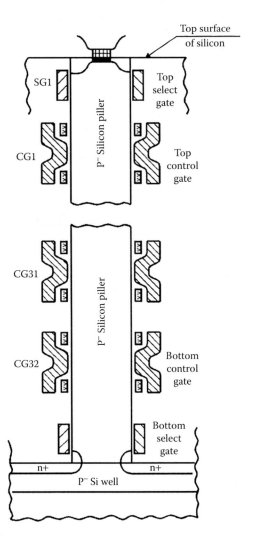

FIGURE 16.34 NAND flash design with vertical channel containing 32 string arranged vertically in a single silicon pillar with buried N+ link diffusion in a P well. (From Bhattacharyya, A., Extendable multilevel SGT-CT NROM and NAND flash devices with enhanced characteristic for multifunctional NVM technology applicable to 20 nm technology node and beyond, Internal Memo, ADI Associates, April 12, 2013.)

Shorter retention and faster carrier transport/trapping-detrapping would yield DRAM like functionality, while slower transport/trapping-detrapping would result longer retention providing NVM functionality. Illustrations of such devices will be covered in Chapters 21 and 23 in Part III of the book.

- Biplanar or multiplanar charge-trapping design: Trapping at different tunneling distances from the silicon interface while maintaining the energy barrier and trapping depth being the same in both cases. The first trapping plane is closer to the silicon charge injecting interface (typically single direct tunneling), whereas the second trapping plane is significantly at a greater distance from the interface changing the transport mode to FN tunneling with distinctly different energetics as explained in Chapter 13. Examples are covered previously in Chapter 13.

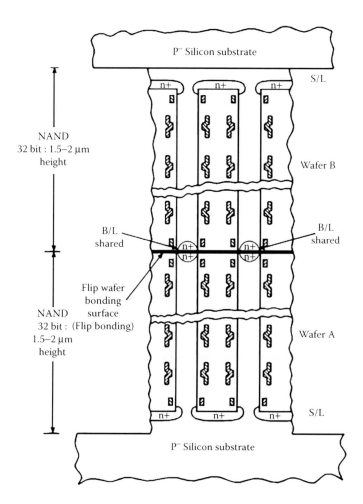

FIGURE 16.35 Vertical flip-chip bonding scheme inside the NAND flash array to achieve 64 bit string high-density vertical channel NAND flash design. (From Bhattacharyya, A., Extendable multi-level SGT-CT NROM and NAND flash devices with enhanced characteristic for multifunctional NVM technology applicable to 20 nm technology node and beyond, Internal Memo, ADI Associates, April 12, 2013.)

16.4.3.4 S-SGT Very High-Density (3D) NAND Design

Several concepts of stacked SGT design could be employed to achieve extremely high bit density (Tb NVM NAND flash) for 20 nm technology node and beyond. These are based on several stack design concepts discussed earlier in Chapter 13 and also in this chapter [41,42,44]. These are as follows:

- Vertically stacking 16 or 32 or 64 bit NAND strings with single level storage cells.
- Vertically stacking 2 or 3 or 4 levels of storage cells employing either split channel NAND design or by multilevel charge storage over diffusion edges ([22] of 2.3).
- Doubling the memory density of either of schemes of (1) and (2) above by sharing the bit lines and bit-line contacts by "flip-bonding" or "flip stacking" twin arrays on top of each other. This concept is illustrated in Figures 16.33 through 16.35.

This leads us to the discussions on current development status of full 3D devices and arrays.

16.5 FULL 3D NV DEVICES AND ARRAYS

Two different schemes of improving NAND memory bit density on silicon chip were evolved during 2006/2007 time frame. One was stackable planar or pseudoplanar (hybrid) NAND arrays from Samsung and Macronix International discussed in Chapter 15. The second was a 3D NAND array structure known as bit cost scalable (BiCS) by Toshiba Corporation [45]. Since then investigations on full 3D NAND cells and arrays have gained momentum and several 3D designs have been published. Some of these have been reviewed in the ITRS road map published in 2011 [46] and by S.W. Park in Flash Memory Summit in 2012 [47]. We will briefly review the progress in this section and highlight some of the relevant attributes.

Proposed 3D NAND structures by Toshiba, Samsung, Hynix, and Micron/Intel Technology are illustrated in Figure 16.36 with key features and issues mentioned.

P-BiCS NAND from Toshiba: [45,48–50]

Details of the 3D BiCS memory features are shown in Figure 16.37. The BiCS cell is conceptually similar to the SSGT cell features from the standpoint of NAND gates surrounding the vertical channel for coupling with surface bit-line contacts and buried source line. It also has several unique features as follows:

1. A low cost simple process of 3D integration called "punch-and-plug" scheme of punching a whole stack of electrode plates and plugging with poly silicon at the same time.
2. A "Macroni" body vertical FET strings as shown in Figure 16.37c and d. The Macroni body of poly silicon plug/substrate features a hollow cylindrical center which gets filled with nitride forming a set of cylindrical pillars or plugs. The vertically fixed Vth select gate devices feature SiN gate dielectric. The surround gate NAND string NVM elements initially featured SiN tunnel dielectric as well as Si-N trapping dielectric with oxide as blocking dielectric (N/N/O) for stack design. The original proposal was subsequently improved by pipe-shaped BiCS [49] for channel contact and a stack design update consisting of a double tunnel layer of SiN plus SiO_2 with SiN trapping layer and SiO_2 blocking layer (N/O/N/O) [50]. The thin Macroni body provides superior subthreshold characteristics compared

	p-BiCS (Toshiba)	TCAT (Samsung)	3D FG (Hynix)	Micron
Structure			CG FG CG	
	Tanaka. H, VLSIT 2007	J. Jang, VLSIT 2009	S. Whang, IEDM 2010	G. Hawk, FMS 2011
Key features	- P+ SONOS Cell	- TANOS Cell	- Floating gate	?
Key issue	- Large cell size - Reliability	- Large cell size - SL resistance	- Process of bit separation - Disturbance	?

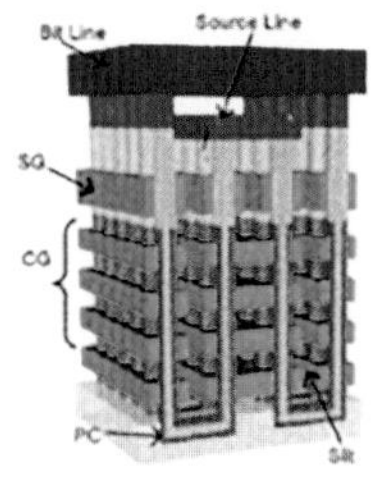

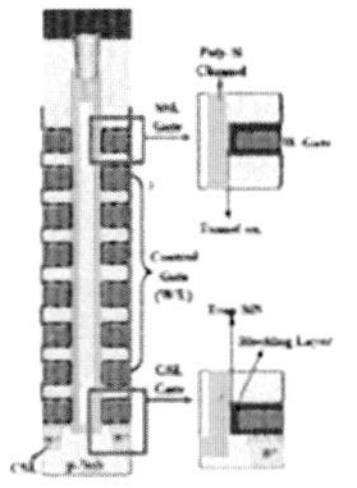

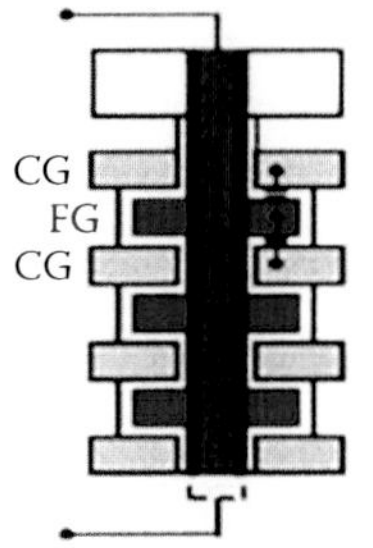

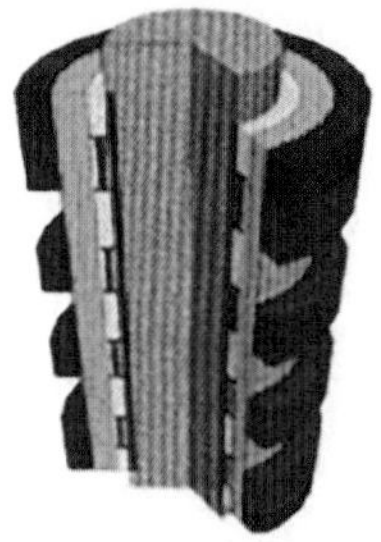

FIGURE 16.36 Proposed 3D NAND structures from leading providers. (From Park, S. W., Prospect for new memory technology, S. K. Hynix, *Flash Memory Summit*, Santa Clara, CA, August, 2012.) (Reproduced from EE Times, August, 2012, courtesy EE Times).

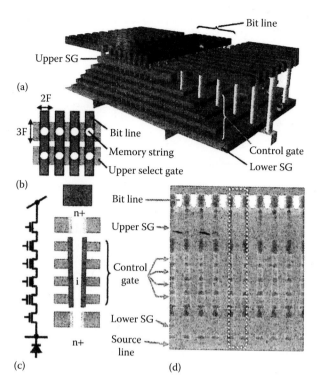

FIGURE 16.37 BiCS features: (a) 3D structure; (b) 2-D Layout, (c) Vertical NAND design; and (d) Vertical-cross-section across (b). (From Tanaka, H., Kido, M., Yahashi, K. et al., Bit cost scalable technology with punch and plug process for ultra high density flash memory, *VLSI technology*, pp. 14–15, [Toshiba] © 2007 IEEE; Fukuzumi, Y., Katsumata, R., Kito, M. et al., Optimal integration and characteristics of vertical array devices for Ultra-High Density, Bit-Cost scalable flash memory, *IEDM*, pp. 449–452. [Toshiba] © 2007 IEEE.)

to the conventional solid polysilicon body as shown in Figure 16.38a from the Id-Vg characteristics of the device and tighter Vth distribution as shown in Figure 16.38b. The concept of Macroni body vertical FET is shown in Figure 16.39. The polysilicon Macroni body is lightly doped or undoped to prevent forming p-n junction within the plug and the FETs work in the depletion mode.

3. For erase operation, hole current is generated by GIDL (gate induced drain lowering) near the lower select gate to raise the body potential required for erasing. Common N+ diffusion is formed on the silicon substrate at the bottom for the source line.

The vertical channel architecture of the 3D NAND array is shown in Figure 16.40 demonstrating the vertical placement of the NAND strings with the word-lines configured in the horizontal planes.

Program-erase characteristics of NAND memory could be achieved at lower voltage levels compared to conventional planar FG-NAND memory. Such characteristics are shown in Figure 16.41: a memory window of +4 V, −1 V (~5 V window) at +14 V at 100 ms write and −13 V, 100 ms erase. Endurance and retention characteristics are shown in Figure 16.42a and b, respectively.

BiCS 128 Gb FLASH memory with vertical 8 bit string was fabricated at 90 nm technology node with 6F^2 cell [45,48–50]. The authors addressed the extendibility of the concept down to 15 nm technology node by means of reducing the thickness of the Macroni body and addressing the concepts of reducing the 6F^2 cell to 4F^2 cell. Extendibility of Macroni body vertical FET Vth variations in 300 mm wafer is analyzed and shown in Figure 16.43. Conceptual proposal of evolution from 6F^2 to 4F^2 memory string has been proposed using double-layered alternate select gate structure. Relative bit cost reduction potential as a function of density enhancement

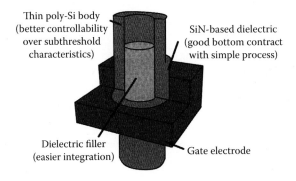

FIGURE 16.38 (a) Id-Vg, and (b) Vth spread characteristics between conventional FET and Macroni-body vertical FET. (From Tanaka, H., Kido, M., Yahashi, K. et al., Bit cost scalable technology with punch and plug process for ultra high density flash memory, *VLSI technology*, pp. 14–15, [Toshiba] © 2007 IEEE.)

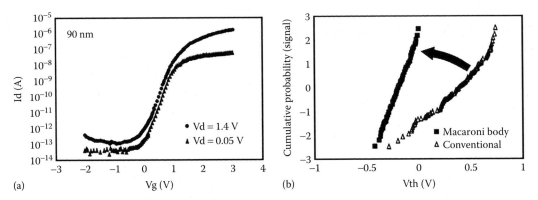

FIGURE 16.39 Concept of macroni body vertical FET. (From Tanaka, H., Kido, M., Yahashi, K. et al., Bit cost scalable technology with punch and plug process for ultra high density flash memory, *VLSI technology*, pp. 14–15, [Toshiba] © 2007 IEEE.)

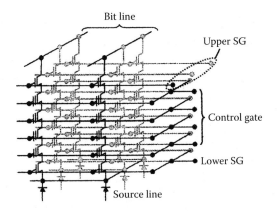

FIGURE 16.40 BiCS vertical channel 3D NAND architecture. (From Tanaka, H., Kido, M., Yahashi, K. et al., Bit cost scalable technology with punch and plug process for ultra high density flash memory, *VLSI technology*, pp. 14–15, [Toshiba] © 2007 IEEE.)

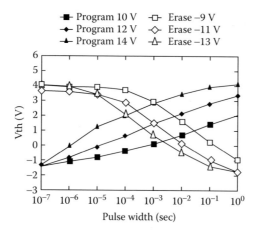

FIGURE 16.41 Program/erase characteristics of BiCS NAND device. (From Tanaka, H., Kido, M., Yahashi, K. et al., Bit cost scalable technology with punch and plug process for ultra high density flash memory, *VLSI technology*, pp. 14–15, [Toshiba] © 2007 IEEE.)

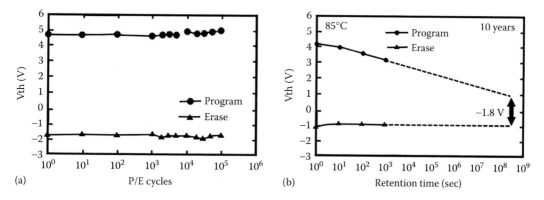

FIGURE 16.42 Endurance and retention of BiCS NAND device. (From Tanaka, H., Kido, M., Yahashi, K. et al., Bit cost scalable technology with punch and plug process for ultra high density flash memory, *VLSI technology*, pp. 14–15, [Toshiba] © 2007 IEEE.)

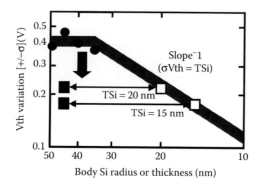

FIGURE 16.43 Extendibility potential and Vth variation reduction of Macroni body. (From Tanaka, H., Kido, M., Yahashi, K. et al., Bit cost scalable technology with punch and plug process for ultra high density flash memory, *VLSI technology*, pp. 14–15, [Toshiba] © 2007 IEEE.)

with increased number of layers in the NAND string as projected in Figure 16.43. A 10-fold cost reduction potential was claimed for 1 Tb memory.

The 3D TCAT-TANOS V-NAND from Samsung: [51–54]: Samsung proposed a terabit cell array transistor (TCAT) technology and vertical cell array architecture incorporating surround-metal gated (W/TaN) vertical TANOS NAND stack for ultra-high density NAND flash memory in 2009. Similar to the Toshiba approach, the vertical NAND adopted polysilicon channel. The fixed Vth select gates incorporated silicon/nitride/oxide/metal design whereas the NAND memory element incorporated TANOS stack design with oxide for tunneling layer. Bulk erase and cell operations were successfully demonstrated. Program disturbance with increasing number of layers were demonstrated to be well contained. In 2013, Samsung announced that it had initiated production of 128 Gb vertical NAND (V-NAND) chip based on charge-trapping (CT-TANOS) cell design with a technology capable of stacking up to 24 layers [53,54]. Samsung claimed a reliability enhancement of 2X to 10X over the planar conventional 128 Gb FG-NAND product announced previously in April 2013 which was using 3-bit multilevel FG cell design. Samsung called the 3D V-NAND technology as "CTF" (charge-trap flash) technology and uses a novel vertical interconnect process to link memory cells. The V-NAND chip being of the same size as the planar chip is expected to be fabricated in relaxed design rules and technology nodes resulting in significantly lower cost/bit.

The 3D FG-NAND Design from Hynix Semiconductor Inc. [43]: S. Whang and coworkers published their 3D "dual-control-gate" S-SGT-NAND flash cell and array architecture in 2010. The cell was called "DC-FC" (Dual-Control-Floating Gate) Cell by the authors. The cell has been introduced earlier in the section on S-SGT and is a surround floating gate design interacting with both the top and bottom control gates (dual CG). The cell structure provides high coupling perhaps at the expense of higher neighboring cell interference.

However, the cell exhibited operation at lower programming and erasing (15 V for programming and 11 V for erasing) when compared with planar FG-NAND cell designs. Large memory window of 9.2 V was demonstrated capable of MLC operation of 2, 3 bit/cell and possibly 4 bit/cell. The authors claimed minimal FG-FG interference of 12 mv/V and believed the design was targeted for 1 Tb density file storage application.

The Micron-Intel 3D NAND cell [55,56]: G Hawk from Micron Technology introduced a 3D NAND cell design jointly being developed with Intel in 2011. The cell design incorporates an annular trapping region in relaxed geometry to store as much as 10,000 electrons and therefore not limited by FEE effect. The NAND flash cell proposal incorporated deep trench technology of Micron used in conventional DRAM. The NAND flash memory array design displayed a 16 layer deep NAND string. At the time of writing this book, the details of the cell and array design were not available.

SAMSUNG 3D 256 Gb V-NAND Flash PRODUCT [57]: At the time, this book was ready for publication (April 2016), Samsung announced their latest 3D V-NAND flash product featuring 256 Gb 3-bit (TLS) multilevel flash with 48 levels of lithography produced at 21 nm technology node. The chip size was 99 mm^2 with a bit density of 2600 Mb/mm^2, compared to Samsung's Planar 64 Gb (TLS) flash product, released in 2015 with a chip size of 86.4 mm^2 with a bit density of 740 Mb/mm^2. The memory cell is a charge-trap device (CT-NVM) in contrast to the planar floating gate device (FG-NVM). Additionally, the NAND memory cell is a variation in vertical channel surround-gate device with polysilicon channel-metal-gate FET. Oxide isolation vertical stud is employed for the polysilicon substrate bit string. Conceptually, the device is similar to the BiCS from Toshiba discussed earlier. For details, the reference is provided [57].

REFERENCES

1. K. Prall, and K. Parat, 25 nm 64 Gb NAND technology and scaling challenges, *IEDM*, 2010, pp. 102–105.
2. H.-T. Lue, Y.-H. Hsiao, K.-Y. Hsieh et al., Scaling feasibility study of planar thin floating gate (FG) NAND flash devices and size effect challenges beyond 20 nm, *IEDM*, Washington DC, 2011, pp. 203–206.

3. T.-H. Hsu, H.-T. Lue, W.-C. Peng et al, A high-performance and scalable FINFET BE-SONOS device for NAND-FLASH memory application, *VLSI Technology, Symposium and Application*, Hsinchu TWN, 2008. pp. 56–57.

4. H. Takato, K. Sunouchi, N. Okabe et al., High performance CMOS surrounding gate transistor (SGT) for ultra high density LSIs, *IEDM*, 1988, pp. 222–225.

5. D. Hisamoto, W.-C. Lee, J. Kedzierski et al., Fin-FET-a self-aligned double-gate MOSFET scalable to 20 nm, *IEEE Trans. Electron Devic.*, 47(12), 2320–2325, 2000.

6. Y. K. Lee, S. K. Sung, J. S. Sim, et al., Multi-level vertical channel SONOS nonvolatile memory on SOI, *IEEE Electron Devic. Lett.*, 23, pp. 664–666, 2002.

7. Y. K. Lee, J. S. Sim, S. K. Sung et al., Multi-level vertical channel SONOS nonvolatile memory on SOI, *VLSI*, 2002, pp. 208–209.

8. T. Endoh, M. Hioki, H. Sakuraba et al., Floating channel type SGT flash memory, ECS, Japan, 199th Fall Meeting, Vol. 99–92, Abstract 1323, Hawaii, October pp. 17–22, 1999.

9. S. H. Lee, J. J. Lee, and J.-D. Choe et al., Improved post-cycling characteristic of FinFET NAND Flash, *IEDM*, San Francisco CA, 2006, pp. 33–36. 56–57.

10. P. Xuan, M. She, B. Harteneck et al., Fin FET SONOS flash memory for embedded applications, *IEDM*, 2003, pp. 609–612.

11. M. Specht, R. Kommling, L. Dreeskomfeld et al., Sub-40 nm tri-gate charge trapping nonvolatile memory cells for high-density applications, *VLSI*, Honolulu HI, 2004, pp. 244–245.

12. A. Bhattacharyya, Scalable high density, high performance hybrid planar-vertical NV memory cells and contactless arrays, Micron Internal Docket: 2004–0942, 2004, U.S. Patents: 7166888 (1/23/07), 7273784 (9/25/07), 7964909 (6/21/2011).

13. A. Bhattacharyya, Extremely high density two-bit-per-cell planar-vertical embedded trap, NVM cell and contactless array, Micron Internal Docket: 2004–0943, 2004, U.S. Patents: 7671407 (3/2/2010), 7838362 (11/23/2010).

14. A. Bhattacharyya, A vertical cross-point high density split channel direct tunnel non-volatile memory, Micron Internal Docket: 2004–0929, August 2, 2004, U.S. Patents: 7365388 (4/29/2008), 7671407 (3/2/2010).

15. A. Bhattacharyya, Defining high speed high endurance scalable NVM device using multimechanism carrier transport and high density structures for contact-less Arrays, Micron Internal Docket: 2004–1008, 9/1/2004; U.S. Patents: 7244981 (7/17/07), 7553735 (6/30/09).

16. L. Masoero, G. Molas, F. Brun et al., Scalability of split-gate charge trapping memories down to 20 nm for low power embedded memories, *IEDM*, Washington DC, 2011, pp. 9.5.1–9.5.4.

17. D. Shum, A. K. Tilke, L. Pescini et al., Highly scalable flash memory with deep trench isolation embedded into high performance CMOS for the 90 nm node and beyond, *IEDM*, Washington DC, 2005, pp. 355–358.

18. D. Lee, F. Tsui, J.-W. Yang et al., Vertical floating-gate 4.5 F^2 split-gate NOR flash at 110 nm node, *VLSI*, 2004, pp. 72–73.

19. J. Song, T. Kim, D. Oh et al., U-shaped FLOTOX cell for 16 G MLC (Multi-Level Cell) NAND flash memory devices, *NVSMW*, 2005 or 2006.

20. M. Specht, U. Dorda, L. Dreeskornfeld et. al., 20 nm tri-gate SONOS memory cells with multiple level operation, *IEDM*, 2004, pp. 1083–1086.

21. T.-H. Hsu, H.-T. Lue, E.-K. Lai et al., A high Speed BE-SONOS NAND-FLASH utilizing the field-enhancement effect of FinFET, *IEDM*, 2007, pp. 913–916.

22. J.-R. Hwang, T.-L. Lee, H.-C. Ma et al., 20 nm gate Bulk-FinFET SONOS flash, *IEDM*, Washington DC, 2005, pp. 161–164.

23. C. Friederich, M. Specht, T. Lutz et al., Multi-level p+ tri-gate SONOS NAND string arrays, *IEDM*, San Francisco CA, 2006, pp. 963–966.

24. J. P. Chiu, Y. L. Chou, H. C. Ma et al., Program charge effect on random telegraph noise amplitude and its device structural dependence in SONOS flash memory, *IEDM*, Baltimore MD, 2009, pp. ISBN 978-1-4, 244-5639-0.

25. C. H. Jan, M. Agostinelli, H. Deshpande et al., RF CMOS technology scaling in High-k/Metal Gate, era for RF SoC (system-on-chip) applications, *IEDM*, San Francisco CA, 2010, pp. 647–650.

26. C.-C. Yeh, C.-S. Chang, H.-N. Lin et al., A low operating power FinFET transistor module featuring scaled gate stack and strain engineering for 32/28 nm SoC technology, *IEDM*, San Francisco CA, 2010, pp. 34.1.1–34.1.4.

27. T.-J. K. Liu, FinFET history, fundamentals and future, 2012 Symposium on VLSI technology short course, June 11, 2012, UC Berkeley.

28. R. Merritt, Intel's FinFETs approach draws fire from rivals, EE Times, 12/13/2012.

29. R. Meritt, GloFo, Samsung in race to 14 nm, EE times, 2/6/13.

30. R. Meritt, IBM, Intel face off at 22 nm, EE Times, 12/10.12.
31. R. Merritt, IBM outlines fab future beyond FinFETs, EE Times, 2/12/13.
32. C. H. Jan, U. Bhattacharyya, R. Brain et. al, A 22 nm SoC platform technology featuring 3-D Tri-Gate and High-k/Metal Gate, optimized for ultra low power, high performance and high density SoC application, *IEDM*, San Francisco CA, 2012, pp. 44–47.
33. S. D. Suk, S.-Y. Lee, S.-M. Kim et al., High performance 5 nm radius twin silicon nanowire MOSFET (TSNWFET) fabrication on bulk si wafer characteristics and reliability, *IEDM*, Washington DC, 2005, pp. 717–720.
34. K. H. Cho, S. D. Suk, Y.-Y. Yeoh et al., Observation of single electron tunneling and ballistic transport in twin silicon nanowire MOSFETs (TSNWFETs) fabricated by top-down CMOS process, *IEDM*, 2006, pp. 543–546.
35. K. H. Yeo, S. D. Suk, M. Li et al., Gate-All-Around (GAA) twin silicon nanowire MOSFET (TSNWFET) with 15 nm length gate and 4 nm radius nanowire, *IEDM*, San Francisco CA, 2006, pp. 539–542.
36. N. Singh, F. Y. Lim, W. W. Fang et al. Ultra-narrow silicon nanowire Gate-All-Around CMOS devices: Impact on diameter, channel-orientation and low temperature on device performance, *IEDM*, San Francisco CA, 2006, pp. 547–550.
37. L. K. Bera, S. Nguyen, N. Singh et al., Three dimensionally stacked SiGe nanowire array and Gate-All-Around pMOSFETs, *IEDM*, San Francisco CA, 2006, pp. 551–554.
38. J. Fu, K. D. Buddhahaju, S. H. G. Teo et al., Trap layer engineered Gate-All-Around vertically stacked twin Si-nanowire nonvolatile memory, *IEDM*, Washington DC, 2007, pp. 79–82.
39. A. Hubert, E. Nowak, K. Tachi et al., A stacked SONOS technology, up to 4 levels and 6 nm crystalline nanowires, with gate-all-around or independent gates (O-Flash), suitable for full 3d integration, *IEDM*, Baltimore MD, 2009, pp. 637–640.
40. A. Bhattacharyya, Extendable multi-level SGT-CT NROM and NAND flash devices with enhanced characteristic for multifunctional NVM technology applicable to 20 nm technology node and beyond, Internal Memo, ADI associates, April 12, 2013.
41. T. Endoh, K. Kinoshita, T. Tanigami et al., Novel ultra high density flash memory with a stacked-surrounding gate transistor (S-SGT) structured cell, *IEDM*, 2001, pp. 33–36.
42. T. Endoh, K. Kinoshita, T. Tanigami et al., Novel ultra-high density flash memory with a stacked-gate-surrounding gate transistor (S-SGT) structured cell, *IEDM*, 2003, pp. 945–951.
43. S. J. Whang, K. H. Lee, D. G. Shin et al., Novel 3D dual control gate with surrounding floating gate (DC-SF) NAND flash cell for 1 Tb file storage application, *IEDM*, 2010, pp. 668–671.
44. A. Bhattacharyya, Integrated surround-gate multifunctional memory device, Micron Internal Docket: 2004–1348, 5/26/2004, Patent Application abandoned.
45. H. Tanaka, M. Kido, K. Yahashi et al., Bit cost scalable technology with punch and plug process for ultra high density flash memory, *VLSI technology*, 2007, pp. 14–15. [Toshiba].
46. ITRS Roadmap: 2011, IEEE.
47. S. W. Park, Prospect for new memory technology, S. K.Hynix, *Flash Memory Summit*, Santa Clara, CA, August, 2012.
48. Y. Fukuzumi, R. Katsumata, M. Kito et al., Optimal integration and characteristics of vertical array devices for Ultra-High Density, Bit-Cost scalable flash memory, *IEDM*, Washington DC, 2007, pp. 449–452. [Toshiba].
49. R. Katsumata, M. Kito, Y. Fukuzumi et al., Pipe-shaped BiCS flash memory with 16 Stacked layers and multilevel cell operation for ultra high density storage devices, *VLSI Tech.*, Kyoto Japan, 2009, pp. 136–137. [Toshiba].
50. Y. Komori, M. Kido, M. Kito et al., Disturbless flash memory due to high boost efficiency on BiCS structure and optimal memory film stack for ultra-high density storage device, *IEDM*, San Francisco CA, 2008, pp. 1–4, DOI. 10-1109. [Toshiba].
51. J. Jang, H.-S. Kim, W. Cho et al., Vertical cell array using TCAT technology for ultra high density NAND flash memory, *VLSI Tech.*, Kyoto, Japan, 2009, pp. 192–193. [Samsung].
52. W. Kim, S. Choi, J. Sung et al., Multilayered vertical gate NAND flash overcoming stacking limit for terabit density storage, *VLSI Tech.*, 2009, pp. 188–189. [Samsung].
53. P. Clarke, 3D NAND production starts at Samsung, EE Times, 8/6/2013. [Samsung].
54. P. Clarke, Samsung confirms 24 layers in 3D NAND, EE Times, 8/9/2013. [Samsung].
55. G. Hawk, 3D NAND flash cell stack, *Micron Technology Flash Summit*, 2012, EE Times, EDA 360 insider, "A look at some genuine 3D NAND cells", courtesy of Micron: 2/19/12. [Micron].
56. P. Clarke, Intel outlines 3-D NAND transition, EE Times, 6/2/2013. [Micron].
57. K. Gibb, Techinsights, First look at Samsung's 48L 3D V-NAND flash, E.E.Times, 4/6/2016.

17 Emerging NVMs and Limitations of Current NVM Devices

CHAPTER OUTLINE

This chapter takes into account the current state of volatile and nonvolatile memory products and device attributes in the context of market and application drivers at present and in foreseeable future. It also outlines the current limitations of such devices. Additionally, it also briefly introduces the emerging nonvolatile memory contenders, namely, phase change memory [PCM (PRAM)], STT-MRAM, and Re-RAM (RRAM). These emerging memories have unique attributes and challenges and are poised to displace conventional silicon based volatile and nonvolatile memories. It is suggested that SUM devices (Part III) has the prospect to meet the challenges against the emerging memories.

The digital world of intra and interconnectivity has touched all aspects of our lives requiring exponential growth in transactions and storage. The storage or memory solutions have essentially taken two forms: The embedded form in close proximity with the processing units for transactional efficiency [1,2,16–18]; and the stand-alone form [3–16] to satisfy volume or capacity requirements for transactional capability. During the last two decades, we have witnessed phenomenal growth of silicon based NVMs in the form of NROM and NAND flash devices and arrays increasingly displacing DRAM at the "performance-end" and magnetic storage media (HDD and Tapes) in the "capacity-end" of the storage hierarchy of the digital system. This has been due to the storage requirements fueled by an ever-increasing growth of portable devices at one end and enterprise/large storage systems at the other end, which is expected to continue in foreseeable future. The important question to address is that how such devices are poised for future. In this section, we will attempt to address such question by examining the limitations of such devices for future market penetration and growth and extendibility requirements to sustain current growth. We will address issues both at the device level and at the subsystem/system level. Additionally, we will briefly outline the enhancement requirements to sustain such growth against new and emerging solutions of other types of NV devices.

Both embedded and stand-alone nonvolatile memories are being used in a diverse range of applications today with a wide range of requirements of capacity, memory subsystem objectives, and device attributes. It is worth noting that the attributes at the core cell/array/device levels impact at the higher functional integration levels such as: (i) at the SSD level for the stand-alone memory which integrates the memory units and the controller functions, and (ii) at the application level of integration whereby cost/power/performance/reliability for certain application is evaluated. We will address the relevant "drivers" at different integration levels impacting the market horizon for the current NVM devices.

17.1 DEVICE LEVEL AND FUNCTIONAL LEVEL ATTRIBUTES OF DRAM, NVMs (SSDs), AND HDD

Tables 17.1 and 17.2 list some of the key attributes of DRAM, NROM, and NAND NVMs and HDDs at device levels [1,2,18] and functional levels, respectively ([3] 1/n [17]).

17.2 THE NVM MARKET HORIZON AND DRIVING FACTORS

The NVM market today could be broadly classified into four major subgroups: (I) "Consumer" level such as smart card and automotive requiring simple well-defined sets of functions [18]; (II) Intelligent "Multifunctional Mobile" level such as cell-phone, tablets, laptops, and PCs with increasing networking and functional capabilities; (III) "Enterprise System" Level with increasing emphasis on performance, data integrity, "IOP" cost, and efficiency and data availability often in concert with DRAM; and (IV) "Cloud/Data Center" level requiring as much as zettabytes (1E21) of storage with increasing demand on cost reduction, higher IOPs, and power reduction and increasing capacity often optimized through combining SSD cache functions with HDD high volume low cost storage. The driving factors at different integration levels are summarized in Table 17.3.

17.2.1 Observations

From the Tables 17.1 and 17.2, the limitations associated with market expansion of silicon based NVM devices and products can be ascertained and potential device enhancements could be suggested for continued market drivers for devices and products.

This is summarized in the following section:

17.2.1.1 Consumer and Portables

Applications requiring consumer and multifunctional portables for both embedded and stand-alone forms of memory where storage requirements could be limited to ~1 terabyte (1E12 byte), primary drivers for expanding market shares are cost/bit, form factor, power, and reliability. In spite of cost/bit advantage, SDD (with moving parts) could not be favored due to power, form factor, and reliability disadvantages. As a result, SSD and associated NROM and NAND flash devices have gained momentum in recent past over HDD options and have firmly established market horizon. This is expected to continue in near future. However, to expand the market horizon, especially in the higher performance

TABLE 17.1

Key Design and Functional Attributes of Current NVMs

		NROM (CT)		NAND Flash (FG)			
	DRAM	SLC	DLC	SLC	DLC	TLC	HDD
Cell Size/bit	6 F2	4–5 F2	2.5 F2	4 F2	2 F2	<1.5 F2	0.5 F2
Read Time	<50 ns	~50 ns	~50 ns	~50 ns	~50 ns	100 ns	2 µs
Write/Erase Time	<10 ns	100 µs/1 ms	100 µs/1 ms	1/0.1 ms	1/0.1 ms	2/0.1 ms	1 µs
Write Energy: J/bit	15	1.00E–16	1.00E–16	2.00E–16	2.00E–16	>2E16	Low
Retention	~60 ms	10 years	10 years	10 years	1–10 years	1 year	10 years
Endurance	>1E16	~1E7	~1E7	1.00E+05	1.00E+04	1.00E+03	1.00E+18
Integration Complexity For Embedded	Highest	Low	Low	High	High	High	NA

TABLE 17.2
Key Functional Attributes of DRAM, NROM-NVM, SSD, and HDD

	DRAM	NROM (CT)		SSD: FG-NAND			HDD
		SLC	DLC	SLC	DLC	TLC	
COST/BIT	Highest	High	Lower	Low	Lower	Lower	0.1 × of TLC-NAND
ACCESS	Random: highest ≪50 ns	Random: high	Lower	Serial: low	Lower	Lower	Serial: lowest ≫ us
CYCLE	Highest ≪30 ns	Slow ~1 ms	Slower	Slow ~10 ms	Slower	Slower	Slowest ~> us
POWER CONSUMED YRLY	High	0.25 × DRAM		~SAME as DRAM			Highest: ~>100 × DRAM
LATENCY	Lowest	High	Higher	High	Higher	Higher	Highest
DATA INTEGRITY:							
Volatility	Volatile	Nonvolatile		Nonvolatile			Nonvolatile
ERROR RATE (UBER)	Low	1.00E-15		1.00E-14			?
ECC Overhead	Low	Low	High	Higher	Highest		?
Retention	<100 ms	10 years	10 years	~1 year	~< 1 year		Longest ≫ 10 years
Endurance (W/E Cycles)	1.00E+16	1.00E+07	>1E5	<1E5	≤3 E4		1.00E+18
DATA UPDATE/AVAILABILITY							
IOP (per sec)	>1E7	>5E6	<5 E5	~5 E4	~3 E4		<2 E2
Power Efficiency IOP/watt	High	NA		~1500/watt			10–15/watt
Cost Efficiency: $/IOP	High (low cost/IOP	NA		0.05–0.15/IOP			Very low: 0.5–2.5/IOP
I/O Channel Throughput	~500 MT/s	~400 MT/s		40–400 MT/s			<1 KT/sec
RELIABILITY	High	High		High			Low

TABLE 17.3

Key Application/Market Drivers of DRAM, NROM-NVM, SSD, and HDD

	DRAM	NROM (CT)		SSD: FG-NAND			HDD
		SLC	DLC	SLC	DLC	TLC	
A: Consumer:							
Cost/Bit	N	N	+	+	++	+++	+++++
Volatility	–––	+	++	+	+	+	+++
Power	–	N	N	N	N	N	–––
Form Factor	+	++	++	+	++	++	––
Reliability	N	+	++	+	+	N	––––
B: Mutifunctional Portables:							
Cost/Bit	N	N	+	+	++	+++	+++++
Volatility	–––	+	++	++	++	+	++++
Power	–	N	N	N	N	N	––––
Form Factor	+	++	++	+	++	++	–
Endurance	+++	+	+	N	–	–	++++
IOP&IO	++	++	++	+	+	N	––––
Reliability	+	+	++	+	+	N	––––
C: Enterprise:							
Cost/Bit	N	N	+	+	++	+++	+++++
Cost/Transaction	N	N	+	+	++	+++	–
Cost/Performance	++++	++	+++	+	N	–	–
Cost/Data Integrity	N	N	N	N	–	–	N
Cost/IOP	+++	++	++	++	+	N	–––
Nonvolatility	–––	++	++	+	+	N	++++
Endurance	+++++	++	++	+	+	N	+++++
Performance	+++++	++	++	+	+	N	––––––
Transaction/sec	+++++	+++	+++	++	++	+	––––––
IOP (Data Alterability)	+++++	+++	+++	++	+	+	––––––
Throughput	+++++	++++	++++	+++	++	+	–––––
Power Consumption	N	++	++	N	N	N	––––––
Reliability	+	++	++	+	N	–	––––––
D: Cloud:							
Cost/Bit	N	N	+	+	++	+++	+++++
Cost/Transaction	N	N	+	+	++	+++	–
Cost/Data Integrity	++++	++	+++	+	N	–	N
Cost/Performance	++++	++	+++	+	N	–	–
Cost/IOP	+++	++	++	++	+	N	––––––
Cost/I-O	+++	++	++	++	+	N	–––––
Capacity	N	+	++	+	++	+++	+++++
Future CGR	N	+	++	++	++	+++	+++
Nonvolatility	–––	++	++	+	+	N	++++
Endurance	+++++	++	++	+	+	N	+++++
Performance	+++++	+++	+++	+++	++	+	–– (15k RPM)
Transaction/sec	+++++	+++	+++	++	++	+	–––––
IOP (Data Alterability)	+++++	+++	+++	++	+	+	–––––
Throughput	+++++	++++	++++	+++	++	+	–––––
Power Consumption	N	++	++	N	N	N	––––––
Reliability	+	++	++	+	N	–	–––––

Market/Application Segments: A: Consumer; B: Multifunctional Portables; C: Enterprise; D: Cloud

arena and further displace DRAM requirements, NVM devices need to enhance programming speed and endurance while maintaining the advantages against DRAM in cost/bit, nonvolatility (10 years), and reliability. This is particularly true for applications favoring DRAM-like performance. Potential solutions to be discussed in Chapter 18 and in part III to fulfill such market need would be FP/CT-based multifunctional devices and SUMS with simultaneous DRAM and NVM functionality.

17.2.1.2 Enterprise

Enterprise memory requirements are increasingly driven to higher performance, higher transactional efficiency, higher data integrity with lower ECC overhead cost and complexity, lower power consumption, and higher reliability for the memory subsystems. The performance (as well as cost/performance), IOP, throughput, transactional cost, and reliability advantages of SDD over HDD have been driving increasing SSD applications in enterprise market displacing HDD in its traditional role. This trend is expected to continue in foreseeable future either in the form of total replacement of HDD or in the form of partial replacement (for lower end cost-sensitive market). It is projected that SSD storage gigabyte will grow nearly 5× between 2011 and 2016 (5 years) for enterprise market [17]. The limitations of current SSD solutions for enterprise market dependent on FG-DLC and FG-TLC devices are (a) ECC overhead cost associated with high UBER; (b) lower than desired programming speed; (c) low endurance; (d) decreasing retention limits (<1 year for TLC); (e) increased cost of overprovisioning, enhanced programming algorithm affecting write latency, system management overhead to ensure data integrity; and (f) limited memory system reliability. Many of these limitations could be overcome through device enhancements through band-engineered FPNVM and CT-NVM stack designs providing MLC multifunctional attributes to be discussed in Chapter 23 and part III. It should be noted that current memory hierarchy involves embedded L1 and L2 SRAMs with 6-device cells (typically >30 F2/bit) with significant on-board or stand-alone DRAM as L3. Both SRAM and DRAM are volatile memories and critical instructions and data have to be off-loaded to and fetched from permanent storage areas from time to time for permanency. Multifunctional embedded and stand-alone NVMs with DRAM functionality coupled with NVM characteristics could be employed through architectural enhancements of memory subsystems in enterprise solutions, thereby potentially reducing traditional SRAM and DRAM contents and providing cost/performance improvements of future systems.

17.2.1.3 Cloud

Historically the most significant driving factors in the past for mass storage had been nonvolatility, cost/bit, and capacity. Such requirements had been well fulfilled by magnetic tapes and HDD for systems requiring mass storage.

As the Internet evolved over the last two decades, the mass storage market has been exploding with ever-growing cloud computing. In such environment, not only the above mass storage attributes are important drivers but also others, namely, performance (transactions/sec), IOP, throughput, power efficiency, reliability, and cost issues associated with all the above additional driving parameters. During the last two decades, NVM FG-NAND flash became increasingly attractive bridging not only the performance gap between the existing DRAM and HDD/Tape in the memory hierarchy but also providing the additional driving attributes mentioned above for the cloud requirements. This is summarized in the above table on clouds. The impressive success of MLC-FG-NAND flash (DLC and subsequently TLC) products driving the technology feature size nodes during the past decade as well as the silicon based memory volume/revenue has much to do with the growth and application of scaled NAND flash devices for the cloud computing market. This growth is expected to continue in near future for at least up to 15 nm technology node. In the meantime, 3D-CT-NAND flash memory cells are being vigorously pursued including the "VNAND," as discussed in the preceding section, to carry the momentum forward beyond the 20 nm technology node.

It has been noted earlier and illustrated in the above table that with increasing number of NAND bit/cell (SLC->DLC>TLC), essential for cost reduction in cloud application, the nonvolatile properties such as memory state margin, retention, endurance, and UBER are progressively degraded limiting the current FG-NAND flash future application and appeal. Cost/bit reduction in proposed 3D-NAND designs is still to be proven for 20 nm technology node and beyond. Additionally, non-silicon based NVM options currently being pursued (e.g., MRAM, discussed later) may prove to be viable contenders in sub-15 nm technology nodes. Therefore, to keep the market momentum continuing, the scaled NAND flash devices not only have to overcome the limitations mentioned above but have to demonstrate cost-effective viability for 4 b/cell planar or nonplanar designs with enhanced device attributes, lower ECC and system overheads, and improved reliability in order to expand market presence and effectively compete against emerging contenders for the cloud computing high capacity market.

17.3   EMERGING CONTENDERS FOR CONVENTIONAL SILICON BASED NVM MEMORIES

Among the silicon based memory cells, FG NAND flash had been successfully scaled to 4 F2/cell and provided MLC capability of 2 and 3 bit/cell operation while trading off retention, endurance, and reliability. Further scalability to ~1F2/cell has been projected in future years as being the ultimate scalability for such device which falls short in capacity when compared to HDD [4, 20]. During the last decade, several nonvolatile memory contenders have emerged with future potential to replace DRAM at the higher performance end and to replace NAND flash and even HDD in the high capacity end [20]. Several of these contenders integrate readily with conventional CMOS technology and consist of two-device memory cell having an FET switch and a bistable resistive or magnetic nonvolatile nonsilicon element for nonvolatility [19–21]. In this book, these contending memories have been classified as "Nonsilicon based" to distinguish them from the traditional silicon based memories such as SRAM, DRAM, NROM, and NAND flash. Detailed and all-inclusive discussion of these memories is out of scope of this book. Readers are being referred to an extended review on the subject entitled, "Alternate Memory Technologies," which was contributed by multiple authors and edited jointly in a significant chapter by G.F. Derbenwick and J.E. Brewer in a recent 2008 publication [22]. The topics included NROM memories, Ferroelectric Memories (FeRAM), Magnetic (MRAM), Resistive Memories, and Chalcogenide (GST)-based Phase Change Memory (PCM or PCRAM). The NROM has been considered in this book as a part of silicon based conventional NVMs and has been extensively discussed. Other emerging memories employ "non-silicon" nonvolatile elements as the integral part of the memory cell. These emerging memories are either in their proto-type phase or in the earlier phase of product life when compared to NROMs and NANDs. However, some of these emerging memories hold considerable promise to displace conventional memories. A brief discussion of selected such memories and their significant features will be discussed here and compared with current memory products to provide a perspective of potential challenges to silicon based memories in future decades.

Memory bit cell scaling of SRAM, DRAM, and NAND is schematically compared with potential nonsilicon based emerging memory solution by D. Eggleston [20] in Figure 17.1a. The 2004 ITRS memory cell requirement road map includes NROM, MRAM, and FeRAM in similar plots as shown in Figure 17.1b. The performance and reliability gap for different applications with the current memory solutions are illustrated in Figure 17.2 [20]. Figure 17.3 illustrates the current memories against several future memory potentials in the write-performance/memory capacity space [21].

It is evident from Figures 17.1 and 17.2 that while conventional NVM devices enjoy density advantages over most of the emerging NVM devices, serious performance disadvantages exist with such devices which could only be fulfilled by volatile SRAM and DRAM cells at the present time. This creates the potentials of emerging memories, namely, MRAM and FeRAM (higher performance options) to replace SRAM and DRAM in the performance space; and PCRAM and RRAM

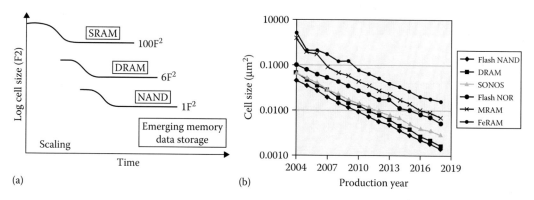

(a)   (b)

FIGURE 17.1   (a) Memory bit cell scaling potentials for (a) SRAM, DRAM, NAND, and emerging NVMs. (Eggleston, D., The impact of emerging memory, *Flash Memory Summit*, Santa Clara, CA, August, 2012.). Reproduced with permission from RAMBUS, (b) ITRS 2004 roadmap for conventional and emerging memory cell size requirements. (ITRS Roadmap 2012, Nonvolatile memory technologies with emphasis on flash, Edited by J.E. Brewer and M. Gill, copyright, *IEEE Inc.*, 2008, p. 17.)

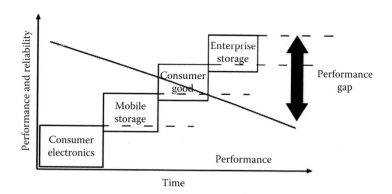

FIGURE 17.2   NAND bit cell trend against application requirements showing performance and reliability gaps. (Eggleston, D., The impact of emerging memory, *Flash Memory Summit*, Santa Clara, CA, August, 2012. Reproduced with permission from RAMBUS.)

(higher capacity options) to replace the capacity space currently held by NROM and NAND flash cells (Figure 17.3). However, since all emerging memories mentioned above involve relatively new materials, challenges related to high-volume manufacturability and reliability need to be met to bring to fruition such potentials.

Figures 17.1 and 17.2 confirm the need of higher capacity for the applications/market horizon as well as enhanced speed and reliability as discussed earlier. The above figures also suggest that such potential could be delivered by the emerging memory solutions as shown in Figure 17.3. We will briefly discuss PRAM (or alternately PC RAM, PCM or phase change memory), MRAM (or alternately MTJ-RAM, PMTJ-RAM, STT-MRAM), and ReRAM (alternately RRAM, MOHJO, Memristor) devices, current status, and prospects. Cell structures for all the above three device types employ a CMOS transistor acting as a switch which couples with a bistable resistive nonvolatile element placed above the transistor between two metal level electrodes. The bistable memory element provides the stable high resistance state (1) and the stable low resistance state (0) when appropriately programmed, thereby creating the nonvolatile memory. However, the fundamental mechanisms involved in changing the resistive characteristics are different in each case altering relevant nonvolatile properties, namely, retention, endurance, latency, programming and reading voltages, power

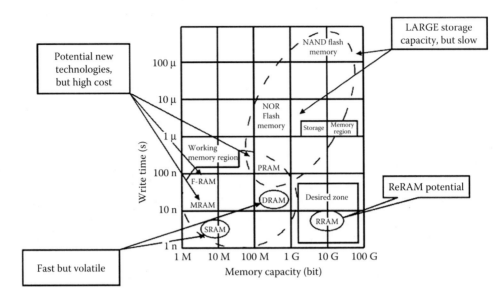

FIGURE 17.3 Write performance versus memory capacity plots for current and emerging memories. (Cleveland, L., Metal oxide hetero junction operation nonvolatile memory [Next Generation of Nonvolatile Memory], 4DS Inc., *Flash Memory Summit*, Santa Clara, CA, August, 2012. Reproduced with permission from 4DS Inc.)

requirements, and reliability. The cells are preferably configured in cross-point arrays to minimize cell area and maximize bit density.

17.3.1 PCM/PRAM

The phase change memory or PCM has been in the market until recently from Micron Technology providing a 10 years of memory retention, >1 E6 endurance at programming speed 2–3 orders of magnitude faster than NAND, and operating (read, write/erase) at low voltage (1.5–3.0 V). The memory product was 120 Mb NVRAM produced at 90 nm feature size and has been withdrawn by the manufacturer recently in favor of FG/FP-NAND citing better scalability of the NAND product. At the core of the PCM device is a chalcogenide glass, GST (Ge2Sb2Te5 or some variations) which switches from stable amorphous high resistance phase to a crystalline low resistance phase when heat or voltage during programming is applied to the material. Extendibility and scalability of PCM devices have been questioned in terms of performance and capacity. We would, therefore, not discuss the device any further.

17.3.2 STT-MRAM

The spin-transfer torque (STT) magnetic tunnel junction (MTJ) memory has received considerable attention in recent years when it was demonstrated that when the magnetic moment is configured to be perpendicular to the silicon substrate (vertical spin vectors), the memory cell could be potentially scaled below 10 nm with significantly lower critical current requirement for switching the direction of the magnetic moment [19]. The MRAM works on the principle of bistable magneto-resistance effects dependent on the direction of magnetic moment which could be altered from high resistance state to low resistance state by altering (programming) the direction of the magnetic moment within the magneto-resistive material. The STT-MRAM has the attractive features of potentially low MTJ switching current, high thermal stability, high speed operation, scalability, long retention (>10 years), high endurance (>1E10), and low power. The memory cell consists of a switching NFET transistor (processed in the front-end-of-line) combined with a STT-MTJ bistable resistor (processed at the back-end-of-line at the metal-interconnect level) in series with the switching transistor. The STT-MRAM is, therefore, CMOS logic technology compatible. However, the MTJ material and stacking structures are complicated, and scalability to smaller dimensions raises issues on stack structure integrity and reproducibility, retention, sense margin, parasitic resistance effects, and degradation of high/low resistance ratio for the memory states. Everspin Technology Inc. of Chandler, Arizona, has thus far offered MRAM product up to 64 Mb implemented in 130 nm technology node. STT-MRAM is a potential NVM replacement of current NAND flash with higher endurance capability as well as a potential replacement of DRAM if the endurance limits could be reliably extended by several orders of magnitude through future innovations.

17.3.3 ReRAM or RRAM

A relatively recent contender of silicon based NVMs, resistive RAM, could potentially be a serious contender of current NVM devices spanning the application possibilities from embedded to stand-alones and of displacing NAND/SDD and HDD in future [19]. The nonvolatile core consists of a thin film of transition metal oxide sandwiched between two metal electrodes which could be switched between stable low resistance and high resistance states at switching currents of <10 uA and at switching voltage of <3 V. The cell can be configured either with a switching transistor similar to PRAM or MRAM or without a switching transistor with multistack cross bar array architecture as a memristor cell [not shown]. The memristor cell displays nonlinear I–V characteristics [19]. The mechanism associated with the change in resistance is reproducible and stable ionic redistribution under the influence of heat or an electric field. ReRAM has the attractive attributes of low power operation with low write energy of ~1E−13 J/bit, scalability with cell size <4F2, low latency with W/E time in the range of ~10 ns, long retention (>10 years), and high endurance (~1E12).

Good nonlinear I–V characteristics of the memristor cell have been demonstrated with characteristic switching curve [19]. ReRAM product development currently at the proto type phase is being actively pursued by several memory companies.

17.4 REQUIREMENTS OF MEMORY ATTRIBUTES FOR FUTURE APPLICATIONS/SYSTEMS

It is anticipated that the applications of memory in forthcoming decades would be driven by accelerated growth in portable and embedded systems globally interconnected through high speed networking, virtualization, and 3D imaging. As a result, all the market sectors identified in Section 17.2 would require enhancement in memory attributes in general and NVM in particular along with interface design and architecture as well as system optimization. From memory standpoint, the winners will be required to have the following attributes:

(a) Compatibility and ease of integration with base line technology and manufacturing; (b) cost, power, and scalability advantages eliminating refresh power and volatility; (c) performance and reliability advantages with built in high IOPs and throughputs; (d) interface compatibility between different memory groups (DDRx/LPDDRx, etc.); and (e) built-in multifunctionality with cost/power/performance trade-offs. Table 17.4 reproduces the ITRS road map for DRAM, NAND, and other emerging nonvolatile memories [19]. A graphical representation of how the current silicon based memories and potential silicon based NVMs (MSUMs) could be positioned along with the emerging nonsilicon based options as discussed above within the context of density/performance-reliability space is schematically shown in Figure 17.4. The multifunctional SUM devices (MSUMs) are discussed in detail in Part III.

TABLE 17.4

ITRS Road Maps for Current and Future Memories

	DRAM	NAND Flash	PCM	FeRAM	STT-MRAM	ReRAM
Feature Size (nm)	36	22	45	180	65	<65
Cell Area	6 F^2	4 F^2	4 F^2	22 F^2	20 F^2	4 F^2, 8 F^2
W/E Time	<10 ns	1/0.1 ms	100 ns	65 ns	35 ns	<10 ns
Retention	64 ms	10 years	>10 years	10 years	>10 years	>10 years
Durability	>1E16	1.00E+04	1.00E+09	1.00E+14	>1E12	1.00E+12
W/E Voltage (V)	2.5	15	3	1.3–3.3	1.8	<1
Read Voltage (V)	1.8	1.8	1.2	1.3–3.3	1.8	<1
Write Energy (J/bit)	4.00E–15	>2E–16	6.00E–12	3.00E–14	2.50E–12	1.00E–13

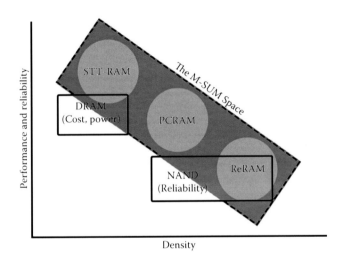

FIGURE 17.4 Conceptual representation of memory options in density/reliability-performance applications space with superimposed MF SUM (part III) and emerging "non-FET-based NVMs."

REFERENCES

1. R. Jha, ITRS roadmap for non-volatile memories, *Flash Memory Summit*, Santa Clara, CA, August, 2012.
2. E. Grochowski, and R.E. Fontana Jr., Future technology challenges for NAND flash and HDD products, *Flash Memory Summit*, Santa Clara, CA, August, 2012.
3. C. Lay, Next great SSD technology, *Flash Memory Summit*, Santa Clara, CA, August, 2012.
4. L.M. Grupp, The bleak future of NAND flash memory, *Flash Memory Summit*, Santa Clara, CA, August, 2012.
5. J.H. Yoon, Future technology challenges for NAND flash and HDD products, *Flash Memory Summit*, Santa Clara, CA, August, 2012.
6. D. Vilfort, Flash implication on system design, *Flash Memory Summit*, Santa Clara, CA, August, 2012
7. T. Rahman, How SSD Fit in different data center applications, *Flash Memory Summit*, Santa Clara, CA, August, 2012.
8. M. Abraham, Comparison of NAND Flash by application Requirement, *Flash Memory Summit*, Santa Clara, CA, August, 2012.
9. D. Dinker, Architectural approaches to explaining flash memory, *Flash Memory Summit*, Santa Clara, CA, August, 2012.
10. H. Huh, C. Jeon, C. Yang et al., A 64Gb NAND Flash memory with 800 MB/sec Synchronous DDR interface, *ISSCC*, 2012, pp. 1–4.

11. N. Shibata, K. Kanda, T. Hisada et al., A 19 nm 112.8 mm 2 g4Gb Multilevel Flash Memory with 400 Mb/sec/pin 1.8v Toggle Mode interface, *ISSCC*, 2012, pp. 159–162.
12. C. Kim, J. Ryu, T. Lee et al., A 21 nm High Performance 64 Gb MLC NAND Flash Memory with 400 MB/s Asynchronous Toggle DDR interface, *ISSCC*, 2012, pp. 981–989.
13. J. Cooke, Flash uses in high-performance platforms, *Flash Memory Summit*, Santa Clara, CA, August, 2012.
14. S. Stead, SSD for storage systems, *Flash Memory Summit*, Santa Clara, CA, August, 2012.
15. J. Scaramuzzo, Smart storage systems for future, *Flash Memory Summit*, Santa Clara, CA, August, 2012.
16. H. Cheng, Optimize your system design with flash memory, *Flash Memory Summit*, Santa Clara, CA, August, 2012.
17. G. Hawk, Flash storage growth trends, *Flash Memory Summit*, Santa Clara, CA, August, 2012.
18. R. Strenz, Embedded Flash technologies and their applications: Status & outlook, *IEDM*, Washington DC, 2011, pp. 211–214.
19. S.W. Park, Prospect for new memory technology, SK Hynix, *Flash Memory Summit*, Santa Clara, CA, August, 2012.
20. D. Eggleston, The impact of emerging memory, *Flash Memory Summit*, Santa Clara, CA, August, 2012.
21. L. Cleveland, Metal oxide hetero junction operation nonvolatile memory (Next Generation of Nonvolatile Memory), 4DS Inc., *Flash Memory Summit*, Santa Clara, CA, August, 2012.
22. ITRS Roadmap 2012, Nonvolatile memory technologies with emphasis on flash, Ed. J. E. Brewer and M. Gill, copyright, *IEEE Inc.*, 2008, p. 17.

18 Advanced Silicon Based NVM Device Concepts

CHAPTER OUTLINE

We have addressed in the previous sections multiple concepts of silicon based NVM device extendibility, current-state-of-the-art in such devices and products, and limitations of current devices and products. We have also briefly explored the future and the current status of "nonconventional" or "non-FET-based" NVM contenders potentially challenging the current dominant roles of NROM and NAND flash products. Additionally, we have reviewed the status of quasi-3 and full 3D NAND flash development efforts, including a brief review of recent development activities, in 3D nanowire-based O-flash memory device and technology. In this section, we will focus on the potential approaches to enhance the conventional NVM device attributes and functionality based on the knowledge and understanding established thus far in the previous sections. This chapter emphasizes advanced and extendable stack designs for silicon based NVM devices and provides illustrations primarily with planar and trench-based devices and arrays. However, it should be assumed that such concepts could be extended to the emerging 3D devices and arrays which include both conventional 3D proposals and Gate-All-Around (GAA)/nanowired (NW) devices and arrays discussed in Chapter 16. The objective would be to reduce and possibly eliminate the major limitations faced by the currently established approaches in cost, performance, and scalability. This section is intended to serve as a prelude to the final part of this book: the development and evolution of silicon based universal memory (SUM) from the currently established knowledge base of technology, devices, and products.

18.1 DEVICE PARAMETER ENHANCEMENT CONSIDERATION

Unique NVM device parameters at stack design and memory cell device level are memory window, Charge retention, Write–Erase Performance or Programming Speed, Power Reduction or Voltage Scalability, Write–Erase Cyclability or Endurance, and Cyclic reliability. Other important parameters such as memory-state integrity, SCE, FEE, and sense-margin are also sensitive to process technology, feature size scalability, programming algorithm, array design, application, and so on. We have discussed both the former group of parameters and the latter group of parameters previously. We will briefly focus on selected stack-design consideration for enhancement of the first group of device parameters not only from application enhancement for conventional NVMs but also from the standpoint of device concepts leading to SUM devices described in Section 2.4 of this book.

18.1.1 Device Parameter Drivers

Keeping in view the base CMOS platform technology compatibility, technology integration, and scaling, we will focus on the extendibility considerations discussed in previous sections for the CT-NVMs toward device parameter drivers and identify the appropriate stack designs for SUM devices.

18.1.1.1 Memory Window and Charge Retention

Critical to large stable memory window is to provide capacity of high-density charge storage with very low leakage. This requires deep offset trapping dielectric such as GaN combined with high-density charge-storage layer such as In-SRN with high electron energy barriers interfacing to reduce leakage to substrate (tunnel dielectric) and gate (blocking dielectric). Conventional device option in the past consisted of Si_2ON_2/In-SRN, to be preferably replaced with GaN/In-SRN. At the expense of added process complexity associated with added dielectric layer with higher barrier energy, OR-SiON or Al_2O_3 may also be considered to further enhance window and reduce leakage (this is discussed in Chapter 13 under "trap-engineering.")

18.1.1.2 Charge Retention and Programming Speed

Critical to simultaneous achievement of fast programming and charge retention is to incorporate asymmetric band-engineered and multilayered tunneling layers associated with direct tunneling of charge transport both during writing and during erasing. The background related to this aspect of device design has been provided in Chapters 12 and 13. Ultra-thin interface layer with silicon substrate needs to be compatible with scaled CMOS FETs from device design perspective. This layer could be best achieved either by an LPCVD OR-SiON layer with controlled oxygen concentration or by plasma nitridation of an ultra-thin layer of SiO_2 as previously discussed. Alternately, ALD techniques may also be considered followed by an appropriate anneal. The multilayered tunnel dielectrics could be tailored by appropriately trading charge retention and programming speed based on application requirements. This is further discussed in considerable detail in this section and in Part III related to SUM devices. While potential application of many Hi-K dielectric films could be feasible, promising dielectric films include HfO_2, La_2O_3, Al_2O_3, HfSiON, HfLaON, and HfTaON. Some of the oxynitrides such as the ones mentioned above could be applicable for blocking layer for NVM devices and simultaneously for the gate dielectric for the scaled FET devices as discussed earlier.

18.1.1.3 Power Reduction

Power reduction is tied with gate stack voltage scalability and transport mechanism. This in turn is reflected into gate stack EOT and employment of high K layers in the gate stack design while fulfilling unique NVM functionality and application requirements. This is further discussed with specific applications later in this section.

18.1.1.4 Endurance and Cyclic Reliability

Critical to endurance and cyclic reliability are the programming peak field, transport mechanism for programming, and dielectric film selection for the stack design. This is also inter-twined with application requirements and programming performance and programming algorithm. End-of-life endurance could only be ensured through DTM devices as explained earlier. Most of the proposed SUM devices in Part III are based on DTM device concepts to meet endurance and cyclic reliability.

18.2 APPLICATION PARAMETER ENHANCEMENT CONSIDERATION

Keeping in view the current and future application requirements discussed in the previous chapter and device extendibility considerations discussed in earlier chapters, this chapter will illustrate concepts of enhanced properties and attributes applicable to future NROMs and NANDs. Enhanced stack design and device/array concepts will be proposed to address a broad range of future applications and market segments within the framework of both embedded and stand-alone memory offerings. We will make the following assumptions in addressing this subject:

1. Enormous growth in portable digital electronics could be anticipated in the next decade requiring a diverse range of embedded NROM and NAND flash memories with enhanced attributes to effectively compete with emerging options.
2. Continued expansion of internet applications will accelerate the requirements of higher performing, cost sensitive, nonvolatile mass-storage systems for enterprise, and database centers displacing lower performing HDD and tape solutions.

We will briefly address the key future application drivers for silicon based embedded and stand-alone NVM devices from device design perspectives and propose cost-effective future device concepts to: (a) maintain and expand application horizon against the scaling challenges and limitations being faced through current approaches; and (b) effectively compete against emerging "non-silicon" contenders such as MRAM and RRAM.

18.3 APPLICATION DRIVERS FOR EMBEDDED AND STAND-ALONE NVMs

Embedded NVMs were customarily employed in the past for code and BIOS applications in systems. In the past decade embedded NVMs have witnessed phenomenal growth in multifaceted applications in consumer electronics and portables. Novel and smart applications continue to emerge and expand market potentials requiring wide range of embedded NVMs. It is estimated that embedded NROM and NAND memory bit requirements may grow as much as 10X within the next few years due to the expanding demands in automotive, smart cards, smart phones, tablets, smart instruments, robotics, medical electronics, safety equipment and even smart toys, and others covering a wide range of memory needs. The memory arrays will be embedded in micro-controllers and microprocessors as ASIC solutions for SOCs, and integrated with MMUs (memory controllers) in SOPs to deliver the desired form factors and memory capacities required for the specific applications. Embedded applications in particular require ease of integration and compatibility with the common platform scaled logic technology. Consequently, the NVM devices and technology features need to be compatible with the Bulk or SOI logic platform and would require minimal additional process complexity to satisfy cost/performance/power objectives. At the same time, NVM device attributes need to be significantly enhanced from the current-state-of-the-art to compete against future nonsilicon NVM contenders. From such considerations, it is appropriate to note that embedded FG-NAND flash cells/arrays would be a disadvantage compared to CT-FP NROM options as discussed earlier in this book in terms of compatibility with the integration schemes and device enhancements. This is particularly true for the range of applications where the embedded memory capacity requirements are relatively smaller. For cost-sensitive embedded applications with lower performance requirements and requiring large memory content, embedded floating plate/conductive plate (FP/CP) NAND-MLC arrays could be justified where cost of memory and market volume are the main drivers.

Stand-alone NROMs and Embedded NROMs, in principle, may require similar enhancements in device attributes and features although compatibility with platform technology would be less of a concern. However, stand-alone NROMs would require enhancements on device and functional attributes to effectively compete with emerging options in performance, power, and compatibility with higher performance scaled CMOS peripheral devices. In our exposition of advanced device concepts, we will treat the embedded and the stand-alone NROMs within similar functional/application groups.

NAND flash technology and products in stand-alone form are optimized for larger capacity memory requirements both for enterprise systems and large database cloud storage applications. To expand into the application space of the HDD and tape arena, cost/bit reduction, higher capacity, and higher endurance would be required. Advanced device concepts of both FP/CP types and CT/FP types providing larger memory window and potential 4 bit/cell MLC capability with stack design to provide higher endurance and lower Vpp could be the preferred future approaches.

18.4 FUNCTIONAL AND ARCHITECTURAL REQUIREMENTS AND GROUPING OF NROMs

Before discussing future embedded and stand-alone NROM device concepts to enable expanded applications and market expansion, we will first examine the potential functional and architectural requirements of the widely diverse NROM market segments and narrow down the device and integration options compatible to both embedded and stand-alone platforms. We will subdivide the potential functional and architectural requirements and associated market segments into four general subgroups: (a) most cost-sensitive low-function applications, "MCLA"; (b) low-cost lower function applications, "LCLA"; (c) cost-performance intermediate function applications, "CPIA"; and (d) high-performance high-function applications, "HPHA." Subsequently, we will narrow down the integration and device options among many potential possibilities discussed in Chapters 13 through 17 based on the author's experience and perspectives. We will also discuss device directions to fulfill future scalability and device enhancements.

The application segments are shown in Table 18.1 for near term technology nodes outlining the potential functional and architectural requirements for NVM-based systems.

18.5 NVM EMBEDDED DEVICE TYPES AND OPTIONS

All application segments described in the table above could, in principle, be served by either embedded and/or stand-alone NROMs (or EEPROM) from the standpoint of capacity requirements although embedded CT NAND may aptly be justified for very large high-end HPHA applications requiring multi-T Bytes of memory. This is believed to be true today and is expected to be true in near future since the memory capacity requirements for most NROM applications may not exceed 1 T Bytes (1E12 Bytes). The majority of above application segments have thus far been served by three types of NOR NVM cells, namely, (1) Split-gate MONOS or SONOS NROM cell, (2) Two-Transistor NOR or NROM cell, and (3) One-transistor EEPROM cell, even though the Split-gate MONOS or SONOS cells (type 1) have been most widely used [1]. Several embedded applications require certain stringent specifications which include:

(a) Higher temperature memory retention and reliability over a broader temperature range of application from 40°C to 175°C;
(b) Built-in ECC requirements to ensure ultra-low fail rate (< 1 ppm over life time, typically 1E-14 with ECC);
(c) High endurance;
(d) Low power consumption;
(e) Higher read performance; and
(f) Scalability to future technology nodes.

Masoero and co-workers investigated scalability of split-gate CT-NVMs and demonstrated that high temperature retention in such devices could be enhanced by incorporating silicon nanocrystals in trapping nitride (SRN) compared to trapping nitride of stoichiometric composition [2]. SRN provides deeper additional traps due to higher band offset, thereby enhancing thermal stability of trapped charges [3]. However, it should be noted that the size, distribution, and composition of silicon nanocrystals are critical in optimizing such properties and at higher silicon content, leakage through such layer could be significantly enhanced degrading retention and memory window (see Section 3.1, Chapter 3 on nanocrystal devices and Chapter 5 on SRN). The investigation of L. Masoero et al., also demonstrated that speed, power, and memory window could be best optimized in split-gate CT devices by selecting programming (writing) through source-side injection and erasing through hole tunneling mechanisms.

TABLE 18.1

Near-End Architectural and Functional Requirements of NROM Devices (Embedded and Stand-Alones) by Application Segments

	MCLA (Most Cost-Sensitive Low Function)	LCLA (Low Cost Low Function)	CPIA (Cost/Perf. Intermediate Function)	HPHA (Hi Perf. Hi Function)
APPS/Functional Range:	Fixed/Limited	Variable/Limited	Significant/Multiple	Large/High-end
MMU (Controller) Rqmnts:	Simple core	Simple core	Multicapability	Sophisticated
−ECC	No	Yes	Yes	Extensive
I/O Requirements:	Small	Larger/Limited	Large	Large
−Throughput	Current Level (CL)	CL	CL	Higher
NVM Requirements:	Code+Data	Code+Data	Code+Data	Code+Data
−Capacity Variable:	Current Level (CL)	Variable	Larger	Largest
−Nonvolatility without Refresh	10 years	10 years	10 years	Lifetime
−Speed: Read/Program	Current Level (CL)	CL	10–100 X	100–1000 X
				DRAM (optional)
Latency	Current Level (CL)	CL	CL	Lower (DRAM-like)
−Power/Voltage	Low	Low	Low	Low to Moderate
−Endurance CL	~1E15	10–100 X CL	1000 × CL	EOL > 1E12
−Reliability	Current Level (CL)	High	High	High
−Arrays	Variable/ASIC Common	Variable/ASIC Common	ASIC and Custom Common	ASIC and Custom Common+ Custom
−Memory Interface	Bulk	Bulk	Bulk	Bulk/SOI
Technology Platform Scalability:				
Feature/Capacity/cost	Yes	Yes	Yes	Yes
Device Enhancements	Yes	Yes	Yes	Yes
Functional Enhancements	Yes	Yes	Yes	Yes

We will focus on device concepts applicable to split-gate CT-NVMs of advanced planar and quasi-planar (dual gate and FinFET SONOS) and trench-based vertical channel configurations to address the requirements of all NROM segments outlined in Table 18.1. Enhancements of device properties and scalability have been extensively discussed in previous sections and many options of device stack designs were illustrated. We will review consideration of (a) integration schemes to minimize process complexity and dielectric developments; (b) programming mechanism options and other stack design considerations to achieve desired device properties for various market/application segments; (c) select specific devices and array concepts based on (a) and (b) above, and (d) illustrate specific examples.

18.6 NVM DEVICE INTEGRATION OPTIONS

Technology integration schemes for enhanced devices have been discussed in Chapter 14. For embedded applications, the integration schemes should be compatible to platform CMOS technology and preferably follows similar and compatible process steps of fabricating CMOS PFET and NFET peripheral devices with least number of additional dielectric films and lithographic steps. Commonality of dielectric films for fixed Vth FET devices and variable Vth NVMs could be sought in the interface layers at the silicon substrate (for bulk) or Silicon/Si-Ge film (for SOI) interface at the bottom end and the work-function controlling gate interface at the top end of the gate insulator stack for all NVMs. Additionally, the high K scalable EOT gate insulator film characteristics for fixed Vth should preferably be similar to that of the charge blocking insulator of the NVM gate stack

(Chapter 14). It would be advantageous if such film could be deposited at the same time for all gate stacks. Similarly, all metal gates could be the same for both FET and NVM devices. Since employment of high K dielectric films for both the FET gates and the blocking layer for the NVM gate stack are assumed to be common, substituted metal gate approach of integration scheme for workfunction control is also assumed [4]. Such integration approach reduces the number of dielectric developments and deposition/photo steps. For NVM device requirements optimized for the different application segments, the simplest stack design could be appropriate for the most cost-sensitive and least demanding segment, namely, MCLA. With increasing requirements of enhanced device properties from LCLA→CPIA→HPHA, the number of dielectric layers and the associated process/integration steps could increase. However, such process/complexity adders could be minimized through selection of appropriate dielectric film options which could be deposited and patterned through a least number of added steps and dielectric film layers and yet achieve the desired device attributes. Such schemes will be specified below.

We will assume that the platform technology integration scheme as discussed in Table 14.1 (Chapter 14) being the basic integration framework for all application segments for planar, dual gate, and FinFET device options. In such framework, gates are defined earlier and source/drain diffusions follow after gate processing. For vertical channel device integration scheme, the buried diffusion has to be placed earlier and require higher thermal budget and typically follows after the trench isolation is formed. However, for very high-performance HPHA requirements with custom designs of device stacks, the SMLK scheme shown in Figure 14.8 could also be considered where gate is formed before the diffusion formation and impurity activation.

18.7   NVM PRODUCT AND STACK DESIGN BASIC CONSIDERATION

Aside from special functionality and integration requirements associated with platform CMOS technology, central to all future NVM products are the requirements of low power/voltage programmability, high reliability, and common memory interface. The latter item is being associated with memory array/controller designs while the other two items are dependent on programming mechanisms and stack design. Advanced stack designs would therefore require higher K dielectrics and lower energy charge transport mechanisms. These are discussed below.

18.8   ADVANCED NVM DEVICES AND ARRAY CONCEPTS

Incorporating high K films in NVM stacks and gate stack formation scheme:

In discussing the advanced device concepts for future applications, we will narrow down the potential options of the number of dielectric film development required from feasibility, device integration and characterization, R&D efforts for incorporating the dielectric film or films into the baseline CMOS technology, reliability qualification, and high yield, high volume production into NVM devices and arrays. It should be recognized that such efforts take considerable time and resources. Furthermore, development and applications of such films should not only be based on the wider applicability among the four subsets of embedded markets, namely, MCLA, LCLA, CPIA, and HPHA but also would fulfill the need for multiple generations of device enhancements belonging to each of the subsets. It has also been discussed before that the selection of such dielectric films in NVM device stack design should preferably be based on minimization of processing (deposition steps, tool requirements, and lithographic steps) parameters to reduce cost and enhance yield/throughput.

It will be assumed in this discussion on dielectric films that all films discussed in Section 1.5 of this book, namely, oxide (SiO_2), nitride (Si_3N_4), oxynitrides of two major subgroups (oxygen-rich and nitrogen-rich): OR-SiON and NR-SiON (e.g., Si_2ON_2), silicon-rich nitrides, especially In-SRN (two-phase consisting of silicon nanocrystals embedded in nitride) will be considered as conventional dielectric films with long history in applications in silicon devices and products. It should be noted that with the exception of films of thermal SiO_2, thermally converted nitridized-oxide (OR-SiON), or plasma nitridized-oxide (SiON), all the films mentioned above could be deposited with desired film characteristics within a single LPCVD tool with appropriate chemistry, thermodynamics, and processing variables. The only higher K dielectric film to be considered "conventional" is Alumina (Al_2O_3) since it had been widely used as DRAM capacitor dielectric and both FG-NAND flash devices and CT-TANOS NROM devices as blocking layer. It should also be noted that In-SRN in conjunction with high K dielectric films play multiple effective roles in advanced stack designs as mentioned in Part I Chapter 4. In terms of defining current state-of-the-art embedded devices, we will only assume stack designs incorporating above conventional dielectric films for the stack as references against advanced embedded and stand-alone devices.

Among the many high K dielectric films discussed in previous sections, the author limits for most NVM applications to only two or three single metal oxides specifically, HfO_2 (Eb = 5.1 ev, K = 24), Pr_2O_3 (Eb = 5.1 eV, K = 31), and/or La_2O_3 (Eb = 4.3 eV, K = 30), all have characteristically low leakage at reduced field and high K values, the last two being lanthanides (see appendix for details). HfO_2 has been much investigated for CMOS gate oxide applications. Three other more complex high K dielectrics are selected: one belonging to oxynitride family, HfSiON (Eb = 6.9 eV, K = 14) (optionally HfAlON with slightly higher K and similar leakage), while others are dual metal oxides, AlHfO (Eb = 6.2 eV, K = 14) and $AlLaO_3$ (Eb = 6.5 eV, K = ~27). These dielectric films have demonstrated strong thermal stability, very low leakage and with proper processing yields, very low trap density, and high breakdown strength. For nitride replacement for charge trapping, either the conventional Si_2ON_2 (Eb = 6.5, K = 7) of the oxynitride film could be considered or higher K, GaN (Eb = 3.39 eV, K = 10) film could be considered. Both films have long development history with deep energy traps and have been widely used in micro-electronics. GaN has the unique characteristics of deep electron energy offset, discussed earlier. Ultra-shallow thin films of most of the above high K films can be readily deposited by sputtering or by LPCVD processing and many of the above by ALD processing at low temperatures.

In integrating multiple layers of high K films for NVM gate stack, the processing approach to be discussed consists of initially defining the active NVM gate stack regions and selectively depositing the tunneling layers and the charge-storage layer (or layers) over those regions. Subsequently, all peripheral CMOS gate stack regions are defined with the "common" gate insulator for the CMOS devices (both NFET and PFET) and the blocking layer for the nonvolatile devices are deposited at the same time followed by gate-interface layer (or layers)/metal gate processing for all device types. Subsequently, the sidewall spacer processing could be carried out for all devices followed by LDD and S/D processing, silicide formation over diffusion, activation anneal, etc. Such process flow ensures reduction of processing and lithographic steps and ease of integration desired for all device subsets. In certain cases, gate insulator step for CMOS devices may precede post surface clean and an ultra-shallow deposition (~1 nm) of interface oxide (SiO_2) before depositing the high K dielectric film. The charge-storage layers (SiON or GaN) with or without a capping In-SRN layers exposed to such treatment remain unaffected by either a low temperature thermal oxidation of exposed silicon substrate or by the ozone treatment of the surface to form the interface oxide (plus low temperature oxidizing anneal if required). Therefore, a potential process flow option for the above scheme at gate insulator processing steps is as follows:

All regions except NVM gate stack region remains protected through an appropriate masking processing:

NVM gate region exposed down to the silicon substrate
↓
NVM tunneling layers deposited (a)
↓
NVM blocking layers deposited (b)
↓
Protective passivation layer deposited (optional, e.g., In-SRN) (c)
↓

Masking stripped selectively removing layers (a), (b), and (c) from all appropriate regions
↓
CMOS gate regions and NVM gate regions selectively defined and exposed (other regions masked)
↓
Ultra-shallow SiO_2 formed over CMOS gate regions after appropriate cleaning of interfaces (optional)
↓
Deposition of CMOS device gate insulator/NVM blocking layer (high K film)
↓
Post deposition anneal (optional)
↓
Metallic interface layer formation (e.g., TaN)
↓
Metal gate formation ("gate first/diffusion later" processing)
↓
Sidewall formation for all gates
↓
Standard process flow afterwards for LDD/S-D, and so on

Selecting transport mechanisms for gate stack device designs:

It has been outlined earlier that future device designs require low power higher performance programming and high endurance. Therefore, low energy transport mechanisms during programming and erasing are key to achieve such goal. Advanced device designs will be based primarily on low energy electron transport (hole transport requires higher energy due to higher effective mass). Most of the proposed stack designs for advanced devices would therefore preferably be DTM based involving. Single or multiple modes of direct tunneling of electrons for tunneling and charge transport. An option for consideration may also include source-side injected electrons for writing which also requires relatively lower energy compared to other mechanisms of electronic charge injection and transport. Block erasing will be envisioned to achieve appropriate tradeoffs between standby retention and erase speed whereby both hole tunneling and electron back tunneling are involved for erasing. Multimode writing and erasing will also be illustrated for achieving higher writing and erasing speed at the expense of relatively higher power and reduced endurance/reliability. Transport mechanisms and stack designs should be applicable to both planar and nonplanar device configurations, including FinFETs and nanowired (e.g., O-flash) devices, even when not explicitly mentioned.

18.9 ADVANCED NVM DEVICES AND ARRAYS: DEVICE STACK AND BAND FEATURES

Before discussing advance devices for future applications, it is worth noting the limitations of conventional CTSONOS devices and variations with single dielectric layer for tunneling. If the tunnel dielectric layer such as SiO_2 (K = 3.9) or OR-SiON (K = 5) is in the direct tunnel regime with thickness < 2 nm (EOT ranging < 1–2 nm), the charge retention is limited to < ~1E4 sec due to reverse tunneling of stored charges even though a device programmed at lower average W/E field would exhibit improved endurance. On the other hand, if the tunnel dielectric thickness is increased to operate in FN tunneling regime with thickness ≥ 5 nm, the scaled SONOS device may satisfy charge retention requirement of 10 years when programmed at an average field of > 10 MV/cm, with programming voltage level typically exceeding 10–12 V, 1 ms (gate stack EOT typically in the range of 12–18 nm) and endurance limited to ~<1 E7 W/E cycles. This is true even when oxide for tunneling and blocking could be replaced by OR-SiON and trapping nitride could be replaced by NR-SiON (e.g., Si_2ON_2) although retention and endurance properties are improved by such structures. When blocking oxide or oxynitride layer is replaced by higher K dielectric layer of alumina (Al_2O_3, K = 10) with TaN gate as in scaled TANOS design, the programming voltage and speed could be optimized around 10 V, 1 ms regime with 10 years of retention and 1E7 W/E endurance (Chapter 13 on Enhanced TANOS) for SLC design with reduced programming power. However, greater reduction of power and enhancements in speed and endurance could not be further achieved while meeting retention requirements. Therefore, further scaling in programming power and NVM attributes required simultaneous employments of multilayered and band-engineered asymmetric internal field-aided lower energy charge transport mechanisms with lower EOT stack design employing higher K dielectric films. Such approach scales programming power (proportional to the square of Vpp), reduces average programming field, and reduces charge leakage during standby, thereby enhancing device attributes and reliability while reducing functional power requirements.

Examples of stack designs for advanced devices are shown in Table 18.2 for the market segments identified in Table 18.1. Progressive generations of enhancements are combined with requirements of dielectric film developments and added complexity in Tables 18.2 through 18.4, respectively, for first, second, and third generations of device enhancements. In Table 18.5, device designs for special

TABLE 18.2
Advanced CT-NVM Stack Designs: First Generation

Application Segment:	MCLA-SLC	LCLA-MLC	CPIA-MLC	HPHA-MLC	NOTE
	(Fixed func., Lo cost)	(Variable func., Lo cost) (multi func., C/P)		(High func., Hi end)	
(a1) Tunnel Layers: First: oxide (SiO_2) or oxygen-rich oxynitride (OR-SiON) for all cases				Si/Insulator Interface	
EOT:	~3 nm	~1 nm	~1 nm	~1 nm	
(a2) Tunnel layer: Second:	nitride $	Hafnia $	Hafnia[a] or La_2O_3 [a] or Si- NC/SRN [a] #	[a] Middle layer	
EOT:	~2 nm	~5 nm	~0.5 nm	~1 nm	$ 2-layer VARIOT
(a3) Tunnel layer: Third:	None	None	Same as a1)	~1 nm for CPIA/HPIA	3-layer DTM # RTB
(a) EOT: a1)+a2)+a3):	**~5 nm**	**~3 nm**	**~2.5 nm**	**~2.5 nm**	
(b) Trapping Layer:	Si_2ON_2 (5 nm)	Si_2ON_2 (5 nm)	Si_2ON_2/SRN^	Si_2ON_2/SRN^	^ Larger W
EOT:	**< 3 nm**	**< 3 nm**	**< 3 nm**	**<3 nm**	
(c) Blocking Layer:	HfSiON or AlHfON or $AlLaO_3$ for FET gate Insulator/Blocking Oxide Selection				
EOT:	~2 nm for HfSiON and AlHfON and ~1 nm for $AlLaO_3$				
(a)+(b)+(c):					
EOT:	**<10 nm**	**<8 nm**	**<~7 nm for CPIA and HPHA**		
Vpp/tpp (Write)	Vpp/tpp (Write)		5.0 V/1 ms 5 V/100 us and 6 V/10 us options for CPIA/HPIA		

Remark: First generation dielectric development may be either HfSiON (preferred for reliability) or HfO_2 or AlHfON for CMOS gate dielectric and embedded NVMs, and can be replaced for others in the table for stack design.

TABLE 18.3
Advanced CT-NVM Stack Designs: Second Generation

Application Segment:	MCLA-MLC	LCLA-MLC	CPIA-MLC	HPHA-MLC
	(Fixed func., lo cost)	(Variable func., Lo cost)	(multi func., C/P)	(High func, Hi end)
(a) Tunnel:	OR-SiON/HfSiON [a]	OR-SiON/La$_2$O$_3$/ORSiON ^	[a] 2-layer Variot	
(b) Trapping	GaN@	GaN+SRN	@ Larger W: 2 b/cell design	
EOT:	2 nm	2 nm		
(c) Block:	HfSiON	LaAlO$_3$		
EOT:	2 nm	1 nm		
(a)+(b)+(c): EOT:	6.5 nm	5 nm		
Vpp/tpp	5 V/1 ms	5V/1–10 us (high perf.) and 3 V/1 ms (cost/perf.)		

Remark: Second Generation dielectric development requires GaN trapping layer and La AlO$_3$ CMOS gate/NVM Blocking layer developments. MCLA with relaxed spec and LSLA with tighter spec could be served with the same stack design. Similarly, CPIA and HPHA are served by the same stack design operating at different programming conditions at different performance and power levels. The scaling reduced the power levels for all market segments. All segments with MLS designs (2 b/cell). The three-layer DTM barrier provides high-performance yet enhanced retention (see Section 3.1.1).

TABLE 18.4
Advanced CT-NVM Stack Designs: Third Generation

Application Segment:	MCLA-MLC	LCLA-MLC	CPIA-MLC	HPHA-MLC
	(Fixed func., lo cost)	(Variable func., Lo cost)	(Multi func., C/P)	(High func., Hi end)
(a) Tunnel:	OR-SiON/LaAlO$_3$ [a]	OR-SiON/HfSiON/La$_2$O$_3$^	[a] 2-layer Variot	
EOT:	<2.0 nm	~1.5 nm	^ 3-layer PBO	
(b) Trapping	GaN	GaN /SRN		
EOT:	2 nm	2 nm		
(c) Block:	LaAlO$_3$	LaAlO$_3$		
EOT:	<1.0 nm	<1.0 nm		
(a)+(b)+(c): EOT:	<5.0 nm	<4.5 nm		

Remark: Second Generation dielectric development requires GaN trapping layer and La AlO$_3$ CMOS gate/NVM Blocking layer developments. MCLA with relaxed spec and LSLA with tighter spec could be served with the same stack design. Similarly CPIA and HPHA are served by the same stack design operating at different programming conditions at different performance and power levels. The scaling reduced the power levels for all market segments. All segments with MLS designs (2 b/cell). The three layer DTM barrier provides high-performance yet enhanced retention (see Chapter 2).

attributes were illustrated for specialized applications and HPHA (High Performance High-function Applications). The stack designs are applicable to both planar and nonplanar devices. Device schematics and array designs are illustrated following the explanation of the tables. Table 18.5 will define stack designs for three special attributes applicable to further enhancements for future applications: (1) multimechanism charge transport design for higher performance, (2) very low power design, and (3) multifunctional stack design which will be followed by subsequent illustrations.

TABLE 18.5

Advanced CT-NVM Stack Designs for Special Attributes[a]

Tunnel	Stack			Transport		
	Trap	Block	EOT	t	Vpp/tpp	Note
OR-SiON (1 nm)/ HfO$_2$(3 nm):	GaN (5 nm)	HfO$_2$/IN-SRN(5/3 nm)	<5 nm	MultiMech.	3V-5V500/ 5 ns	V.Hi Perf.
OR-SiON/Si$_2$ON$_2$/HfO$_2$ (0.5/1/2 nm)	GaN/HfO$_2$ (2/4 nm)	LaAlO$_3$(7 nm)	<4 nm	DT-PBO	2 V/1 ms	V.Lo Pwr^
OR-SiON/HfO$_2$/ Si$_2$ON$_2$(tr-a)/Al$_2$O$_3$/ GaN(tr-b)[1/1.5/3/2/5]		LaAlO$_3$[7 nm]	6.5 nm	MultiMech.	2.5–5 V// 50 ns/1 ms	"MF"

"MF": DRAM plus NVM functionality
[a] See text for details
^ Ref. 8.11 (V.low power)

18.9.1 Advanced NVM Devices: Explanations and Illustrations of Devices and Band Features

Except for the very high-performance device stack design outlined in Table 18.5, all advanced device designs mentioned above are tunneling based NROM ("TBNROM") devices of split channel designs [3] for enhancement mode low power operation most suited for embedded applications. Stack designs are applicable to not only planar configurations but also back gated [5], FinFET and vertical channel [6,7], and (see previous section) nonplanar devices. Additionally, most of the design combines voltage scaling, multiple direct tunneling (DTM), and band-engineered charge transport with appropriate multilayered high K films for energy efficiency, scalability, reliability, and enhancements in device characteristics [8–11]. The RTB-DTM options were employed for HPHA applications in examples in Table 18.2 [12–15] to facilitate MLC operability. However, most device options illustrated would yield 2 bit/cell functionality due to larger memory window derived through multilayered tunneling with asymmetric energy barriers combined with deep energy/deep offset trapping layers such as GaN replacing traditional nitride layer. Furthermore, 4 bit/cell designs could be feasible for CPIA and HPHA applications by reversing current flow and applying appropriate sensing scheme utilizing localized trapping (15.7.1 on enhanced CT-MNSC devices). Enhanced device performance could be achieved by incorporating device design through enhanced gate injection [16] and invoking multimechanisms for charge transport during programming [9,17] through band engineering and by designing device stack with PBO-DTM tunneling [17], the latter approach provides simultaneously higher performance and higher endurance at lower power. PBO-DTM designs also provide enhanced scalability and lower voltage programming, thereby lowering operational power [8,9,11,18]. Finally, DTM design concepts could be applied to provide both DRAM and NVM functionality by simultaneously storing charges at different tunneling distances to create multifunctional memory cells [9,18]. The selection of high K dielectric films for the above embedded device illustrations were driven by: (a) minimization of high K film development for simultaneous applications in scaled CMOS peripheral device gate insulators and as blocking layer for NVM devices, (b) minimization of deposition and lithography steps for integration into embedded technology platforms, and (c) reliability of both fixed Vth and NVM devices.

Figure 18.1a and b shows the schematic cross-sections of advanced one-device NROM cell and splitgate NROM cell respectively following the process schemes discussed above. Figure 18.1c and d shows the corresponding PFET and NFET scaled peripheral devices respectively in future technology.

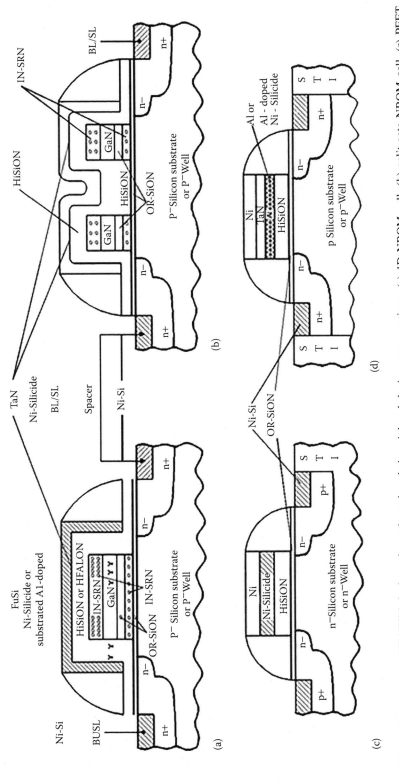

FIGURE 18.1 Advanced CT-NVM device schematic and scaled peripheral device cross-sections: (a) 1D-NROM cell, (b) split-gate NROM cell, (c) PFET, and (d) NFET.

Most of the advanced CT-NVM stack designs illustrated in Tables 18.2 and 18.5 are based on:

(a) band-engineered stack designs with multilayered direct tunneling (single or multiple direct tunneling) for writing; plus,
(b) asymmetric tunneling barrier for erasing; plus,
(c) deep level trapping coupled; plus,
(d) very low leakage high K dielectric film for charge blocking in combination with high work-function gate.

In order to achieve high retention and eliminate back injection preventing stored charge compensation. It has been explained in Part I in multiple chapters and in Part II in Chapter 13 on band engineering that the direct tunnel mode of electron transport provides energy efficiency, high programming performance, and least damage to the dielectric media during charge transport promoting device reliability and high endurance. Furthermore, two types of multilayer tunnel barriers were emphasized: the two-layered and triple-layer VARIOT barriers and triple-layer PBO barriers (Figures 13.8 and 13.14). The RTB (Resonant Tunnel Barrier) example as noted in Table 18.2 could also be considered, in principle, a variation of VARIOT barrier with dual direct tunneling energy states. Both VARIOT and PBO barriers provide internal field-aided electron transport during writing enhancing writing speed and both provide asymmetric high energy barriers reducing reverse tunneling. The VARIOT barriers promote and extend "no-transport" regime (Figures 13.8 and 13.12) at lower field with steeper J versus V characteristics, thereby preventing charge leakage to the substrate and yielding devices with larger memory window and longer retention. The PBO-DTM barriers provide maximum scalability with the best trade-offs in power, window, retention, and endurance. Band diagrams for the triple-layer VARIOT and PBO barriers from the stack designs illustrated in the above tables are shown in Figures 18.2 and 18.3, respectively.

Within the constraints of high K dielectric development, integration, and applications into all types of embedded devices, power reduction considerations have been a key factor particularly in embedded NVM designs. Appropriate high K applications to lower stack EOT and therefore $\pm$Vpp have been illustrated in the above tables to conserve power for the desired performance targets. Two-layer direct tunneling with higher K lower band gap second layer of VARIOT and PBO barriers could be effective in lowering stack EOT and reduce programming power by lowering Vpp and enhancing internal field-aided transport of electrons (writing) and holes (erasing). In similar ways, further reduction in power could be achieved through three-layer direct tunnel PBO stack design to achieve $\pm$2 V, 1 ms W/E programmability as exemplified in the stack design for very low power in

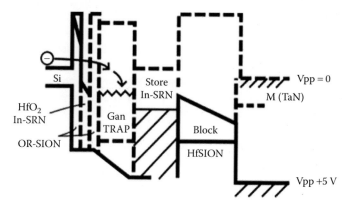

FIGURE 18.2 Band features for triple-tunnel-layer VARIOT consisting of OR-SiON/HfO$_2$ or IN-SRN/ OR-SiON (~2.5 nm EOT) with GaN+IN-SRN storage (~2.0 nm EOT) and HfSiON Blocking (~2 nm EOT) for the device of Figure 8.1 showing electron transport by multiple direct tunneling in bold during writing and dotted during initial state of no applied potential.

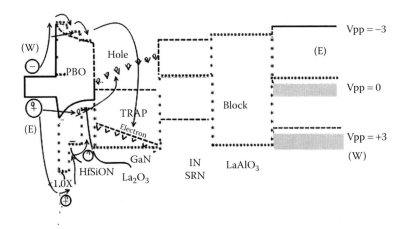

FIGURE 18.3 Illustration of band features during writing (Vpp: +3 V) and erasing in bold (Vpp: −3 V) for a triple-layer PBO stack design consisting of < 1 nm of SiO_2/2 nm of HfSiON/2.5 nm of La_2O_3 (EOT~1.5 nm) with GaN+IN-SRN charge storage and $LaAlO_3$ blocking demonstrating field-aided tunneling during writing and erasing of electrons and holes respectively from silicon substrate. Only the band modulations of tunneling and trapping layers are depicted.

Table 18.5 [8] with OR-SiON/Si_2ON_2/HfO_2 three-layer ultra-thin PBO tunneling structure whereby the band features would be similar to the one shown in Figure 18.3, where the device is further scaled with the stack EOT ~< 4 nm.

Two other special attributes for NVM NROM device designs are illustrated in Table 18.5. The first one is an example of very high-performance device derived from multimechanism carrier transport using a two-layer band-engineered tunneling structure. The tunnel layers consist of an ultrathin OR-SiON layer of 1 nm with an over layer of 3 nm thick HfO_2. During writing, higher energy electrons are supplied from source-side injection along with simultaneous transport of lower energy electrons from the silicon substrate by direct tunneling. Erase operation involved hot hole transport from silicon due to high field from the diffusion-substrate depletion layer combined with cooler hole transport from the gate by enhanced field emission.

For the split channel configuration, the control gate (or the word-line) wraps around the central fixed Vth element (separate select gate in series) [16] schematically shown in Figure 18.4a, and writing and erasing are schematically illustrated by the band diagrams of 18.4b [19]. The array layout scheme for the planar device is shown in Figure 18.5 [16]. Such stack design could be incorporated in FinFET and in vertical channel devices for embedded applications with similar device attributes for sub-15 nm nodes.

The second special attribute is related to the stack design to provide multifunctionality within the same cell, specifically the DRAM and NVM functionality built within the same cell. Such cell design will be called multifunctional cell or MF cell. Multiple variations of MF cells will be discussed in detail in Part III within the framework of SUM. One such cell for embedded application is illustrated in Table 18.5. The cell contains two trapping regions separated by a large band gap high K dielectric film of Al_2O_3. The first trapping dielectric of Si_2ON_2 is placed only 2.5 nm from the silicon/insulator interface separated from the interface by 1nm of OR-SiON and 1.5 nm of HfO_2 (or alternately La_2O_3 not shown in the table) to form a two-layer PBO/VARIOT barrier for electron transport. The second trapping layer of GaN is placed at 7.5 nm from the interface for charge storage with 10 years of retention for NVM functionality.

The first trap filling into Si_2ON_2 is very fast and is accomplished at significantly lower Vpp (Vpp1) due to the band structure and field-aided direct electron tunneling during "NV-DRAM" writing and enhanced direct hole tunneling during "NV-DRAM" erasing. Typically, "NV-DRAM" programming operation takes place at ±2.5 V, 50 ns, yielding an initial memory window of >1.5 V for fast sensing with retention > 100 sec and is required to be refreshed the memory state either

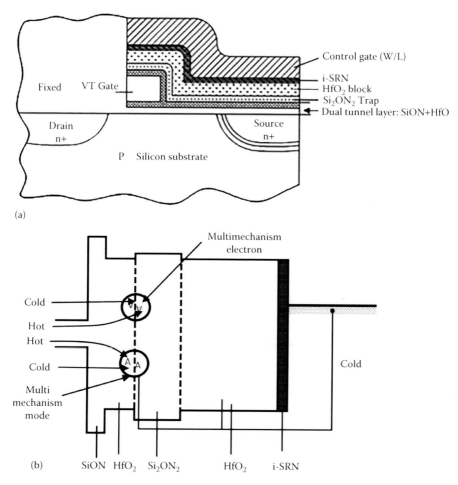

FIGURE 18.4 Band-engineered high-performance embedded device incorporating multimechanism carrier transport: (a) split-gate device schematic cross section, (b) multimechanism electron transport for writing and multimechanism hole transport for erasing. (From Bhattacharyya, A., A novel low power non-volatile memory and gate stack, ADI Associates Internal Docket, [MIDD45]; U.S. Patent: 7612403 [12/2]; Oh, C. W., Kim, N. Y., Choi, Y. L., et al., 4-bit double SONOS memories [DSMs] using single level and multi-level cell schemes, *IEDM,* 2006, pp. 967–970.)

within every 100 sec or the memory state to be stored permanently into the high retention NVM part of the same cell. The NV-DRAM part of the cell provides >> 1E12 (end-of-life) endurance. The NVM part of the cell operates on FN tunneling mode at higher programming voltage ±5 V to ±6 V, 1 ms to yield a memory window of > 4 V; >> 1E7 endurance and > 10 years of retention at operational condition. The stack EOT would be ~6.5 to 7 nm and the blocking layer of 7 to 10 nm of LaAlO$_3$ is employed with TaN gate to provide very low leakage. The device should preferably be a split-gate device operating at all enhancement mode for low power operation. The band diagram of the device is shown in Figure 18.6a reflecting the DRAM operating condition of +2.5 V (DRAM writing) and in Figure 18.6b reflecting the NVM operating condition of +5 V (NVM writing). J.H. Yi disclosed a two-transistor multiple direct tunnel memory [20] with FG structure demonstrating DRAM functionality at ±5 V with write speed of 100 ns at retention of 200 sec. It should be noted, however, that the traditional DRAM product today operates with 1, 2 V on bit-line and nearly 3 V on word-line while providing access time of < 5 ns. This implies that NV-DRAM would face considerable challenge in meeting competitive operational power requirement even though the refresh power for traditional DRAM could be higher.

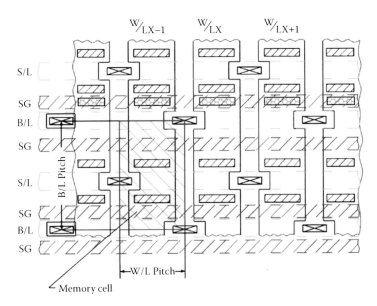

FIGURE 18.5 Array layout of the high-performance split-gate device sharing bit lines and source lines between adjacent cells. (From Bhattacharyya, A., A novel low power non-volatile memory and gate stack, ADI Associates Internal Docket, [MIDD45]; U.S. Patent: 7612403 [12/2].)

18.10   SCALABLE AND NONPLANAR NROMS

We have thus far discussed scalable planar NROM devices and stack designs and pointed out that most stack designs could be considered for nonplanar devices. In the previous section, we have reviewed nonplanar and 3D devices. In this subsection, we will discuss some of the concepts of advanced scalable and nonplanar NROMs in embedded and stand-alone environments and will explore some of the apparent advantages of higher density, feature size scalability, and immunities against SCE, FEE, and parasitic effects. We will include planar front back-gated devices, trench-based vertical channel devices, and FinFETs in this subsection for embedded and stand-alone applications. It should be implicit in addressing nonplanar devices that the special advantages outweigh the additional complexity in integrating such device proposals.

1. Dual-Plane or Double-Gate NROM: For non-FinFET technology both bulk and SOI, double-gated and/or back-side NROM implementation with advanced stack design could be considered for cost-effective solution for diverse applications [5,21]. The device structures are similar to that shown in Figure 15.30a and b, respectively, for bulk and SOI implementation options with stack designs outlined in Tables 18.2 and 18.5. Typically, back-gate NVM would have density advantage over the split-gate device with performance advantages over the planar split-gate implementation at the expense of additional process complexity. Additionally, 2 bit/cell implementation could also be achieved for the back gate with the select fixed Vth device as the front gate device. Other possibilities include 4 bit/cell DSM cell implementations [22], 2 bit/cell MNSC implementation [23], and 4 bit/cell MNSC implementation [24].

2. FinFET Embedded NROM: In recent years, Trigate FinFET has been the FET devices of preference for the CMOS platform technology and for many embedded applications [25]. FinFET-TANOS devices demonstrated superior device characteristics compared to planar TANOS and has been reviewed in the previous section. Advanced stacks as suggested in Tables 18.2 and 18.5 could be incorporated into the FinFET technology down to sub-15 nm node to derive superior NROM-NVM devices.

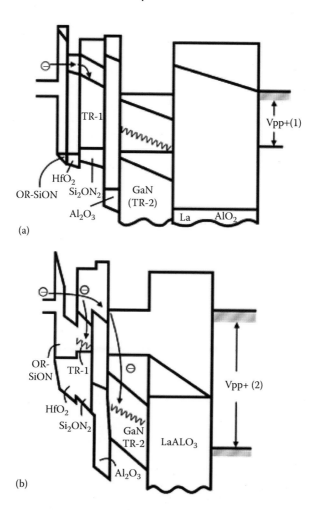

FIGURE 18.6 Embedded multifunctional device band features: (a) DRAM writing, (b) NVM writing. (From Bhattacharyya, A., A very low power memory sub-system with progressive non-volatility, ADI Associates Internal Docket, 8/2/2004; U.S. Patents: 7276760 [10/02/2007], 7385245 [06/10/2008]. [MIDD39] MF; Bhattacharyya, A., Advanced split gate cell design for embedded applications Internal Mem, ADI Associates, June 10, 2013.)

3. TRENCH-BASED Vertical Channel Embedded and Stand-alone NROMs: Trench-based vertical channel CT-NROM devices have been discussed in Section 16.1 in considerable detail as potential solutions for feature size scaling and removing the constraints of FEE and SCE. Cell and Array designs suggested [5–7] could be readily applicable with advanced stack designs proposed in Tables 18.2 and 18.5 to provide advanced embedded NVM devices and arrays to satisfy the diverse market segments for the future technology nodes.

18.11 MLS AND DENSE NROM DESIGN CONCEPTS: BOTH PLANAR AND NONPLANAR

18.11.1 Planar MLS Potentials

CT-NROM devices when programmed by means of hot carrier injection exhibit localized trapping with intrinsic capability of 2 bit/cell MNSC type design discussed in Chapter 15, Section 15.6.1 [5,8,9,17,19,26–28]. Using VGA architecture, memory density of $3F^2$ to $4F^2$ could be achieved for such

NROM devices. However, this bit density comes at the expense of reduced endurance, reliability, and higher power. On the other hand, normal planar DTM cell [8] of the type shown in Figure 18.1a, operating as 1 b/cell design could yield memory bit density of ~6F^2 cell. For low cost low power embedded applications high bit density would be highly desirable. For planar configurations, this could be feasible by the split-gate cell design of the type as shown in Figure 18.1b whereby charges could be selectively trapped over the diffusion regions of the split-gate NROM isolated by the fixed Vth element at the center of the device acting both as the select device and controlling the low Vth state of the memory cell. By reversing the roles of bit-line and source line and sensing the current flow, 2 bit/cell capability could be readily achieved. The array layout for such cell is shown in Figure 18.7. Such a layout could yield an 8F^2 design with 4F^2/bit for a DLS cell as mentioned above. However, with appropriate stack design, the charge trapping could be significantly enhanced by programming over each diffusion region of the cell enlarging the memory window capable of four memory states over each node. This would provide 4 bit/cell capability per cell yielding bit density of 2F^2/bit for embedded and stand-alone devices. Further enhancements of each node with TLS capability cannot be ruled out. Therefore, even with a scheme of relaxed design rules (longer channel length NVM devices) to address feature size scaling issues such as FEE and SCE, MLS planar NROM cells could be envisioned to achieve low cost and high yield potentials without trading off important NVM device characteristics such as endurance, power, and reliability. Many planar NVM designs with enhanced attributes described for 1 bit/cell operation could be modified with the above approach of multibit per cell design further enabling embedded and stand-alone planar NVM devices for application extendibility and cost/performance point of view. Additionally, most planar array designs could be reconfigured into trench-based planar/vertical channel array configurations in contactless VGA architecture with buried diffusion lines to overcome feature size-induced scalability limitations.

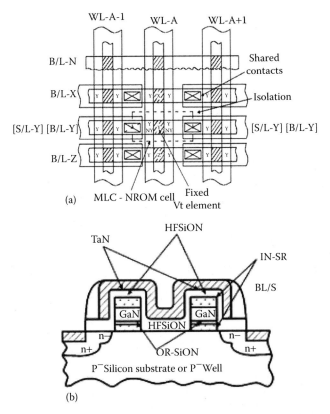

FIGURE 18.7 Array layout for planar DTM split-gate NROM-MLS cell with fixed Vth select gate at the center. (From Oh, C. W., Kim, N. Y., Choi, Y. L., et al., 4-bit double SONOS memories [DSMs] using single level and multi-level cell schemes, *IEDM,* 2006, pp. 967–970.)

18.11.2 NonPlanar Cell Designs and MLS Potentials

Both stand-alone and embedded technology features integration of both logic and memory elements. In nonplanar configurations, such designs add more process complexity when compared to the planar version of NROM technology. At the same time, embedded nonplanar technology is required to be cost and complexity sensitive to be effective in responding to the application requirements. Therefore, we will limit our discussion on nonplanar designs to FinFET and trench-based vertical channel device options only since the integration complexity is limited for such nonplanar options when compared to 3D options such as Gate-all-around or full 3D configurations discussed in Chapter 16.

The split-gate planar device concept discussed in subsection A above could be extended to the FinFET devices by similarly incorporating double layer overlapping gate designs whereby the middle polysilicon fixed Vth gate is placed at the center of the silicon fin acting as the select gate regulating the low Vth state of the NVM device. The overlapped metal gate (e.g., TaN/W) will be the control gate of the device. The fixed Vth device could have the same oxide/high K blocking layer gate insulator structure as the peripheral n-type FinFET fixed Vth device following the process sequences described elsewhere [27]. Potential trench-based vertical channel MLS concepts for embedded NROM will be described below.

Vertical channel MLS designs can take advantage of scaled feature size and MLS cell designs thereby providing denser memory arrays and cost-effectiveness in spite of added process complexity. An example of a trench-based split-gate design consisting of two NROM devices within a trench with capability of providing two or more bits of memory per device (a total of 4 or 6 or 8 bits per trench) could be envisioned and is shown in Figure 18.8 [27]. The two devices are linked by a common buried n+ linked diffusion, but capable of operating independently. The fixed Vth gates for the center of the channel are also common and formed in similar scheme as the surround gate and contacted out of the array. The surface diffusions act interchangeably as both bit-line and source line (reverse channel current mode) and have self-aligned contact scheme. The word-lines overlap the fixed Vth gate vertically and are separated by the polyplug oxidized for isolation. The vertical poly-plug at the center of the trench contacts the buried link-diffusion and is contacted outside the array. The array layout is shown in Figure 18.9. The surface n+ diffusions could be shared between neighboring trenches to enhance bit density (not shown).

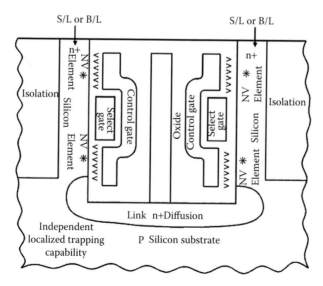

FIGURE 18.8 Vertical channel split-gate NROM cell features within a single trench. (From Bhattacharyya, A., Extending planar and non-planar DTM device and array concepts to achieve high bit density thereby expanding market application base for embedded and stand-alone NVM memories, ADI Associates, Internal Memo, July, 2013.)

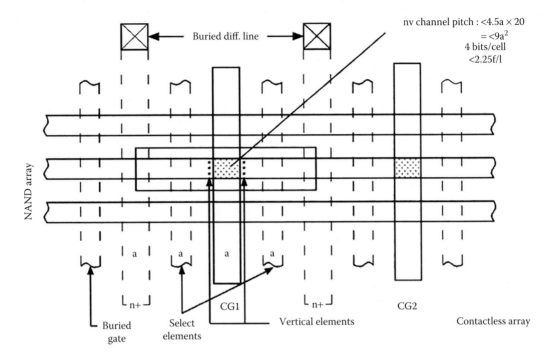

FIGURE 18.9 Array Layout of Vertical channel NROM cell design of Figure 18.8. (From Bhattacharyya, A., Extending planar and non-planar DTM device and array concepts to achieve high bit density thereby expanding market application base for embedded and stand-alone NVM memories, ADI Associates, Internal Memo, July, 2013.)

Depending on the MLS levels, bit density of $3F^2$/bit and significantly higher bit density could be feasible.

18.12 ADVANCED NAND DESIGN CONCEPTS: PLANAR AND NONPLANAR

18.12.1 Embedded NAND Flash

Embedded NAND has the potential of providing low cost/bit solution for applications requiring large amount of embedded memory [3,6–8,17,21,27]. For such applications, one of the preferred approaches could be to take advantage of feature size scaling and associated fundamental challenges to consider vertical channel NAND devices and arrays with minimum process complexity. The author has proposed several variations in trench-based vertical channel NAND concepts [6,7,22]. Other planar options [3,8,27] could also be reconfigured into vertical channel embedded NAND flash devices while preserving some of the key enhanced device attributes. However, from the practical embedded NAND flash device and product point of view, the conventional and process-compatible stack design with feature size scalability is considered to be the preferred approach in the form of FPNAND devices. One stack design for such NAND could consist of ultra-thin oxide or ORSiON combined with HfSiON for tunneling/Si_2ON_2-SRN or preferably TaN (conductive storage medium for larger window for DLC capability: [28] for charge storage and HfSiON for charge blocking). Metal gate technology could be applicable for such design with compatible work-function control for the PFET and NFET CMOS devices with common Ni-FuSi/TaN gate processing [29,30]. Additionally, the N+ and P+ diffusions and "SALI" (self-aligned local interconnects) contacts could be silicides with Ni-Si for controlling device series resistances and contact resistances. For embedded applications, the common device elements thus could be summarized as follows:

NFET: SiON (1 nm)/HfSiON (8 nm)/Ni-Fusi/TaN (with or without Al or Pt (I.I) doping for work-function control, Gate (EOT~3 nm)

PFET: Same as above (with or without an additional Ni I.I (ion implant) for work-function control, Gate (EOT~3 nm); NVM: SiON/HfSiON/TaN/HfSiON/NiFuSi/TaN: (1/5/6) (TiN)/8/10 (TiN): Gate (EOT~5 nm)

Such integration approach as above not only minimizes the complexity in embedded technology enhancing cost reduction and yield but also provides at the same time scaled relatively high-performance devices with enhanced attributes and reliability. The NAND flash device targets are expected to be as follows:

Vpp: ± 5 V, 1 ms; Memory window: > 5 V (DLC Capability); Retention: 10 years, Endurance > 1E7 W/E

While the CMOS devices could be considered either as planar conventional or FinFET depending on the base platform technology, the embedded NAND flash could optionally be of trench-based vertical channel to provide high bit density. The schematic cross-sections of NFET, PFET, and vertical channel NAND flash are shown in Figures 18.10a, b and 18.11, respectively. Vertical channel could be configured with a pair of NAND devices per trench linking with other devices through buried N+ diffusion and surface N+ diffusion for the entire string. The fixed Vth select devices could be formed with vertical channel as shown.

A contactless array layout for the vertical channel NAND flash is shown in Figure 18.12. With DLS capability per cell and with two cells per trench as shown in Figure 18.11b, a bit density of ~1.5 F2/bit could be feasible for embedded applications.

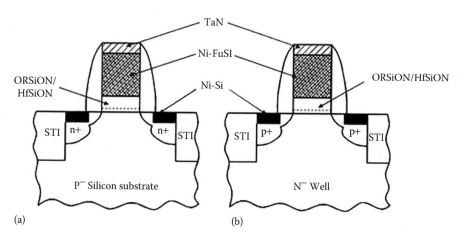

FIGURE 18.10 CMOS device schematics in advanced embedded NAND flash Technology: (a) NFET, (b) PFET. (From Kim, Y. H., Cabral, Jr. C., Zarfar, S., et al., Systematic study of work function engineering and scavenging effect using NiSi alloy FUSI metal gates with advanced gate stacks, *IEDM,* 2005, pp. 657–660.)

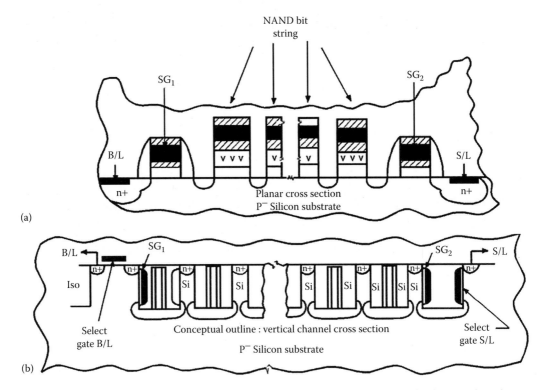

FIGURE 18.11   NAND flash string schematic cross section options in Embedded Technology (a) planar layout, and (b) vertical channel layout with two NVM devices per trench, note that select gate FETs are also of vertical channel.

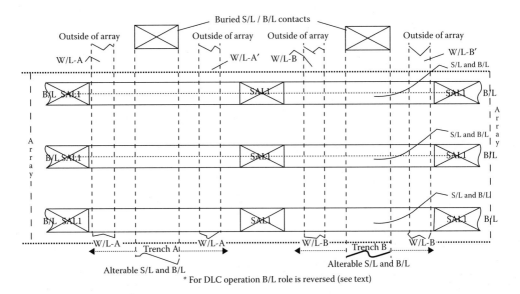

FIGURE 18.12   Embedded NAND flash contactless array layout [27].

18.12.2 Advanced Stand-Alone NAND Flash

It has been stated earlier that the stand-alone NAND flash drives the NVM memory bit volume with CGR of 70% and is expected to continue in foreseeable future driven by continued cost reduction and feature size scaling. However, it is not clear if such momentum could continue beyond 15 nm technology node. Key challenges for continued momentum include: (a) cost-effective demonstration and integration of full 3D NAND flash product design proposals; (b) 4 b/cell MLC designs with low RBER and low ECC overhead cost; (c) enhancements in endurance and retention of memory states; and (d) improved product reliability. From device perspectives we have noted that progress has been made in terms of changing the NAND flash device from FG type to CT-TANOS type [31–34] and conductive plate/floating plate (CP/FP) type [36] demonstrating scalability down to 15 nm technology node. Additionally, multilevel planar and nonplanar integration of high-density NAND flash arrays and functionality has been demonstrated up to 64 and 128 Gb levels [35–38]. However, cost implication of stand-alone nonplanar NAND flash arrays is yet to be determined. The devices proposed for the above-referenced NAND flash arrays have large memory windows with potential TLC capabilities and endurance > 1E5 W/E.

18.12.2.1 Cyclability

An advanced stand-alone NAND flash stack design based on trench-based vertical channel technology is proposed here in the form of contactless array layout as of Figure 18.12 to provide $\leq$ 1.5 F2/bit density, which would have further device capability and enhancements. The FET and NVM devices will be integrated with similar philosophy proposed for embedded NAND flash design as above. The FET and the NVM stack designs would be altered as listed below [28].

NFET: SiON (1 nm)/LaAlO$_3$ (8 nm)/SiON (1 nm)/ "Ni-Fusi/TaN (10 nm)"-(Si/Metal Gate) (with or without Al or Pt (I.I) doping for work-function control, (EOT~2.5 nm)

PFET: Same as above (with or without an additional Ni I.I (ion implant) for work-function control, (EOT~2.5 nm)

NVM: SiON(1 nm)/La$_2$O$_3$(7 nm) (tunneling)/TaN(6 nm) + SRN(6 nm) or GaN(5 nm) + SRN (6 nm)(charge storage)/LaAlO$_3$(8 nm)/ SiON(1 nm) (charge blocking)/NiFuSi/TaN(10 nm): (1/5/6 nm)(TiN)/8/10 nm (TaN): (EOT~3.5–5 nm)

It should be noted that the silicon/insulator interface layer is common and the gate insulator/gate for the FET devices are compatible to the NAND flash blocking layer/gate design. The design provides high-performance FET devices at VDD = 2 V and cost/performance design for NAND flash with high endurance and low programming power. The expected device attributes are listed below:

Memory window: >>8 V good for TLC design
Programming conditions: Vpp+- ±4 V, 1 ms Memory Retention: 10 years (EOL Window >> 5 V Memory Endurance: >> 1 E7 cycles
Low RBER/Low ECC Requirements

With the trench-based vertical channel design of one-transistor or one-and-a-half transistor cell, the NAND flash should be scalable to 10 nm node with advanced lithography yielding ~ 1 F2/bit memory using TLC cells and contactless array layout.

18.13 ADVANCED NANOCRYSTAL DEVICE CONCEPTS

Silicon nanocrystals or nanodots (NC or ND) in nitride in the form of SRN layers have been well characterized in the past [39] and potential advantages and limitations in applications in NVM devices have been demonstrated. This has been discussed in Part I, Chapters 2 and 4 in this book.

During the past decade, considerable R&D activities were conducted with Si-NC, Ge-NC, and a variety of metal NCs with high work functions (W, Rh, Pd, Pt, etc.) embedded in oxide or nitride or other dielectric films as trapping centers for potential CT-NVM devices with the expectation of applications in enhanced NROM and NAND NVMs.

Most publications were limited to capacitors and transistors and considerable understanding of quantum dot devices and discrete trapping centers were derived from such investigations. This has been briefly summarized in Part I Section 3.2.4. Dielectric films containing high-density metal nanodots (MNDs) of 2–3 nm diameter was used as charge retention layer demonstrating high endurance [40]. Thus far productization of NC-NVMs has not been successful due to integration related difficulties and metal NC contamination issues. Lately, N. Ramaswamy et al. [34] reported a comparative study of integrated NAND flash arrays with Pt-NC and Rh-NC "Charge-trap-flash" (NCCTF) devices and conventional oxide/nitride (optimized composition) CTF devices (called "SN-CTF") and conductive charge-storage CTF (which they called "Conductive floating gate" device which in our terminology would be called floating plate [FP] or conductive floating plate [CFP] device) [32]. Their investigation showed no apparent advantages of NC-CTF over FP or CFP device on NAND applications especially regarding SILC related device reliability issues. Their investigations also showed that while memory window Vth variation from cell to cell could be minimized with higher density of metal NCs (>5 E12/cm^2) and large memory window for TLC could be achievable, such device is prone to lateral charge redistribution related Vth instability. So far, this has been the most comprehensive study on NC-CT NAND flash devices and arrays and associated technology demonstrating future challenges with such devices within the framework of the selection of the NAND cell designs. Within the framework of single cell SLC NROM designs, lateral redistribution of charges may not be a critical issue.

NC-NVMs are attractive since such devices provide planar charge confinements and potential deep energy discrete charge-trapping centers with high work-function metal-embedded NC-devices. Since charges do not propagate perpendicular to the plane of confinement, with appropriate band engineering, the devices could provide significant enhancements in programming speed and endurance. By trapping charges in discrete planes at different distances from silicon substrate and gate, multilevel memory devices could be achieved providing MLC capability. Additionally, with appropriate stack design, retention of charges could be modulated in multilayered band-engineered stack designs to simultaneously achieve high-speed DRAM functionality and lower speed NVM functionality within the same NVM cell. These attributes provide attractive possibilities for NC-based NROM cells in providing multilevel and multifunctional cell design concepts. Furthermore, NC-based trapping and charge-trapping dielectrics could be appropriately combined to yield enhanced NVM attributes.

A variety of NC-based multilevel and multifunctional NVM stack designs have been proposed with and without charge-trapping dielectric medium to achieve advanced device attributes [16,18,24,26,27,41–47]. Some of these concepts will be discussed in Chapter 3 of this book where the multifunctional NVM devices based on charge-trapping dielectrics (CTNVMs) will be reviewed in detail. References on others are provided at the end of the section.

18.14   OTHER ADVANCED MLC NVM DEVICE CONCEPTS

We have discussed several advanced MLC device concepts [27] both for NROM (4 bit/cell) and NAND flash devices in Chapter 7. Since such concepts could be applicable for appropriate application segments discussed here, no further discussion will be made here.

18.15   MULTIFUNCTIONAL NVM DEVICES

By appropriate device stack design, retention of memory states in NVM devices could be varied by many orders of magnitude. Additionally, charge transport rate could be altered by at least 4 to 5 orders of magnitude by a combination of tunneling mode and barrier engineering. This provides opportunity to achieve multifunctional memory characteristics and thereby expand the application potential of NVM devices. This is the basis of many of the SUM proposals. An illustration of an NROM device to provide simultaneously DRAM and NVM functionality has been discussed earlier and an example for such a stack design is provided in Table 18.5. Multifunctional NVM devices will be discussed in detail in Part III on SUM.

REFERENCES

1. R. Strenz, Embedded flash technologies and their applications: Status and outlook, *IEDM*, Washington DC, 2011, pp. 211–214.
2. L. Masoero, G. Molas, F. Brun et al., Scalability of split-gate charge trap memories down to 20 nm for low power embedded memories, *IEDM*, Washington DC, 2011, pp. 215–218.
3. A. Bhattacharyya, K. Prall, and L. C. Tran, Band engineered multi-gated non-volatile memory with enhanced attributes, ADI associates internal Docket, July 10, 2004; U.S. Patents: 7279740 (10/9/07), 7749848 (7/6/2010), 8063436 (11/22/11). [MIDD36].
4. C. S. Park, B. J. Cho, L. J. Tang et al., Substituted aluminum metal gate on high K dielectric for low work-function and fermi-level pinning free, *IEDM*, San Francisco CA, 2004, pp. 299–302.
5. A. Bhattacharyya, Novel back-side trapped non-volatile memory, ADI Associates Internal Docket, U.S. Patents: 7851827 (12/14/10), 8058118 (11/15/2011).
6. A. Bhattacharyya, Scalable high density, high performance hybrid planar vertical NV memory cells and contactless arrays, ADI Associates Internal Docket, 2004; U.S. Patents: 7166888 (1/23/07), 7273784 (9/25/07), 7964909 (6/21/2011). [MIDD42].
7. A. Bhattacharyya, Extremely high density two-bit-per-cell planar-vertical. Embedded trap NVM cell and contactless array, ADI Associates Internal Docket, 2004; U.S. Patents: 7671407 (3/2/2010), 7838362 (11/23/2010).
8. A. Bhattacharyya, A vertical cross-point high density split channel direct tunnel non-volatile memory, ADI Associates Internal Docket, August 2, 2004; U.S. Patents: 7365388 (4/29/2008), 7671407 (3/2/2010). [MIDD 37].
9. A. Bhattacharyya, A very low power memory sub-system with progressive non-volatility, ADI Associates Internal Docket, 8/2/2004; U.S. Patents: 7276760 (10/02/2007), 7385245 (06/10/2008). [MIDD39] MF.
10. A. Bhattacharyya, A vertical cross-point high density split channel direct tunnel non-volatile memory, ADI Associates Internal Docket, August 2, 2004; U.S. Patents: 7365388 (4/29/2008), 7671407 (3/2/2010). [MIDD 40].
11. A. Bhattacharyya, A novel low power non-volatile memory and gate stack, ADI Associates Internal Docket, U.S. Patent: 7612403 (12/22/2009).
12. S. Kim, S. J. Baik, C. Huo et al., Robust multi-bit programmable flash memory using a resonant tunnel barrier, *IEDM*, 2005, p. 881.
13. A. Bhattacharyya, Scalable integrated logic/NV memory technology with associated gate stack structure and process, ADI Associates Internal Docket, May 23, 2004; U.S. Patent: 7544990 (06/09/2009).
14. A. Bhattacharyya, Enhanced multi-bit non-volatile memory using resonant tunnel barrier, ADI Associates Internal Docket, August 18, 2005; U.S. Patents: 7482651 (01/27/2009), 7867850 (01/11/2011).
15. A. Bhattacharyya, Enhanced RTB devices, internal memo, ADI Associates, Unpublished, August 2005.
16. A. Bhattacharyya, Defining high speed high endurance scalable NVM device using multi-mechanism carrier transport and high density structures for contact-less arrays, ADI Associates Internal Docket, 9/1/2004; U.S. Patents: 7244981 (7/17/07), 7553735 (6/30/09), 8159875 (4/17/12).
17. A. Bhattacharyya, A novel low power non-volatile memory and gate stack, ADI Associates Internal Docket; U.S. Patent: 7612403 (12/2).
18. A. Bhattacharyya, Advanced scalable nano-crystal device and process, ADI Associates Internal Docket, November 19, 2004, [MIDD 47], Redefined: Scalable multi-functional and multi-level nano-crystal NVM, U.S. Patents: 7476927, 01/13, 09; 7759715, 07/30/2010; 7898022, 2011; 8242554, 08/14/2012.
19. A. Bhattacharyya, "Advanced split gate cell design for embedded applications" Internal Mem, Adi Associates, June, 10, 2013, Unpublished.
20. A. Bhattacharyya, "Novel Band-Engineered Nano-Crystal NVM device Utilizing enhanced Gate injection", ADI Associates Internal Docket, Dec. 4, 2004; U.S.Patent 7629641, 12/22/2009.
21. A. Bhattacharyya, A novel high density SONOS NAND, ADI Associates Internal Docket, July 2005, U.S. Patent: 7829938 (11/9/2010).
22. C. W. Oh, N. Y. Kim, Y. L. Choi et al., 4-bit double SONOS memories (DSMs) using single level and multi-level cell schemes, *IEDM*, San Francisco CA, 2006, pp. 967–970.

23. R. Ranica, A. Villaret, P. Mazoyar et al., A new 40 nm SONOS structure based on backside trapping for nano-scale Memory, *IEEE Transactions. on Nanotechnology*, 4(5), 2005, pp. 581–587.

24. A. Bhattacharyya, A novel 4 bit/cell dual channel MNSC device, Internal Memo, ADI Associates, September 2007 (Unpublished).

25. C. H. Jan, U. Bhattacharya, R. Brain et al., A 22 nm SoC platform technology featuring 3-D tri-gate and highk/metal gate, optimized for ultra-low power, high performance and high density SoC applications, *IEDM*, 2012, pp. 44–47.

26. G. Zhang, W. S. Hwang, S. M. Bobade et al., Novel ZrO_2/Si_3N_4 dual charge storage layer to form step-up potential wells for highly reliable multi-level cell application, *IEDM*, Washington DC, 2007, pp. 83–86.

27. A. Bhattacharyya, Extending planar and non-planar DTM device and array concepts to achieve high bit density thereby expanding market application base for embedded and stand-alone NVM memories, ADI Associates, Internal Memo, July 2013.

28. Y. H. Kim, C. Cabral, Jr., S. Zarfar et al., Systematic study of work function engineering and scavenging effect using NiSi alloy FUSI metal gates with advanced gate stacks, *IEDM*, Washington DC, 2005, pp. 657–660.

29. A. Lauwers, A. Veloso, T. Hoffmann et al., CMOS integration of dual work function phase controlled NI FUSI with simultaneous silicidation of NMOS (NiSi) and PMOS (Ni-rich silicide) gates on HfSiON', *IEDM*, Washington DC, 2005, pp. 661–664.

30. Y. Park, J. Choi, C. Kang et al., Highly manufacturable 32 Gb multi-level NAND flash memory with 0.0098 um 2 cell size using TANOS (Si-Oxide-Al_2O_3-TaN) cell technology, *IEDM*, 2006, pp. 29–32.

31. Z. Huo, J. Yang, S. Lim et al., Band engineered charge trapping layer for highly reliable MLC flash memory, *VLSI Symp.*, Kyoto, Japan, 2007, pp. 138–139.

32. C.-H. Lee, S.-K. Sung, D. Jang et al., A highly manufacturable integration technology for 27 nm 2 and 3 bit/cell NAND flash memory, *IEDM*, San Francisco CA, 2010, pp. 5.1.1–5.1.4.

33. Y. Q. Wang, D. Y. Gao, W. S. Hwang et al., Fast erasing and highly reliable MONOS type memory with HfO_2 high-k trapping layer and Si_3N_4/SiO_2 tunneling stack, *IEDM*, San Francisco CA, 2006, pp. 971–974.

34. N. Ramaswamy, T. Graetinger, G. Puzzilli et al., Engineering a planar NAND cell scalable to 20 nm and beyond, *IEDM*, Monterey MD, 2013 [Micron].

35. J. Hwang, J. Seo, Y. Lee et al., A middle-1X nm NAND flash memory cell (MOX-NAND) with highly manufacturable integration technologies, *IEDM*, 2011, pp. 199–202.

36. Peter Clarke, 3D NAND production starts at Samsung, EE Times, 8/6/2013. [Samsung].

37. Peter Clarke, Samsung confirms 24 layers in 3D NAND, EE Times, 8/9/2013. [Samsung].

38. A. Bhattacharyya, Defining high speed high endurance scalable NVM device using multi-mechanism carrier transport and high density structures for contact-less arrays, ADI Associates Internal Docket, 9/1/2004; U.S. Patents: 7244981 (7/17/07), 7553735 (6/30/09).

39. A. Bhattacharyya, R. Bass, W. Tice et al., Physical and electrical characteristics of LPCVD silicon rich nitride, ECS Fall Meeting, October 1984, New Orleans.

40. M. Takata, S. Kondoh, T. Sakaguchi et al., New non-volatile memory with extremely high density metal nano dots, *IEDM*, New York, 2003, 22.5.1–22.5.4.

41. A. Bhattacharyya, High density nano dot non-volatile memory, ADI Associates Internal Docket, July 2005, U.S. Patent: 7690265 (4/15/2010).

42. A. Bhattacharyya et al., A germanium nano-crystal high performance NVM using graded germanium-silicon carbide insulator, ADI Associates Internal Docket, November 29, 2004, "Combined volatile and nonvolatile memory device with graded composition," U.S. Patent: 77680062 (8/3/2010).

43. A. Bhattacharyya, High performance multi-level non-volatile memory device, ADI Associates Internal Docket, December 4, 2004, U.S. Patent: 7429767 (9/30/08), 7579242 (8/25/09), 8129213 (3/6/12).

44. A. Bhattacharyya, Discrete trap nonvolatile multi-functional memory device, ADI Associates Internal Docket, August 11, 2005, "Discrete trap nonvolatile device for universal memory chip Application," U.S. Patent: 7436018 (10/14/08), 7786516 (10/14/10).

45. A. Bhattacharyya, Novel metal nano-crystal non-volatile device and process scheme ADI Associates Internal Docket, November 2006, Unpublished.

46. A. Bhattacharyya, Scalable multilevel cell nano-crystal device with reduced threshold dispersion, ADI Associates Internal Docket, U.S. Patent: 7898850 issued 3/1/2011, titled: memory cells, electronic system, methods of forming memory cells, and methods of programming memory cells.

47. A. Bhattacharyya, Memory cell comprising DRAM nanoparticles and NVM nanoparticles, original docket titled "Advanced scalable Nano-crystal NVM device and Process," ADI Associates Internal Docket, August 2005, MIDD 75: US Patent: 7759715 (7/30/2010), 8193568 (6/5/2012).

Part III

SUM: Silicon Based Unified Memory

INTRODUCTION

The final section of this book explores and suggests the direction of future developments in silicon based nonvolatile memory (NVM) through the evolutionary concepts of what is generally termed "Silicon Based Unified Memory" or SUM.

WHY SILICON BASED UNIFIED MEMORY: SUM?

Silicon based digital technology has created this new age of revolution in developmental achievement unprecedented in human history touching every aspect of human life and activities in the merely five decades of its existence. Scaling of silicon based CMOS technology has been the cornerstone of such success for the past four decades of technology integration providing unit functions of logic and memory in unceasing miniaturized form at ever-increasing performance with decreasing cost and energy usage. This is expected to continue at least for the next two decades since the CMOS FET devices and technology have successfully been scaled and mass produced from micrometer dimensions to nanometer dimensions in such a short time. The silicon based NVM of FET-based devices is also expected to continue the unforeseen growth momentum of 70% CGR for at least next 5 years and possibly for longer period of time. The continued momentum beyond 15 nm technology node would require not only feature size scalability but also 3D structures with enhanced device attributes and functionality. The challenge associated with integration, mass fabrication, and economic justification cannot be overstated. Such development also requires close coupling with innovative memory storage and system architecture, associated integration, and "fault-tolerant"/ "reliability-enhancement" schemes necessary to expand application spectrum, while continuing the momentum of reduction of cost/memory bit and energy requirements. In spite of impressive progress in the current forms of NAND and NROM NVM-FET products, these devices and products face serious challenges against emerging "non-silicon" based NVMs such as R-RAMs and M-RAMs which could displace the current application space of DRAMs and NROMs/NANDs in terms of density, reliability, power, and performance.

The limitations of current silicon FET-based NVMs and the prospects of emerging nonsilicon based NVMs have been discussed in Chapters 17 and 18 of the preceding part. The extendibility potentials of current silicon FET-based NVMs are also illustrated. However, from the system perspective, the scope of current and future silicon based NVMs, even if successfully scaled to the end of the ITRS road map is relatively constrained and limited. For example, such an approach would not be able to fulfill a "power-failure-free," unlimited capacity, "unconstrained-cost-performance" system design without battery backup, and application of the current memory hierarchical approach. Additionally, current silicon based NVMs require significantly higher programming voltage and therefore, constrained to consume significantly higher active power. Current and future system designs are therefore constrained by the significant usage of low density high performance volatile SRAMs, the ever-increasing requirements of working memory bits supplied by volatile DRAMs, and the extensive large database storage requirements of undesirably lower performance yet cost-effective HDDs. Additionally, current approaches are severely constrained by extensive ECC requirements for data integrity and "DRAM-driven" standardization for storage hierarchy.

The "Silicon based Unified Memory" (SUM) devices address the challenge to overcome some of the above constraints in current system design in potentially overcoming the limitations of performance, power and endurance of the current NROMs and NAND flash devices, both at the higher performance-end, and at the capacity end. For the higher-performance application end, the proposed higher density unifunctional (USUM) memory cells and arrays could potentially replace the SRAMs and NROMs; while for general applications, multifunctional URAM and MSUM memory cells and arrays could potentially replace the current memory hierarchy consisting of DRAMS, NANDs, and HDDs. This is further elaborated in Chapter 19. Since the proposed USUM, URAM, and MSUM devices are based on the current silicon based CMOS platform technology and associated integration schemes, all logic and memory devices are unified under the same technology framework. Thus, the term, "Silicon based Unified Memory" or SUM.

FUNDAMENTAL CONCEPT BEHIND SUM DEVICE & TECHNOLOGY FRAMEWORK

The unified concept behind all SUM and URAM devices is the evolution of NVM cells in defining relative permanency of digital memory states by integrating active silicon devices with a charge storage (electrons or holes) medium.

The active device is typically an FET and optionally a gate controlled p-i-n or an n-i-p diode or thyristor. The charge storage medium is typically a single-layer or a multilayer dielectric film stack with or without the combination of a unipotential storage medium such as i-SRN (or IN-SRN). It should be emphasized that such an approach is evolutionary and scalable both in terms of material development and technology integration capability of microelectronics and nano electronics. Therefore, SUM and URAM devices utilize the progressive learning of the present-state-of-the-art in silicon based integrated technology. Such learning has been extensively discussed by reviewing the progress in NROM and NAND NVMs in the previous parts: Part I and Part II of this book.

The evolutionary progression of silicon based NVMs into SUM has been mentioned in the pursuit of device and application enhancements in several places in earlier parts of this book as a lead in to this final part of this book. Such evolution has been facilitated by the development of ultrathin high K films in multilayered and band-engineered gate insulator stack designs and their successful incorporation and integration into scaled metal gate CMOS technology. This replaced the traditional roles of SiO_2 and Si_3N_4 films in NVM gate insulator stack designs in silicon gate CMOS technology. Other NVM device concepts utilized multilevel storage through planar and nonplanar (e.g., vertical channel) structures to overcome device scaling limitations. Yet additional device concepts emerged by integrating different tunneling modes built into the gate insulator stack designs to provide multifunctionality within the same memory cell by trading off memory retention and endurance. SUM device concepts and proposals are, therefore, presented as evolutions of current silicon based devices and NVMs with compatibility with the CMOS logic devices and technology integration schemes and as natural progression of silicon based technology and integration schemes to gain momentum in market acceptance and application by effectively exploiting the following advantages:

1. Established and mature manufacturing base
2. Single transistor cell concepts
3. Demonstrated multifunctionality
4. Demonstrated MLC capability
5. Compatibility with platform CMOS logic devices and technology
6. Demonstrated enhancements in performance, nonvolatility, and endurance

Thus, the framework of SUM:

SUM will be based on the NVMs discussed in Part I and Part II of this book. As stated above, SUM is defined in terms of a broad memory concept applicable to all types of digital hardware, systems, and subsystems. Such memory concept covers the domain ranging from the simplest consumer products, unifunctional, and multifunctional portables to a diverse set of computing systems and large-scale storage systems for local networking, communication, and Internet. SUM definition is multidimensional in the sense that SUM devices and memory cells are proposed to address simultaneously from the standpoint of system perspective, application perspective, as well as from memory device and technology perspective. System and application perspectives of SUM devices are beyond the scope of this book and are expected to be covered in depth in other future book or books. System perspectives are briefly introduced in this section in the context of memory hierarchy, whereas application perspectives are also briefly discussed in the last chapter. This part of this book is intended to focus primarily on the emerging memory device concepts and associated perspectives based on the six elements of advantages stated above. We have briefly included integration and application perspectives of SUM cells and devices at the end of this final part of this book. We are cognizant of the fact that the success of future SUM devices as covered in this part will be dependent on not only of the stated technology and device advantages at the memory cell/device level as above, but also on traditional and innovative applications. This also implies innovations and applications in traditional and innovative system architectures and associated firmware developments in forthcoming decades. We expect that SUM devices would serve as catalyst towards innovative and energy efficient memory system architecture and novel system designs as well as promote innovative applications.

ORGANIZATION AND CONTENT OF PART III

The conventional and emerging memories were briefly discussed earlier in Part II Chapter 17, reviewing the limitations of conventional NVMs against the emerging memories. In Part II, Chapter 18, we highlighted the need of extendibility of conventional devices leading to the development and emergence of silicon based "Unified Memory" devices: "THE SUM DEVICES" for this part of this book.

Chapter 19 briefly describes the different SUM device concepts and perspectives and classifications highlighting the broad range of applicability and functionality that such devices provide. This includes background of SUM and URAM cells and definitions of USUM, FSUM, and URAM cells. Potential application advantages and memory subsystem architectural advantages were also briefly discussed as rationale for the development SUM devices, technologies, and products for the forthcoming decades. Intended device attributes of SUM devices (including "URAM" devices discussed in Chapter 23) were compared with other memory devices highlighting the advantages of SUM devices over others. Merits of SUM devices in providing broader range of applicability and functionality compared to conventional volatile and NVMs such as SRAM, DRAM, NROM, NAND flash, and HDD as well as emerging NVMs such as RRAM, PRAM, and MRAM are noted.

Chapter 20 briefly outlines application of technology learning from Part I and Part II of this book into the future SUM devices, stack designs, and technology integration concepts. This includes selection and employment of dielectric films and gate stack designs for the SUM devices as well as employment of compatible integration schemes. This chapter discusses technology and device interplay, introduces band engineering for multilayered tunneling and trapping, and presents multiple examples of stack designs for SUM devices.

Chapter 21 describes band engineering for device design considerations for SUM devices. This includes multiple illustrations of band engineering for both USUM and FSUM devices.

Chapter 22 describes several potential options of USUM memory cells and arrays. These include cell concepts involving a single charge storage layer integrated into an active silicon device as well as charge-trapping nonvolatile (CTNV) FET-based memory cells of planar and vertical channel configurations with extendable stack designs discussed in Part II.

Chapter 23 describes exclusively a range of multifunctional cell concepts to functionally replace conventional volatile memory cells in system hierarchy. This includes integrated, DRAM-NVRAM cells, SRAM-DRAM-NVRAM cells, URAM cells, and others.

Chapter 24, the final chapter, is a brief introduction to functional integration, packaging, and potential application of SUM devices.

This part of this book is largely conceptual. Since Part III mostly focuses on future device concepts and primarily at pre-prototype phase, the R&D efforts and characteristic data for such devices are limited. The intent of Part III is to serve as a catalyst for the development and realization of SUM devices and products.

19 SUM Perspective, Device Concepts, and Potentials

CHAPTER OUTLINE

SUM devices were conceived to overcome limitations of current silicon based memory cells and associated hierarchical requirements of storage devices for current and future digital systems. Such limitations have been discussed in previous parts of this book and fundamental approaches to overcome such limitations have been identified. In this chapter, we define and discuss perspectives for SUM devices, the conceptual background for such devices as well as applicability, and functionality of these devices for current and future digital systems.

19.1 SUM DEVICE PERSPECTIVE: OBJECTIVE, APPLICABILITY, AND FUNCTIONALITY

19.1.1 SUM Objective

From applicability perspective, SUM devices need to cover the broadest performance spectrum of memory applicability in fulfilling digital operations from 50 operations per sec to 500 million operations per sec providing memory bandwidth of nearly 7 orders of magnitude. From electronic systems functional and memory requirement perspective, it would be highly desirable if the information for the working memory and the information to be stored permanently could be contained at the same memory cell. From cost, throughput, memory organization, and system efficiency (latency) perspective, it would be equally desirable if a significant portion of cache and main memory could be integrated at required performance within the same high bit density memory array eliminating multiple levels of (lower bit density) cache memories from the memory hierarchy. From memory device perspective as well as technology integration/process complexity perspective, similar desirability could be attained if the memory stack design provides process compatibility and commonality with peripheral logic devices. At the same time, it would be highly desirable from device perspective to achieve higher bit density and to provide compatible NROM and NAND memory arrays while yielding appropriate power/performance/density trade-offs. Such memory device concept could not only have wider applicability but also could have universal appeal. This is the objective that defines NVM-SUM. Thus, the SUM device concepts are intended to serve multidimensional perspectives.

19.1.2 SUM Applicability

SUM applicability is intended to span the entire range of digital electronics. At the very low end, it is expected to be an integral part of a silicon based system-on-chip (SOC) solution fulfilling the roles of BIOS, code, and data simultaneously, while directly interacting with the built-in registers and caches associated with the processor or controller through a common bus. Examples of such applications could be smart card, automotive electronic systems, and even electronic toys and low-end smart phones. At the mid-range, SUM devices could play the role of code (BIOS and instructions in NROM arrays), working memory with dual functionality of DRAM and NVRAM (multifunctional NROM arrays), as well as file storage functions (NAND SDDs replacing HDDs). The entire memory system in mid-range could be silicon based and may or may not contain traditional DRAMs. Such applicability could be in a diverse form of portables such as smart phones, tablets, and laptops. In the high end such as Enterprise systems and Clouds, SUM-based high capacity NAND-memory systems could be the key solutions toward providing high throughput data storage and retrieval as well as memory archival systems in conjunction with HDDs and tapes. Additionally, in complex three-dimensional (3D) graphics, bio-imaging, gaming systems, and video processing, SUM devices could serve as the primary source of memory.

19.1.3 SUM Functionality

We will categorize SUM devices into two groups: *uni-functional SUM ("USUM") and multi-functional SUM ("MSUM")*. Both USUM and MSUM devices could have wide range of attributes in terms of power/performance characteristics to meet the application requirements of the system or subsystem. The USUM memory cell and array could provide a range of functionality in terms of latency, programming speed, endurance, and retention with multiple memory arrays within a single chip designed to provide unique attributes from each memory arrays. Alternately, multiple USUM memory chips could be connected to the same memory bus each with unique attributes. The MSUM memory cell and array, however, would have the capability of simultaneously providing traditional DRAM-like functionality (lifetime endurance, lower latency, faster programming, and shorter retention) and traditional NROM/NAND functionality built into the same memory cell. The MSUM memory could, therefore, replace the need of separate DRAM for working memory and SDD-memory for storage in a variety of systems. It should be noted that unlike SRAM and DRAM cells, where charge storage are associated with silicon capacitive element, the charge storage of all types of NVMs is either in dielectric media or in conductive and semiconductive media isolated by surrounding dielectric media. USUM and MSUM cells could be both planar (2D) and nonplanar (3D) types and may operate by different charge transport mechanisms.

19.2 SUM DEVICE CONCEPTS AND CLASSIFICATIONS

SUM devices are broadly classified into two groups distinguished from each other from the standpoint of memory cell functionality, yet unified from the standpoint of technology integration and scalability within each other and within the framework of silicon based logic technology. These two groups are unifunctional SUM devices or USUMs and multifunctional SUM devices or MSUMs. The URAM device groups are considered as a subset of the MSUM family of memory devices. The SUM devices are all nonvolatile and are proposed to replace conventional memories in functionality for all digital systems as well as the conventional memory hierarchy for the digital systems within a unified memory architecture and technology framework.

19.2.1 USUM and MSUM/URAM Definitions

USUM devices by our definition are designed to provide a defined set of device attributes within the framework of a specific memory array. A multiplicity of arrays each with defined set of attributes may be employed to fulfill the system functional objectives as normally conceived through the conventional memory hierarchy. Such concept was proposed earlier in the memory scheme conceptually similar to those discussed in Figure 14.4 in Part II, Chapter 14 for the SOC integration scheme. Several USUM memory arrays of different memory device types could be employed based on common platform CMOS technology to cover the entire system functionality requirements from BIOS, code, working memory, and file storage functions. While cost-performance and capacity objectives within each type of memory arrays could be significantly different, together, such approach meets system objective while unified under a single technology platform.

   MSUM devices, however, by our definition are designed to provide within the cell multiple functionality of data or instruction storage with different nonvolatility attributes, simultaneously. Such memory cell could functionally replace, for example, traditional SRAM and DRAM memory cells, or DRAM and flash (NROM/NAND) memory cells, or multiple storage devices such as SDD and HDD through

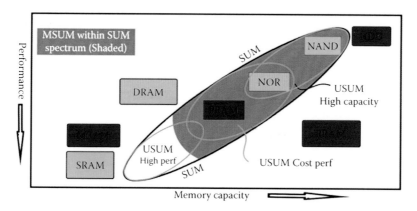

FIGURE 19.1 Application spectrum of potential USUM, MSUM/URAM, and conventional volatile and non-volatile memories within the memory hierarchy.

MSUM arrays. Therefore, in principle MSUM arrays could address the memory hierarchy or the major part of the performance/memory capacity spectrum as shown shaded in Figure 19.1. Furthermore, the MSUM devices could potentially replace the density/reliability-performance applications space defined for the proposed RE-RAMs, PC-RAMs, and STT-RAMs as suggested in Figure 17.4 in Part II. In early 2004, and through 2004–2007, several such memory cells were proposed. Recently, similar memory concepts have been proposed, especially, by the investigators from KAIST International, which the associated authors called "Unified RAM" or "Universal RAM" or URAM. URAMs in this book are discussed as a subset of MSUMs based on their multifunctional attributes. Different proposed versions of USUM and MSUM cells and arrays are described in Chapters 22 and 23, respectively.

19.2.2 Historical Background of SUM-RAM Cells

Aside from conventional SRAM and DRAM and NV-DTM RAM cells, several other volatile RAM concepts emerged during 1999 to 2002 based on (i) floating body charges in PDSOI substrate and (ii) depletion layer charge storage in bulk substrate. The former one was based on an FET device and gave rise to "capacitor-less DRAM" memory cells [1,2]. The "capacitor-less DRAM" evolved into a NVDRAM cell in 2002 with enhanced attributes [10]. The depletion layer charge storage in bulk silicon was based on a thyristor device and gave rise to TRAM [3,4] devices, and to an even higher performance device called GL-TRAM [5]. During 2002, several 1T NV-DRAM cells were proposed including a memory cell concept on SOI substrate with extended retention based trapping in on charge a trapping dielectric within the floating body of a PDSOI substrate [6]. Subsequently, several NV-RAM cells were proposed during 2002 through 2004 for both SOI and bulk silicon based technology utilizing:

1. An integrated Schottky Diode-Floating-Plate NVM device with the diode as the charge source (bulk version: [6])
2. Floating body charge source of an SOI-FET with an integrated trapping layer: (SOI version: [7]); combined Bulk and SOI version: [8]
3. Utilizing the negative differential resistance (NDR) mechanisms of diodes and thyristors and storing charges in the intrinsic regions of gate controlled P-i-N or N-i-P diodes [8,9]
4. Charge storage in the gated p region of PNPN thyristor [9,10]

Additionally, direct tunnel memory based CT-NVM cells were also proposed for USUM applications [11–13]. These memory cells formed the basis of USUM devices to provide functionalities of conventional SRAM, DRAM, and NVRAMs implementable within a common silicon based platform technology. Details of such cell concepts and operational schemes will be described in Chapter 22.

Several cell concepts also emerged from extended NVM devices during 2004 and subsequently, which functionally integrated SRAM, DRAM, and conventional NVM devices within the framework of unit memory cell [14–21]. These types of memory cells were called multifunctional silicon based unified NVM memory cells or MSUM cells to differentiate such devices from the USUM cells.

Lately, there has been renewed interest in conductive floating-gate MSUM devices similar to those proposed in earlier years employing IN-SRN in the gate stack designs to derive multilevel NVM cell designs and multifunctional cell designs [4,11,22,23]. Multilayered conducting floating gate/floating plate structures in the gate stack design to achieve simultaneous functionality of DRAM and NVM as previously conceived [4,11] had been demonstrated for planar flash technology, utilizing TaN as the conductive floating plate [20,23]. This type of MSUM devices will also be briefly outlined in Chapter 23.

In recent years, several similar cells were proposed by investigators from KAIST and other organizations. These multifunctional cells also form the basis of several MSUM devices and will be reviewed in Chapter 23 along with the above-mentioned MSUM cells. Historically, J-W Han proposed a capacitor-less multifunctioning DRAM/NVRAM cell in 2007 and extended such device concepts in 2007–2008-time frame into buried n-well [24,25]. Han demonstrated multifunctionality using buried $Si_{1-y} Cy$ substrates as well as into Six Ge_{1-x} substrates [25]. While all such devices displayed DRAM functionality with <50 ns programming speed, the memory retention characteristics were limited. The above devices were of both planar and of FinFET configurations. FinFET multifunctional memory cell with a SONOS stack was demonstrated to provide both conventional NVM device characteristics and DRAM functionality within the same single device FinFET memory transistor [26]. This was called URAM which will be elaborated in Chapter 23.

The relative roles of USUM and MSUM/URAM devices are schematically shown in Figure 19.1. The roles of conventional SRAM, DRAM, NROM, and NAND are also shown as well as the emerging MRAM, PCRAM, and RERAM devices for comparison. Some of the emerging devices have the potential to replace traditional silicon based volatile and nonvolatile memories.

19.3 SUM DEVICES AND ARRAYS IN MEMORY HIERARCHY

SUM devices are intended to be applicable to the entire spectrum of the memory hierarchy. Some of the intended applications would be to minimize or replace the functionality of conventional volatile DRAM at the cost/performance end of the memory hierarchy and conventional nonvolatile HDD/TAPE on the other end of the memory hierarchy.

We have taken the liberty of defining a "non-rigorous" definition of roles of various traditional memory levels in replacing the functionality of each currently employed conventional memory cells from the standpoint of memory sub-system and memory hierarchy throughout this part of the book. In discussing the functional replacements of conventional SRAMs, DRAMs, NROMs, SDDs and HDDs and so on, by USUM and MSUM memory cells and arrays, we may not imply exact replacement of parameters of individual memory cells of conventional memory types and associated hierarchy with a corresponding attributes of certain SUM memory cell types to be discussed in this part of the book. Therefore, a certain proposed USUM memory cell may not have the exact replication of attributes when compared to a conventional SRAM memory cell defined by a certain technology generation. Similarly, a certain MSUM memory cell with multiple functionality replacement roles of conventional DRAM memory cell/array parameters and NAND cell/array parameters may not exactly replicate the characteristics of specific DRAM cells/arrays and NAND cells/arrays. However, when appropriately employed, SUM memory cells and arrays would replicate the intended memory system objectives and in turn, the intended parameter objectives of the digital systems measured in terms of energy consumption, throughput, cost, performance and durability.

In discussing this part of the book, we broadly and often overlappingly defines the roles the USUM and MSUM memory cells and arrays play in replacing traditional memory hierarchy of L1, L2, L3, and L4 in digital system design. From conceptual understanding of replacement of conventional memory hierarchy, we will assume the following broad definition of functionality:

L1 memory: Fastest memory throughput performing the execution of CPU
L2 memory: Cache implementation
L3 memory: Working memory functional implementation
L4 memory: High-capacity storage functionality of data, operating systems, and applications

Our non-rigorous above definition implies that the USUM memory cells and arrays discussed in this part could be considered as functional replacements for conventional SRAM, DRAM, NROM, and NAND/HDD memories. Therefore, specific USUM memory cells/arrays proposed may functionally replace either L1 or L2, or L3 or L4.

The proposed MSUMs, being only "FET-based" charge-trapping memory cells, could largely be considered as functional replacements simultaneously for DRAM, NROM, and NAND/HDD memories. However, faster versions of some MSUM memory cells may also replace the conventional SRAM cells depending upon application objectives.

19.3.1 USUM

How USUM devices fit into the memory hierarchy is conceptually illustrated in Figure 19.2. Multiple device concepts have emerged which could efficiently displace traditional, (a) 6-device SRAM cell for L1/L2; (b) 2-device DRAM cell for L3 (main working memory); (c) NAND/HDD permanent large capacity storage (L4); and additionally, (d) NROM for code. Such a design for memory requirements is expected to be achieved by a unified approach of NV-CT USUM memory arrays, each containing devices with well-defined stack EOTs with progressively higher bit density through appropriate USUM memory cell selection. The USUM memory cells and associated stack designs including the selected charge trapping dielectric layer/layers to be employed are expected to be compatible for all memory arrays. Additionally, the integration scheme is also expected to be fully compatible to the general CMOS technology integration scheme. The band-engineered NV-CT-DTM devices had been at the heart of some of the USUM device concept and will be discussed in detail in this part.

19.3.2 MSUM/URAM

MSUM/URAM devices provide flexibility in memory hierarchy and memory architecture by providing multiple functionality within a memory cell in addition to enhancing memory capacity.

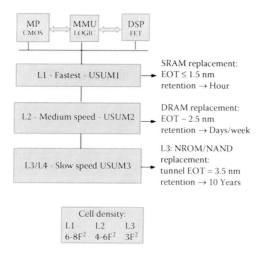

FIGURE 19.2 CMOS technology compatible NV-CT-DTM-USUM memory arrays in a memory sub-system with performance/capacity/non-volatility trade-offs covering the conventional memory hierarchy. (Bhattacharyya, A., Scalable flash/NV structures and devices with extended endurance, *ADI Associates internal publication*, August 2001. U.S. Patents: 7012297 [3/14/2006], 7250628 [7/31/2007], 7400012 [7/15/2008], 7750395 [7/6/2010].)

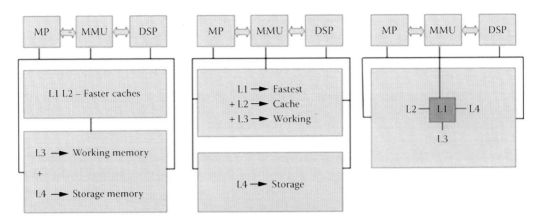

FIGURE 19.3 Conceptual illustrations of CMOS technology compatible NV-CT-MSUM memory arrays in a Memory Subsystem merging the roles of conventional Memory Hierarchy: Options: (a) L1 and L2 + L3 and L4, (b) L1 + L2 + L3 and L4, (c) L1 + [L2 + L3 + L4].

Through MLC capability with varying performance levels, such cell can replace the traditional functionality of (a) SRAM and DRAM, or (b) DRAM and NROM, or (c) NROM and NAND, or (d) NAND and HDD, and so on. This could be achieved within the framework of a single FSUM cell thereby providing unique application flexibility for tailored system design as well as for diverse system applications of memory subsystems. Such capability is not currently available in traditional memory hierarchical framework. This is conceptually illustrated in embedded memory subsystem options in Figure 19.3a–c where multiple levels of memory arrays are merged.

19.4 COMPARATIVE ATTRIBUTES OF SUM VERSUS OTHER MEMORIES

SUM device attributes could be compared with silicon based conventional memory devices such as DRAM, NROM, NAND, as well as "non silicon" based contending and emerging memories such as HDD, PCM, RRAM, and MRAM.

Table 19.1 lists relevant device attributes of SUM devices against DRAM and NAND at one end and PCM, RRAM, and HDD at the other end. It should be noted that SUM device characteristics

TABLE 19.1
Comparative Attributes of SUM and Various Other Types of NVM

Parameter	DRAM	NAND	SUM	PCM	RRAM	HDD
Cell size	$\geq 6F^2$	$<4F^2$	$\ll 4F^2$	$4F^2$	$6{-}10F^2$	$0.5F^2$
Cell structure	1T+1C	1T	1T	1T+1R	1T+1R	None
Read time (ns)	<10	50	2–50	20–50	20–50	2000
W/E time (ns)	<10	1E5–1E6	<5–1E6*	100	20–50	1000
Retention (s)	~0.06	1.00E+08	>10/>1E9*	1.00E+08	1.00E+08	>1E8
Endurance (W/E cycles)	1.00E+16	1E3–1E5	>1E9/>1E15*	1.00E+12	1.00E+12	1.00E+16
W/E volts (V)	2.5	>12	2.5/8*	1.5–3.0	1.5–3.0	3.0–5.0
Read volts (V)	1.5	1.5–2.0	1.5–2.0	1.2	<1.0	3.0–5.0
W/E energy (J/bit)	4.00E-15	>2E-16	<1E-16	6.00E-12	1.00E-13	<1E-15
Max density (Gb/in²)	<100	150–550	300–500	200 approx.	200 approx.	750–1000

*Based on CT-NVM Ultimate Parameter Objectives of advanced multi-functional SUM devices with both DRAM and NVM functionality.

have the potential to compete against some of the emerging contenders such as RRAM with the advantages mentioned earlier.

19.5   APPLICATION ADVANTAGES OF SUM DEVICES AND TECHNOLOGY

SUM device advantages from Table 19.1 are summarized as follows:

Nonvolatility
Potential reliability
Lowest W/E energy
Density ~0.5x HDD
One device cell
Multiple functionality
Enhanced endurance
High density
Comparable read performance
Scalability
Manufacturability
Multilevel capability
Progressive W/E performance
Competitive cost/bit
CMOS technology compatibility

Perhaps the greatest potential advantages of SUM devices when compared with other emerging devices are CMOS technology compatibility, competitive cost/bit for bulk technology options, areal density, manufacturability, power advantages at system level, and potential reliability. Additionally, such devices are expected to provide system level design flexibility. The remaining sections of Part III are expected to provide the background toward such advantages.

REFERENCES

1. S. Okhonin, M. Nagoga, J. M. Sailese et al., A SOI capacitor-less 1T DRAM concept, *IEEE Intl. SOI Conference*, Durango CO, 2001, pp. 153–154.
2. P. Fazan, S. Okhonin, M. Nagoga et al., Capacitor-less 1-Transistor DRAM, *IEEE Intl. SOI Conference*, Williamsburg VA, 2002, p. 10.
3. F. Nemati, and J. D. Plummer, A novel thyristor-based SRAM cell (T-RAM) for high speed, low voltage, giga-scale memories, *IEDM*, Washington DC, 1999, pp. 283–286.
4. A. Bhattacharyya, R. S. Bass, and G. D. Grise, Non-volatile memory cell having silicon-rich nitride charge trapping layer, IBM Internal Technical Reports: 1984; U.S. Patent: 4870470 (09/26/1989).
5. A. Bhattacharyya, Gated lateral thyristor-based radom access memory cell (GLTRAM), *ADI Associates internal publication*, June, 2002, U.S. Patent: 7042027 (5/9/2006), 746054 (11/25/2008). V.
6. A. Bhattacharyya, A novel one-device non-volatile DRAM device concept and structure, *ADI Associates internal publication*, May 2002, U.S. Patent: U.S.6903969 (6/7/2005), 7130216 (10/31/06), 7485513 (02/03/2009).
7. A. Bhattacharyya, One transistor SOI non-volatile random access memory cell, *ADI Associates internal publication*, July 22, 2002, U.S. Patent. 6888200 (5/3/2005), 6917078 (7/12/2005), 7184312 (2/27/2007), 7339830 (03/04/2008), 7440317 (10/21/2008).
8. A. Bhattacharyya, High performance one-transistor memory cell, original title: High performance capacitor less one transistor memory devices in bulk and SOI technology, *ADI Associates internal publication*, July 2, 2003, U.S. Patent: 7432562 (10/07/2008), 7625803 (12/01/2009), 7968402 (06/28/2011), 81255003 (02/28/2012).
9. A. Bhattacharyya, SOI non-volatile (Self Refreshed) two-device SRAM using floating body charge, *ADI Associates internal publication*, May 2003, U.S. Patent, 7224002 (05/20/2007).

10. A. Bhattacharyya, Memory cell with trenched gated thyristor, (original Title: A cross point stable memory cell using trenched gated-thyristor), *ADI Associates internal publication*, April 30, 2004, U.S. Patents: 7145186 (12/05/2006), 7440310 (10/21/2008).

11. A. Bhattacharyya, Scalable flash/NV structures and devices with extended endurance, *ADI Associates internal publication*, August 2001, U.S. Patents: 7012297 (3/14/2006), 7250628 (7/31/2007), 7400012 (7/15/2008), 7750395 (7/6/2010).

12. A. Bhattacharyya, Embedded trap direct tunnel NVM and "A vertical channel cross point high density split channel direct tunnel NVM" and "A very low power memory sub-system with progressive non-volatility," *ADI Associates internal publication*, July–August, 2004, U.S. Patents: 7276760 (10/02/2007), 7365388 (04/29/2008), 7385245 (06/10/2008).

13. A. Bhattacharyya, Integrated two-device multiple configuration alterable ROM/SRAM and Scalable integrated logic/NV memory technology with associated gate stack structure and process, *ADI Associates internal publication*, June 2004, U.S. Patents: 7208793 (04/24/2007), 7544990 (06/09/2009), 7728350 (06/01/2010).

14. A. Bhattacharyya, Integrated surround-gate multifunctional memory device, *ADI Associates internal publication*, 5/26/2004.* Patent Application abandoned.

15. A. Bhattacharyya, Extendable multi-level SGT-CT NROM and NAND flash devices with enhanced characteristics for multi-functional NVM technology applicable to 20 nm technology node and beyond, Internal Memo, ADI associates, April 12, 2013.

16. A. Bhattacharyya et al., Integrated DRAM-NVRAM multilevel memory cell, *ADI Associates internal publication*, May 5, 2004, U.S. Patents: 7158410 (01/02/2007), 7349252 (03/25/2008), 7379336 (05/27/2008), 7403416 (07/22/2008), 7417893 (08/26/2008), 7459740 (12/02/2008), 7457 159 (11/25/2008).

17. A. Bhattacharyya, Surround gate multi-functional memory chip: Device concept, technology and chip architecture, *ADI Associates internal publication*, (December 2004).

18. A. Bhattacharyya, Multi-functional memory device with graded composition insulator stack, *ADI Associates internal publication*, Patent Application submitted December 5, 2004, U.S. Patents: 7525149 (04/28/2009), 77680062 (08/03/2010).

19. A. Bhattacharyya, Discrete trap non-volatile device for universal memory chip application, *ADI Associates internal publication*, May 29, 2005 (MIDD 52), U.S. Patents: 7436018 (10/14/2008), 7476927 (01/13/2009), 7786516 (10/14/10), 8143657 (03/27/2012), 8242554 (08/14/2012).

20. A. Bhattacharyya, Embedded trap nano-crystal multi-level-cell device for DRAM and universal memory, Updated Title: "Embedded rap Nano-crystal Multi-level-cell Device Universal Memory", *ADI Associates internal publication*, U.S. Patent. 7759715 (07/30/2010), 8193568 (06/05/2012).

21. A. Bhattacharyya, Extendable multi-level SGT-CT NROM and NAND flash devices with enhanced characteristics for multi-functional NVM technology applicable to 20 nm technology node and beyond, Internal Memo, ADI associates, (Unpublished) April 12, 2013.

22. N. D. Spigna, D. Schinke, S. Jayanti et al., A novel double floating-gate nified memory device, *VLSI ans Sustem-on-Chip*, 2012, pp. 53–58.

23. B. Sarkar, N. Ramanan, S. Jayanti et al., Dual floating gate unified memory MOSFET with simultaneous dynamic and non-volatile operation, *IEEE Electron Device Letters*, 35, pp. 48–50, 2014.

24. J.-W. Han, S.-W. Ryu, C. Kim et al., A Unified-RAM (URAM) cell for multi-functioning capacitor-less DRAM and NVM, *IEDM*, 2007, pp. 929–932. [KAIST-URAM].

25. J.-W. Han, S.-W. Ryu, S. Kim et al., Energy band engineered Unified-RAM (URAM) for multi-functioning 1TDRAM and NVM, *IEDM*, San Francisco CA, 2008. [KAIST–URAM].

26. D.-I. Moon, J.-S. Oh, S.-J. Choi et al., Multi-functional universal device using a band-engineered vertical structure, *IEDM*, Washington DC, 2011 pp. 551–554. [KAIST 3D: SRAM+DRAM+NVRAM].

20 SUM Technology

CHAPTER OUTLINE

This chapter covers three important technology topics related to the design of SUM devices. These are (a) the preferred selection of dielectric films applicable to SUM devices, (b) technology integration considerations related to the fabrication of SUM devices, and (c) associated stack design considerations related to achieving the desired device properties. These topics are closely coupled and applicable for all CT-SUMs: The USUM as well as the FSUM devices are based on charge trapping and charge storage within dielectric media. It is also to be noted here that band-engineering considerations are also closely coupled to the above three topics of SUM technology defining charge storage and transport characteristics. While the topic of band engineering for such devices will be exclusively discussed in Chapter 21, band energetics will be considered in this chapter for stack design, dielectric selection, and technology integration standpoint. The chapter is limited in scope and selective as explained below.

20.1 CONSIDERATION AND SELECTION OF DIELECTRIC FILMS FOR SUM DEVICES

Cell concepts for SUM devices are discussed in Chapters 22 and 23, respectively, for USUM and MSUM devices. All MSUM cells and several USUM cells discussed in the above chapters are FET-based. These memory cells operate similar to the traditional FET-based NVMs discussed in Parts I and II of this book requiring multiple layers of dielectric films for satisfying the functions of tunneling/trapping-storage/blocking for the gate stack nonvolatility requirements. However, several proposed USUM cells operate on floating bulk silicon body and/or partially/fully depleted SOI substrate where memory states are created either by field effects or by NDR effects. For nonvolatile operations, these cells store charges directly in a trapping layer embedded within the silicon floating body. Although trapping dielectric layers in such devices could either be a single layer of Si_3N_4 (nitride) or GaN and optionally, such trapping layer could be sandwiched between two layers of IN-SRN (injector SRN to enhance charge transport and storage), the integration schemes are different. Such integration schemes are explained in relevant References 1,2 and will not be elaborated here. This section focuses on consideration and preferential selection of dielectric films for gate stack design for "conventional" USUM and MSUM band-engineered FET-based memory cells.

It should also be pointed out that the scope of this section as well as sections on technology integration and stack design are also limited primarily in the context of "Charge-Trapping Dielectric-based NVMs (CT-NVMs)," although similar devices could employ nanocrystal-based charge storage or a combination of nanocrystal-based and trapping-layer-based charge storage, or even floating-plate-based charge storage schemes. Technology integration in the latter types of charge storage schemes is relatively more complex and only referenced.

20.1.1 Dielectric Layer Selection for Silicon Interface: Preferred Film: OR-SiON

In recent years, scaled FET device technology employs higher K dielectric films at the silicon interface. For reasons of compatibility with the FET devices employed for designing peripheral logic circuitry and NV-stack design structures, it is suggested that all SUM devices should have an ultrathin layer of OR-SiON in the thickness range of 1–1.5 nm. Such a dielectric layer passivates the silicon interface effectively, replacing SiO_2 with a somewhat higher K value and simultaneously provides a desirable energy barrier for NVM devices. It has been discussed in Chapter 13 that the optimum silicon–insulator interface barrier energy should be in the range of 5.5–7.5 eV for advanced NVM devices from the standpoint of writing performance and stored charge retention. The OR-SiON layer assumed for many of the USUM and MSUM devices discussed below provides an interface band gap of 7.2 eV using a composition of N/N+O of ~0.18 with atomic concentration of oxygen in the range of 50–60% in the film (see Chapter 4 for details). The above OR-SiON layer could be fabricated in multiple ways, either by LPCVD or by ALD or by thermal/plasma oxidation/nitridation techniques followed by appropriate plasma anneal. Unless specifically stated, all SUM device design assumes the above common silicon interface.

20.1.2 Insulator-Metal-Interface (IMI) Layer Selection for Metal Gate Interface: Preferred Film → TiN or TaN

Metal gate interface layer is a critical technology item for both FET device design and NVM device gate stack design. This has been explained both in Parts I and II of this book. This is equally applicable for SUM device and technology. The preferred layers are either TiN or TaN with characteristics of compatibility and desired higher work function requirements for advanced CMOS technology. Therefore, for SUM technology, such interface layer should be assumed unless specific options are discussed.

20.1.3 Charge Trapping and Charge Storage Layers for SUM Technology: Preferred Films Are: Si_3N_4, or Si_2ON_2, or GaN or a Combination

Charge trapping dielectric films have been reviewed both in Part I (Chapter 4) and Part II (Chapter 12) of this book. While Si_3N_4 films have been widely employed for charge trapping in gate stack designs due to CMOS process compatibility and stability, it had been demonstrated that Si_2ON_2 films and other NR-SiON films are similarly process compatible and stable and are superior in terms of providing improved retention and memory window. A shorter list of charge trapping dielectric films for SUM technology is suggested in Table 20.1. An attractive charge trapping film (widely employed in electronics but not significantly yet in NVM devices) is GaN for its characteristics negative conduction band offset with respect to silicon, desirable band gap, and significantly higher trap density ($>>1E13/cm^2$). Although trap depth has not been reported, limited device data

TABLE 20.1
Selected Charge Storage and Charge Trapping Layers for Sum Technology

Dielectric	K	Φb-e/SiO_2 eV	Trap Depth/Capture Cross-Section
Si_3N_4^	7	1.1	1.1/1E–13
Si_2ON_2^	7.5	0.8	1.6/1E–13
AlN	10	0.4(?)/4.4	NA
Ta_2O_5	26	3.1	2.7/NA
GaN#^	10	3.65	1.6/NA

^ Provides Chemical Passivation; # negative band offset.

showed excellent retention at room temperature and suggests deeper intrinsic trap depth in GaN (see Chapter 12 for details). Suggested charge trapping layers for most SUM devices, therefore, have been either of nitride or Si_2ON_2 or GaN for single trapping layer or a combination of nitride or Si_2ON_2 and GaN for multitrapping layered stack designs. Additionally, GaN provides chemical passivation similar to Si_3N_4 and Si_2ON_2. Selected charge trapping dielectric films for SUM devices are provided in Table 20.2.

20.1.3.1 Role of I-SRN (IN-SRN) for Charge Storage of SUM Devices

Most higher performance SUM devices are either derivatives of DTMs (direct tunnel memories) or employ band-engineered PBO stack designs (Chapter 13) to enhance internal field aided carrier transport. An over layer of I-SRN not only facilitates the confinements of charges as "charge reservoir," but also enhances the erase performance. Since I-SRN could be deposited along with Si_3N_4 or Si_2ON_2 as a single step process, and patterning processes are similar, I-SRN has often been incorporated in SUM devices to enhance device characteristics.

20.1.4 Blocking Layer Selection for SUM Stack and CMOS FET Gate Insulator for SUM Technology

To facilitate gate stack design, overall technology integration, and to contain processing cost, it is highly desirable to select the blocking layer for the SUM stack to be compatible and complementary to the CMOS FET gate stack dielectric film. It has been discussed at length in Chapter 12 that many of the requirements for scaled FET gate dielectric properties and blocking layer properties of FET-based NVMs are equivalent. These properties are: trap free dielectric, extremely low conductivity, higher K values to achieve lowest possible EOT, chemical passivity against processing contaminants, thermal and structural stability, high breakdown strength and compatibility with appropriate metallic interface for work-function control. It had been previously suggested that several oxynitrides with larger band gap (≥ 6 eV) fit above demanding requirements. While OR-SiON films (band gap: 7.2 eV) have most of the desired attributes, the K value is low (K $\cong$ 5.0). Band characteristics of high K films for general NVM technology compatible for scaled FET gate dielectric application are listed in Table 20.2.

Among the above multiple options, three dielectric films are particularly selected due to their stability and reliability. These are HfSiON, HfTaON, and HfLaON. Of the above three dielectric films, HfSiON has been most extensively investigated and integrated into the scaled CMOS technology. For advanced SUM stacks designed to achieve larger memory window for MLC capability, a dual

---

TABLE 20.2
FET Gate Dielectric and Blocking Dielectric Commonality

Parameter	Low Leak, Low EOT, Low Trap Density, High Structural/Thermal Stability, Gate Metal Compatibility						
	OR-SiON[a]	La$_2$O$_3$#	HfSiON[a]	HfAlON[a]	HfTaON[a]	HfLaON[a]	HfLaSiON[a]
K	5.0	30	14	16^	18^	20	14^
Ub-e/Si	3.0	2.3	3.0	2.8^	1.8^	2.9^	2.6^
Ub-h/Si	3.1	0.9$	2.8	3.0^	3.1^	2.0^	2.4^
Ub	7.2	4.35	6.9	6.8^	6.0	6.0^	6.0

^ Values estimated.

[a] All a Oxy-nitride Films demonstrate High Temperatures, High Reliability and TaN Metal Gate Compatible (includes La$_2$O$_3$).

TABLE 20.3

Preferred Hi-K Blocking Dielectric Films Compared with OR-SiON and Al$_2$O$_3$

Films:	OR-SiON	Al$_2$O$_3$	HfSiON	HfTaON	HfLaON
K:	5	10	14	18	20
Φb (eV)	7.2	8.7	6.9	6.0	6.0
Φb-e/Si (eV)	3.0	2.75	3.0	1.8^	2.9^
Φb-h/Si (eV)	3.1	4.85	2.8	3.1^	2.0^

^ Estimated value.

blocking layer is often considered (Chapter 22). In such cases, a secondary blocking under layer of Al$_2$O$_3$ is incorporated to provide larger band gap interface to the stored charge to enhance memory window and retention. The electronic properties of SUM blocking layers are listed below along with the OR-SiON as reference dielectric layer in Table 20.3.

All the above amorphous oxynitride thin films have been demonstrated to have high temperature structural stability, low leakage, high reliability, and TiN or TaN interface compatibility.

20.1.5 Tunneling Dielectric for SUM Technology

Fundamentals: Consideration and selection for tunneling dielectric films for SUM devices are complex. Since tunneling-based electronic transport is electrode dependent, electrons and holes are provided either from the silicon substrate or from the gate. For normal mode NVM devices, stack design is based on silicon being the primary source of carriers whereas for reverse mode, the gate being the primary source. Unless explicitly stated, we will be discussing the normal mode SUM designs compatible with CMOS FET devices in stack designs as stated earlier. Additionally, the SUM device design considerations as exemplified here involve electron injection and transport (and not hole) as primary means to provide desirable attributes toward n-channel NVMs. Since silicon/insulator interface and insulator/gate interface are critical parameters in device design and technology compatibility associated with such stack designs is critical, the insulator selections for design examples will be bound by such constraints.

Let us critically examine the driving requirements for the CT-SUM tunneling layer (or layers) in device stack designs based on the above constraints. For high-performance SUM devices, electron transport from silicon to the trapping sites through the tunneling dielectric is required to be most energy efficient, yet the reverse transport must be minimal for nonvolatility. This implies the transport need to be asymmetric when the external field is removed. Therefore, the dielectric medium (or media) between the charge storage region and either silicon or gate should exhibit extremely low conductivity at stored-charge-induced internal field (and temperature) regardless of the transport mechanism. Additionally, stored charges are required to be removed and returned to the substrate at preferably equally high speed (during erase) when a reverse external field is applied. It has been pointed out in Chapter 13 that a single layer dielectric would be extremely limited to fulfill such contradictory requirements especially against the constraints stated above. Hence, the need for band engineering which has been discussed in that chapter. From Band-engineering considerations, a three-layered-tunneling structure design provides most flexibility in providing the desired charge transport for high-performance/cost-performance/high density SUM stack designs. However, several proposals have been outlined here with minimum number of tunneling layers meeting application requirements while reducing integration complexity.

It should also be noted that the band characteristics of most dielectric films except for few, for example, AlN, La$_2$O$_3$, Si$_3$N$_4$, and Pr$_2$O$_3$, have high barrier energy for holes when compared to electrons with reference to silicon (see Chapter 12). Additionally, holes move slower compared to electrons within dielectric medium due to higher effective mass. Therefore, for faster transport, electrons would be preferred compared to holes in SUM stack designs.

Design Considerations for Selection of Tunnel Layers: In designing tunneling dielectric layers of CT-SUM normal mode devices, the following considerations will be attempted to balance trade-offs between performance parameters and nonvolatility, while assuming "Normal mode" device design requirements:

1. Primarily electron transport back and forth from silicon substrate to the storage centers.
2. Thinnest tunneling media to promote direct tunneling (single or multiple) with internal-field-aided transport for speed-driven designs; intermediate tunneling distance to charge storage using direct tunneling for performance/retention trade-offs associated with cost/performance designs; and longer tunnel distance with enhanced internal-field-aided F.N. Transport for high-density stack designs with EOL retention.
3. To meet such requirements, DTM, PBO, RTB, VARIOT, and CRESTED tunnel barrier designs would be appropriately considered (see Chapter 13).
4. Furthermore, for speed/retention high-performance designs, retarding field layer within DTM devices will be considered which is conceptually similar to "double tunnel junction" devices (see Chapter 13) [3].

Reverse mode SUM devices will be briefly mentioned at the end of this chapter for completeness of concepts. Similar general considerations also apply for reverse mode SUM device stack designs where the active carrier source is the gate and not the silicon substrate.

Some of the tunneling layer concepts and associated band characteristics for USUM and MSUM devices are illustrated below. Additional details are presented in Chapter 21.

20.1.5.1 For USUM Device

CT-USUM Tunnel-Layer Design and Selection for Performance-Driven Stack (with built-in retarding field):

An example of the triple-layered USUM tunnel layer design is considered here. This consists of 1–1.5 nm of OR-SiON/1.0 to 1.5 nm In-SRN/and varying thickness of a hafnium-based oxynitride film, for example, HfTaON with overall tunnel layer EOT optimized for each arrays to meet performance objectives of L1, L2, and L3, respectively. To minimize integration and processing complexity, the same oxynitride layer should also be the blocking layer as well as the FET gate insulator. This approach aids in processing of gate stack insulating layers for the peripheral FETs and NVMs. Additionally, the trapping/charge storage design for the arrays should also preferably be identical. The band diagrams at VG = 0 and Vg = +Vpp are shown in Figure 20.1 left and right, respectively, with GaN trapping layer. It should be noted that the thickness of the In-SRN layer defines the silicon nanocrystal diameter. Electrons are transported from silicon to the nanocrystals through the interface SiON by direct tunneling first and subsequently from the nanocrystals to the GaN trapping centers through the HfTaON layer. For L1 and L2 applications, the thickness for the HfTaON layers is adjusted for direct tunneling (e.g., 1 to 1.5 nm for L1 array, and 2.0 to 2.5 nm for L2 array, respectively). For L3 application, the thickness of HfTaON could be ~5 nm to ensure EOL retention with modified FN tunneling mode operative for electron transport. The tunnel layers illustrate PBO barrier with internal field aided electron transport enhancing programming speed. Charges stored in nanocrystal traps provide retarding field analogous of "Coulomb Blocade," reducing back tunneling. Figure 20.1 (right) illustrates the band bending and electron transport when a positive potential for writing is applied at the gate.

A second option of built-in retarding field design could be provided by generating fixed negative charges through incorporation of two ultrathin layers of SiO_2/Al_2O_3 (or OR-SiON/Al_2O_3) as exemplified in Figure 13.32. Such a double layer could be introduced in place of the In-SRN layer as

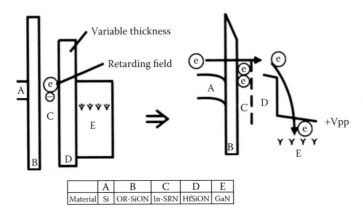

	A	B	C	D	E
Material	Si	OR-SiON	In-SRN	HfSiON	GaN

FIGURE 20.1 Band characteristics of CT-USUM triple-layered retarding field tunnel layer design consisting of OR-SiON/In-SRN/HfTaON for L1 applications: Left: Vg = 0; Right: Vg = +Vpp (writing).

discussed above for the high-performance USUM device as well as for the cost/performance multifunctional MSUM devices (see Section 23.2.2). This second approach would enhance retention at the expense of programming speed.

20.1.5.2 For MSUM Device

20.1.5.2.1 OPTION 1: Performance-Driven CT-MSUM Tunnel Layer Design (Both Electron and Hole Transport, no retarding field)

An example of CT-MSUM tunnel layer design (double or triple-layered) for simultaneous L1 and L2 performance capability within one cell is illustrated here. The first two tunnel dielectric layers B and C (Figure 20.2 left figure) consists of 1 nm each of OR-SiON and La_2O_3 films followed by a 2.5 to 3.0 nm of nitride (Layer D) for trapping. To further enhance retention for the L1 device, a 1 nm layer of Al_2O_3 (third tunnel layer, not shown) could be placed between the La_2O_3 (Layer C) and the trapping nitride layer (Layer D) which would create negative fixed charge at the interface of the two layers (not shown) providing built-in retarding field for the trapped electrons in nitride reducing reverse tunneling. The double-layered version (without the Al_2O_3 layer) provides multiple PBO direct tunneling of both electrons and holes (due to low barrier energy for holes in La_2O_3) providing high speed transport during writing and erasing. Layer E is either a thicker layer of La_2O_3 or HfSiON (3–5 nm) as shown in Figure 20.3, followed by a GaN (layer F) of 5 nm for charge trapping for longer retention.

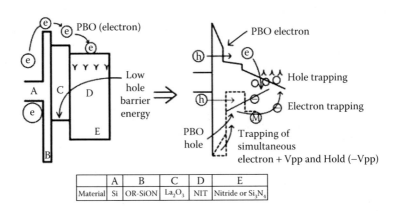

	A	B	C	D	E
Material	Si	OR-SiON	La_2O_3	NIT	Nitride or Si_3N_4

FIGURE 20.2 Band characteristics of CT-MSUM tunnel layers for simultaneous L1/L2 applications: Left: Vg = 0; Right: Vg = +Vpp (writing).

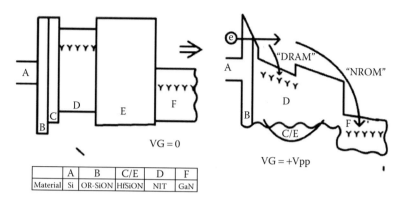

	A	B	C/E	D	F
Material	Si	OR-SiON	HfSiON	NIT	GaN

FIGURE 20.3 Band characteristics of CT-MSUM tunnel layers for L2/L3 application: Left: Vg = 0; Right: Vg = +Vpp (writing).

When Vpp is applied for short duration (30 to 50 ns), traps in nitride would be effective. At a somewhat higher Vpp and longer duration (100 to 300 ns), GaN traps will be filled. Accordingly, the cell would be multifunctional with speed/retention trade-offs built within each cell. An additional over-layer (Layer G, not shown) could be placed to provide a charge storage layer of In-SRN of high conductivity (5 nm) and the blocking/FET dielectric layer could consist of ~7 to 10 nm of HfSiON or other oxynitrides such as HfTaON or Hf LaON. The memory cell could operate at +5 to 6 V (Vpp) with EOL endurance. The band diagrams for the tunnel layers at Vg = 0 and Vg = +Vpp are illustrated in Figures 20.2 left and right, respectively. The B/C tunnel layers in the example below form double-layer PBO barrier.

20.1.5.2.2 OPTION 2: Cost-Performance CT-MSUM Tunnel Layer Design Example with Simultaneous L2 and L3 Capability (Electron Transport Only)

A potential design approach for simultaneous L2 and L3 capability may involve the first two tunnel layers B and C consisting of ultrashallow 1 nm each layers of OR- SiON and HfSiON (or HfLaON), the former being the common interface layer as before. However, both layers have similar Φb-e/ Si of 3 eV. The first trapping layer is also a 3 nm thick Si_3N_4 (layer D) followed by a thicker tunnel layer of 5 nm of HfSiON (or HfLaON) as shown in Figure 20.3, layer E, below. The deep trapping layer of GaN (layer F) is deposited on top of the thicker tunnel layer of HfSiON or HfLaON. The blocking layer/FET gate insulator layer would also be a layer of 8 to 10 nm of HfSiON (or HfLaON). The layer C (HfSiON) has higher K value and, therefore, the electrostatic field due to charges stored in nitride during retention is lower compared to layer B, lowering the rate of charge loss to the substrate compared to a single 2 nm thick OR-SiON layer. Alternately with an added 1 nm layer of La_2O_3 (three-layer barrier instead of two), a classical VARIOT barrier would be formed. In either options, while the nitride trapping provides DRAM like functionality through multiple direct tunneling, the GaN trapping through modified FN tunneling provides NROM like performance and longer term retention for L3 NVM applications.

Depending on the final stack EOT, the cell provides simultaneous L2 (DRAM) and L3 (NROM) performances in the Vpp range of +6.5–7 V at programming speeds of <100 ns (L2) and <1 ms (L3), respectively, with retention levels of >1000 sec and >1E9 sec, respectively. The device operates primarily by transport of electrons since energy barriers associated with holes are significantly higher. The DRAM part is designed for EOL endurance, while the NROM part is designed for extended endurance (>1E9 w/E cycling). Figure 20.3 left figure and right figure illustrates the band diagrams for the tunnel layers at Vg = 0 and Vg = +Vpp, respectively. With or without the ultrathin La_2O_3 tunnel layer between SiON and HfSiON tunnel layers, the device characteristics are similar to those with VARIOT tunnel barrier.

20.1.5.2.3 OPTION 3: Electron-Only CT-MSUM Crested Tunnel Barrier Design with Multifunctionality (DRAM/NROM Functionality)

A third tunneling option based on crested tunnel barrier could also be considered to provide multifunctional MSUM devices. In general, a well-balanced crested barrier could provide better charge retention for direct tunnel devices by providing larger electron barrier for the charges stored within close proximity. An example of such tunnel barrier could consist of ultrathin layers of HfAlO/Al$_2$O$_3$/HfAlO or HfO$_2$/Al$_2$O$_3$/HfO$_2$ or similar triple layer design where Al$_2$O$_3$ could be replaced by SiO$_2$ or OR-SiON. Each layer may be of thickness in the range of 1–1.5 nm for electron transport by multiple direct tunneling. The processing for the interface layer is important and required to be such as to provide well bonded and high temperature stable silicon/insulator interface and appropriate inert anneal to ensure the stability of the interface. The significance of the silicon/insulator interface preparation, especially when the first high K ultrathin layer is a single oxide (e.g., HfO$_2$ or Pr$_2$O$_3$) or bi-metal oxide (e.g., HfAlO), has been discussed and illustrated elsewhere [4,5]. When the first trapping layer and subsequent stack structures are considered, the approach could be identical to the example of Option 2 as mentioned earlier. The multifunctional capability and the device characteristics could be similar to the Option 2 above, providing relatively longer DRAM retention and somewhat slower programming speed. The band characteristics of such option are exemplified in Figure 20.4.

Several other DTM-based band engineered tunnel dielectric structures have been discussed previously in the context of multifunctional device stack design in Chapter 13 [6,7].

Aside from integration complexity, nanocrystal tunnel layer designs provide attractive tunnel layer options by themselves [8–12] as well as in combination of discrete charge trapping dielectric layers mentioned above [10]. However, the discussion on nanocrystal-based SUM devices will be limited at this time in this book.

It could be noted that aside from the silicon interface layer of OR-SiON, multiple high K tunnel dielectric layers have been identified in USUM and MSUM device examples depending on the functionality requirements. Dielectric tunnel layers such as HfSiON, HfLaON, and HfTaON and Al$_2$O$_3$ have been previously discussed as blocking layers as well as scaled FET dielectric layers. Two single metal oxides HfO$_2$ (K = 25, band gap: 5.7 eV) and La$_2$O$_3$ (K = 30, band gap: 4.3 eV) were also identified. Both HfO$_2$ and La$_2$O$_3$ are characterized as high K, low leakage dielectric films appropriately suited as gate insulator for scaled FET devices and had been extensively investigated. It has also been noted earlier that La$_2$O$_3$ is one of the few high K dielectric film with low hole barrier energy with reference to silicon, suited for faster hole transport. From the standpoint of carrier transport by tunneling form silicon substrate, high K dielectric films with low leakage and band gap in the range of 5–7 eV would be most desirable as tunnel layers in developing enhanced NVM device characteristics. It should also be noted that all the above high K oxy-nitride films exhibit band gaps within the above desirable range.

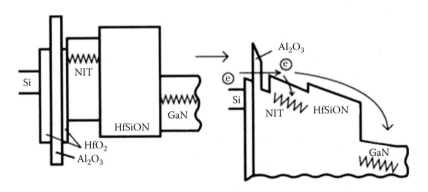

FIGURE 20.4 Crested tunnel barrier equivalent of stack design example of Figure 21.6b (Chapter 21): Triple layered (HfO$_2$/OR-SiON/HfO$_2$) DRAM with deeper trap NROM functionality.

20.2 INTEGRATION SCHEME FOR SUM TECHNOLOGY

20.2.1 SUM Technology and Device Interplay

The objective of SUM technology is to seamlessly replace current silicon based volatile memories such as SRAM and DRAM arrays as well as nonvolatile memories such as NROMs and NAND flash arrays with SUM arrays to functionally substitute current memory hierarchy of L1 or L2 or L3 functions more cost-effectively and provide simultaneously enhanced applicability and power-performance advantages. Such objective requires technology compatibility with the current and future scaled CMOS-FET technology for processor, MMU, I/O, and other system logic functions at the same time. This also implies that required writing speed, nonvolatility (retention), endurance, and reliability of memory elements are met and exceeded for functional replacements of conventional memory devices and arrays.

USUM devices achieve such objectives by ensuring compatibility not only with the basic CMOS technology integration schemes with minimal additional processing steps but also with an integrated approach of CMOS FET devices and NVM stack designs incorporating multiple layers of dielectric films to achieve desired logic and NVM properties with minimal additional processing steps and with high degree of commonality [13]. MSUM devices and arrays further such possibilities by built-in multiple memory functions within each memory cell with minimal added process complexity. A schematic representation of the various device cross-sections with stack designs is illustrated for a case example for FET (example NFET) and L1, L2, and L3 devices in Figure 20.5a–d, respectively, which is further explained in the section on gate stack designs. This is reflected in the device integration scheme below for DTM-based USUM devices as the case example.

20.2.2 Technology for Planar and 3D SUM Devices and Arrays

It is worth reiterating here that SUM technology is an evolutionary extension of scaled CMOS technology. Both planar (2D) and 3D SUM devices are compatible with scaled CMOS technology, and fabrication of such devices is assumed to follow similar schemes as CMOS bulk and/or SOI technology. It should be assumed that in general, technology integration approaches, process flow, array configurations, and other related considerations for SUM devices are expected to be similar to that of

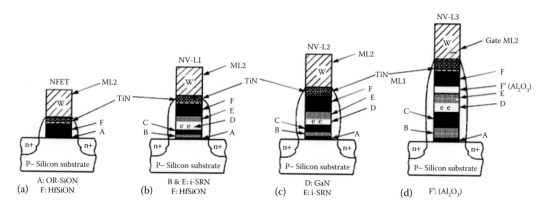

FIGURE 20.5 Schematic cross-sections of the FET Logic (NFET) device and GBE–CT-DTM USUM devices with common stacking layers identified: (a) NFET logic device, (b) L1-USUM, (c) L2-USUM, and (d) L3-USUM devices. Note that interface layers and gate layers are common to all devices. (Bhattacharyya, A., "Embedded trap direct tunnel NVM" and "A Vertical Channel Cross Point High Density Split Channel Direct Tunnel NVM" and "A Very Low Power Memory Sub-system with Progressive Non-volatility", ADI Associates internal publication July–August, 2004, U.S. Patents: 7276760 [10/02/2007], 7365388 [04/29/2008], 7385245 [06/10/2008].)

current and future CMOS technology nodes. Other than the unique features of USUM and MSUM memory cells and associated stack designs, SUM technology should be compatible to scaled CMOS technology and could be readily incorporated into CMOS platform technology. Integration schemes identified and discussed in Part II of this book (Chapter 14) and the general process flow outlined for extendible NVM technology in that chapter should be readily applicable for SUM technology. Advanced metal gate CMOS technology as discussed above should be assumed as the framework for SUM devices and technology. Therefore, no further elaboration is necessary for SUM technology. It is important to point out here that CMOS 3D technology as it evolves would be equally applicable to SUM devices and associated stack designs. This includes devices with gate-all-around configurations and nanowire devices as discussed in Part II, Section 16.3.2. However, it would be appropriate to examine the common elements in SUM gate stack designs and peripheral device gate stack design to optimize the process integration schemes and minimize processing complexity.

20.2.3 Common Elements of USUM and MSUM

Common features for USUM and MSUM stack designs in combination with the peripheral CMOS FET devices are:

1. CT-NVM FET devices within high-performance, cost/performance, and high density array schemes to be compatible to the scaled CMOS FETs [13] such that the silicon/insulator interface dielectric film and the blocking high K dielectric film as well as gate interface layer/M1 (first metal) are identical for process integration and compatibility (see Chapter 14 for details).
2. Memory system architecture incorporating multiple arrays with progressive nonvolatility within the same chip and technology [6] are compatible.
3. DTM-based single device or merged two-device cells with band engineering and high K stack designs to optimize performance, endurance, and density [14] are applicable.
4. Both planar and nonplanar cell structures for scalability, capacity, and performance enhancements [14–18] are also applicable.

While several such devices have been referenced and discussed previously in specific contexts, the significance of such devices to provide SUM cells for future applicability will be addressed below in this section.

20.3 STACK DESIGNS FOR SUM DEVICES

20.3.1 Guiding Device Concepts for Stack Design for "CT-DTM" USUM

Key considerations in device and stack designs are as follows:

1. Ensuring the peak programming field to be low enough for all device types to achieve EOL endurance for L1 and L2 device types and very high endurance for L3 device types
2. Multiple direct tunneling (double tunnel junction structure, Chapter 13) DTM device with two layer PBO electron tunneling for low energy electron transport to enhance programming speed at reduced Vpp
3. Large band offset high K GaN trapping layer combined with i-SRN conductive charge storage medium to achieve simultaneously large memory window and charge retention
4. Progressive tunnel EOT adjustments for L1 to L2 to L3 device types to achieve appropriate programming speed/retention trade-offs
5. Low leakage high K blocking layers combined with high work-function gate electrode (TaN) eliminating undesirable charge injection and compensation from the gate

6. Last but not the least, selection of high K layers such as HfSiON dielectric film for structural stability, extremely high (>= 20 MV/cm) breakdown strength, and low intrinsic charge trapping characteristics to enhance FET and NV device reliability

The above items are key elements for consideration of USUM device characteristics and stack designs. It should be mentioned that multiple other stack designs are feasible to satisfy the above key considerations for CT-DTM USUM stack designs and could be derived from illustrations provided in Part II of this book.

20.3.2   Stack SUM Devices Design Examples of "CT-DTM" USUM

Gate stack design complementary to the above figures is listed in Tables 20.4 and 20.5.
    Key processing commonalities are as follows:

1. Same silicon/ insulator (Si-I) interface layer, same blocking insulator/FET gate dielectric layer, and same gate layers for all devices.
2. Same large band offset trapping layer (GaN) combined with i-SRN charge storage layer to maximize charge storage and enhance memory window and retention. I-SRN serves as highly conductive charge storage medium for all NV devices
3. With the exception of Al_2O_3 (which could be deposited by sputtering), all films, namely, i-SRN, ORSiON, HfSiON, and GaN could be deposited in common LPCVD tools with multilayer processing capability.
4. The tunneling and trapping layers are all common for all NVM devices except the thickness of HfSiON in the tunnel layer for each device type. An integration scheme could be adopted which would first define only the NVM arrays and subsequently forming interface layer, the i-SRN ultrathin layer (~1 nm Si-nanocrystal/nitride two-phase layer for double

TABLE 20.4
Detailed Stack Design **Attributes of Devices of Figure 20.5

**Metal Gate in All Cases Either Aluminum (Al) or Tungsten (W)

Device Type	Dielectric Layers (DL) and Attributes							Gate and Stack Attributes		
	Si-I/Tu1 DL-A	SI1 DL-B	Tu2 DL-C	Tr1 DLD	SI2 DL-E	B11 DL-F	IMI	MIS EOT	OTHERS	
Logic NFET	OR-SiON					HfSiON	TaN	~2.5 nm	Vdd = 1.5 V	
	1–1.5 nm					6 nm				
L1-SUM-NVM	OR-SiON	i-SRN	HfSiO N	GaN	i-SRN	HfSiON	TaN	~6 nm	Double Jct. Design	Tunnel
	1–1.5 nm	1 nm	2.0 nm	5 nm	5 nm	6 nm			Tunnel EOT <2 nm	
L2-SUM-NVM	OR-SiON	i-SRN	HfSiO N	GaN	i-SRN	HfSiON	TaN	~7 nm		
	1–1.5 nm	1 nm	3–4 nm	5 nm	5 nm	6 nm			Tunnel EOT <2.5 nm	
L3-SUM-NVM	OR-SiON	i-SRN	HfSiO N	GaN	i-SRN	AL$_2$O$_3$ + HfSiON	TaN	~9.5 nm	Enhanced Blocking	
	1–1.5 nm	1 nm	6–8 nm	5 nm	5 nm	4 nm + 6 nm			Tunnel EOT ~3.3 nm	

TABLE 20.5
GBE-CT-DTM USUM Device Parameter Objectives

		NFET VT = 0.3V		
	L1-SUM-NVM	L2-SUM-NVM	L3-SUM-NVM	
W/E: Vpp/Time	±4.0 V/30–50 ns	±5.0 V/1 μs	±7.5 V/1 ms	
WINDOW: INITIAL/EOL:	2 V/1 V	>3.0 V/>1.5	>6.0 V/>4.0 V*	* Double-layer Blocking
RETENTION:	Hours	Days		>> 10 years Electron Trapping/Storage
ENDURANCE:	EOL#	EOL#		>> 1E7 W/E# Band Engg./Lo Peak Field
(Peak W/E Field: MV/CM)	<7.5	<7.5	<7.8	

tunnel junction), and the 2 nm HfSiON layer for all NVM devices. Subsequently, L1 array region is protected, while additional thickness of HfSiON is deposited over the defined regions of L2 and L3 arrays. The process is repeated by protecting L1 and L2 regions, and additional required HfSiON is selectively deposited in the defined region of L3 array.

The process step then is followed by common deposition of trapping layers over all NVM device stacks. Subsequent processing steps are described below:

The FET device region could be defined and interface layer can be formed thermally, while I-SRN protects all NVM device regions. This step will be followed by selective deposition of sputtered Al_2O_3 (low temperature processing) only over L3 device regions on L3 arrays followed by common deposition of final HfSiON layer (blocking layer for NV devices and gate insulator for FET devices) to complete the stack designs for all devices. Finally, TaN and metal gate deposition and definition are deposited for all devices.

20.3.3 High-Performance Stack Options for CT-DTM USUM Device Design (L1 Applications)

Aside from the above example, several other device and stack options could also be considered for CT-DTM USUM devices to achieve higher performance based on band-engineered stack designs (Chapter 13) at the expense of some additional process complexity. Two of the options are worth illustrating.

The first one is based on progressive band offset (PBO) device and stack design incorporating an additional high K dielectric ultrathin films characterized by very low leakage as the direct tunneling medium (e.g., HfTaON or HfLaON each with K = 20 and leakage nearly 2 orders of magnitude lower than HfSiON (K = 14) and 5 orders of magnitude lower than same thickness of SiO_2 film). This will be called as PBO-DTM USUM.

The second one is based on multimechanism carrier transport to achieve very fast writing [13]; this will be termed as MMCT USUM.

These are discussed in the next chapter on band engineering for SUM devices.

20.3.4 Stack Design Examples of CT-MSUM Devices

Table 20.6 lists some examples of the stack design schemes for CT-MSUM cells with specific combinations of dielectric layers based on the band engineering discussed in Chapter 13 and the band diagrams discussed earlier in this chapter for the selected dielectric films for blocking, trapping, and tunneling. For most of the examples, the interface dielectric layer would be an ultrathin layer of OR-SiON for interface stability common with the FET device and the blocking layer would be common to the FET gate insulator and as discussed earlier. The gates for both logic and memory devices

TABLE 20.6

Examples of Stack Design Attributes of CT-MSUM Devices: CMOS Technology: SUB 50 nm Node: Power Supply: 5.0 V[a]

GROUP 1

Dielectrics	DL-B	DL-C	DL-D	DL-E	DL-F	sDL-G	DL-H	DL-I	STACK EOT	NOTE
Mode	Tu	Tr1	Tu	Tr2	Tu	Tr3	SL	BL	BL	
Thickness	1.0 nm/1.5 nm	3 nm	3 nm	3.5 nm	3.5 nm	3.5 nm	3.5 nm	3.5 nm	7 nm	
1.1: L2/L3	—/La_2O_3	Si_3N_4	La_2O_3	GaN			In-SRN	HfLaON[a]	6.8 nm	PBO
1.2: L2/L3	in-SRN^	Si_3N_4	La_2O_3	GaN			In-SRN	HfLaON[a]	~7.0 nm	^Retarding Field
1.3: L3/L4	# HfTaON	Si_3N_4	La_2O_3	GaN			In-SRN	HfTaON	~7.6 nm	Variot
1.4: L3/L4	La_2O_3	Si_3N_4	La_2O_3	GaN			In-SRN	HfSiON	~7.6 nm	PBO
1.5: L3/L4	La_2O_3/HfLaON	HfLaON	HfLaON	GaN			In-SRN	HfLaON	~7.3 nm	Variot
1.6: L2/L3/L4	—/La_2O_3	La_2O_3	La_2O_3	GaN	La_2O_3	GaN	In-SRN	HfLaON[a]	~9.5 nm	PBO/DTM
1.7: L2/L3/L4: in-SRN/La_2O_3	Si_3N_4	Si_3N_4	La_2O_3	Si_3N_4	La_2O_3	GaN	In-SRN	La_2O_3[b]Al_2O_3	~10.0 nm	Enh. Window

[a] or Hf SiON; ^Retarding Field; # or La_2O_3; [b] Fixed −ve Charge.

GROUP 2: No OR-SiON Passivation Layer. Passivated Silicon Interface through Surface Treatment

Dielectrics	DL1/2/3	DL-4	DL-5	DL-6	DL-7	DL-8	DL-9	STACK EOT	NOTE
Mode	Tunnel	Tr1	SL1	Tr2	SL2	BL1	BL2		
Thickness: nm	1.5/2.0/2.0	3 nm	3.5 nm	3.5 nm	3.5 nm	3.5 nm	7 nm		
2.1: L2/L3	HfO_2/OR-SiON/HfO_2	Si_3N_4	In-SRN	HfO_2^^	In-SRN	La_2O_3	^Al_2O_3	~10.5 nm	CRESTED
2.2: L2/L3	Si_3N_4/OR-SiON/Si_3N_4	Si_3N_4	In-SRN	GaN	In-SRN		HfSiON	~10.5 nm	CRESTED
2.3: L1/L2	HfO_2/OR-SiON/HfO_2/LaAlO3O2	Si_3N_4*	In-SRN	GaN	In-SRN	La_2O_3	In-SRN	~<5 nm: 2-layer	CRESTED

The Stack below Belongs to Stack Design with Multifunctional Characteristics plus MLC Capability for L3 NROM/NAND Devices

2.4: L3/MLC-L4   Change stack of 1.4 with Blocking layers consisting of La_2O_3, 2 nm/Al_2O_3, 3 nm/HfSiON, 7 nm: ~ 9 nm EOT; And,

2.5: L2/L3/MLC-L4   Similarly change stack of 1.7 as 3 above for the blocking layers: ~10.5 total EOT.

^^or GaN Fixed negative charge.

* Thickness optimized for performance/retention/functionality.

[a] Abbreviations: Dielectric Layers: DL; Tunneling Layer: Tu; Trapping Layer: Tr; Storage Layer: SL; Blocking Layer: BL.

[b] Silicon/Insulator Interface Layer SI: OR-SiON (1.0–1.5 nm); Insulator/Metal Interface Layer: TaN.

would also be the same and assumed to be TaN for the insulator/gate interface for the work-function requirement. The FET gate characteristics would be assumed to be similar to the ones shown for the USUM device/technology options.

Stack designs for four CT-MSUM device types are described in the following table based on tunneling layer designs, functionality integrations, and multilevel storage capability.

Type 1 is based on PBO or VARIOT barrier tunneling schemes amenable to provide internal field-aided enhanced electron transport during writing. First two options of this type are examples of MSUM stack designs to functionally replace conventional SRAM and DRAM simultaneously within each memory cell. The third, fourth, and fifth examples are designed to replace slower L2 and L3 functions (NROM and DRAM functionality) within the memory hierarchy. The sixth and seventh options demonstrate examples of simultaneously replacing L1, L2, and L3 functionality within the framework of a single memory cell. It is significant to recognize that multiple band-engineered dielectric layers are involved both in tunneling and charge storage in order to derive both multifunctionality and to achieve desired parameters for nonvolatility. It is also important to note that the peak programming field associated with the stack design is 8 MV/cm in order to achieve the required endurance for L1, L2, and L3 functionality. The requirements for L3 functionality were defined to achieve 10 years of memory retention and endurance $\geq$1E9 W/E cyclability, significantly superior to current NVM limits (see Table 20.7: MSUM device parameter objectives, for details).

TABLE 20.7

CT-MSUM Device Parameter Objectives

DEVICE TYPE GROUP 1	1.1 Thru 1.5	1.6 and 1.7	
Design Point: (Peak Field)	8 MV/cm	8 MV/cm	
Stack EOT	~7.5 nm	~10 nm	
Programming: Vpp	±6 V–6.5 V	±8.0 V	±8.0 V
	L2-MSUM	L3-MSUM	L4-MSUM
PROGRAMMING: Tpp	30–50 ns	~300 ns	~1 ms
WINDOW: INITIAL/EOL	>1.5/>0.75 V	>2.5/>1.5 V	>4 V/>2.5 V
RETENTION	>100 sec	>3000 sec	>>10 Years
ENDURANCE	EOL	EOL	>>1E9 W/E
DEVICE TYPE GROUP 2	**2.1 and 2.2**		**2.3**
Design Point: (Peak Field)	~9 MV/cm		
Stack EOT	~10.5 nm		~5 nm
Programming: Vpp	±9.0 V		±4.5 V
	L3-MSUM	L4-MSUM	L2-MSUM
PROGRAMMING: Tpp	~300 ns	~1 ms	!50 ns
WINDOW: INITIAL/EOL	>2.5/>1.5 V	>4 V/>2.5 V	>1.5/>0.75 V
RETENTION	>10,000 sec	≫10 Years	>100 sec
ENDURANCE	EOL	>1E8 W/E	EOL
DEVICE TYPE GROUP 3 & 4	**3 (L3/L4)**	**4 (L2/L3/L4)**	
Design Point: (Peak Field)	8 MV/cm	8 MV/cm	
Stack EOT	~9.0 nm	~10.5 nm	
Programming: Vpp	±8.0 V	±9.5 V	
	L2-MSUM	L3-MSUM	L4-MSUM
PROGRAMMING: Tpp	30–50 ns	~300 ns	~1 ms
WINDOW: INITIAL/EOL	>1.5/>0.75 V	>2.5/>1.5 V	>6 V/>4 V
RETENTION	>100 sec	>3000 sec	≫10 Years
ENDURANCE	EOL	EOL	≫1E9 W/E

Note: Memory: L2 Cache; L3: Working; L4: File.

Type 2 is based on crested barrier tunneling schemes intended to optimize performance, retention, and functionality with appropriate trade-offs. It should be noted that unlike Type 1, this type of design does not incorporate OR-SiON as the Si-I interface layer. Instead silicon surface is treated using a passivation scheme to quench the interface states. Consequently, lower band gap dielectric films such as HfO_2 or Si_3N_4 could be employed as the first tunnel layer to derive appropriate charge transport characteristics to simultaneously provide L2 and L3 functionality (examples 2.1 and 2.2) and L1 and L2 functionality (example 2.3). It may be noted that higher performance in the example of stack design of 2.3 is based on simultaneous electron and hole transport during writing and erasing by employing HfO_2 at the silicon interface and La_2O_3 plus In-SRN (i-SRN) at the gate interface whereby both silicon and gate are active in providing carriers during programming.

Type 3 and Type 4 provide unique examples of multifunctional and multilevel capabilities combined. Type 3 is an example of memory cell providing large memory window for MLC-L3 functionality combined with L2 performance capability. It combines the stack design of type 1 example 4 with triple layered blocking consisting of 2 nm of La_2O_3, 3 nm of Al_2O_3, and 7 nm of HfSiON to provide significantly enhanced window and retention required for MLC-L3 functionality. Type 4 is a further enhancement based on Type 1 example 7 to provide MLC-L3 capability in similar ways while simultaneously providing L1 and L2 (and L3) functionality. Some of the above examples are also illustrated in the next chapter (Chapter 21) on band engineering. Details of cell and array functionality are described in Chapter 23.

20.3.5 Key Design Concepts for MSUM Stacks

1. L1 and L2 performance achieved by direct tunneling (PBO or VARIOT or CRESTED barriers), whereas L3 by either multiple direct tunneling or by modified FN tunneling.
2. For direct tunneling options involving both electrons and holes as participating careers, faster performance could be achieved aided by internal field (PBO). Such dielectric options are: La_2O_3, Si_3N_4, and HfLaON. These films are characterized by ≤ 2 eV barrier energy for holes against silicon. Since most other dielectric films exhibit higher barrier energy for holes, designs require effective and enhanced electron transport as primary charge carriers to meet performance objectives.
3. Design points are at lower peak field to ensure EOL endurance for L1 and L2.
4. Incorporation of a "retarding field" layer either via (a) an In-SRN layer (stack option 1.2) through coulomb blockade effect or (b) an Al_2O_3 layer (stack option 1.7) through fixed negative charge effect to enhance retention.
5. Multiple trapping planes are introduced at different tunnel distances to achieve retention targets for L1, L2, and L3; while nitride trapping provides faster detrapping for L1 erasing, deeper offset GaN or HfO_2 trapping provides appropriate erase performances for L2 and L3. Example of integrated duel-offset trapping plane design (nitride/GaN) is also provided.
6. Dual blocking layer (Al_2O_3 plus HfSiON) with or without negative fixed charge (ref.11B) provides larger memory window required for MLC-L3 design.

20.3.6 Parameter Objectives for MSUM Devices

Parameter objectives for the CT-MSUM devices of all above four types are derived from the following considerations:

1. Functionality Considerations: The objectives for MSUM devices have been to achieve attributes to replace conventional L1, L2, and L3 functionality from overall system performance perspective. To achieve such goal through CT-MSUM devices, design objectives were set to achieve: (1) programming speeds of 30–50 ns, 300 ns, and 1000 ns, respectively,

for L1, L2, and L3 performance replacements; and (2) end-of-life endurance requirements for replacement of L1 and L2 functions combined with >>1E9 W/E cycle capability for L3 functions. It has been assumed that such goal will replace conventional memory hierarchy without adversely impacting overall system performance for most future digital systems with additional benefits of power, cost, and design flexibility. The key stack design considerations have been band-engineered DTM-based designs to achieve item (1) and with peak-field design point <=8 MV/cm to meet item (2) requirement. For some group 2 devices, with thicker EOT stacks, the peak-field design point was raised to ~9 MV/cm to achieve the required performance–retention–endurance trade-offs, since lower band gap higher K dielectrics (HfO_2 and Si_3N_4) and crested barrier designs were employed for such stacks which provide higher reliability at W/E conditions.

2. Nonvolatility Considerations: MSUM cells simultaneously provide >100 sec, >3000 sec, and >10 years for memory state retentions while emulating, respectively, L1, L2, and L3 memory functionality. Such attributes significantly reduce refresh requirements of memory states and ensure nonvolatility and memory state stability and system overhead associated with ECC requirements. To ensure such objectives, i-SRN layers were appropriately introduced to provide retarding field reducing back-tunneling of electrons, and a combination of deep-offset trapping layer (e.g., GaN) with i-SRN charge storage scheme was incorporated. Nonvolatility was further enhanced with multiple blocking layer scheme using Al_2O_3 which provides high barrier energy coupled with fixed interface negative charges enhancing memory state retention.

3. Memory State Stability, Design Points, and MLC Considerations: Both enhanced endurance and nonvolatility as discussed in (a) and (b) provide memory state stability for MSUM cells. Asymmetric barrier engineering, deep offset trapping, extended charge storage, and fixed interface negative charge designs provide drastically reduced charge leakage of stored charges during stand-by state. Such design approach yields memory state stability as well as large memory window for MLC capability for MSUM cells. Additionally, as mentioned in item (a) the peak field operational point during programming is significantly reduced compared to conventional NVM devices to achieve stress induced stability and enhanced reliability of the memory states for MSUM designs.

Table 20.7 summarizes the device parameter objectives for the MSUM devices.

It is important to recognize from the above examples that SUM technology provides a unified approach toward silicon based advanced nonvolatile memory device stack designs and complementary to the scaled CMOS technology platform.

REFERENCES

1. A. Bhattacharyya, R. S. Bass, and G. D. Grise, Non-volatile memory cell having silicon-rich nitride charge trapping layer, IBM Internal Technical Reports: 1984, U.S. Patent: 4870470 (09/26/1989).
2. A. Bhattacharyya, One transistor SOI non-volatile random access memory cell, ADI Associates internal publication, July 22, 2002, U.S. Patent: 6888200 (5/3/2005), 6917078 (7/12/2005), 7184312 (2/27/2007), 7339830 (03/04/2008), 7440317 (10/21/2008).
3. Y.-K. Choi, T.-W. Han, S. Kim et al., High speed flash memory and 1T-DRAM on dopant segregated Schottky barrier (DSSB) FinFET SONOS device for multi-functional SoC applications, *IEEE, IEDM*, San Francisco CA, 2008.
4. A. Bhattacharyya, "Lanthanide Dielectrics with controlled interfaces", ADI Associates Internal Memo, March 2007, U.S. Patents: 7662693 (2/10), 7956426 (6/7/2011)
5. A. Bhattacharyya, Promising high-k mixed oxy-nitride films containing at least one rear-earth metal, for example, LaTaON or LaHfON for electronic applications, ADI Associates internal publications (Unpublished), 2006; U.S. Patent Publications: a) HfLaON films: 20090294924 (12/03/09) and b) TaLaON films: 20090236650 (9/24/09) by L. Forbes, K.Y. Ahn, and A. Bhattacharyya.

6. A. Bhattacharyya, Scalable flash/NV structures and devices with extended endurance, ADI Associates internal publication, August 2001, U.S. Patents: 7012297 (3/14/2006), 7250628 (7/31/2007), 7400012 (7/15/2008), 7750395 (7/6/2010).

7. A. Bhattacharyya, A novel low power battery operated non-volatile memory device and associated gate stack, ADI Associates internal publication, September 4, 2004, (MIDD 45), U.S. Patent: 7612403 (11/03/2009).

8. L. Forbes, High density single transistor vertical DRAM gain cell, Internal Publications, Micron Internal Publications (Unpublished) August 2002 and January 2004.

9. A. Bhattacharyya, Advanced scalable nano-crystal device and process, ADI Associates internal publication, December 2004, (MIDD 47), U.S. Patents: 7759715 (07/30/2010), 7898022 (03/01/2011.

10. A. Bhattacharyya, *Future developments of SONOS technology and devices and Stack designs leading to "Silicon-based Universal Memory"*, Technical exchanges by A. Bhattacharyya, ADI associates to M. Durcan, Micron Technology, Boise, Idaho, April 2005 and 05/25/2005.

11. A. Bhattacharyya, Multi-functional memory device with graded composition insulator stack, ADI Associates internal publication, Patent Application submitted December 5, 2004, U.S. Patents: 7525149 (04/28/2009), 77680062 (08/03/2010).

12. A. Bhattacharyya, High performance multi-level ET/DT NVM Device, ADI Associates internal publication, December 5, 2004, U.S. Patents: 7429767 (09/28/2008), 7553735 (06/2009), 7579242 (08/25/2009), 8159875 (04/17/2012).

13. A. Bhattacharyya, "Embedded trap direct tunnel NVM" and "A Vertical Channel Cross point High Density Split Channel Direct Tunnel NVM" and "A Very Low Power Memory Sub-system with Progressive Non-volatility", ADI Associates internal publication July–August, 2004, U.S. Patents: 7276760 (10/02/2007), 7365388 (04/29/2008), 7385245(06/10/2008).

14. A. Bhattacharyya, Integrated two device multiple configuration alterable ROM/SRAM and "Scalable integrated logic/NV memory technology with associated gate stack structure and process", ADI Associates internal publication, June 2004, U.S. Patents: 7208793 (04/24/2007), 7544990 (06/09/2009), 7728350 (06/01/2010).

15. A. Bhattacharyya, A novel stack design concept for charge trapping NVM cell with band engineered tunneling and trapping layers providing large memory window and 3 bit/cell MLC capability, ADI associates, internal memo, August 2007 (unpublished).

16. A. Bhattacharyya, Back-Sided trapped non-volatile device, ADI Associates internal publication, July 12, 2004, U.S Patent Publication 257324, December 11, 2008, U.S. Patents: 7749848 (7/6/2010) and 7851827 (12/14/2010).

17. A. Bhattacharyya, A novel 4 bit/cell dual channel MNSC device, ADI associates, internal memo, September 2007 (Unpublished).

18. A. Bhattacharyya, Integrated surround-gate multifunctional memory device, ADI Associates internal publication, 5/26/2004.* Patent Application abandoned.

21 Band Engineering for SUM Devices

CHAPTER OUTLINE

The preceding chapter introduced the significance of band energetics in selecting the appropriate dielectric layers for stack designs for SUM devices and in particular multilayered tunnel dielectrics and trapping schemes to achieve L1, L2, and L3 functionality requirements. This chapter illustrates several examples of band engineered USUM and MSUM device concepts based on the learning derived from Chapter 13 to achieve desired device parametric and functionality objectives. We will provide two examples of band engineered USUM devices for performance enhancement: the first one is called Progressive Band Offset Direct Tunnel Memory for USUM ("PBO-DTM-USUM") and the second one is called Multimechanism Carrier Transport associated USUM ("MMCT-USUM"). We will also illustrate several examples of Band Engineering to achieve multifunctionality for MSUM devices and discuss relative merits of such concepts.

21.1 BAND ENGINEERING FOR USUM DEVICES

21.1.1 PBO-DTM USUM Device

The stack design is similar to that of the L1-SUM-NVM device discussed in Table 20.4 in Chapter 20 (the advanced double tunnel junction [DJT] DTM with OR-SiON/In-SRN/HfSiON tunnel layers). The design is also similar in terms of the trapping/storage and blocking layers and in the selection of the interface layers for the silicon substrate and gate design. The unique difference is in the tunneling layer design which in its simpler form consists of a double-layered combination of 1 to 1.25 nm of OR-SiON (Φ_{b-e} = 2.5 ev, K = 5, Eb = 7.3 ev) and an ultra-thin layer of HfTaON (Φ_{b-e} = 0.5 ev, K = 20, Eb ~ 4,5 ev) of 3 to 6 nm providing a PBO tunnel barrier. Note that in this design HfTaON is replacing the HfSiON layer employed for the DTJ device design. Additionally, as stated before, HfTaON or HfLaON provides significantly lower conductivity compared to HfSiON enhancing device properties and reduces back-tunneling of stored charges to silicon substrate. The thickness of HfTaON or HfLaON would be dependent on speed/retention trade-off objective. Due to the large band offset associated with HfTaON, direct tunneling of electron transport to the trapping dielectric of GaN is extremely fast, aided by the internal field induced by the PBO barrier. The same holds for the HfLaON option. Successive band offset and deep trapping into GaN provide excellent charge retention minimizing the back tunneling of electrons to the substrate. Leakage of electrons through the blocking HfSiON (optionally, HfTaON or HfLaON) layer to the gate is also minimized due to the high barrier energy of the HfSiON blocking layer characterized by low leakage. Other integration approaches are similar as mentioned earlier for L2 and L3 arrays. The band diagram of the USUM stack is shown in Figure 21.1. The difference between the previously mentioned DTJ device and this PBO device is shown in Figure 21.2, where the band diagram for the two devices are superimposed with the difference in the tunneling design of the DTJ device (shaded) for comparison. The PBO device is expected to be significantly faster in performance when compared with the DTJ device.

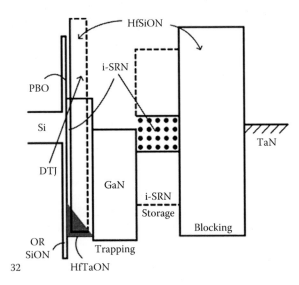

FIGURE 21.1 Band diagram comparison between PBO-DTM-USUM (solid line) and DTJ-DTM. USUM (dotted line) device and stack design concepts.

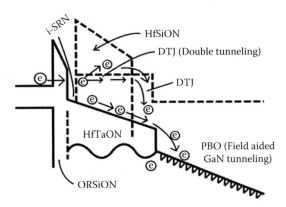

FIGURE 21.2 Difference in electron transport between PBO: Solid line and DTJ: Dotted line during writing for devices illustrated in Figure 21.1.

21.1.2 Multimechanism-Carrier-Transport (MMCT) USUM Device

This type of CT-SUM devices is programmed utilizing both source-side injection and direct tunneling, simultaneously, for writing and erasing. During writing, electrons are transported to the trapping centers by means of simultaneous transport of hot electrons from source-side injection as well as "cooler" electrons by direct tunneling from the substrate [1]. During erasing, hot holes are injected upward from to the trapping layer due to high field at the diffusion/substrate depletion layer, and simultaneously "cooler" holes are injected from the gate electrode downward to the trapping layer by quantum mechanical tunneling. Source-side hot carrier injection had been known to be relatively energy efficient compared to drain-induced hot carrier injection. The simultaneous dual mode (both gate and substrate acting as carrier source) and dual mechanism of carrier injection (quantum mechanical tunneling and hot carrier injection), for both writing and erasing, significantly improves rate of charge transport and thereby programming speed. Programming speed could be enhanced by greater than four orders of magnitude through this approach compared to the Fowler Nordheim tunneling mechanism used in conventional NVM devices. Multimechanism-carrier-transport (MMCT) device design could readily achieve sub-hundred nano-sec device programming performance, yet the transport could be sufficiently energy efficient without seriously compromising endurance and reliability. The device source lines and bit lines are engineered to obtain high-depletion field at the p-substrate /n+ junction. The MMCT device design could also be implemented for both NROM and NAND arrays and could be configured either in planar or vertical channel memory cell designs and associated memory array designs. The MMCT device concept could be extended for PBO-DTM and DTJ-DTM tunneling structures for speed enhancements especially for L1 and L2 applications.

To facilitate source-side injection of electrons for the CT-SUM (NFET) device, the N+ source/ drain diffusion is surrounded with a p-type (boron doped) halo region at the gate edge. The halo doping is engineered to achieve a high-depletion field at the substrate/N+ diffusion junction. High field at the gate edge enhances the edge field during writing enhancing hot electron injection into the tunnel dielectric from the source region while simultaneously "cooler" electrons tunnel from the substrate through the tunnel insulator by direct tunneling. During erase operation, dual mechanisms are also simultaneously operative. A layer of iN-SRN is deposited at the gate interface to enhance hole injection from the gate during erase. At the same time, hot holes are injected from silicon substrate due to the high field at the diffusion substrate depletion layer. The blocking dielectric layer could be either La_2O_3 (Φb-h = 0.9 ev, K = 30) or HfO_2 (Φb-h = 3.1 eV, K = 24), although the former would be preferred due to lower Φb-h for enhanced hole injection from the gate during erasing. The iN-SRN layer at the top serves also as a passivation layer preventing oxygen and contaminant diffusion during process integration promoting compositional integrity and reliability.

A typical gate stack design for a planar one-device or merged-gate (memory element) MMCT SUM memory cell containing above features for L1 application may consist of:

Tunnel layer: OR-SiON/HfSiON (K = 14, J/Jox = 1E-3), or
HfTaON (K = 20, J/Jox = 1.5E-4), or HfLaON (K = 20, J/Jox = 1E-5), EOT ~ 1.5 nm
Trapping/Storage layer: 5 nm GaN/5 nm i-SRN, EOT ~ 2 nm
Blocking Layer: 10 nm La_2O_3/5 nm i-SRN, EOT ~ 1.5 nm
Total stack EOT: ~5 nm; Physical Thickness: ~23 nm; Vpp: ±<5 V, 10–30 ns, Vdd: <2.5 V

Memory retention is expected to be in hours with EOL endurance and EOL memory window ≥1 V. The unique cell features are shown for NROM cross-section in Figure 21.3 and NAND cross-section

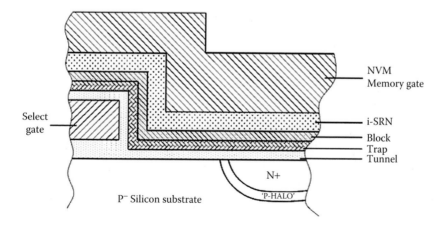

FIGURE 21.3 MMCT-USUM planar merged gate NROM device schematic cross-section with P-Halo and top i-SRN for MMCT operation. (From Bhattacharyya, A., Integrated two-device multiple configuration alterable ROM/SRAM and "Scalable Integrated Logic/NV Memory Technology with Associated Gate Stack Structure and Process"; ADI Associates internal publication, June 2004, U.S. Patents: 7208793 [04/24/2007], 7544990 [06/09/2009], 7728350 [06/01/2010].)

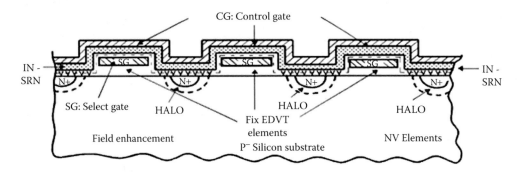

FIGURE 21.4 MMCT planar merged gate NAND schematic cross-section with Halo and i-SRN. (From Bhattacharyya, A., Integrated two-device multiple configuration alterable ROM/SRAM and "Scalable Integrated Logic/NV Memory Technology with Associated Gate Stack Structure and Process"; ADI Associates internal publication, June 2004, U.S. Patents: 7208793 [04/24/2007], 7544990 [06/09/2009], 7728350 [06/01/2010].)

in Figure 21.4, respectively. As stated earlier, MMCT SUM devices could also be implemented for not only for L1/ L2, but also for L3 and L4 applications with appropriate stack designs to meet functionality requirements for such memory arrays.

21.2   BAND ENGINEERING FOR MSUM DEVICES

Key design concepts and features in proposing selection of dielectric films for MSUM devices have been exemplified in Tables 20.3 and 20.4 in the previous chapter. We will further illustrate such concepts through the progression of band engineering over the past decade. The limitations of oxide-based NVM device in scalability, reliability, and in achieving "DRAM-like" functionality were reviewed by C. Hu in 1996 ([1], Chapter 4). Since the turn of the century, replacement of oxide with OR-SiON and other higher K dielectric options (Chapter 12) for the FETCMOS technology provided the impetus to enhance programming performance as well as endurance (key requirements for multi-functionality) for the NVM devices. Fundamental NVM device design approach for extending endurance by reducing peak programming field in the stack design (Chapter 5) and DTM stack design concepts and band-engineering approach for device parameter enhancements have been discussed in considerable detail in Chapter 13. These elements formed the basis of MSUM concepts.

21.2.1   Background and Historical References

Extending DRAM capability into NVM technology was first proposed by in 1989 (Chapter 6) and afterward in early 2004 [7]. Several multifunctional CT-NVM devices and arrays were proposed with built-in DRAM and NVRAM capability based on DTM with PBO barriers [2–4], with CRESTED barrier [5–8], and combined CRESTED/VARIOT barrier [7,8] in 2004 and 2005. Additional concepts included embedded nanocrystal-based trapping [4,9,10], combining both discrete trapping (trapping dielectric) and embedded trapping (nanocrystal-based trapping) [3,5–8]. Most of the concepts involved charge trapping in multiple planes at different distances from the silicon substrate to derive the appropriate speed versus retention trade-offs. However, concept of uniplanar trapping with distinctly different trapping depths consisting of successive monolayers of two trapping dielectric films (e.g., Si_3N_4 and GaN) was also introduced for multifunctional device characteristics [8]. Additionally, application for multifunctional DTM devices was presented containing negative fixed charge to enhanced retention [4,11]. DTM design was central to significantly enhance programming performance [11]. For some higher performance designs, direct tunneling of both electrons and holes was employed [5,7,12], whereas for other designs PBO and VARIOT direct tunneling of electrons was employed to achieve high programming speed through internal field aided transport [2,3]. CRESTED barrier

DTM designs and combination of VARIOT and CRESTED barrier designs were adopted to achieve desired trade-offs between speed and retention [7,8]. All the above multifunctional concepts form the basis of MSUM devices as mentioned above including MLC and nonplanar device design concepts discussed in Part II of this book. Multiple examples of extending NVM device properties such as memory window, retention, and endurance through employment of band engineering and high K dielectric films are provided in Chapter 13. Following similar approach stack design could yield an appropriate performance/nonvolatility trade-off to achieve, for example, the "DRAM" functionality. Such an example is illustrated in 21.2.2. The stack design could be extended to integrate both the DRAM functionality and the NVM functionality. This is how the MSUM devices are proposed.

21.2.2 Band Diagram Illustrations for MSUM Devices

Four CT-MSUM device types outlined in the stack design Table 20.3 are illustrated here. As stated previously, Type 1 is based on PBO or VARIOT barrier tunneling schemes; Type 2 is based on CRESTED barrier tunneling schemes.

Type 3 and Type 4 provide multifunctional and multilevel capabilities combined.

The DRAM functionality through a two-layered CRESTED barrier DTM design is illustrated below. Several selected illustrations of band diagrams are shown below to demonstrate the concepts leading to CT NVM devices (discrete trap and nanocrystal-based embedded trap) evolve into multifunctional CT-NVM devices and subsequently into CT-FSUM device as illustrated in Figure 21.5.

21.2.3 PBO Multifunctional CT-NVM → PBO-CT-MSUM: 1.1 of Table 20.3

Figure 21.6a shows the band diagram for the multifunctional CT-NVM device [2], whereas Figure 21.6b shows the same for the CT-MSUM device 1.1 with similar PBO band barrier characteristics for high-performance electron transport. Important differences to note are as follows: (a) tunnel layers for MSUM device are thinner (both OR-SiON and La_2O_3) and selection of La_2O_3 as the second tunnel layer (layer B) to ensure both electron and hole tunneling for faster write/erase performance, (b) DRAM trapping with Si_3N_4 layer (layer C) to enhance detrapping due to shallower

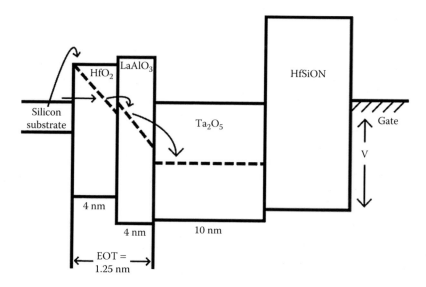

FIGURE 21.5 Band diagram illustration of a proposed NV-DRAM stack design using a two-layered CRESTED barrier and high K dielectrics. (From Bhattacharyya, A., Discrete trap non-volatile device for universal memory chip application, ADI Associates internal publication, May 29, 2005 (MIDD 52), U.S. Patents: 7436018 [10/14/2008], 7476927 [01/13/2009], 7786516 [10/14/10], 8143657 [03/27/2012], 8242554 [08/14/2012].)

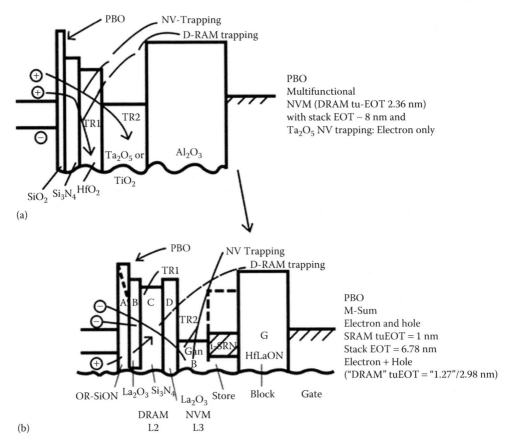

FIGURE 21.6 Device evolution from multifunctional CT-NVM to CT-MSUM device with L2 (DRAM) and L3 (NVM) functionality using DTM-PBO barriers: (a) multifunctional NVM [2], (b) MSUM with Stack Design from Table 20.3.

trap depth compared to HfO_2 trapping of the previous case, thereby aiding faster erasure for DRAM performance, and (c) incorporation of I-SRN as charge storage layer to enhance memory window for the high NVM memory state. It should also be noted that at programming Vpp $\pm \geq 5.5$ V, the effective tunnel distance during "DRAM" operation would be ~1.27 nm (calculated) due to barrier thinning and not the physically derived EOT of ~3 nm.

21.2.4 VARIOT-CT-MSUM 1.3 of Table 20.3

A functionally similar to the above MSUM device is illustrated in Figure 21.7, where the high programming performance is derived from the VARIOT barrier using only single carrier electron transport for both writing and erasing. High hole barrier for all the relevant tunnel layers prevents hole tunneling from the silicon substrate, whereas the VARIOT barrier aids in high speed electron transport due to barrier thinning. This is shown schematically in the band diagram where ultra-thin layers of OR-SiON/HfTaON/OR-SiON provide the VARIOT barrier with sharp rise in electron fluence with applied voltage during writing and erasing. It should be noted that VARIOT barrier provides barrier thinning. The MSUM design shows multiple direct tunneling first through A and then through C while gaining internal field along the conduction band of layer B during writing. The effective tunnel distance becomes equal to the thickness of layer A and a fraction of the thickness of layer C during writing. The reverse situation happens during erase. The device is designed to simultaneously provide NV-DRAM (L2) and NROM (L3) functionality.

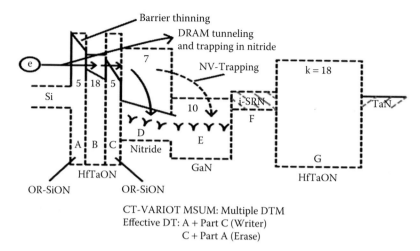

CT-VARIOT MSUM: Multiple DTM
Effective DT: A + Part C (Writer)
C + Part A (Erase)

FIGURE 21.7 Band diagram of CT-SUM example of VARIOT tunnel barrier designed for L2 and L3 functionality with nitride and GaN trapping for DRAM and NROM, respectively (Table 20.3).

21.2.5 CRESTED Barrier CT-MSUM (Table 20.3) → CRESTED Barrier-CT-MSUM 2.1 of Table 20.3

CRESTED barrier NVM designs are often considered for cost/performance multifunctional applications to achieve both DRAM-like attributes and NVM attributes of NROM and NAND with extended nonvolatility and endurance. CRESTED barrier designs could be considered for either single source carrier type (silicon substrate for normal mode device or gate for reverse mode device as described in Part II, Chapter 13) or dual source carrier type where both silicon substrate and gate could be active in providing electrons or holes. While most designs employ silicon substrate as active carrier source: examples of MSUM devices of 2.1 and 2.2 in Table 20.3, both silicon and gate could also be considered as active carrier sources as shown in the example of 2.3 in the same table. We will illustrate devices with silicon carrier source first as proposed in option 2.1. We will follow up with dual carrier source device design for NVM as well as MSUM devices for the stack design example of 2.3 illustrated in Table 20.3.

Figure 21.8 illustrates the band diagram for the MSUM device of 2.1. Similar to the multilevel CT-NVM device of CRESTED barrier type discussed in Chapter 13 of Part II. The device is based

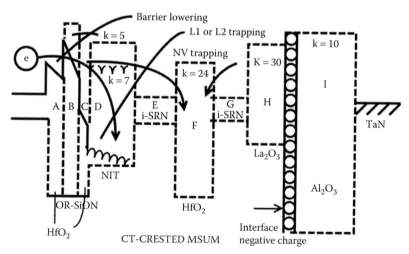

FIGURE 21.8 Band diagram of the CRESTED barrier MSUM device of option 2.1 with Bi-planar trapping and negative fixed charge design.

on electron transport only due to high energy barrier design for holes. The device is a two-layered DT tunneling media consisting of HfO_2 and OR-SiON of ultra-shallow thickness. During writing or erasing, barrier height of OR-SiON is lowered as shown for writing in the band diagram. The thickness of the tunnel layers of HfO_2 and OR-SiON determines the programming and erasing speed and thereby the functional objectives of either L1 or L2. The trapping layer for such functionality is the nitride layer as shown. The second trapping layer placed further from the silicon/insulator interface is a layer of HfO_2 for providing larger window EOL retention for NROM or NAND device. The in-SRN layers at both sides (layers E and G) of the HfO_2 trapping layer (layer F) provide charge storage on one hand, enhancing memory window, and on the other hand, enhancing transport during writing and erasing improving programming speed. The negative fixed charges generated at the La_2O_3/Al_2O_3 interface due to the post deposition anneal after depositing the blocking layers (La_2O_3 and Al_2O_3) minimizes charge loss to the gate and provides larger window for the NROM/NAND element to enable MLC design. The retention enhancement of internal charge generated stack design will be further discussed.

21.2.6 CRESTED Barrier Multifunctional CT-NVM [40] → CRESTED Barrier-CT-MSUM 2.3 of Table 20.3: Dual Carrier Source Design

The CRESTED barrier MSUM example is derived from the CT-NVM design of the two-layered tunneling version discussed in Chapter 13 with reference to CT-MLC-NVM designs (Figure 21.8). Both silicon substrate and the gate provide electronic carriers for programming and charge trapping. The example is a six-layered gate stack design for L1 and L2 applicability whereby the speed and retention could be traded off by varying the thickness of the first trapping nitride layer (layer C) and the blocking La_2O_3 layer (layer E). The device could be further extended by adding additional trapping plane of GaN for example, and an additional dielectric layer of La_2O_3 between the double-layered trapping plane shown here to add NV functionality (L3) either SLC or MLC (not shown). The above device could have an EOT ~<5 nm and operable at ±Vpp ≤3.5 V. Depending on the device design and functional objectives, the stack with L3 capability could be operable in the range of ±6 to 8 V of Vpp for stack design ranging from 7 nm to 10 nm EOT. Figure 21.9 illustrates the band diagram, whereas Figures 21.10 and 21.11 show the schematic cross-section of a planar device and the band bending during writing and erasing of the device, respectively.

The band engineered FET-based USUM and MSUM cells are described in Chapters 22 and 23, respectively.

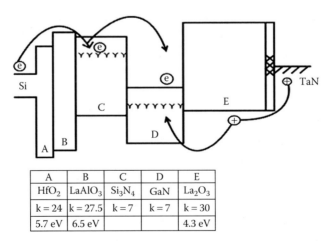

A	B	C	D	E
HfO_2	$LaAlO_3$	Si_3N_4	GaN	La_2O_3
k = 24	k = 27.5	k = 7	k = 7	k = 30
5.7 eV	6.5 eV			4.3 eV

FIGURE 21.9   CRESTED MSUM device of option 2.3 (see text): L1 and L2 applicability with merged dual energy trapping layers. (From Bhattacharyya, A., High performance multi-level ET/DT NVM device, ADI Associates internal publication, December 5, 2004, U.S. Patents: 7429767 [09/28/2008], 7553735 [06/2009], 7579242 [08/25/2009], 8159875 [04/17/2012].)

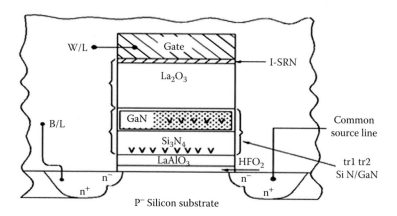

FIGURE 21.10 Schematic cross-section of the device of Figure 21.9. (From Bhattacharyya, A., High performance multi-level ET/DT NVM device, ADI Associates internal publication, December 5, 2004, U.S. Patents: 7429767 [09/28/2008], 7553735 [06/2009], 7579242 [08/25/2009], 8159875 [04/17/2012].)

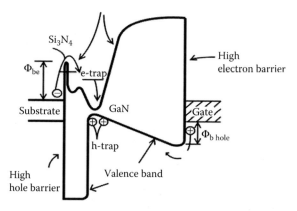

FIGURE 21.11 Band bending during programming and charge trapping for device of Figure 21.10. (From Bhattacharyya, A., Farnworth, W., and Farrar, P. F., High density nano-dot nonvolatile memory, ADI Associates internal publication, [MIDD 48], Title changed to: "Extremely Hi Density [Terabit] nano dot memory", U.S. Patent office Publication, 2010090265, dated 04/15/2010.)

REFERENCES

1. A. Bhattacharyya, Integrated two-device multiple configuration alterable ROM/SRAM and "Scalable Integrated Logic/NV Memory Technology with Associated Gate Stack Structure and Process"; ADI Associates internal publication, June 2004, U.S. Patents: 7208793 (04/24/2007), 7544990 (06/09/2009), 7728350 (06/01/2010).
2. A. Bhattacharyya, A novel low power battery operated non-volatile memory device and associated gate stack, ADI Associates internal publication, September 4, 2004, (MIDD 45), U.S. Patent: 7612403 (11/03/2009).
3. A. Bhattacharyya, Integrated surround-gate multifunctional memory device, ADI Associates internal publication, 5/26/2004.* Patent Application abandoned. (Unpublished).
4. A. Bhattacharyya, Novel band-engineered nano-crystal non-volatile memory device utilizing enhanced gate injection, ADI Associates internal publication, Dec. 5, 2004, U.S. Patent. 7629641 (12/22/2009).
5. A. Bhattacharyya, Advanced scalable nano-crystal device and process, ADI Associates internal publication, December 2004, (MIDD 47), U.S. Patents: 7759715 (07/30/2010), 7898022 (03/01/2011).
6. A. Bhattacharyya, "Multi-functional memory device with graded composition insulator stack", ADI Associates internal publication, Patent Application submitted December 5, 2004, U.S. Patents: 7525149 (04/28/2009), 77680062 (08/03/2010).

7. A. Bhattacharyya, High performance multi-level ET/DT NVM device, ADI Associates internal publication, December 5, 2004, U.S. Patents: 7429767 (09/28/2008), 7553735 (06/2009), 7579242 (08/25/2009), 8159875 (04/17/2012).

8. A. Bhattacharyya, Discrete trap non-volatile device for universal memory chip application, ADI Associates internal publication, May 29, 2005 (MIDD 52), U.S. Patents: 7436018 (10/14/2008), 7476927 (01/13/2009), 7786516 (10/14/10), 8143657 (03/27/2012), 8242554 (08/14/2012).

9. A. Bhattacharyya, Surround gate multi-functional memory chip: Device concept, technology & chip architecture, ADI Associates internal publication, December 2004.

10. A. Bhattacharyya, Future developments of SONOS technology and devices and stack designs leading to Silicon-based universal memory, Technical exchanges by A. Bhattacharyya, ADI associates to M. Durcan, Micron Technology, Boise, Idaho, April 2005 and 05/25/2005.

11. A. Bhattacharyya, "Scalable flash/NV structures and devices with extended endurance", ADI Associates internal publication, August, 2001, U.S. Patents: 7012297 (3/14/2006), 7250628 (7/31/2007), 7400012 (7/15/2008), 7750395 (7/6/2010).

12. Y. Q. Wang, D. Y. Gao, W. S. Hwang et al., Fast erasing and highly reliable MONOS type memory with HfO_2, high -k trapping layer and Si_3N_4/SiO_2 tunneling stack, *IEEE, IEDM,* 2006, pp. 973–976.

13. A. Bhattacharyya, W. Farnworth, and P. F. Farrar, High density nano-dot nonvolatile memory, ADI Associates internal publication, (MIDD 48), Title changed to: "Extremely Hi Density (Terabit) nano dot memory", U.S. Patent office Publication, 2010090265, dated 04/15/2010.

22 Uni-Functional SUM
The USUM Cells and Arrays

CHAPTER OUTLINE

USUM cells are all nonvolatile memory cells. However, nonvolatility in these cells is derived from two distinct approaches. The first approach requires embedding a charge trapping layer and functionally integrating such a layer with the active device such as an FET (bulk or SOI), or a diode (p-i-n or n-i-p), or a thyristor (pnpn or npnp). The second approach relies on an extension of DTM-based NVM cells through band-engineered, multilayered gate stack designs. USUM cells operating within the framework of the first approach are defined and labeled by the associated active device and designed to functionally replace a conventional memory exclusively such as SRAM or DRAM or NROM or NAND. DTM-based USUM cells are also designed to functionally replace the above-mentioned conventional memories only through modifications of their stack designs, all being FET-based and, therefore, not exclusive in the same sense as the first type of NVMs. We shall initially discuss the USUM cells of the first type. Subsequently, we will discuss DTM-based USUM cells with the background of stack designs and band diagrams already provided in Chapters 20 and 21.

It should be appropriate to mention the cell density and performance implications of the proposed USUM cells to be discussed in this chapter intended to functionally replace the conventional memories. The high performance USUM cells of the first approach are based on either gate-controlled diode or gate-controlled thyristor with cell density in the range of ~8 F^2 and are intended for functional replacement of conventional SRAMs of cell density in the range of 50–100 F^2. The cost performance USUM cells of the same type are based on floating body SOI or bulk single transistor ("capacitor-less") design and are intended to functionally replace DRAM (typically 8–10 F^2 in cell density) with cell density of ~6 F^2. The DTM-based USUM cells of the second type with MLC (2 bit per cell) capability [1–3] for functional replacement of either L1 or L2 or L3 or L4 could achieve cell density ~4 F^2.

22.1 THE FB RAM USUM CELL

22.1.1 The FB RAM Cell: (The Floating Body-Charge Trapping USUM RAM Cell)

The author had proposed floating plate nonvolatile memory devices using a gate insulator stack containing silicon-rich insulators [4–6]. In this proposal, carriers generated in the floating body of the PD-SOI transistor are trapped by providing an appropriate trapping layer such as silicon-rich nitride (SRN) placed in the X-Y plane of the floating body. The process of fabricating such a device will be explained later. The trapped-charge state could be neutralized by generating and injecting charges of opposite polarity: for example, holes by electrons and vice versa. The charged state and the neutralized state create the binary memory state of the memory. Since the energy barrier for the trapped state could be in the order of 1 eV, charged state (similarly, the neutralized state) is retained

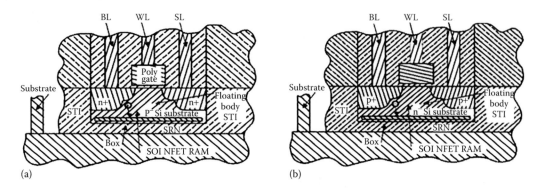

FIGURE 22.1 Schematic cross-section for 1T USUM RAM cells: (a) The NFET SOI version and (b) the PFET SOI version. (From Bhattacharyya, A. et al., Non-volatile memory cell having silicon-rich nitride charge trapping layer, IBM Internal Technical Reports, U.S. Patent: 4870470 (09/26/1989), 1984; Bhattacharyya, A., One transistor SOI non-volatile random access memory cell, *ADI Associates internal publication*, U.S. Patent: 6888200 [5/3/2005], 6917078 [7/12/2005], 7184312 [2/27/2007], 7339830 [03/04/2008], 7440317 [10/21/2008], July 22, 2002.)

providing long retention and nonvolatility. The sensing is carried out by measuring the difference in channel conductance of the cell transistor between the charged state and the neutralized state to define memory states of "1" or "0."

The schematic cross-sections for the 1-T USUM RAM cells are shown in Figure 22.1a and b for n-SOI-FET and p-SOI-FET devices, respectively. The detailed discussion will be confined to the NFET version while a complementary approach should be assumed for the PFET version. The difference from the standard PD-SOI transistor is the presence of the trapping layer at the box–body interfaces as shown in Figure 22.1a and b. The composition of the trapping SRN layer is tailored to achieve an appropriate trap-energy depth and trap density which in turn determines difference in transistor channel conductance between "1" and "0" state and the charge-retention/charge-neutralization specifications and characteristics [4,6].

22.1.1.1 Mode of Body-Charge Generation

Body charge could be generated either by high field impact ionization at the drain edge (FET mode) or by relatively low field parasitic bipolar (lateral n-p-n) mode [7]. The latter mode (lateral n-p-n mode) is preferred because of improved reliability for the write operation. To generate holes in the body of the NFET transistor (p-type silicon), a shorter negative drain pulse is superimposed on a negative gate pulse. As the gate pulse returns to ground, the substrate is concurrently pulsed negative. This causes the generated holes to drift toward the trapping layer (vertical drift field) and get trapped raising the body potential positive and enhancing the transistor channel conductance (threshold is reduced and remains reduced). The state achieved is write "1" and remains stable. To neutralize the holes (write "0"), electrons are generated in the body by forward biasing the drain–body (n ± p) diode and creating a vertical drift field to push the generated electrons toward the trapping layer. This is achieved by pulsing the drain negative, followed by a positive substrate pulse to overlap the drain pulse while keeping the gate at constant low positive potential. Electrons drift to the trapped insulator and neutralize the positive trapped charges, thereby reducing the body potential and raising the threshold of the transistor. This is write "0" and remains stable. The transistor channel conductance is reduced at this state. The write "1" pulsing scheme and illustration of hole generation and trapping are shown in Figure 22.2a and b, respectively. The write "0" pulsing scheme and illustration of hole detrapping plus electron and hole charge neutralization during write "0" are shown in Figure 22.3a and b, respectively.

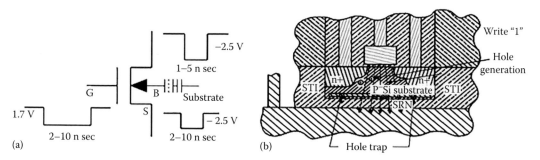

FIGURE 22.2 Pulsing scheme (a) and schematic of hole generation and trapping in the floating body (b) of the memory cell. (From Bhattacharyya, A., One transistor SOI non-volatile random access memory cell, *ADI Associates internal publication*, U.S. Patent: 6888200 [5/3/2005], 6917078 [7/12/2005], 7184312 [2/27/2007], 7339830 [03/04/2008], 7440317 [10/21/2008], July 22, 2002.)

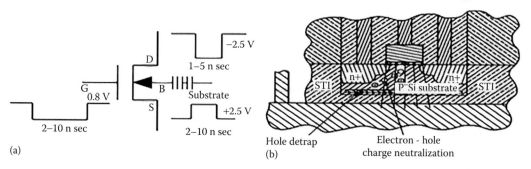

FIGURE 22.3 (a) Pulsing scheme for Write "0" and (b) Schematic of the hole detrapping and electron-hole charge neutralization in the floating body of the memory cell. (From Bhattacharyya, A., One transistor SOI non-volatile random access memory cell, *ADI Associates internal publication*, U.S. Patent: 6888200 [5/3/2005], 6917078 [7/12/2005], 7184312 [2/27/2007], 7339830 [03/04/2008], 7440317 [10/21/2008], July 22, 2002.)

22.1.1.2 Reading the Memory Cell

For sensing the stable "1" and "0" states, either a DRAM-like approach of using a reference cell with a current mode differential sense amplifier scheme or a SRAM-like direct cell-current sense amplifier scheme could be adopted depending on the cell design and performance specification. This is illustrated in Figure 22.4a and b for DRAM-like and SRAM-like sensing schemes, respectively. The potential characteristics of gate (V_g), drain (V_d), and substrate during read "1" and read "0" and corresponding currents (I_s) through the memory device are illustrated in Figure 22.5.

Table 22.1 describes the operational scheme for the capacitor-less DRAM cell [6].

22.1.1.3 Processing of FB RAM Device Structure

Device processing for the structures shown in Figure 22.1 involves standard PD-SOI CMOS device processing with an appropriate adjustment of channel-tailor implant for as processed FET threshold adjustment. The special processing steps involved for the creation of box–body interface trapping layer could be as follows:

1. Standard processing steps through isolation (STI).
2. Block mask to define/open NFET-active region.
3. Ion implant silicon, ammonia (NH_3), and hydrogen (optional) with appropriate energy and concentration to achieve the desired refractive index after post-processing anneal [4].

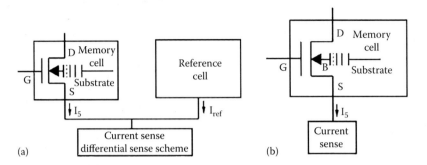

FIGURE 22.4   (a) DRAM cell sensing scheme and (b) SRAM cell sensing scheme of the memory cell. (From Bhattacharyya, A., One transistor SOI non-volatile random access memory cell, *ADI Associates internal publication*, U.S. Patent: 6888200 [5/3/2005], 6917078 [7/12/2005], 7184312 [2/27/2007], 7339830 [03/04/2008], 7440317 [10/21/2008], July 22, 2002.)

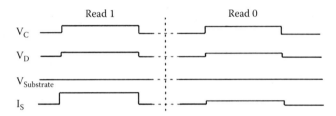

FIGURE 22.5   Pulsing scheme and sensing current is for Read "1" and Read "0" for the memory cell. (From Bhattacharyya, A., One transistor SOI non-volatile random access memory cell, *ADI Associates internal publication*, U.S. Patent: 6888200 [5/3/2005], 6917078 [7/12/2005], 7184312 [2/27/2007], 7339830 [03/04/2008], 7440317 [10/21/2008], July 22, 2002.)

TABLE 22.1
Operations for Single Transistor Capacitor-Less NVDRAM Cell

Operation	Bit Line	Word Line	Substrate	Remarks
Write "1"	−2.5 V	−1.7 V	−2.5 V	Holes are generated in the body and are trapped
	1–5 ns	2–10 ns	2–10 ns	in the trapping layer. Vr is reduced by 200 mV.
Write "0"	−2.5 V	0.8 V	2.5 V	Electrons are generated in the body and
	1–5 ns		2–10 ns	neutralize the trapped holes. Vr returns to
				original value.
Half-Select Cells	0.3 V	As above	As above	No change.
Read "1"	0.3 V	0.8 V	Gnd	Current is 2–3 orders of magnitude higher.
Read "0"	0.3 V	0.8 V	Gnd	Current is lower. Device threshold is designed
				to put the device in subthreshold operation for
				Read "0" operation.

Source: Bhattacharyya, A., One transistor SOI non-volatile random access memory cell, *ADI Associates internal publication*, U.S. Patent 6888200 (5/3/2005), 6917078 (7/12/2005), 7184312 (2/27/2007), 7339830 (03/04/2008), 7440317 (10/21/2008), July 22, 2002.

Ammonia could be replaced by active nitrogen, while silicon could be replaced by other active silicon sources (e.g., silane and dichlorosilane).

4. Post-implant inert anneal (optionally, RTA, or inert plasma anneal in nitrogen).

5. Standard PD-SOI CMOS fabrication steps.

22.2 THE GDRAM USUM CELL

The gated diode charge storage nonvolatile USUM RAM cell stems from the concept of storing charges at the intrinsic node of an integrated gated diode within the memory cell following the proposal discussed in references [8,9]. The GDRAM cell could store charge either in a gated P-i-N diode or in a gated N-i-P diode built-in within the memory cell. A brief description of the concept and cell operation is outlined below.

The memory cell consists of a switching transistor (either NFET or PFET) integrated in series at the floating node with a P-i-N or N-i-P diode. The cell is implementable in bulk silicon technology, in FinFET technology as well as in SOI technology. The memory state of the cell is determined by the charge stored at the intrinsic region of the diode altering the potential of the floating node. In either implementation of the diode (P-i-N or N-i-P), the diode is gate controlled to significantly enhance the programing performance (write/erase speed) and significantly reduce the standby power of the memory states.

In the bulk silicon implementations, the diode could be configured either laterally or vertically. Therefore, in principle, one could consider a total of 16 variations of such cell options. In the planar silicon substrate, the vertical implementation could be achieved by forming a silicon pillar. Lateral implementation may provide faster performance at increased process complexity and cell size [9]. The diode functions as a negative differential resistance (NDR) device similar to those of P-N-p-N thyristor or N-P-n-P thyristor. The thyristor implementation of the memory cell will be discussed later on. The cell is sensed in similar ways as conventional DRAM by turning on the switching transistor and measuring the change in bit-line potential related to the memory state associated with the floating node potential (high or low) which is the common node to both the transistor and the diode as shown in the above figures. However, the sensing speed is significantly faster than that of a DRAM due to the large difference in current level of the forward ("on" state) and reverse ("off" state) of the gated diode. Figure 22.6 illustrates the schematic cross-section of the memory cell consisting of an NFET with both an ungated diode version (Top Figure: P-i-N or N-i-P) as well as the

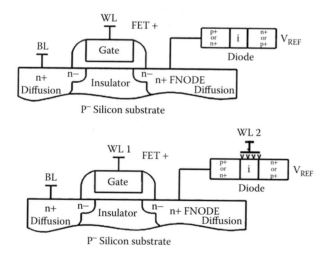

FIGURE 22.6 The conceptual schematics of the diode charge storage USUM RAM cell: Top: switching transistor (NFET) integrated with ungated diode (P-i-N or N-i-P); Bottom: switching transistor (NFET) integrated with gated (WL2) diode (P-i-N or N-i-P). (From Bhattacharyya, A., High performance one-transistor memory cell, original title: High performance capacitor less one transistor memory devices in bulk and SOI technology, *ADI Associates internal publication*, U.S. Patent: 7432562 [10/07/2008], 7625803 [12/01/2009], 7968402 [06/28/2011], 81255003 [02/28/2012], July 2, 2003; Bhattacharyya, A., SOI non-volatile (self refreshed) two-device SRAM using floating body charge, *ADI Associates internal publication*, U.S. Patent: 7224002 [05/20/2007], May, 2003.)

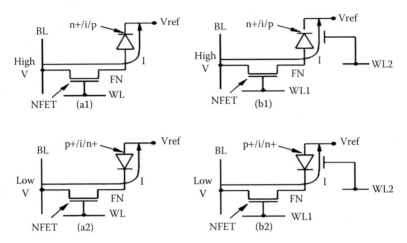

FIGURE 22.7 Circuit schematics of diode charge storage USUM RAM cells; (a1): ungated N-i-P diode option, (a2): ungated P-i-N diode option, (b1): gated N-i-P diode option, (b2): gated P-i-N diode option. The WL2 is tied with the gate of the diode. (From Bhattacharyya, A., High performance one-transistor memory cell, original title: High performance capacitor less one transistor memory devices in bulk and SOI technology, *ADI Associates internal publication*, U.S. Patent: 7432562 [10/07/2008], 7625803 [12/01/2009], 7968402 [06/28/2011], 81255003 [02/28/2012], July 2, 2003.)

gate-controlled diode version (Bottom Figure: P-i-N or N-i-P). The top version of the memory cell being an ungated diode storage cell requires operationally a single gate node WL as shown in the top figure for the switching transistor. However, the programming performance is slow and power consumption is high. The bottom version of the memory cell consists of an access gate WL1 and the memory gate WL2, the latter providing the gate-controlled diode characteristics with significantly superior power or performance.

The corresponding circuit schematics are shown in Figures 22.7 showing options: (a1) [for N-i-P], (a2) [for P-i-N], (b1) [for N-i-P], and (b2) [for P-i-N], respectively, when integrated with an NFET. Similarly, four other cell options and associated circuit schematics are feasible when a PFET switching transistor is employed and integrated with the above four diode options (not shown).

22.2.1 The Diode I-V Characteristics

The diode I-V characteristics are shown in Figure 22.8 for P-i-N and N-i-P diodes, respectively. The I-V curves for gate-controlled diodes are shown solid, while those without gate control are shown dotted. During writing of the memory cell, the built-in diode is switched in the forward-biased mode ("on" state), while the gate control limits the holding current at that state. During switching to the "off" state, gate-enhanced switching facilitates the stored charge removal through enhanced electron-hole recombination and drift of the excess stored charge, thereby enabling significantly faster switching of the diode. Additionally, by limiting the holding current at the "on" state of the gate-controlled diode, standby power is reduced. It should be noted that gated diode lowers the break over voltage and enhances the stability of the standby "1" and "0" memory states [9,10].

The gated diode charge storage RAM cell could be configured either with a vertical diode or with a lateral diode as noted earlier. Schematic cross-sections of a memory cell with a gated N+-i-P+ vertical diode are shown in Figure 22.9a, while a memory cell with a gated P+-i-N+ vertical diode is shown in Figure 22.9b. A cell layout scheme for either of the above memory cells with shared bitline contact schemes between adjacent cells is illustrated in Figure 22.9c. Two layers of polysilicon could be employed for the word line 1 (polysilicon 1, for the switching FET element) and word line 2 (polysilicon 2, for the diode gate), respectively, as shown. The first-level metal, M1, could be used

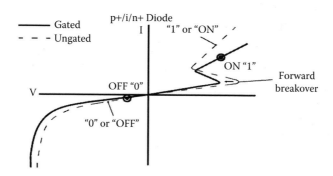

FIGURE 22.8 NDR/I-V characteristics of gated (solid curve) and un-gated (dotted curve) built-in P-i-N diode (a) and N-i-P diode (b) of the USUM RAM cell in bulk silicon. (From Bhattacharyya, A., High performance one-transistor memory cell, original title: High performance capacitor less one transistor memory devices in bulk and SOI technology, *ADI Associates internal publication*, U.S. Patent: 7432562 [10/07/2008], 7625803 [12/01/2009], 7968402 [06/28/2011], 81255003 [02/28/2012], July 2, 2003.)

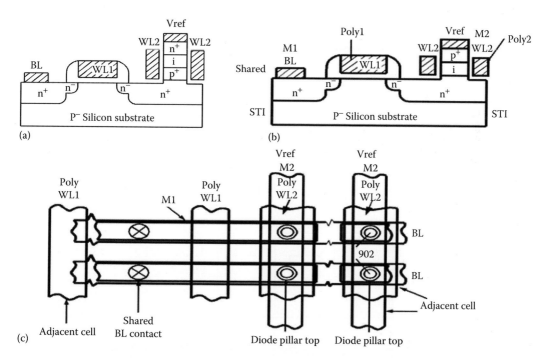

FIGURE 22.9 (a) Schematic cross-section of a gated-diode charge storage RAM cell with vertical N+-i-P+ gated diode. (From Bhattacharyya, A., High performance one-transistor memory cell, original title: High performance capacitor less one transistor memory devices in bulk and SOI technology, *ADI Associates internal publication*, U.S. Patent: 7432562 [10/07/2008], 7625803 [12/01/2009], 7968402 [06/28/2011], 81255003 [02/28/2012], July 2, 2003; Bhattacharyya, A., SOI non-volatile (self refreshed) two-device SRAM using floating body charge, *ADI Associates internal publication*, U.S. Patent: 7224002 [05/20/2007], May, 2003.); (b) and (c) Schematic cross-section of a gated-diode charge storage RAM cell: (b) with vertical P+-i-N+ gated diode; (c) Cell layout scheme with shared bit-line contact with double polysilicon word lines and metal 1-bit line and Metal 2 Vref Line. (From Bhattacharyya, A., High performance one-transistor memory cell, original title: High performance capacitor less one transistor memory devices in bulk and SOI technology, *ADI Associates Internal Publication*, U.S. Patent: 7432562 [10/07/2008], 7625803 [12/01/2009], 7968402 [06/28/2011], 81255003 [02/28/2012], July 2, 2003.)

for the bit lines. The second-level metal M2 connects the salicided metal pads on top of the diode of each cell and runs orthogonal to M1.

The M2 lines are connected to the common Vref node outside the array. The above layout could yield ~12 F^2 cell compared to a typical SRAM cell of >50 F^2. Denser cell schemes with lateral diode configurations placed over the switching transistor have also been illustrated [9].

22.2.2 Cell Operation

Cell operational schemes of memory cells of Figure 22.9a with gated N+-i-P+ diode and Figure 22.9b with gated P+-i-N+ diode are described in Figures 22.10 and 22.11, respectively.

22.2.2.1 Memory Cell 22.9 (a): Gated N+-i-P+ Diode

The pulsing scheme of word lines 1 and 2, bit line, and the diode reference node is illustrated below for write "1," read "1," write "0," and read "0."

For a Vdd (power supply voltage) of 2.5 V, the pulse potential levels of WL1 could be 2.5 V, of BL could also be 2.5 V, the reference potential at the diode gate could be 0.8 V, and the WL2 potential could be 1.5 V. WL2 is pulsed for both write 1 and write 0 operations, while bit line is pulsed only for write 1 as shown. During write 1, WL2 assists in the reverse breakdown of the n+/p+ junction, while the raised potential of the n+ floating node capacitively couples the p+ cathode of the diode, forward biasing the diode. The diode forward-biased "on" mode is also enhanced by the word line 2 pulse. Charges get stored at the intrinsic region of the diode. During write 0, the bit line is held low, and the diode remains reverse biased. The word line 2 pulse assists in switching the diode transition from on to off completely within the short duration of the WL2 pulse. At the steady state (bit-line potential = 0 V), the floating node n+ potential assumes the steady state potential (=0 V) of the bit line when the cell state is written "0". The state of the cell is read by turning the access transistor on and sensing the bit-line potential. Therefore, when the cell is written "1," bit-line potential is raised during reading the cell. When the cell is written 0, bit-line potential remains low when the cell is read. The cell performance is dependent on the diode turn-off speed (transition from write 1 to write 0) which is dramatically enhanced by the gate-assisted turn-off characteristics of the diode by enhancing the removal of stored charge from the intrinsic region of the diode. Since the stored

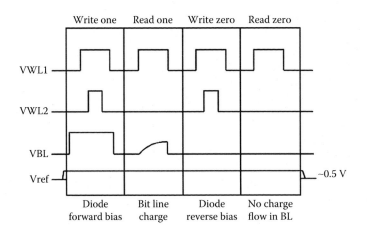

FIGURE 22.10 Pulsing schemes for memory cell with gated N+-i-P+ diode. (From Bhattacharyya, A., High performance one-transistor memory cell, original title: High performance capacitor less one transistor memory devices in bulk and SOI technology, *ADI Associates internal publication*, U.S. Patent: 7432562 [10/07/2008], 7625803 [12/01/2009], 7968402 [06/28/2011], 81255003 [02/28/2012], July 2, 2003; Bhattacharyya, A., SOI non-volatile (self refreshed) two-device SRAM using floating body charge, *ADI Associates internal publication*, U.S. Patent: 7224002 [05/20/2007], May, 2003.)

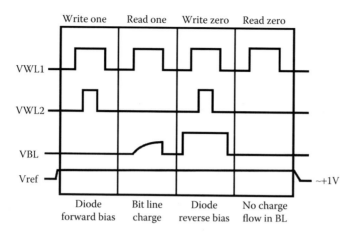

FIGURE 22.11 Pulsing schemes for memory cell with gated P+-i-N+ diode. (From Bhattacharyya, A., High performance one-transistor memory cell, original title: High performance capacitor less one transistor memory devices in bulk and SOI technology, *ADI Associates internal publication*, U.S. Patent: 7432562 [10/07/2008], 7625803 [12/01/2009], 7968402 [06/28/2011], 81255003 [02/28/2012], July 2, 2003; Bhattacharyya, A., SOI non-volatile (self refreshed) two-device SRAM using floating body charge, *ADI Associates Internal Publication*, U.S. Patent: 7224002 [05/20/2007], May, 2003.)

charge volume is significantly lower in the lateral device, the diode is expected to be turned off faster. It has been noted earlier that the standby leakage current of the diode at the steady state is also reduced by the word line 2 excitation.

Cell operational schemes of memory cells of Figure 22.9b with gated P+-i-N+ diode are described below.

22.2.2.2 Memory Cell 22.9 (b): Gated P+-i-N+ Diode

The pulsing scheme of word lines 1 and 2, bit line, and the diode reference node is illustrated below for write "1," read "1," write "0," and read "0."

The node potentials could be similar to those of the previously discussed cells. While WL2 is pulsed for both write 1 and write 0 operations, bit line is pulsed only for write 0 which is reverse of the previous (N+-i-P+ gated diode) cell operation. During write 1 the bit line is held low, the diode being forward biased enhanced by the word line 2 gating. Charges get stored at the intrinsic region of the diode and the n+ floating node (cathode) potential is raised. During write 0, the bit line is held high, and the diode remains reverse biased. The word line 2 pulse assists in switching the diode transition from on to off completely within the short duration of the WL 2 pulse during write 0. The access transistor discharges the floating node and the potential returns to ground. During read 1, the access device is turned on and the floating bit-line potential rises as the charges flow out from the floating n+ node to the bit line. During read 0, the n+ floating node is at steady state of ground and the potential of the floating bit line does not change. Note that the reference potential in this case is nearly 1 V and significantly higher than the previous case to aid forward biasing of the diode.

22.2.3 SOI Technology Implementation of the Diode Charge Storage RAM Cell

The above RAM cell has a unique advantage in implementation in PD-SOI technology since the cell could operate effectively without requiring gate control of the diode, thereby yielding a higher density memory cell at high performance and at lower power.

Charges can be generated in the floating body of the SOI transistor either by impact ionization when the transistor is operating in FET mode or by parasitic lateral bipolar action (n-p-n for SOI-NFET and p-n-p for SOI-PFET) when operating in bipolar mode. Both excess electrons and holes could be generated in such manner altering the floating body potential. In SOI-FET implementations, floating body-generated excess charges are utilized to enhance the p/i/n or n/i/p diode performance and charge storage for the proposed memory cells, thereby eliminating the requirement of gate control for the diode. This provides process simplicity and increased density and can be implemented using standard SOI-CMOS technology. The memory cell would consist of a SOI transistor whereby the gate is tied to the word line, one of the diffusions is tied to the bit line, while the other is floating and integrates the p/i/n or the n/i/p diode. The other end of the diode is tied to a reference potential. The schematic cross-sections of the SOI implementations of the vertical diode charge storage RAM cells are shown in Figures 22.12a with P+-I-N+ diode and 22.12b with N+-i-P+ diode, respectively. The cell can be laid out as polysilicon 1 for the word line, Metal 1 for shared bit line, and either a polysilicon 2 or a Metal 2 for the Vref line (which could also be shared) as shown in Figure 22.13, yielding a ≤8–10 F² cell size. Cell options with integrated lateral diode and processing schemes have been described in reference [9].

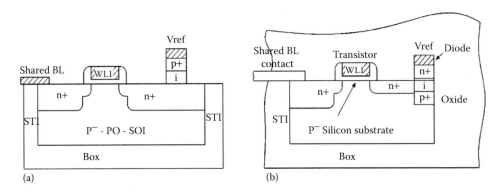

(a) (b)

FIGURE 22.12 Cross-sectional schematics of SOI diode charge storage RAM cells: (a) with P+-i-N+ diode and (b) with N+-i-P+ diode. (From Bhattacharyya, A., SOI non-volatile [self refreshed] two-device SRAM using floating body charge, *ADI Associates internal publication*, U.S. Patent: 7224002 [05/20/2007], May, 2003.)

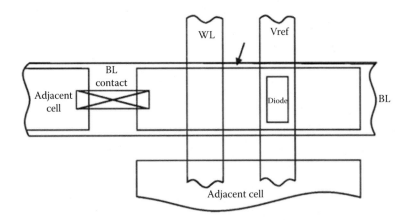

FIGURE 22.13 Layout scheme of cells of Figure 22.12. (From Bhattacharyya, A., SOI non-volatile [self refreshed] two-device SRAM using floating body charge, *ADI Associates internal publication*, U.S. Patent: 7224002 [05/20/2007], May, 2003.)

22.2.4 SOI Diode Charge Storage RAM Cell Operation

The cell operation pulsing scheme for the SOI diode charge storage RAM cell with ungated N+-i-P+ diode as shown in Figure 22.7 a1 without requiring WL2 is shown in Figure 22.14. Detailed explanation is also provided (Figure 22.14).

The steady state reference potential (Vref) for the diode is set around 0 to +0.5 V to reduce diode leakage and standby current. This could also be the scheme during write "0" and read "0" operation. However, during write "1" and read "1," the reference potential is set at −0.8 V for forward biasing the diode. The word line WL1 potential could be set to the power supply voltage, typically to 2.0–2.5 V. To write "1" into the cell, word line is pulsed up and subsequently bit line is pulsed negative for a short duration to generate excess electrons into the body by forward biasing the drain–body junction of the access transistor. Concurrent to the word line pulse, the diode node is pulled down to around −0.8 V to −1.0 V to forward bias the diode. Charges are collected at the n+/i diode region due to the reference potential and the cell gets written "1." To read "1," the bit line is precharged to be high, reference potential is held to ground and the word line is turned on. Flow of charge from the floating node is sensed by the potential drop in the bit line as shown. To write "0" into the cell, both word line and bit line are pulsed up as shown, while the reference potential is held to ground. SOI floating body, as well as the "P+" pocket of the diode, gets capacitively coupled to positive potential due to the word line transition. Consequently, excess hole generation in the body raises the potential of the p-pocket of the diode, setting the intrinsic region to the reference potential (Vref at 0 V), thereby writing "0" into the cell. To read 0, word line is turned on and reference potential is held to ground. Bit-line potential stays up and unchanged reading "0."

For the SOI diode charge storage RAM cell with ungated P+-i-N+ diode of Figure 22.12a, the operational pulsing scheme is appropriately modified, although writing into the cells is similar [9]. The steady state reference potential of the P+-i-N+ diode is set at ground level at the standby state. The word line pulsing for reading and writing "1" and "0" is similar to that in the previous example. For writing "1", while the word line and bit-line pulsing are similar to the previous case, the reference potential is reversed (pulsed positive) to forward bias the P+-i-N+ diode. Charges are

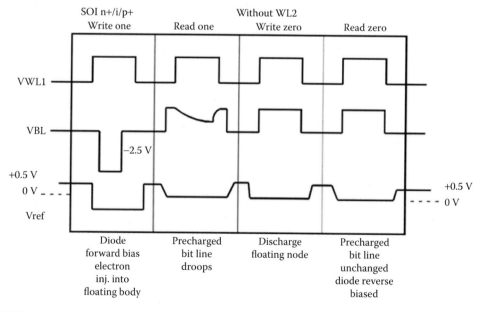

FIGURE 22.14 The pulsing scheme for the ungated (N+-i-P+) SOI diode charge storage RAM cell. (From Bhattacharyya, A., SOI non-volatile [self refreshed] two-device SRAM using floating body charge, *ADI Associates internal publication*, U.S. Patent: 7224002 [05/20/2007], May, 2003.)

collected at the N+/I node reflecting the reference potential and written "1" state is achieved. For reading "1," bit line is set floating and the reference potential is set to ground. As word line is turned on, charges flow from the floating node charging up the bit line, whereby read "1" is sensed. The diode node is kept at ground during write "0," remains reverse biased while charges are removed from the floating node writing the "0" state. Read "0" is similar as before.

22.2.5   Process Considerations for the Gated P-i-N or N-i-P Diode Charge Storage USUM RAM Cell Bulk Versions

These devices could be fabricated using conventional bulk silicon CMOS technology. The process sequence consists of: (a) defining and etching the diode pillar region on a starting intrinsic or very lightly doped p-silicon substrate; (b) implanting the bottom P+ (with reference to the N+-i-P+ diode) or the bottom N+ (with reference to the P+-i-N+ diode) part of the diode using an oxide cap protection on top of the silicon pillar; (c) forming the polysilicon 1 word line WL1; (d) forming the polysilicon 2 word line WL2 gates using anisotropic polysilicon etch after gate oxidation; (e) implanting source–drain diffusion regions and n+ doped polysilicon gate; (f) planarization; (g) forming diode anode by ion implantation through the contact hole. The depth of the intrinsic region is controlled by the appropriate thermal anneal following final ion implantation process. The lateral diode version of the process has been explained in reference [9]. SOI Versions: These devices could be fabricated using standard SOI silicon CMOS technology by following process sequences (a) through (g) similar to the one outlined above.

22.3   THE GTRAM USUM CELL

The name "GTRAM USUM Cell" types stand for the gated thyristor-based charge storage USUM RAM cells [1,10,11]. Instead of gated diodes, these cell types employ gated thyristors for nonvolatile charge storage. The concept of gated thyristor-based charge storage cells in SOI and bulk technology is as follows. This will be described in section 22.4.1.

The TRAM [10] and GLTRAM [1] cells have volatility characteristics similar to conventional SRAMs even though such cells have been considerably denser. The USUM version of such cell was proposed to be a vertically integrated gated thyristor cell in bulk implementation featuring two word lines, in principle similar to the previously discussed diode cell, while the SOI implementation is a single word line based on floating body-induced charge trapping similar to that discussed in the diode version of the USUM cell. One of the SOI versions of the cell will be discussed first [10]. Similar cells could be laid out in cross-point arrays and implementable both in SOI and in bulk technology [11]. This will be described later on in this section.

22.3.1   High Performance SOI Nonvolatile Two-Device SRAM Using Floating Body Charge

The concept consists of a single switching floating body NFET transistor with a laterally integrated thyristor integrated at the common floating node. The floating N+ node of the access transistor (switching transistor) is also the cathode of the thyristor (NPNP) and forms the memory storage node. The memory states are defined by the NDR effects of the lateral-gated PNPN thyristor thus formed. The thyristor operation is gate controlled for fast turn-on and turn-off to provide fast memory performance. To provide nonvolatility, a trapping layer is incorporated within the floating body of the active region of the SOI device. The trapping layer is held either at a positive potential by tapping excess holes injected into the floating body or at negative potential by injecting excess electrons into the floating body. Negatively charged trapping layer will enhance the stability of memory state "0," whereas positively charged trapping layer will enhance the stability of state "1" during standby. The

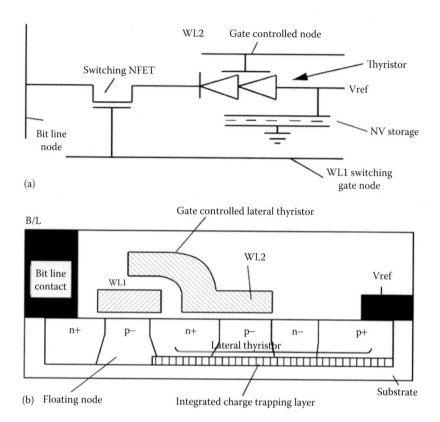

(a)

(b) Floating node   Integrated charge trapping layer   Substrate

FIGURE 22.15 NV two-device floating body charge high performance cell: (a) Circuit schematic and (b) Cross-section. (From Bhattacharyya, A., Memory cell with trenched gated thyristor, (original Title: A cross point stable memory cell using trenched gated-thyristor), *ADI Associates internal publication*, U.S. Patents: 7145186 [12/05/2006], 7440310 [10/21/2008], April 30, 2004.)

memory states are "written" or programmed by turning on the lateral thyristor along with the access transistor to charge or discharge the floating node. The memory states are "read" in a standard manner by turning on the access transistor and sensing the potential of the bit line. Concurrent to the memory cell programming operation, the charging or discharging of the trapping layer is performed for nonvolatile storage of the memory state. The device, therefore, could perform as both a standard SRAM memory and/or a nonvolatile memory. Additionally, the concept provides simultaneously high density and high SRAM-like performance bridging the application requirements of both DRAM and SRAM. Ref. [10] describes the process of fabrication of the cell in detail. The circuit schematic and the cross-section of such an SRAM cell is shown in Figure 22.15a and b, respectively.

22.4 THE CPRAM USUM CELL

22.4.1 The CPRAM Cell: (The Cross-Point Thyristor-Based Charge Storage RAM Cell)

Figure 22.16 shows the schematic cross-section of the cross-point memory device on an SOI substrate [2].

Similar to a conventional DRAM device, the bit-line contacts the drain node of the access transistor, while the word line is connected to the gate node of the access transistor. The n+ storage node of the memory device is simultaneously the source node of the access transistor and the cathode of the vertical thyristor placed on the trench of the memory device. At the bottom of the trench is a p+ region formed by ion implantation and subsequent thermal anneal to connect to the substrate

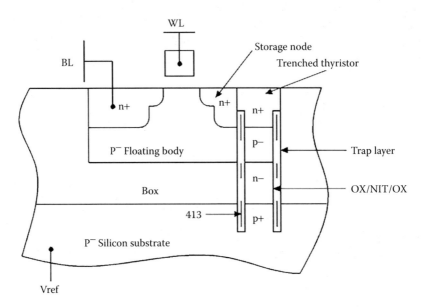

FIGURE 22.16 Schematic cross-section of the thyristor-based cross-point charge trapping cell. (From Bhattacharyya, A., Embedded trap direct tunnel NVM and A vertical channel cross point high density split channel direct tunnel NVM and A very low power memory sub-system with progressive non-volatility, *ADI Associates internal publication*, U.S. Patents: 7276760 [10/02/2007], 7365388 [04/29/2008], 7385245 [06/10/2008], July–August, 2004.)

below the buried oxide (BOX) region and forms the anode of the thyristor. The p+ anode nodes of all memory cells get tied to the bottom substrate which is held at a reference potential +V ref which is typically around 0.6–1.0 V. This arrangement eliminates anode contact for individual memory cell aiding memory density. Along the vertical side wall of the trench is an insulator stack consisting of thin layers of oxide/oxynitride/oxide which separates the floating body of the access transistor from the vertical thyristor and acts as the gate insulator as well as the trapping layer (oxynitride). The floating body acts as the gate for the thyristor as will be explained later on. This arrangement eliminates the requirement of a second word line for gating the thyristor as originally proposed by Nemati and Plummer [12]. Figure 22.17 shows the schematic cross-section for the TRAM memory cell shown in Figure 22.16. The equivalent circuit schematic for the proposed cross-point cell of Figure 22.16 is shown in Figure 22.18a and compared with the corresponding circuit schematic of Figure 22.18b for the TRAM cell. It should be noted that the floating body not only gates the thyristor (p-/n- region) but

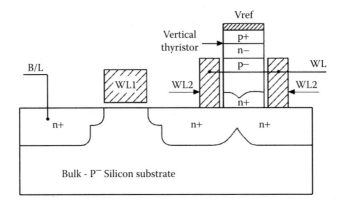

FIGURE 22.17 The TRAM cell. (From Nemati, F., and Plummer, J. D., A novel thyristor-based SRAM cell (T-RAM) for high speed, low voltage, giga-scale memories, *IEDM*, p. 283 © 1999 IEEE.)

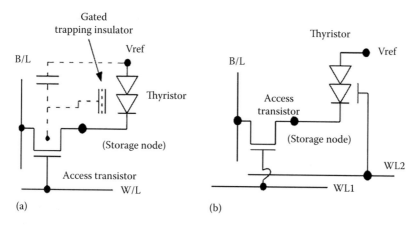

FIGURE 22.18 Circuit schematic comparisons of the (a) thyristor-based cross-point charge trapping cell and (b) TRAM cell. (From Bhattacharyya, A., Gated lateral thyristor-based random access memory cell [GLTRAM], *ADI Associates internal publication*, U.S. Patent: 7042027 [5/9/2006], 746054 [11/25/2008], June, 2002; Bhattacharyya, A., Memory cell with trenched gated thyristor, (original title: A cross point stable memory cell using trenched gated-thyristor), *ADI Associates internal publication*, U.S. Patents: 7145186 [12/05/2006], 7440310 [10/21/2008], April 30, 2004.)

also capacitively couples to the p+ anode of the thyristor (via BOX capacitance), thereby providing an enhanced effect on the switching "on" and "off" of the thyristor. This is similar to SOI version of diode charge storage and the thyristor-based cells explained earlier in detail for the operation of such memory.

The memory cell can be readily fabricated in SOI technology as well as in bulk technology. The key process steps consist of: the trench formation and sidewall insulator formation, followed by a p+ implant at the bottom of the trench, the thyristor region is epitaxially grown to yield either a lightly doped p-silicon or n-silicon. Subsequently either the n-region or the p-region is formed by counterdoping. Finally, the access transistor with the associated n+ regions is fabricated using standard technology approach. A typical cell layout is shown in Figure 22.19. The memory cell is formed at the cross-point of the polysilicon word line and the metal bit line.

For bulk implementation of the above cell, isolated floating bodies required for the cell could be achieved by creating either (a) a localized buried oxide isolation for the floating p-body or (b) a buried diffusion isolation for the p-floating body. Such processing schemes are illustrated in Figure 22.20a–c for oxide isolation and Figure 22.21a and b for diffusion isolation, respectively.

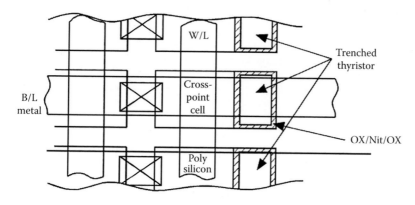

FIGURE 22.19 An example of layout of the cross-point memory cell. (From Bhattacharyya, A., Embedded trap direct tunnel NVM and A vertical channel cross point high density split channel direct tunnel NVM and A very low power memory sub-system with progressive non-volatility, *ADI Associates internal publication*, U.S. Patents: 7276760 [10/02/2007], 7365388 [04/29/2008], 7385245 [06/10/2008], July–August, 2004.)

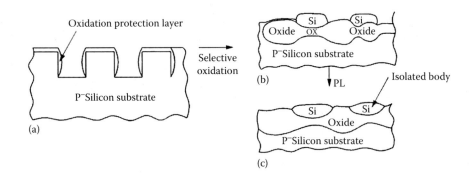

FIGURE 22.20 Oxide isolation scheme for the floating p-silicon body.

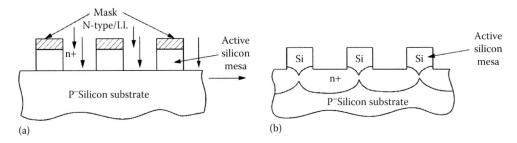

FIGURE 22.21 Diffusion isolation scheme for the floating p-Silicon body.

The oxide isolation scheme involves the following major steps: (1) define and etch into the silicon substrate, active cell regions (silicon pillars) for the memory cell; (2) protect the side wall and top with hard oxidation-resistant mask as shown in step A of Figure 22.20a, step A; (3) low temperature wet oxidation with postoxidation N_2 anneal to form bottom and sidewall oxide as shown in Figure 22.20b; and finally (4) surface planarization (chem.-mech-polish: CMP) to achieve oxide isolated active silicon islands for device processing (22.20c).

The diffusion isolation scheme involves the following major steps:

1. Define and etch into the silicon substrate, active cell regions (silicon pillars or mesas) for the memory cell similar to the previous option; (b) provide implant mask on the top surface and sidewall (if needed) of all active regions followed by n+ (phosphorus) ion implant of the bottom surface of the silicon substrate as shown in Figure 22.21a, step A.
2. Appropriate post-implant annealing in inert environment to drive the dopant vertically and laterally to form diffusion isolation as shown in Figure 22.21b, step B.

22.4.2 Cell Operation

Operational wave forms of the memory device during read, write, and standby are shown in Figure 22.22. This will be explained in detail in this section. The fundamental operation of the device is similar to those explained in [9–11]. The NDR effects of the thyristor are utilized to achieve the two stable memory states. These two memory states are defined as "0" for the low conduction high impedance OFF state of the thyristor and "1" for the high conduction low impedance ON state of the thyristor. Writing or storing "1" implies raising the potential of the floating n+ node by putting positive charges into the node and subsequently sensing ("read 1") the transfer of the stored positive charges to the floating bit line. Conversely, writing or storing "0" implies discharging the potential of the floating n+ node to 0 so that bit-line potential remains unchanged ("read 0") on

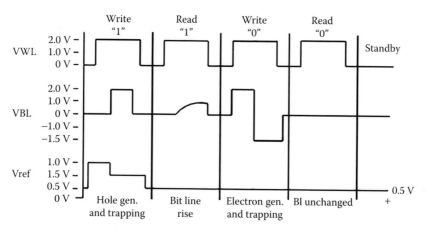

FIGURE 22.22 Operational waveform of the cross-point thyristor-based USUM cell. (From Bhattacharyya, A., Embedded trap direct tunnel NVM and A vertical channel cross point high density split channel direct tunnel NVM and a very low-power memory subsystem with progressive nonvolatility, *ADI Associates internal publication*, U.S. Patents: 7276760 [10/02/2007], 7365388 [04/29/2008], 7385245 [06/10/2008], July–August, 2004.)

subsequent sensing. To write "1" to the device, the thyristor is forward biased by raising the reference potential sufficiently positive and subsequently pulsing the word line to Vdd. As the word line is pulsed high, the floating body capacitively couples and gates the thyristor turning the thyristor in the high conduction mode. Bit line is subsequently pulsed up, putting the access transistor into saturation and raising the potential of the floating node to Vdd-(Vt+Vref). Thus "1" is stored into the device. During this process, excess holes are generated in the floating body and get trapped into the trapping layer. The effect of the trapped positive charge, during standby, is to compensate any loss of positive charge of the floating node due to recombination effects, thus enhancing the stability of the storage "1" state. To write "0" to the device, Vref is held to ground, bit line is pulsed up followed by the word line so that the access device conducts and discharges the floating node. The thyristor is switched to reverse bias and as before, the word line pulse capacitively couples the floating body and gates the thyristor speeding up the turn-off speed of the thyristor. Subsequently, the bit line is pulsed negative to forward bias the n+/p diode of the access transistor and to inject excess negative charge into the floating body. The negative charging of the trapping layer has the effect of maintaining the thyristor as well as the n+/p- junction strongly reverse biased, during standby, enhancing the stability of the "0" state. The device and the trapping layer act in a similar fashion to the device proposed in reference [9].

The operational algorithm for writing and reading of the cross-point cell is summarized in Figure 22.23.

22.5 THE FET USUM CELLS

The stack design and band engineering of FET-based USUM cells has already been discussed in chapters 20 and 21 respectively. These cells are advanced nonvolatile memory cells and labeled by band-engineered gate stack designs employing higher K dielectric multilayered films for tunneling, charge trapping and storing, and charge blocking functions to enhance memory device properties in terms of MLC capability, programming performance, endurance, and nonvolatility. Stack designs and band engineering requirements to provide L1, L2, and L3 functionality have already been discussed. These USUM cells are direct tunnel memory-based cells and employ enhanced direct tunneling modes for higher performance charge transport and enhanced charge trapping/storage modes combined with effective charge-retention mode for nonvolatility and MLC capability. These

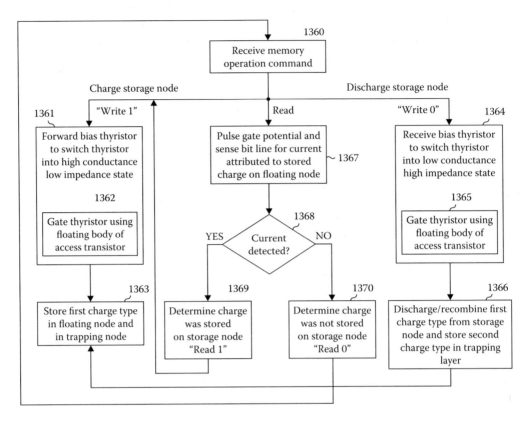

FIGURE 22.23 Operational algorithm for writing and reading of the cross-point high performance NV USUM cell of 22.4. (From Bhattacharyya, A., Embedded trap direct tunnel NVM and A vertical channel cross point high density split channel direct tunnel NVM and A very low power memory sub-system with progressive non-volatility, *ADI Associates internal publication*, U.S. Patents: 7276760 [10/02/2007], 7365388 [04/29/2008], 7385245 [06/10/2008], July–August, 2004.)

memory cells are generalized as charge trapping direct tunnel memory (CT-DTM) cells. Specific memory cells are often distinguished by differing concepts of direct tunneling charge transport as, for example, Direct-tunnel-Jct DTM (DTJ-DTM), progressive-band-offset DTM (PBO-DTM).

These memory cells are assumed to complement conventional memory cells such as SRAM, DRAM, NROM, and NAND in terms of functionality, read–write schemes, operability, and technology compatibility with advanced main-line CMOS technology. Additionally, these cells could be configured both as planar and as nonplanar (including vertical channel and 3D) cells as outlined in chapters 15, 16 and 18 for advanced silicon based NVM cells. In terms of array design layouts, these USUM cells follow similar approaches. Additionally, these memory cell concepts should be applicable to vertical channel 3D-nanowired arrays with gate-all-around configurations for future 10 nm design rules and beyond the ITRS road map. Therefore, discussion on the memory cells and arrays of CT-DTM USUM cells will be limited in this chapter.

22.5.1 Configurations and Operations of Memory Cells and Arrays for CT-NVM USUM Devices

CT-NVM USUM devices could be envisioned as one device and alternately a merged two-device NROM memory cell as well as a NAND flash cell element similar to other conventional NVM devices. These devices could be configured to be planar, FinFET, planar-vertical, and vertical-channel devices. We will briefly present some of options and possibilities of such memory cells and arrays in this subsection.

Planar one-device CT-NVM USUM NROM Cells: The planar single transistor USUM NROM cell could follow similar approach to that of the embedded trap DTM cell described in Part I with reference to VGA layout. Such a cell could be implemented either as an SLC (1 bit per cell) or as a DLC (2 bit per cell) utilizing localized trapping in GaN over the source and drain regions (see Section 2.2 in Chapter 2). Depending on the design objectives, the DLC cell could be applicable for both L2 and L3 applications nearly doubling the bit density (~3 F × F per bit). To improve the bit density, the source line could be buried and shared between two adjacent memory cells and may only be contacted with metal lines at specific intervals within the array to reduce series resistance. This scheme is shown in both Chapters 16 and 18 with reference to array layout examples.

22.5.2 Planar Merged Two-Device (Split-Gate) CT-NVM USUM NROM Cells

The single device conventional NROM cell is illustrated in Figure 22.24a [2,3,13]. Such a cell does not provide "over-erasure" protection discussed earlier. For performance-driven memory, the merged two-device USUM cells (or split-gate cells) follow configurations and layout considerations previously described in Chapters 15 and 18. For such memory cells with a fixed Vth access gate element in series with one or more NVM element(s) between the source and drain could be more desirable. It has been discussed in earlier parts of this book that split-gate cells would provide "over-erasure" protection (fixed Vth element defining the "erase state"), enhanced memory states stability, lower leakage and power requirements [12]. The split-gate planar USUM NROM cells could be configured with multiple variations outlined as follows:

1. Poly 1 gate for fixed Vth while overlapped TaN/Poly 2 NV element as shown in Figure 22.24b.
2. The reverse configuration with TaN/Poly 1 gate for NVM element with overlapped Poly 2 gate for fixed Vth as shown in Figure 22.24c.
3. Fixed Vth element (Poly 1) at the center of the channel with two-NV elements with common memory gate as shown in Figure 2.25a.
4. Fixed Vth element at the center of the channel with two different memory gates at either side (source diffusion and drain diffusion) while both overlapping the fixed Vth element

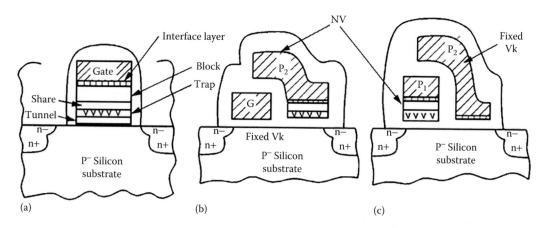

FIGURE 22.24 Planar CT-USUM NROM cell configurations: (a) One-device cell; (b) Two-device split gate with P2 memory gate; and (c) Two-device split gate with poly one memory gate. (From Bhattacharyya, A., Embedded trap direct tunnel NVM and A vertical channel cross point high density split channel direct tunnel NVM and A very low power memory sub-system with progressive non-volatility, *ADI Associates internal publication*, U.S. Patents: 7276760 [10/02/2007], 7365388 [04/29/2008], 7385245 [06/10/2008], July–August, 2004.)

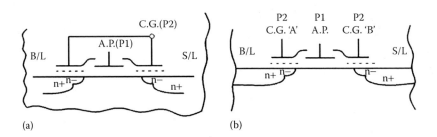

(a)   (b)

FIGURE 22.25 Planar CT-USUM NROM cell configurations: (a) 2D- split gate with access gate at the center and common memory gate [Dual Gated Cell] and (b) 2D- split gate with access gate at the center and two separate memory gates [Triple-gated Cell]. (From Bhattacharyya, A., Embedded trap direct tunnel NVM and A vertical channel cross point high density split channel direct tunnel NVM and A very low power memory sub-system with progressive non-volatility, *ADI Associates internal publication*, U.S. Patents: 7276760 [10/02/2007], 7365388 [04/29/2008], 7385245 [06/10/2008], July–August, 2004.)

(Poly1) as shown in Figure 22.25b. While the second and third options of the cell configurations could be used for L1 applications as SLC cells, the fourth and fifth options could be used for L2 and L3 applications providing MLC capabilities (2 bit per cell and even 3 bit per cell [8 memory states for 28(E)]) with appropriate stack designs as exemplified in Table 20.2. It should also be noted that with appropriate boron "Halo implant" and i-SRN gate interface, multiple mechanisms could be operative to enhance write or erase performances of USUM cells (MMCT designs).

22.5.2.1 Planar/Vertical and Vertical-Channel Single Device and Merged Two-Device (Split-Gate) CT-NVM USUM NROM Cells

The advantages of trench-based vertical channel and planar-vertical channel configurations in device scalability have been discussed in Part II, Chapter 6 of this book. Several illustrations of cells and array layouts have been provided for CT-NROM and NAND devices. Such configurations as well as FinFET and other nonplanar configurations (e.g., SGT) are equally applicable for CT-NVM USUM cells for NROM and NAND arrays to satisfy L1, L2, and L3 applications[27]. We will limit our discussion here with some illustrations of (a) planar-vertical NROM cell with access gate being planar in P1 and shared overlapping memory gate in P2, providing two NROM memory cells per trench either in SLC or DLC stack designs enhancing density and scalability. The design is illustrated in Figure 22.26a and also compared with an all vertical-channel design for the dual gate cells with common buried source line further improving density as illustrated in Figure 22.26b. It should be noted that these

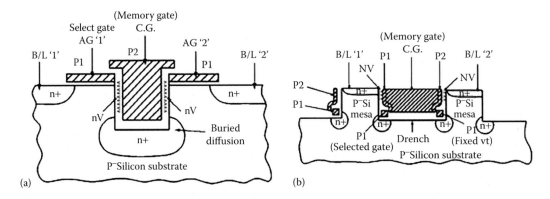

(a)   (b)

FIGURE 22.26 Planar-vertical and vertical-channel equivalent split-gate NROM USUM cells: (a) Planar-vertical and (b) Vertical channel.

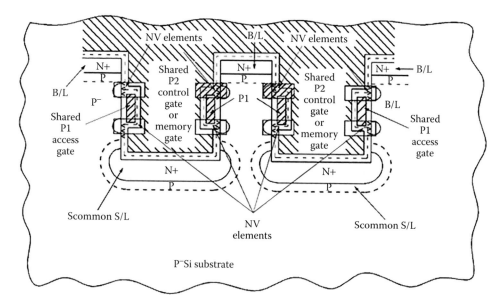

FIGURE 22.27 Vertical-channel configuration of NROM USUM cell of Figure 22.25 with shared WL, access gate, and buried SL. (From Bhattacharyya, A., Embedded trap direct tunnel NVM and A vertical channel cross point high density split channel direct tunnel NVM and A very low power memory sub-system with progressive non-volatility, *ADI Associates internal publication*, U.S. Patents: 7276760 [10/02/2007], 7365388 [04/29/2008], 7385245 [06/10/2008], July–August, 2004.)

cells are vertical-channel equivalents of the planar cell shown in Figure 22.24b. A vertical-channel configuration of the NROM USUM cell of Figure 22.25 is also shown in Figure 22.27 whereby both the fixed Vth access gate (P1) and the memory gate (P2) are shared between two NROM cells within a single trench. With appropriate array layout (see Chapter 16), and self-aligned contact schemes as well as shared bit-line approach with neighboring cells, such schemes could potentially yield <6 F2 per bit high performance SLC USUM design for L1 and <3 F2 per bit DLC design for L2 implementable in sub 20 nm CMOS technology nodes.

22.5.2.2 CT-NROM USUM Array Layout

In general, CT-NROM USUM array layouts are similar to those CT-NROM NVMs discussed in Chapter 16 of Part II of this book. For example, the array layout of the vertical-channel NROM USUM device of Figure 22.27 could be similar to those shown in Figure 16.5.

A typical conceptual array layout for the planar version of the MMCT USUM split-channel device is shown in Figure 22.28. The layout is in VGA architecture. Bit line is shared between adjacent memory cells to improve layout density. The fixed Vth select gate (P1) is placed at the center of the channel (Figure 22.25) and runs along the Y-direction parallel to bit-line and source-line directions. The word lines (P2) run along the X-direction as shown overlapping P1 select gates. The trapping regions and the stacks for the USUM NV memory elements at both diffusion edges are shown shaded underneath the word line 1 only. The bit lines and the word lines are contacted over the isolation regions to improve the array layout. The isolation in the X-direction is not shown in the picture which should be added in the word pitch.

22.5.2.3 Operational Schemes of CT-NROM USUM Cells

The operational schemes for most CT-NROM USUM cells are similar to those of CT-NROM NVM cells discussed in Part II. For one-device USUM cells based on DTM gate stack designs (see Table 3.3), while the voltage drop across the multilayered tunnel layers are kept sufficiently high (~1/3 of Vpp) during programming to ensure high electron fluence for high speed, the peak field is

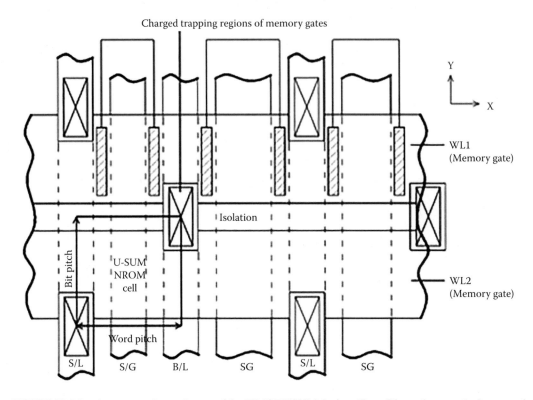

FIGURE 22.28 A conceptual array layout of the MMCT USUM device. (From Bhattacharyya, A., Integrated two-device multiple configuration alterable ROM/SRAM and Scalable integrated logic/NV memory technology with associated gate stack structure and process, *ADI Associates internal publication*, U.S. Patents: 7208793 [04/24/2007], 7544990 [06/09/2009], 7728350 [06/01/2010], June, 2004.)

kept below 7.5 MV/cm, nearly one-third of breakdown field to ensure dielectric integrity and device reliability. Additionally, the band-engineered tunnel design provides low enough storage field and high barrier energy to ensure low leakage to the substrate and gate to provide long retention. Low energy internal field aided transport provides EOL endurance without any device degradation. In general, the half-select devices are kept at nearly half (or lower) of the Vpp level to ensure neither program disturb nor read disturb. For L1 applications, the high Vth (written state) is targeted at +2 V, while the low Vth state (erased state) is targeted at 0 V (initial window, W = 2 V). The EOL window is targeted to be >1 V. Vread is targeted at +1 V. For L2 applications, similar considerations or operability are employed except the stack design yields larger memory window, longer retention when programmed at somewhat larger Vpp with longer duration. For L3 applications requiring EOL retention (>10 years), performance and endurance are traded against retention and stack designs are modified to achieve enhanced Fowler–Nordheim tunneling or multiple direct tunneling at lower field for writing yet asymmetric reverse tunneling and high energy barrier to minimize charge loss during storage and larger energy requirement for erasing. Consequently, Vpp and duration are raised for programming a larger EOT stack design to achieve larger memory window with reduced endurance. However, peak programming field is still maintained to be in low enough range compared to conventional NVM-NROM to ensure at least three orders of magnitude enhancement in retention for the USUM L3 devices. Additionally, the stack design provides larger memory window to enable multilevel storage for enhancing bit density and reduce cost per bit.

	Source Line	Bit Line	Select Gate	Program Gate	Vt(PG)
WRITE	+3.5 V	0 V	1.0/0 V	+4.5 V	+2.0 V
ERASE	+4.5 V	0 V	+3.5 V	+1.0 V	−1.5 V
READ	0 V	1.5 V	+1.0 V	+0.5 V	

FIGURE 22.29 Typical operational conditions for MMCT-USUM-NROM cells. (From Bhattacharyya, A., Integrated two-device multiple configuration alterable ROM/SRAM and Scalable integrated logic/NV memory technology with associated gate stack structure and process, *ADI Associates internal publication*, U.S. Patents: 7208793 [04/24/2007], 7544990 [06/09/2009], 7728350 [06/01/2010], June, 2004.)

For split-gate cells, the fixed Vth element is designed to target a Vth $\cong$ 0.5 V. This sets the Vth low state of the device regardless of the "erased" state of the memory element of the USUM cell. The stack design considerations are essentially similar to the above one-device USUM objectives with appropriate modification of memory window.

For split-gate high performance cells employing multiple carrier transport mechanisms (MMCT SUM cells), a typical operational condition is shown in Figure 22.29 to ensure both direct tunneling and hot carrier injection modes to be operative. Stack design considerations have been explained earlier for MMCT devices.

22.5.3 High Density CT-NAND USUM CELLS

Multiple concepts of high density CT-NAND-USUM cells could be envisioned based on the planar and nonplanar CT-NVM devices and arrays discussed in Part II, Chapters 5 and 6 before [3,13–26]. These include stacked planar single crystal and TFT devices and arrays [13,14]; multi-bit high performance arrays [15,16]; backside FETs and DSM-based MLC devices and arrays [17–20], as well as SGT-based devices and arrays [21–24]. These will not be discussed any further. A very high density split-channel device-based vertical-channel NAND USUM cell is discussed in Section 22.5.4 which has not been mentioned before.

22.5.4 Very High Density Vertical-Channel CT-NAND USUM Configurations

Vertical channel trench-based NAND designs have been briefly illustrated in Chapter 16 of Part II earlier [25,27]. The dual-gated and the triple-gated planar cell designs illustrated in Figure 22.28 could be reconfigured into vertical-channel SLC, DLC, and even TLC CT-NAND-USUM designs to significantly enhance bit density for L3 applications. Each trench could potentially provide four pairs of NAND string of DLC (16 bits) per trench, extending into 8 pairs and 16 pairs (with deeper trench) to potentially provide 32 bits and 64 bits, respectively, per trench. Since trench technology has been matured over the decades for DRAM applications, it is the author's contention that cost-effective CT-NAND USUM solutions for standalone high density USUM NVMs through the above approach could be viable in future decades when compared with full 3D-NAND approaches which appear to be more complex in high volume implementation. The dual-gated and the triple-gated (in DLC implementation could be 4 bits per cell: a total of 32 bits per trench) four pair NAND configurations are illustrated in Figure 22.30a and b, respectively.

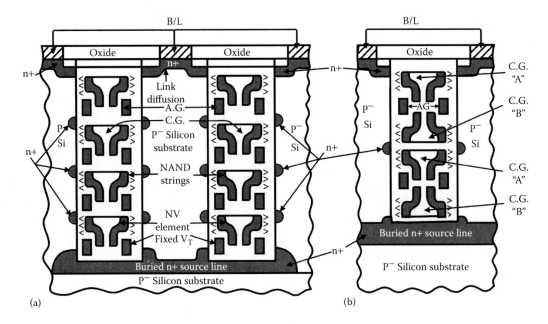

FIGURE 22.30 Vertical channel CT-NAND USUM configuration concepts: (a) Four-pairs of dual-gated DLC-NAND cells: 16 bits/trench and (b) four-pairs of triple-gated DLC-NAND cells: 32 bits/trench. (From Bhattacharyya, A., Extendable multi-level SGT-CT NROM and NAND flash devices with enhanced characteristics for multi-functional NVM technology applicable to 20 nm technology node and beyond, Internal Memo, ADI associates, April 12, 2013.)

REFERENCES

1. A. Bhattacharyya, Memory cell with trenched gated thyristor, (original Title: A cross point stable memory cell using trenched gated-thyristor), *ADI Associates internal publication*, April 30, 2004, U.S. Patents: 7145186 (12/05/2006), 7440310 (10/21/2008).
2. A. Bhattacharyya, Embedded trap direct tunnel NVM and A vertical channel cross point high density split channel direct tunnel NVM and A very low power memory sub-system with progressive non-volatility, *ADI Associates internal publication*, July–August, 2004, U.S. Patents: 7276760 (10/02/2007), 7365388 (04/29/2008), 7385245 (06/10/2008).
3. A. Bhattacharyya, Scalable high performance non-volatile memory cells using multimechanism carrier transport, *ADI Associates internal publication*, November, 2004, U.S. Patent: 7244981 (07/17/2007), 7553735 (06/30/09).
4. A. Bhattacharyya, R. S. Bass, and G. D. Grise, Non-volatile memory cell having silicon-rich nitride charge trapping layer, IBM Internal Technical Reports: 1984; U.S. Patent: 4870470 (09/26/1989).
5. A. Bhattacharyya, A novel one-device non-volatile DRAM device concept and structure, *ADI Associates internal publication*, May, 2002, U.S. Patent: U.S.6903969 (6/7/2005), 7130216 (10/31/06), 7485513 (02/03/2009).
6. A. Bhattacharyya, One transistor SOI non-volatile random access memory cell, *ADI Associates internal publication*, July 22, 2002, U.S. Patent: 6888200 (5/3/2005), 6917078 (7/12/2005), 7184312 (2/27/2007), 7339830 (03/04/2008), 7440317 (10/21/2008).
7. A. Bhattacharyya, Scalable flash/NV structures and devices with extended endurance, *ADI Associates internal publication*, August, 2001, U.S. Patents: 7012297 (3/14/2006), 7250628 (7/31/2007), 7400012 (7/15/2008), 7750395 (7/6/2010).
8. A. Bhattacharyya, High performance one-transistor memory cell, original title: High performance capacitor less one transistor memory devices in bulk and SOI technology, *ADI Associates internal publication*, July 2, 2003, U.S. Patent: 7432562 (10/07/2008), 7625803 (12/01/2009), 7968402 (06/28/2011), 81255003 (02/28/2012).
9. A. Bhattacharyya, SOI non-volatile (self refreshed) two-device SRAM using floating body charge, *ADI Associates internal publication*, May, 2003, U.S. Patent: 7224002 (05/20/2007).

10. F. Nemati, and J. D. Plummer, A novel thyristor-based SRAM cell (T-RAM) for high speed, low voltage, giga-scale memories, *IEDM*, Washington DC, 1999, pp. 283–286.

11. A. Bhattacharyya, Gated lateral thyristor-based random access memory cell (GLTRAM), *ADI Associates internal publication*, June, 2002, U.S. Patent: 7042027 (5/9/2006), 746054 (11/25/2008).

12. A. Bhattacharyya, Integrated two-device multiple configuration alterable ROM/SRAM and Scalable integrated logic/NV memory technology with associated gate stack structure and process, *ADI Associates internal publication*, June, 2004, U.S. Patents: 7208793 (04/24/2007), 7544990 (06/09/2009), 7728350 (06/01/2010).

13. S.-M. Jung, J. Jang, W. Cho et al., Three-dimensionally stacked NAND flash memory technology using stacking single crystal Si layers on ILD and TANOS structure for beyond 30 nm node, *IEDM*, 2006, pp. 37–40.

14. E.-K. Lai, H. T. Lue, Y.-H. Hsiao et al., A multi-layer stackable thin-film transistor (TFT) NAND-type flash memory, *IEDM*, San Francisco CA, 2006, pp. 41–44.

15. H. T. Lue, S.-Y. Wang, E. K. Lai et al., A novel P-channel NAND-type flash memory wit 2 bit/cell operation and high programming throughput (9 > 20 Mb/sec), *NVSMW*, 2005.

16. A. Bhattacharyya, A novel stack design concept for charge trapping NVM cell with band engineered tunneling and trapping layers providing large memory window and 3bit/cell MLC capability, ADI associates, internal memo, August 2007 (unpublished).

17. R. Ranica, A. Villaret, P. Mazoyer et al., A new 40 nm SONOS structure based on backside trapping for nano-scale memory, *IEDM* (*"ST Microelectronics"*), 2004, pp. 99–102.

18. A. Bhattacharyya, ref. 5.52: A. Bhattacharyya, Back-sided trapped non-volatile device, *ADI Associates internal publication*, July 12, 2004; U.S Patent Publication 257324, December 11, 2008, U.S. Patents: 7749848 (7/6/2010) and 7851827 (12/14/2010).

19. C. W. Oh, N. Y. Kim, S. H. Kim et al., 4-bit double SONOS memories (DSMs) using single level and multi-level cell schemes, *IEDM*, San Francisco CA, 2006, pp. 967–970.

20. A. Bhattacharyya, A novel 4 bit/cell dual channel MNSC device, ADI associates, internal memo, September 2007 (Unpublished).

21. A. Bhattacharyya, A novel low power battery operated non-volatile memory device and associated gate stack, *ADI Associates internal publication*, September 4, 2004, (MIDD 45), U.S. Patent: 7612403 (11/03/2009).

22. A. Bhattacharyya, Integrated surround-gate multifunctional memory device, *ADI Associates internal publication*, May 26, 2004.* Patent Application abandoned.

23. S. J. Whang, K.-H. Lee, D. G. Shin et al., Novel 3D dual control gate with surrounding floating gate (DC-SF) NAND flash cell for 1 Tb file storage application, *IEDM*, 2010, pp. 668–671; and S. J. Whang, K.-H. Lee, D. G. Shin et al., Novel 3-dimensional dual control-gate surrounding-gate(DC-SF) NAND flash cell for 1Tb file storage application, *IEDM*, San Francisco CA, 2010, pp. 29.7.1–29.7.4. [Hynix]

24. A. Bhattacharyya, Very high density trench based dual-gated and triple gated CT-USUMNAND designs and implementation concepts, Internal Memo, ADI Associates, April 9, 2013.

25. A. Bhattacharyya, Extendable multi-level SGT-CT NROM and NAND flash devices with enhanced characteristics for multi-functional NVM technology applicable to 20nm technology node and beyond, Internal Memo, ADI associates, April 12, 2013.

26. R. Ohba, Y. Mitani, N. Sugiyama et al., 35 nm floating gate planar MOSFET memory using double junction tunneling, *IEEE NVSMW*, Washington DC, 2005.

27. A. Bhattacharyya, Embedded trap direct tunnel NVM and A vertical channel cross point high density split channel direct tunnel NVM and A very low power memory sub-system with progressive non-volatility, *ADI Associates internal publication*, July–August, 2004, U.S. Patents: 7276760 (10/02/2007), 7365388 (04/29/2008), 7385245 (06/10/2008).

23 Multifunctional SUM
The MSUM Cells and Arrays

CHAPTER OUTLINE

MSUM cells provide simultaneously multifunctional memory attributes within the framework of a single memory cell. Such multifunctionality could be employed to meet requirements of conventional SRAM, DRAM, NROM, and NAND memory array requirements currently provided through different cell and array designs with and without the CMOS logic process/integration technology compatibility. We discuss in this chapter several cell and array design concepts which could potentially replace conventional memories by providing simultaneously two or more traditional memory cells and array functions as mentioned above while providing desired nonvolatility. Additionally, we provide here several MSUM cell examples to achieve traditional DRAM and NAND/NROM functionality following the principles of extendibility discussed in Part II of this book for extendable silicon based NVMs.

DRAM technology not only made an incredible stride and established a pivotal role as working memory and main memory in digital systems but also drove the technology scaling for over three decades. However, data retention in DRAM has been in the order of milliseconds and requires frequent refresh and the data often needs to be backed up using slower "nonvolatile" memories such as NANDs and HDDs reducing system performance and consuming additional energy in the process. There has always been a desire to seek development of an alternative to DRAM with equivalent performance with desired nonvolatility. Thus far, silicon FET-based nonvolatile memories did not meet such possibility due to very limited endurance and slower performance (slower access and cycle time due to orders of magnitude slower programming speed).

Two important developments provided the pathways toward solving the limitations on speed and performance. The first one is related to the development of energy efficient DTM memories whereby significantly higher speed (by nearly four orders of magnitude) electronic charge transport could be achieved at modest average field; and the second one is related to the understanding of the relationship between the energy transfer of electronic charges during writing and erasing and the generation of stress-induced defects and subsequent degradation in interfacing dielectric films. It was clearly established that this second aspect adversely affects the dielectric film integrity, causes enhanced leakage (and eventual breakdown), and limits the device endurance. The application of band-engineered multilayered DTM-based devices enable electronic charge transport at significantly lower energy levels during writing and erasing, transferring little or no energy to the interfacing dielectric film, thereby ensuring high endurance and reliability of the device. Several SUM devices are built on such concept to provide unlimited endurance while providing DRAM-like performance. Stack designs and band engineering for multiple FET-MSUM DTM NVM devices have been discussed and identified in previous chapters. We have discussed different MSUM cells and arrays covering the above approaches and others in section 23.1 and subsequent sections in this Chapter.

We shall first discuss a type of MSUM cell which integrates a vertical channel DRAM gain cell with a traditional NVM cell. Subsequently, we shall describe several band-engineered DTM-based CT-MSUM cells as suggested above. We shall also reference recently published multilayered conductive floating-gate/floating-plate MSUM stack designs demonstrating simultaneous DRAM/NVRAM functionality. The chapter will finally cover URAM MSUM cells.

23.1 INTEGRATED DRAM-NVRAM MULTILEVEL AND MULTIFUNCTIONAL MSUM CELL

Concept: The original concept is based on integrating a single transistor vertical memory DRAM gain cell [1] with a shared vertical gate floating-plate device [2,3] to provide not only enhanced charge storage for the DRAM part of the cell but also provide high endurance NVRAM capability with high density and multilevel nonvolatile storage capability. The cell could also be used as elements of FPGA as well as elements for alterable/reconfigurable logic. The MSUM cell is built on the above concept with the band-engineered multilevel nonvolatile element (NVRAM) which further enhances performance, endurance, and application of such cells. The cell is multifunctional in the sense that the DRAM part could be used as a regular working memory and with appropriate algorithm the data or instruction could be permanently stored. Additionally, with multilevel capability, such cell could also store BIOS or appropriate code to carry out other functions such as reconfigurable logic or PLA functions. The original concept of the memory cell and the added enhancements are described below.

The MSUM cell: A vertical "capacitor-less" DRAM gain cell had been previously proposed by L. Forbes [1] which is a vertical channel switching FET formed on one of the side wall of the silicon mesa integrated with a floating buried diffusion node and a body polysilicon gated MIS capacitor formed on the opposite side of the silicon mesa. The MSUM cell replaces the MIS capacitor with a nonvolatile element containing a TaN-Metal-gate with blocking layer/trapping-charge storage layer or layers and appropriate band-engineered tunnel layer (or layers) integrated with the DRAM switching transistor through the buried diffusion node. Furthermore, the nonvolatile memory control gate formed inside the trenched silicon substrate is also shared with the neighboring MSUM cell. A schematic cross-section of a pair of the MSUM cell is shown in Figure 23.1a and the equivalent electrical circuit for the memory cell is shown in Figure 23.1b respectively. It should be noted that p-body of the cell is made isolated from the p-substrate in the bulk silicon implementation of the cell. In the SOI implementation, a buried-oxide layer isolates the p-floating body from the substrate. The SOI embodiment is shown in Figure 23.2. The stack designs for the

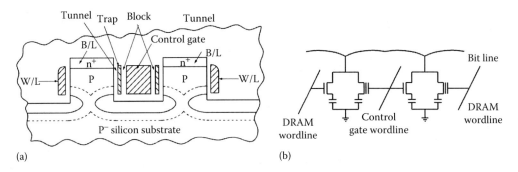

(a)   (b)

FIGURE 23.1 Integrated DRAM-NVRAM MSUM cell in bulk silicon: (a) Schematic cross-section and (b) equivalent circuit. (From Bhattacharyya, A. et al., Integrated DRAM-NVRAM multilevel memory cell, *ADI Associates internal publication*, U.S. Patents: 7158410 [01/02/2007], 7349252 [03/25/2008], 7379336 [05/27/2008], 7403416 [07/22/2008], 7417893 [08/26/2008],7459740 [12/02/2008], 7457159 [11/25/2008], May 5, 2004.)

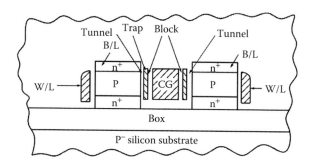

FIGURE 23.2 SOI implementation of integrated DRAM-NVRAM MSUM cell: Schematic cross-section. (From Bhattacharyya, A. et al., Integrated DRAM-NVRAM multilevel memory cell, *ADI Associates internal publication*, U.S. Patents: 7158410 [01/02/2007], 7349252 [03/25/2008], 7379336 [05/27/2008], 7403416 [07/22/2008], 7417893 [08/26/2008], 7459740 [12/02/2008], 7457159 [11/25/2008], May 5, 2004.)

TABLE 23.1

Operational Conditions for MSUM Cells for Multilevel Storage

V-B/L[a]	V-W/L[a]	Vsl[a]	V-CG[a]	Remark
2.5	2.5	Gnd	Gnd	DRAM-Write "0"
Gnd	0.8	−2.5	Gnd	DRAM-Write "1"
Float	0.8	Gnd	Gnd	DRAM-read "0" (vt + 0.5)
Float	0.8	Gnd	Gnd	DRAM-read "1"[a]
2.5	2.5	Gnd	−2.5	[b]NVRAM-Write "0"
Gnd	0.8	−2.5	+2.5	[b]NVRAM-Write "1"
Float	0.8	Gnd	Gnd	[b]NVRAM-Write "A0"
Float	0.0	Gnd	1.2	[b]NVRAM-Read "B0", Vt = 1
Float	0.8	Gnd	Gnd	[b]NVRAM-Read "A1"[a]
Float	0.0	Gnd	1.2	[b]NVRAM-Read "B1"
Gnd	Gnd	+9[a]	(Vx − cg) (10 ms)	NVRAM-Write "1x"[c]
Gnd	Gnd	+9[a]	Gnd	NVRAM-write "0x"@

[a] B/L potential unchanged, none to little charge transfer.

[b] Virtual 2-bit store.

[c] Electron trapping, Vt (1x) >2 V.

memory element of the cell follows a similar scheme as previously proposed in GBE-CT-DTM USUM cell or cells (Table 20.5). This is provided in Table 23.1 as shown below to tailor different targets of NVM properties with i-SRN floating charge storage plate as originally proposed [3, 7, 9]. Additionally, the NV element could be either SLC or MLC depending on the memory window design and retention targets.

23.1.1 Multifunctionality

When the control gate is held to ground, the cell works like a DRAM similar to the one explained in Reference 8 of Chapter 22, with an important difference: When the device is in high conductance (write "0"), some of the excess hole charge in the floating body (generated by impact ionization) tunnels through the trapping layer and gets trapped. Consequently, the device conductance is further increased creating a "fat 0." Conversely, when electrons are generated in the floating body to create the lower conductance state (write "1"), some of the excess electrons gets trapped into the trapping layer with the consequent effect of yet lower conductance, "fat 1." Thus, the effect of the trapping layer would be to improve the logic separation of the DRAM memory state and/or state retention.

For NVRAM operation, the control gate is pulsed to a negative potential concurrent to pulling up to Vdd, both word line and bit line driving the access device to saturation. Strong lateral field between the floating body and the control gate drives excess holes generated into the body to tunnel through the tunnel oxide and gets trapped. Due to this hole trapping, the adjacent body potential is raised positive and held permanently until the trapped state is discharged (by trapping electrons). This is nonvolatile "zero" state and can be sensed readily (Read "0") by turning the access device word line up and sensing the current through the bit line. To write the nonvolatile "one" state, the control gate is pulsed positive concurrent to forward biasing either the drain-body (Figure 23.1a) or the source-body (Figure 23.2) diodes to inject excess electrons into the floating body. The trapping layer traps excess electrons and gets raised to a permanent negative potential. As a result, the access device Vt is raised and the device does not conduct during Read "1". The device remains in the nonvolatile "one" state until the trapping layer electrons are neutralized by injecting holes.

For multilevel NVRAM operation, the above nonvolatile "0" and "1" states can be addressed or read either by the word line of the access device (corresponding to the Vt-wl "0" and Vt-wl "1," respectively) or by the control gate device (corresponding to the Vt-cg "0" and Vt-cg "1," respectively). For the same degree of charge storage in the trapping layer, the control gate Vts would be significantly different than those of access device Vts and, therefore, bilevel addressing could be achieved and the device achieves virtual "dual-bit" storage for the same written state. Additional multilevel nonvolatile storage could be achieved, by directly storing increasing density of charges (e.g., electrons) into the trapping layer by programming the control gate conventionally with increasing programming voltages to generate multiple high Vt states (multiple Vt-cg "1" s) and addressing by means of both word line and control gate and establishing appropriate sensing schemes to separate all levels of storage states.

For field programmable operation as well as for alterable logic applications, appropriate electron charge density or hole charge density could be stored into the trapping layer by programming via the control gate converting the device into a PROM. The device could, therefore, be used not only for updating BIOS but also as "ALTERABLE SWITCH," at appropriate logic nodes for FPGA and reconfigurable logic applications.

23.1.2   Operational Attributes

The above MSUM cells require significantly lower field across the dielectric stack for programming via the control gate for L1, L2, or L3 implementations dependent on stack EOT and stack design as exemplified in Table 23.1. Consequence of the lower programming field is the increased endurance capability for such devices and scalability both in geometry and voltages. This has been discussed earlier. Example of operational voltages is provided below for a relatively simple gate insulator stack consisting of 4.5 nm SiO_2 as tunnel insulator, 6 nm Silicon-Rich-Nitride [14] as trapping dielectric, and 6.5 nm SiO_2 as charge blocking layer (EOT ~12 nm) requiring a programming voltage below ~ 9 V. By appropriate gate insulator stack design employing higher K gate insulators, the programming voltage level could be further scaled as explained earlier.

The above logic separation between "1x" and "0x" could be used for multilevel store; the programming voltage, Vx-cg could be altered to Vy-cg to create a different logic level separation "1y" and "0y" for multilevel store. Similar approach could also be employed for PROM "1" and "0" writing.

It should be noted that the operational characteristics are dependent on the MSUM stack design for the NV element. Therefore, for a band-engineered L3 stack design of Table 23.1 with a stack structure of, for example, 2.5 nm OR-SiON + 1.5 nm i-SRN + 6 nm HfSiON (tunnel)/5 nmGaN + 5 nm i-SRN (Trap and storage)/4 nm Al_2O_3 + 6 nm HfSiON (Block); with an effective EOT ~ 9 nm, would require Vpp 7.5 V, 1 ms for programming to achieve >6.0 V memory window and DLS capability with enhanced NVM attributes. Additionally, the switching NFET device operation is predicated with 2.5 nm EOT and Vdd = 1.5 V as compared to the above MOS-FET example of EOT of 4.5 nm and Vdd = 2.5 V. Therefore, the operational levels of V-BL, V-CG, etc., would change accordingly.

23.1.3   Array Layout

Array layout of the above cell could be considered to be similar to that of vertical channel CT-NROM USUM array layout of the cell shown in Figure 22.27 in the previous chapter. This in turn has been discussed in Chapter 16, Part II in the context of array layout for the vertical channel "shared word line" NROM devices. The above FSUM device shares the control gate between two adjacent FSUM cells analogous to the shared word line for the NROM cell discussed in Part II. Here the layout of the DRAM word lines would be equivalent to those of vertical channel access gates for the NROM array discussed in Part II. Therefore, no further elaboration of the array layout is made.

23.2   THE BAND-ENGINEERED DTM MSUM CELLS AND ARRAYS

Discussed previously in the context of stack design (Chapter 20) and band engineering (Chapter 13), these MSUM cells provide built-in multifunctionality within each cell. In principle, functionality of SRAM, DRAM, and NROM could all be achievable within an MSUM cell through appropriate stack design and band engineering. Such capability makes MSUM cells very attractive for future applications. We have provided several stack design examples of such cells in Chapter 20 and illustrated band diagrams for several device design options in Chapter 21. We shall briefly describe cell functionality and array design concepts in this chapter. It should also be mentioned that novel and unique functionality could be achieved through integrating both logic and memory functions within such memory cells, thereby creating potentially "smart memory subsystems" by appropriate design of programmable logic arrays using MSUM devices.

Separately, we shall briefly outline band-engineered nanocrystal-based MSUM device design concepts and as well as reverse mode MSUM device design concepts.

23.2.1   General Considerations for DTM MSUM Cell Design

Four CT-MSUM cell types distinguished by the respective stack design have been described in Chapters 20 and 21 with reference to stack design and band engineering, respectively. To reiterate the cell types, Type 1 is based on PBO or Variot barrier-tunneling schemes, Type 2 is based on crested barrier tunneling schemes, and Type 3 and Type 4 provide multifunctional and multilevel capabilities combined.

In defining MSUM cells of Type 3 and Type 4, the following additional considerations were made:

1. The lowest Vth of the memory state would be the reference state with maximum stability through the end of life. This is best achieved through the split-channel device design where the lowest Vth state is defined the threshold of the fixed Vth (typically NFET) CMOS FET design.
2. The writing a multifunctional MSUM cell for SRAM or DRAM or NROM/NAND functionality is always preceded by a "FULL ERASE" to ensure the device starts from the reference low Vth state.

The L1 and L2 functionality programming conditions are separated and uniquely defined from NROM/NAND programming and memory window spectrum to ensure clean and unambiguous sensing of such memory states as distinct ones in comparison to the NROM/NAND SLC and MLC memory window without the possibilities of overlap and RBER.

Many options may be considered in providing multifunctionality and MLC operability for band-engineered CT-MSUM device families. Stack design examples of such memory cell types were listed in Table 20.3 and device parametric objectives for such cell types were shown in Table 20.4. Multilayered tunnel barrier of different types combined with single and/or multilayered trapping dielectric were considered to derive multifunctionality and MLC capability. Options 1.1, 1.2, and 2.3 of Table 20.3, for example, provide simultaneous SRAM and DRAM (L1/L2) functionality, while options 1.3, 1.4, 1.5, 2.1, and 2.2 provide simultaneous DRAM and NROM [or NAND] (L2/L3) functionality. Finally, options 1.6 and 1.7 were examples to provide simultaneous multifunctionality of SRAM, DRAM, and NROM [NAND] within a single memory cell. MLC operability was also noted in several options including, for example, option 2.4.

Several band diagrams were also illustrated for CT-MSUM device families in Chapter 21. These include PBO tunnel layer design in Figure 21.6 (stack design option 1.1), Variot tunnel layer design in Figure 21.7 (stack design option 1.3), and crested tunnel barrier designs in Figures 21.8 and 21.9 (stack design options 2.1 and 2.3, respectively), with potential MLC capabilities.

Additional design considerations for MLC NROM and NAND applications are discussed below.

23.2.2 MSUM Cell Band Characteristics for NROM and NAND Applications

Two MSUM device types are mentioned with MLC NVM capability for NROM/NAND applications in Table 20.3, one with built-in L2 functionality (DRAM) defined as Group 3, while the other one with built-in L1 and L2 functionality (SRAM and DRAM) defined as Group 4, CT-MSUM devices. Both are of DTM-PBO/Variot tunneling design to take advantages of barrier thinning and internal field-induced transport enhancement of charge carriers. However, as mentioned above, in the context of crested barrier MSUM designs, MLC NROM/NAND designs could also be designed with crested barrier tunneling structures with a different set of performance/retention trade-offs taking advantage of barrier lowering.

Some of the key design considerations for Group 3 and Group 4 devices are (a) large memory window for MLC capability; (b) virtually no leakage to the gate during standby; (c) large sensing margin between the submemory states to clearly distinguish the states at the EOL time frame at worst case operating conditions; (d) well separated memory states from SRAM/DRAM types of states dictated by significantly faster programming transitions and the general considerations highlighted in the beginning of the section should be addressed into the CT-MSUM stack designs.

Band diagram of Group 4, CT-MSUM device is illustrated in Figure 23.3. The two-layered PBO barrier contains an integrated ~1–1.5 nm in-SRN layer defining the size of nanocrystals. This layer provides the required retarding field for the first set of charges trapped into the first nitride (layer C) to meet the objectives of speed and retention for the L1 functionality. The second nitride layer (layer E)

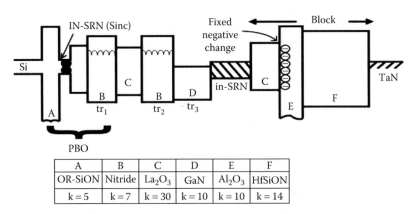

A	B	C	D	E	F
OR-SiON	Nitride	La_2O_3	GaN	Al_2O_3	HfSiON
k = 5	k = 7	k = 30	k = 10	k = 10	k = 14

FIGURE 23.3 Band diagram for CT-MSUM device Type 4 with MLC capability.

TABLE 23.2

Operational Conditions for DLS NROM (or NAND) MSUM Cells

NROM	L1	L2	
Vth Reference: "0" or "00" V	0.5 ± 0.2	0.5 ± 0.2	0.5 ± 0.2
Vth "1" or "11": INI(initial)/EOL V	2.0 ± 0.35	3.0 ± 0.5	7.0 ± 1.5
Vth: "01" V		4.5 ± 0.9	
0.25 V	0.75 V		
Sense "0"/"00" (off) and "on"	0.75 V	1 V	0.75 V, 1 V
Sense "1"	1.5 V	2.25 V	2.75/3.5/4.5/5.25/5.5[a]

[a] Sense "10"/"01"/"11" (current sensing and algorithmic sorting).

separated by a well-defined tunneling dielectric layer of La_2O_3 defines the memory window for the L2 functionality and required retention. While the programming characteristics for L1 functionality is defined by direct tunneling of electrons (writing) and holes (erasing), the programming characteristics for the L2 functionality are defined by additional tunneling only of electrons into the second nitride trapping layer (writing) and detrapping and tunneling back of electrons from the second nitride layer from the trapping centers to the silicon substrate. The resulting programming characteristics for L1 and L2 functionality are distinct and separate and can be sensed appropriately to provide the required functionality and distinctly different sense margins and memory window.

The multilevel NROM (or NAND) element operates at higher programming voltage level and longer time duration shown in Table 23.2; for example, Vth "1" is 3.0V for single level storage whereas Vth "11" is 7.0 V for 2-bit storage (DLS, as shown in Table 23.2). The two memory states uniquely define the DLS characteristics of the NROM element, and are significantly higher to those of L1 and L2 elements. Examples of various Vth states for the MSUM device are given with appropriate sense margins to illustrate the operability of such device (Table 23.2).

The multilevel NROM (or NAND) element operates at higher programming voltage level and longer time duration. The two memory states uniquely define the DLS characteristics of the NROM element, for example, are significantly higher to those of L1 and L2 elements. Examples of various Vth states for the MSUM device are given below with appropriate sense margins to illustrate the operability of such device (Table 23.2).

23.2.3 CT-MSUM Cell and Array Configurations

Both planar and nonplanar MSUM cells could be envisioned similar to those of planar and nonplanar CT-NVM devices discussed in Part II of this book. It is important to recognize that the split gate cell designs have distinct advantages to meet programming and sensing requirements for the MSUM designs especially when designed to meet simultaneously L1, L2, and MLC NROM/NAND requirements as well as low-power embedded requirements [16]. For higher performance requirements, VGA NROM array configuration is preferred. A typical layout and circuit schematics are shown in Figure 23.4a and b, respectively [10] for the RAM array. For serial access, conventional NAND layout and schemes described earlier apply as well [8]. Vertical channel adaptations of MSUM cells and arrays including surrounding gate NROM and NAND cells [4,13] should follow similar approach as discussed in Part II in the planar and nonplanar sections.

It is the author's perspective that although full 3D cells for NVM technology are being actively pursued in recent time for sub 20 nm technology node, cost-effectiveness for such approach is not clear. Vertically stacked planar cells and trench-based vertical channel cells and arrays can be cost-effective especially with MLC NROM and NAND arrays could prove to be cost-effective especially with MLC NROM and NAND arrays. USUM and FSUM designs in stacked planar and vertical

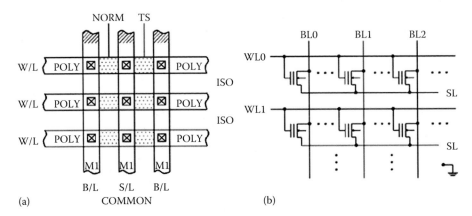

FIGURE 23.4 Typical layout and circuit schematic for MSUM NROM arrays: (a) Layout and (b) circuit schematic. (From Bhattacharyya, A., Discrete trap non-volatile device for universal memory chip application, ADI Associates internal publication, U.S. Patents: 7436018 [10/14/2008], 7476927 [01/13/2009], 7786516[10/14/10], 8143657 [03/27/2012], 8242554 [08/14/2012], May 29, 2005.)

channel configurations have the potential to effectively cover the entire performance-capacity spectrum of the memory requirements for the industry in foreseeable future.

23.3 OTHER CT-MSUM DEVICES

Several other CT-MSUM devices could be proposed based on band engineering and replacement of specific trapping dielectrics by either (a) graded composition of multidielectrics [8], or (b) nanocrystal-induced trapping centers [5–7,10–12], or (c) reverse mode device designs employing different trapping media [6,10]. Novel NVM devices with DRAM and NROM/NAND multifunctional capability based on graded composition of SiC and GeC with embedded nanocrystal trapping planes have also been proposed [8]. Such devices may also be considered for MSUM applications.

Multifunctional CT-NVMs based on nanocrystal (NC)-based trapping centers have been proposed as early as 2004 [4–6,9]. Examples of combining nanocrystal planes (embedded trap) with charge trapping dielectrics (discrete traps) to achieve desired MSUM devices have also been proposed [7,10]. Subsequently, all NC-based multifunctional and multilevel NROM/NAND designs were also proposed [11–13]. It should be apparent that many concepts of extendibility of CT-NVMs could be applicable to CT-MSUM devices through judicious combinations of trapping dielectric and trapping centers to enhance the application potentials of SUMs in future timeframe.

23.3.1 Nanocrystal-Based CT-MSUM Device

A nanocrystal-based CT-MSUM device with capability of multilevel DRAM (DLC-DRAM) combined with multilevel NROM (DLC-NROM) within the same memory cell will be reviewed here for illustration [12].

Other nanocrystal-based CT-MSUM devices are provided in the references mentioned above. A normal mode version of the device will be described here. The device could be configured and operated in the reverse mode which has also been described in detail in the reference and will not be discussed here. The device embraces the following features:Direct Tunneling is assumed for efficient carrier transport. Band-engineering concepts elaborated in

Chapter 13, Part II forms the basis of stack design to achieve the desired functionality.

For the DRAM functionality: An ultrathin layer of high K dielectric film with characteristic lower energy barrier interfacing the silicon substrate provides direct tunnel high-performance transport of electrons to the trapping region. Multi-level storage is enabled by incorporating trapping layers with two well- defined work functions and   trapping charges in two different trap depths induced by nanocrystals incorporated in dieletric medium. In this manner, 4 memory states for DLC-DRAM could be achieved. Reverse tunneling is contained to achieve retention of >100 sec. The design provides infinite endurance for the DRAM element.

For the NVM functionality: Two additional planar stacks are incorporated separated by larger tunnel distance containing appropriate nanocrystals for NVM trapping to provide DLC capability such that a high energy dielectric barrier separates the DRAM element from the NVM element to provide appropriate isolation between the two elements. The NVM element contains higher work function metal nanocrystals of progressively larger sizes to reduce quantum confinement and to enhance retention with progressively larger physical distance from the substrate to achieve distinct and multiple memory states with low Vth dispersion through planar coulomb blockade [7]. The NVM element could emulate the DRAM memory states through appropriate program algorithm or alternately could operate as a nonvolatile memory only. The NVM element features a double-layered PBO barrier which provides faster electron transport and yet asymmetric tunnel barrier [10] to provide many orders of magnitude enhanced retention than conventional SONOS. Other features include high K blocking and different programming schemes for DRAM element and NVM element to ensure different programming characteristics and charge trapping and detrapping for DRAM and NVM elements. For the NVM element, tunnel distance, trap density (nanocrystal size and distribution), and trap depths (nanocrystal work function) are optimized to achieve the desired memory states and programming conditions.

Figure 23.5 shows the schematic cross-section of the gate stack of a one-transistor planar memory cell (NROM) consisting of an ultrathin (~2 nm) layer of HfO_2 tunnel layer for the DRAM element with an ultrathin (~1.5 nm) SiO_2 layer containing both W and Pt nanocrystals with work functions of 4.6 eV (Wf1) and 5.3 eV (Wf2), respectively. The stack design contains of three layers of HfSiON films with embedded nanocrystals of either Platinum or Tungsten. The thickness of the HfSiON layer defines the size of the nanocrystal diameters of W and/ or Pt. As the Figure 23.5 suggests, successive HfSiON layers of ≥3.5 nm (second layer) and >= 4.5 nm (third layer) is deposited over the first plane of nanocrystals as defined by the thickness of the specific HfSiON layers. The final trapping layer contains a larger diameter (~5 nm) of Pt nanocrystals as shown. The blocking layer is also a film of HfSiON of thickness ranging from 7 nm to 10 nm. A TaN gate is considered for such device. The larger W and Pt nanocrystal

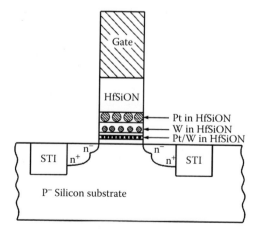

FIGURE 23.5   Schematic cross-section of a single transistor nanocrystal-based CT-MSUM device. (From Bhattacharyya, A., Embedded trap nano-crystal multi-level-cell device for DRAM and universal memory, Updated Title: Embedded trap nano-crystal multi-level-cell device universal memory, *ADI Associates internal publication*, U.S. Patent. 7759715 [07/30/2010], 8193568 [06/05/2012].)

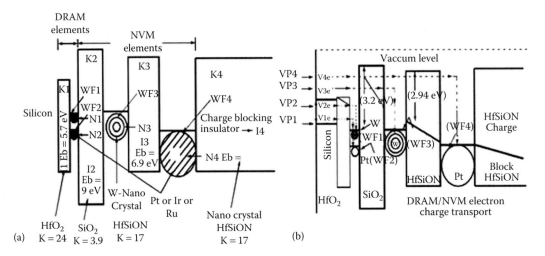

FIGURE 23.6 Band diagrams for the device of Figure 23.5: (a) Vg = 0 and (b) at different Vpp for MLS capability of DRAM and NVM elements. (From Bhattacharyya, A., Embedded trap nano-crystal multi-level-cell device for DRAM and universal memory, Updated Title: Embedded trap nano-crystal multi-level-cell device universal memory, *ADI Associates internal publication*, U.S. Patent. 7759715 [07/30/2010], 8193568 [06/05/2012].)

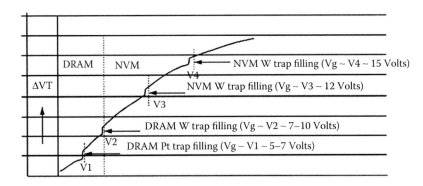

FIGURE 23.7 Change in Vth during writing at different Vpps (Vgs) for the device of Figure 23.5. (From Bhattacharyya, A., Embedded trap nano-crystal multi-level-cell device for DRAM and universal memory, Updated Title: Embedded trap nano-crystal multi-level-cell device universal memory, *ADI Associates internal publication*, U.S. Patent. 7759715 [07/30/2010], 8193568 [06/05/2012].)

containing layers (the second and the third trapping layers) provide trapping centers for the NVM elements. Figure 23.6a and b shows the band diagrams for the stack at Vg = 0 and Vg = Vp1/Vp2/Vp3/Vp4, respectively, while Figure 23.7 shows the change in Vth versus Vpp characteristics for the device during writing. Since nanocrystal CT device characteristics are strongly dependent on the work function, size, and distribution of nanocrystals, it should be noted that significant work function difference needs to exist between the nanocrystals embedded between the first tunnel layer (HfO$_2$) and the second tunnel layer (SIO$_2$), the sizes of nanocrystals for the NVM trapping layers, and the separation between the trapping planes (3.5 nm to prevent charge flow between the trapping layers by direct tunneling) to achieve the desired programming and retention objectives.

23.3.1.1 Nanocrystal Size, Trapping Characteristics, and Estimated Memory Window (Shift in Vth due to Trap Filling)

Fundamentals of nanocrystal-based charge trapping NVMs have been discussed in Part I, Chapter 3. Nanocrystal work function and size determine the nature of the quantum well of the

TABLE 23.3

Relationship between Nanocrystal Size and NVM Parametrics

Average Nanocrystal (NC) Size (nm)	Planar Density (N/cm²)[a]	Charge Density	
		1e/NC Coulomb/cm²	2e/NC Coulomb/cm²
1 nm	4.94E+12	7.90E–07	1.56E–06
2 nm	3.31E+12	5.30E–07	1.06E–06
3 nm	2.37E+12	3.80E–07	7.60E–07
4 nm	1.76E+12	2.80E–07	5.70E–07
5 nm	1.38E+12	2.20E–07	4.40E–07
6.5 nm	1.00E+12	1.60E–07	3.20E–07

[a] Spacing between nearest neighbors = 3.5 nm.

nanocrystal-induced trapping sites, and the number of charges filling each of such traps determines the NVM device characteristics [11]. Therefore, work function, size, and distribution of nanocrystals are key parameters in such device design. Techniques of employing uniform distribution of nanocrystals using self-assembly have been published over a decade back. Based on uniform planar distribution of nanocrystals and assuming 100% trap filling of one electron or two electrons per trap per nanocrystal, required current density for such trap filling process during writing as a function of nanocrystal size could be computed. This is shown in Table 23.3. In such computation, an average nanocrystal spacing of 3.5 nm is assumed to avoid direct tunneling of charges between nearest neighboring trapped charge centers within the plane.

As an approximate guide for device stack design, location of the trapping plane and gate stack EOT are important parameters in determining the effect of such charge trapping on memory window (change in threshold of the device induced by trap filling). For gate stack designs of 10 nm EOT (approximate Vpp ~±10V) and 20 nm EOT (approximate Vpp ~±20V), with the same set of assumptions as shown in the table, the memory windows could be computed for different size of nanocrystals. This is also shown in the following table. It should be apparent that while smaller size nanocrystals would provide higher trap density and memory window in the range of 1.1–2.3 V (see the case for EOT = 10 nm, two electrons per nanocrystal trapping, 3 nm nanocrystal size, and 3.5 nm spacing), the retention, for the larger memory window case is limited by the presence of 2 electrons per trap. Larger nanocrystal size would provide better retention with lower memory window. Therefore, CT devices strictly based on nanocrystal-induced trapping considerations have limited trade-offs between memory window and retention. This limitation could be overcome by multilayered nanocrystal memory designs [11]. Nanocrystal-based multiplanar memory designs have the advantage of reduced Vth dispersion between memory states and well-defined charge centroids and consequently high endurance. MSUM designs which combine the nanocrystal trapping with dielectric trapping and selected high trap density deep offset dielectric films have the advantages of both. Such devices have been proposed to achieve multifunctional and MLC capability simultaneously [7,10–12].

Gate Trapping ΔVT		Memory Window (ΔVT) for Nanocrystal Size					
Distance		1 nm	2 nm	3 nm	4 nm	5 nm	6.5 nm
10 nm	1e/NC	2.33	1.59	1.16	0.86	0.69	0.51
	2e/NC	4.65	3.17	2.31	1.73	1.38	1.03
20 nm	1e/NC	4.65	3.17	2.31	1.73	1.38	1.03
	2e/NC	9.23	6.24	4.51	3.35	2.65	1.95

23.3.1.2 Operational Scheme for Nanocrystal-Based MSUM Device of Figure 47 and Associated MLC Memory States

Four different programming levels: ±Vp1, ±Vp2, ±Vp3, and ±Vp4 are required to establish both MLC capabilities of DRAM and NVM elements [12]. The lower ones: ±Vp1 and ±Vp2 are associated with DRAM functionality of trapping and detrapping of electronic charges associated with N1 and N2 for DRAM functionality, whereas the higher ones ±Vp3 and ±Vp4 are associated with tapping and detrapping of N3 and N4 for NVM functionality. The common reference state with the lowest Vth is associated with the "erased state" for all elements whereby all carriers are removed from N1/N2/N3 and N4. This is schematically shown in Figure 23.8 where all memory states are defined. It should be noted that Vp1 < Vp2 < Vp3 < Vp4.

23.3.1.3 DRAM MLC States of 11D/10D/01D and Reference State 00

When +Vp1 is imposed, first the deeper trap associated with N2 gets filled creating the written memory state of [10 D] (DRAM) by direct tunneling and setting up a reverse field due to "planar coulomb blockade" which results in minimal filling of N1 and requires a higher programming voltage level of +Vp2 to create a higher written state of [11 D]. All the memory states of DRAM functionality could be erased to the reference level of [00] with −Vp2. By imposing +Vp1 followed by −Vp1, the DRAM memory state of partial detrapping of charges from N2 occurs establishing the memory state of [01 D].

23.3.1.4 NVRAM MLC States of 11NV/10NV/01NV and Reference State 00

When +Vp3 is imposed, carrier trapping takes place in N3 along with N1 and N2. This establishes the written state of [10 NV]. A higher programming voltage is required to establish trapping in N4 due to longer tunneling distance. Therefore, a +Vp4 is imposed to enable trapping in N4 and consequently the trap filling of N4 establishes the highest written memory states of [11 NV] whereby all nanocrystal-based trapping levels are filled. When −Vp4 is imposed, all traps are emptied and

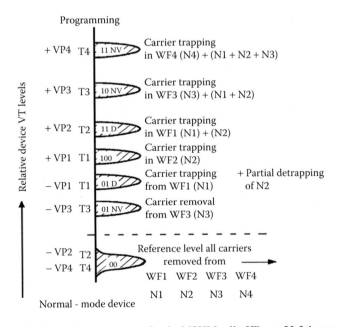

FIGURE 23.8 Programming and memory states for the MSUM cell of Figure 23.5 demonstrating MLC capabilities of DRAM and NVM. (From Bhattacharyya, A., Embedded trap nano-crystal multi-level-cell device for DRAM and universal memory, Updated Title: Embedded trap nano-crystal multi-level-cell device universal memory, *ADI Associates internal publication*, U.S. Patent. 7759715 [07/30/2010], 8193568 [06/05/2012].)

the reference erased state of [00] gets re-established. When +Vp4 is followed by Vp3, carriers are removed from N1, N2, and N3 resulting in the lowest written memory state of [01 NV].

23.3.2 Reverse Mode CT-MSUM Devices

So far we have discussed the normal mode CT-MSUM devices with silicon substrate being the primary source of carriers and charge trapping closer to the silicon/insulator interface to achieve highest Vth shift (and therefore the memory window) per unit electronic charge trapping [6,7,10,12]. However, if the reverse mode NVM is considered, the silicon/insulator interface could be made passive with the consequence of improved interface stability and reliability. Reverse mode NVMs have been discussed briefly in Part II. SUM device concepts could be extended to reverse mode operations to potentially enhance reliability of such devices. A highly reliable multifunctional reverse mode device has been proposed by the author in early 2005 [10]. Band diagram of one such device is shown in Figure 23.9. An HfSiON tunnel layer is sandwiched between two laminates: one interfacing the TaN or N+ Silicon gate and the other interfacing the trapping dielectric layer of TiO_2. The blocking dielectric interfacing the silicon substrate is another layer of HfSiON film as shown in the figure. Each laminate consists of multiple ultrathin layers of alternate films of nitride and HfO_2 formed by the ALD process. The laminate at the gate enhances electron and hole injection due to band bending and could be considered as a tunnel-enhancing layer, whereas the laminate at the trapping layer interface provide multienergy trapping and can be considered as a retention-enhancing layer in combination with the single energy trapping centers of TiO_2. The two laminates and the tunneling layer of HfSiON form the "reverse" crested barrier. With proper optimization of the laminates and thicknesses of the tunneling and blocking HfSiON layers as well as the TiO_2 layer, an MSUM device could be envisioned to functionally replace both traditional DRAM and SRAM devices. Many options of dielectric films could be combined to provide such schemes for both forward and reverse mode of operations of such MSUM devices. This is illustrated in Figure 23.9.

A relatively simpler stack design for multifunctional reverse mode MSUM device is shown in Figure 23.10 [6] consisting of a double-layered tunneling dielectrics, one each for DRAM and NVM performance and a double-layered trapping dielectrics with shallower (nitride) trapping layer for DRAM functionality and deeper trapping layer (GaN) for NVM functionality (retention). The gate

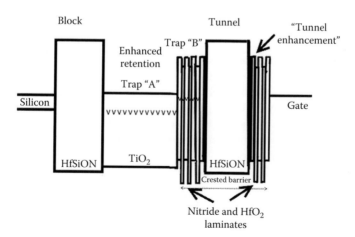

FIGURE 23.9 Band diagram of a novel reverse mode MSUM device with HfSiON tunnel layer sandwiched between laminates of HfO_2/Si_3N_4 forming crested tunnel barrier. (From Bhattacharyya, A., Discrete trap non-volatile device for universal memory chip application, *ADI Associates internal publication*, (MIDD 52), U.S. Patents: 7436018 [10/14/2008], 7476927 [01/13/2009], 7786516[10/14/10], 8143657 [03/27/2012], 8242554 [08/14/2012], May 29, 2005.)

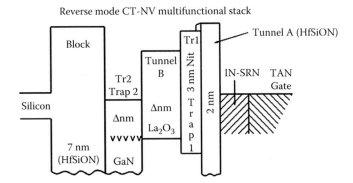

FIGURE 23.10 Band diagram for a double tunnel layer and a double trapping layer multifunctional reverse mode MSUM device. (From Bhattacharyya, A., Novel band-engineered nano-crystal non-volatile memory device utilizing enhanced gate injection, *ADI Associates internal publication*, U.S. Patent. 7629641 [12/22/2009], December 5, 2004.)

stack consists of Si/HfSiON/GaN/La$_2$O$_3$/Si$_3$N$_4$/HfSiON/in-SRN/TaN (gate) as shown in the figure below. The in-SRN layer interfacing the gate provides enhanced charge injection from the gate to achieve the desired speed objectives.

23.3.2.1 Multilayered Conductive Floating/Floating-Plate MSUMs

The very first NVM cell was proposed to be a metal (conductive) floating-gate NVM device by Khang and Sze in 1967 (Figure 2.1). Subsequently, multiple NVM device had been proposed in earlier years employing highly conductive IN-SRN with or without conventional silicon floating gate in the gate stack designs to derive multilevel NVM cell designs and multifunctional cell designs [10,12,14,15]. In recent years, there has been renewed interest in conductive floating-gate/floating-plate devices to address scalability. Multilevel NVM flash technology using conductive floating gate has been discussed in Part II enabling 128 Gb Flash NAND product at 20 nm technology node [16]. Instead of IN-SRN, ultrathin layers of TaN were employed by P. Blomme et al. demonstrating reduction in parasitic side-wall coupling effects [17] and also by S. Jayanti et al. [18] in 2010. Scalability of TaN as conductive floating gate down to 1.0 nm thickness was demonstrated by S. Jayanti and coinvestigators in the same year [18]. This has been briefly outlined in Chapter 15.

The background work has led to the recent demonstration of multilayered conductive floating gate (floating plate) MSUM device feasibility, demonstrating simultaneous functionality of DRAM and NVM of the memory cell discussed briefly below.

MSUM device functionality was demonstrated by incorporating conductive floating gates (plates) at different distances from the silicon interface and appropriate charge storage in the floating-gate layers through programming, by N. D. Spigna and coworkers in 2012 [19]. Their MIS gate stack used Pd gate as well as dual layer stack of Pd floating gates separated by interfloating-gate high K dielectric films. The gate stack structure consisted of n-type silicon substrate/SiO$_2$ tunnel dielectric/ bottom Pd floating gate (FG1)/interfloating-gate HfO$_2$ dilelectric/top Pd floating gate (FG2)/thicker HfAlO blocking dielectric/and finally the Pd control gate. The authors demonstrated ~300 mV flatband shift at ±7.5 V during "dynamic" mode pulsing for 200 ms followed by refresh at ±10 V (500 ms) as indicative of DRAM functionality with memory retention in the range of ~10 sec. The NVM capability was demonstrated by programming in the range of ±17 to ±20 V for ~500 ms with memory window (flatband shift) in the range of 4–7 V. NVM retention of 10 years and endurance of 1E5 W/E cycle were also demonstrated.

Follow-on work by the same R&D group as above confirmed the dual DRAM/NVRAM functionality in transistor structure by replacing Pd floating gate with TaN floating gate and Pd control gate with W-TaN control gate [20].

23.4   URAM AND OTHER MULTIFUNCTIONAL SILICON BASED MEMORIES

A brief review of other silicon based multifunctional memories developed and published recently is being made here. Some of these concepts could be classified within the general definition of MSUM devices. Regardless of the terminology, such emergence of multifunctional devices is expected to widen the application frontiers of the nonvolatile memories in future years.

23.4.1   The URAM and "Universal" Devices from KAIST

The URAM cells were mentioned at the beginning of the section from the historical context [21–23]. J.-W. Han in 2008 demonstrated DRAM functionality in a buried Si-Ge substrate FinFET device utilizing hole charge storage in the body of the substrate, while the ONO gate stack of the same device performed as a conventional CT-NVM device with ~3 V memory window and 10 years of retention programming at 12 V at 100 μsec. The DRAM retention was in the order of microsec.

A vertical channel multifunctional "Universal" novel device was proposed by D.-H. Moon et al. [23] also from KAIST in 2011. A 1T cell was conceived out of each silicon pillar with tailored silicon doping of the vertical channel with a bottom and top regions of N+ for source and drain, respectively. A cross-sectional TEM image of the FET pillar and the doping characteristics are shown in Figure 23.11a and b, respectively. The device gate wraps around each silicon pillar similar to the SGT structure (see Part II) with a band-engineered gate stack design consisting of O1-N1-O2-N2-O3 dielectric layers and Ti-TiN-W gate. SEM image of the silicon pillar array and a cross-sectional image of the gate stack are shown, respectively, in Figure 23.12a and b. Depending on the operational conditions, the device can exhibit single Vth normal FET characteristics at the lowest drain potential as well as a "steep-slope" FET (SSFET) at higher drain potential; the latter characteristics could be employed for low-power operations with an ON-OFF current ratio of 1E7. The SSFET feature results from hole generation and accumulation at the middle of the FET channel induced by weak impact ionization due to the inhomogeneously tailored channel doping mentioned above. At higher drain potential, the drain voltage versus drain current characteristics are altered from stronger impact ionization to the BJT mode establishing hysteresis between forward and reverse characteristics of the FET device. This phenomenon was utilized to achieve a volatile bithreshold memory state applicable for DRAM/SRAM functionality. The hole accumulation during weak impact ionization and the transition from impact ionization to the BJT mode for the

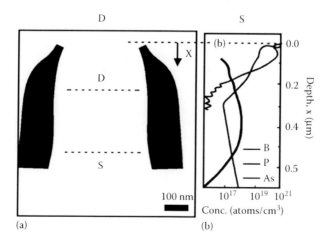

FIGURE 23.11   Multifunctional universal memory: (a) Cross-sectional TEM image of the 1T cell and (b) doping profile of the vertical channel of a silicon pillar. (From Moon, D.-I. et al., Multi-functional universal device using a band-engineered vertical structure, *IEDM*, p. 551 © 2011 IEEE.)

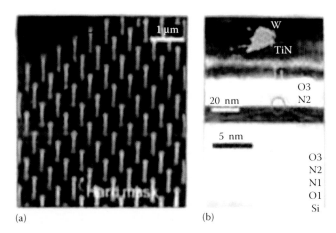

FIGURE 23.12 Multifunctional universal memory: (a) Tilted SEM image of silicon pillar array and (b) NV gate stack layers. (From Moon, D.-I. et al., Multi-functional universal device using a band-engineered vertical structure, *IEDM*, p. 551 © 2011 IEEE.)

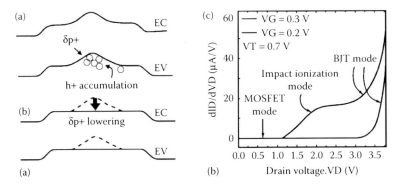

FIGURE 23.13 Multifunctional universal memory: (a) Channel doping effect on band energy and hole accumulation during weak impact ionization and (b) Id-Vd slope characteristics of the vertical channel FET due to transition from Impact Ionization mode to BJT mode at higher Vdd for forward and reverse bias of the device. (From Moon, D.-I. et al., Multi-functional universal device using a band-engineered vertical structure, *IEDM*, p. 551 © 2011 IEEE.)

device are shown, respectively, in the band diagram of Figure 23.13a and the change in Id-Vd slope characteristics in Figure 23.13b.

As mentioned above, wide hysteresis generated due to the parasitic BJT mode of the device was used for providing the bithreshold memory states for the DRAM/SRAM functionality of the 1T memory device at higher Vdd. The operational bias conditions for programming, reading, and standby states are shown in the table in Figure 23.14a, and the transient measurement of the source current during read "0" and read "1" for a 5 ns programming for writing and 5 ns programming for erasing is shown in Figure 23.14b demonstrating an attractive sense margin of 40 μA/μm for a relatively simpler sensing scheme. The Vd and Vg ranges for the DRAM/SRAM operations were demonstrated and the stability of the memory states after 1E16 W/E cycling was also demonstrated.

The binary memory state stability and retention of 1 sec was demonstrated by optimizing doping profile of the channel and the Vd range of operation.

The NVM functionality characteristics of the NOR and NAND memory cells are shown in Figure 23.15a and b, respectively. The NOR writing and erasing involved channel hot electron

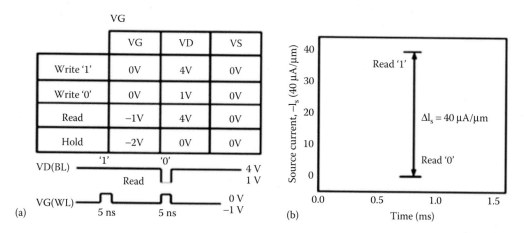

FIGURE 23.14 Multifunctional universal memory: (a) Operational conditions to achieve DRAM/SRAM Bithreshold functionality and (b) transient measurement of the DRAM/SRAM cell demonstrating write "1" and write "0" operations at ns programming and resulting sense margin between the memory states of nearly 40 μA/μm between read "1" and read "0". (From Moon, D.-I. et al., Multi-functional universal device using a band-engineered vertical structure, *IEDM*, p. 551 © 2011 IEEE.)

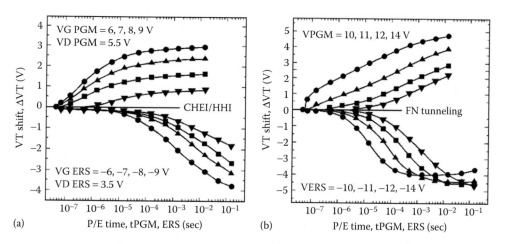

FIGURE 23.15 Multifunctional universal memory: NVM operations for (a) NOR cell using CHEI for writing and HHI for erasing and (b) NAND cell using F–N tunneling for both writing and erasing. (From Moon, D.-I. et al., Multi-functional universal device using a band-engineered vertical structure, *IEDM*, p. 551 © 2011 IEEE.)

injection (writing) and hot hole injection (erasing) while for the NAND, F–N tunneling was employed. Depending on the operational conditions, initial memory windows in the range of 2–3 V and retention of >1E8 sec at 300K were demonstrated. Additionally, P/E cyclic endurance of >1E4 was also demonstrated. Retention and endurance characteristics of the NAND device are shown in Figure 23.16a and b, respectively.

By appropriate operational conditions, the 1T multifunctional universal memory cell could be envisioned to simultaneously perform as NOR, NAND, SRAM/DRAM, as well as conventional/low-power MOSFET device within the framework of a unified memory technology. Such a scheme of bias map and memory hierarchy concept is shown in Figure 23.17a and b, respectively.

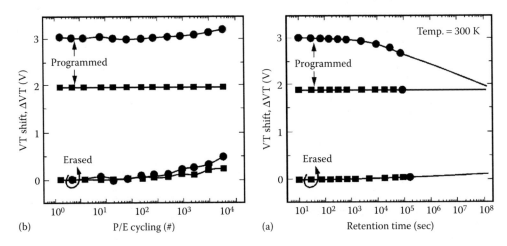

FIGURE 23.16 Multifunctional universal memory: NVM characteristics for NAND cell: (a) Retention at 300K and (b) endurance. (From Moon, D.-I. et al., Multi-functional universal device using a band-engineered vertical structure, *IEDM*, p. 551 © 2011 IEEE.)

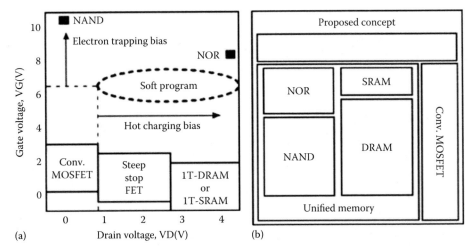

FIGURE 23.17 Multifunctional universal memory: (a) Bias scheme and (b) memory hierarchy. (From Moon, D.-I. et al., Multi-functional universal device using a band-engineered vertical structure, *IEDM*, p. 551 © 2011 IEEE.)

23.4.2 Other MSUM/UNIFIED Memory Concepts

Most multifunctional and unified memory concepts have emerged during the past 10 years [23–27]. It is anticipated that some of these concepts will come to product fruition soon, particularly those in the pursuit of terabit level capacity with enhanced endurance. A. Bhattacharyya and coworkers proposed a planar very high density CT-NVM with multifunctional DRAM/NVM functionality based on charge trapping within a single nanodot per memory cell [24]. Combining (a) "nano-nvm" device concepts similar to the above one and possibly others with (b) the vertical channel concepts developed by D.-I. Moon et al. [23] and associated integration scheme and (c) emerging multi-level nanowire/surround gate structures and devices discussed in Chapter 16, Part II [references 6,8–10,39], multiple MSUM devices and arrays could be envisioned. Such device and technology could potentially drive the NVM technology for the future generations beyond the ITRS road map. The advantages of these concepts are obvious. These concepts overcome scalability limitations both at feature-size and memory cell levels by simultaneously combining multifunctional CT-based DRAM/NROM/NAND devices and low-power FET devices for logic circuitry within the framework of 1T device concepts and single memory technology integration scheme [25]. Therefore, potentials for high performance, high capacity, low power, low cost/bit solutions are feasible for future memory subsystems based on the generalized SUM concepts. Additionally, as suggested elsewhere, such memory developments act as catalyst for novel digital systems, devices, and applications so far not available.

Two other recent concepts of multifunctional devices are referenced at the end of this chapter to draw reader's attention. One was called the "BT-RAM" cell by T. Sukiyaki and coinvestigators from Sony who published their work in 2007 [26], and the other one by Y.-K. Choi et al. from KAIST in 2008 [27].

REFERENCES

1. L. Forbes, High density single transistor vertical DRAM gain cell, Internal Publications, August 2002 and January 2004.
2. A. Bhattacharyya, Scalable flash/NV structure, *ADI Associates internal publication*, October 2001, U.S. Patents: 6743681 (06/01/2004), 6998667 (02/14/2006), 7528043 (05/05/2009).
3. A. Bhattacharyya et al., Integrated DRAM-NVRAM multilevel memory cell, *ADI Associates internal publication*, May 5, 2004, U.S. Patents: 7158410 (01/02/2007), 7349252 (03/25/2008), 7379336 (05/27/2008),7403416 (07/22/2008), 7417893 (08/26/2008), 7459740 (12/02/2008), 7457159 (11/25/2008).
4. A. Bhattacharyya, Surround gate multi-functional memory chip: Device concept, technology & chip architecture, *ADI Associates internal publication*, December 2004.
5. A. Bhattacharyya, Advanced scalable nano-crystal device and process, *ADI Associates internal publication*, December 2004, (MIDD 47), U.S. Patents: 7759715 (07/30/2010), 7898022 (03/01/2011).
6. A. Bhattacharyya, Novel band-engineered nano-crystal non-volatile memory device utilizing enhanced gate injection, *ADI Associates internal publication*, December 5, 2004, U.S. Patent. 7629641 (12/22/2009).
7. A. Bhattacharyya, High performance multi-level ET/DT NVM device, *ADI Associates internal publication*, December 5, 2004, U.S. Patents: 7429767 (09/28/2008), 7553735 (06/2009), 7579242 (08/25/2009), 8159875 (04/17/2012).
8. A. Bhattacharyya, Multi-functional memory device with graded composition insulator stack, *ADI Associates internal publication*, Patent Application submitted December 5, 2004, U.S. Patents: 7525149 (04/28/2009), 77680062 (08/03/2010).
9. A. Bhattacharyya, Transitional strategy from floating gate to floating plate and to charge trapping technology and devices and development of NV-DRAM, Internal technical communications from ADI Associates to K. Prall, cc: M. Durcan, Micron Technology, Boise, Idaho, USA, and March 31, 2004.
10. A. Bhattacharyya, Discrete trap non-volatile device for universal memory chip application, *ADI Associates internal publication*, May 29, 2005 (MIDD 52), U.S. Patents: 7436018 (10/14/2008), 7476927 (01/13/2009), 7786516(10/14/10), 8143657 (03/27/2012), 8242554 (08/14/2012).
11. A. Bhattacharyya, Scalable multi-level cell nano- crystal device with reduced threshold dispersion, *ADI Associates internal publication*, U.S. Patent, 8228743 (07/24/2012).
12. A. Bhattacharyya, Embedded trap nano-crystal multi-level-cell device for DRAM and universal memory, Updated Title: Embedded trap nano-crystal multi-level-cell device universal memory, *ADI Associates internal publication*, U.S. Patent. 7759715 (07/30/2010), 8193568 (06/05/2012).
13. A. Bhattacharyya, Extendable multi-level SGT-CT NROM and NAND flash devices with enhanced characteristics for multi-functional NVM technology applicable to 20nm technology node and beyond, Internal Memo, ADI associates, (Unpublished) April 12, 2013.
14. A. Bhattacharyya, R. S. Bass, and G. D. Grise, Non-volatile memory cell having silicon-rich nitride charge trapping layer, IBM Internal Technical Reports: 1984; U.S. Patent: 4870470 (09/26/1989).
15. A. Bhattacharyya, Scalable flash/NV structures and devices with extended endurance, *ADI Associates internal publication*, August 2001, U.S. Patents: 7012297 (3/14/2006), 7250628 (7/31/2007), 7400012 (7/15/2008), 7750395 (7/6/2010).

16. N. Ramaswamy, T. Graetinger, G. Puzzilli et al., Engineering a planar NAND cell scalable to 20nm and beyond, *IEDM*, Monterey CA, 2013 [Micron].

17. S. Jayanti, X. Yang, R. Suri et al., Ultimate scalability of TaN metal floating gate with incorporation of high-K blocking dielectric for Flash memory applications, *IEDM*, San Francisco CA, 2010, pp. 5.3.1–5.3.4.

18. P. Blomme, M. Rosmaulan, A. Cacciato et al., Novel dual layer floating gate structure as enabler of fully planar flash memory, *VLSI Technology*, Honolulu HI, June 2010.

19. N. D. Spigna, D. Schinke, S. Jayanti et al., A novel double floating-gate nified memory device, *VLSI ans Sustem-on-Chip*, 2012, pp. 53–58.

20. B. Sarkar, N. Ramanan, S. Jayanti et al., Dual floating gate unified memory MOSFET with simultaneous dynamic and non-volatile operation, *IEEE Electron Dev. Lett.*, 35, 48–50, 2014.

21. J.-W. Han, S.-W. Ryu, C. Kim et al., A unified-RAM (URAM) cell for multi-functioning capacitor-less DRAM and NVM, *IEDM*, 2007, pp. 929–932 [KAIST-URAM].

22. J.-W. Han, S.-W. Ryu, S. Kim et al., Energy band engineered unified-RAM (URAM) for multi- functioning 1TDRAM and NVM, *IEDM*, San Francisco CA, 2008 [KAIST-URAM].

23. D.-I. Moon, J.-S. Oh, S.-J. Choi et al., Multi-functional universal device using a band-engineered vertical structure, *IEDM*, Washington DC, 2011, pp. 551–554 [KAIST 3D: SRAM+DRAM+NVRAM].

24. A. Bhattacharyya, W. Farnworth, and P. F. Farrar, High density nano-dot nonvolatile memory, *ADI Associates internal publication*, (MIDD 48), Title changed to: Extremely Hi Density (Terabit) nano dot memory, U.S. Patent office Publication, 2010090265, dated 04/15/2010.

25. (a) A. Bhattacharyya, MSUM device with extremely high density of 1T DRAM and NV memory based on vertical channel silicon nano-pillars or nano-wires, Internal memo, ADI associates, December 14, 2012. Also (b) A. Bhattacharyya, Extendable Multi-level SGT-CT NROM and NAND Planar and 3D (including nano-wired) Flash Devices with enhanced characteristics for multi-level and multi-functional NVM technology applicable to 20nm technology node and beyond, Internal Memo, ADI associates, April 12, 2013.

26. T. Sukiyaki, M. Nakamura, M. Yamagita et al., Extremely low-voltage and high speed operation bulk thyristor SRAM/DRAM (BT-RAM) cell with triple selective epitaxy layer (TEL) bulk thyristor RAM, *IEDM 2007*, Washington DC, pp. 933–936.

27. Y.-K. Choi, T.-W. Han, S. Kim et al., High speed flash memory and 1T-DRAM on dopant segregated Schottky barrier (DSSB) FinFET SONOS device for multi-functional SoC applications, *IEEE, IEDM*, San Francisco CA, 2008.

24 SUM Functional Integration, Packaging, and Potential Applications

CHAPTER OUTLINE

This final chapter will briefly explore functional integration potentials at the chip and packaging levels. The chapter briefly mentions potential impacts of SUM devices for future system designs and memory system architecture, leading to application potentials of SUM devices in current and future digital systems .

24.1 SUM AND NVM INTEGRATION PERSPECTIVE: CHIP AND PACKAGING

Future technology integration schemes both in chip and package levels for the NVM products have been briefly covered in Chapter 14. SUM is an evolutionary progression of silicon based NVM technology which has already adopted concepts of band engineering and high K dielectric films for NROM and NAND products. SUM chip-level integration has the unique advantage of successfully incorporating and integrating advances in scaled CMOS technology features from SOC platform technology for the past and present technology nodes [1]. This trend is expected to continue in foreseeable future down to 7 nm technology node [12] and beyond. It is also expected that NVM technology in general and SUM in particular will take advantages of evolving packaging technology of integrating multiple chips into cost-effective packaging schemes through silicon visas (TSV) and embedded multidie interconnect bridge (EMIB) and others (mentioned earlier) providing SIP-SUM (system in package) products with enriched flexibility and functionality as well as enhanced capability.

24.2 INTEGRATION AT SILICON TECHNOLOGY/CHIP LEVEL

SUM devices discussed in Part III are primarily derivatives of conventional CT-based NVM devices (e.g., NROM and NOR and NAND flash) and secondarily CMOS bulk and FET based. All devices proposed are implementable in bulk and SOI silicon technology and require scaled CMOS FET peripheral circuits for the memory arrays similar to those of current silicon based NVM products such as NROM and NAND. Similar to the currently scaled silicon based NVM devices, SUM products at silicon level would be either in the form of embedded memory or in the form of stand-alone memory with evolutionary scaled CMOS circuitry for memory functionality as well as built-in processors or controllers or memory management units (MMUs) to fulfill the functionality and application objectives of the memory chip. Other than the unique memory cell designs and device stack designs, SUM devices should be considered as evolutionary derivatives of current silicon based NVM devices. Therefore, all basic constraints of scaling and integration at silicon level apply to SUM devices. Additionally, in order to ensure broad and diverse range of applicability for SUM, the following integration compatibility should be assumed:

1. Both Bulk [1] and equivalent SOI platform technology compatibility with planar and non-planar (e.g., FinFET) devices and circuits with on-chip power management capability.
2. Compatibility with ITRS road map of technology and device scaling.

3. Unique device and operational requirements of SUM could be readily incorporated into the mainstream CMOS technology. These include multilayered high K dielectric films for the gate stack with work-function defined metal gate, on-chip higher voltage generation from low Vss power supply for programming, neighboring cell parasitic coupling reduction, and unique nonplanar device integration, etc. [2–10].

24.3 INTEGRATION AT PACKAGING LEVEL

Recent advances in packaging schemes have significantly broadened the application base of NVM products. Various high-density SIM (HDS) are being currently employed to provide SIP functionality for a variety of portable devices incorporating DRAM, flash, and EEPROM devices. For higher density memory requirements, hybrid memory cubes (HMCs) and TSV inter-chip connection schemes have been successfully employed for chip-to-chip integration [11,12]. These have been briefly discussed in Chapter 14 in Part II of the book before Chip-to-Chip functional integration through TSV vastly improves functionality through power reduction and enabling higher IOP capability. Such chip-to-chip connections could be both planar and vertical as illustrated in Figure 24.1. Another cost-effective chip to chip interconnection scheme known as "EMIB" has been disclosed recently by Intel Corporation [12]. Cross-section and EMIB package prototypes are illustrated in Figures 24.2 through 24.4, respectively.

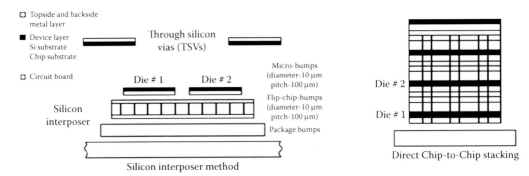

FIGURE 24.1 TSV packaging scheme of interconnecting planar chips and vertically stacked chips. (From Bansal, S., 3D-IC is now real: Wide I/O driving 3D-IC TSV, *Flash Memory Summit*, Santa Clara, CA, August 2012; Courtesy of Cadence Inc., Staunton, VA.)

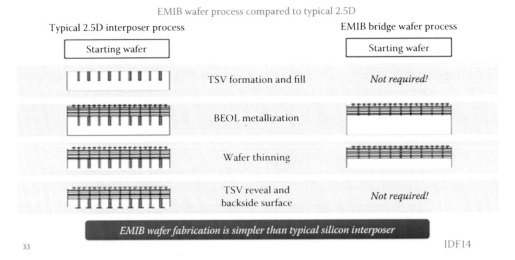

FIGURE 24.2 EMIB fabrication comparison with PRIOR TSIV technology. (From Merritt, R., Intel opens door on 7nm, foundry, *EE Times*, September 11, 2014; Courtesy of Intel Corporation, Santa Clara, CA.)

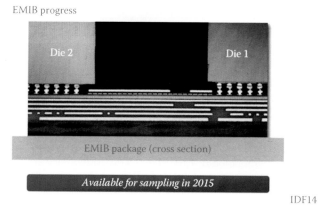

FIGURE 24.3 Cross-section of EMIB package applicable for potential SUM chips. (From Merritt, R., Intel opens door on 7nm, foundry, *EE Times*, September 11, 2014; Courtesy of Intel Corporation, Santa Clara, CA.)

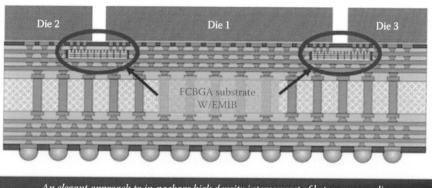

FIGURE 24.4 Interconnecting diverse-functional chips by EMIB packaging scheme. (From Merritt, R., Intel opens door on 7nm, foundry, *EE Times*, September 11, 2014; Courtesy of Intel Corporation, Santa Clara, CA.)

SUM chips incorporating such packaging schemes have the potential of providing a diverse set of customized current and future applications, but also through multifunctionality and enhanced capacity have the potential of vastly broadening the application base.

24.4 INTEGRATION AT LARGE SYSTEM LEVEL

Multiple terabytes of SSDs containing NAND flash chips are being increasingly employed in large enterprise systems and even larger capacity as caches (with HDD) for cloud storage systems (file servers/transaction processors) and distributed data centers all over the world. In both of these groups of applications, high IOPs are extremely desirable for performance and efficiency as well as higher endurance than currently available without compromising EOL retention. Additionally, low UBER and lower ECC requirements and associated overhead will be highly desirable to achieve reliability objectives and to minimize cost and power. USUM arrays with 4 bits/cell capability appropriately packaged and architected for high throughput combined with MSUM arrays to provide effective cache capability for large database memory sub-systems with MMUs and ECC controllers in advanced TSV/HMS packaging scheme could be envisioned to provide large system level integration solutions.

24.5   INTEGRATION AT FUNCTIONAL AND ARCHITECTURAL LEVEL

We have thus far discussed integration characteristics of SUM at the hardware level from conventional system architecture perspectives. Simultaneously available multifunctional capability within a memory cell at different performance range has the potential of providing PLA (programmable logic array) capability, fuzzy logic capability, associative memory capability, and "smart system" capability not considered previously using conventional NVM products. NROMs are widely used for code applications. MSUM NROMs could be considered and configured for "smart" code update applications as well as what may be called "self-adaptive," updatable logic, and system applications. Many software and random logic applications could be integrated into SUM arrays to provide enhanced functionality not considered until now. A combination of USUM and MSUM memory arrays could be considered to integrate software, logic, and memory functions through novel memory and system architecture to provide unique functional enhancements. One possibility is to design a smart memory controller built into the memory array to achieve required ECC capability to enhance data integrity. Another possibility is to achieve "smart" I/O and enhanced throughput capability built into the memory architecture especially associated with large relational database applications, networking, and routing Internet data streams.

SUM devices could potentially serve as catalyst to provide unified advanced memory systems and sub-systems for current and future applications of digital electronics. SUM devices could facilitate developments of advanced digital systems and memory sub-systems to simultaneously provide the following:

a. Enhanced memory density per package,
b. Cost reduction per function,
c. Enhancement in data throughput,
d. Reduction in overall Energy Consumption, and,
e. Enhancement in memory durability and system reliability.

It has been noted by Richter [13] that for portable applications employing multi-core microprocessors, memory sub-systems have successfully replaced DRAM-HDD combination by introducing micro-controller with SDD (NAND flash) architecture and achieved equivalent data bandwidth and system performance. Current employment of low latency SSDs based on either fast NAND with direct page loading design or enhanced VG NOR flash have already demonstrated enhanced performance, energy efficiency and reliability [13]. Application of USUM and FSUM devices with optimized and advanced memory architecture have the potential to advance digital system capability by multiple orders of magnitude in system attributes.

SUM memory based system optimization and advanced digital system development is beyond the scope of this book. Such subject matter is for future publications. However, it should be worth mentioning that SUM-driven system architecture development could achieve significant advancement in system attributes and could lead to universal and wider (current and future) applications in digital systems in the following areas [14]:

1. "SUM SYSTEM-ON-CHIP" (SSOC),
2. "SUM-driven SELF ADAPTIVE SYSTEMS" (SdSAS),
3. "SUM-driven PARALLEL PROCESSING SYSTEMS" (SdPPS),
4. SUM-driven RELATIONAL & GENERAL DATA-BASED SYSTEMS" (SdRDS & SdGDS),
5. "SUM-driven PARALLEL PROCESSING SYSTEMS" (SdPPS),
6. "SUM-driven COMMUNICATION SYSTEMS" (SdCS), and
7. Many other system applications not mentioned above.

It is conceivable, therefore, that SUM technology may vastly create new and novel applications in data processing, communications, networking, and imaging while enhancing the existing ones.

24.5.1 SUM and NVM Application Perspective

It may not be farfetched to assume that SUM may replace the entire traditional memory concepts of SRAM, DRAM, NROM, NAND, and HDD sub-groups into a unified SUM product sets covering the performance/capacity spectrum of the memory requirements for the entire silicon based digital products. It may also be conceivable that SUM devices may inspire new system architecture, new functions, as well as new applications and development of a broad range of "self-adaptive" smarter digital devices. The driving forces for such transition are as follows:

1. Compatibility with silicon-CMOS-based digital electronics.
2. Compatibility with silicon based foundry capability for high volume production.
3. Compatibility with advanced multichip packaging schemes.
4. Power reduction and voltage scaling.
5. Cost/bit reduction.
6. Enhanced performance in system and sub-system levels.
7. Enhanced flexibility, feature ability, and functional capability (enhanced IOPs).
8. Enhanced reliability (high endurance, low UBER, etc.).
9. Expanding the current trend of replacing DRAM-HDD applications with NVM (SUM)-SSD applications and SSD-HDD-based large storage systems applications to enriched architecture memory systems with SUM.
10. Potential for novel future SUM applications: for example, in self-adaptive digital systems utilizing SUM-based PLAs and other possibilities through innovative applications in smart systems, robotics, space, biomedicine, genetics, etc.

We will examine first the obvious application space for the SUM devices by displacing current memory products. Subsequently, we will speculate potentials for advanced applications through innovations.

24.6 CURRENT NVM AND DRAM MARKET: POTENTIAL SUM APPLICATIONS

From the standpoint of worldwide semiconductor market in 2012 of a total of $313 B, DRAM and flash markets have been $32.1B (10.25%) and $32.4B (10.35%), respectively. Since then flash has overtaken DRAM revenue growing at a CGR of 70% per year. This trend is expected to continue for at least next 5 years. About 88% of total NVM revenue comes from NAND flash products associated primarily with enterprise and cloud applications, whereas NOR flash and NROM products provided a revenue of ~$ 4 B in 2012. NOR flash and NROM products are primarily used for code applications for a diverse set of products from the very low end to the high end, whereas the NAND products are primarily applied for data storage. However, SSDs may contain both NAND flash and NROMs (embedded in the controller) for data and code, respectively. The current NVM products are either embedded or stand-alone chips. The embedded NOR flash or NROMs are widely used for code and low data storage requirements such as smart cards, automotive, RFID, consumer electronics, as well as a diverse set of applications for code and BIOS covering the low-end, mid-range, and high-end systems. Typically, the memory requirements are less than few GB. This is shown in Table 24.1. The SSD NAND flash applications may cover a wide range of system performance from consumer to enterprise to cloud data center applications requiring memory capacity ranging from 4 GB to 100s of TB (1E12B = 1TB). This is shown in Table 24.2.

TABLE 24.1

Current NOR Flash and NROM Applications

Low End	Mid-Range	High End
Automotive	Smart phones	DSL servers
Smart card	Tablets	Modem
Camera	Computers	Router
Industrial	Laptops	Switches
Cameras/toys	LCD monitors	LANs
General electronics	Printers	High-end phones, Low-end phones, Tablets, etc
Low-end phones, tablets, etc.	Desktops PCs	High-end portables
Low-end printers	GPS	PCIe SSD, etc.
	Bluetooth	

TABLE 24.2

SSD: NAND Flash Applications

Consumer/Client	Enterprise/Cloud/Data Center
Low performance (4–16 GB)	
High performance plus high endurance	High performance plus high endurance
Smart phone/tablets (16–32 GB)	Multiple TB
Medium performance (32–64 GB) high IOPs	High IOPs
High performance (256 GB)	

24.7   ADVANCED APPLICATIONS

At the current time frame, future novel applications of SUM devices are speculative. With innovative memory and system architectures, with novel firmware and code developments, with advanced "self-adaptive" fuzzy logic capability, with "smart" networking systems and sub-systems (Hubs, Routers, etc.), and with vastly expanding imaging capability, sky could be the limit of applications of SUM devices and technology. Therefore, it could be imagined that SUM devices would find applications not only in traditional market space as mentioned above but also in a vast range of advanced applications in the following market sectors:

Biogenetics, bioimaging, biomedicine
Space (communications, data analysis, imaging)
Robotics
Drones and aeronautical vehicles
Smart transportation and traffic regulation systems
Green energy developmental systems
Weather prediction and materiology
Advanced agriculture and water conservation
Educational supplements, smart toys, and others

It has been discussed earlier that SUM devices and products cover the entire memory spectrum and have the potential to expand the application market by not only replacing the DRAM and increasing memory bits from HDD but also creating additional novel applications beyond the traditional market space. It is, therefore, conceivable that SUM revenue could exceed $100.00 B within a decade.

REFERENCES

1. C. H. Jan, U. Bhattacharyya, R. Brain et al., A 22nm SoC platform technology featuring 3-D tri-gate and high-k/metal gate, optimized for ultra low power, high performance and high density SoC application, *IEDM*, San Francisco CA, 2012, pp. 44–47.
2. A. Bhattacharyya, Integrated surround-gate multifunctional memory device, *ADI Associates Internal Publication*, May 26, 2004. Patent Application abandoned.
3. A. Bhattacharyya, Multi-functional memory device with graded composition insulator stack, *ADI Associates Internal Publication*, Patent Application submitted December 5, 2004, U.S. Patents: 7525149 (04/28/2009), 77680062 (08/03/2010).
4. S. J. Whang, K.-H. Lee, D. G. Shin et al., Novel 3D dual control gate with surrounding floating gate (DC-SF) NAND flash cell for 1 Tb file storage application, *IEDM*, 2010, pp. 668–671; S. J. Whang, K.-H. Lee, D. G. Shin et al., Novel 3-dimensional dual control-gate surrounding-gate (DC-SF) NAND flash cell for 1Tb file storage application, *IEDM*, San Francisco CA, 2010, 29.7.1–29.7.4. [Hynix]
5. Y. H. Kim, C. Cabral, E. P. Gusev et al., Systematic study of work function engineering and scavenging effect using NiSi alloy FUSI metal gates with advanced gate stacks, *IEDM*, Washington DC, 2005, pp. 657–660.
6. H. Tanaka, M. Kido, K. Yahashi et al., Bit cost scalable technology with punch and plug process for ultra high density flash memory, *VLSI Technology*, Kyoto, Japan, 2007, pp. 14–15.
7. J. Jang, H. S. Kim, W. Cho et al., Vertical cell array using TCAT technology for ultra high density NAND flash memory, *VLSI Technology*, Kyoto, Japan, 2009, pp. 192–193.
8. W. Kim, S. Cho, J. Sung et al., Multilayered vertical gate NAND flash overcoming stacking limit for terabit density storage, *VLSI Technology*, 2009, pp. 188–189.
9. N. Ramaswamy, T. Graetinger, G. Puzzilli et al., Engineering a planar NAND cell scalable to 20nm and beyond, *IEDM*, Monterey CA, 2013 [Micron].
10. S. Jayanti, X. Yang, R. Suri et al., Ultimate scalability of TaN metal floating gate with incorporation of high-K blocking dielectric for flash memory applications, *IEDM*, San Francisco CA, 2010, pp. 5.3.1–5.3.4.
11. S. Bansal, 3D-IC is now real: Wide I/O driving 3D-IC TSV, *Flash Memory Summit*, Santa Clara, CA, August, 2012.
12. R. Merritt, Intel opens door on 7nm, foundry, *EE Times*, September 11, 2014.
13. D. Richter, "Flash Memories, Economic Priciples of Performance, Cost and Reliability Optimization", Springer Series in Advanced Microelectronics, Springer, 2014, pp. 172–174.
14. A. Bhattacharyya, "Potential of SUM-driven Advanced Digital System Architecture and Application Horizon", ADI Internal Publication (Unpublished), March 2017.

Conclusion

Silicon based unified memories (SUMs) have great potential for next decade and beyond in providing unified nonvolatile memory products for the digital systems. The SUMs are evolutionary device concepts based on advanced CT-NVM devices and other charge-trapping device concepts and fully compatible to the scaled CMOS technology. The CT-NVM-based SUMs incorporate band engineering and high K dielectrics, as well as DTM based highly efficient charge transport, in the gate stack designs and provide options to the replacements of traditional SRAM, DRAM, NROM, NAND flash, and SSD memory hierarchy existing today. SUM devices could be effectively configured in conventional planar 2D arrays, in trench-based vertical channel arrays, in full 3D configurations as well as nano-wired-gate-all-around arrays providing multilevel multifunctional high capacity high performance NVMs. SUM technology could be readily integrated at system-on-chip level as well as in advanced packaging levels to provide a broad range of diversified applications for the current and future digital electronics. SUM devices have the potential to provide a unified memory system covering the entire performance/capacity spectrum of the memory hierarchy. Additionally, SUM devices have the potential to act as catalyst to provide novel memory subsystems with simultaneous enhancements in Memory Density per package, cost reduction per function, significant enhancement in data throughput, reduction in energy consumption, and enhancement in memory and system durability. Therefore, SUMs have expanded market potentials not only for the current applications but also for the enhanced and innovative futuristic applications in the ever-expanding digital universe.

THE PURPOSE OF THIS BOOK IS TO FULFILL THE DREAM OF MAKING SUM A DRIVING FORCE FOR DIGITAL MEMORY UNIVERSE AND FUTURE DIGITAL SYSTEMS OF TOMORROW.

Appendix: Rare Earth Metal-Based Future Dielectric Thin Films for SUM Devices

The significance of ultrathin higher dielectric constant thin dielectric films for the gate stack design of silicon FET-based NVM devices has been elaborated in Chapters 4, 12, and 20 in Parts I, II, and III respectively. In Chapter 12, several rare earth metal-based dielectric films have been discussed. These included single metal oxides, namely, La_2O_3 and Pr_2O_3; mixed metal oxides and oxynitrides, namely, $LaAlO_3$ and HfLaON. We have also discussed the properties of HfLaSiON in Chapter 20 with nearly symmetric band offset (interfacing silicon) for gate dielectric and blocking layer application. Rare earth metal oxides are popularly called as *Lanthanides*, Lanthanum (La) being the lowest atomic number of the group in the periodic table and La_2O_3 is the lightest molecular weight among lanthanides. Development of rare earth metal-based ultrathin dielectric films has been relatively recent, motivated to seek ultimate scalability for the FET devices. Fundamental investigation of most lanthanides demonstrated excellent thermal stability, high k values (>>10), low conductivity, and acceptable range of ultimate breakdown strength (>4MV/CM) in the EOT range of 1–3 nm. These attributes are attractive options to replace SiO_2 and other ultrathin dielectric films for sub-20 nm technology nodes.

Lanthanides are also attractive for silicon based NVM applications since these oxides provide lattice parameter matching with silicon substrate (least interfacial stress) and attractive range of band gap (5–6 eV). Most investigated lanthanides are La_2O_3, and Pr_2O_3; the latter one exhibits very low leakage and larger band gap. Advantages of lanthanides such as Pr_2O_3 outweigh the challenges associated with technology integration and manufacturability.

This appendix of rare earth metal-based dielectric films for NVMs is extracted from a technical presentation on the subject given by the author to Micron Technology, Boise, Idaho in 2007 [8]. The objective would be to foster future R&D efforts to enable SUM device development and application utilizing the unique advantages of rare earth metal-based high K dielectric films for NVM device stack design.

It should be noted that rare earth metals not only form silicide, and oxide, with elements Si, and O respectively; these metals could also form carbide with C and nitride with N (and oxynitride with O and N) with the potential of application as high K charge trapping dielectric films besides the previously discussed application as tunneling and blocking dielectric films. However, more work is needed in the development of such films for charge trapping applications.

The appendix is an extremely brief synopsis of the following topics related to rare earth metal-based ultrathin insulators. A limited bibliography is provided at the end for more intensive review and covers the development only up to the year 2007. The reader is encouraged to consult more recent developments in the field. The topics covered are (A) General Introduction and Special Attributes of Lanthanides; (B) Process Options for Lanthanide Ultrathin Films; (C) Structure of Thin Dielectric Films (Lanthanides); (D) Interface Characteristics with Silicon and SiO_2; (E) Materials and Device Characteristics; (F) Challenges in Integration with Silicon Technology; and (G) Potential Solutions.

A.1 GENERAL INTRODUCTION AND SPECIAL ATTRIBUTES OF LANTHANIDES

Figure of merit for common single metal oxides, namely, Al_2O_3, ZrO_2, HfO_2, Ta_2O_5, and TiO_2 has been reviewed in Table 3.3 [1–9]. Even with a K value of 80, the figure of merit for TiO_2 is merely 1.25 when compared to HfO_2 of 1.7 due to significantly higher conductivity of the former dielectric film. By comparison, the figure of merits of La_2O_3 and Pr_2O_3 is both 3.5 due to significantly lower conductivity with K value around 30 (K = 25 for HfO_2). This alone makes the lanthanides more attractive for tunneling and blocking dielectric films for gate stack design with the desired band gap for advanced NVMs and especially for SUM devices. It should be noted that such films are also attractive for MIS capacitor applications in general. Another attractive feature of lanthanides is their lattice compatibility with silicon aiding to provide stress-free interface between silicon substrate and the insulating film. However, it should also be noted that most lanthanides may readily form silicide at the silicon interface at elevated processing temperature and, therefore, requires appropriate interface passivation discussed in Sections A.4 and A.7. Another potential advantage when compared to ZrO_2 or HfO_2 is the relatively higher thermal stability of lanthanides from the standpoint of CMOS technology integration. Multiple process options exist for producing quality films for lanthanides such as MBE, E-beam, MOCVD, and ALD. Ultrathin Pr_2O_3 films are specifically more appealing due to their extremely low leakage and lower reactivity with silicon and will be covered in greater detail below.

Table A.1 provides the published characteristics of lanthanide films.

TABLE A.1
Published Attributes of Rare Earth Metals and Lanthanides

Lanthanides: Key Parameters							
Rare earth metals:	La	Pr	Nd	Sm	Gd	Dy	Er
Atomic number:	57	59	60	62	64	66	68
Lanthanides:	La_2O_3	Pr_2O_3	Nd_2O_3	Sm_2O_3	Gd_2O_3	Dy_2O_3	Er_2O_3
Mol. weight:	325	330	336	348	361	373	382
Thermal stability:	Yes	Yes	Yes	Yes	Yes	Yes	Yes
Reactivity with H_2O:	Hi	HI	Med	Med	Low	Low	N/A
Silicide formation	No	No	Yes	?	No	?	?
Interface-oxide (Si)	Yes	Yes	Yes	Yes	Yes	Yes	Yes
Silicate	Stable	Stable	N/A	N/A	N/A	N/A	N/A
Alumnate	Stable	N/A	N/A	N/A	N/A	N/A	N/A
Dielec. Bkdn: MV/cm	3.3–5.6	3.4–5.0	High	Low	Low	High	?
K: (SC):	30	31	14	N/A	22	N/A	N/A
K: (Amorphous)	?	11–24	12–13	11–16	14–15	14–16	?
K: (Poly/ALD)	?	10–15	10.4	10.0	8.9	8.4	10.0
K-Silicate		21–22					
Leakage	Med	Low	Low	Low	Low	Med	?
Eb (eV)	4.3	4.9–5.2	5.8		5.2		

N/A: Not Available.

Pr_2O_3 is thus far the most investigated. High K, Very Low Leakage, High Dielectric, Higher thermal stability with stable silicates and possibly aluminates are positive attributes.

A.2 ULTRATHIN LANTHANIDE FILMS OVER SILICON SUBSTRATE: PROCESS OPTIONS

Ultrathin lanthanide films can be deposited by a variety of techniques. During the earlier years prior to 2000, primary deposition processes were e-beam evaporation and MBE under high vacuum conditions [5,10–21].

Due to the hygroscopic nature of several lanthanide films and varying reactivity with silicon, most film processing was achieved at ultra high vacuum (UHV) MBE (often multi chamber) systems either by e-beam evaporation or by pulsed laser evaporation of a source or by laser sputtering a target material. After the turn of the century, advanced ALD and MOCVD multichamber tools were made available. Both MOCVD and ALD techniques have been updated with liquid injection capability in recent years (LI-MOCVD and LI-ALD). Key features of various process options are listed in Table A.2.

A.3 STRUCTURE OF THIN FILMS OF LANTHANIDES ON SILICON SUBSTRATE

Crystallographic structures of all lanthanides grown over silicon substrate had been shown to be similar [5,22,23,42]. Epitaxial growth has been demonstrated dependent on the crystallographic plane of the silicon substrate exhibiting close lattice match with silicon substrate. Crystal structures are dependent on pressure, temperature, substrate characteristics, and processing mode. Generally, two forms of crystallographic structures are exhibited, the hexagonal La_2O_3 structure and the cubic structure similar to either Mn_2O_3 or CaF_2 cubic crystal structure. This is outlined below.

1. Hexagonal La_2O_3 structure, for example:
   Pr_2O_3 films grown on Si (111) at ultra high vacuum (e-beam),
   or
   Pr_2O_3 films sputter-grown on polycrystalline TiNx at 400°C (substrate temperature, 1E-8 mbar),
   or
   Pr_2O_3 films grown by MOCVD on Si (100) at high substrate temperature of 750°C
2. Cubic Mn_2O_3 or Cubic CaF_2 structure; for example:
   Pr_2O_3 films grown on Si (001) at ultra high vacuum (e-beam) at high substrate temperature (>600°C)

TABLE A.2
Process Features

Key Process Environment	Source	Substrate Temperature	Post Deposition Anneal and Film
1A) High vacuum 10E-10-10E-7mbar e-beam	Pr_6O_{11} crystala (W or Mo or Ti-boat)	RT-725°C Si (001), Si (111)	Anneal ranging 600–8900°C, 5 min in N_2 and/or RTA 900°C in N_2 cubic single crystal Pr_2O_3 (100) hexagonal single crystal Pr_2O_3 (111) (IN-situ polysilicon Cap deposition)
1B) High vacuum pulsed laser deposition	Pr_6O_{11}	RT-900°C Si (001)	Single crystals of Pr_2O_3 and Pr_6O_{11} with silicate interface
1C) High vacuum laser sputtering: 4E-7 torr	$LaAlO_3$ crystals	700°C Si (001)	High quality single crystal $LaAlO_3$ without interface layer
2A) Hot-wall, Low pressure MOCVD	Pr(tmhd)3 Precursor in Argon (170°C)	400–650°C and 750°C a-silicate →	a-SiO_2/a-silicate/a-Pr_2O_3 a-SiO_2/a/nc silicate/Poly-Pr_2O_3 Poly-Pr_2O_3 on anneal in O_2 hexagonal Pr_2O_3
2B) Liquid inj. MOCVD	Pr[N9SiMe3)2]3 + H_2O	n-Si (001) 200–400°C Anneal 900–1000 →	a-silicate:[Pr9.33$(SiO_4)_6O_2$] crystalline silicate

Crystallographic structures have been investigated by either the X-ray diffraction pattern of the film formed or by X-ray scan studies and observing the associated peaks along with the silicon substrate peaks (see references 5, 22 and 42).

A.4 INTERFACE CHARACTERISTICS OF SILICON-LANTHANIDE AND SILICON-SiO$_2$-LANTHANIDE FILMS

For ultrathin dielectric films grown over silicon with or without the presence of a native oxide (SiO$_x$), the interface characteristics at the monolayer level are of critical importance from the standpoint of device characteristics and reliability [16–18,24–26,31]. The basic mechanism of the chemical bonding and the electronic exchange in forming hetero-oxide with rare earth metals has been investigated by several authors and the need of interface passivation for effective integration of ultrathin lanthanide films for silicon devices had been identified. The associated mechanism of oxygen-exchange, silicate formation, and hygroscopic characteristics for lanthanide films is discussed below.

- Lanthanides react readily with silicon with or without the presence of native oxide (SiO$_2$) to form stable silicate even at room temperature at very low partial pressure of oxygen and –OH (UHV condition). Both oxygen-exchange and silicon-exchange could take place simultaneously at the interface region via a mixed Si-O-M phase without an SiO$_2$ interface layer. This reactivity would result in the formation of an active silicate phase. For Si/epitaxial Pr$_2$O$_3$ as well as for Si/native SiO$_2$/epitaxial Pr$_2$O$_3$ systems, a mixed stable interface of Si-O-Pr silicate would be readily formed irrespective of the thickness of Pr$_2$O$_3$ film (even at high growth temperature, e.g., >600°C).
- Oxygen diffuses readily through both amorphous and crystalline lanthanides. Depending on the partial pressure of oxygen and –OH ion and temperature, the silicon/insulator inter-face may form a SiO$_2$ layer (amorphous) with amorphous or crystalline silicate phase with an over-layer of crystalline or amorphous lanthanide structure. For Pr$_2$O$_3$/Si-O-Pr/SiO$_2$/Si system thus formed, oxygen and silicon exchange could take place at the silicate phase and such intermixing gets enhanced at vacuum (UHV) annealing at elevated temperature at the expense of SiO$_2$ layer.
- Lanthanides react with water and readily absorbs –OH ion, which is accumulated at the silicon/lanthanide interface. The larger the ionic radius and lower the electro-negativity, the more reactive the lanthanide is with water. Therefore, among Pr$_2$O$_3$, Sm$_2$O$_3$, Gd$_2$O$_3$, and Dy$_2$O$_3$, Pr$_2$O$_3$ is most reactive with water, whereas Dy$_2$O$_3$ is least reactive. Due to this strong hygroscopic nature, an oxygen-rich interface would readily form for Si/Pr$_2$O$_3$ sys-tem. Such interfacial oxide/silicate would readily grow with higher annealing temperature. Consequently, electrical characteristics would be significantly affected.
- Understanding and control of stoichiometry of silicon/lanthanide as well as lanthanide/metal interfaces are critical elements to control and reproduce electrical and physical char-acteristics and requires optimized integration scheme for reproducible device properties.
- Pr$_2$O$_3$/Si interface layer composition and bonding characteristics (Si-O-Pr) and hygroscop-icity have been investigated in-situ (in UHV) by several investigators: Refs [12,25,28,39], using core level X-ray photoelectron spectroscopy (XPS), Fourier transform IR spectros-copy (FTIR), and high-resolution transmission electron microscopy (HRTEM).

A.5 MATERIAL AND DEVICE CHARACTERISTICS OF ULTRATHIN LANTHANIDE FILMS

The following highlights the properties of ultrathin lanthanide dielectric films formed over silicon substrate, and related properties, as well as, associated MOS/FET device characteristics [26–41]:

- Film properties strongly dependent on processing, interface composition, and subsequent annealing: for example, K values for Pr_2O_3 could be strongly dependent on processing variables as given in Table A.3.
- Film interface and consequently film properties strongly dependent on oxygen and water content (–OH) in the film and in the processing environment. Interface oxidation due either to annealing in oxidizing ambient or due to water absorption would lower the effective dielectric constant of the film and adversely increase EOT.
- Superior dielectric properties achieved for epitaxial films compared to amorphous films. However, if interface could be stabilized, even amorphous films exhibit improved characteristics. Both epitaxial films and amorphous films show excellent leakage characteristics. Leakage current density for different films with EOT = 1.4 nm is shown in Table A.4. Current density versus applied field characteristics of single crystal Pr_2O_3 film is shown in Figure A.1.
- Polycrystalline Pr_2O_3 MIM capacitor on TiNx/Si (001) to control silicate formation with metal (Aluminum) counter electrode could achieve enhanced charge storage density suitable for RF and IC applications. Due to the interface silicate formation, the EOT and the effective dielectric constant values are expected to be sensitive to thickness especially for ultrathin films. For amorphous lanthanides, this is illustrated in Table A.5.
- Preliminary n-MOS device had been fabricated with e-beam deposited Pr_2O_3 dielectric film on Si (001) substrate, in-situ polysilicon capped to prevent oxygen and –OH absorption at the silicon interface, with RIE SF_6/Cl_2 poly silicon gate etch and 900°C, 10 sec TRA for S/D activation, followed by forming gas anneal for 10 minutes at 425°C. VT offset by 1 V due to high density of fixed negative charge and sub-Vt slope of 145 mV/sec due to high density of interface states degrade device characteristics. However, FET functionality was demonstrated.

TABLE A.3
Pr_2O_3 Dielectric Constant: Process and Structure Sensitivity

Single Crystal	Crystalline	Polycrystal	a-PrOx	a-Silicate
Pr_2O_3	Silicate	Pr_2O_3	MOCVD	ALD
30	~22–25	~15	~11–24	10–16[a]

[a] With high temp. 800°C N_2 anneal K increased to ~16–20.

TABLE A.4
Leakage Current Density of Ultrathin Pr_2O_3 Compared to SiO_2 and Metal Oxides

Material	Jg (A/cm²) @ Vg = 1V
Pr_2O_3	$5 * 10^{-9}$
HfO_2	$1 * 10^{-4}$
ZrO_2	$5 * 10^{-4}$
SiO_2	~1
SiO_2 (3 nm)	~10^{-4}

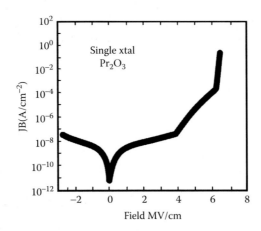

FIGURE A.1   Typical J versus E characteristics of single crystal Pr_2O_3. (From Jeon, S., and Hwang, H., *J. Appl. Phys.*, 93, 6393, 6395, 2003.)

TABLE A.5
Dielectric Constant of Thick and Thin Amorphous Lanthanides

Material	Epsilon (ε) Thin	Epsilon (ε) Thick	Do (Interface)	EOT/J (A/cm² @ Va ~ 1V)
Pr_2O_3	15	15.2	10.7	13A/5.4E-4
Nd_2O_3	12.9	13.2	8.2	12.4A/1.1E-4
Sm_2O_3	11.4	15.7	5.5	12.2A/9.86E-5
Gd_2O_3	13.9	15.3	10	13A/3E-3
Dy_2O_3	14.3	16	12	13.3A/2.3E-5
PrTiO	23		6	10.2A/2.9E-3

A.6   APPLICATION OF LANTHANIDES IN SI-BASED TECHNOLOGY: ASSOCIATED CHALLENGES

Major challenges of single metal oxide lanthanides are associated with the following [42–44]:

1. Interface control with silicon due to strong reactivity related to oxygen and –OH ion exchanges at the interface
2. High diffusivity of oxygen through the bulk M_2O_3 or MO_2 structures of lanthanides
3. Enhanced oxygen and water absorption in amorphous films
4. Degraded electrical properties and unstable interface associated with amorphous films
5. Higher substrate temperature and UHV requirements to achieve quality films
6. Integration challenges to achieve contamination-free quality film within the context of conventional silicon technology
7. Relatively limited process window with known precursors for obtaining films using CVD, MOCVD, and ALD techniques for low cost production

Due to the above limitations, the films reported thus far exhibit:

1. Active silicate interface with or without SiO_2 formation with silicon
2. Unacceptable interface state density: >1E12/cm²
3. Unacceptable negative fixed charge: ~1E12/cm²
4. Unacceptably high residual hydrogen and carbon in films deposited at lower temperature and atmospheric pressure (e.g., by ALD techniques)

A.7 PROPOSED POTENTIAL SOLUTIONS

1. Require interface passivation, TiNx showed promise [3,31,44]
2. Require in-situ capping of lanthanide with polysilicon [15]
3. Other approaches for effective interface passivation and encapsulation against oxygen, –OH, and contaminants while processing with MOCVD methods [11,18,31,37,43,44]
4. Optimization of interface treatment and anneal to reduce and stabilize interfaces and to reduce interface density and fixed negative charges [10,31,43–45]

A.8 CONCLUSION

Advantages of lanthanides outweigh the challenges: Relatively high K values. Low trap density, very low leakage, relatively more stable silicates and aluminates makes such dielectric attractive solutions near-end for RF and DRAM capacitors. More work is needed to find practical solutions for gate insulator for FET and FLASH applications. These should be forthcoming.

REFERENCES

1. K. J. Hubbard, and D. G. Seblom, Thermodynamic stability of binary oxides in contact with silicon, *J. Mater. Res.*, 11(11), 2757–2776, 1996.
2. H. J. Osten, J. P. Liu, P. Gaworzewski et al., High-k gate dielectric with ultra-low leakage current based on praseodymium oxide, *IEEE, IEDM*, San Francisco CA, 2000, pp. 653–656.
3. S. Jeon, K. Im, H. Yang et al., Excellent electrical characteristics of lanthanide (Pr, Nd, Sm, Gd, and Dy) oxide and lanthanide-doped oxide for MOS gate dielectric applications, *IEDM*, Washington DC, 2001, pp. 471–474.
4. J. W. McPherson, J. Kim, A. Shanware et al., Trends in the ultimate breakdown strength of high dielectric-constant materials, *IEEE Trans. Electron Dev.*, 50, 1771–1778, 2003.
5. H. J. Osten, J. P. Liu, H.-J. Mussig, and P. Zaumseil, Epitaxial, high-K dielectrics on silicon: the example of praseodymium oxide, *Mat. Res. Soc. Symp.*, 744, 991–994, 2003.
6. A. Bhattacharyya, Selection and consideration of high K dielectric for replacement of SiO_2 for future gate insulator for logic and memory (FLASH/DRAM) applications, ADI Associates Internal Micron Memo, January 6, 2006 (Unpublished).
7. A. Bhattacharyya, Considerations and status of high K dielectric capacitor technology for micro and nano electronics, ADI Associates Internal Micron Memo, June 1, 2006 (Unpublished).
8. A. Bhattacharyya, A review on high K rear-earth metal oxides focussing on Pr_2O_3, *ADI Associates, Invited Presentation at the Micron Technology*, Boise, ID, March, 2007.
9. O. Sneh, R. B. Clark-Phelps, A. R. Londergan et al., Thin film atomic layer deposition equipment for semiconductor processing, *Thin Solid Films*, 402, 248–261, 2002.
10. A. Nakajima, H. Ishii, T. Kitade, and S. Yokoyama, Atomic-layer-deposited ultrathin Si-Nitride gate dielectrics - A better choice for sub-tunneling gate dielectrics, *IEEE, IEDM*, 2003, pp. 7803–7806.
11. K. Endo, and T. Tatsumi, Metal organic atomic layer deposition of high-k gate dielectric using plasma oxidation, *Jpn J. Appl. Phys.*, 42, 1685–1687, 2003.
12. D. Wolffarm, M. Ratzke, M. Kappa et al., Pulsed laser deposition of thin Pr_xO_y films on Si(100), *Mater. Sci. Eng. B*, 109, 24–29, 2004.

13. R. L. Nigro, V. Raineri, and C. Bongiorno, Dielectric properties of Pr_2O_3 high-k film grown by metal organic chemical vapor deposition on silicon, *Appl. Phys. Lett.*, 83, 129–131, 2003.

14. H. C. Aspinall, J. Gaskell, P. A. Williams et al., Growth of praseodymium oxide thin films by liquid injection MOCVD using a novel praseodymium alkoxide precursor, *Chem. Vap. Deposition*, 9, 235–238, 2003.

15. H. J. Osten, E. Bugiel, and A. Fissel, Epitaxial praseodymium oxide: A new high-K dielectric, *Mater. Res. Soc., Symp. Proc.*, 744, M1.5.1–M1.5.10, 2003.

16. J. Paivasaari, M. Putkonen, and L. Niinisto, A comparative study on lanthanide oxide thin films grown by atomic layer deposition, *Thin Solid Films*, Vol. 472, 1–7, 2004.

17. K. Kukli, M. Ritala, T. Pilvi et al., Evaluation of a praseodymium precursor for atomic layer deposition of oxide dielectric films, *Chem. Mater.*, 16, 5162–5168, 2004.

18. R. J. Potter, P. R. Chalker, T. D. Manning et al., Deposition of HfO_2, Gd_2O_3 and PrO_x by liquid injection ALD techniques, *Chem. Vap. Deposition*, 11, 159–169, 2005.

19. H. C. Aspinall, J. Gaskell, P. A. Williams et al., Growth of praseodymium oxide and praseodymium silicate thin films by liquid injection MOCVD, *Chem. Vap. Deposition*, 10, 83–89, 2004.

20. H. C. Aspinall, J. Gaskell, Y. F. Loo et al., Growth of gadolinium oxide thin films by liquid injection MOCVD using a new gadolinium alkoxide precursor, *Chem. Vap. Deposition*, 10, 301–305, 2004.

21. R. L. Nigro, R. G. Toro, G. Malandrino et al., Effects of deposition temperature on the microstructural and electrical properties of praseodymium oxide-based films, *Mater. Sci. Eng., B*, 118, 117–121, 2005.

22. R. L. Nigro, R. G. Toro, G. Malandrino et al., Effects of the thermal annealing processes on praseodymium oxide based films grown on silicon substrates, *Mater. Sci. Eng., B*, 118, 192–196, 2005.

23. K. Y. Ahn, L. Forbes, and A. Bhattacharyya, M-I-M capacitors using Pr_2O_3 dielectric: Invention disclosure, MARI, micron technology, November 2006. U.S.P.C. Application No. 20080268605, dated October 30, 2008 entitled: Capacitor and methods with praseodymium oxide insulators, U.S. Patent: 9,231,047 (2016).

24. S. C. Choi, M. H. Cho, S. W. Whangbo et al., Epitaxial growth of Y_2O_3 films on Si(100) without an interfacial oxide layer, *Appl. Phys. Lett.*, 71, 903–905, 1997.

25. D. Schmeisser, The Pr_2O_3/Si(001) interface, *Mater. Sci. Semicond. Process.*, 6, 59–70, 2003.

26. D. Schmeisser, P. Hoffmann, and G. Beucken, Electrical properties of the interface formed by Pr_2O_3 growth on Si(001), Si(111) and SiC(0001) surfaces, *J. Phys.: Condens. Matter*, 16, S153–S160, 2004.

27. B. W. Busch, W. H. Schulte, E. Garfunkel, and T. Gustafsson, Oxygen exchange and transport in thin zirconia films on Si(100), *Phys. Rev.*, 62, 290–293, 2000.

28. Z. M. Wang, J. X. Wu, Q. Fang, and J.-Y. Zhang, Photoemission study of high-k praseodymium silicate formed by annealing of ultrathin Pr_2O_3 on SiO_2/Si, *Thin Solid Films*, 462–463, 118–122, 2004.

29. A. Fissel, Z. Elassar, O. Kirfel et al., Interface formation during molecular beam epitaxial growth of neodymium oxide on silicon, *J. Appl. Phys.*, 99, 074105, 2006.

30. X. Guo, W. Braun, B. Jenichen, and K. L. Ploog, Reflection high-energy electron diffraction study of molecular beam epitaxy growth of Pr_2O_3 on Si(001), *J. Cryst. Growth*, 290, 73–79, 2006.

31. A. Bhattacharyya, Scalable logic and FET DRAM technology using PrO_x/Pr-silicate/Pr-aluminates or other Lanthanides/Silicates/Aluminates as gate insulators and storage capacitor dielectric films; and Processing Lanthanides for silicon with controlled interface, ADI Associates Internal memo, March 18, 2007: U.S. Patent Publications: 20090090950 (03/26/2009), 20100109068 (05/06/2010), 201100291237 (12/01/2011) and 20120205720 (07/19/2012); Multiple U.S. Patents.

32. J. P. Chang, M. L. Steigerwald, R. M. Fleming et al., Thermal stability of Ta_2O_5 in metal-oxide-metal capacitor structures, *Appl. Phys. Lett.*, 74, 3705–3707, 1999.

33. J. Kwo, M. Hong, A. R. Kortan et al., High k gate dielectrics Gd_2O_3 and Y_2O_3 for Silicon, *Appl. Phys. Lett.*, 72, 130–132, 2000.

34. J. A. Gupta, D. Landheer, J. P. McCaffrey, and G. I. Sproule, Gadolinium silicate gate dielectric films with sub-1.5 nm equivalent oxide thickness, *Appl. Phys. Lett.*, 78, 1718–1720, 2001.

35. J. Kwo, M. Hong, A. R. Kortan et al., Properties of high k gate dielectrics Gd_2O_3 and Y_2O_3 for Si, *J. Appl. Phys.*, 89, 3920–3927, 2001.

36. S. Jeon, and H. Hwang, Electrical and physical characteristics of $PrTi_xO_y$ for metal-oxide-semiconductor gate dielectric applications, *Appl. Phys. Lett.*, 81, 4856–4858, 2002.

37. Y. Morisaki, T. Aoyama, Y. Sugita et al., Ultra-thin ($T_{eff}^{inv} = 1.7nm$) poly-Si-gated $SiN/HfO_2/SiON$ high-k stack dielectrics with high thermal stability (1050C), *IEEE, IEDM*, 2002, pp. 861–864.

38. M. Youm, H. S. Sim, H. Jeon et al., Metal oxide semiconductor field effect transistor characteristics with iridium gate electrode on atomic layer deposited ZrO_2 high-k dielectrics, *Jpn J. Appl. Phys.*, 42, 5010–5013, 2002.

39. S. Jeon, and H. Hwang, Effect of hygroscopic nature on the electrical characteristics of lanthanide oxides (Pr_2O_3, Sm_2O_3, Gd_2O_3, and Dy_2O_3), *J. Appl. Phys.*, 93, 6393, 2003.
40. M.-H. Cho, D. W. Moon, S. A. Park et al., Enhanced thermal stability of high-dielectric Gd_2O_3 films using ZrO_2 incorporation, *Appl. Phys. Lett.*, 84, 678–680, 2004.
41. D. Schmeisser, and H.-J. Muessig, Stability and electronic properties os silicates in the system SiO_2-Pr_2O_3-Si(001), *J. Phys. Condens. Matter*, 16, 153–160, 2004.
42. C. Wenger, J. Dabrowski, P. Zaumseil et al., First investigation of metal-insulator-metal (MIM) capacitor using Pr_2O_3 dielectrics, *Mater. Sci. Semicond. Process.*, 7, 227–230, 2004.
43. M. Suzuki, M. Tomita, T. Yamaguchi et al., Ultra-thin (EOT~3A) and low leakage dielectrics of La-aluminate directly on Si substrate fabricated by high temperature deposition, *IEDM*, 2005, pp. 445–448.
44. A. Bhattacharyya, Properties and applications of silicon oxynitride films (technical Memo, IBM, 1976): Unpublished.
45. A. Bhattacharyya, Promising high-k mixed oxy-nitride films containing at least one rear-earth metal, for example, LaTaON or LaHfON for electronic applications, ADI Associates internal publications (Unpublished), 2006; U.S. Patent Publications: a) HfLaON films: 20090294924 (12/03/09) and b) TaLaON films: 20090236650 (9/24/09) by L. Forbes, K.Y. Ahn, and A. Bhattacharyya.

Index

Note: Page numbers followed by f and t refer to figures and tables, respectively.

1D-NROM cell, 378f
1 Gb NROM design, 131f
1 Gb SLC device, 101, 101f
1 Mb EEPROM SONOS device, 20
1-T USUM RAM cells, 434, 434f
2 bit/cell SNOS device, 111f
2-layer crested barrier
    with bi-planar level trapping, 245–246
    with single/dual-energy level trapping, 245
3 bit/cell program-verify, 286f, 287f
3D-CT-NAND flash memory cells, 359
3D FG-NAND Design, 352
3D NV devices and arrays, 348–352, 350f
3D SONOS flash memory, 340, 340f
3D S-SGT-FG-NAND structure, 343, 344f
3D V-NAND technology, 352
4 bit/cell CT-NROM, 281
4 Gb SLC NAND chip, 128f
4 Mb EEPROM NAND chip, 123–124, 124f
4 Mbit EEPROM, 122
8 Gb chip MLC NAND device, 101, 102f
8 Kb AROS, 19
8 Kb EAROM, 14
16 GB MLC NAND flash memory, 9f, 10
16 Kb EEPROM, 120
32 bit NAND string, 127

A

ACEE (array contactless EEPROM) cell, 121–122
Advanced CT-NVM
    device, 378f
    stack designs, 379
        first generation, 375t
        second generation, 376t
        for special attributes, 377t
        third generation, 376t
Advanced MLC NVM device, 390
Advanced nanocrystal device, 389–390
Advanced NVM device, 372–382
    and array concepts, 372–374
    stack and band features, 374–382
Advanced stand-alone NAND flash, 389
"Air-gap" technology, 282–283
Al_2O_3. *See* Alumina (Al_2O_3) film
Alterable Read Only Storage (AROS) cell, 119, 120f
Alternate Memory Technologies, 360
$Al/TiO_2/Si_3N_4/Si$ gate stack characteristics, 187f
Alumina (Al_2O_3) film, 178–180, 198, 373
    mean time to failure, 180f
    nonstoichiometric silicate formation, 179
    stress stability, 179
    ultra-thin, leakage characteristics, 179, 179f
Aluminates, 188
    of lanthanides, 188–191

Amorphous dielectric, 40
Amorphous metal oxide thin films, 176–178
Anode Hole Injection (AHI), 53
Anode Hydrogen Release (AHR), 53
AROS (Alterable Read Only Storage) cell,
    119, 120f
Array contactless EEPROM (ACEE) cell, 121–122
Asymmetric band gap-engineered (ABE) DTM, 247

B

Ballistic Fowler–Nordheim tunneling, 52
Band
    diagram
        CT-SONOS devices, 216
        FG devices, 214–216, 215f
    gap, 39
        insulators, 52
Band-engineered DTM MSUM cell, 463–466
    band characteristics for NROM and NAND
        applications, 464–465
    CT-MSUM cell and array configurations,
        465–466
    design consideration, 463–464
Band-engineered multiple direct-tunnel high K device,
    247
Band engineering, 196, 219
    application for NVM device attributes, 233–247
        CT/NC/DT NVMs, 236–237
        endurance enhancement, 246–247
        memory retention enhancement, 240–247
        memory window enhancements, 237–240
        programming speed enhancement, 237
        property enhancements, 237–240
        trap engineering, 240
        Vpp lowering & programming time and power,
        233–235
    design options, 220–221
    for direct tunnel multilayer tunnel dielectrics,
        231–233
        double-layer, 231–232
        triple-layer, 232, 233f
    DT NVM, 247–253
        band-engineered multiple direct-tunnel high K
        device, 247
        BE-SONOS DTM, 249–250
        DTJ DTM, 248–249, 248f
        PBO-DTM, 251–253
        RTB DTM, 250–251
    in MFC NVM, 259–260
    in MLC NVM, 256–259
    stack
        design, 220
        requirements to enhanced retention,
        243–244

Band engineering (*Continued*)
  SUM devices, 423–431
    MSUM, 426–431
    USUM, 423–426
  thicker multilayer tunnel dielectrics, 221–230
    double-layer dielectrics, 221–224
    progressive band offset, 230
    triple-layer dielectrics, 226–230
  using tunnel dielectric films
    multilayer, 218–220
    single-layer, 217–218
Band-to-band
  direct tunneling, 103
  tunneling, 220
Barrier
  energy asymmetry, 219
  lowering, 223, 226, 226f
  offset effect, 223, 226, 228
  thinning, 222, 222f, 226, 428
Bentchkowsky, D. Frohman, 18–19
BE-SONOS, 303
BE-SONOS DTM, 249–250
  band diagrams, 249, 250f
  characteristics, 250, 250f
  retention enhancement, 251f
BiCS. *See* Bit cost scalable (BiCS)
Binary state(s), 3–4
  DRAM cell, 6
Biplanar/multiplanar charge-trapping design, 346
Bit cost scalable (BiCS), 348, 349f
  3D NAND architecture, 350f
  endurance and retention, 351f
  program/erase characteristics, 351f
Bits, 3
Blocking dielectric film, 197
Blocking oxide/OR-SiON, 107–108
Body charge, 434
Box–body interface trapping layer, 435
BT-RAM cell, 477
Built-in Operating System (BIOS), 7
Bulk-controlled mechanism, 42

C

Capacitive coupling ratio, 99
Capacitor charge density, 40
Capacitor-less DRAM cell, 399
Carrier transport, 90
CB (coulomb blockade), 22
Cell-to-cell coupling (CCC), 282–284
Charge
  capture probability, 196
  confinement, 40
  high K dielectric film for
    blocking, 199, 207–208
    CT, 199, 204–207
    tunneling, 198
  holding capacity, 40
  retention characteristics, 85f
  storage capacity, 102
  transport characteristics, 40
  trap NVM cells and arrays, 130–134
    disturbs in, 138
  trapping/detrapping, 40

Charge injecting SRO (CI-SRO) film, 68
Charge-trap flash (CTF) technology, 352
Charge-trapping (CT) devices, 11
  drain disturb transient characteristics, 157f
  nonplanar multibit/cell vertical channel,
    325–329
    trench-based vertical channel CT NROM,
      325–327, 326f
    vertical channel DTM-NROM, 327–329,
      328f, 329f
    vertical channel FG-NAND, 329
  radiation responses, 156–158
  voltage scaling, 169–170
Charge-trapping flash (CTF) memory cells, 293
Charge-trapping (CT) memory cell
  advantage over FG memory cell, 18
  reliability issues, 151–156
    characteristics, 151
    data retention loss, 153–155
    drain disturb, 156
    hot-carrier-induced CT cells, 153
    STS, 155–156
    thicker tunnel oxide cells, 153
    ultrathin tunnel oxide cells, 151–152
Charge trapping SRNs (CT-SRNs) film, 66, 77
Chemical vapor deposition (CVD) nitride, 55–56
CI-SRO (charge injecting SRO) film, 68
CMOS-field-effect transistor (FET) technology, 4
Complementary metal–oxide–silicon (CMOS) technology,
  3, 271
  challenges, 271
Conductive FG memory cell, 291f, 293
Conductivity mechanisms, dielectric, 45–49
  bulk-controlled Poole–Frenfel, 45, 45f
  electrode-controlled quantum mechanical, 46
    Direct Tunneling/modified Fowler–Nordheim,
      46–47
    enhanced Fowler–Nordheim, 48, 49f
    Fowler–Nordheim, 47, 47f
Control gate (CG), 36–37
Conventional CT devices, 18–20
  CT-NVM cell designs, 19–20
    pass gate memory cell, 20
    TMS, 19–20
  MNOS/MXOS/MONOS/SNOS/SONOS, 18
  MNOS, MXOS to SONOS CT devices, 18–19
Conventional dielectric films, 51
  J *versus* E plots, 76–77, 76f
  for NVM device, 51–70
    Si_3N_4, 55–56
    SiO_2, 51–55
    SiONs, 56–64
    SRIs, 64–70
Conventional NVM device
  limitations, 161
  scaling challenges, 159–161
    FEE, 159–160
    limitations and challenges, 160–161
    SCE, 159
    vertical scalability, 160
  stack designs, limitations, 216–217
Conventional silicon based NVMs memories,
  360–363
"Correction before coupling" (CBC) reprogram, 296

Coulomb blockade (CB), 22
CPRAM USUM cell, 445–449
    operation, 448–449
Crested barrier, 226
    2-layer
        with bi-planar level trapping, 245–246
        with single or dual-energy level trapping, 245
        with deep-offset trapping, 245
        with high K dielectrics, 229–230
        tunnel, 226
    vs. Variot, 228, 229f
CRESTED barrier MSUM device, 429–430, 429f
Cross-point thyristor-based charge storage RAM cell,
        445–449, 446f
    operation, 448–449
        waveform, 449f
    vs. TRAM cell, 447f
CT. *See* Charge-trapping (CT) devices
CT-DTM USUM, 414–416
CT-MLC
    multilayer NVM stack designs
        DCSL cell, 303–305
        DSM cell, 305–306
    triple trapping layer stack design, 301f
CT-MSUM device(s), 416–419, 466–472
    cell and array, 465–466
    crested barrier tunneling schemes, 419
    nanocrystal-based, 466–471
        for DRAM functionality, 466–467
        DRAM MLC states, 470
        for NVM functionality, 467
        NVRAM MLC states, 470–471
        operational scheme, 470
        single transistor, 467–468, 467f
        size, trapping characteristics and memory window,
            468–469
    parameter objectives, 418t
    PBO/VARIOT barrier tunneling schemes, 418
    reverse mode, 471–472
        band diagram, 471f
        multilayered conductive floating/floating-plate, 472
    tunnel layer
        band characteristics, 410f, 411f
        cost-performance, 411
        electron-only CT-MSUM crested tunnel barrier,
            412
        performance-driven, 410–411
CT-NAND USUM cell
    high density, 455
    very high density vertical-channel, 455, 456f
CT-NC device, 11, 102
CT-NROM SONOS device, 144
CT-NROM USUM cells, 453–455
CT-NVM cell designs, 19–20
    pass gate memory cell, 20
    TMS, 19–20
CT-NVM devices
    charge loss
        and leakage paths, 81–84
        mechanisms, 83–84
    stack design, 102–116
        blocking oxide/OR-SiON, 107–108
        elements and device properties, 103–108
        NC, 115–116

ONO stack design and SONOS device properties,
        108–110
    SONOS stack designs, 110–115
    trapping layer, 106–107
    tunneling oxide, 103–106
    tunneling modes, 104f
    writing and erasing process, 73
CT-NVM USUM devices, 450–451
CT-SONOS device, 102–103
    band diagram, 216
CT-SRNs (charge trapping SRNs) film, 66, 77

D

DCSL. *See* Dual-charge storage layer (DCSL)
Defect centers/traps, 40
DEIS EAROM device, 68
Dennard, R. H., 6
Device scalability of conductive FG cells, 292f
Dielectric
    breakdown, 40
    conductivity mechanisms. *See* Conductivity
        mechanisms, dielectric
    constant (K), 40
    films
        carrier transport in, 42
        conductivity, 45
        electronic properties of, 39–41
        multilayer, carrier transport mechanisms
            for, 49
    strength, 40
Dielectric films
    for NVM device, 51–70
        Si_3N_4, 55–56
        SiO_2, 51–55
        SiONs, 56–64
        SRIs, 64–70
    for SUM devices, 405–412
        blocking layer selection, 407–408
        charge trapping and charge storage layers,
            406–407
        CMOS FET gate insulator, 407–408
        IMI layer for metal gate interface, 406
        I-SRN role for Charge Storage, 407
        silicon interface, 405–406
        tunneling dielectric, 408–412
Diffusion isolation scheme, 448, 448f
Digital memories, 3–4
Digital systems, memory hierarchy in, 9, 9f
DINOR (divided bit-line NOR) cell, 17, 128–129
Direct tunnel device, 220
Direct tunneling, 46–47
Direct tunneling memory (DTM), 29–31, 30f, 90. *See also*
        Direct tunnel NVM (DT NVM) devices
    ETDTM, 30–31, 31f
    improved, 30–31
    limitations, 30
    process flow, 30f
    schematic cross section, 30f
Direct tunnel multilayer tunnel dielectrics, band
        engineering, 231–233
    double-layer, 231–232
        modified, 232
    triple-layer, 232, 233f

Direct tunnel NVM (DT NVM) devices. *See also* Direct tunnel memory (DTM)
  band engineering for, 247–253
    band-engineered multiple direct-tunnel high K device, 247
      BE-SONOS DTM, 249–250
      DTJ DTM, 248–249, 248f
      PBO-DTM, 251–253
      RTB DTM, 250–251
  based multifunctional NVMs, 259f
Discrete traps device, 11
Disturbs, 309
  CT (NROM) devices and arrays, 138
  FG devices and arrays, 134–136
    program, 136, 136f
    read, 134–135, 135f
    minimization/mitigation, 136–138
    read disturb, 136–137
Divided bit-line NOR (DINOR) cell, 17, 128–129
DLS NAND product, 312
Double data rate (DDR) interface, 296
Double-Gate NROM, 382
Double-layer
  dielectrics, 221–224
    band options, 221f, 224t
    characteristics, 224t
    current–voltage characteristics, 224–225
    high K high barrier/low K low barrier, 223–224
    high K low barrier/low K high barrier, 223
    low K high barrier/high K low barrier, 222–223
    low K low barrier/high K high barrier, 224
  direct tunnel dielectrics, 231–232
Double SONOS memory (DSM) cell, 305–306
Double tunnel junction (DTJ), 232, 303
Drain disturb
  CT device and array, 38
  FG device and array, 136, 136f
    mitigation/minimization, 137–138
  reliability issue
    in CT cell, 156
    in FG devices, 150
DRAM. *See* Dynamic random access memory (DRAM)
DSM (double SONOS memory) cell, 305–306
DTJ. *See* Double tunnel junction (DTJ)
DTJ DTM, 248–249, 248f
DTM. *See* Direct tunnel memory (DTM)
DT NVM. *See* Direct tunnel NVM (DT NVM) devices
Dual-charge storage layer (DCSL), 303
  energy band diagram, 304f
  memory, 303f, 304f, 305f
Dual-Control-Floating Gate (DC-FC) Cell, 352
Dual-electron injector structure (DEIS) cell, 21
Dual-Plane NROM, 382
"Dual-pulse" channel, 296
Dynamic random access memory (DRAM), 6, 7f, 459
  functional attributes, 357t
  market, 482–483
    drivers, 358t
  memory, 4f

E

EAROM (electrically alterable ROM), 14
E-beam evaporation, 28

ECC solutions for SDD, 317f
EEPROM (electrically erasable and programmable read-only memory), 8
Effective oxide thickness (EOT), 40, 52
  impact on memory window, 75
Electrically alterable ROM (EAROM), 14
Electrically erasable and programmable read-only memory (EEPROM/E²PROM), 8
Electrically programmable read-only memory (EPROM), 8
Electrode compatibility, 44
Electrode-controlled quantum mechanical tunneling mechanisms, 46
  Direct Tunneling/modified Fowler–Nordheim, 46–47
  enhanced Fowler–Nordheim, 48, 49f
  Fowler–Nordheim, 47, 47f
Electron
  affinity, 39
    dielectric, 43
  detrapping mechanism, 79
Electronegativity, 39
Embedded multidie interconnect bridge (EMIB) packing scheme, 480–481, 481f
Embedded NAND, 386
Embedded NAND flash device, 386–389, 387f, 388f
Embedded NROMs, 369
Embedded NVMs, 269, 369–370
Embedded Trap DTM (ETDTM), 30–31, 31f
Embedded traps device, 11
Endurance, 86–92, 90t
  charge fluence, dielectric conductance, and QBD, 89–90
  definition, 86
  and depend types, 90
  MXOS device with high, 91–92, 92f
  and peak field, 87–89
  requirement for high, 91
  and retention, 309
Enhanced CT-MNSC devices
  band-engineered MNSC device, 309–310
  dual-gated enhanced CT-MNSC device, 310–312
enhanced Fowler–Nordheim (F–N) tunneling, 42, 48, 49f, 220, 247, 249
Enhancement current density, 48
Equivalent SiO₂ oxide thickness (EOT), 40, 42
Erase, 307–308
Erase-Saturation, 26
Erasing process, 3
Erratic bit phenomenon, 148, 148f
Error correction code (ECC), 294
Etch rate, 61
ETDTM (Embedded Trap DTM), 30–31, 31f
ETOX cell, 122, 128
ETOX NVM cell, 16
Extendable S-SGT NV device, 342–343
Extrinsic defects, 174
Extrinsic traps, 196

F

Fast erase cells, 150
FB RAM USUM cell, 433–436
  body-charge generation, 434, 435f
  memory cell reading, 435, 436f
  structure, processing of, 435–436

Feature size scalability, 167
FEE (few-electron effects), 159–160, 323
FEOL (front-end-of-line) level, 269
FET. *See* Field-effect transistor (FET)
FET USUM cell, 449–456
  CT-NAND USUM
    high density, 455
    very high density vertical-channel, 455, 456f
  CT-NVM USUM devices, 450–451
  planar merged two-device CT-NVM USUM NROM
      cells, 451–455
    CT-NROM USUM cells, 453–455
    planar/vertical and vertical-channel,
        452–453
Few-electron effects (FEE), 159–160, 323
FG. *See* Floating-gate (FG) devices
FG-FG coupling, 100f
FG-MLC
  cell designs, 284–287
  extendibility challenges, 287
    advanced technology and improved program
        algorithm, 289
    controlling read-current reduction, 290
    STI/sidewall fringing field and cell geometry
        effects, 289–290
    sub-lithographic patterning, 288
FGNAND. *See* Floating-Gate NAND (FGNAND)
FGNAND flash
  array circuit, 331f, 360
  devices, 97–102
    FG-FG coupling, 99, 100f
    floating poly gate, GCR, and STI, 97–102
    GCR *vs.* floating-gate height, 99, 100f
    NAND cell and array, 97, 98f
    Vth dispersion, 100, 101f
  products, 281
FG-NAND SLC devices, 284
FG-NVM device, 13–18, 14f
  band diagram, 78f
    and charge loss modes for, 81–82
  charge transfer and storage in, 72–73, 72f
    erasing process, 72–73, 72f
    writing process, 72, 72f
  current components, 77–78
    band diagram-based assessment, 78–79
  DINOR cell, 17
  endurance characteristics, 87, 88f
  ETOX/FLOTOX cell, 16
  evolution chronology, 15–16
  FAMOS, 13–14
  in first two decades, 17t
  HIMOS cell, 16
  NAND cell, 17–18
  SAMOS, 14
  SAMOS 8 Kb EAROM, 14–15
  SIMOS, 15
  stack design with PBO, 230f
  technology, process flow of, 144
  transistor memory cell, 13, 14f
  tunnel oxide thickness limit, 80–81
  window and retention enhancement, 81
Field-effect transistor (FET)
  device scaling, 173
  gate-insulator, 176

Field-induced traps, 196
FinFET
  technology and devices, 329–336
  transistor, 324
FinFET Embedded NROM, 382
FinFET SONOS devices, 112f, 330–331
  flash, 329–330
  programming characteristics and memory
      window, 333f
FinFET TANOS device, 333f
Flash devices, 8
Floating body-charge trapping USUM RAM Cell,
    433–436
Floating-gate avalanche injection MOS (FAMOS) device,
    13–14
Floating-gate (FG) cell, 99, 99f
  reliablilty issues, 148–150
    data retention degradation, 149–150
    drain disturb, 150
    erratic bit phenomenon, 148, 148f
    fast erase cells, 150
    transconductance degradation, 149
Floating-gate (FG) devices, 11
  band
    bending, 215f
    diagram, 214–216, 215f
    engineering in, 233–234, 238
  drain disturb transient characteristics, 157f
  energy band diagram, 37, 37f
  parasitic coupling, 282
  radiation responses, 156–158
  slimming, 288
  voltage scaling, 168–169
Floating-gate flash NVM cells, 119–122
  ACEE, 121–122
  ETOX, 122
  FLOTOX, 120, 121f
Floating-Gate NAND (FGNAND), 167
  devices, 386
Floating-gate NOR cells and arrays, 128–129
  DINOR memory cell and VGA architecture,
      129
Floating plate (FP) devices, 11
FLOTOX, 120, 121f
  cell, 16
Fowler–Nordheim (FN) tunneling, 24, 42,
    47, 47f
  write and erase mode, 120
Fringing field effect (FFE), 324
Front-end-of-line (FEOL) level, 269

G

GAA NV devices, 336–340
GAA-SONOS nanowire device, 336
GaN, 373
Gate-all-around (GAA) device, 295
Gate coupling ratio (GCR), 99, 281
  for MLC NAND design, 284
gated N+-i-P+ diode, 440–441, 440f
gated P+-i-N+ diode, 441, 441f
Gate-first process process integration scheme, 183,
    191, 210
Gate oxide capacitance, 52

Gate stack, NVM device
   attributes, 36–38
      CG, 36–37
      charge blocking layer, 36
      charge storage layer, 36
      energy band diagram, 37–38, 37f
      interface A, 36
      interface B, 36
      interface C, 36
      interface D, 36
      tunnel insulator, 36
   design
      challenges, 272
      characteristics, 272
   dielectric elements, reliability issues, 147–148
   layers and interfaces, 35
   material, 44–45
GCR (gate coupling ratio), 99, 281
GDRAM (gated diode charge storage nonvolatile) USUM
      cell, 437–445, 437f
   in bulk silicon implementations, 437
   circuit schematics, 438f
   diode I-V characteristics, 438–440, 439f
   operation, 440–441
      gated N+-i-P+ diode, 440–441, 440f
      gated P+-i-N+ diode, 441, 441f
   SOI diode charge storage RAM cell, 443–444, 443f
   SOI technology implementation, 441–442, 442f
      SOI-FET, 442
Germanium nanocrystal device characteristics, 29
GL-TRAM, 399
GTRAM (gated thyristor-based charge storage) USUM
      cell, 444–445

H

Hafnia (HfO$_2$) film, 180–184
   advantages, 182–183
   band diagram, 194, 194f
   capacitors, band diagram, 195, 195f
   doped polysilicon gate stack, 180, 181f
   interface engineering, 183–184
   leakage characteristics, 180, 181f
   SILC characteristics, 182, 183f
   silicate formation, 180
Hafnium and zirconium
   bimetal oxides and aluminates of, 188
   nitrides and oxynitrides of, 191–192
Hafnium silicon oxynitride (HfSiON) film, 192–193
   band diagram, 194, 194f
   FET device characteristics, 193
   I-V
      characteristics, 192, 192f
      stress stability, 193, 193f
Half-select cells, 135
Halo doping, 425
Hard disk drives (HDD)
   functional attributes, 357t
   market drivers, 358t
HDS (high-density SIM), 268, 481
HfLaSiON film, 191
HfO$_2$. *See* Hafnia (HfO$_2$) film
HfSiON. *See* Hafnium silicon oxynitride (HfSiON) film
High-density SIM (HDS), 268, 481

High K dielectic films
   crested barrier with, 229–230
   vs. conventional SiO$_2$, 217–218
High K dielectric films, NVM
   Al$_2$O$_3$, 178–180
   aluminates of lanthanides, 188–191
      LaAlO$_3$, 189–191
   band diagram comparison, 194–195
   for charge
      blocking, 199, 207–208
      CT, 199, 204–207
      tunneling, 198
   common, 175–176, 195–197
      blocking, 197
      trapping, 196–197
      tunnel, 196
      unique functional requirements, 195
   device design objectives and dielectric selection
      options, 208–209
   FET gate-insulator requirements, 176
   future perspective, 209–210
   for gate stack design, 199–201
   general requirements, 174–175
   hafnium and zirconium
      bimetal oxides and aluminates of, 188
      nitrides and oxynitrides of, 191–192
   HfO$_2$, 180–184
   HfSiON, 192–193
   historical perspective, 173–174
   integration in silicon based CMOS NVM technology,
      210
   leakage
      characteristics, 174f
      and retention, 203–204
   leakage characteristics for tunnel application,
      203–204
   Ta$_2$O$_5$, 188
   thin amorphous films, 176–178
   TiO$_2$, 187
   for tunneling, 201–202
   ZrO$_2$, 184–187
High-performance dual-mode NVM device, 260–261, 261f
High Vth memory retention state, 214
HIMOS cell, 16
Hopping mechanism, 45
Hot-carrier-induced CT cells, 153
Hybrid Memory Cube (HMC) integration scheme, 268
Hynix semiconductors, 288

I

IBM R&D groups, 19
Incremental step-pulse programming-mode (ISPP), 80
Inhibit Channel-Coupling-Reduction (ICCR), 296
Inhibit scheme, 125
injector-SRN (IN-SRN/I-SRN), 11, 66, 68–69
   layer, 327
Insulator stack structure, 25
Integrated DRAM-NVRAM MSUM cell, 460–463, 460f
   multifunctionality, 461–462
   SOI implementation, 460–461, 461f
Interface engineering, 183–184, 235
Interlayered band diagram, 302f
Internal Schottky Mechanism, 45

Interpoly dielectric (IPD), 78, 226
    layer, 36
    ONO
        charge blocking layer, 284
        leakage characteristics, 78–79, 79f
Intrinsic defects, 174
Intrinsic trap density, 196
IPD. *See* Interpoly dielectric (IPD)

K

Kahng, D., 13, 119
K effect, 222, 222f, 224, 228

L

Lanthanides, 188–189
Lanthanum aluminate (LaAlO₃) film, 189–191,
    190f
    band diagram, 194, 194f
    leakage current characterization, 189
Likharev, K. K., 226
LOCOS isolation, 143
Log TDDB, 54
Lo/Hi barrier band diagram, 231f
LPCVD nitride, 55–56

M

Mask programming process, 7
MATHS, 198
Memory
    arrays, 3
    cells, 3
    FET device, 13
    hierarchy in digital systems, 9, 9f
    levels for 1-bit/cell (SLC), 284f
    retention. *See* Retention
    states, 71
    window (W), 71–75
        charge transfer and storage in FG, 72–73, 72f
        generation, 73–75
        Vpp and EOT impact on, 75
        vs. charge transport, trapping, 75t
Memory management units (MMUs), 265
Memristor cell, 363
Merged Gate cell, 20
Metal(s)
    gate interface layer, 406
    nanocrystal device characteristics, 28–29
    oxide figure of merit, 43t
Metal–alumina–oxide–silicon (MAOS) substrate, 12
Metal–insulator–silicon (MIS) capacitors, 6, 20
    dielectric film in, 42
Metal–nitride–oxide–silicon (MNOS) substrate, 12
Metal–oxide–silicon field effect transistor (MOSFET), 13
    device evolution, 324f
Metal–silicon oxynitride–oxide–silicon (MXOS)
    device(s), 63
        band diagram, 91, 92f
        enhanced charge retention, 85–86
        with high endurance, 91–92, 92f
    substrate, 12
MFC. *See* Multifunctional cell (MFC)

MHTHS, 198
Micron-Intel 3D NAND cell, 352
Micron Technology, 295, 352, 362
MIS capacitor. *See* Metal–insulator–silicon (MIS)
    capacitors
MLC (multilevel cell), 281
MMCT USUM. *See* Multimechanism-Carrier-Transport
    (MMCT) USUM
MNOS CT device, 18–19
MNSC (multiple node storage cell), 297
Modified direct tunneling, 219
Modified double-layer direct tunnel barrier, 232
modified Fowler–Nordheim tunneling, 46–47
MONOS device
    band diagrams, 81–82, 82f
    retention time modeling, 84
    stability and retention, 83–84
    vs. MOXOS, 113f
Most cost-sensitive low-function applications (MCLA),
    372
MOXOS device, 113f
MRAM, 362
MSUM. *See* Multifunctional SUM (MSUM) cell
MSUM device, 398–399
    band engineering, 426–431
        background and historical references, 426–427
        band diagram, 427
        CRESTED barrier CT-MSUM, 429–430
        CRESTED barrier multifunctional
            CT-NVM, 430
        PBO multifunctional CT-NVM to PBO-CT-MSUM,
            427–428
        VARIOT-CT-MSUM, 428, 429f
    features, 414
    memory array, 401–402, 402f
    parameter objectives, 419–420
    stacks, 419
    tunnel layer of, 410–412
MSUM NROMs, 482
    arrays, 466f
Muler, R. G., 15
Multibit per cell (MLC) NVM, 256–259
Multibit S-SGT device, 343–347
    SGAA-NROM/SGT-NROM, 344, 345f
    SGT multifunctional NROM, 344–346
    S-SGT very high-density (3D) NAND design, 347
    surround gate floating gate, 343, 344f
Multifunctional cell (MFC), 380
    NVM, band engineering in, 259–260
Multifunctional CT-NVM device, 427, 428f
Multifunctional NVMs, 344–345
    devices, 390–391
Multifunctional SUM (MSUM) cell, 459–477
    band-engineered DTM, 463–466
        band characteristics for NROM and NAND
            applications, 464–465
        CT-MSUM cell and array configurations,
            465–466
        design consideration, 463–464
    CT-MSUM devices, 466–472
        cell and array, 465–466
        nanocrystal-based, 466–471
        reverse mode, 471–472
    functionality, 472

Multifunctional SUM (MSUM) cell (*Continued*)
   integrated DRAM-NVRAM, 460–463
      array layout, 463
      for field programmable operation, 462
      multifunctionality, 461–462
      for multilevel NVRAM operation, 462
      for NVRAM operation, 462
      operational attributes, 462–463
      SOI implementation, 460–461, 461f
   operational conditions, 461t
Multifunctional universal memory, 473–476, 473f, 474f,
      475f, 476f
Multi-layered SXOS stack, band diagram, 108, 109f
Multilayer tunnel dielectric films, band engineering in,
      218–220
   for direct tunnel, 231–233
      double-layer, 231–232
      triple-layer, 232, 233f
   thicker. *See* Thicker multilayer tunnel dielectrics
Multilevel cell (MLC), 281
Multilevel NVM devices, 281–282
Multimechanism-Carrier-Transport (MMCT) USUM, 416
   device, 424–426, 426f
      array layout, 454f
Multiple node storage cell (MNSC), 297
Multiple node storage devices (MNSD), 297
Multiple trap energy, 302
Multiple trapping layer, 300
MXOS. *See* Metal–silicon oxynitride–oxide–silicon
      (MXOS)
MXOS CT devices, 19

N

NAND device
   process flow, 145–146, 146f
   SASTI process flow in, 143
NAND EEPROM, 126
   device, cross-sectional view, 142f
NAND flash, 369
   applications, 483t
   devices, band engineering in, 234
   memory, 8, 17, 160
NAND flash cell, 122–128
   4 Mb NAND EEPROM, 124–126
      erase inhibit, 125
      erase operation, 124
      NAND cell inhibit scheme, 125
      program inhibit, 125–126
      read operation, 125
      writing and programming operation,
         124–125
   floating gate, 123–124
      progress in, 126–128
NAND memory cell, 17–18, 282
Nanocrystal-based CT-MSUM device, 466–471
   for DRAM functionality, 466–467
   DRAM MLC states, 470
   for NVM functionality, 467
   NVRAM MLC states, 470–471
   operational scheme, 470
   single transistor, 467–468, 467f
   size, trapping characteristics and memory window,
      468–469

Nanocrystal CT-NVM devices, 20–29
   history, 20–21
   nanocrystal device characteristics, 25
      germanium, 29
      metal, 28–29
      silicon, 26–28
   nanocrystal NVM devices, 24–25
   nanocrystal physics and charge trapping, 21–23
Nanocrystals (NCs), 11
   charge trapping, 11
   CT devices, stack design, 115–116
   device, 11
   device characteristics, 25
      germanium, 29
      metal, 28–29
      silicon, 26–28
   NVM devices, 24–25
Nanodots (NDs), 21–23
   density effect on threshold shift, 23, 23t
   device, 11
Nano-Wired (NW) FET (TSNWFET) devices, 336–340
Narrow width effect (NWE), 324
NC-NVMs, 390
NCs. *See* Nanocrystals (NCs)
NDs. *See* Nanodots (NDs)
Negative differential resistance (NDR)
   device, 437
   effects, 448
NFET, 378f, 387
Nitride, 55–56, 106–107, 199
   band structure, 107f
   charge trapping, 11
   electron and hole trapping, 59–60
   films, 55
      nonstoichiometric, 56
   I *versus* $E^{0.5}$ plots, 58f
   traps creation, 55–56
   vs. oxynitride, MIOS trapping, 59–60
Nitride NOR, 110, 130
Nitrogen-rich-SiON (NR-SiON), 57
   advantage, 63
   films, 204
"Non-airgap" option, 283
Nonplanar NVM devices, 324–325
Nonplanar "wrap-around" cell, 282
Nonsilicon based memories, 360
Nonstoichiometric nitride films, 56
Nonstoichiometric silicate (Alx Siy Oz) formation, 179
Nonvolatile memory (NVM), 4, 265–279
   cell and arrays, 119
      charge trap NVM, 130–134
      disturbs and mitigation, 134–138
      floating-gate flash NVM, 119–122
      floating-gate NOR, 128–129
      NAND flash, 122–128
   conventional silicon based, emerging contenders,
      360–363
      PCM/PRAM, 362
      ReRAM/RRAM, 363
      STT-MRAM, 362–363
   current, 482–483
      functional attributes, 356t, 357t
   devices. *See* NVM device(s)
   EPROM, EEPROM, and E^2PROM, 7–8

FET, 10f
fundamental concepts in, 10
future applications/systems, memory attributes for, 363, 363f, 363t
gate stack
high K films in, 372–374
transport mechanisms, 374
market horizon and driving factors, 356, 358t, 359–360
cloud/data center level, 356, 359–360
consumer level, 356
enterprise, 356, 359
intelligent multifunctional mobile level, 356
observations, 356, 359–360
masking processing, 373–374
NROM and NAND flash memories, 8–9
product and stack design basic consideration, 372
with progressive nonvolatility, 255, 256f
ROM, 7
SRN for, 11
technology, 141
integration. *See* Technology integration, NVM
Normalized CCC ratio *versus* NAND cell technology scaling node, 283f
Normal mode NVM device, 95
NOR NVM cells, 370
NROM, 169, 482
flash
devices, 102
memory, 8
scalable and nonplanar, 382–383
Dual-Plane/Double-Gate, 382
FinFET Embedded, 382
TRENCH-BASED Vertical Channel Embedded and Stand-alone, 383
NROM device(s), 130
advancement in, 131–132
cells and arrays, 130
disturbs in, 138
cross-section, 130f
general operations, 131
near-end architectural and functional requirements, 371t
process flow and integration scheme, 144–145
product
potential and challenges, 134
scalability road map, 133f
scaling and product progression, 133
NROM-NVM
market drivers, 358t
split-channel cell, 254–255, 254f
NROM USUM cells
split-gate, 452f
vertical-channel configuration, 453f
NR-oxynitride, 106–107
band structure, 107f
NR-SiON. *See* Nitrogen-rich-SiON (NR-SiON)
NVM. *See* Nonvolatile memory (NVM)
NVM device(s)
advanced, 372–374
application parameter enhancement consideration, 368–369
charge retention and programming speed, 368
CT-SRN films, 69–70
direct tunnel, band engineering in, 247–253

electrostatic potential in, 76
embedded, 369–371
application drivers for, 369–370
endurance and cyclic reliability, 368
groupings and nomenclature, 11–12
historical progression, 13–31
IN-SRN in, 68–69
integration options, 371–372
memory window and charge retention, 368
naming conventions for, 12t
normal mode, 95
NROMs, functional and architectural requirements, 370
parameter driver, 367–368
power reduction, 368
and product attributes, 161t
properties, 71–92
endurance and its traits, 86–92
retention, 75–86
window, 71–75
reliability, 147–158
reverse mode, 95
stack
cross-sections, 35f
design, 95–116
stand-alone, application drivers for, 369–370
Vth memory states stability, 76
NVM-unique technology integration features, 142–144
high-voltage generation, transmission, and circuit requirements, 142
process flow and integration scheme, 144
SASTI scheme, 143

O

"Off" memory state, 3
Off-stoichiometric nitride, 65
"On-Chip-Randomizer" scheme, 296
"On" memory state, 3
ONO. *See* Oxide-nitride-oxide (ONO)
OR-SiON. *See* Oxygen-rich-SiON (OR-SiON)
Over-erasure protection, 451
Oxide Impact Ionization (OII), 53
Oxide isolation scheme, 448, 448f
Oxide-nitride-oxide (ONO)
IPD, 226
charge blocking layer, 284
leakage characteristics, 78–79, 79f
stack design, 108–110, 331
Oxygen-rich-SiON (OR-SiON), 57, 173
advantage, 63
for silicon interface in SUM device, 405–406
Oxynitride, 56. *See also* Silicon oxynitrides (SiONs)
electron and hole trapping, 60
refractive index and, 59t
trap depth in, 59

P

Pass gate memory cell, 20
PBO. *See* Progressive band offset (PBO) barrier
PBO-DTM. *See* Progressive band offset DTM (PBO-DTM) device
PBO-DTM USUM, 416

PCT. *See* Planar charge trapping (PCT)
P/E memory window degradation, 289f
PFET, 378f, 387
    device, 19
p+ gated tri-gate SONOS device
    disturb characteristics, 336f
    retention characteristics, 336f
Phase change memory (PCM) device, 362
Phase-state low electron-number drive random access
        memory (PLEDM) cell, 160
Planar and wrap-around floating-gate flash designs, 290–291
    GCR reduction yielding smaller memory window, 293
    high K IPD and metal gate for planar conductive FG
        cell, 291–293
    STI edge-fringing field effect, 293–294
Planar charge trapping (PCT)
    characteristics, 24
    MLC devices, 298–299
        advanced CT-MLC stack design, 300–302
        multiple trapping stack layer, 300
        single trapping stack layer, 299
        stack design options for planar CT-MLC
            devices, 299
        three-bit/cell storage, 302–303
    MNSC devices and extendibility, 306–307
        disturbs, 309
        endurance and retention, 309
        erase, 307–308
        programming (writing), 308
        sensing (reading), 308
Planar CT-USUM NROM cell configurations, 451f, 452f
Planar DTM split-gate NROM-MLS cell, 384f
Planar floating gate
    device options, 292t
    MLC NAND flash devices, 282–284
Planar MLC FG-NAND flash/SSD limitations and
        scalability issues, 316
    ECC and controller mitigation, 316–318
    endurance and overprovisioning, 318
    MLC NAND data integrity issues, 318–319
Planar MLC NVM devices
    future, 312–315
    multiplanar stackable NAND devices and technology,
        315–316
Planar MLS potentials, 383–384
Planar MNSC device, 306
Planar one-device CT-NVM USUM NROM cells, 784
Polysilicon control gate, 283
Poole–Frankel mechanism, 42, 45, 59
Potential energy barriers, 37–38
PRAM device, 362
Pre-silicon gate technology, 119
Process sensitivity, dielectric film, 44
Program disturb
    CT device and array, 138
    FG device and array, 136, 136f
Programming (writing), 308
    and erasing characteristics, 101, 102f
    the memory cell, 3
Program-verify algorithm, 285
Progressive band offset (PBO) barrier, 230, 230f
    with deep-offset trapping, 244–245
Progressive Band Offset Direct Tunnel Memory for
        USUM (PBO-DTM-USUM) device, 423, 424f

Progressive band offset DTM (PBO-DTM) device,
        251–253
    stack options, 253t
Pulsed nucleation layer (PNL) method, 28, 115–116
"Punch-and-plug" scheme, 348

Q

QBD, 54
Quad-NROM memory, 306
Quad SPT (QSPT) technique, 288, 288f, 296
Quantum confinement (QC), 22
Quantum mechanical
    direct tunneling, 29
    tunneling mechanisms, 46f
        electrode-controlled, 46

R

Random telegraph noise (RTN), 330
Read disturb
    CT device and array, 138
    FG-device and array, 134–135, 135f
        mitigation, 136–137
Reading mode, memory, 3
Read-only memory (ROM), 7
ReRAM device, 363
Residual cell-to-cell coupling (RCCC), 296
Resonance tunneling, 42
Resonant tunnel barrier (RTB), 232
    DTM device, 250–251
        advanced, 253t
    retention characteristics, 252f
Resonant Tunneling, 45
Rested tunnel barrier, 28
Retention, 75–86
    attributes, 84
    charge-loss mechanisms, 155f
    CT NVM device, charge loss and leakage paths,
        81–84
    definition, 75
    enhanced charge retention, MXOS devices,
        85–86
    enhancement through band engineering, 240–247
    FG-NVM devices
        charge loss and leakage paths, 77–78, 78f
        current components, band diagram-based
            assessment, 78–79
        tunnel oxide thickness limit, 80–81
        window and retention enhancement, 81
    loss, 153–155, 216
    retention, charge transport, and dielectric conductivity,
        76–77
    tunnel oxide charge loss characteristics, 79–80
Reverse crested barrier, 471
Reverse direct tunneling, 219
Reverse mode CT-MSUM devices, 471–472
    band diagram, 471f
    multilayered conductive floating/floating-plate, 472
Reverse mode NVM device, 95, 260
ROM (read-only memory), 7
Rossler, B., 15
RRAM device, 363
RTB. *See* Resonant tunnel barrier (RTB)

S

SAMOS 8 Kb EAROM, 14–15
SAMOS EAROM device, 14
SASTI (self-aligned shallow trench isolation) integration
        scheme, 143, 281
Scalable two-transistor memory (STTM) cell, 160
SCE (short-channel effect), 159, 323–324
Self-aligned-double patterning lithography, 296
Self-aligned local interconnect (SALI), 325
Self-aligned shallow trench isolation (SASTI) integration
        scheme, 143, 281
Semiconductors, work functions, 29f
Sense margin, 285
Sensing (reading), 308
SGCL-DRAM cell, 342
SGT. *See* Surrounding gate transistor (SGT)
SGT-NROM device, 344, 345f
SGT-NVM devices
    device characteristics, 343t
    stack design options, 343t
Shallow trench isolations (STIs), 37, 97
Short-channel effect (SCE), 159, 323–324
Si_3N_4, 55–56
Sidewall image transfer (SIT), 287
SILC. *See* Stress-induced leakage current (SILC)
Silicate, 179
Silicide, 176
Silicon
    based digital memories
        definition, 3
        DRAM, 6, 7f
        SRAM, 5–6, 5f
        volatile, 4
    based NVM technology
        CMOS, high K films integration in, 210
        feature size scalability in, 167
    gate, 296
    nanocrystal
        CT-MOIOS NVM cell, 21f
        device characteristics, 26–28
Silicon based unified memory (SUM) devices, 29, 397
    advantages, 403
    applicability, 397–398
    application, 482–483
        advanced, 483–484
        perspective, 482
    and arrays in memory hierarchy, 400–402
    attribute comparison, 402–403, 402f
    concepts and classifications, 398–400
    dielectric films for, 405–412
        blocking layer selection, 407–408
        charge trapping and charge storage layers, 406–407
        CMOS FET gate insulator, 407–408
        IMI layer for metal gate interface, 406
        I-SRN role for Charge Storage, 407
        silicon interface, 405–406
        tunneling dielectric, 408–412
    functionality, 398
    intergration at
        functional and architectural level, 482
        large system level, 481
        packaging level, 480–481
        silicon technology/chip level, 479–480

MSUM, 398–399
    and NVM integration perspective, 479
    objective, 397
    perspective, 397–398
    stack designs for, 414–420
        CT-DTM USUM, 414–416
        CT-MSUM, 416–419
    SUM-RAM Cells, 399–400
    technology, integration scheme, 413–414
        and device interplay, 413
        MSUM stacks, 419
        objective, 413
        for planar and 3D SUM devices and arrays,
            413–414
        USUM and MSUM, common features, 414
    URAM, 399
    USUM, 398
Silicon carbide (SiC), 39
Silicon-gate CMOS process technology, 141–142
Silicon–oxide–nitride–oxide–silicon (SONOS), 11
    device, 19
        band diagram, 81–82, 82f
        blocking oxide thickness, 108
        device properties, 108–110
        endurance of, 87
        FinFET, 112, 112f
        nitride thickness in, 107
        with oxynitride, 112–115
        programming characteristics, 110, 110f
        reliability issue, 151–156
        retention time modeling, 84
        split-channel, 111f
        stability and retention, 83–84
        stack design, 110–112
        voltage scaling, 169–170
    stack, 337
        band diagram, 108, 109f
Silicon oxynitrides (SiONs), 56–64, 191
    advantage, 63
    applications in NVM devices, 63–64
    for electronic applications, 56
    etch rate, 61, 61f
    I *versus* $E^{0.5}$ plots, 58f
    properties, 60–62
        breakdown strength and, 62, 63t
        dielectric constant and, 62
        refractive index and, 60–62
    Si, O, N phase diagram and compositions, 56, 56f
    trap depths, 58
    trapping characteristics, 57–59
    vs. nitride, MIOS trapping, 59–60
Silicon-rich insulators (SRIs), 64–70
    attributes, 64
    IN-SRN films, 68–69
Silicon-rich nitrides (SRNs), 11, 20–21, 64–70
    annealed, characteristics of, 65t
    current *vs.* field characteristics, 67, 67f
    electron-trapping characteristics, 66f
    hole-trapping characteristics, 67f
    single-phase, 65
    two-phase, 65–68
Silicon-rich oxides (SROs), 20–21, 64–70
Single-layer tunnel dielectric films, 217–218
Single level cell (SLC), 103, 126, 255

Single metal multiple high K (SMLK) integration scheme, 272
    CMOS FET gate stack and NVM gate stack, 273–276, 274f, 275f, 276t
    well and isolation schemes, 272–273
Single-phase SRNs, 65
Single transistor capacitor-less NVDRAM cell, 436t
Single trapping layer, 299
Si-NVMs, 281
SiO_2. *See* Thermal oxide (SiO_2) film
SiO_2 film, 217–218
SiO_2 tunnel dielectric film, reliability issues, 148–156
    CT cell issues, 151–156
        characteristics, 151
        data retention loss, 153–155
        drain disturb, 156
        hot-carrier-induced CT cells, 153
        STS, 155–156
        thicker tunnel oxide cells, 153
        ultrathin tunnel oxide cells, 151–152
    floating-gate cell issues, 148–150
        data retention degradation, 149–150
        drain disturb, 150
        erratic bit phenomenon, 148, 148f
        fast erase cells, 150
        transconductance degradation, 149
SiONs. *See* Silicon oxynitrides (SiONs)
SIP. *See* System-in-package (SIP) integration scheme
Six-transistor CMOS SRAM cell, 5f
SLC (single level cell), 103, 126, 255
SMLK. *See* Single metal multiple high K (SMLK) integration scheme
SNOS
    device, 20
    programming characteristics, 104f
SOC. *See* System-on-chip (SOC) integration scheme
SOI
    diode charge storage RAM cell, 443–444, 443f
    nonvolatile two-device SRAM, 444–445
    technology implementation, 441–442, 442f
    wafers, 331
SONOS. *See* Silicon–oxide–nitride–oxide–silicon (SONOS)
SONOS FET-NVM device, 306
SONOS-MLC NAND, 334f
SONOS/MONOS devices, 297
SONOS NAND arrays, 331
SONOS-nanowire device, 337, 339f
SONOS-tri-gate device, 332f
Spacer patterning technology (SPT), 288
Spin-transfer torque (STT) MTJ memory, 362
Split-gate
    CT-NVMs, 370–371
    device, 103, 110
    NROM cell, 378f
    planar USUM NROM cells, 451
Split Gate cell, 20
Split Gate Device Cell, 155f
SRAM (static random access memory), 5–6, 5f
SRIs. *See* Silicon-rich insulators (SRIs)
SRNs. *See* Silicon-rich nitrides (SRNs)
SROs. *See* Silicon-rich oxides (SROs)
S-SGT. *See* Stacked-SGT (S-SGT)

S-SGT-NAND cells, programming and disturb characteristics, 341, 341f
Stack design, NVM device, 95–116
    CT device, 102–116
        elements and device properties, 103–108
    FG-NAND flash devices, 97–102
        FG-FG coupling, 99, 100f
        floating poly gate, GCR, and STI, 97–102
        GCR *vs.* floating-gate height, 99, 100f
        NAND cell and array, 97, 98f
        Vth dispersion, 100, 101f
Stacked Gate Avalanche Injection MOS (SAMOS) device, 14
Stacked Gate Injection MOS (SIMOS) device, 15
Stacked-SGT (S-SGT), 340
    NAND flash memory array, 340
    SEM photograph, 340, 341f
Stand-alone
    NROMs, 369
    NVMs, 269, 369–370
State-of-the-art in planar MLC FG-NAND flash device, 295–296
    charge storage, 297
    PCT MLC/MNSC flash devices, 296–297
Static random access memory (SRAM), 5–6, 5f
Steady-state Fowler–Nordheim tunneling, 52
"Steep-slope" FET (SSFET), 473
Stress-free insulating film, 56
Stress-induced leakage current (SILC), 24, 53, 79, 149–150
STS (sub-threshold slope degradation), 155–156
STT-MRAM device, 362–363
Sublithographic patterning, 287
Substrate doping effect (SDE), 324
Sub-threshold slope (SS), 289, 337
Sub-threshold slope degradation (STS), 155–156
SUM device. *See* Silicon based unified memory (SUM) devices
Surround gate NV devices, 340–347
    extendable S-SGT, 342–343
    multibit S-SGT, 343–347
    S-SGT and GAA device, 341–342
Surrounding gate transistor (SGT), 324
    devices, 342f
        stack/gate cross-section, 343f
System-in-package (SIP) integration scheme, 265–267, 266f, 267f
System-on-chip (SOC) integration scheme, 267–268, 267f
Sze, S. M., 13, 119

T

Ta_2O_5 (tantalum oxide) film, 188
Tail cells, 149–150, 150f
TaN/alumina/nitride/oxide/silicon (TANOS), 303
    charge-trapping memory cell, 173
    device, 37
        memory retention in, 241–242
TANOS. *See* TaN/alumina/nitride/oxide/silicon (TANOS)
TANOS-follow-on band-engineered devices, 242–243
TANOS-FOLLOW-ON devices, 303
TANOS-NAND FinFET device, 331
Tantalum oxide (Ta_2O_5) film, 188

Technology integration, NVM
    devices/arrays challenges and, 277–279
        NAND flash, 277–278
        NOR flash and NROM, 278–279
    FEOL, 269–271, 270t
    interconnect and packaging levels, 265–268
        functional integration, 268, 268f
        SIP, 265–267, 266f, 267f
        SOC, 267–268, 267f
    memory level, 269
        embedded NVMs, 269
        stand-alone NVMs, 269
    schemes, 271–276
        CMOS platform technology, 271
        gate stack design, 271–272
        planar implementation, 271–272
    transition and integration challenges, 276–277, 277t
Technology scaling, 173
Terabit cell array transistor (TCAT) technology, 352
Texas Instrument Corporation, 121
Thermal cross-shift (TCS), 296
Thermal oxide (SiO$_2$) film, 51–55
    carrier transport and energy dissipation, 51–52
    defect generation and oxide degradation, 53–54
    oxide reliability, 54
Thicker multilayer tunnel dielectrics, 221–230
    double-layer dielectrics, 221–224
        band options, 221f, 224t
        characteristics, 224t
        current–voltage characteristics, 224–225
        high K high barrier/low K low barrier, 223–224
        high K low barrier/low K high barrier, 223
        low K high barrier/high K low barrier, 222–223
        low K low barrier/high K high barrier, 224
    PBO, 230, 230f
    triple-layer dielectrics, 226–230
        crested barrier, 226
        Variot barrier, 226–228
Thicker tunnel oxide cells, 153
Thin dielectric films, properties, 38–43
    bulk and interface defects and charge trapping, 41
    charge transport, 42
    electronic properties, 39–41
        band gap and electron affinity, 39–40
        charge confinement and charge transport
            characteristics, 40
        charge holding capacity/dielectric constant, 40
        dielectric breakdown/dielectric strength, 40
    figure of merit, 42–43
        metal work function and electron affinity, 43
    physical, chemical, and thermal stability, 38–39
Three-layer barrier engineering, 235
Time to breakdown (TDDB), 54
Titanium oxide (TiO$_2$) film, 176, 187
TLC-NAND flash products, 160
TMS (tri-gate memory cell), 19–20
TRAM cell, 446f
    vs. CPRAM USUM cell, 447f
Transconductance, 149
Trap-assisted direct tunneling, 42, 45, 45f
Trap engineering, 240
Trapezoidal barrier, 103–104
Trapezoidal energy barrier, 216, 220

Trapping dielectric, 11
    film, 196–197, 207
Traps, 41, 53
Trench-based vertical channel CT NROM, 325–327, 326f
TRENCH-BASED Vertical Channel Embedded and
    Stand-alone NROMs, 383
Triangular energy barrier, 103, 215, 220
Trigate FinFET, 382
Tri-gate memory cell (TMS), 19–20
Trigate SONOS devices, 332f
Tri-gate transistor, characteristics, 335f
Triple-layer dielectrics, 226–230
    crested barrier, 226
        with high K dielectrics, 229–230
    Variot barrier, 226–228
        vs. crested, 228, 229f
Triple-layer direct tunnel barrier, 232, 233f
Triple-layer SiON stack design, 108–110
Triple-tunnel-layer VARIOT, 379f
Triple-well CMOS NVM structure, 142f
TSNWFET (Twin Silicon Nanowire MOSFETs), 336
T-SRN/TR-SRN, 11
Tunnel
    current density, 46
    dielectric film, 196
        multilayer, 218–220
        single-layer, 217–218
    engineering, 196, 298
    insulator engineering, 28
    oxide, 282
        charge loss characteristics, 79–80
Tunnel-based NVM devices, 95
Tunneling
    characteristics, 201t, 202t
    high K dielectric films for, 201–202
    probability (Pt), 46
Tusunoda, K., 30
Twin Silicon Nanowire MOSFETs (TSNWFET), 336
Two-phase SRN, 65–68
    electrical characteristics, 66–68

U

Ultra-thin dielectric film. See High K dielectric films,
    NVM
Ultrathin triple-layer VARIOT asymmetric tunnel layers,
    298
Ultrathin tunnel oxide cells, 151–152
Uncorrectable bit error rate (UBER), 316
Unified RAM (URAM), 399
Uniform node storage cell (UNSC), 297
Uni-functional SUM (USUM) cells, 433–456
    CPRAM, 445–449
        operation, 448–449
    FB RAM cell, 433–436
        body-charge generation, 434, 435f
        memory cell reading, 435, 436f
        structure, processing of, 435–436
    FET, 449–456
        CT-NAND USUM, 455–456
        CT-NVM USUM, 450–451
        planar merged two-device CT-NVM USUM
            NROM cells, 451–455

Uni-functional SUM (USUM) cells (*Continued*)
   GDRAM, 437–445
      diode I-V characteristics, 438–440, 439f
      operation, 440–441
      SOI diode charge storage RAM cell, 443–444
      SOI technology implementation, 441–442
   GTRAM, 444–445
Uni-functional SUM (USUM) device, 398, 413
   band engineering for, 423–426
      MMCT, 424–426, 426f
      PBO-DTM USUM, 423, 424f
   CT-DTM, 414–415
   features, 414
   memory array, 401, 401f
   tunnel layer of, 409–410
Universal RAM (URAM), 399
   cell and "universal" devices from KAIST, 473–476
USUM. *See* Uni-functional SUM (USUM) cells

V

Variable oxide thickness dielectric barrier, 226
Variot barrier, 226–228, 249
   vs. crested, 228, 229f
VARIOT-CT-MSUM, 428, 429f
Vertical channel
   DTM-NROM, 327–329, 328f, 329f
   FG-NAND, 329
   MLS designs, 385
   split-gate NROM cell, 385f
Vertical EOT scalability. *See* Voltage scaling/scalability, NVM
Vertical flip-chip bonding scheme, 347f
Vertical NAND (V-NAND) chip, 352
Virtual ground array (VGA) architecture, 129, 129f, 145f
   1 Gb NROM design in, 131f
V-NAND (vertical NAND) chip, 352

Volatile memories, 4
   DRAM, 6, 7f
   SRAM, 5–6, 5f
Volatile silicon based digital memories, 4
Voltage scaling/scalability, NVM
   CT device, 169–170
   floating-gate devices, 168–169
   history, 168
   overview, 167–168
   recent developments in, 168–169
   SONOS, 169–170

W

Word length, memory cell, 3
Work function, 43
Wrap-around cell, 282
"Wrap-around" memory cell, 288
Write–erase cycling, 83–84, 149
Write–saturation, 26
Written standby state, 214

Z

Zirconia (ZrO_2) film, 184–187
   advantages, 184
   band diagram, 184f
   cross-sectional TEM micrographs, 185f
   J-V and C-V characteristics, 184–185, 185f
   leakage characteristics, 184, 185f
   processing techniques for, 184
   properties and characteristics, 184
   Pt/ZrO_2/p-Si structure, 186f
   SILC characteristics, 185, 186f
   transport mechanisms in, 186
Zirconium nitride (ZrSiON), 192
ZrO_2. *See* Zirconia (ZrO_2) film